ELECTRIC CIRCUIT ANALYSIS

ELECTRIC CIRCUIT ANALYSIS

Second Edition

S. A. Boctor

Chairman
Department of Electrical Engineering
Ryerson Polytechnic Institute

PRENTICE HALL, Englewood Cliffs, NJ 07632

Library of Congress Cataloging-in-Publication Data

BOCTOR, S. A.
    Electric circuit analysis/S. A. Boctor.—2nd ed.
      p.   cm.
    Includes index.
    ISBN 0-13-251141-X
    1. Electric circuit analysis.   I. Title.
TK454.B584   1992
621.319′2—dc20   91-16330
                                 CIP

Editorial/production supervision
    and interior design: *Julie Boddorf*
Cover design: *Ben Santora*
Cover photo: *Dominique Sarraute/The Image Bank*®
Prepress buyers: *Mary McCartney and Ilene Levy*
Manufacturing buyer: *Ed O'Dougherty*
Acquisitions editor: *Holly Hodder*
Editorial assistant: *Cathy Frank*

Printed in the United States of America

10 9 8 7 6 5 4 3 2 1

ISBN   0-13-251141-X

PRENTICE-HALL INTERNATIONAL (UK) LIMITED, *London*
PRENTICE-HALL OF AUSTRALIA PTY. LIMITED, *Sydney*
PRENTICE-HALL CANADA INC., *Toronto*
PRENTICE-HALL HISPANOAMERICANA, S.A., *Mexico*
PRENTICE-HALL OF INDIA PRIVATE LIMITED, *New Delhi*
PRENTICE-HALL OF JAPAN, INC., *Tokyo*
SIMON & SCHUSTER ASIA PTE. LTD., *Singapore*
EDITORA PRENTICE-HALL DO BRASIL, LTDA., *Rio de Janeiro*

To
Beatrice and Saba
Sylvia and David
The memory of my father

and to
The memory of Dr. N. Bakhoum,
the man who taught me how to think.

CONTENTS

3 SIMPLE RESISTIVE CIRCUITS   53

4 PRACTICAL SOURCES   120

8 CAPACITANCE AND TRANSIENTS IN *RC* CIRCUITS 271

9 INTRODUCTION TO MAGNETISM 329

13 POWER IN AC CIRCUITS 543

14 RESONANCE IN AC CIRCUITS 572

15 LOOP AND NODE ANALYSIS OF GENERAL AC CIRCUITS 613

16 NETWORK THEOREMS FOR AC CIRCUITS   666

17 MUTUALLY COUPLED CIRCUITS AND TRANSFORMERS   698

18 THREE-PHASE CIRCUITS   745

PREFACE

This text provides freshmen students with the necessary foundation in electric circuit analysis. Since many colleges and universities have, for some time, offered such a course to first-year engineering and engineering technology students, this book treats the subject material based primarily on algebra and trigonometry. The treatments of all the circuit analysis foundation topics are carried out in a complete and thorough manner. Any new mathematical concept is introduced and treated in conjunction with the corresponding circuit analysis topic. Those topics that need calculus concepts are introduced with a clear physical meaning. For the sake of rigor and completion, derivations based on calculus are provided in the appendices of the corresponding chapters.

The topics dealt with in this book introduce students to the basic concepts and tools needed for further education in the fields of electrical engineering technology, including electronics, electrical apparatus and machines, and advanced networks and systems studies.

The objective of this text is to lay the foundation for the methodologies of circuit analysis through a clear understanding of the basic concepts. Each chapter introduces the required basic concepts in a clear and complete manner to achieve learning through understanding. The thinking ability is developed through applying the concepts learned to numerous general solved examples at the end of each chapter, stressing analytical reasoning. The emphasis is on the why and how before getting involved in the usual algebraic manipulations. This philosophy is followed throughout the text to show that in practical situations a number of ideas and topics are really needed in solving such problems.

In this edition, each chapter begins with a summary of the topics' objectives and a glossary of the important and newly introduced terms is provided at the end of the chapter. Also, more than 220 additional problems of the drill type are provided in order that the student may gain confidence and practice in the new concepts introduced in each chapter, before tackling the more challenging problems on the topics treated. The drill problems are designed specifically to be directly applicable to the new concepts introduced, and complete answers to all the odd-numbered drill problems are provided at the end of each chapter. The challenge problems are marked with roman numerals so that a(ii) would indicate a higher degree of difficulty. Both the drill and the challenge problems are categorized as to the section to which they apply. In each chapter a number of problems appropriate as exercises for computer-aided analysis or programming are marked with an asterisk. The book now provides more than 600 problems without undue repetition. Answers to selected problems are given at the end of the book. Complete solutions are provided in the instructor's solution manual. Instructors are also provided with a transparency manual, additional assignments, and test problem sets manual with complete solutions and a Pspice manual with demonstration disks. A companion laboratory manual is also available.

Briefly, the book deals with the basic concepts of electrical circuits and the methods used to analyze them. It stresses the fundamental understanding of Ohm's law, KVL, and KCL. In the first seven chapters, specific analysis tools are introduced and applied to dc resistive circuits. These include voltage- and current-division principles, network theorems, and the general techniques of loop and node analysis. In particular, Chapter 4 deals with practical voltage and current sources and their duality, together with the design of practical voltage dividers. Graphical interpretations and techniques based on the $V-I$ characteristics of electrical devices are also stressed in this chapter.

Electrical measurement instrumentations are dealt with in Chapter 7. Even though meter-movement instrumentations are becoming more obsolete, they are examined here to stress the loading-effect problems and to illustrate practical applications of circuit analysis concepts. Also, introductory examination and operation of the oscilloscope is treated there. This topic may, however, be covered after Chapter 11.

In Chapters 8 and 10, capacitance and inductance are defined in terms of the electric and magnetic field phenomena, in terms of the appropriate physical variables and in terms of the corresponding differential forms of Ohm's law. RC and RL circuit transients are treated comprehensively through numerous solved examples. Chapter 9 provides an introduction to magnetism and simple magnetic circuits, leading to Faraday's and Lenz's laws. This coverage is sufficient to explain the electromagnetic phenomena, ac voltage generation, and self- and mutual induction.

The second half of the book is devoted to topics on the analysis of ac circuits. Through experience, this practice has been shown to be quite effective. The treatment here is, however, quite distinct from that of many other books. The time domain and phasors are stressed throughout, from simple circuits to more complicated ones. The relationship between time-domain waveforms (oscillograms) and the corresponding phasor representations and manipulations is treated extensively through numerous examples. From the beginning, the concept of average power is developed and used.

After developing the basic concepts of ac circuit analysis in Chapters 11 and 12, a generalized treatment of power in ac circuits and the resonance phenomena are covered in Chapters 13 and 14. Chapters 15 and 16 demonstrate that the analysis techniques developed for dc circuits do apply in ac circuits, keeping in mind that one now deals with complex numbers instead of real numbers. Chapters 17 and 18 deal with some specialized topics that are quite useful in many fields of electrical engineering technology.

Chapter 19 deals with SPICE, one of the most commonly used, flexible, and powerful computer-aided analysis and design software packages. Through demonstrating how an electric circuit can be described and how a certain type of analysis and a certain output can be requested, the usefulness of such a CAD tool becomes clear, especially when dealing with complicated circuits. The familiarity with SPICE gained through this study permits the reader to use other available software packages and facilitates using the other aspects of SPICE which extend beyond the scope of this text.

All symbols, diagrams, and units used in this text follow the IEEE recommended standards and the International System of Units (SI). The subjects covered can be treated quite comfortably in two academic semesters. However, it is left to the discretion of the teaching faculty to select the appropriate topics from the material presented in order to construct the option suitable for their own curriculum.

The author is indebted to his colleagues at Ryerson, in particular J. Van Arragon, who helped shape the text in its present form. The contributions made by S. Prabhu and M. T. Ghorab to the laboratory manual and the instructor's solution manual are gratefully acknowledged. The valuable constructive comments and encouragements of Professors A. W. Avtgis, L. H. Hardy, J. L. Brice, D. B. Beyer, M. Boyle, and J. N. Tompkin, who reviewed the manuscript, are gratefully appreciated. Thanks are also due to Miss C. Gervais for typing the manuscript and to the editorial staff at Prentice Hall, in particular, Holly Hodder, Julie Boddorf, and Alice Barr, who proposed and implemented this new and highly effective production of the second edition.

S. A. Boctor

ELECTRIC CIRCUIT ANALYSIS

1 | BASIC CONCEPTS

OBJECTIVES

- ■ Familiarity and understanding of the basic quantities encountered in electric circuits, in particular the electric charge, current, voltage, and power.
- ■ Ability to interrelate the basic electrical quantities with each other and comprehend their fundamental physical definitions.
- ■ Familiarity with the positively defined direction of the electrical quantities, their symbols, and the units used to measure these quantities.
- ■ Introduction to the basic sources of electrical energy.
- ■ Ability to make simple calculations involving electrical energy, power, and efficiency.

INTRODUCTION

This chapter introduces the reader to the basic quantities that will be encountered in studying electric circuits. The definitions of these quantities and the symbols and units used to measure them are examined, together with the interrelationships between these quantities. Also, the conventions that relate to the positively defined direction of the basic electrical quantities are clarified.

The symbolic letters used to indicate an electrical quantity can be in either capital or lowercase form. Capital letters usually indicate constant or uniform quantities; the corresponding lowercase forms are used to indicate the instantaneous value of such quantities, implying that these quantities are functions of time (vary with time).

Electricity is a form of energy that has been known to mankind for many centuries. However, the laws and relations that govern its fundamentials, uses, and conversions to other forms of energy (such as heat, light, sound, or mechanical energies) have been discovered only during the last 300 years. Many conventions are retained from the earlier historical concepts that were accepted as this branch of science was developed. However, recent international agreements have been concluded to adopt uniform conventions and standards, in particular because of the tremendous impact and use of electricity and electronics in almost all of our modern everyday activities. This varies from home appliances and power tools to industrial machines and equipment and from wired and wireless communication systems to computers and calculators. In fact, all kinds of laboratories dealing with physical sciences, production and manufacturing, medical sciences, and education are stocked with various types of electrical and electronic measuring equipment.

A standard systems of measurements is of prime importance to producers, manufacturers, sales companies, and users of electrical and electronic components and equipment. The units used to describe the magnitude of an electrical quantity must conform to this standard in order that everyone understands what the measurement means. In 1960, the General Conference on Weights and Measures adopted the International System of Units (see Section 1.2), abbreviated SI, based on the MKS metric system.

The major emphasis in this book is on the methods of analysis of electric circuits. An electric circuit is formed by connecting the terminals of a number of electric elements, using conducting wires. The electric elements include electric sources of energy and loads, including resistors, capacitors, inductors, and possibly semiconductor electronic devices. Many of these elements are discussed and treated in detail in the following chapters. Usually, the interconnection of elements to form an electric circuit (or network) is done in a prescribed manner in order to perform a desired function. *Analysis* of electric circuits refers to the computations required to determine one or more of the unknown electrical quantities (to be defined in this chapter) associated with one or more of the elements constituting the electric circuit.

Almost all of the mathematical manipulations introduced and used in this book to analyze electric circuits can easily be performed using a hand-held scientific calculator. Figure 1.1 shows some of the most suitable scientific calculators (including Sharp, Hewlett-Packard, Texas Instruments, etc.). When the complexity of the electrical or electronic circuit increases or the number of electrical quantities to be determined is quite large, there is justification for

MODEL EL-506P
"THIN MAN"™ WALLET-SIZE
WITH 56 SCIENTIFIC FUNCTIONS, TEXT
Extra full-featured scientific calculator with Memory Safe Guard™

- Trigonometric functions (sin, cos, tan) and their inverses.
- Hyperbolic functions (sinh, cosh, tanh) and their inverses.
- Hexadecimal, octal, binary.
- Rectangular/polar coordinate conversions.
- Exponential (base 10 and base e) and their inverses (logarithms).
- 3 angle modes (degree/radian/grad).
- Power (y^x) and its inverse.
- Complex number calculations.
- Mean, sum, and standard deviation.
- 10-digit LCD with scientific notation expression.
- 15 levels of parentheses and 4 pending operations.
- Independently accessible 3-key memory.
- Percent key.
- Automatic Power Off (APO)™.
- Wallet and batteries included.
- 2-23/32"(W) × 9/32"(H) × 5-1/32"(D).

(a)

Figure 1.1 (a) Sharp Scientific Calculator Model EL506P; (b) Sharp Scientific Computer Model EL5500II; (c) HP-11C Scientific Calculator; (d) Texas Instruments Scientific Calculator Model T155-II. [(a), (b) Courtesy of Sharp Electronics Corporation; (c) courtesy of Hewlett-Packard; (d) courtesy of Texas Instruments Inc.]

MODEL EL-5500-II
PORTABLE SCIENTIFIC COMPUTER
WITH DUAL KEYBOARD, WITH TEXT
Two machines in one: Scientific Calculator, BASIC Computer.

- Ability to use statistical information and scientific functions in BASIC mode.
- 59 scientific functions, including hexadecimal.
- 19 statistical functions, including linear regression.
- 3534 BASIC program steps.
- 16-character dot-matrix display.
- Ability to store and recall 18 separate programs under different labels.
- 26 data memories for storing values in BASIC mode.
- 18 instant BASIC instructions for fast programming.
- Includes slip-on hardshell cover, application book, and batteries.
- Optional CE-126P Printer, with Cassette Interface for storing programs (cassette recorder not included).
- 6-11/16"(W) × 3/8"(H) × 2-27/32"(D).

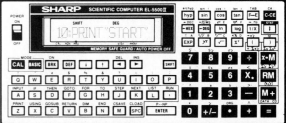

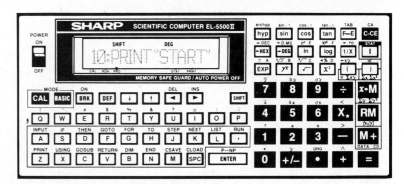

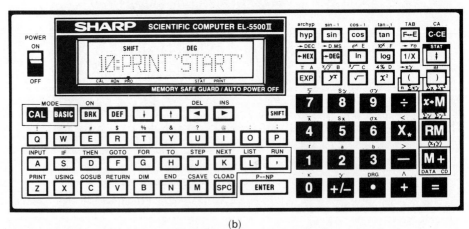

(b)

Figure 1.1 *(cont.)*

(c)

(d)

Figure 1.1   (*cont.*)

programming the solution and using personal computers or mainframe computers to aid in the circuit analysis. An example of such a computer-aided circuit analysis program is discussed in Chapter 19, together with examples of its use in different types of network analysis. Some of the end-of-chapter problems [designated with an asterisk (*)] can be used to practice such computer-aided circuit analysis or any other suitable programming language.

1.2 THE INTERNATIONAL SYSTEM OF UNITS

The adoption of the International System of Units (SI) is intended to allow scientists in all fields of physics and engineering to publish the results of their investigations in a manner that can be understood, verified, and made use of by other workers in all countries. It is also intended to expand and facilitate the international market in manufactured goods among countries, especially when such goods and machinery are made to similar specifications and are compatible with each other.

The SI system is based on the MKS units (meter-kilogram-second) for distance, weight, and time, rather than the cgs units (centimeter-gram-second) or the British system, which uses foot-pound-second units. The MKS system was preferred over the cgs system because the former has more practical unit sizes. Also, the MKS system of units is based on the decimal number system (the metric system), which has 10 as its base. This fact favored it over the British system, which has many conversion factors (e.g., 3, 12, 100, 1000, 5760, etc.) between the various sizes of each measuring unit.

The three basic units mentioned above, together with the coulomb as the basic unit for the charge and the Kelvin or Celsius degrees for temperature, are the principal units from which almost all other physical quantities can be derived. In other words, the newton for force, watts for power, joules for energy, amperes for electric current, volts for electric voltages, ohms for electric resistances, and so on, can all be derived or expressed in terms of the meter, kilogram, second, and coulomb. All the units used in this text are SI units. They are defined and interrelated with each other as and when they are introduced in the study of the subject matter.

Because physical quantities vary tremendously in size, one may be confronted with having to write leading and trailing zeros, as in

$$0.000\,003\,3\,\text{farad} \quad \text{(a capacitor size)}$$

or

$$2\,000\,000.0\,\text{hertz} \quad \text{(a frequency size)}$$

This is usually inconvenient, confusing, and wasteful of space. To overcome this, the SI system of units uses *unit prefixes* based on the powers of 10 and replaces such powers of 10 by symbols. It should always be remembered that these prefixes are not part of the unit but integral parts of the numerical value of the physical quantity and should be manipulated as such. Representing a number in terms of a power-of-10 multiplier is usually referred to as *scientific notation*: for example,

$$2\,300\,500.00 = 2.3005 \times 10^6 \quad \text{(in scientific notation)}$$

or

$$0.000\,030\,7 = 3.07 \times 10^{-5} \quad \text{(in scientific notation)}$$

In addition and subtraction, it is better to use the same power-of-10 multiplier for all the numbers involved in the operation. However, in multiplication, the exponents of the different powers of 10 for each number are added algebraically, while in division the exponents of the different powers of 10 for the numbers in the denominator are subtracted algebraically from the exponents of the powers-of-10 multipliers for the numbers in the numerator. For example:

(1) $\quad 0.007 + 0.00035 - 0.0101 = 7.0 \times 10^{-3} + 0.35 \times 10^{-3} - 10.1 \times 10^{-3}$

$$= -(10.1 - 7.35) \times 10^{-3}$$

$$= -2.75 \times 10^{-3}$$

(2) $\quad\quad\quad\quad (2.2 \times 10^5) \times (3.0 \times 10^{-2}) \times (1.9 \times 10^3)$

$$= (2.2 \times 3.0 \times 1.9) \times 10^{5-2+3}$$

$$= 12.54 \times 10^6 = 1.254 \times 10^7$$

(3) $$\frac{6.8 \times 10^3 \times 7.5 \times 10^{-6}}{1.5 \times 10^{-2} \times 8.0 \times 10^4} = \frac{6.8 \times 7.5}{1.5 \times 8.0} \times 10^{3-6-(-2+4)}$$
$$= 4.25 \times 10^{-5}$$

A summary of the most commonly used prefixes of the SI system is given in Table 1.1. As mentioned before, these prefixes replace the corresponding powers of 10 by symbols. The names used to refer to these symbols are also indicated in the table.

TABLE 1.1

Power of 10	Prefix name	Symbol
10^{12}	tera	T
10^{9}	giga	G
10^{6}	mega	M
10^{3}	kilo	k
10^{-2}	centi	c
10^{-3}	milli	m
10^{-6}	micro	μ
10^{-9}	nano	n
10^{-12}	pico	p
10^{-15}	femto	f

EXAMPLE 1.1

Convert the following values to scientific notation and numerical equivalents.
(a) 10.5 MW.
(b) 5.7 μs.

Solution
(a) 10.5 MW = 10.5 × 10⁶ W = 10 500 000.00 W.   *[handwritten: 1.05 ×10⁷ W]*
(b) 5.7 μs = 5.7 $\times 10^{-6}$ s = 0.000 005 7 s.

EXAMPLE 1.2

Convert the following numbers to scientific notation and use SI prefixes to represent the equivalent values.
(a) 0.000 25 A.
(b) 500 000.0 V.

Solution
(a) 0.000 25 A = 2.5 $\times 10^{-4}$ A = 0.25 $\times 10^{-3}$ A = 0.25 mA.
(b) 500 000.0 V = 5.0 $\times 10^5$ V = 500.0 $\times 10^3$ V = 500.0 kV.

1.3 ELECTRIC CHARGE

Modern experimental investigations of the physical nature and properties of elements (e.g., hydrogen, oxygen, copper, iron, lead, silicon, etc.) have revealed that each atom has a positively charged nucleus, consisting of positively charged protons and neutral neutrons. Around the nucleus, negatively charged electrons revolve in specific orbits. Thus the atomic structure resembles planets traveling around the sun. The charge on one electron is equal to the charge on one proton, even though the mass of a proton is 1836 times that of an electron. When an atom is in the "neutral" state, the number of electrons is equal to the number of protons. Each element has a specific number of such electrons (and protons); for example, hydrogen has one electron, oxygen has eight electrons, copper has 29 electrons, and so on. The physical and chemical properties of the various elements are a function of their atomic structure and the number of electrons in their atoms.

About 200 years ago, Charles A. Coulomb, a French physicist, discovered that the force between charged bodies is proportional to the product of the charges and inversely proportional to the square of the distance between them. Unlike charges produce a force of attraction between them, while like charges result in a repulsion force. The complex forces acting on the electrons in each of the allowable orbits will not be examined here. However, a simpler notion would be to consider that the centrifugal force resulting from the rotational motion of an electron around the nucleus of an atom compensates for the attraction force that tries to pull the electron into the positively charged nucleus.

The orbits of the electrons are at different distances from the nucleus and each corresponds to a specific amount of energy. The outermost electrons (farthest orbits) have the weakest binding forces. Their energy levels are referred to as *valence bands*. Adding a little energy to these electrons, through heat or light, permits them to escape the binding force of the parent atom. Metals such as copper, aluminum, silver, gold, and so on, constitute a group of elements characterized by the fact that each of their atoms has one or two electrons in the valence band. These electrons are very loosely attached to the parent atom. Even at room temperature, they gain enough energy to escape, becoming *free electrons*.

Once an atom loses one or more electrons, it becomes a positively charged ion since the net charge on it is mostly positive. In metals these positively charged parent atoms are stationary, constituting the crystalline structure of the metal. In gases and liquids, positively charged ions can move. The liberated free electrons in metals do, in fact, move through the crystalline structure created by the stationary parent atoms, but with a net average displacement (with respect to time) of zero, because of the random statistical nature of such motion.

Under the influence of an external energy source (e.g., a battery), as shown schematically in Fig. 1.2, the free electrons will acquire energy and will be driven by this externally generated force toward the positive terminal of the energy source (marked +). The net displacement of electrons corresponds to the transfer of charge from one position to another.

The path taken by an electron, as shown in Fig. 1.2, takes this electron from the valence band of one atom into that of another atom. As it moves, it leaves behind a space that used to be occupied by it. We can also think of charge transfer as a situation in which the electrons are stationary but an electron deficiency (like a bubble), referred to as a *hole*, having an equivalent charge as the electron but positive ($h^+ \triangleq e^-$), is moving in the direction opposite to the electron motion, as indicated in Fig. 1.2. This equivalent representation resembles the flow of water molecules in one direction or bubbles moving in the opposite direction. Thus even though in reality, charge transfer in metals is due to the motion of negatively charged electrons in one direction, this is exactly equivalent to the flow of positively charged holes in the opposite direction. The

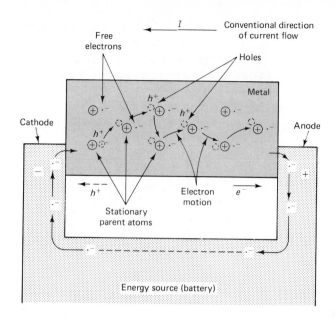

Figure 1.2

concept of holes might be thought of as artificial in this case; however, in semiconductors (elements such as silicon and germanium), these positively charged current carriers (h^+) do, in fact, also exist. Modern physics has actually proven that a hole behaves as though it has a mass slightly higher than that of the electron! Electrons and holes are called current carriers.

In Fig. 1.2 the energy source (battery) is acting like a pump, drawing electrons from one side of the metal conductor (the side in contact with the positive terminal) and supplying the same number of electrons to the conductor from the other side (the side in contact with the negative terminal). The conductor is thus always neutral at any given time; that is, the total amount of positive charge equals the total amount of negative charge, even with such charge transfer.

The symbol used for the quantity of charge is Q or q. The smallest amount of an electrical charge is that of an electron ($1\,e^-$) or of the equivalent hole ($1\,h^+$). This charge is, however, too small to be used in practice as the unit for charge. Instead, the practical unit for charge is the *coulomb,* with the symbol C, according to the SI system of measurements. The unit charge of 1 coulomb is

$$1\,\text{C} = \text{charge on } 6.242 \times 10^{18}\,\text{electrons}$$

or

$$1\,e^- \triangleq 1\,h^+ = \frac{1}{6.242 \times 10^{18}} = 1.602 \times 10^{-19}\,\text{C}$$

EXAMPLE 1.3

What is the charge of 10 million hole current carriers?

Solution   The amount of charge can be obtained by multiplying the number of hole current carriers by the charge of each hole, that is

$$Q = 10 \times 10^6 \times 1.602 \times 10^{-19} = 1.602 \times 10^{-12} \quad \text{C} = 1.602\,\text{pC}$$

Note that 10^{-12} is replaced by the letter p (for pico; see Table 1.1 for the SI prefixes). The p is not part of the unit C (for coulomb) but part of the numerical value of the quantity. This remark applies to all prefixes for all quantities and *should never* be forgotten.

1.4 ELECTRIC CURRENT

The net displacement (flow or motion) of current carriers (electrons or holes) through a cross-sectional area of a conductor, such as a copper wire, is called the *electric current*, or simply the *current*. The symbol used for the magnitude of the current is I or i, from the French word "intensité."

As discussed before, any piece of conducting metal has a large number of free electrons which continuously move in random, with a net zero displacement in any given direction. Only when an external force acts on the current carriers, as would be exhibited by applying an external energy source (e.g., a battery) to the conductor, will the current carriers be forced to move mainly in one direction. As indicated in Fig. 1.2, the electrons are attracted to the positive terminal of the source, which acts as a pump, transporting them back (through chemical reaction) toward the negative terminal of this source, again to continue their drift along the conductor. This constitutes the continuous motion of the charge described as the electric current. The motion of current carriers in a wire resembles the flow of water molecules in a pipe or a stream. Motion is always a displacement with respect to time.

Definition. *Current* is the rate of flow of charge.

Rate refers to variation with respect to time. Clearly, if a certain amount of charge Q did flow through a cross section of a conductor in a time t, then the smaller t is, the faster is the motion (i.e., the higher must be the current). In analogy with the flow of water molecules, one usually says that a stronger current exists when the water particles are flowing faster (i.e., in a shorter period of time). This indicates that mathematically, current I is expressed as

$$I = \frac{Q}{t} \qquad (1.1)$$

When 1 C of charge flows through a cross section of a conductor in 1 s, one unit of current results, called 1 ampere or 1 A. Thus

$$1\,\text{A} = 1\,\text{C/s} \qquad (1.1a)$$

The unit of current is the ampere, named in honor of the French physicist André Ampère, who was the first to define this unit in terms of the electromagnetic phenomenon, as explained in Chapter 9.

If the charge flowing in the conductor is not constant but varies with time (i.e., q), current is defined as

$$i \cong \frac{\Delta q}{\Delta t} \quad \text{A} \qquad (1.2)$$

where Δq is the small amount of charge flowing through a cross section of the conductor in a time Δt. As Δt tends to zero,

$$i = \lim_{\Delta t \to 0} \frac{\Delta q}{\Delta t} = \frac{dq}{dt} \quad A \tag{1.2a}$$

Current is thus the derivative of charge with respect to time. This calculus definition of the electric current is presented here for the sake of completeness. In talking about current in electric circuits or in elements of an electric circuit, we always say current "in" or "through" the element.

Historically, early conceptions about current were based on the belief that the current carriers were positively charged. Only after the discovery of electrons did scientists realize that these negatively charged current carriers are the actual cause of charge transfer (i.e., current) in metallic conductors. However, during the past 40 years, after the discovery of transitors and the tremendous exploitation of semiconductors in the electronic industry, scientists realized that there are, in fact, two types of current carriers, electrons and holes. To eliminate such confusion, the universally accepted convention is that *the direction of flow of electric current is the direction in which the positive current carriers* (e.g., holes) *would flow*. This convention is indicated in Fig. 1.2.

In many situations in electric circuits, neither the magnitude nor the direction of the current in a certain element may be known. In such cases, we assume a certain reference direction of the current flow in this element. Based on this assumption, the mathematical solution of the circuit relations (explained in Chapter 3) may result in a positive or a negative value for this unknown current. A positive value for current indicates that the current is flowing in the assumed direction; a negative value indicates that the current is flowing in the direction opposite to that assumed. As indicated in Fig. 1.3, a current I flowing in one direction is exactly the same as $-I$ flowing in the opposite direction. In other words, reversing the direction of the current flow corresponds to reversing its sign (from + to −, or vice versa).

When the magnitude and direction of the current flowing in an element of an electric circuit do not vary with time, as indicated in Fig. 1.4(a), the current is

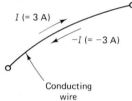

Figure 1.3

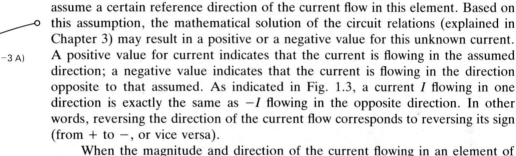

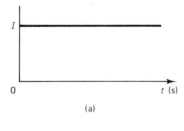

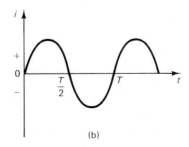

Figure 1.4 (a) dc current; (b) ac current.

referred to as *direct current* (dc). On the other hand, if the current in a certain electric circuit has a continuously varying magnitude with time and changes its direction of flow (say, every $T/2$ seconds), as shown in Fig. 1.4 (b), the current is referred to as *alternating current* (ac). In many situations, current may in general be a function of time, $i(t)$.

■ EXAMPLE 1.4

If the charge flowing in a certain element is 7.5 C for 0.5 min, what is the current in this element?

Solution Since $I = Q/t$ and $t = 0.5 \, \text{min} = 30 \, \text{s}$,

$$I = \frac{7.5}{30} = 0.25 \, \text{A}$$

■ Note that Q must be expressed in coulombs and t in seconds.

■ EXAMPLE 1.5

The current in a certain element was measured to be 5 A. What is the time in which 4 mC of charge flows through this element?

Solution Since the current is the rate of flow of charge,

$$I = \frac{Q}{t}$$

then

$$t = \frac{Q}{I} = \frac{4 \times 10^{-3}}{5} = 0.8 \times 10^{-3} = 0.8 \, \text{ms}$$

Again note that 4 mC is 4×10^{-3} C because m is a part of the numerical value ■ of the charge, representing 10^{-3} (i.e., milli).

1.5 VOLTAGE OR ELECTRIC POTENTIAL DIFFERENCE

As explained above, a source of energy (e.g., a battery) must be used to provide the required work (or energy) necessary to cause the motion of current carriers. Current, which reflects the transfer of charge with respect to time, will then result. The work provided by the source initiates the forces that make the charges move and it is thus exhibited as the momentum of the moving charges. The battery (or in general, the energy source) is thus referred to as the *electromotive force* (EMF).

Energy is the capacity to do work. Work is done by spending stored energy. Work and energy are measured in *joules,* abbreviated J, according to the SI system of units. Potential energy is one form of energy reflecting the capacity of doing work. When a mass is placed on a table, it has a potential energy ($= mg \times d$ where m is the mass, g the gravitational acceleration, and d the distance from ground) because when the mass is allowed to fall, it will give up this amount of energy to the ground. There are many types of electrical energy sources. All of them convert various forms of energy to electrical potential energy difference, usually referred to by the symbol E or e. This conversion could be from chemical (batteries), mechanical (generators), light (photovoltaic cells), or thermal heat (thermocouples) sources. Various types of commercial batteries are shown in Fig. 1.5.

The electric energy source creates potential energy difference between two points, which are the terminal contacts of the source, analogous to two planes at different heights from the ground (d_1 and d_2) each having a different potential energy difference with respect to ground, and with a potential energy difference between them. The symbolic representation of a battery which produces a constant (with respect to time) potential energy difference between its terminals of value E is shown in Fig. 1.6. The positive (+) terminal represents *the higher potential point* and the negative (−) terminal represents *the lower potential point*. These potential energy levels need not be positive or negative numbers, but could be numbers such as 5 and 2 or −1 and −3, where the first number is higher than the second number. The higher potential point (higher number) is thus always denoted the positive (+) terminal; for example, the level 5 is the + terminal and

Figure 1.6

(a)

Figure 1.5 (a) Various types of commercial dry cells; (b) cross section of a zinc chloride cell; (c) cross section of an alkaline cell; (d) cross section of lithium miniature cell. (Courtesy of Eveready–Union Carbide.)

"EVEREADY" No.1250 BATTERY

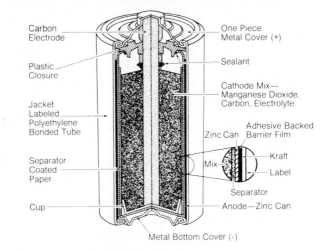

Carbon Electrode

Plastic Closure

Jacket Labeled Polyethylene Bonded Tube

Separator Coated Paper

Cup

One Piece Metal Cover (+)

Sealant

Cathode Mix— Manganese Dioxide, Carbon, Electrolyte

Zinc Can

Adhesive Backed Barrier Film

Mix

Kraft

Label

Separator

Anode—Zinc Can

Metal Bottom Cover (-)

CUTAWAY OF SUPER HEAVY DUTY ZINC CHLORIDE CELL

(b)

"EVEREADY" NO. E95 BATTERY

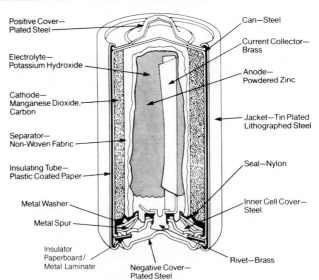

Positive Cover— Plated Steel

Electrolyte— Potassium Hydroxide

Cathode— Manganese Dioxide, Carbon

Separator— Non-Woven Fabric

Insulating Tube— Plastic Coated Paper

Metal Washer

Metal Spur

Insulator Paperboard/ Metal Laminate

Negative Cover— Plated Steel

Can—Steel

Current Collector— Brass

Anode— Powdered Zinc

Jacket—Tin Plated Lithographed Steel

Seal—Nylon

Inner Cell Cover— Steel

Rivet—Brass

CUTAWAY OF CYLINDRICAL ENERGIZER ALKALINE CELL

(c)

14

Figure 1.5 *(cont.)*

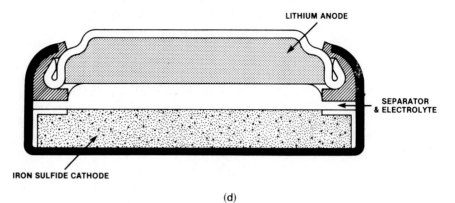

EVEREADY.

1.5 VOLT LITHIUM IRON SULFIDE
MINIATURE CELL CONSTRUCTION

LITHIUM ANODE

SEPARATOR
& ELECTROLYTE

IRON SULFIDE CATHODE

(d)

Figure 1.5 (*cont.*)

the level 2 is the − terminal. The long and short lines in Fig. 1.6 represent the positive and negative terminals, respectively. The potential energy difference E is also referred to as the *voltage difference* (or simply *voltage*) between the two points. It will be defined properly shortly. Throughout this book, the +, − notation will be used to denote the voltage polarities (i.e., the higher and lower potential points, respectively).

Conventionally, current is the flow of positively charged current carriers (charge Q). Through the load shown in Fig. 1.7 (the box representation), the positive charges will be attracted to the negative side of the battery. Thus the current direction in the load will be from the positive toward the negative terminal of the battery, in agreement with the schematic diagram of Fig. 1.2. Figure 1.7 represents a simple electric circuit consisting of a source and a load. Each is a two-terminal device. One terminal of the source and one terminal of the load are connected together; the other terminals of each are also connected together, either directly or using conducting wires. These two terminals are

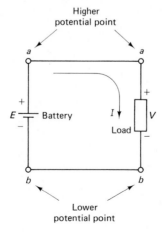

Figure 1.7

denoted a and b in Fig. 1.7. The potential energy difference between them is a unique value. The electric potential energy difference, or voltage, across the load is usually denoted V or v. In the case shown here

$$E = V = V_{ab} \qquad (1.3)$$

V_{ab} means the voltage or potential differences of point a (the first subscript) with respect to point b (the second subscript).

Once the positive charge carriers, constituting the electric current I, reach the negative side of the battery ($-$ or point b), they will be forced to move from this lower potential point toward the higher potential point ($+$ or point a) inside the battery. Because the positive terminal of the battery would repel the positive charge Q as it is transferred inside the battery, work has to be spent against the repulsion force. This work is produced from the chemical reaction, converted into potential energy. The positive charge Q is said to have *gained* potential energy. The motion of the charge inside the battery is said to experience a *potential rise*. When this charge moves in the load, it is transferred from the higher potential point (terminal a) toward the lower potential point (terminal b). It is then said that the charge experiences a *potential drop*. In the load, the energy gained by the charge from the battery is now lost to the load and converted to heat (as in stove heating elements) or to light (as in light bulbs).

The analogy with the water system discussed earlier provides more clarication. As a mass of water (resembling positive charges) moves downhill in a pipe, under the influence of gravity, it loses potential energy since it moves from higher to lower levels. A pump (resembling the battery) must then be used to lift the water mass uphill again, thus raising the potential energy of the water mass. The energy provided by the pump usually comes from another source of energy.

Definition. The *voltage*, or *potential energy difference*, between two points in an electric circuit is the amount of energy (or work) required to move a unit charge. In mathematical terms,

$$\boxed{V \text{ or } E \text{ (electric potential difference)} = \frac{W \text{ (work or energy)}}{Q \text{ (the charge)}}} \qquad (1.4)$$

The unit of voltage is the *volt*, abbreviated V, named in honor of Antonio Volta, the Italian scientist who invented the first battery. When the work done to transfer 1 C of charge between two points is 1 J, the voltage between the two points is 1 V:

$$1\,V = 1\,J/1\,C \quad (1 \text{ joule/coulomb}) \qquad (1.4a)$$

When we talk about voltage, it is always stated that the voltage is "between" two points or "across" the terminals of an element, never in or through an element. Voltage is always the difference in potential energy (per unit charge) *between two points*.

Following the polarity convention for voltage introduced above, let point a be 5 V higher in potential than point b. We then write

$$V_{ab} = 5\,V$$

that is, point a, with respect to point b, is higher in potential by 5 V.

It is also correct to say that point b is lower in potential than point a by 5 V, that is, the voltage of point b, with respect to point a, is

$$V_{ba} = -5 \text{ V}$$

Thus

$$V_{ab} = -V_{ba} \qquad (1.5)$$

In words, *reversing the subscripts of the voltage necessitates changing the sign in order not to change the meaning of the voltage.* This situation is represented in Fig. 1.8. The same element is connected between the same two points. In case 1, with b as the reference, a is higher by 5 V; this is the meaning of $V_{ab} = 5$ V. In case 2, with a as the reference, b is lower by 5 V; this is the meaning of $V_{ba} = -5$ V. Obviously, both cases mean the same thing.

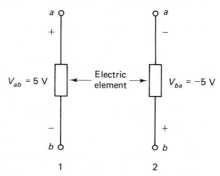

Figure 1.8

The discussion above indicates that in a manner similar to the conclusions reached about current and its assumed reference direction, we can also assume a reference direction for an unknown voltage and then solve the circuit relations (as shown in Chaper 3). If the answer is a positive number, it means that the assumed reference direction is correct. If the answer is a negative number, it means that the polarities of the assumed reference direction must be reversed to obtain a positive voltage measurement.

In any load in an electric circuit, current must enter the element from the higher-voltage terminal and leave this element from the lower-voltage terminal, as shown in Fig. 1.9. The voltage and current reference directions shown in this figure are said to be *associated*. This is the universally accepted convention.

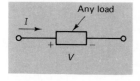

Figure 1.9

EXAMPLE 1.6

If a dry-cell chemical reaction needed 55 J of work to produce 50 C of charge separation between the battery terminals, what is the terminal voltage of this battery?

Solution   Based on the definition of voltage in eq. (1.4),

$$V = \frac{W}{Q} = \frac{55 \text{ J}}{50 \text{ C}} = 1.1 \text{ V}$$

EXAMPLE 1.7

A current of 0.3 A flows through a filament and releases 9.45 J of heat energy in 5 s. What is the voltage across the filament?

Solution Here the work done by the charge as it flows in the element is known, but the amount of charge is unknown. However, the current I and time t are given. Thus as

$$I = \frac{Q}{t}$$

then

$$Q = It = 0.3 \times 5 = 1.5\,\text{C}$$

Now applying Eq. (1.4) yields

$$V = \frac{W}{Q} = \frac{9.45}{1.5} = 6.3\,\text{V}$$

Electrical energy sources are produced in many types. They are broadly classified as one of the following:

1. *Voltage sources:* providing a fixed specified voltage across its terminals, independent of the load connected to it
2. *Current source:* providing a fixed specified current flow from its terminals, independent of the load connected to it

More detailed discussion of these types of sources is given in Chapter 4. It should be mentioned, however, that current sources are in general constructed from voltage sources.

Sources are also classified as dc or ac sources. Dc sources produce a constant output independent of time, while ac sources produce an output that varies continuously, in a prescribed manner, as a function of time. The source polarities of ac sources also reverse continuously every specified amount of time. The circuit symbols used for the various types of electrical energy sources are shown in Fig. 1.10. Variable-voltage laboratory power supplies provide an adjustable dc output voltage by conversion (through rectification) of the ac voltage from the power outlets. Some familiar laboratory-type power supplies are shown in Fig. 1.11.

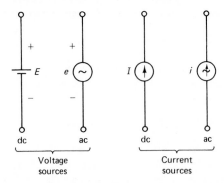

Figure 1.10

Batteries can store only a specified amount of energy. This is the amount of work provided by the chemical reaction to sustain a constant potential energy difference across the battery's terminals. Since energy and charge are linearly

Figure 1.11 Laboratory type, test equipment, and system power supplies. (Courtesy of Lambda Electronics, Division of VEECO Inc.)

related through Eq. (1.4), battery manufacturers prefer to specify batteries in terms of their charge capacity Q. When this charge capacity "dries up," the source can no longer sustain the specified voltage across its terminals. As the charge Q can be expressed, using Eq. (1.1), by

$$Q = It \quad C$$

a more practical unit for the charge capacity of batteries is in terms of current × time, or ampere-hour (abbreviated Ah). The lifetime of a battery can then easily be obtained by dividing its ampere-hour capacity by the constant amount of current that is expected to be drawn from it. Hence

$$\text{lifetime (h)} = \frac{\text{ampere-hour capacity (Ah)}}{\text{average current drawn (A)}} \quad (1.6)$$

1.5   Voltage or Electric Potential Difference

EXAMPLE 1.8

A battery can store 1.2 MJ of chemical energy. It has a terminal voltage of 6 V. If it is to deliver 4 A continuously, what is the lifetime of this battery in hours?

Solution To deliver the energy stored in it at a constant voltage of 6 V, the battery should separate a total charge of Q, where

$$V = \frac{W}{Q}$$

Therefore,

$$Q = \frac{W}{V} = \frac{1.2 \times 10^6}{6} = 0.2 \, \text{MC} = 200,000 \, \text{C}$$

But 1 C is 1 A·s = $\frac{1}{3600}$ Ah. Therefore,

$$Q = \frac{200,000}{3600} = 55.556 \, \text{Ah}$$

Now using Eq. (1.6) gives

$$\text{battery lifetime} = \frac{\text{Ah capacity}}{\text{current}} = \frac{55.556}{4} = 13.89 \, \text{h}$$

1.6 ELECTRIC POWER AND ENERGY

When a person raises a certain mass a fixed distance from the ground, he or she would feel more tired if the mass were lifted very fast than if it were raised slowly. Even though the same amount of work (or energy) is done (or spent) in both cases, in one case the time taken to do this work is short, while in the other case it is longer. The rate of spending energy is very important in all practical physical applications. This quantity is called the power. The symbol used for power is P or p.

Definition. *Power* is the rate of doing work.

The word *rate* implies "with respect to time." Thus mathematically,

$$\boxed{P = \frac{W}{t}} \tag{1.7}$$

When 1 joule of energy (or work) is spent in 1 second, a unit of power results. The unit of power is called the *watt*, abbreviated W, named in honor of James Watt, who invented the steam engine. Thus

$$1 \, \text{W} = 1 \, \text{J/s} \tag{1.7a}$$

Since work and time have been identified as influencing the definitions of electric voltage and current, power can be related to these quantities as shown below. Multiplying the numerator and denominator by Q yields

$$\boxed{P = \frac{W}{t} = \frac{W}{Q}\frac{Q}{t} = VI \qquad \text{W}} \tag{1.8}$$

Thus

$$1\,\text{W} = 1\,\text{V} \times 1\,\text{A} \tag{1.8a}$$

If the voltage and the current are functions of time, the power will also be a function of time:

$$p = vi \quad \text{W} \tag{1.9}$$

This definition of power is quite general. The power in any electrical element can be obtained by multiplying the voltage across the element by the current flowing in it.

For practical applications, the watt is usually a small unit. One practical unit of power that is still being used extensively (from the British system of units) is the *horsepower*, abbreviated hp:

$$1\,\text{hp} = 746\,\text{W}$$

Another practical unit of power used extensively for appliances and industrial machinery is the *kilowatt*, abbreviated kW:

$$1\,\text{kW} = 1000\,\text{W}$$

Also, 1 hp = 0.746 kW.

It is important to differentiate between the absorption of power (as in electric loads) and the generation or delivering of power (as in the electrical energy sources). Referring to the discussion of voltage and Fig. 1.7, when current flows in a load element from the higher potential point (+) to the lower potential point (−), energy is spent by the charge. Thus the load is absorbing power. This situation is shown in Fig. 1.12(a). This same situation is also observed when charging a source, because then the source is also absorbing power.

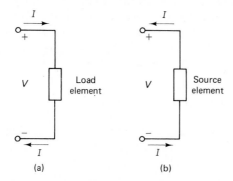

Figure 1.12 (a) Load absorbing power; (b) source delivering power.

On the other hand, when current flows in a source from the lower potential point (−) toward the higher potential point (+), the electric charges are gaining energy. Here the source is generating power or delivering power to other elements in the electric circuit. This situation is shown in Fig. 1.12(b).

EXAMPLE 1.9

What is the power delivered from a 30 V source if the current supplied by this source is 2 A (leaving its positive terminal)?

Solution   The power delivered by the source is

$$P = VI$$
$$= 30 \times 2 = 60\,\text{W}$$

EXAMPLE 1.10

What is the current flowing in a 60 W, 120 V lamp?

Solution   The lamp is a load, absorbing energy at the rate of 60 W in this example. Since

$$P = VI \qquad \text{then} \qquad I = \frac{P}{V}$$

Therefore,

$$I = \frac{60}{120} = 0.5 \, \text{A}$$

Electric energy is spent at a rate indicated by the power, according to Eq. (1.7). Electrical energy sources supply this energy to produce the potential energy difference (voltage) required to produce the current flow (transfer of charge), according to Eq. (1.4). The symbol for the electrical energy is W or w and its unit is the joule, abbreviated J. Using either of these equations, the energy can be expressed as

$$W = P \times t \quad \text{J} \tag{1.10a}$$
$$= V(I \times t) \qquad \text{[using Eq. (1.8)]}$$
$$= VQ \quad \text{J} \tag{1.10b}$$

The joule is $1 \, \text{W} \cdot \text{s}$, according to Eq. (1.7a). This unit is quite small for measuring the total energy used by households or manufacturing companies, especially over a long period of time. Electrical utilities charge for the expenditure of energy based on a much larger unit of energy called the kilowatthour, abbreviated kWh. It is the total energy spent for 1 h at the constant rate of 1 kW (i.e., 1000 W) of power. Thus

$$1 \, \text{kWh} = 1000 \, \text{W} \times 3600 \, \text{s} = 3.6 \times 10^6 \, \text{W} \cdot \text{s} = 3.6 \times 10^6 \, \text{J}$$

EXAMPLE 1.11

At 6 cents per kilowatthour, what is the cost of leaving a 60 W lamp on for 5 days?

Solution   The power (rate of using energy) is given: 60 W or 0.06 kW. Also the time is known: 5 days = $5 \times 24 = 120$ h; therefore,

$$W = P \times t = 0.06 \, \text{kW} \times 120 \, \text{h} = 7.2 \, \text{kWh}$$
$$\text{cost} = 7.2 \times 6 = 43.2 \, \text{cents}$$

Efficiency η (the Greek lowercase letter eta).   Many electrical components that convert energy from electrical to mechanical or heat form, or vice versa, for example, motors, do not perform such a conversion without some internal losses. Power is lost in overcoming friction. It is dissipated as unuseful heat emission. Thus the input energy (or power, P_{in}) is usually larger than the useful output energy (or power, P_{out}).

The efficiency is defined as

$$\eta\,(\%) = \frac{P_{out}}{P_{in}} \times 100 \tag{1.11}$$

and is usually expressed as a percentage. Obviously, it is always less than 100%. Electric motors are rated on the basis of their output power. Equation (1.11) can be used to calculate the motor's efficiency if the input power is also known, or to calculate the input power if the efficiency is known.

EXAMPLE 1.12

What is the input power of a 2 hp motor whose efficiency is 80%?

Solution   Since

$$\eta\,(\%) = 80 = \frac{P_{out}}{P_{in}} \times 100 = \frac{2 \times 746}{P_{in}} \times 100 \qquad \text{(note: 1 hp} = 746\,\text{W)}$$

then

$$P_{in} = \frac{2 \times 746}{0.8} = 1865\,\text{W}$$

The input power (1865 W) is clearly larger than the output power, which is $2 \times 746 = 1492\,\text{W}$. The amount of power lost internally in the motor is thus $1865 - 1492 = 373\,\text{W}$.

PROBLEMS

Section 1.3

(i)   **1.1.**   Find the charge in coulombs of 10^{16} electrons.

(i)   **1.2.**   What is the charge in coulombs of 5×10^{19} holes?

(i)   **1.3.**   How many electrons would constitute a charge of $-4.5\,\text{mC}$?

Section 1.4

(i)   **1.4.**   What is the current in a wire if 5×10^{20} electrons pass through in 30 s?

(i)   **1.5.**   What is the amount of charge that passes by an arbitrary point in a wire in 2.5 ms if the current was measured to be $40\,\mu\text{A}$?

(i)   **1.6.**   What is the time required to accumulate a charge of 80 mC on a plate if the current flow was observed to be 2 A?

Section 1.5

(i)   **1.7.**   What is the amount of energy gained by 10 C of charge when passing through a 9 V battery?

(i)   **1.8.**   If a battery provides 80 J of work to 16 C of charge, what is the terminal voltage of the battery?

(i)   **1.9.**   Find the amount of charge that can be separated by a 1.5 V battery if the energy produced by the chemical reaction is 7.5 J.

(ii)   **1.10.**   An element carrying a current of 10 mA dissipates 5 J of energy in 20 s. What is the voltage across the element?

(i) **1.11.** What is the power absorbed by an element if 3 C of charge passes through it in 6 s and the voltage across the element is 12 V?

(ii) **1.12.** An important energy unit in physics is the electron volt (eV). It is the energy gained or released by an electron in moving through a potential difference of 1 V. Express this unit in joules.

(i) **1.13.** What is the power absorbed by an element when the voltage across it is 12 V and the current through it is 2.5 A? What is the energy released by this element in 1 min?

(i) **1.14.** What is the energy in kilowatthours if the current flowing in a 120 V appliance is 2 A and the appliance was turned on for 30 min?

(ii) **1.15.** A 6 V battery can supply 0.2 A of current for 5 h. What is the amount of charge delivered? What is the total amount of work done by this battery?

(i) **1.16.** Find the current drawn by a 2 kW appliance connected to a 240 V source.

(ii) **1.17.** A battery is rated at 80 Ah and has a lifetime of 20 h. What is the rated current? What power can it deliver if its terminal voltage is 6 V? What is the total amount of energy stored in this battery?

(ii) **1.18.** The energy supplied by a source is 1.2 kWh when connected to a load for 15 min. If the current flow is 3 A, what is the voltage across the load?

(i) **1.19.** A 6 hp motor with a 90% efficiency is operated from a 120 V source. What is the input power to the motor and the current it draws?

(i) **1.20.** If the motor in problem 1.19 is operated for 1.8 h, what is the cost when it is operated at a charge of 7.5 cents/kWh?

GLOSSARY

Ac Source: An electric energy source that produces an output voltage or current that varies with time in a prescribed manner.

Ampere (A): The SI unit of current measurement. One ampere is the rate of flow of 1 C of charge in 1 s.

Ampere-hour (Ah): A practical unit of charge capacity of batteries. One ampere-hour is equivalent to 3600 C.

CGS System: The system of units using the centimeter, gram, and second as the fundamental units of measurement.

Conductor: A material, usually metallic, that has a very large number of free current carriers. It thus allows the flow of electric current with minimal voltage applied.

Coulomb (C): The SI unit of electric charge. It is equal to the charge carried by 6.242×10^{18} electrons.

Dc Current Source: An electric source of energy that provides a fixed current level to any electric load connected to it. Its value does not vary with time.

Dc Voltage Source: An electric source of energy that provides a fixed voltage level to any electric load connected to it. Its value does not vary with time.

Direct Current (dc): The electric current whose value does not change with time.

Efficiency: The percentage or fraction of the input power that is converted by the electric device into a useful output power. It is usually expressed as the ratio of output to input powers.

Electromagnetic Phenomenon: The interrelationship between electricity and magnetism.

Electron: The particle with the smallest negative electric charge that rotates in a specific orbit around the nucleus of an atom.

Free Electron: An electron that is unassociated with any particular atom and is therefore free to move through the crystalline lattice structure of the element.

Hole: An empty space created in the valence band of an atom due to the escape of an electron that becomes a free electron.

Horsepower (hp): A unit of measurement of power in the British system of units: 1 hp = 746 W.

Joule (J): A unit of measurement of energy (or work) in the SI system of units.

Kelvin (K): A unit of measurement of temperature in the SI system of units: K = 273 + °C.

Kilogram (kg): A unit of measurement of weight in the SI system of units: 1 kg = 1000 g.

Kilowatt (kW): A unit of measurement of power in the SI system of units: 1 kW = 1000 W.

Kilowatthour (kWh): A unit of measurement of energy in the SI system of units: 1 kWh = 3.6×10^6 J.

Meter (m): A unit of measurement of distance in the SI system of units: 1 m = 100 cm.

MKS System: The system of units using the meter, kilogram, and second as the fundamental units of measurement.

Negative Ion: An atom that has gained extra electron(s) in its valence band.

Neutron: An uncharged particle found in the nucleus of an atom.

Newton (N): A unit of measurement of force in the SI system of units.

Nucleus: The structure in the center of an atom that contains both the protons and the neutrons.

Positive Ion: An atom that has lost electron(s) from its valence band.

Potential Difference: The difference in potential energy between two points in an electric circuit or in a mechanical system.

Potential Energy: The energy possessed by a particle or a mass due to its position.

Pound (lb): A unit of measurement of mass (weight or force) in the English system of units.

Proton: The particle with the smallest positive charge found in the nucleus of an atom.

Power: The rate of doing work.

Scientific Notation: The method of expressing numerical values through the use of powers of 10.

Second (s): The unit of measurement of time in all of the systems of units.

Semiconductor: A material that is neither a good conductor nor a good insulator. It has a relatively small number of free electrons. Its conductivity is controlled by the addition of other impurity atoms to the semiconductor material.

SI System: The system of units adopted by the IEEE and the international scientific community as the standard fundamental units of measurement.

***SI Unit Prefixes*:**   The symbols (and their names) used to designate specific powers of 10 in the SI system of units.

***Valence Band*:**   The energy levels corresponding to the allowable electron orbits that are farthest away from the nucleus of an atom.

***Volt (V)*:**   The SI unit of measurement of the potential energy difference between two points in an electric circuit. If 1 J of energy is spent to move a charge of 1 C between two points, the difference in potential between the two points is said to be 1 V.

***Voltage Cell*:**   An electrical energy source that converts chemical, heat, or light energy into electrical energy.

***Watt (W)*:**   The SI unit of measurement of power. One watt is equivalent to the rate of change of energy when 1 J is spent or generated in 1 s.

2 | RESISTANCE AND OHM'S LAW

OBJECTIVES

- Understanding the physical phenomena that create and affect the electrical resistance.
- Becoming so familiar with Ohm's law and its application to the extent that it becomes "second nature."
- Recognizing the limits of maximum safe operating conditions of a resistor, and the interrelationship between power, voltage, and current.
- Understanding the graphical representation of the voltage-current relationship for linear and nonlinear elements, and the use of the power hyperbola.
- Familiarity with the various types of practical resistors and the color coding system.

When current carriers (free electrons) flow in a conducting material (a metal) under the influence of an external energy source such as a battery, they accelerate, gaining momentum and hence kinetic energy. There is a great probability that these electrons may collide with the stationary atoms forming the crystalline structure of the material. On collision, the current carriers lose their momentum, and their kinetic energy is transferred to the stationary atoms. Energy conversion thus takes place, with the kinetic energy converting into heat, as in electric stove elements, or light, as in incandescent lamps. After each collision, the electrons again accelerate, gain momentum and kinetic energy, and the process is repeated.

This collision process is the reason behind the property of the material which is exhibited as opposition to current flow. This property is called the *electrical resistance* of the material and is denoted by the symbol R.

Recent research studies have investigated the electrical properties of a class of material compounds that belong to the family of ceramics (such as lithium barium fluoride, calcium titanate, yttrium barium copper oxide, etc). Such materials can exhibit superconductivity at temperatures in the range 25 to 100 K. Superconductors have zero resistance. This phenomenon can be explained in the simplest manner through the realization that engineered flaws in the crystalline structure of the material can eliminate the atomic lattice vibrations and hence reduce the collision probability and resistance to zero. Once the results of these investigations are implemented into practical devices, it is predicted that a new evolution of electrical engineering applications of superconductors will result, parallel to that caused by the discovery and exploitation of semiconductors.

2.1.1 Factors Affecting Resistance

Every material has a specific number of free electrons that may take part in the process of charge transfer (i.e., the electric current). Also, each material could have a different crystalline structure, some more dense than others. Thus each material will oppose the flow of electric current differently. This particular characteristic of the material is referred to as the *specific resistivity* of the material and has the symbol ρ (the Greek lowercase letter rho).

The physical dimensions of the material also affect its resistance. The longer the length of the material, *l,* as shown in Fig. 2.1, the more would be the probability of collisions, hence the larger the resistance becomes. The larger the

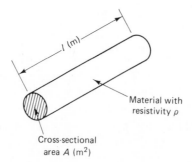

Cross-sectional area A (m^2)

Figure 2.1

cross-sectional area A of the material, the easier it is for electrons to flow through it, hence the lower the electric resistance becomes.

Thus the resistance R of a material is directly proportional to its length and inversely proportional to its cross-sectional area. The constant of proportionality is the specific resistivity of this material. In mathematical form,

$$R = \rho \frac{l}{A} \quad \Omega$$

(2.1)

The unit of resistance is the ohm and has the symbol Ω (the Greek capital letter omega). The relationship between the ohm and the basic electrical quantities defined in Chapter 1 is derived in Section 2.2.

In Eq. (2.1), l is in meters and A is in square meters; thus

$$\rho = R(\Omega) \times \frac{A(\mathrm{m}^2)}{l(\mathrm{m})}$$

has the unit $\Omega \cdot \mathrm{m}$. Table 2.1 gives the specific resistivity ρ for different materials at room temperature (20°C).

Silver is the best *conductor*; however, copper is almost as good and a lot cheaper. Almost all good-conducting wires are made of copper. Aluminum is also cheap and lightweight but is not as good a conductor as copper. A conductor made of aluminum with the same dimensions as a conductor made of copper would be 1.65 times larger in resistance. As Eq. (2.1) indicates, materials with low resistivity result in a lower value of resistance than materials with a higher value of resistivity. *Insulators* provide very high resistances and thus allow an insignificant amount of current to flow. Such materials are used to form the insulating coating of conductors (copper wires) so that they can be grouped together without current flow between them. They are also used in making capacitors, as shown in Chapter 8. Insulators are used to support power transmission lines and telephone lines.

TABLE 2.1

Material	Resistivity ($\Omega \cdot \mathrm{m}$)
Conductors	
Silver	1.6×10^{-8}
Copper	1.7×10^{-8}
Aluminum	2.8×10^{-8}
Constantan	49×10^{-8}
(copper-nickel alloy)	
Semiconductors	
Carbon	4×10^{-5}
Germanium	0.45
Silicon	2500
Insulators	
Paper	10^{10}
Mica	5×10^{11}
Glass	10^{12}

Germanium and silicon are neither good conductors nor good insulators. They are called *semiconductors*. They are used primarily in the production of diodes, transistors, and all types of integrated circuits. Their resistivity can be carefully controlled by adding the required amount of other elements, usually aluminum, gallium, phosphorous, zinc, or arsenic. Carbon is not a good conductor and it is also not useful as a semiconductor (even though it is classified as such). It is used primarily to produce commercial resistor elements.

EXAMPLE 2.1

Find the resistance at 20°C of
(a) A copper wire 1 m long with a circular cross-sectional area of diameter 2 mm.
(b) A constantan wire 2.5 m long with a circular cross-sectional area of diameter 1 mm.

Solution Using Eq. (2.1) and the resistivity data given in Table 2.1, the resistance of each wire can be calculated, provided that the area A is known.

$$\text{(a)} \quad A = \frac{\pi d^2}{4} = \frac{\pi}{4}(2 \times 10^{-3})^2 = 3.14 \times 10^{-6}\,\text{m}^2$$

Thus

$$\text{resistance of the copper wire} = \rho\frac{l}{A}$$

$$= 1.7 \times 10^{-8} \times \frac{1.0}{3.14 \times 10^{-6}}$$

$$= 5.4 \times 10^{-3}\,\Omega$$

$$\text{(b)} \quad A = \frac{\pi d^2}{4} = \frac{\pi}{4}(1 \times 10^{-3})^2 = 7.85 \times 10^{-7}\,\text{m}^2$$

and

$$\text{resistance of the constantan wire} = \rho\frac{l}{A}$$

$$= 49 \times 10^{-8} \times \frac{2.5}{7.85 \times 10^{-7}}$$

$$= 1.56\,\Omega$$

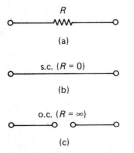

Figure 2.2 (a) Resistor; (b) conducting or short-circuit wire; (c) open-circuit or open switch.

In circuit diagrams, a resistive element is represented symbolically as shown in Fig. 2.2(a). The zigzag line emphasizes opposition to current flow. Two special cases of resistance occur quite often in electic circuits. A short circuit (s.c.) is the case corresponding to $R = 0\,\Omega$. Symbolically, this situation corresponds to a conducting wire, as shown in Fig. 2.2(b). The resistance of a short copper wire (see Example 2.1) is very small and can easily be approximated to zero. An open circuit (o.c.) is the case corresponding to $R = \infty\,\Omega$. Symbolically, this situation corresponds to "no path" between the circuit points. It is represented as an open switch (i.e., no connection) between the points, as shown in Fig. 2.2(c).

2.1.2 Conductance

Conductance is the ability of a material to allow electric current to flow. It is therefore just the opposite of resistance; in other words, conductance is the inverse of resistance. The symbol for conductance is G. The SI unit for conductance is the siemens, abbreviated S. Another widely accepted unit of conductance is the *mho*, which is "ohm" spelled backward and has the symbol ℧ (inverted omega). Thus

$$G = \frac{1}{R} \quad \text{S or } ℧ \tag{2.2}$$

From Eq. (2.1),

$$G = \frac{1}{\rho} \frac{A}{l} = \sigma \frac{A}{l} \quad \text{S} \tag{2.3}$$

where

$$\sigma = \text{specific conductivity} = \frac{1}{\rho} \quad \text{S/m} \tag{2.3a}$$

Since ρ has the unit $\Omega \cdot m$, σ has the unit $\Omega^{-1} \cdot m^{-1} = S/m$, as specified above.

2.1.3 Effect of Temperature on Resistance

As the temperature of a metallic conductor is increased, the parent atoms constituting the crystalline structure of the material will acquire a larger amount of kinetic energy, exhibited in the form of vibrations around their average positions (called *phonons* of energy). This increase in the vibration activity of the parent atoms makes the collision process of the current carriers more probable. Thus in general, the resistance of a metallic conductor (copper, aluminum, etc.) increases with a rise in temperature.

For poor conductors (e.g., carbon) and semiconductors (e.g., germanium, silicon), a rise in temperature usually results in the liberation of more electrons from their parent atoms as these electrons acquire sufficient amounts of energy to enable them to overcome the binding force with the nucleus of the atom. An increase in the number of available current carriers means that the material is becoming more conductive. This can more than make up for the increase in collisions between the current carriers and the vibrating atoms. The overall result is that a rise in temperature would make these materials more conductive (i.e., less resistive). In other words, the resistance decreases with temperature.

In some materials (e.g., constantan), the two processes described above may be in approximate balance. This means that the possible increase in resistance due to the increase in the collisions between the current carriers and the vibrating atoms cancels out with the possible increase in conductivity due to the liberation of more current carriers. The resistance of such materials is therefore minimally affected by changes in temperature.

A typical graphical representation of the variation of the resistance (R) of a metallic wire with the corresponding variations in temperature (T) is shown in Fig. 2.3. The resistance of any material is zero at the absolute zero temperature ($0\,\text{K} = -273°\text{C}$). If the approximately linear portion of the behavior curve is extended (the dashed line), it will intersect with the temperature axis at a point called the *inferred zero temperature*, denoted by T_0 (which is numerically a negative number).

The approximately linear portion of the behavior curve can be described mathematically as follows:

$$\text{slope} = \frac{R_2}{T_2 - T_0} = \frac{R_1}{T_1 - T_0}$$

Therefore,

$$R_2 = R_1 \frac{T_2 - T_0}{T_1 - T_0} = R_1 \frac{T_1 - T_0 + T_2 - T_1}{T_1 - T_0} \tag{2.4a}$$

$$= R_1[1 + \alpha(T_2 - T_1)] \tag{2.4b}$$

where

$$\alpha = \frac{1}{T_1 - T_0}$$

$$= \text{temperature coefficient of resistance, } °\text{C}^{-1} \tag{2.4c}$$

Note that in the numerator of Eq. (2.4a), the term $T_1 - T_1(=0)$ is added, having no effect on the ratio expression. Usually, T_1 is taken to be the room temperature (i.e., 20°C), while T_2 could be any other temperature at which the value of the resistance, R_2, is to be determined. The temperature coefficient of resistance, α, is positive for metal conductors, resulting in an increase in the resistance as the temperature rises. α is negative for poor conductors, resulting in a decrease in the resistance as the temperature rises. When α is very small (almost zero), there is minimal change in the value of the resistance as the temperature changes. Table 2.2 indicates typical values of α (at room temperature) for different materials.

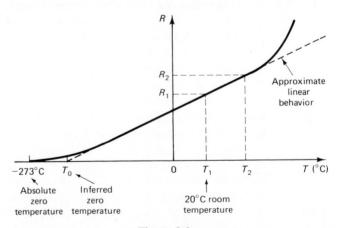

Figure 2.3

TABLE 2.2 Temperature Coefficient of Resistance (α) at 20°C

Material	α (°C^{-1})
Silver	0.003 8
Copper	0.003 93
Aluminum	0.003 91
Tungsten	0.005
Carbon	−0.000 5
Constantan	0.000 008

■ EXAMPLE 2.2

Find the resistance of a tungsten filament at 200°C if it has a 10 Ω resistance at room temperature (20°C).

Solution From Table 2.2 the temperature coefficient of resistance (α) for tungsten is 0.005. Applying Eq. (2.4b) and noting that T_1 is 20°C and T_2 is 200°C, we have

$$R_{200°C} = R_{20°C}[1 + \alpha(200° - 20°)]$$
$$= 10(1 + 0.005 \times 180) = 19\ \Omega$$

2.2 OHM'S LAW

The transfer of current carriers through a resistive element necessitates the expenditure of energy, in the form of work done by an external electrical energy source such as a battery, to overcome the opposition that the resistive element exhibits due to the collision mechanism discussed in Section 2.1. This implies that the higher the value of the resistance (R), the more work has to be done. As work per unit charge is the voltage (V), then:

$$V \propto R \tag{2.5}$$

that is, voltage is directly proportional to the value of the resistance.

Also, for a specific value of a given resistive element, the higher the voltage, the more the kinetic energy of the charges being transferred through this element. This reflects as faster charge transfer. Since the amount of charge transferred per unit time is the electric current (I), then

$$V \propto I \tag{2.6}$$

that is, voltage is directly proportional to the value of the resulting current flow. Combining relations (2.5) and (2.6) gives

$$V \propto RI$$

Thus

$$V = kRI$$

Since the unit of R has not yet been defined specifically, the constant of proportionality k in the equation above can be combined with R to define the resistance value, or it can be taken as unity without any loss of generality. Hence

$$V = RI$$ (2.7a)

This relationship was discovered by the German physicist Georg Ohm and it is referred to as *Ohm's law*. It is a basic law in electric circuits. It can also be written in any one of the following forms:

$$R = \frac{V}{I}$$ (2.7b)

$$I = \frac{V}{R}$$ (2.7c)

or in terms of the conductance G (=1/R), Eq. (2.7c) can also be expressed as

$$I = GV$$ (2.8)

From Eq. (2.7b) it is clear that the unit of the resistance (R), which was called the ohm (Ω) in Section 2.1, is in fact a derived unit, related to the voltage and current units through

$$R(\Omega) = \frac{V(V)}{I(A)}$$

that is,

$$1\,\Omega = 1\,V/A$$ (2.9)

All the forms of Ohm's law in Eqs. (2.7) and (2.8) are important and are suitably applied in different situations in circuit analysis.

A resistive element (resistance) is an electrical load. The conventional associated reference directions of current flow and voltage drop for an electrical resistance are indicated in Fig. 2.4. Current flows from the higher potential terminal toward the lower potential terminal. This conforms with the associated reference directions of current and voltage, developed in Chapter 1.

Refer to Fig. 2.2 and use Ohm's law:

1. For $R = 0$, that is, a short circuit or connecting wire, with any finite amount of current flow,

$$V = IR = I \times 0 = 0\,V$$ (2.10a)

2. For $R = \infty$, that is, an open circuit or open switch with any finite voltage across the terminals,

$$I = \frac{V}{R} = \frac{V}{\infty} = 0\,A$$ (2.10b)

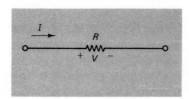

Figure 2.4

To measure the current flowing in a resistance, an ammeter is connected to the resistance as shown in Fig. 2.5. The ammeter monitors the current flowing in it, which is the same as the current flowing in R if the voltmeter is not connected or if the voltmeter is ideal (i.e., $R_V = \infty$) because then no current will be flowing in the voltmeter! The ammeter is a polarized instrument. This means that the current must enter the instrument from the positive side (marked +) in order to obtain an upscale deflection of the pointer. An ideal ammeter has zero resistance (i.e., $R_A = 0$). The voltage across the resistance is measured by connecting a voltmeter across the terminals of the resistive element as shown in Fig. 2.5. The voltmeter monitors the voltage across its own terminals, which is the same as the voltage across the resistance. It is also polarized. Thus its positive terminal must be connected to the higher potential point of the resistance (marked +) in order to obtain an upscale deflection of the pointer. An ideal voltmeter has an infinite resistance (i.e., $R_V = \infty$). The value of the resistance R can then easily be obtained by applying Ohm's law [Eq. (2.7b)]:

$$R = \frac{V}{I} = \frac{\text{voltmeter reading}}{\text{ammeter reading}}$$

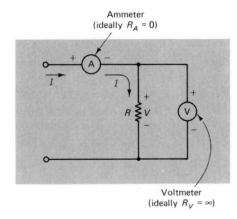

Ammeter
(ideally $R_A = 0$)

Voltmeter
(ideally $R_V = \infty$)   **Figure 2.5**

EXAMPLE 2.3

What is the resistance of a lamp if a 6 V battery results in a 100 mA current flow in it?

Solution From Ohm's law,

$$R = \frac{V}{I} = \frac{6}{100 \times 10^{-3}} = 60\ \Omega$$

EXAMPLE 2.4

What is the voltage drop across a 560 Ω resistance when the current flowing in it is 20 mA?

Solution From Ohm's law,

$$V = IR = 20 \times 10^{-3} \times 560 = 11.2\ \text{V}$$

EXAMPLE 2.5

If the voltage across a $2\,k\Omega$ resistor is $8\,V$, what is the current flowing in the resistor?

Solution From Ohm's law,

$$I = \frac{V}{R} = \frac{8}{2 \times 10^3} = 4\,mA$$

2.3 POWER DISSIPATED (OR ABSORBED) IN A RESISTANCE

The external energy source provides work to the current carriers to overcome the resistance of the resistive element. This work is converted through the collision mechanism into other forms of energy, usually heat (as in physical resistors, stove elements, or heaters) or light (as in lamps).

Broadly speaking, resistors can be used for one of the following functions:

1. To limit or control the amount of current flow
2. To achieve energy conversion from electrical to heat, light, or mechanical energies.
3. To connect electrical components [i.e., connecting wires; here the resistive elements (wires) should have very small resistance, ideally of zero value].

As discussed in Chapter 1, the rate of energy absorption is the electrical power (P) dissipated in a resistance (R). Power is defined as

$$P = VI \quad W \tag{2.11}$$

where V is the voltage (in volts) across the terminals of the resistor and I is the current (in amperes) flowing in this resistor.

From Ohm's law, $I = V/R$. Thus substituting in Eq. (2.11), we have

$$P = V\frac{V}{R} = \frac{V^2}{R} \quad W \tag{2.11a}$$

Also from Ohm's law, $V = IR$. Thus substituting in Eq. (2.11) yields

$$P = IRI = I^2R \quad W \tag{2.11b}$$

All three forms of power in Eq. (2.11) are equivalent. Any one may be more useful in certain situations, as will be noted throughout.

Resistors have to dissipate the heat energy generated internally due to the collision mechanism. The larger the physical size of the resistor, the more surface area it has and could therefore dissipate the heat energy faster than a smaller size resistor. In general, practical resistors cannot dissipate the heat energy generated faster than a certain maximum rate. Resistors are therefore classified according to their maximum power rating, referred to as P_{max}. Once the power dissipated in the resistor exceeds this value, it usually burns out. Knowing the value of a resistor and its power rating, it is easy to deduce the maximum possible voltage and current that can be safely applied to it without destruction, using Eq. (2.11a) and (2.11b), respectively:

$$V_{\text{max}} \text{ (maximum safe voltage across } R) = \sqrt{P_{\text{max}}R} \qquad (2.12)$$

and

$$I_{\text{max}} \text{ (maximum safe current through } R) = \sqrt{\frac{P_{\text{max}}}{R}} \qquad (2.13)$$

Note that

$$\frac{V_{\text{max}}}{I_{\text{max}}} = \sqrt{P_{\text{max}}R}\sqrt{\frac{R}{P_{\text{max}}}} = R$$

that is, Ohm's law is satisfied at this limiting condition.

The energy dissipated in any resistance, or the equivalent work done by the externally applied electrical source, is

$$W = P \times t = V \times I \times t \quad \text{J} \qquad (2.14)$$

EXAMPLE 2.6

Calculate the power dissipated in each of the resistors in Examples 2.3, 2.4, and 2.5.

Solution In Example 2.3:

$$P = VI = 6 \times 100 \times 10^{-3} = 0.6 \text{ W}$$

$$\left(\text{also} = I^2R = \frac{V^2}{R} \text{ with } R = 60 \, \Omega \right)$$

In Example 2.4:

$$P = I^2R = (20 \times 10^{-3})^2 \times 560 = 0.224 \text{ W}$$

$$\left(\text{also} = \frac{V^2}{R} = VI \text{ with } V = 11.2 \text{ V} \right)$$

In Example 2.5:

$$P = \frac{V^2}{R} = \frac{8 \times 8}{2000} = 0.032 \text{ W} = 32 \text{ mW}$$

$$(\text{also} = I^2R = VI \text{ with } I = 4 \text{ mA})$$

EXAMPLE 2.7

A 2 W, 5 kΩ resistor is available. Find the maximum safe voltage and current that can be applied to this resistor without damage.

Solution Since

$$P_{\text{max}} = \frac{V^2_{\text{max}}}{R}$$

then

$$V_{max} = \sqrt{P_{max}R} = \sqrt{2 \times 5000} = 100\,\text{V}$$

Also,

$$P_{max} = I_{max}^2 R$$

Therefore,

$$I_{max} = \sqrt{\frac{P_{max}}{R}} = \sqrt{\frac{2}{5000}} = 0.02\,\text{A} = 20\,\text{mA}$$

Note that

$$\frac{V_{max}}{I_{max}} = \frac{100}{0.02} = 5000\,\Omega = 5\,\text{k}\Omega$$

satisfying Ohm's law.

EXAMPLE 2.8

Find the energy used by a 20 Ω heater element operating from a 120 V source for 2 h.

Solution

$$\text{Heater power} = \frac{V^2}{R} = \frac{120 \times 120}{20} = 720\,\text{W} = 0.72\,\text{kW}$$

Therefore,

$$\begin{aligned}
\text{energy used by the heater} &= P \times t = 0.72 \times 2 = 1.44\,\text{kWh} \\
&= 720 \times 2 \times 60 \times 60 \\
&= 5.184 \times 10^6\,\text{J}
\end{aligned}$$

EXAMPLE 2.9

Find the resistance of each of the following elements.
(a) When a 10 V source is applied to the element, it dissipates 0.25 W.
(b) The element dissipates 50 mW when the current flowing in it is 5 mA.

Solution (a) As $P = V^2/R$, then

$$R = \frac{V^2}{P} = \frac{10 \times 10}{0.25} = 400\,\Omega$$

(b) As $P = I^2R$, then

$$R = \frac{P}{I^2} = \frac{50 \times 10^{-3}}{(5 \times 10^{-3})^2} = 2\,\text{k}\Omega$$

Here

$$\begin{aligned}
V = IR \quad &\text{(from Ohm's law)} \\
&= 5 \times 10^{-3} \times 2 \times 10^3 = 10\,\text{V}
\end{aligned}$$

and

$$VI = 10 \times 5 \times 10^{-3} = 50\,\text{mW} \quad \text{(check)}$$

2.4 V–I (OR I–V) CHARACTERISTICS OF RESISTORS

When a resistor obeys Ohm's law,

$$V = RI$$

it is called a *linear resistor*. This is because the plot of V versus I is a straight line, as shown in Fig. 2.6. The equation of a straight line is

$$y = mx$$

Therefore, with the V variable represented by the vertical axis y, and the I variable represented by the horizontal axis x, the V–I characteristics of a linear resistor are represented (as shown in Fig. 2.6) by a straight line passing through the origin, with a slope m equal to the value of the resistance R. The higher the value of the resistance, the steeper would be the slope.

Ohm's law for such linear resistors can also be written in the form

$$I = GV = \frac{1}{R}V$$

Therefore, if I is represented as the vertical variable y and V is represented as the horizontal variable x, the equation above resembles

$$y = mx$$

which again is the equation of a straight line passing through the origin, with a slope m equal to the value of $G(=1/R)$. The I–V characteristics of linear resistors are as shown in Fig. 2.7. Here the higher the resistance, the smaller would be the slope, as indicated in the figure.

The power dissipated in any resistance is

$$P = VI$$

By choosing some suitable values of V (within the axis range) and calculating the corresponding values of I such that the product VI is a constant, equal to a certain power dissipation, the resulting set of (V, I) or (I, V) points when represented in Figs. 2.6 or 2.7 and connected by a curve corresponds to what is known as the

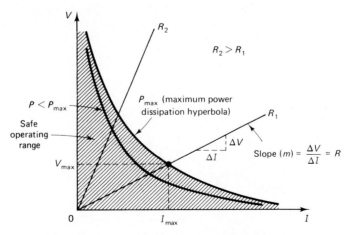

Figure 2.6   V–I characteristics of linear resistors.

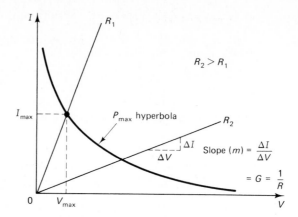

Figure 2.7 *I–V* characteristics of linear resistors.

power hyperbola. When the value of P is replaced by P_{max}, the maximum power dissipation of a resistor (or any element in general), the resulting curve will obviously represent the maximum power dissipation hyperbola, as shown in Figs. 2.6 and 2.7. This curve sets the limits for maximum safe operation. The V–I (or I–V) segment of the characteristics, to the left of the P_{max} hyperbola, corresponds to safe operating conditions. The segment to the right of the P_{max} hyperbola should not really exist because there the resistor is physically damaged (burned). The point of intersection of the maximum power dissipation hyperbola and the V–I (or I–V) characteristics of the element determines the values of the maximum safe voltage (V_{max}) and maximum safe current (I_{max}) allowable for such an element.

Some elements (e.g., diodes, lamps) do not obey Ohm's law. Their V–I or I–V characteristics can be represented by an equation of the form

$$I = kV^n + b \qquad (2.15)$$

where n is usually not equal to 1. The constant b may or may not be zero. Such electrical components are called *nonlinear resistors.* Their characteristics are usually curves, as shown in Fig. 2.8, not straight lines. Here the meaning of the term "resistance" is not related directly to any of the parameters in Eq. (2.15). At any specific operating point (a fixed value of V or I) two definitions of

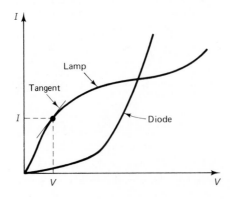

Figure 2.8 *I–V* characteristics of nonlinear elements.

Chapter 2   Resistance and Ohm's Law

resistance can be derived:

$$R_{dc} = \text{dc resistance}$$

$$= \text{static resistance} = \frac{V}{I} \tag{2.16a}$$

and

$$r_{ac} = \text{ac resistance}$$

$$= \text{dynamic resistance} = \frac{dV}{dI} \tag{2.16b}$$

$$= \text{slope of the tangent to the } V\!-\!I \text{ curve at the operating point}$$

The values resulting from these two definitions of resistance are not constant but are different at different operating points. They are both useful in different situations, as will become apparent through a study of electronic components and circuits.

■ EXAMPLE 2.10

The $I\!-\!V$ characteristics shown in Fig. 2.9 correspond to a resistance of value R. Draw on the same graph the $I\!-\!V$ characteristics for a resistance of value
(a) $2R$.
(b) $\frac{1}{2}R$.

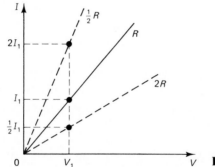

Figure 2.9

Solution A line can be determined completely if two points on the line are known. The $I\!-\!V$ characteristics of a linear resistor is a straight line passing through the origin. Therefore, only one more point has to be known. For the characteristics of the given resistor R, assume that a voltage V_1 is applied to it. This will result in a current flow I_1, as shown in Fig. 2.9, where

$$\frac{V_1}{I_1} = R \qquad \text{or} \qquad I_1 = \frac{V_1}{R}$$

(a) Now allowing the same voltage V_1 to exist across the $2R$ resistor, the current will be

$$\frac{V_1}{2R} = \tfrac{1}{2}I_1$$

Thus the point $(V_1, \tfrac{1}{2}I_1)$ is a point on the linear characteristics, and the line representing the resistor $2R$ can be drawn.

(b) Similarly, allowing the same voltage, V_1, to exist across the $\frac{1}{2}R$ resistor, the current will be

$$\frac{V_1}{\frac{1}{2}R} = 2\frac{V_1}{R} = 2I_1$$

Therefore, the point $(V_1, 2I_1)$ is a point on the linear characteristics, and the line representing the resistor $\frac{1}{2}R$ can be drawn. ∎

2.5 PRACTICAL RESISTORS

There are many types, sizes, and shapes of practical resistors. Some physical resistors are shown in Fig. 2.10. Resistors are basically specified by

1. Their value, which is either printed or color coded on the cylindrical body of the resistor. Resistors range in value from a fraction of an ohm to many megohms.
2. Their power rating or wattage, which is usually printed on the resistor body or identified by the size of standard resistors. The power rating ranges from a fraction of a watt to many tens of watts.

Resistors are either fixed or variable. Of the *fixed resistors,* the most common type is the *carbon-composition resistor,* shown schematically in Fig. 2.11. The values of these resistors depend on the relative proportion of the carbon granules and the insulating powder material, which are heat pressed to produce the cylindrically shaped body of these resistors. Wire leads are embedded in the ends and the resistor's body is coated with insulating material. The larger these carbon resistors are, the higher is the power rating. Standard power ratings for this type of resistors are $\frac{1}{8}$, $\frac{1}{4}$, $\frac{1}{2}$, 1, and 2 W.

Carbon-film and *metal-film resistors* are two other important types of fixed resistors. These resistors are made by depositing a thin layer of carbon or metal on the surface of a ceramic insulating rod. After attaching the end leads, the body of the resistor is covered with an insulating coating. Metal-film resistors are much more accurate, stable, and less noisy than carbon-type resistors. They are, however, more expensive. By the way, *noise* is any undesired electrical signal. Almost all electrical components produce noise.

Thin- and *thick-film resistors* are produced in a similar manner. Here resistive inks are deposited as a film on a ceramic surface and shaped as spiral or zigzagging lines. Integrated-circuit resistors are produced from semiconductor materials. They are integral components of the chip, which may include a large number of other electronic components, such as diodes and transistors.

Another important type of fixed resistor is the *wire-wound resistor*. These are produced by winding an accurately dimensioned metal wire on a ceramic core and then coating it with an insulating jacket. These resistors have many advantages. They can dissipate more power than the other types and hence have the highest power ratings, up to many tens of watts. They can be made very precise, with an accuracy tolerance of better than 1%, sometimes as low as 0.01%. They are much more stable with variation of temperature than are other

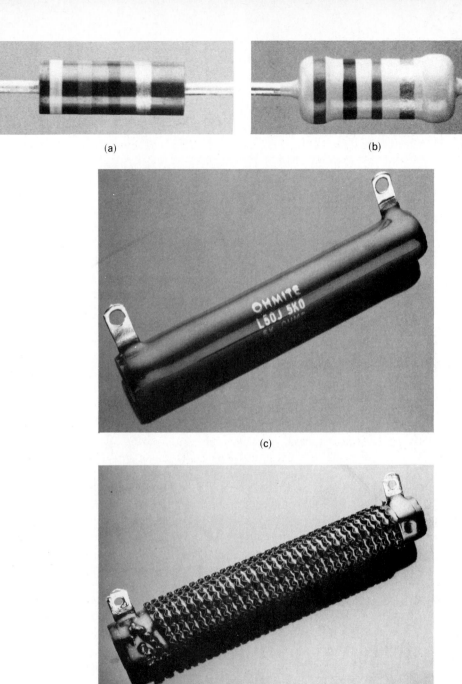

(a)

(b)

(c)

(d)

Figure 2.10 (a) Carbon-composition resistor; (b) carbon-film resistor; (c), (d) wire-wound resistors; (e) chip resistors; (f) metal-film resistors. [(a)–(d) Courtesy of Ohmite Manufacturing Co.; (e), (f) courtesy of Allen-Bradley Co.]

(e)

(f)

Figure 2.10 (*continued*)

types. Obviously, they are also more expensive than the common carbon-composition resistors.

Variable resistors are constructed in a manner similar to that of the wire-wound resistors. However, in the variable construction, the wound wire is not insulated. A movable contact arm makes a wiping contact with the bare wire. Figure 2.12 shows some typical variable resistors. The circuit symbols for these components are shown in Fig. 2.13. Usually, variable resistors have three

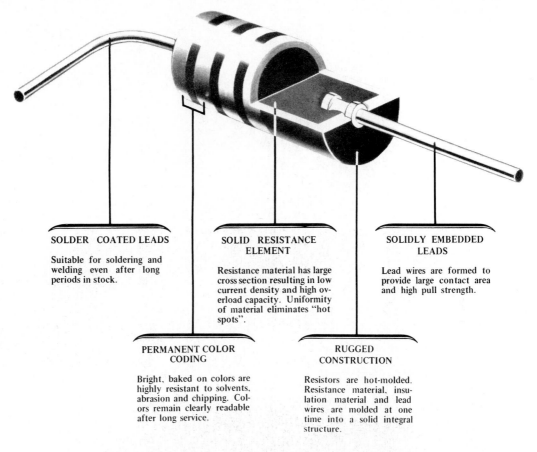

SOLDER COATED LEADS

Suitable for soldering and welding even after long periods in stock.

SOLID RESISTANCE ELEMENT

Resistance material has large cross section resulting in low current density and high overload capacity. Uniformity of material eliminates "hot spots".

SOLIDLY EMBEDDED LEADS

Lead wires are formed to provide large contact area and high pull strength.

PERMANENT COLOR CODING

Bright, baked on colors are highly resistant to solvents, abrasion and chipping. Colors remain clearly readable after long service.

RUGGED CONSTRUCTION

Resistors are hot-molded. Resistance material, insulation material and lead wires are molded at one time into a solid integral structure.

Figure 2.11 Carbon-composition resistor (hot-molded). (Courtesy of Allen-Bradley Co.)

terminals. The wire is wound between terminals a and c (i.e., the resistance between points a and c is the resistance of the whole wire). The movable contact is terminated at point b. Depending on the position of this movable arm, the resistance between points a and b will be variable, an adjustable fraction k (0.0 to 1.0) of R. When used in this manner, the variable resistor is called a *rheostat*. The same component can be used to construct a variable-voltage source when connected to a fixed voltage source, E, as shown in Fig. 2.14. The output voltage between points a and b is a fraction of E, kE, where k is adjustable depending on the position of the movable contact arm b. The element is then called a *potentiometer* or *pot*. It is an example of voltage dividers, discussed in Chapters 3 and 4.

Other types of variable resistors, such as *decade resistor boxes,* could have any required value, say between 0.0 and 999,999 Ω by selecting the positions of dial contact switches. Each of the different powers of 10 dials can be adjusted independently. The box usually contains standard resistors, which are connected based on the position of the appropriate dial switch.

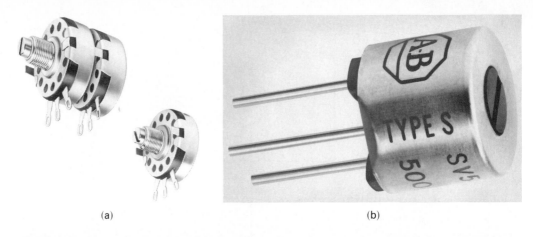

(a)

(b)

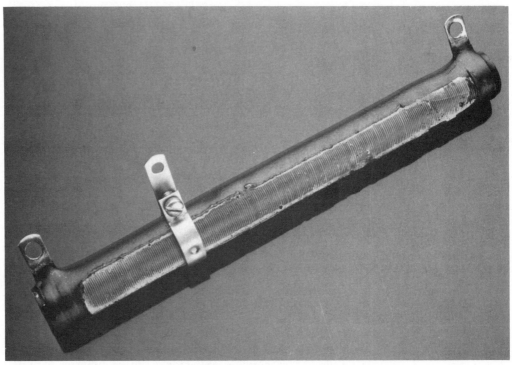

(c)

Figure 2.12 (a), (b) Panel-type variable resistors (potentiometers); (c) adjustable resistor. [(a), (b) Courtesy of Allen-Bradley Co.; (c) courtesy of Ohmite Manufacturing Co.]

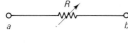

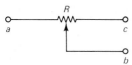

Figure 2.13 Rheostat.

2.5.1 Standard Resistance Values and Tolerance

The number of different-value resistors that are commonly manufactured are limited. Table 2.3 shows the standard values that have been agreed to. Note that in each column the two significant digits of the numerical values of resistors are the same. The difference between the columns is the power-of-10 multiplier.

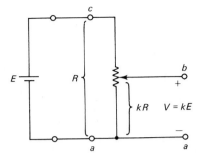

Figure 2.14 Potentiometer.

These tabulated values are called the *nominal values* since the actual values of the resistors are probably not the same but very close to these nominal values. The variations about these nominal values are called the *tolerances* and are usually expressed in percentages. For carbon-composition resistors, the common tolerances are 5, 10, and 20%, indicating that the actual value of the resistor can vary from the nominal value by ±5, ±10, and ±20%. For example, a 680 Ω, 5% resistor would vary from the nominal value by up to $680 \times 0.05 = 34$ Ω. Thus the actual value of such a resistor could be anywhere from $680 - 34 = 646$ Ω to $680 + 34 = 714$ Ω. Similarly, one can calculate the range of any other resistor value for the tolerance specified.

TABLE 2.3 Standard Values of Commercially Available Resistors: 5% Tolerance[a]

Ohms (Ω)					Kilohms (kΩ)		Megohms (MΩ)	
0.10	**1.0**	**10**	**100**	**1000**	**10**	**100**	**1.0**	**10.0**
0.11	1.1	11	110	1100	11	110	1.1	11.0
0.12	**1.2**	**12**	**120**	**1200**	**12**	**120**	**1.2**	**12.0**
0.13	1.3	13	130	1300	13	130	1.3	13.0
0.15	**1.5**	**15**	**150**	**1500**	**15**	**150**	**1.5**	**15.0**
0.16	1.6	16	160	1600	16	160	1.6	16.0
0.18	**1.8**	**18**	**180**	**1800**	**18**	**180**	**1.8**	**18.0**
0.20	2.0	20	200	2000	20	200	2.0	20.0
0.22	**2.2**	**22**	**220**	**2200**	**22**	**220**	**2.2**	**22.0**
0.24	2.4	24	240	2400	24	240	2.4	
0.27	**2.7**	**27**	**270**	**2700**	**27**	**270**	**2.7**	
0.30	3.0	30	300	3000	30	300	3.0	
0.33	**3.3**	**33**	**330**	**3300**	**33**	**330**	**3.3**	
0.36	3.6	36	360	3600	36	360	3.6	
0.39	**3.9**	**39**	**390**	**3900**	**39**	**390**	**3.9**	
0.43	4.3	43	430	4300	43	430	4.3	
0.47	**4.7**	**47**	**470**	**4700**	**47**	**470**	**4.7**	
0.51	5.1	51	510	5100	51	510	5.1	
0.56	**5.6**	**56**	**560**	**5600**	**56**	**560**	**5.6**	
0.62	6.2	62	620	6200	62	620	6.2	
0.68	**6.8**	**68**	**680**	**6800**	**68**	**680**	**6.8**	
0.75	7.5	75	750	7500	75	750	7.5	
0.82	**8.2**	**82**	**820**	**8200**	**82**	**820**	**8.2**	
0.91	9.1	91	910	9100	91	910	9.1	

[a] **Boldface** figures are 10% values.

Table 2.3 shows the values of the 5% tolerance standard resistors. Each second value in the table (boldface figures) represents those standard resistors available in 10% tolerance. Also, each fourth value in the table represents those standard resistors available in 20% tolerance (e.g., 10, 15, 22, 33, 47, 68, 100, etc.). For example, a 560 Ω resistor could be bought either as a 5% or a 10% tolerance component, but not as a 20% tolerance component. Clearly, the lower the tolerance specification, the more expensive the component would be because of the selection procedures required.

2.5.2   Color Coding

When the body of the physical resistor is not sufficiently large to print the numerical value of the resistor on it, it is usually color coded. This is common for carbon-composition resistors and some smaller wire-wound resistors.

Three or four color bands are printed on the body of the resistor, with the first band closer to one end of the resistor. The fourth band, whether present or not, specifies the tolerance of the resistor. If this band is gold, it specifies 5% tolerance; silver specifies 10% tolerance; if no band is present, the tolerance is 20%.

The colors of the first and second bands specify the first and second digits of the numerical value of the resistor, while the color of the third band specifies the power-of-10 multiplier. The color bands are always read left to right starting from the end that has the band closest to it. Table 2.4 indicates the color coding used.

TABLE 2.4   Color Coding for Carbon-Composition Resistors

Color	First-band digit	Second-band digit	Third-band multiplier
Black	0	0	$10^0 = 1$
Brown	1	1	$10^1 = 10$
Red	2	2	$10^2 = 100$
Orange	3	3	$10^3 = 1000$
Yellow	4	4	$10^4 = 10,000$
Green	5	5	$10^5 = 100,000$
Blue	6	6	$10^6 = 1,000,000$
Violet	7	7	$10^7 = 10,000,000$
Gray	8	8	$10^8 = 100,000,000$
White	9	9	$10^9 = 1,000,000,000$
Gold			0.1
Silver			0.01

EXAMPLE 2.11

What is the nominal value and tolerance of a resistor having bands colored in the order orange, white, yellow, and silver?

Solution The band colors are matched with the corresponding digits as follows:

orange   white   yellow (power-of-10 multiplier)   silver (tolerance)
  3   9 10^4 10%

Thus the element has a value of $39 \times 10^4 = 390 \, k\Omega$, with a 10% tolerance.

EXAMPLE 2.12

A resistor has three colored bands only. They are in the order brown, green, and gold. What is the value and tolerance of this resistor?

Solution Since the fourth band is not present, the element must have 20% tolerance. Matching the color bands with the corresponding digits:

brown   green   gold (power-of-10 multiplier)
  1   5   0.1

Thus the element value = $15 \times 0.1 = 1.5 \, \Omega$

PROBLEMS

$4.329 \text{ of } \times 10^{-13}$

Section 2.1

(i) **2.1.** Find the resistance of a 120 m copper wire which has a circular cross section with a diameter of 0.25 cm. Assume that this conductor is at room temperature.

(i) **2.2.** A 30 m constantan wire at room temperature has a circular cross section with a diameter of 0.1 cm. What is its resistance?

(i) **2.3.** A carbon rod has a length of 2 cm. What should be its diameter to result in a conductance of 0.001 S at room temperature?

(i) **2.4.** What is the length of a 10 Ω aluminum wire at 20°C if its diameter is known to be 0.15 cm?

(i) **2.5.** Calculate the resistance of each of the elements in Problems 2.1–2.4 at
  (a) 150°C.
  (b) −10°C.
  Consider room temperature to be 20°C.

(i) **2.6.** A certain resistance was measured to be 30 Ω at 20°C and 40 Ω at 95°C. Find the temperature coefficient of this resistance and its inferred zero temperature.

Sections 2.2 and 2.3

(i) **2.7.** The current flowing in a 15 Ω resistor is 20 mA. What is the voltage drop across this resistance and the power dissipated in it?

(i) **2.8.** The voltage across a 0.002 S conductance is 50 V. What is the current flowing in this element? What is the value of the resistance of this element? How much power is dissipated in it?

(i) **2.9.** What is the resistance of an element that allows 17 mA of current to flow in it when a 6 V battery is connected across its terminals? Calculate the power dissipated in this element.

(i) **2.10.** What is the voltage of a source that produces 0.1 A of current flow in a 40 Ω resistor? What will be the current if this source is applied to a 4 kΩ resistor?

(i) **2.11.** Find the value of the current and the power in a 10 kΩ resistor if the voltage of the source applied to it is 12 V.

(i) **2.12.** Find the current in a 20 kΩ resistor if the power dissipated in it is 200 mW. What is the voltage across this resistor?

(i) **2.13.** The power dissipated in a 7.5 kΩ resistor is 3 W. What is the voltage across this resistor? What is the current through this element?

(i) **2.14.** Find the maximum safe current and voltage for each of the following elements.
   (a) A resistor of value 200 Ω and maximum power rating 0.5 W.
   (b) A resistor of value 1 kΩ and maximum power rating 0.225 W.
   (c) A conductance of value 0.02 S whose maximum power rating is 0.25 W.

(i) **2.15.** What is the value of the current flowing in a 100 W lamp if the applied source has a voltage of 125 V? What is the resistance of this lamp?

(i) **2.16.** Find the resistance of and the current flowing in a 2 kW heater when the voltage of the applied source is 110 V.

(ii) **2.17.** How long should a 75 Ω heater be plugged in to a 120 V source to produce 15 kJ of heat energy?

(ii) **2.18.** Complete the following table for each of the components indicated.

$R(\Omega)$	$G(S)$	$V(V)$	$I(A)$	$P(W)$
		25		1.0
	0.025		0.4	
300				0.75
		100	0.02	

(ii) **2.19.** Given the four resistors
   (1) 100 Ω, 0.25 W rating   (2) 1000 Ω, 0.1 W rating
   (3)   25 Ω, 2 W rating   (4) 500 Ω, 0.125 W rating
   which element has the maximum safe voltage and which has the maximum safe current?

(i) **2.20.** In Problem 2.19 find the element that has the least safe voltage and the element that has the least safe current.

Section 2.4

(i) **2.21.** Draw the I–V characteristics of each of the following resistors on the same graph.
   (a) A 200 Ω resistor.
   (b) A 500 Ω resistor.
   Also draw the maximum power dissipation hyperbola corresponding to $P_{max} = 0.15$ W. From your graph find the maximum permissible voltage and current for each element.

Section 2.5

(i) **2.22.** Find the resistance range of
   (a) A 120 Ω 10% tolerance component.
   (b) A 2.7 kΩ 5% tolerance component.

(ii) **2.23.** The two resistors in Problem 2.22 had a current flow of 50 mA. What is the range of the voltage across each? If a battery of 10 V is applied to each of them in turn, what is the range of current flowing in each?

(i) **2.24.** Find the nominal resistance value and the tolerance of the following carbon-composition resistors having (in order) the color bands indicated.

    (a) Green, blue, orange, silver.
    (b) Brown, black, red, gold.
    (c) Brown, green, yellow.
    (d) Red, violet, gold, silver.
    (e) Orange, white, silver, gold.
    (f) Red, red, orange.
    (g) Brown, red, black, silver.

GLOSSARY

Absolute Zero Temperature: The temperature (273°C) at which all molecular motion ceases.

Color Coding: A method of identifying the value of an element and its tolerance. For resistors, color bands around the body of the element are used for such coding.

Conductance (G): The characteristic property of an element that indicates the ease with which electric current may be allowed to flow in the element. It is measured in siemens (S).

Dynamic Resistance: The slope of the tangent of the V–I characteristics of an element at a specified point on the curve.

Noise: The undesirable signals existing in an electrical system. All electrical components and elements generate a certain amount of noise.

Nonlinear Element: An element whose V–I characteristics correspond to a curve. Such elements do not obey the linear form of Ohm's law.

Ohm (Ω): The SI unit of measurement of a resistance. One ohm is equivalent to 1 V of applied potential difference to the resistance per ampere of current flow.

Ohm's Law: The relationship between the applied voltage and the resulting current flow in an electric element.

Open Circuit: A condition equivalent to an infinite value of the resistance, corresponding to a zero value of the current flow.

Potentiometer: A three-terminal device, made of a variable resistance, that can be used to provide a variable level of potential difference between two of its terminals.

Power Hyperbola: The curve connecting the (V, I) points that correspond to the relationship $V \cdot I =$ constant. The constant is the value of power dissipated by the electric component.

Power Rating: The maximum safe value of power dissipation in an electric element, beyond which physical destruction of the element could occur.

Resistance (R): A measure of the opposition of an element to the flow of the electric current carriers. It is measured in ohms.

Resistivity (ρ): The electric resistance of a sample of a material whose length is 1 m and whose cross-sectional area is $1\,\text{m}^2$. It is a specific property of a given material.

Rheostat: A three-terminal device constituting a variable resistance between two of its terminals.

Short Circuit: A condition equivalent to a zero value of the electric resistance between two points. The corresponding voltage difference between these two points is also zero.

Siemens (S):   The SI units of the conductance. One siemens is equivalent to the flow of 1 A of electric current per 1 V of potential difference.

Static Resistance: The value of the resistance of an element obtained by dividing the voltage (V) by the current (I) at a specific point on the $V-I$ characteristics of the element.

Superconductivity: The property of a material that corresponds to a zero value of the electric resistance: the ability of the material to allow the flow of the electric current carriers even with zero potential energy difference across the terminals of the material.

Temperature Coefficient of Resistance (α):   The fractional variation of the value of a resistance per unit change of the temperature (1°C) of the resistance. A positive value of α corresponds to an increase of the value of the resistance as the temperature rises, while a negative value of α corresponds to a decrease of the value of the resistance as the temperature rises.

3 | SIMPLE RESISTIVE CIRCUITS

OBJECTIVES

- Understanding the meaning of the topological terms used to describe electric circuits and the various types of interconnections of electric circuit elements.
- Understanding the meaning of circuit ground, and the voltages referenced to ground.
- Becoming so familiar with both KVL and KCL and their circuit application to the extent that they become "second nature".
- Ability to analyze various series circuits, parallel circuits and series-parallel circuits using Ohm's law, KVL and KCL, to obtain any required electrical quantity (voltage, current or power), or any unknown circuit element (resistor or source) value.
- Introduction to the Wheatstone bridge and its use as an accurate method of determining the unknown value of a resistor.

In this chapter analysis techniques for simple circuits made of resistors and dc sources are developed and applied to many basic electric circuit configurations. It is important to remember that voltages and currents in any part of such circuits are constant with respect to time, as the term "dc" implies.

Elements in an electric circuit can be interconnected in various possible ways and topologies. Before discussing the analysis techniques that would be more suitable for specific interconnections, some basic definitions have to be examined.

A *node* in an electric circuit is the point at which two or more components are connected together. These points are usually marked with dots. The circuit in Fig. 3.1 has the nodes *a, b, c, d,* and *e.* These points could also be considered as "solder points" in the circuit. In general, a point or a node in an electric circuit specifies a certain potential energy level (or voltage level). Since connecting wires used to connect one terminal of an element to another terminal of a second element are usually considered to have zero resistance, the potential difference (or voltage) across such connecting wires is zero ($IR = I \times 0 = 0$). Thus connecting wires are essentially at the same potential level as the node to which they are connected. Therefore, these connecting wires are also considered to be part of the node, as, for example, nodes *c* and *e* in Fig. 3.1. The dashed lines indicate that the node in question includes all the connecting wires tied to it. Later, the difference between a principal and a secondary node will be examined.

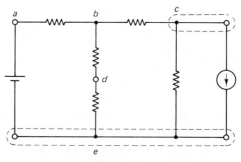

Figure 3.1

A *branch* is simply any element connected between two nodes. These elements could be sources, resistances, or other types of elements, as will be discussed later. It can easily be deduced that the circuit in Fig. 3.1 has seven branches, five resistive branches, and two branches containing sources.

A *loop* is any closed path in an electric circuit. For example, the closed paths *abdea* and *bcedb* in the circuit in Fig. 3.1 constitute two loops. Similarly, the outside closed path *abcea* is also a loop. A *mesh* is a special type of loop that does not have a closed path in its interior. The first two loops identified above are also meshes, but the third one is not.

3.2 KIRCHHOFF'S CURRENT LAW

To be able to analyze electric circuits and thus determine the voltages and currents at different parts of the circuit, Ohm's law by itself is not sufficient, except for the very basic single-load circuits. With the introduction of the two

Kirchhoff's laws, a large number and various types of electric circuits can be analyzed. In other chapters (e.g., Chapter 5) more general techniques are studied.

Kirchhoff's current law (KCL) can be stated in either of the following forms:

1. The algebraic sum of currents at a node (or, in general, any closed surface in an electrical network) is zero. Here currents entering and currents leaving the node (or the surface) must be assigned opposite algebraic signs.
2. The sum of currents entering a node (or a closed surface) must equal the sum of currents leaving this node (or the closed surface).

This law is plausible because at any given time any conductor and indeed any circuit element in general is always neutral. The amount of charge entering the element must be the same as the amount of charge leaving the element because the element as a whole cannot create or destroy any amount of this charge. Even elements that can store charge (as shown later when the capacitor is examined) have equal and opposite sign quantity of charge stationed at two different parts (or sides) of the element, but as a whole the element is still neutral at every instant of time. This is also true for sources. Since nodes are, in fact, infinitesimally small conductors, and closed surfaces contain neutral elements at all times, KCL makes sense.

To show how KCL would be applied, consider the portion of a network shown in Fig. 3.2. For the node (or the closed surface indicated with the dashed line), let the currents entering the node be considered positive; then obviously, those leaving the node should be considered negative. The application of the first form of KCL results in

$$I_1 + I_2 + (-I_3) + (-I_4) = 0$$

or by rearranging the terms, we have

$$I_1 + I_2 = I_3 + I_4$$

which is actually the second form of KCL. In general, the algebraic sum of currents at any node is

$$\sum_{i=1}^{n} I_i = 0 \qquad (3.1a)$$

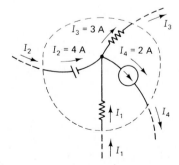

Figure 3.2

or, equivalently,

sum of currents entering a node = sum of currents leaving this node (3.1b)

Since I_1 is the only unknown in the example of Fig. 3.2, then using the numerical values given yields

$$I_1 + 4 = 3 + 2$$

Therefore,

$$I_1 = 1\,\text{A}$$

In many situations, neither the magnitude nor the direction of many currents in a given circuit may be known. Current directions would then be assumed arbitrarily based on one's best judgment. The solution of the appropriate KCL expressions will then result in the numerical values and signs of these unknown currents. If the resulting sign of the current being determined is positive, this implies that the assumed direction was the correct one, as in the example of Fig. 3.2. If, on the other hand, the resulting sign of the current being determined is negative, this only means that the proper direction of this particular current flow is opposite to that assumed (i.e., this current flow direction should be reversed).

■ EXAMPLE 3.1

Find the unknown current I in the circuit portion shown in Fig. 3.3.

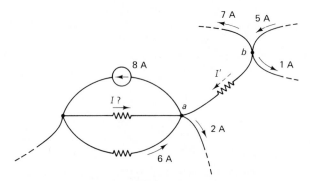

Figure 3.3

Solution   Here it is more appropriate to start with node b since only one unknown branch current is involved with this node, while at node a two branch currents are unknown. The current I' is assumed to flow from b to a, as shown in Fig. 3.3. Applying KCL at node b gives us

$$5 = 7 + 1 + I'$$

sum of currents entering = sum of currents leaving

Therefore,

$$I' = 5 - 8 = -3\,\text{A}$$

Now apply KCL at node a:

$$I + 6 + I' = 8 + 2$$

sum of the currents entering = sum of the current leaving

Therefore,

$$I = 10 - 6 - I' = 4 - (-3) = 7\,\text{A}$$

From these results one can easily conclude that the real direction of flow of the current I is as assumed in Fig. 3.3, while the real direction of flow of the current I' is from a to b, opposite that assumed in Fig. 3.3. The magnitude of I' is still 3 A, as determined.

3.3 KIRCHHOFF'S VOLTAGE LAW

When the current (direction of flow of the positively charged current carriers) flows in an element from the lower potential point (or negative terminal) toward the higher potential point (or positive terminal), as, for example, in the case of current flow through a voltage source, a voltage (or potential energy) rise is encountered. Voltage drop is encountered when the current flows in an element from the higher potential terminal toward the lower potential terminal, as, for example, in the case of current flow through a load element. Kirchhoff's voltage law (KVL) can be stated in either of the following two equivalent forms:

1. The algebraic sum of all voltages around any closed loop in an electric circuit is zero. Here a voltage rise may be taken with a positive sign, while a voltage drop would then be taken with a negative sign, or vice versa.

2. In any closed loop,

    sum of voltage rises = sum of voltage drops.

The first form of the KVL can be expressed mathematically as

$$\sum_{i=1}^{n} V_i \text{ (around any closed loop)} = 0 \qquad (3.2)$$

In applying KVL to a closed loop in an electric circuit, one chooses a certain starting point (or node) and traces (or goes around) the loop in a chosen direction, either clockwise or counterclockwise, and ends at the same starting point. To write KVL for the loop: Whenever an element is encountered (in the direction chosen) from its positive terminal first, its voltage is taken with a positive sign, while the voltage of the element that is encountered (in the direction chosen) from its negative terminal first is taken with a negative sign. In the context of KVL, the reference to closed loop means that the tracing of the elements' voltages around the loop should start and end at the same point (close the path) even if the loop is physically open (i.e., contains an open circuit).

The application of KVL can be illustrated by considering the closed loop shown in Fig. 3.4. Node (or point) a is chosen as the starting (and end) point and the direction of tracing the voltages around the loop is chosen to be the clockwise direction. The element voltages E, V_1, V_2, V_3, and V_4 are encountered in order. The potential of the terminals encountered first for each element are $-$, $+$, $-$, $-$, and $+$, respectively. Thus the first form of KVL provides the following equation:

algebraic sum of the voltage around any closed loop = 0

$$-E + V_1 - V_2 - V_3 + V_4 = 0 \qquad (3.3a)$$

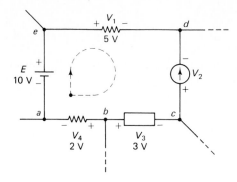

Figure 3.4

or by rearranging the terms in Eq. (3.3a), we obtain

$$\underbrace{E + V_2 + V_3}_{} = \underbrace{V_1 + V_4}_{} \tag{3.3b}$$

sum of voltage rises = sum of voltage drops

which is the equivalent second form of KVL. In this second form, the voltage rises and voltage drops follow the definition above, assuming a hypothetical current flowing in the direction of the arrow shown dashed in Fig. 3.4.

It should be emphasized that the expression one would obtain for KVL for this loop, starting and ending at any other point, whether in a clockwise or counterclockwise direction, is exactly the same as Eq. (3.3b) (try it!). This implies that KVL is unique for each loop, independent of the starting point or the direction chosen (except for the equivalent of multiplying the entire equation by -1). The procedure outlined above for writing KVL for any loop is strongly recommended. This will help establish this law as "second nature."

In the example illustrated in Fig. 3.4, the voltage V_2 across the current source is unknown, while the other voltages are all given. Since neither the magnitude nor the polarities of this voltage (V_2) are known, the polarities shown in Fig. 3.4 are assumed. Substituting the numerical values for the voltages in Eq. (3.3b) results in

$$10 + V_2 + 3 = 5 + 2$$

Therefore,

$$V_2 = -13 + 7 = -6 \text{ V}$$

The negative sign in the answer means that the proper polarities for the voltage difference V_2 are opposite those assumed. If the polarities are reversed, the answer for V_2 will be positive, indicating that point d is in fact higher in potential than point c.

KVL is also plausible since voltages are in fact potential energy differences between two points (or levels). Thus starting at a point in a loop and tracing the voltage rises and drops around the loop to come back to the same starting point means that the overall potential energy difference encountered is zero (i.e., the potential energy difference between the starting and end points, which is exactly the same point).

■ EXAMPLE 3.2

Find the voltage V_S across the open switch (an open circuit) in the circuit shown in Fig. 3.5.

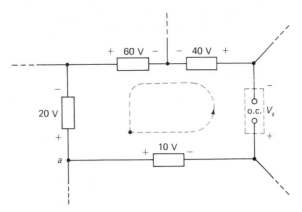

Figure 3.5

Solution   The polarities for the voltage difference across the switch terminals, V_S, are assumed as shown in Fig. 3.5. Even though the circuit is physically open circuited, the loop is still traced as a closed path, with the open switch considered as just another element.

Following the procedure outlined above for writing KVL, this loop will be traced starting and ending at point a, in a counterclockwise direction as indicated. The voltages encountered are in turn $+10$, $+V_S$, $+40$, -60, and -20; thus from KVL,

$$+10 + V_S + 40 - 60 - 20 = 0$$

Therefore,

$$V_S = 80 - 50 = 30 \text{ V}$$

The positive answer indicates that the assumed polarities for V_S are properly (or correctly) chosen.

EXAMPLE 3.3

Find the unknown voltages in the circuit shown in Fig. 3.6.

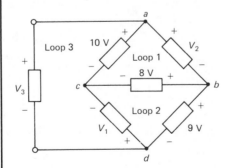

Figure 3.6

Solution   The polarities for the unknown voltages V_1, V_2, and V_3 are assumed as indicated in Fig. 3.6. KVL could then be written for each of the three loops, starting at point c and tracing the voltages for each loop in a clockwise direction. As outlined in the discussion above:

For loop 1:

$$-10 + V_2 + 8 = 0$$

Therefore,

$$V_2 = 10 - 8 = 2 \text{ V}$$

For loop 2:

$$-8 + 9 + V_1 = 0$$

Therefore,

$$V_1 = -9 + 8 = -1 \text{ V}$$

For loop 3:

$$-V_1 - V_3 + 10 = 0$$

Therefore,

$$V_3 = 10 - V_1 = 10 - (-1) = 11 \text{ V}$$

The assumed polarities for the voltages V_2 and V_3 are properly chosen, while the polarities for the voltage V_1 should be reversed for V_1 to be a positive number.

3.4 SERIES CIRCUITS

Elements are said to be connected in series if one (and only one) terminal of an element is connected to one terminal of another element. There should be no other terminal of a third element connected to the common node b between the first two elements. The elements in Fig. 3.7(a) satisfy these conditions and are therefore referred to as *connected in series*. In Fig. 3.7(b), a terminal of a third element is connected to the common node b, and in Fig. 3.7(c) the two terminals of the two elements are connected to each other (i.e., there are two common

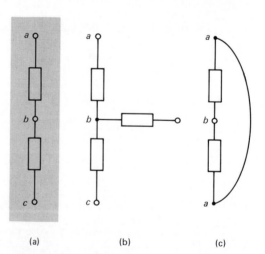

(a) (b) (c)

Figure 3.7 (a) Series; (b), (c) nonseries.

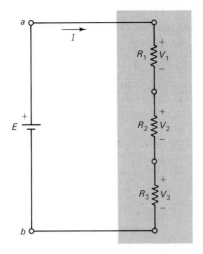

Figure 3.8

nodes, not one). Thus the connections of elements in Fig. 3.7(b) and (c) are not series connections.

An example of a series resistive circuit is shown in Fig. 3.8. It contains three resistors, R_1, R_2, and R_3, and a voltage source E, all connected in series. In such a series circuit only two elements are connected to each node. According to KCL, this type of connection implies that the current in each of the series-connected elements is the same, because this one current, I, enters and leaves each node. This is, in fact, another definition (or a special property) of the series circuit: *Components are said to be connected in series if the same current flows in all of them.*

Let the current flowing from the source E be denoted I. As this current flows through each of R_1, R_2, and R_3, a voltage difference will be exhibited across each element in accordance with Ohm's law and the associated polarity (or direction) convention discussed in Chapter 1. These voltages, denoted V_1, V_2, and V_3, are all voltage drops, as the current will be flowing from the higher potential terminal toward the lower potential terminal of each element. The polarities of each of these voltage drops are therefore as indicated in Fig. 3.8. Applying Ohm's law to each resistive element yields

$$V_1 = IR_1 \tag{3.4}$$

$$V_2 = IR_2 \tag{3.5}$$

$$V_3 = IR_3 \tag{3.6}$$

To analyze this circuit, KVL can be written for the closed loop starting and ending at, say, point b and going clockwise (in the direction of the current flow). Then

$$-E + V_1 + V_2 + V_3 = 0 \quad \text{(algebraic sum of voltages} = 0) \tag{3.7}$$

or upon rearranging the terms,

$$E = V_1 + V_2 + V_3 \quad \text{(sum of voltage rises} = \text{sum of voltage drops)} \tag{3.8}$$

Substituting for V_1, V_2, and V_3 in Eq. (3.8), using Ohm's laws [Eqs. (3.4), (3.5), and (3.6)], gives us

$$E = IR_1 + IR_2 + IR_3$$
$$= I(R_1 + R_2 + R_3)$$
$$= IR_T = IR_{eq} \tag{3.9}$$

where

$$R_T = R_{eq} = R_1 + R_2 + R_3 \tag{3.10}$$

R_T refers to the total resistance of the circuit, which can also be referred to as the *equivalent resistance* of the circuit. For a circuit containing n number of resistors, Eq. (3.10) can be extended to

$$\boxed{R_T = R_{eq} = R_1 + R_2 + R_3 + \cdots + R_n} \tag{3.11}$$

If all n resistors are equal (say R),

$$R_T = R_{eq} = R + R + R + \cdots + R = nR \tag{3.12}$$

It is clear from Eq. (3.11) that the total resistance of a series circuit is greater than the largest of the series-connected resistors.

Equation (3.9) can be interpreted as Ohm's law of a simple, single-load circuit, as shown in Fig. 3.9. Comparing the circuits of Figs. 3.8 and 3.9, it can be concluded that a number of series-connected resistances, between terminals a and b of a source, can be replaced by a single resistance, R_T (or R_{eq}), between the same two terminals of the source, such that the same source, E, drives the same current, I, in both circuits. The circuit in Fig. 3.9 is called the *equivalent circuit* to that shown in Fig. 3.8. The equivalent circuit is not usually a real circuit but a mathematically equivalent model of the original circuit that permits one to analyze the original circuit in a simpler and quicker manner. The equivalent-circuit concept applies not only to the series circuit but also to many other types of circuits, as shown in this chapter and in many other parts of the text.

Because the current is the same in all the series-connected components, every voltage in the circuit can be found once the value of the current is determined. Applying the equivalent-circuit concept, we first determine

$$R_T = R_1 + R_2 + R_3 + \cdots$$

Then from Ohm's law,

$$I = \frac{E}{R_T}$$

This value of I, the source current in the equivalent circuit, is the same as that in

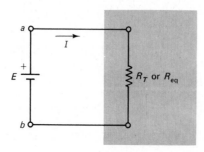

Figure 3.9 Equivalent-circuit representation.

the original circuit. Thus

$$V_1 = IR_1 \qquad V_2 = IR_2 \qquad V_3 = IR_3 \quad \text{etc.}$$

Note that if the value of one resistance in a series circuit is changed, R_T and I will also change. Hence all the voltages in the circuit will also consequently change.

3.4.1 The Voltage-Division Principle

In the series circuit discussed above, one can conclude from Eqs. (3.4), (3.5), and (3.6) that

$$\frac{V_1}{V_2} = \frac{IR_1}{IR_2} = \frac{R_1}{R_2} \tag{3.13}$$

Similarly,

$$\frac{V_1}{V_3} = \frac{R_1}{R_3} \qquad \frac{V_2}{V_3} = \frac{R_2}{R_3} \quad \text{etc.} \tag{3.14}$$

This means that the ratio of the voltages across any two elements connected in series is the same as the ratio of their connecting resistances.

Also, by considering the ratio of the voltages in Eqs. (3.4), (3.5), and (3.6) to the total source voltage E expressed in Eq. (3.9), we can conclude that

$$\frac{V_1}{E} = \frac{IR_1}{IR_T} = \frac{R_1}{R_T} = \frac{R_1}{R_1 + R_2 + R_3}$$

Therefore,

$$\boxed{V_1 = E \frac{R_1}{R_1 + R_2 + R_3}} \tag{3.15}$$

Note that $E/(R_1 + R_2 + R_3)$ is, in fact, I. Thus Eq. (3.15) is another form of Ohm's law. Similarly,

$$\boxed{V_2 = E \frac{R_2}{R_1 + R_2 + R_3}} \tag{3.16}$$

and

$$\boxed{V_3 = E \frac{R_3}{R_1 + R_2 + R_3}} \tag{3.17}$$

Equations (3.15), (3.16), and (3.17) are referred to as the *voltage-division principle*. In effect, this principle states that if a voltage exists between two nodes with series-connected elements between these two nodes, this voltage divides between the elements by the ratio of the resistance of each element to the total resistance of all the elements connected between the two nodes. The largest portion of the voltage will appear across the largest resistance.

■ EXAMPLE 3.4

Find the current I and the voltage across each element in the circuit shown in Fig. 3.10. Also find the voltage between nodes c and b, V_{cb}.

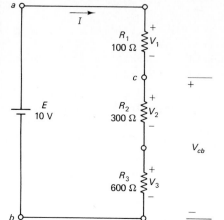

Figure 3.10

Solution   Following the procedure described above, we obtain

$$R_T = R_1 + R_2 + R_3$$
$$= 100 + 300 + 600 = 1000 \, \Omega$$

Thus

$$I = \frac{E}{R_T} = \frac{10}{1000}$$
$$= 0.01 \, \text{A} = 10 \, \text{mA}$$

Therefore, from Ohm's law for each component,

$$V_1 = IR_1 = 0.01 \times 100 = 1 \, \text{V}$$
$$V_2 = IR_2 = 0.01 \times 300 = 3 \, \text{V}$$
$$V_3 = IR_3 = 0.01 \times 600 = 6 \, \text{V}$$

Note that

$$V_1 + V_2 + V_3 = 1 + 3 + 6 = 10 \, \text{V} = E$$

which satisfies KVL for this closed loop.

The voltage V_{cb} is the voltage across elements R_2 and R_3, whose combined resistance is $300 \, \Omega + 600 \, \Omega = 900 \, \Omega$. Therefore,

$$V_{cb} = I(R_2 + R_3) = 0.01 \times 900 = 9 \, \text{V}$$
$$= V_2 + V_3 = 3 + 6 = 9 \, \text{V}$$

The voltages in this example could also be determined without having to find the current I first, through the application of the voltage-division principle. Thus

$$V_1 = E \frac{R_1}{R_1 + R_2 + R_3} = 10 \times \frac{100}{100 + 300 + 600} = 10 \times \frac{100}{1000} = 1 \, \text{V}$$

$$V_2 = E \frac{R_2}{R_1 + R_2 + R_3} = 10 \times \frac{300}{100 + 300 + 600} = 10 \times \frac{300}{1000} = 3 \, \text{V}$$

$$V_3 = E \frac{R_3}{R_1 + R_2 + R_3} = 10 \times \frac{600}{100 + 300 + 600} = 10 \times \frac{600}{1000} = 6 \, \text{V}$$

and

$$V_{cb} = E \frac{R_2 + R_3}{R_1 + R_2 + R_3} = 10 \times \frac{300 + 600}{100 + 300 + 600} = 10 \times \frac{900}{1000} = 9 \text{ V}$$

EXAMPLE 3.5

Determine the values of the three resistors in the circuit shown in Fig. 3.11, given that the total resistance of the circuit is 50 Ω.

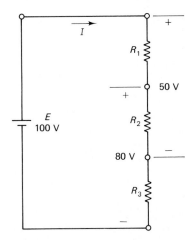

Figure 3.11

Solution The value of a resistor can be determined by applying Ohm's law, provided that the voltage across it and the current flowing in it are known. The current I in this series circuit can be found since E and R_T ($= R_1 + R_2 + R_3$) are given. Here

$$I = \frac{E}{R_T} = \frac{100}{50} = 2 \text{ A}$$

This current flows in all three series-connected resistors. Also, from the circuit diagram.

$$V_1 + V_2 = \text{total voltage across resistors } R_1 \text{ and } R_2$$
$$= 50 \text{ V}$$

and

$$V_2 + V_3 = \text{total voltage across resistors } R_2 \text{ and } R_3$$
$$= 80 \text{ V}$$

From KVL,

$$E = V_1 + V_2 + V_3$$
$$100 = 50 + V_3$$

Therefore,

$$V_3 = 50 \text{ V}$$

and

$$100 = V_1 + 80$$

Thus,

$$V_1 = 20 \text{ V}$$

Now, again using KVL,

$$100 = 20 + V_2 + 50$$

Therefore,

$$V_2 = 100 - 70 = 30 \text{ V}$$

Since all the elements' voltages and the current are now determined, then applying Ohm's law to each element yields

$$R_1 = \frac{V_1}{I} = \frac{20}{2} = 10 \,\Omega$$

$$R_2 = \frac{V_2}{I} = \frac{30}{2} = 15 \,\Omega$$

$$R_3 = \frac{V_3}{I} = \frac{50}{2} = 25 \,\Omega$$

As a check, note that $R_1 + R_2 + R_3 = 10 + 15 + 25 = 50 \,\Omega = R_T$, as given.

■ EXAMPLE 3.6

Find I and V_{ab} in the circuit shown in Fig. 3.12. Also determine the power in the $10 \,\Omega$ resistor and the 5 V and 8 V sources.

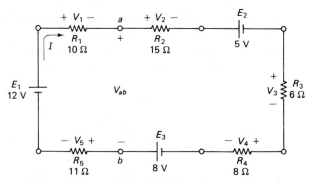

Figure 3.12

Solution   Let us start by assuming that the current I will flow in this series circuit in the direction indicated in Fig. 3.12. The voltage drops V_1, V_2, V_3, V_4, and V_5 across each of the resistors will have the polarities indicated in the figure, according to the associated V–I polarity (direction) convention.

To apply KVL to this closed loop, point b is picked as the starting and ending point and the direction of tracing the loop is counterclockwise. Therefore,

$$+E_3 - V_4 - V_3 - E_2 - V_2 - V_1 + E_1 - V_5 = 0$$

(algebraic sum of voltages = 0). Rearranging the terms in the equation above gives

$$E_1 + E_3 - E_2 = V_1 + V_2 + V_3 + V_4 + V_5 \qquad (3.18a)$$

Using Ohm's law for each resistor yields

$$E_1 + E_3 - E_2 = IR_1 + IR_2 + IR_3 + IR_4 + IR_5$$
$$= I(R_1 + R_2 + R_3 + R_4 + R_5) \qquad (3.18b)$$

or

$$E_T = IR_T \qquad (3.18c)$$

where

$$E_T = E_1 + E_3 - E_2 = 12 + 8 - 5 = 15\text{ V}$$

and

$$R_T = R_1 + R_2 + R_3 + R_4 + R_5 = 10 + 15 + 6 + 8 + 11 = 50\,\Omega$$

Therefore,

$$I = \frac{E_T}{R_T} = \frac{15}{50} = 0.3\text{ A}$$

The rearrangement of terms in Eqs. (3.18a) and (3.18b) indicates that interchanging the location of the elements and sources while preserving the polarity directions does not affect KVL or the value of the current or the voltage across any element in a series circuit. Figure 3.13(a) represents the circuit described by Eqs. (3.18a) and (3.18b). This circuit is equivalent to the original circuit in Fig. 3.12. It is obtained by interchanging the location of the elements through combining the sources together and the resistors together. Finally, Eq. (3.18c) describes the simplest equivalent circuit which is shown in Fig. 3.13(b).

Going back to the original circuit, after I is determined, KVL can be applied to the closed loop (or path) to the left of nodes a and b, to determine V_{ab}. Starting and ending at point b and again tracing this loop in a counterclockwise direction, we have

$$-V_{ab} - V_1 + E_1 - V_5 = 0$$

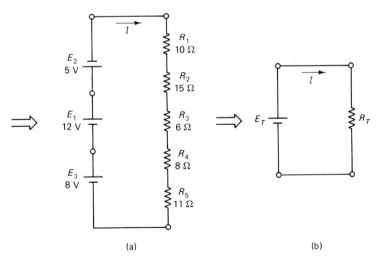

(a) (b)

Figure 3.13 Equivalent circuits.

Therefore,

$$V_{ab} = E_1 - V_1 - V_5$$
$$= E_1 - IR_1 - IR_5$$
$$= 12 - 0.3 \times 10 - 0.3 \times 11 = 12 - 3 - 3.3 = 5.7 \text{ V}$$

The same result would have been obtained if the clockwise direction was chosen or if the loop to the right of nodes a and b was considered.

The power in the 10 Ω resistor is

$$I^2 R_1 = (0.3)^2 \times 10 = 0.9 \text{ W}$$
$$= V_1 I = 3 \times (0.3)$$
$$= 0.9 \text{ W} \quad \text{(absorbing)}$$

The power in the 5 V source is

$$E_2 I = 5 \times 0.3 = 1.5 \text{ W} \quad \text{(absorbing)}$$

This source is absorbing power because the current is flowing in it from the positive terminal, as if it were a load. In reality, this source is being "charged."

The power in the 8 V source is

$$E_3 I = 8 \times 0.3 = 2.4 \text{ W} \quad \text{(delivering)}$$

This source is delivering power because the current is flowing out of it from its positive terminal.

EXAMPLE 3.7

What is the maximum voltage that can be applied to the series combination of resistors shown in Fig. 3.14 without exceeding the power rating of any of them?

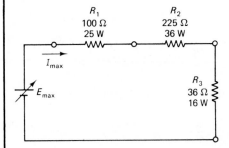

Figure 3.14

Solution The maximum power rating of each resistor is specified in Fig. 3.14. Since the current must be the same in this series connection of resistors, the maximum current-carrying capability of each resistor should first be determined. Use Eq. (2.13):

For R_1:

$$I_{1\,\text{max}} = \sqrt{\frac{P_{1\,\text{max}}}{R_1}} = \sqrt{\frac{25}{100}} = 0.5 \text{ A}$$

For R_2:

$$I_{2\,max} = \sqrt{\frac{P_{2\,max}}{R_2}} = \sqrt{\frac{36}{225}} = 0.4 \text{ A}$$

For R_3:

$$I_{3\,max} = \sqrt{\frac{P_{3\,max}}{R_3}} = \sqrt{\frac{16}{36}} = 0.667 \text{ A}$$

Clearly, the maximum safe current that could flow in all three elements without damaging any of them is

$$I_{max} = 0.4 \text{ A}$$

that is, R_2 is the weakest link in this connection. Therefore,

$$E_{max} = I_{max}R_T = I_{max}(R_1 + R_2 + R_3)$$
$$= 0.4(100 + 225 + 36) = 0.4 \times 361 = 144.4 \text{ V}$$

EXAMPLE 3.8

In the circuit shown in Fig. 3.15, P_1 (the power dissipated in R_1) is 200 W, V_2 (the voltage across R_2) is 50 V, and R_T (the total resistance of the circuit) is 10 Ω. Find

(a) R_3.

(b) The energy supplied by the force, E, in 1 week.

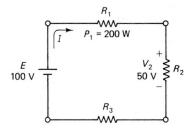

Figure 3.15

Solution   (a) As E and R_T are both given,

$$I = \frac{E}{R_T} = \frac{100}{10} = 10 \text{ A}$$

To determine R_3, the voltage across it, V_3, must first be calculated. V_3 can be obtained through the application of KVL provided that V_1 is also known. Thus V_1 has to be determined first. Since $P_1 = V_1 I$,

$$V_1 = \frac{P_1}{I} = \frac{200}{10} = 20 \text{ V}$$

Now from KVL,

$$E = V_1 + V_2 + V_3$$
$$100 = 20 + 50 + V_3$$

Therefore,

$$V_3 = 30\,\text{V}$$

Thus by applying Ohm's law for resistor R_3, we have

$$R_3 = \frac{V_3}{I} = \frac{30}{10} = 3\,\Omega$$

(b) The power supplied by the source is P_T, where

$$P_T = EI = 100 \times 10 = 1000\,\text{W} = 1\,\text{kW}$$

Therefore, the energy supplied by this source in 1 week is

$$W_T = P_T \times t = 1000\,\text{W} \times 7 \times 24 \times 3600\,\text{s}$$
$$= 604.8 \times 10^6 = 604.8\,\text{MJ}$$

or

$$W_T = 1\,\text{kW} \times 7 \times 24\,\text{h} = 168\,\text{kWh}$$

EXAMPLE 3.9

Find E in the series circuit shown in Fig. 3.16 provided that P_1 (the power dissipated in resistor R_1) is 100 W, V_2 (the voltage across resistor R_2) is 40 V, R_3 is 5 Ω, and the total resistance of the circuit, R_T, is 10 Ω.

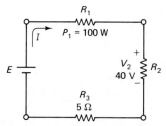

Figure 3.16

Solution Here the current I, the voltage V_1, and the voltage V_3 cannot be found directly. E is related to the voltage drops in the circuit through KVL, that is,

$$E = V_1 + V_2 + V_3$$

To find E, one possible technique is to express each undetermined term in KVL in terms of E, resulting in an equation in terms of one unknown. This can then be solved algebraically to determine the unknown value of E. Here

$$I = \frac{E}{R_T} = \frac{E}{10}$$

$$V_1 = \frac{P_1}{I} = \frac{100}{E/10} = \frac{1000}{E}$$

$$V_3 = IR_3 = \frac{E}{10} \times 5 = \frac{E}{2}$$

while V_2 is given (40 V). Thus substituting in KVL gives

$$E = \frac{1000}{E} + 40 + \frac{E}{2}$$

Multiplying both sides of the equation by $2E$ yields

$$2E^2 = 2000 + 80E + E^2$$
$$E^2 - 80E - 2000 = 0$$
$$(E - 100)(E + 20) = 0$$

Therefore,

$$E = 100\text{ V} \quad (\text{or } -20\text{ V})$$

The negative answer is rejected because it cannot satisfy KVL.

3.4.2   Ground Potential and Single-Subscripted Voltages

Voltages always reflect potential differences between any two points, or nodes, in an electric circuit. They can be denoted according to the name of the element between the two points; for example, V_1 or V_{R_1} is the voltage across element R_1 as shown in Fig. 3.17. They can also be denoted according to the symbols given to the two points; for example, V_{AB} (in Fig. 3.17) is the voltage of point A (how high or how low the potential energy is at A) with respect to point B. This is called *double-subscripted voltage*.

In many circuits it may be convenient to choose one point in the circuit as the reference point (or node) and refer all voltages in this circuit to that common reference. If this node is also assigned the reference voltage value of 0 V, some points in the circuit will have voltages above it (i.e., positive) and some other points in the circuit will have voltages below it (i.e., negative). This point is usually referred to as the circuit *ground*. It is usually recognized in electric circuits by attaching the symbol $\perp$ to it and denoting it by the letter G. Any point in a circuit can be chosen to be the reference point. In a circuit with only one voltage source, if the negative side of the battery is taken to be the reference point, all the other nodes in the circuit will have positive voltages with respect to it because they are higher in potential that this node. This is easy to visualize since current always flows from the higher potential point toward the lower potential point.

When the voltages at the different nodes are referred to this ground (or common reference) point, the double-subscripted voltages V_{AG}, V_{BG}, V_{CG}, and so on, can be written as V_A, V_B, V_C and so on. This convention means that dropping the second subscript (i.e., *single-subscripted voltages*) always refers the potentials of the points indicated to ground (or point G).

In practice, for example in domestic electrical appliances, this reference point is usually connected to the physical ground (which, by the way, is also the zero potential energy level in mechanics) through the water pipes. This terminal is the third (rounded) prong in power outlet sockets. Because physical ground is considered to be the sink of all electric currents, with a potential of 0 V always, it is clear that for safety reasons the chassis of the appliance should be the ground terminal.

On the other hand, in low-current low-voltage circuits, the ground terminal (usually the chassis) need not be physically connected to the real ground. This

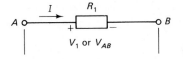

Figure 3.17

point is still considered to be the 0 V (or reference) node, similar to the arbitrary choice of a reference point, or origin, on an axis.

It is very important to recognize that introducing the ground connection to an electric circuit only establishes a reference voltage point; it does not affect or change the application of any of the circuit relations (e.g., Ohm's law or Kirchhoff's laws).

Consider the circuit example in Fig. 3.18(a). The ground (or reference) node is chosen to be the point of connection of elements R_3 and R_4 as shown. Ohm's law and Kirchhoff's laws apply as usual. For simplicity we could think of ground as the reference point on a linear voltage scale as shown in Fig. 3.18(b). Because current flows from point C (higher potential) to point G (lower potential), V_C is positive. Similarly, point B is higher in potential than C and point A is higher in potential than B, as shown in Fig. 3.18(b). The difference in potential between the different nodes is the same as the voltage differences across the corresponding elements. Point G, however, is higher in potential than point D, as the current continues to flow toward the negative side of the battery. Thus V_D is negative. As Fig. 3.18(b) indicates, E is still the sum of the voltage drops V_1, V_2, V_3, and V_4 (KVL). The choice of the reference node does not change any of the circuit relations.

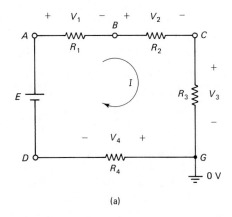

(a)

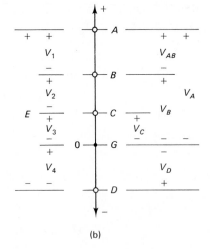

(b)

Figure 3.18

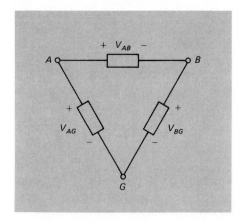

Figure 3.19

If any three points in an electric circuit are considered, as for example the closed loop shown in Fig. 3.19, the application of KVL results in

$$-V_{BG} - V_{AB} + V_{AG} = 0$$

Therefore,

$$\boxed{V_{AB} = V_{AG} - V_{BG} = V_A - V_B}$$

(3.19)

If point G is renamed any other name, for example O, then Eq. (3.19) still holds:

$$\boxed{V_{AB} = V_{AO} - V_{BO} = V_{AO} + V_{OB}}$$

(3.20)

In Equation (3.20), the fact that the exchange of the subscripts changes the sign of the voltage ($V_{BO} = -V_{OB}$) is used, as discussed in Chapter 1. The interpretation of Eq. (3.19) can easily be understood by considering the pictorial linear scale of Fig. 3.18(b).

EXAMPLE 3.10

In the circuit shown in Fig. 3.20, V_P is 26 V; resistor R dissipates 345.6 kJ in 24 h. Find the values of E and R.

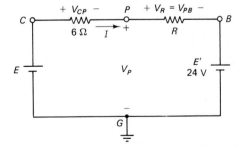

Figure 3.20

Solution The voltage across the resistor R, V_R, is $V_R = V_{PB} = V_P - V_B$, according to Eq. (3.19). But $V_B = E' = 24$ V and V_P is given to be 26 V; therefore,

$$V_R = 26 - 24 = 2 \text{ V}$$

This means that the current I will flow from P to B as shown. To find the value of I, the energy dissipated in R can be expressed as

$$W_R = P_R \times t = V_R I t$$

Therefore,

$$345.6 \times 10^3 = 2 \times I \times 24 \times 3600$$

$$I = 2\,\text{A}$$

$$R = \frac{V_R}{I} = \frac{2}{2} = 1\,\Omega$$

Noting that

$$V_C = E \qquad \text{and} \qquad V_{CP} = V_{6\Omega} = I \times 6 = 2 \times 6 = 12\,\text{V}$$

and again using Eq. (3.19), we have

$$V_{CP} = V_C - V_P$$

Therefore,

$$V_C = E = V_{CP} + V_P = 12 + 26 = 38\,\text{V}$$

3.5 PARALLEL CIRCUITS

Elements are said to be *connected in parallel* if their two terminals are connected to the same two nodes; that is, each element in the parallel connection of elements shares its two terminals with every other element in the connection.

An example of a parallel circuit is shown in Fig. 3.21(a). Here the elements R_1, R_2, R_3, and E have the same two common terminals, nodes a and b. Nodes a and b could be the solder points of the terminals of the elements or they could be the terminals of the source, as shown in Fig. 3.21(b). This figure could actually represent how one may connect this parallel circuit in the laboratory. However, Fig. 3.21(a) is the recommended practice of drawing parallel electric circuit configurations. Note that the nodes contain all short-circuit wires connected to them.

Since the voltage between two nodes is a finite specific quantity, all the elements in the parallel connection must have the same voltage across each of them. This is, in fact, another definition (or a special property) of the parallel circuit: *Elements are said to be connected in parallel if the same voltage exists across each of them.*

In the circuit of Fig. 3.21(a), the elements R_1, R_2, and R_3 have the same voltage drop V ($= E$) across each of them. The source current, denoted I_T, is entering node a. It will then split to three currents, I_1, I_2, and I_3, each flowing in its corresponding component, as shown in the figure. These are called the *branch currents*. The three branch currents then combine at node b to form I_T, which then flows into the negative terminal of the source.

From Ohm's law for each of the parallel branches,

$$V = E = I_1 R_1$$
$$= I_2 R_2$$
$$= I_3 R_3 \qquad\qquad (3.21)$$

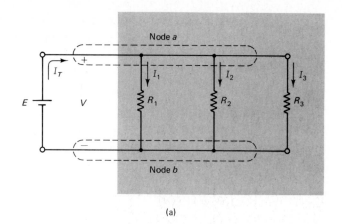

Node a

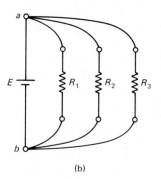

(a)

(b)

Figure 3.21

Applying KCL at node a or node b gives us

$$I_T = I_1 + I_2 + I_3 \tag{3.22}$$

Substituting for each of the branch currents in terms of the voltage V, using Eq. (3.21) yields

$$
\begin{aligned}
I_T &= \frac{V}{R_1} + \frac{V}{R_2} + \frac{V}{R_3} \\
&= V\left(\frac{1}{R_1} + \frac{1}{R_2} + \frac{1}{R_3}\right) \\
&= V(G_1 + G_2 + G_3) \\
&= VG_T = \frac{V}{R_T}
\end{aligned}
\tag{3.23a}
$$

Therefore,

$$V = I_T R_T \tag{3.23b}$$

Here the conductance G of any branch is $1/R$, R being the resistance of the branch. Also,

$$G_T = \text{total conductance of the parallel branches}$$

$$= G_1 + G_2 + G_3$$

$$= \sum_{i=1}^{n} G_i = \text{sum of the } n \text{ branch conductances}$$

$$= \frac{1}{R_T} \tag{3.24}$$

R_T is the equivalent, or total, resistance of the parallel circuit. Equation (3.24) can also be expressed as

$$\boxed{R_T = \frac{1}{G_T} = \frac{1}{G_1 + G_2 + G_3 + \cdots} = \frac{1}{1/R_1 + 1/R_2 + 1/R_3 + \cdots}} \tag{3.25}$$

Equation (3.23b) can be interpreted as Ohm's law of a simple single-load circuit, as shown in Fig. 3.22. This circuit is the equivalent circuit to that shown in Fig. 3.21. This means that any number of parallel resistors connected between two nodes, such as nodes a and b, can be replaced by a single resistance, equal to R_T, such that the current entering or leaving the nodes in the original and in the equivalent circuits is the same. Again, it should be emphasized that this equivalent circuit is only a mathematical model that simplifies the analysis of the circuit.

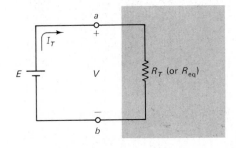

Figure 3.22

The special case of only two resistances in parallel, shown in Fig. 3.23, is important because of its frequent occurrence. Here Eq. (3.25) can be simplified as follows:

$$\boxed{R_{eq} = \frac{1}{1/R_1 + 1/R_2} = \frac{R_1 R_2}{R_1 + R_2} = \frac{\text{product of the two resistances}}{\text{sum of the resistances}}} \tag{3.26}$$

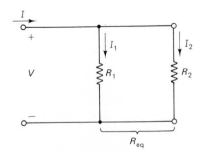

Figure 3.23

Chapter 3   Simple Resistive Circuits

Also, from Fig. 3.23 and Eqs. (3.21) and (3.23b),

$$V = I_1 R_1 = I_2 R_2 = I R_{eq} \tag{3.27}$$

3.5.1 The Current-Division Principle

As a result of Ohm's law for each of the parallel-connected branches [Eqs. (3.21)], we can conclude that

$$\frac{I_1}{I_2} = \frac{V/R_1}{V/R_2} = \frac{R_2}{R_1} = \frac{G_1}{G_2} \tag{3.28a}$$

Similarly,

$$\frac{I_2}{I_3} = \frac{R_3}{R_2} = \frac{G_2}{G_3} \tag{3.28b}$$

and

$$\frac{I_3}{I_1} = \frac{R_1}{R_3} = \frac{G_3}{G_1} \tag{3.28c}$$

Thus the ratio of the currents in two parallel branches is the same as the ratio of the two branch conductances. Also, using Eqs. (3.23a) and (3.25), we have

$$\boxed{I_1 = \frac{V}{R_1} = VG_1 = \frac{I_T}{G_T} G_1 = I_T \frac{G_1}{G_1 + G_2 + G_3 + \cdots}} \tag{3.29a}$$

and similarly,

$$\boxed{I_2 = I_T \frac{G_2}{G_T} = I_T \frac{G_2}{G_1 + G_2 + G_3 + \cdots}} \tag{3.29b}$$

$$\boxed{I_3 = I_T \frac{G_3}{G_T} = I_T \frac{G_3}{G_1 + G_2 + G_3 + \cdots}} \quad \text{etc.} \tag{3.29c}$$

Equations (3.29) constitute the *current-division principle*. The importance of these relationships is in indicating how the total current entering a node, I_T, is split between the parallel branches connected to that node. Clearly, the portion of this total current in a certain branch is the same as the ratio of this branch conductance to the total conductance of all the parallel branches.

A simple form of the current-division principle can be obtained for the special case of two parallel-connected resistances (Fig. 3.23). Here, using Eqs. (3.26) and (3.27), we have

$$\boxed{I_1 = I\frac{R_{eq}}{R_1} = I\frac{1}{R_1} \frac{R_1 R_2}{R_1 + R_2} = I\frac{R_2}{R_1 + R_2}} \tag{3.30a}$$

and

$$\boxed{I_2 = I\frac{R_{eq}}{R_2} = I\frac{1}{R_2} \frac{R_1 R_2}{R_1 + R_2} = I\frac{R_1}{R_1 + R_2}} \tag{3.30b}$$

Thus the current in resistance R_1 is the total current entering the node, times the resistance of the opposite branch, R_2, divided by the sum of the two branch resistances. The second branch current can be obtained similarly.

3.5.2 Special Properties of Parallel Circuits

A. If all n branches have equal resistances, R ($=1/G$), then from Eqs. (3.24) and (3.25),

$$G_T = nG = n\frac{1}{R}$$

or

$$R_T = \frac{R}{n} \tag{3.31a}$$

For two equal resistances, of value R in parallel (i.e., $n = 2$),

$$R_T = R_{eq} = \frac{R}{2} \tag{3.31b}$$

B. When the parallel branches are connected to a constant-voltage source (ideal), then as Eqs. (3.21) indicate, the variation in any branch resistance affects only the current and the power in this branch; it does not affect the current and the power in any of the other parallel branches. For this reason, house wiring is always configured as a parallel circuit.

C. The addition of extra branches in parallel increases the value of G_T, as can easily be concluded from Eq. (3.24). This results in a reduction of the total (or equivalent) resistance of the parallel circuit, R_T (or R_{eq}), since R_T is $1/G_T$.

D. Increasing the value of any of the parallel-connected resistances, R_1 or R_2 or $R_3, \ldots$, reduces the denominator in Eq. (3.25); this results in an increase in R_T (or R_{eq}).

E. When the numerator and denominator in Eq. (3.25) are both multiplied by R_1, or R_2, or R_3, then R_T can be expressed as:

$$R_T = \frac{R_1}{1 + R_1/R_2 + R_1/R_3 + \cdots} \tag{3.32a}$$

$$= \frac{R_2}{1 + R_2/R_1 + R_2/R_3 + \cdots} \tag{3.32b}$$

$$= \frac{R_3}{1 + R_3/R_1 + R_3/R_2 + \cdots} \tag{3.32c}$$

Since in any of the forms in Eqs. (3.32), the denominator is greater than unity, R_T is smaller than any of the parallel-connected resistances; in fact, it is smaller than the smallest of these resistances.

F. Referring to the case of two parallel-connected resistances; if, say, R_2 is equal to zero (i.e., a short-circuit wire), then from Eq. (3.26),

$$R_{eq} = 0$$

and from Eq. (3.27),

$$V = 0$$

Also, from the current-division principle, [Eq. (3.30)],

$$I_1 = 0$$

and

$$I_2 = I$$

This means that short circuiting any resistance makes its value redundant to the rest of the circuit and results in a zero voltage across it, similar to any zero resistance (short-circuit) situation. Also, no current flows in the resistance that is shorted out (R_1 in this case), and all of the circuit's current, I, flows in the zero-resistance (short-circuit wire) branch. Thus the action of short circuiting any resistance effectively removes that resistance from the rest of the circuit since its value does not affect the determination of the currents or the voltages in the rest of the circuit.

G. Also in the case of the two parallel-connected resistances, if $R_1 > R_2$, then from Eqs. (3.30),

$$I_2 > I_1$$

that is, the larger portion of the node current, I, will flow in the branch that has the smaller resistance. This is also true for the general case of any number of parallel-connected resistances.

■ EXAMPLE 3.11

Using KCL, find the unknown currents, I_1, I_2, and I_3 in the circuit shown in Fig. 3.24.

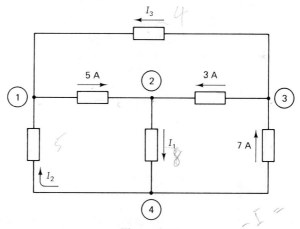

Figure 3.24

Solution   At node 1, two currents are unknown. However, at nodes 2 or 3, only one current is unknown. Thus starting with node 2, we have

$$I_1 = 5 + 3 = 8\,\text{A}$$

At node 3,

$$7 = 3 + I_3$$

Therefore,

$$I_3 = 4\,\text{A}$$

Now, at node 1,

$$I_2 + I_3 = 5$$

Therefore,

$$I_2 = 5 - I_3$$
$$= 5 - 4 = 1 \text{ A}$$

As a check, at node 4,

$$I_1 = 8 = I_2 + 7 = 1 + 7$$

(confirming that the answers above are correct).

EXAMPLE 3.12

Find the voltage V_{ab} across the parallel branches in Fig. 3.25 and then calculate the value of the currents I_1, I_2, and I_3.

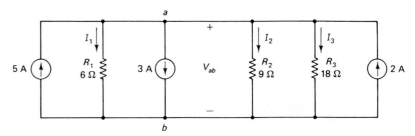

Figure 3.25

Solution Using the equivalent-circuit techniques, first the three resistors can be replaced by one resistor, R_{eq}, where

$$R_{eq} = \frac{1}{\frac{1}{6} + \frac{1}{9} + \frac{1}{18}} = \frac{18}{3 + 2 + 1} = 3 \,\Omega$$

and the net current source entering node a is

$$I_S = 5 + 2 - 3 = 4 \text{ A}$$

The simple equivalent circuit shown in Fig. 3.26 is thus obtained. Applying Ohm's law to this circuit yields

$$V_{ab} = I_S R_{eq} = 4 \times 3 = 12 \text{ V}$$

The same result can be obtained by applying KCL to node a in the original

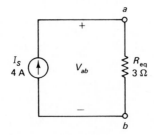

Figure 3.26

circuit:

$$5 + 2 = 3 + I_1 + I_2 + I_3 = 3 + \frac{V_{ab}}{6} + \frac{V_{ab}}{9} + \frac{V_{ab}}{18}$$

$$V_{ab}(\tfrac{1}{6} + \tfrac{1}{9} + \tfrac{1}{18}) = 7 - 3$$

$$V_{ab}\frac{3 + 2 + 1}{18} = 4$$

Therefore,

$$V_{ab} = \frac{4 \times 18}{6} = 12 \text{ V}$$

Now applying Ohm's law to each of the three resistive branches yields

$$I_1 = \frac{V_{ab}}{R_1} = \frac{12}{6} = 2 \text{ A}$$

$$I_2 = \frac{V_{ab}}{R_2} = \frac{12}{9} = 1.333 \text{ A}$$

$$I_3 = \frac{V_{ab}}{R_3} = \frac{12}{18} = 0.667 \text{ A}$$

EXAMPLE 3.13

The power dissipated in R_3 in the circuit shown in Fig. 3.27 is 20 W. Find the value of R_1 and the power dissipated in it, P_1.

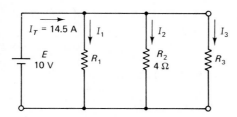

Figure 3.27

Solution   The power in R_3 is

$$P_3 = 20 = EI_3 = 10 \times I_3$$

noting that the voltage across each branch is E. Then

$$I_3 = \frac{20}{10} = 2 \text{ A}$$

Therefore,

$$R_3 = \frac{E}{I_3} = \frac{10}{2} = 5 \ \Omega$$

Using Ohm's law for R_2 gives us

$$I_2 = \frac{E}{R_2} = \frac{10}{4} = 2.5 \text{ A}$$

From KCL applied at the top node of the parallel branches in Fig. 3.27, we have

$$I_T = I_1 + I_2 + I_3$$

Therefore,

$$I_1 = I_T - (I_2 + I_3) = 14.5 - (2.5 + 2) = 10 \, \text{A}$$

$$R_1 = \frac{E}{I_1} = \frac{10}{10} = 1 \, \Omega$$

and

$$P_1 = EI_1 = 10 \times 10 = 100 \, \text{W}$$

EXAMPLE 3.14

Find R so that the equivalent resistance for the two parallel resistances shown in Fig. 3.28 is 5 kΩ. If I_T is 10 mA, find the current in R and the voltage V.

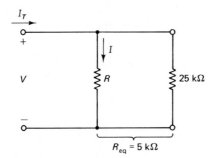

Figure 3.28

Solution The equivalent resistance of two parallel-connected resistances is given by Eq. (3.26). Here

$$5 \, \text{k}\Omega = \frac{R \times 25 \, \text{k}\Omega}{R + 25 \, \text{k}\Omega}$$

$$R + 25 \, \text{k}\Omega = 5R$$

Therefore,

$$R = \frac{25 \, \text{k}\Omega}{4} = 6.25 \, \text{k}\Omega$$

The current-division principle can then be used to obtain the current in R. Here

$$I = I_T \frac{25 \, \text{k}\Omega}{25 \, \text{k}\Omega + 6.25 \, \text{k}\Omega} = 10 \times 10^{-3} \times \frac{25}{31.25} = 8 \, \text{mA}$$

The voltage V is the same across each of the parallel branches. Thus

$$V = IR = I_T R_{eq}$$
$$= 8 \, \text{mA} \times 6.25 \, \text{k}\Omega = 10 \, \text{mA} \times 5 \, \text{k}\Omega = 50 \, \text{V}$$

As a check, the current in the 25 kΩ resistor is

$$10 \, \text{mA} - 8 \, \text{mA} = 2 \, \text{mA}$$

Therefore,

$$V = 2 \, \text{mA} \times 25 \, \text{k}\Omega = 50 \, \text{V} \quad \text{(as above)}$$

EXAMPLE 3.15

Find the maximum safe value of the current source I_S that can be applied to the parallel resistances shown in Fig. 3.29, without causing a breakdown in any of these resistances.

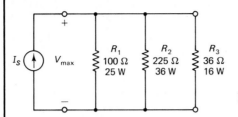

Figure 3.29

Solution   Since the voltage is the same across each of these parallel resistors, the maximum voltage that each resistance can sustain without breakdown has to be found first. Then the smallest of these voltages would be the one chosen, V_{max}, to ensure safe operation of all of them. As shown in Chapter 2:

For R_1:

$$V_{1\,max} = \sqrt{P_1 R_1} = \sqrt{25 \times 100} = 50\ \text{V}$$

For R_2:

$$V_{2\,max} = \sqrt{P_2 R_2} = \sqrt{36 \times 225} = 90\ \text{V}$$

For R_3:

$$V_{3\,max} = \sqrt{P_3 R_3} = \sqrt{16 \times 36} = 24\ \text{V}$$

Obviously, V_{max} should be 24 V. Here R_3 is the weakest link in this parallel-resistances combination.

As V_{max} is now determined, the maximum safe value of the source current follows by applying KCL and Ohm's law:

$$I_S = \frac{V_{max}}{R_1} + \frac{V_{max}}{R_2} + \frac{V_{max}}{R_3} = V_{max}\left(\frac{1}{R_1} + \frac{1}{R_2} + \frac{1}{R_3}\right)$$

$$= 24\left(\frac{1}{100} + \frac{1}{225} + \frac{1}{36}\right)$$

$$= 24 \times 0.0422 = 1.01\ \text{A}$$

3.6 | SERIES–PARALLEL CIRCUITS

Practical electric circuits are seldom of the simple series or simple parallel forms. The circuit shown in Fig. 3.30 is an example of such practical circuits. Parts of such circuits can be identified as belonging to either the series and/or the parallel forms. For this reason they are referred to as *series–parallel circuits*.

In this section we introduce some analysis techniques that are used to solve for the unknown electrical quantities in such circuits; in Chapter 5 we introduce more formal and general analysis techniques to handle any dc electrical network (a network is a generalized circuit).

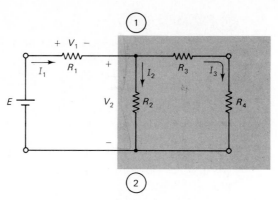

Figure 3.30

In the circuit example shown in Fig. 3.30, the resistance elements R_3 and R_4 have the same current, I_3, flowing in each of them. They are therefore in series and can be replaced by a single resistance, R_5 ($= R_3 + R_4$), as shown in the first equivalent (simplified) circuit in Fig. 3.31(a). R_5 replaces R_3 and R_4 between nodes 1 and 2. The current flowing in R_5 is still I_3.

Note that the voltages between the nodes and the currents in the different branches remain the same as in the original circuit, according to the properties (definition) of equivalent circuits.

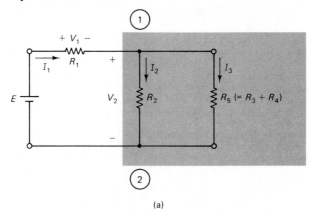

(a)

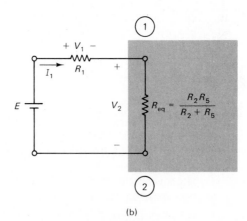

(b)

Figure 3.31

84

Now, in the circuit of Fig. 3.31(a), we can easily see that R_2 and R_5 are in parallel between the nodes 1 and 2, with the voltage drop V_2 (between nodes 1 and 2) being the same across each of these two parallel branches. Thus R_2 and R_5 can be replaced by a single equivalent resistance:

$$R_{eq} = \frac{R_2 R_5}{R_2 + R_5}$$

between nodes 1 and 2, as shown in the second equivalent circuit in Fig. 3.31(b). The voltages between the nodes that still appear in the equivalent circuit (i.e., V_1 and V_2, and hence I_1) are the same as in the original circuit of Fig. 3.30.

This last equivalent circuit in Fig. 3.31(b) is now easy to identify as a simple series circuit, which can be solved as discussed in Section 3.4. Here

$$I_1 = \frac{E}{R_1 + R_{eq}} \tag{3.33}$$

$$V_1 = I_1 R_1 \tag{3.34}$$

$$V_2 = I_1 R_{eq} \tag{3.35}$$

and

$$E = V_1 + V_2 \tag{3.36}$$

Once V_2 is obtained, we can go back to the original circuit to calculate

$$I_2 = \frac{V_2}{R_2} \tag{3.37}$$

and

$$I_3 = \frac{V_2}{R_5} = \frac{V_2}{R_3 + R_4} \quad \text{[as seen from Fig. 3.31(a)]} \tag{3.38}$$

The method described above is called the *equivalent-circuit technique for analyzing series-parallel circuits*. This technique is very simple to apply when all the resistance values are known.

On the other hand, when some of the resistance values are unknown, we must use the algebraic technique for analyzing series–parallel circuits. This technique is based on writing KVL, KCL, and Ohm's law for the various parts of the original (or the equivalent) circuit, in terms of the unknown quantity (or variable). These equations are then solved algebraically to determine the value of the unknown. For example, in the case of the circuit shown in Fig. 3.30 [or Fig. 3.31(a)],

$$E = V_1 + V_2 \quad \text{(KVL)} \tag{3.39}$$

$$I_1 = I_2 + I_3 \quad \text{(KCL)} \tag{3.40}$$

$$V_1 = I_1 R_1 \quad \text{(Ohm's law)} \tag{3.41}$$

$$V_2 = I_2 R_2 \quad \text{(Ohm's law)} \tag{3.42}$$

$$= I_3 R_5 = I_3(R_3 + R_4) \quad \text{(Ohm's law)} \tag{3.43}$$

In many cases, both techniques can be applied simultaneously to different parts of the circuit. The following examples illustrate such analysis procedures.

EXAMPLE 3.16

Find I_S, V_1, and V_2 in the circuit shown in Fig. 3.32.

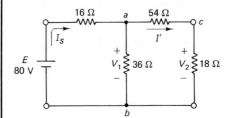

Figure 3.32

Solution   As explained above in regard to the equivalent-circuit technique, we can proceed by replacing the 54 Ω and 18 Ω resistors (which are in series) by an equivalent 72 Ω resistor, as shown in Fig. 3.33(a). Note that the voltage drop V_2 cannot be recognized in Fig. 3.33(a) because node c has disappeared in this equivalent circuit.

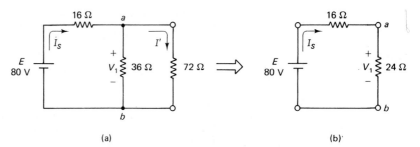

Figure 3.33

In Fig. 3.33(a), the 36 Ω and 72 Ω resistors are in parallel. Their single resistance equivalent is

$$\frac{36 \times 72}{36 + 72} = \frac{36 \times 72}{108} = 24 \ \Omega$$

connected across nodes a and b, as shown in the second equivalent circuit shown in Fig. 3.33(b). V_1 is still the voltage between nodes a and b and I_S is still the current from the voltage source E, in both the original circuit and the equivalent circuit in Fig. 3.33(b). Analyzing this last simple series circuit provides

$$I_S = \frac{80}{16 + 24} = \frac{80}{40} = 2 \ \text{A}$$

and

$$V_1 = 2 \times 24 = 48 \ \text{V}$$

Going back to the original circuit in Fig. 3.32, using the value of V_1 obtained above yields

$$I' = \frac{V_1}{54 + 18} = \frac{48}{72} = 0.667 \ \text{A}$$

and thus

$$V_2 = I' \times 18 = 12 \text{ V}$$

EXAMPLE 3.17

For (a) the ladder circuit shown in Fig. 3.34(a) and (b) the lattice circuit shown in Fig. 3.34(b), find R_T (or the input resistance, R_{in}) if
(i) The terminals a and b are left open circuited.
(ii) The terminals a and b are shorted (i.e., connected together with a short-circuit wire).

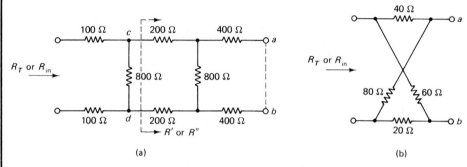

(a)　　　　　　　　　　　　　　(b)

Figure 3.34

Solution　(a) For the ladder circuit in Fig. 3.34(a):
(i) When terminals a and b are left open, the two $400\,\Omega$ resistors have no effect, since no current can flow through this part of the circuit. Therefore, the resistance to the right of nodes c and d is

$$R' = 200 + 800 + 200 = 1200\,\Omega$$

Now R' is in parallel with the $800\,\Omega$ resistance between nodes c and d. Hence

$$R_T = R_{in} = 100 + \frac{800 \times 1200}{800 + 1200} + 100 = 200 + 480 = 680\,\Omega$$

(ii) When terminals a and b are shorted, the two $400\,\Omega$ resistors are in series (i.e., equivalent to $800\,\Omega$), which is now in parallel with the $800\,\Omega$ resistance. Therefore, the resistance to the right of nodes c and d is

$$R'' = 200 + \frac{800 \times 800}{800 + 800} + 200 = 200 + 200 + 400 = 800\,\Omega$$

This equivalent resistance R'' is now in parallel with the $800\,\Omega$ resistor between nodes c and d. Thus in this case,

$$R_T = R_{in} = 100 + \frac{800 \times 800}{800 + 800} + 100 = 200 + 400 = 600\,\Omega$$

(b)　For the lattice circuit in Fig. 3.34(b):
(i) When terminals a and b are left open, the resistance seen at the input (R_T or R_{in}) is made of two parallel branches, one being the $60\,\Omega$ in series with the

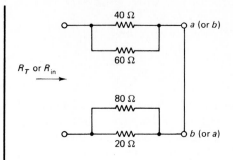

Figure 3.35

20 Ω and the other being the 40 Ω in series with the 80 Ω. Therefore,

$$R_T = R_{in} = \frac{(40 + 80) \times (60 + 20)}{(40 + 80) + (60 + 20)}$$

$$= \frac{120 \times 80}{120 + 80} = 48\,\Omega$$

(ii) When terminals a and b are connected together (i.e., shorted), the circuit can be redrawn as shown in Fig. 3.35. Nodes a and b are actually the same point. It is therefore possible to recognize that the 40 Ω and 60 Ω resistors are actually in parallel, and so are the 80 Ω and 20 Ω resistors. Thus,

$$R_T = R_{in} = \frac{40 \times 60}{40 + 60} + \frac{80 \times 20}{80 + 20}$$

$$= 24 + 16 = 40\,\Omega$$

EXAMPLE 3.18

Obtain the values of I_1 and I_2 in the circuit shown in Fig. 3.36.

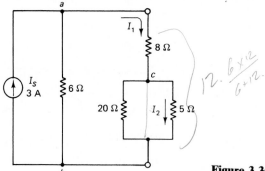

Figure 3.36

Solution To simplify the circuit, one can easily observe that the 5 Ω and 20 Ω resistors, which are in parallel, can be replaced by a single resistor, R_{eq}:

$$R_{eq} = \frac{5 \times 20}{5 + 20}$$

$$= \frac{100}{25} = 4\,\Omega$$

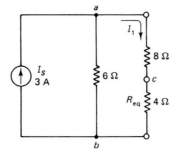

Figure 3.37

as shown in Fig. 3.37. Now, this last equivalent circuit is a two-branch parallel circuit; one is $6\,\Omega$ and the other is $8 + 4 = 12\,\Omega$. The current I_1 in the $8\,\Omega$ resistor can then be found using the current-division principle:

$$I_1 = I_s \frac{6}{6 + 12} = 3 \times \frac{6}{18} = 1.0 \text{ A}$$

Going back to the original circuit in Fig. 3.36, I_1 then splits at node c between the two parallel branches, the $5\,\Omega$ and $20\,\Omega$ resistors. Again using the current-division principle, we have

$$I_2 = I_1 \frac{20}{20 + 5} = 1 \times \frac{20}{25} = 0.8 \text{ A}$$

EXAMPLE 3.19

Find R_1 in the circuit shown in Fig. 3.38, given that the total power delivered by the source is 75 W.

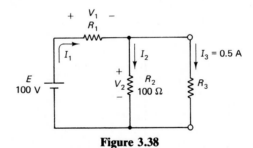

Figure 3.38

Solution The algebraic technique has to be used for this circuit since two of the resistors are unknown. Here the source power is

$$P_T = E \times I_1$$
$$75 = 100 \times I_1$$

Therefore,

$$I_1 = 0.75 \text{ A}$$

Since $I_1 = I_2 + I_3$ (from KCL),

$$I_2 = I_1 - I_3$$
$$= 0.75 - 0.5 = 0.25 \text{ A}$$

Thus the voltage across the two parallel resistors, R_2 and R_3, is

$$V_2 = I_2 R_2 = 0.25 \times 100 = 25 \text{ V}$$

Now applying KVL to the loop containing the source, V_1 can be obtained:

$$E = V_1 + V_2$$

Therefore,

$$V_1 = E - V_2 = 100 - 25 = 75 \text{ V}$$

Applying Ohm's law to the resistor R_1 yields

$$R_1 = \frac{V_1}{I_1} = \frac{75}{0.75} = 100 \text{ } \Omega$$

EXAMPLE 3.20

In the circuit shown in Fig. 3.39, V_A is -60 V. Find the value of the unknown resistor R_1.

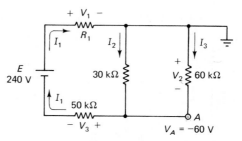

Figure 3.39

Solution   As indicated, point A is below ground voltage (0 V) by 60 V (since $V_A = -60$ V). Therefore, V_2, the voltage across the two parallel branches, the 30 kΩ and 60 kΩ resistors, is 60 V. Application of Ohm's law to these two parallel resistors provides the values of the currents I_2 and I_3. Here

$$I_2 = \frac{60}{30 \text{ k}\Omega} = 2 \text{ mA}$$

and

$$I_3 = \frac{60}{60 \text{ k}\Omega} = 1 \text{ mA}$$

Therefore, from KCL,

$$I_1 = I_2 + I_3 = 2 \text{ mA} + 1 \text{ mA} = 3 \text{ mA}$$

Before applying KVL to the loop containing E in order to determine V_1, the voltage drop V_3 can be found from Ohm's law:

$$V_3 = I_1 \times 50 \text{ k}\Omega = 3 \text{ mA} \times 50 \text{ k}\Omega = 150 \text{ V}$$

As

$$E = V_1 + V_2 + V_3$$

then

$$V_1 = 240 - (60 + 150) = 30 \text{ V}$$

Applying Ohm's law to the resistor R_1 gives us

$$R_1 = \frac{V_1}{I_1} = \frac{30}{3 \text{ m}} = 10 \text{ k}\Omega$$

EXAMPLE 3.21

If I_4 is $\frac{1}{3}I_1$, in the circuit of Fig. 3.40, prove that V_4 is 40% of E.

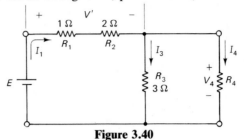

Figure 3.40

Solution As I_4 is $\frac{1}{3}I_1$, then from KCL,

$$I_3 = I_1 - I_4 = I_1 - \tfrac{1}{3}I_1 = \tfrac{2}{3}I_1$$

To obtain the ratio of V_4 to E, each will be expressed in terms of I_1, noting that V_4 is the voltage across the two parallel resistors R_3 and R_4. Thus

$$V_4 = I_3R_3 = \tfrac{2}{3}I_1 \times 3 = 2I_1$$

and

$$V' = (1 + 2)I_1 = 3I_1$$

Therefore, using KVL, we have

$$E = V' + V_4 = 3I_1 + 2I_1 = 5I_1$$

and

$$\frac{V_4}{E} = \frac{2I_1}{5I_1} = 0.4$$

that is, V_4 is 40% of E, as required.

EXAMPLE 3.22

Find the current I in the circuit shown in Fig. 3.41.

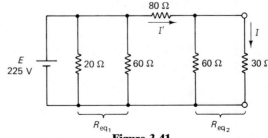

Figure 3.41

Solution   The circuit can first be simplified as shown in Fig. 3.42 with

$$R_{\text{eq}_1} = \frac{20 \times 60}{20 + 60} = \frac{20 \times 60}{80} = 15\ \Omega$$

and

$$R_{\text{eq}_2} = \frac{60 \times 30}{60 + 30} = \frac{1800}{90} = 20\ \Omega$$

The source voltage E is clearly across the two parallel branches, R_{eq_1} and $(80 + R_{\text{eq}_2})$. The current in the 80 Ω resistor, I', can then be obtained easily from Ohm's law:

$$I' = \frac{E}{80 + R_{\text{eq}_2}} = \frac{225}{80 + 20} = 2.25\ \text{A}$$

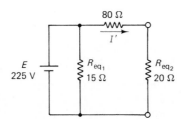

Figure 3.42

Going back to the original circuit in Fig. 3.41 and applying the current-division principle yields

$$I = I' \frac{60}{60 + 30} = 2.25 \times \frac{60}{90} = 1.5\ \text{A}$$

EXAMPLE 3.23

Find R_4 in the circuit shown in Fig. 3.43.

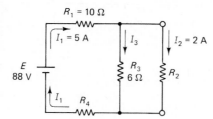

Figure 3.43

Solution   From KCL, I_3 can be obtained:

$$I_3 = I_1 - I_2$$
$$= 5 - 2 = 3\ \text{A}$$

Writing KVL for the loop containing E gives us

$$E = I_1R_1 + I_3R_3 + I_1R_4$$
$$88 = 5 \times 10 + 3 \times 6 + 5 \times R_4$$

Therefore,

$$R_4 = \tfrac{1}{5}(88 - 50 - 18) = 4\ \Omega$$

EXAMPLE 3.24

In the circuit shown in Fig. 3.44, $P_1 = 2P_2 = 4P_3$, indicating the ratios of the power dissipated in each component. Find R_1, R_2, R_3, and E.

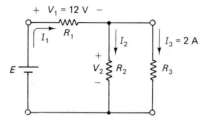

Figure 3.44

Solution   To find the element values, the voltages and currents in the entire circuit have to be determined. R_2 and R_3 are in parallel, with V_2 being the voltage across each. As $2P_2 = 4P_3$, then

$$\cancel{V_2}I_2 = 2\cancel{V_2}I_3$$

and

$$I_2 = 2I_3 = 2 \times 2 = 4\ \mathrm{A}$$

From KCL,

$$I_1 = I_2 + I_3 = 4 + 2 = 6\ \mathrm{A}$$
$$P_1 = I_1V_1 = 6 \times 12 = 72\ \mathrm{W}$$

Therefore,

$$P_2 = \tfrac{1}{2}P_1 = 36\ \mathrm{W}$$
$$P_3 = \tfrac{1}{4}P_1 = 18\ \mathrm{W}$$

But $P_2 = I_2V_2$; therefore,

$$V_2 = \frac{P_2}{I_2} = \frac{36}{4} = 9\ \mathrm{V}$$

As a check, $P_3 = V_2I_3 = 9 \times 2 = 18\ \mathrm{W}$, as above.

Now that all the voltages and currents in the circuit are determined, then from Ohm's law for each component,

$$R_1 = \frac{V_1}{I_1} = \frac{12}{6} = 2\ \Omega$$

$$R_2 = \frac{V_2}{I_2} = \frac{9}{4} = 2.25\ \Omega$$

$$R_3 = \frac{V_2}{I_3} = \frac{9}{2} = 4.5\ \Omega$$

Also by applying KVL to the loop containing E,

$$E = V_1 + V_2 = 12 + 9 = 21\ V$$

EXAMPLE 3.25

Find R_3 in the circuit shown in Fig. 3.45.

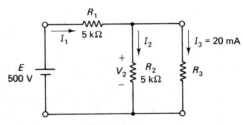

Figure 3.45

Solution   To obtain R_3, V_2 (the voltage across it) has to be found. For this circuit one can write KCL and KVL as follows:

$$I_1 = I_2 + I_3$$

and

$$E = I_1 R_1 + V_2$$

KCL can be expressed in terms of V_2 as follows:

$$I_1 = \frac{V_2}{5000} + 20 \times 10^{-3}$$

Substituting this value of I_1 in KVL gives us

$$500 = \left(\frac{V_2}{5000} + 20 \times 10^{-3}\right) \times 5000 + V_2$$

$$= V_2 + 100 + V_2 = 2V_2 + 100$$

Therefore,

$$V_2 = \frac{1}{2}(500 - 100) = 200\ V$$

and

$$R_3 = \frac{V_2}{I_3} = \frac{200}{20 \times 10^{-3}} = 10 \times 10^3 = 10\ k\Omega$$

EXAMPLE 3.26

If P_3 in the circuit shown in Fig. 3.46 is 160 W, find R_2 and R_3.

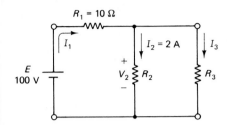

Figure 3.46

Solution Here also V_2 must be determined before R_2 and R_3 can be calculated. Since

$$P_3 = V_2 I_3$$

then

$$I_3 = \frac{160}{V_2}$$

expressing I_3 in terms of V_2. KCL can be written in terms of V_2 as follows:

$$I_1 = I_2 + I_3 = 2 + \frac{160}{V_2}$$

and KVL gives

$$E = I_1 R_1 + V_2$$
$$100 = I_1 \times 10 + V_2$$

Substituting for I_1 in KVL above yields

$$100 = 10 \times \left(2 + \frac{160}{V_2}\right) + V_2$$

$$= 20 + \frac{1600}{V_2} + V_2$$

This algebraic equation can be solved to find V_2. Multiplying both sides of this equation by V_2 results in

$$(100 - 20)V_2 = 1600 + V_2^2$$
$$V_2^2 - 80V_2 + 1600 = 0$$

or

$$(V_2 - 40)^2 = 0$$
$$V_2 = 40 \text{ V}$$

Thus

$$R_2 = \frac{V_2}{I_2} = \frac{40}{2} = 20 \ \Omega$$

Since

$$P_3 = \frac{V_2^2}{R_3}$$

then
$$R_3 = \frac{V_2^2}{P_3} = \frac{(40)^2}{160} = 10 \ \Omega$$

EXAMPLE 3.27

Find E in the circuit of Fig. 3.47, given that P_3 is 120 W.

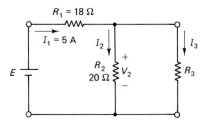

Figure 3.47

Solution Since from KVL,

$$E = I_1R_1 + V_2$$
$$= 5 \times 18 + V_2 = 90 + V_2$$

V_2 has to be determined first. But

$$P_3 = V_2I_3$$

Therefore,

$$I_3 = \frac{120}{V_2}$$

and

$$I_2 = \frac{V_2}{20}$$

Now from KCL,

$$I_1 = I_2 + I_3$$
$$5 = \frac{V_2}{20} + \frac{120}{V_2}$$

This equation can then be solved to determine V_2. Multiplying both sides by $20V_2$ gives us

$$100V_2 = V_2^2 + 2400$$
$$V_2^2 - 100V_2 + 2400 = 0$$
$$(V_2 - 40)(V_2 - 60) = 0$$

Therefore,

$$V_2 = 40 \text{ V or } 60 \text{ V (both solutions are acceptable)}$$

Thus

$$E = 130 \ V \text{ or } 150 \text{ V} \qquad \text{from KVL above}$$

EXAMPLE 3.28

Determine the maximum safe value of the voltage source, E_{max}, that can be applied to the three-resistor circuit shown in Fig. 3.48.

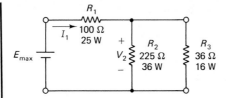

Figure 3.48

Solution Since R_2 and R_3 are in parallel, the maximum safe value of V_2 should first be checked; here

$$V_{R_2(\text{max})} = \sqrt{P_2 R_2} = \sqrt{36 \times 225} = 90 \text{ V}$$

$$V_{R_3(\text{max})} = \sqrt{P_3 R_3} = \sqrt{16 \times 36} = 24 \text{ V}$$

Therefore, V_2 should not exceed 24 V, to prevent the breakdown of R_3.
For $V_{2\,\text{max}} = 24$ V,

$$I_1 = I_2 + I_3 = \frac{V_{2\,\text{max}}}{R_2} + \frac{V_{2\,\text{max}}}{R_3}$$

$$= \frac{24}{225} + \frac{24}{36} = 0.106 + 0.667 = 0.773 \text{ A}$$

However,

$$I_{1\,\text{max}} = \sqrt{\frac{P_1}{R_1}} = \sqrt{\frac{25}{100}} = 0.5 \text{ A}$$

Thus I_1 cannot exceed this value of 0.5 A; in this case R_1 is obviously the weakest link. Hence

$$V_2 = I_{1\,\text{max}} \frac{R_2 R_3}{R_2 + R_3}$$

$$= 0.5 \times \frac{225 \times 36}{225 + 36} = 0.5 \times 31.034 = 15.517 \text{ V}$$

There will be no breakdown in R_2 or R_3 (i.e., both are operating safely). Thus

$$E_{\text{max}} = I_{1\,\text{max}} R_1 + V_2$$
$$= 0.5 \times 100 + 15.517 = 65.517 \text{ V}$$

EXAMPLE 3.29

In the circuit shown in Fig. 3.49, V_P is -16 V. Find R and E.

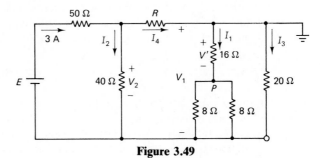

Figure 3.49

Solution Here the potential at point P is 16 V below the ground potential (0 V). Therefore,

$$V' = 16 \text{ V}$$

and

$$I_1 = \frac{V'}{16} = \frac{16}{16} = 1 \text{ A}$$

The overall resistance of the branch containing point P is

$$16 + \frac{8 \times 8}{8 + 8} = 16 + 4 = 20 \text{ }\Omega$$

Therefore,

$$V_1 = 1 \times 20 = 20 \text{ V}$$

Thus

$$I_3 = \frac{V_1}{20} = \frac{20}{20} = 1 \text{ A}$$

and from KCL,

$$I_4 = I_1 + I_3 = 1 + 1 = 2 \text{ A}$$

Also, from KCL applied to the top node of the 40 Ω resistor,

$$3 = I_2 + I_4$$

Therefore,

$$I_2 = 3 - 2 = 1 \text{ A}$$

and

$$V_2 = I_2 \times 40 = 40 \text{ V}$$

The voltage across the resistance R is therefore V_R:

$$V_R = V_2 - V_1 = 40 - 20 = 20 \text{ V}$$

Thus

$$R = \frac{V_R}{I_4} = \frac{20}{2} = 10 \text{ }\Omega$$

and applying KVL to the loop containing E gives us

$$E = 3 \times 50 + V_2 = 150 + 40 = 190 \text{ V}$$

3.7 WHEATSTONE BRIDGE

When an element is connected between points a and b in the circuit shown in Fig. 3.50, this element is said to *bridge* the two branches. This particular circuit is called the *Wheatstone bridge* and it is very useful in measurement systems. It provides the most accurate technique for measuring the value of a resistance. The galvanometer shown, G, is a sensitive current-measuring device (it usually has a very low resistance). The galvanometer has a center-scale null position, to indicate current flow from a to b, or vice versa. Galvanometers are also referred to as *null detectors*. R_g is a guard resistance used to protect the galvanometer against excessive current flow. In general, the bridge circuit cannot be analyzed using the series or the parallel equivalent techniques because with the bridge arm

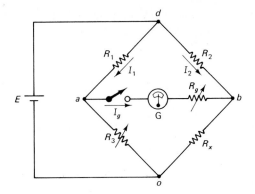

Figure 3.50

in place, no two elements are either in series or in parallel. The general network analysis techniques discussed in Chapter 5 are usually used to analyze such networks. However, if no current flows in the bridge arm, or if the bridge arm is considered as an ideal voltmeter (with infinite resistance), the circuit can easily be analyzed as a series-parallel circuit.

The situation corresponding to zero current flow in the bridge arm, which usually has a finite resistance whether large or small, is called the *balance condition* of the bridge. Here the galvanometer shows zero deflection (which is also called the *null condition*). It corresponds to

$$V_{ab} = I_g R_g = 0 \text{ V} \tag{3.44}$$

Under the condition that no current is flowing in the bridge arm (i.e., $I_g = 0$), I_1 flows in R_1 and R_3, while I_2 flows in R_2 and R_x (see Fig. 3.50). Thus R_1 and R_3 are in series, as are R_2 and R_x. The two series branches are in parallel with the voltage source E. Therefore,

$$\begin{aligned} V_{ab} &= V_{ao} - V_{bo} \quad [\text{see Eq. (3.20)}] \\ &= I_1 R_3 - I_2 R_x \\ &= \frac{E}{R_1 + R_3} R_3 - \frac{E}{R_2 + R_x} R_x \end{aligned} \tag{3.45}$$

At the null or balance condition,

$$V_{ab} = 0$$

Thus

$$\frac{R_3}{R_1 + R_3} = \frac{R_x}{R_2 + R_x}$$

Inverting each side of the equation above yields

$$\frac{R_1}{R_3} + 1 = \frac{R_2}{R_x} + 1$$

Therefore,

$$\frac{R_1}{R_3} = \frac{R_2}{R_x} \tag{3.46a}$$

This null condition is called the *touching-arm ratio condition*. It can also be written as

$$\frac{R_2}{R_1} = \frac{R_x}{R_3} \tag{3.46b}$$

or as

$$R_1 R_x = R_2 R_3 \quad \text{(opposite-arm product condition)} \tag{3.46c}$$

The balance condition above can also be written as

$$R_x = \frac{R_2}{R_1} R_3 \tag{3.47}$$

It indicates that the value of an unknown resistance (e.g., R_x) can be determined in terms of standard resistors (e.g., R_1, R_2, and R_3). R_2/R_1 is called the *multiplier ratio* and it is usually varied in multiples of 10 (i.e., 10^{-2}, 10^{-1}, 1, 10, 100, etc.), while R_3 is a standard decade resistor. By adjusting the multiplier ratio and R_3 until the bridge balance is achieved, Eq. (3.47) can be used to determine the value of the unknown resistance, R_x. Note that as the balance condition is approached, less current flows in the bridge arm. The resistance R_g could be reduced to increase the sensitivity of the null detector. When the balance is reached, R_g could be reduced to zero and the null detector would read zero current.

■ EXAMPLE 3.30

In the bridge circuit shown in Fig. 3.51, find R_x under the following conditions.
(a) $V_{ab} = 0$.
(b) $V_{ab} = 3\,\text{V}$.
(c) $V_{ab} = -4\,\text{V}$.

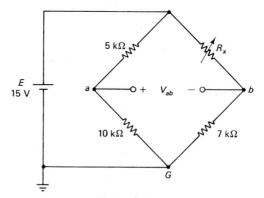

Figure 3.51

Solution (a) When $V_{ab} = 0$, the bridge is in the balance condition; therefore,

$$\frac{R_x}{5\,\text{k}\Omega} = \frac{7\,\text{k}\Omega}{10\,\text{k}\Omega}$$

$$R_x = 5\,\text{k}\Omega \times \frac{7\,\text{k}\Omega}{10\,\text{k}\Omega} = 3.5\,\text{k}\Omega$$

(b) In general, with a voltmeter connected between points a and b,

$$V_{ab} = V_a - V_b$$

$$= \frac{E}{5\,k\Omega + 10\,k\Omega} \times 10\,k\Omega - \frac{E}{7\,k\Omega + R_x} \times 7\,k\Omega$$

$$= \frac{150}{15} - \frac{105\,k\Omega}{7\,k\Omega + R_x} = 10 - \frac{105\,k\Omega}{7\,k\Omega + R_x}$$

With $V_{ab} = 3\,V$,

$$3 = 10 - \frac{105\,k\Omega}{7\,k\Omega + R_x}$$

$$\frac{105\,k\Omega}{7\,k\Omega + R_x} = 7$$

$$7\,k\Omega + R_x = 15\,k\Omega$$

Therefore,

$$R_x = 8\,k\Omega$$

(c) Using the same expression as above and substituting $V_{ab} = -4\,V$ yields

$$-4 = 10 - \frac{105\,k\Omega}{7\,k\Omega + R_x}$$

$$\frac{105\,k\Omega}{7\,k\Omega + R_x} = 14$$

$$7\,k\Omega + R_x = 7.5\,k\Omega$$

Therefore,

$$R_x = 0.5\,k\Omega$$

EXAMPLE 3.31

If $V_A = 2\,V$ in the bridge circuit shown in Fig. 3.52, find R_x.

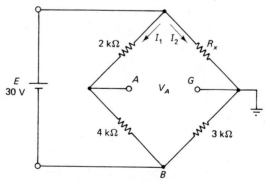

Figure 3.52

Solution   Here

$$V_{AB} = V_A - V_B$$

$$= V_{AG} - V_{BG} = V_{AG} + V_{GB}$$

with

$$V_{AB} = I_1 \times 4\,\text{k}\Omega$$

$$= \frac{30}{2\,\text{k}\Omega + 4\,\text{k}\Omega} \times 4\,\text{k}\Omega = 20\,\text{V}$$

and

$$V_{GB} = I_2 \times 3\,\text{k}\Omega = \frac{30}{R_x + 3\,\text{k}\Omega} \times 3\,\text{k}\Omega$$

Thus,

$$20 = 2 + \frac{90\,\text{k}\Omega}{R_x + 3\,\text{k}\Omega}$$

$$R_x + 3\,\text{k}\Omega = \frac{90\,\text{k}\Omega}{18} = 5\,\text{k}\Omega$$

Therefore,

$$R_x = 2\,\text{k}\Omega$$

DRILL PROBLEMS

Section 3.2

3.1. Obtain the currents I_2 and I_4 in the network shown in Fig. 3.53, using KCL.

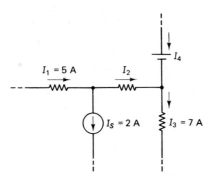

Figure 3.53

3.2. Determine the magnitude and direction of each of the currents I_3, I_5, and I_6 in the network shown in Fig. 3.54, using KCL.

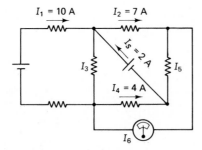

Figure 3.54

Section 3.3

3.3. Using KVL, determine the voltages V_3 and V_{ab} in the network shown in Fig. 3.55.

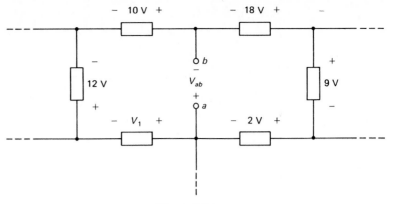

Figure 3.55

3.4. Obtain, using KVL, the voltages V_1 and V_2 in the network shown in Fig. 3.56.

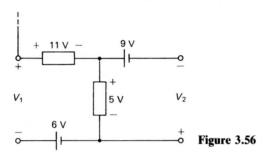

Figure 3.56

Section 3.4

3.5. In the network shown in Fig. 3.57, E is 15 V, R_1 is 6 Ω, R_2 is 10 Ω, and R_3 is 14 Ω. Obtain the total resistance seen by the source, R_T, the current, I, and the voltages V_1, V_2, V_3, and V_{ab}.

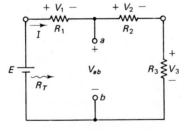

Figure 3.57

3.6. In a network similar to that shown in Fig. 3.57, the resistors have the same values as in Drill Problem 3.5. The source voltage has been changed such that the resulting current, I, is now 0.2 A. Find the value of E and the new values of V_1, V_2, V_3, and V_{ab}.

3.7. In a network similar to that shown in Fig. 3.57, E is 24 V, I is 0.1 A, R_1 is 120 Ω, and R_2 is 80 Ω. Find R_3. Determine the value of the voltages V_1, V_2, and V_3 and the amount of power dissipated in each resistor. Calculate the amount of power supplied by the source and check the validity of the principle of conservation of energy.

3.8. In a network similar to that shown in Fig. 3.57, I is 0.4 A, R_1 is 10 Ω, V_2 is 10 V, and P_3 (the power dissipated in R_3) is 7.2 W. Calculate the values of R_2, R_3, E, V_1, V_3, and V_{ab}.

3.9. Use the voltage-division principle to recalculate the values of V_1, V_2, V_3, and V_{ab} in the circuit described in Drill Problem 3.5.

3.10. In the circuit shown in Fig. 3.58, calculate the value of I and then determine the voltages V_a, V_b, V_c, and V_1.

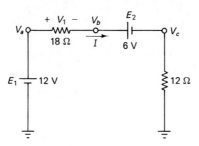

Figure 3.58

3.11. If the two sources E_1 and E_2 in the network shown in Fig. 3.58 are readjusted so that now E_1 is 6 V and E_2 is 12 V, recalculate the values of I, V_a, V_b, V_c, and V_1. What would you conclude about the new direction of I and the polarities of V_1 in this case?

Section 3.5

3.12. In the parallel circuit shown in Fig. 3.59, E is 12 V, R_1 is 30 Ω, R_2 is 60 Ω, R_3 is 120 Ω, and R_4 is 40 Ω. Find the total (or equivalent) resistance, R_T, as seen by the source. Calculate the values of I_T, I_1, I_3, and I_4 and prove the validity of KCL.

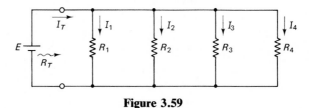

Figure 3.59

3.13. In a parallel circuit similar to that shown in Fig. 3.59, what should be the value of R_4 so that the total resistance becomes 16 Ω? Here R_1, R_2, and R_3 have the same values as in Drill Problem 3.12. If the current in R_4 is to be 0.25 A, find the required voltage of the power supply and calculate the new value of I_T in this case.

3.14. If the voltage source in Drill Problem 3.12 is replaced by a 0.5 A current source, use the current-division principle to calculate I_1, I_2, I_3, and I_4. Obtain the voltage across the terminals of the current source in this case.

3.15. Determine the values of R_T, I_T, I_1, and I_2 in the circuit shown in Fig. 3.60 if E is 30 V, R_1 is 100 Ω, and R_2 is 300 Ω.

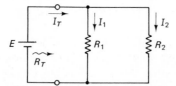

Figure 3.60

3.16. If R_1, in a circuit similar to that shown in Fig. 3.60, is 200 Ω, what should be the value of R_2 so that R_T is 160 Ω? If the current in R_1 is required to be 0.5 A, what will be the voltage of the necessary power supply in this case? Also calculate the values of I_2 and I_T.

3.17. In a circuit similar to that shown in Fig. 3.60, R_1 is 60 Ω and E is 9 V. The power dissipated in R_2 is 2.7 W. Obtain the values of R_2, I_1, I_2, and I_T in this case.

3.18. In a circuit similar to that shown in Fig. 3.60, R_1 is 20 Ω and the power dissipated in it is 20 W. Find the required value of R_2 so that the total current from the source, I_T, is 3 A. Obtain the values of E, I_1, and I_2 in this case and calculate the power provided by the source and dissipated in R_2. Does the principle of conservation of energy apply in this case?

3.19. Use the current-division principle to calculate the values of I_1, I_2, I_3, and I_4 in the circuit shown in Fig. 3.61.

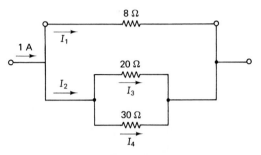

Figure 3.61

Section 3.6

3.20. Analyze the series–parallel circuit shown in Fig. 3.62 to obtain I_1, I_2, I_3, V_1, and V_2 if E is 16 V, R_1 is 80 Ω, R_2 is 200 Ω, and R_3 is 300 Ω.

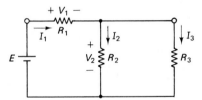

Figure 3.62

3.21. In a circuit similar to that shown in Fig. 3.62, R_1 is 50 Ω, R_2 is 100 Ω, and R_3 is 400 Ω. If V_2 is 5 V, calculate the values of I_1, I_2, I_3, E, and V_1 in this case.

3.22. In a circuit similar to that shown in Fig. 3.62, R_1 is 20 Ω, R_2 is 80 Ω, and R_3 is 120 Ω. If I_3 is 0.1 A, calculate I_2, V_2, I_1, V_1, and E in this case.

3.23. In the circuit examined in Drill Problem 3.22, E is to be changed so that I_1 is now 0.2 A. Find the required value of E in this case and calculate the new values of I_2, I_3, and V_2.

Section 3.7

3.24. In a balanced Wheatstone bridge (i.e., at the null condition), as shown in Fig. 3.50, calculate the values of R_x under the following observations:
   (a) $R_1 = R_2 = 10$ Ω and $R_3 = 350$ Ω
   (b) $R_1 = 10$ Ω, $R_2 = 100$ Ω, and $R_3 = 475$ Ω
   (c) $R_1 = 1000$ Ω, $R_2 = 10$ Ω, and $R_3 = 7942$ Ω

PROBLEMS

Note: The problems below that are marked with an asterisk (*) can be used as exercises in computer-aided network analysis or as programming exercises.

Section 3.2

(i)   **3.1.** Find the unknown currents in the circuit portions shown in Fig. 3.63.

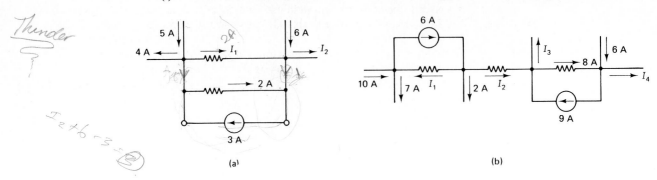

(a)   (b)

Figure 3.63

Section

(i)   **3.2.** Find the value of the current source I and the voltage source E in the circuit shown in Fig. 3.64.

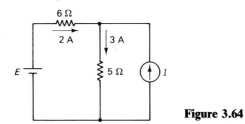

Figure 3.64

Section 3.3

(i)   **3.3.** Apply KVL to find the unknown voltages in the circuit shown in Fig. 3.65.

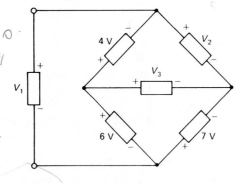

Figure 3.65

Section 3.4

(i)   **3.4.** Find the value of each of the unknown resistances R_1 and R_2 in the circuit shown in Fig. 3.66.

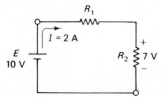

Figure 3.66

(i) **3.5.** Three resistors of value 10 Ω, 40 Ω, and 70 Ω are connected in series with a 12 V source. Find the total resistance of the circuit, R_T, the current in the circuit, I, and the voltage drop across each of the three resistors.

(i) **3.6.** Find the current I and the voltage V in the circuit of Fig. 3.67.

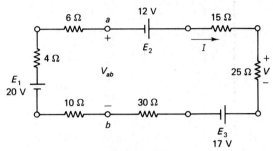

Figure 3.67

(i) **3.7.** In the same circuit shown in Fig. 3.67, calculate the voltage drop V_{ab}, the power dissipated in the 6 Ω and 15 Ω resistors, and the power supplied by the 12 V source.

(ii) **3.8.** In the circuit shown in Fig. 3.68, the voltage drop across R_1. V_1, is 12 V, while the power dissipated in R_2, P_2, is 160 mW, and $P_3 = 2P_2$ Find the values of R_1, R_2, and R_3. Also find the power dissipated in R_1, P_1.

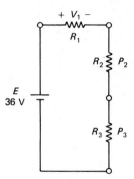

Figure 3.68

(i) **3.9.** What is the maximum voltage that can be applied to the series connection of two resistors, a 200 Ω, 8 W resistor and a 300 Ω, 3 W resistor? Which of the two resistors is the weaker link?

(i) **3.10.** When the switch is open in the circuit shown in Fig. 3.69, the current supplied by E is 1 mA. When the switch is closed, the current supplied by E is 10 mA. Find the values of E and R.

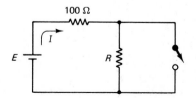

Figure 3.69

(ii) **3.11.** When the applied source has a voltage of 100 V in the circuit shown in Fig. 3.70, the source current is 10 A, while the voltage drops across R_1 and R_2 are 20 V and 30 V, respectively. Find R_3, R_T, and the power in R_3.

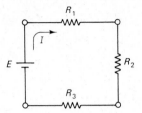

Figure 3.70

(i) **3.12.** The I–V characteristics of three resistors, R_1, R_2, and R_3, are shown in Fig. 3.71.

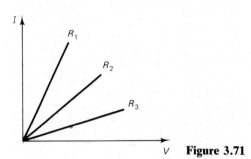

Figure 3.71

Superimpose on the same graph the I–V characteristics of the total resistance of R_1, R_2, and R_3 in series.

(ii) **3.13.** In the circuit shown in Fig. 3.72, E is 100 V, R_1 is 10 Ω, and P_2 is 160 W. Find the value(s) of R_2.

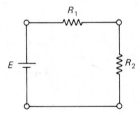

Figure 3.72

Section 3.5

(i) **3.14.** The four resistors, $R_1 = R_2 = 100\ \Omega$, $R_3 = 200\ \Omega$, and $R_4 = 300\ \Omega$, are connected in parallel. What is the value of the total (or equivalent) resistance of this circuit?

(i) **3.15.** When the parallel circuit described in Problem 3.14 is connected in parallel with a 6 A current source, what is the voltage across the combination of resistors? What is the current in each of them?

(i) **3.16.** What is the current in each of the parallel resistors shown in Fig. 3.73? What is the source's current?

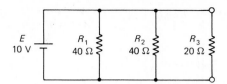

Figure 3.73

(i) **3.17.** Find R_T for the parallel resistors combination shown in Fig. 3.74.

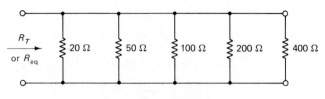

Figure 3.74

(i) **3.18.** In the parallel circuit shown in Fig. 3.75, find R_2. What is the power dissipated in

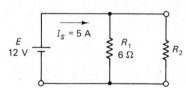

Figure 3.75

R_1 and R_2 and the power supplied by the source? Does the conservation of power hold in this circuit?

(ii) **3.19.** Applying KCL to the circuit shown in Fig. 3.76, find V_{ab}, I_3, I_4, and I_5.

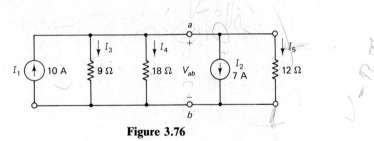

Figure 3.76

(i) **3.20.** In the circuit of Problem 3.19, what is the amount of power supplied (or dissipated) by each component? Prove that the conservation of power holds for this circuit.

(i) **3.21.** Three resistors R_1, R_2, and R_3 are connected in parallel to a voltage source. An increase of R_2 results in an increase of
   (a) Current through R_1.
   (b) Voltage across R_2.
   (c) Power in R_3.
   (d) None of the above.

(i) **3.22.** In the parallel circuit shown in Fig. 3.77, $G_1 = 100\,\text{mS}$, $G_2 = 100\,\text{mS}$, $G_3 = 200\,\text{mS}$, and $I_T = 10\,\text{A}$. Find the value of E and the total power dissipated in this parallel load.

Problems

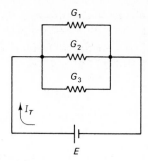

Figure 3.77

(i) **3.23.** The $I-V$ characteristics of three resistors, R_1, R_2, and R_3, are shown in Fig. 3.78. Superimpose on the same graph the $I-V$ characteristics of the equivalent resistance of the three resistors in parallel.

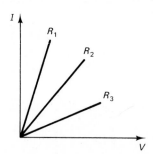

Figure 3.78

(ii) **3.24.** Three resistors are connected in parallel across a 100 V source. The first resistor dissipates 5.0 W, the second resistor draws 33.33 mA, and the third resistor has a resistance of 6 kΩ. Find the equivalent resistance of the circuit and the amount of the current flowing in the voltage source.

(ii) **3.25.** In the circuit shown in Fig. 3.79, find the magnitude and the direction of the current in
   (a) Line AB.
   (b) Line BC.

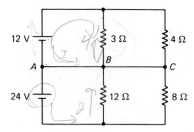

Figure 3.79

Section 3.6

(i) **3.26.** If $R_1 = 10\,\Omega$, $R_2 = 6\,\Omega$, and $R_3 = 30\,\Omega$ in the circuit shown in Fig. 3.80, find the value of E if I is 2 A. What is the value of the voltage drop V?

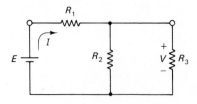

Figure 3.80

(i)   **3.27.** In a circuit similar to that shown in Fig. 3.80, E is 10 V and I is 0.5 A. Find R_3 if R_1 is 4 Ω and R_2 is 20 Ω. What is the value of the voltage drop V?

(i)   ***3.28.** Find the value of the current I, the voltage V, and the power dissipated in the 4 Ω resistor in the circuit shown in Fig. 3.81.

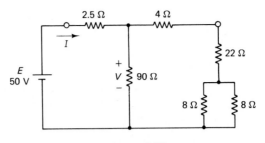

Figure 3.81

(ii) ***3.29.** Through the network simplifications using the equivalent resistance concepts, find the value of the voltages V_1 and V_2 in the circuit shown in Fig. 3.82.

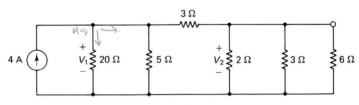

Figure 3.82

(ii) ***3.30.** Find the value of the currents I_1 and I_2 in the circuit shown in Fig. 3.83.

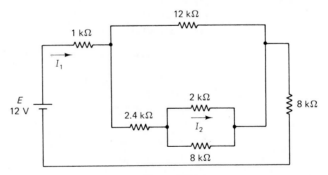

Figure 3.83

(i)   ***3.31.** Analyze the circuit shown in Fig. 3.84, to calculate I_1, I_2, and V.

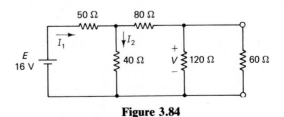

Figure 3.84

(ii) **3.32.** In the series circuit shown in Fig. 3.85, E is 36 V, I is 2 A, while V_1/V_2 is 2 and V_2/V_3 is 3. Find the values of the three resistors R_1, R_2, and R_3.

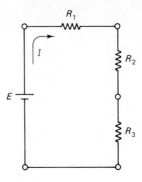

Figure 3.85

(i) ***3.33.** Using current division to obtain I_1 and I_2 and voltage division to obtain V_2 and V_3, analyze the circuit shown in Fig. 3.86. Start by obtaining I and V_1.

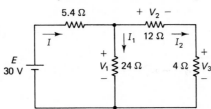

Figure 3.86

(i) **3.34.** If V_A in the circuit shown in Fig. 3.87 is 16 V, find the value of E.

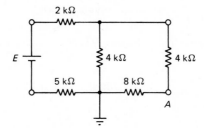

Figure 3.87

(i) **3.35.** In Fig. 3.88, the potential of point A, V_A, is -60 V. Find
    (a) The value of E.
    (b) The power dissipated by the 5 kΩ resistor.

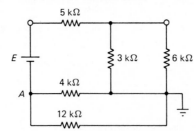

Figure 3.88

(i) **3.36.** In the series–parallel circuit shown in Fig. 3.89:
    (i) A decrease in R_2 results in a decrease of
        (a) The voltage across R_2.
        (b) The voltage across R_1.
        (c) The current through R_2.
        (d) None of these.

(ii) An increase in R_1 results in an increase of
  (a) The power dissipated by R_2.
  (b) The current in R_3.
  (c) The voltage across R_1.
  (d) None of these.

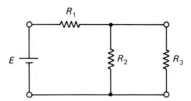

Figure 3.89

(ii) 3.37. In the circuit shown in Fig. 3.90:

$$R_1 = 16\,\Omega \text{ with a power rating of } 9\text{ W}$$
$$R_2 = 4\,\Omega \text{ with a power rating of } 16\text{ W}$$
$$R_3 = 3\,\Omega \text{ with a power rating of } 12\text{ W}$$
$$R_4 = 6\,\Omega \text{ with a power rating of } 6\text{ W}$$

What is the maximum value of E, E_{max}, that can be safely applied to the circuit without exceeding the rating of any of the resistors? Identify the weakest link in this circuit.

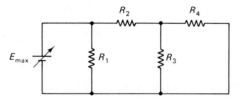

Figure 3.90

(ii) 3.38. For a series–parallel circuit similar to that shown in Fig. 3.89, the values in the following table are known. Fill in the blanks in this table.

Component	$R\,(\Omega)$	$V\,(\text{V})$	$I\,(\text{A})$	$P\,(\text{W})$
R_1				
R_2	60			240
R_3				
Total		300		900

(i) 3.39. In the circuit shown in Fig. 3.91, find the value of R and the power dissipated in it.

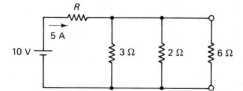

Figure 3.91

(ii) *3.40. Find the current in the 5 Ω resistor in the circuit shown in Fig. 3.92.

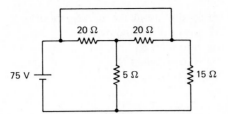

Figure 3.92

(ii)   **3.41.**   Find the value of R in the circuit of Fig. 3.93.

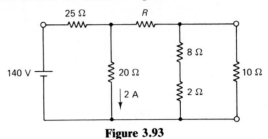

Figure 3.93

(ii)   **3.42.**   Find the values of E and R in the circuit shown in Fig. 3.94.

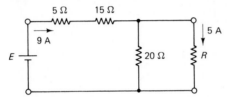

Figure 3.94

Section 3.7

(i)   **3.43.**   Figure 3.95 shows a Wheatstone bridge under balanced conditions. Find R_x with $R_1 = 150\ \Omega$, $R_2 = 30\ \Omega$, and $R_3 = 10\ \Omega$.

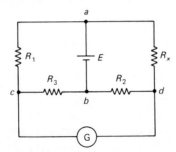

Figure 3.95

(ii)   **3.44.**   Figure 3.96 shows another form of Wheatstone bridge. Find R_x if the conditions for bridge balance were achieved with $R_1 = 10\ \Omega$, $R_2 = 60\ \Omega$, and $R_3 = 120\ \Omega$.

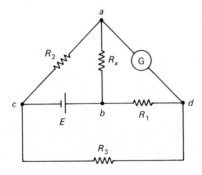

Figure 3.96

(i) ***3.45.** In the Wheatstone bridge shown in Fig. 3.97, find
  (a) V_A.
  (b) V_B.
  (c) V_{AB}.

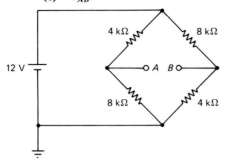

Figure 3.97

General Problems

(ii) **3.46.** In the circuit shown in Fig. 3.98, the current through R_4 is 2 A. The voltages across R_1 and R_2 were measured to be 10 V and 12 V, respectively. The applied source has a voltage of 34 V, while R_3 is 15 Ω. Find the values of the resistances R_1, R_2, and R_4.

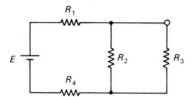

Figure 3.98

(ii) **3.47.** In the circuit shown in Fig. 3.99, I_T is 12 A, I_3 is 8 A, P_2 is 96 W, and R_1 is 6 Ω. Find E.

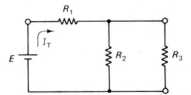

Figure 3.99

(ii) **3.48.** In a series–parallel circuit similar to that shown in Fig. 3.99, E is 100 V, I_T is 2 A, P_1 is 120 W, and $R_1 = R_2$. Find the value of R_3.

(i) **3.49.** In the circuit of Fig. 3.100, V_A is 20 V. Find E and I_T.

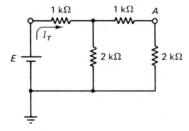

Figure 3.100

(i) ***3.50.** In the circuit shown in Fig. 3.101, find
  (a) The voltage across the 60 Ω resistor.
  (b) The current through the 40 Ω resistor.

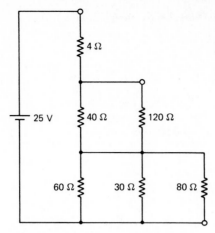

Figure 3.101

(ii)   **3.51.**   The voltage across the 4 kΩ resistor in the circuit shown in Fig. 3.102 is 40 V. Find
      (a) E.
      (b) V_{AB}.

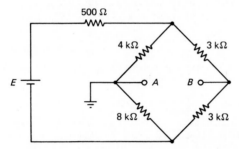

Figure 3.102

(ii)   **3.52.**   The power delivered by E in the circuit of Fig. 3.103 is 160 W. Find the value of
      R.

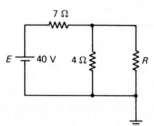

Figure 3.103

(i)   **3.53.**   Using KVL and KCL, find E and I in the circuit shown in Fig. 3.104.

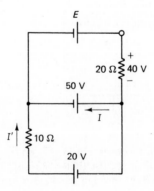

Figure 3.104

(ii) ***3.54.** For the circuit shown in Fig. 3.105, find

    (a) $I_{5\Omega}$.
    (b) $P_{8\Omega}$.
    (c) $V_{2\Omega}$.
    (d) I_{AB}.
    (e) I_{9V}.

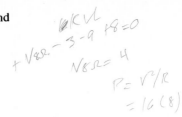

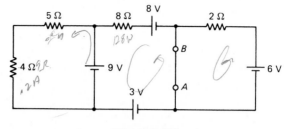

Figure 3.105

(ii) **3.55.** Find the value of the resistance R in the circuit shown in Fig. 3.106 if the ammeter reads 6 A.

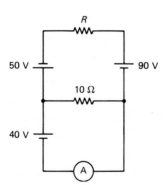

Figure 3.106

(ii) ***3.56.** In the circuit of Fig. 3.107, find

    (a) The current in the 20 V source.
    (b) The power in the 100 V source.
    (c) The power in the 30 V source.

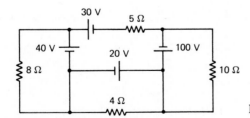

Figure 3.107

(ii) **3.57.** In the circuit shown in Fig. 3.108:

    (a) Find R and $I_{2\Omega}$.
    (b) What should be the new value of R so that when point A is grounded in addition to the existing ground, the 8 A current in the 10 V source remains unchanged?

Problems

117

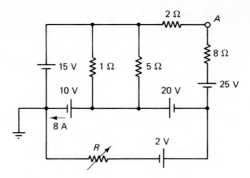

Figure 3.108

(i) **3.58.** Two resistors, R_1 and R_2, with a ratio of $R_2/R_1 = 3$ and a 120 V source are available.

    (a) What is the voltage across each resistor when they are connected in series with the source?

    (b) What is the value of each of the resistors if the total current drawn from the 120 V source is 6 A when the two resistors are connected in parallel with the source?

(ii) **3.59.** In the circuit shown in Fig. 3.109, V_2 is 8 V, I_3 is 1 A, and the power supplied by the source is 60 W. Find R_2, R_4, I_1, and E.

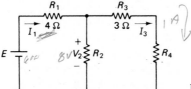

Figure 3.109

(ii) **3.60.** In the circuit shown in Fig. 3.110, find

    (a) V_B.

    (b) E.

    (c) The magnitude and direction of the current in the 20 Ω resistor.

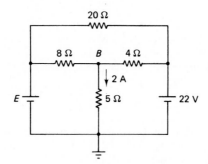

Figure 3.110

GLOSSARY

Branch: A portion of an electrical network consisting of the element connected between two nodes.

Bridge Balance Condition: The condition existing in a bridge circuit at which no current flows in the bridge arm.

Ground: The reference node in an electric circuit whose voltage is considered to be 0 V. This node may or may not be connected to the actual physical ground (earth).

Loop: A closed path in an electric circuit. The starting and end points for tracing the path are in effect the same node.

KCL: Kirchhoff's current law, which states that the algebraic sum of the currents entering and leaving a node in an electric circuit is zero.

KVL: Kirchhoff's voltage law, which states that the algebraic sum of the voltage rises and drops around a closed path in an electric circuit is zero.

Mesh: A special case of a closed path in an electric circuit. A mesh is a loop that does not contain any branches.

Node: The point of connection (junction) of two or more elements (branches).

Null Detector: A sensitive instrument, usually a galvanometer, with a center-scale null position, used to indicate and help in arriving at the situation corresponding to zero current flow in the bridge arm.

Parallel Connection: The method of connection of a number of elements between two nodes in an electric circuit such that all the parallel-connected elements have the same voltage across each.

Series Connection: The method of connection of a number of elements in an electric circuit such that two elements have only one common node between them and no third element is connected to that common node. Series-connected elements each have the same current flow.

Wheatstone Bridge: A bridge circuit used mainly for measurement of the value of a resistance (or an impedance) of an element.

ANSWERS TO DRILL PROBLEMS

3.1. $I_2 = 3$ A, $I_4 = 4$ A

3.3. $V_1 = 5$ V, $V_{ab} = 7$ V

3.5. $R_T = 30\,\Omega$, $I = 0.5$ A, $V_1 = 3$ V, $V_2 = 5$ V, $V_3 = 7$ V, $V_{ab} = 12$ V

3.7. $R_3 = 40\,\Omega$, $V_1 = 12$ V, $V_2 = 8$ V, $V_3 = 4$ V, $P_1 = 1.2$ W, $P_2 = 0.8$ W, $P_3 = 0.4$ W, $P_T = 2.4$ W $= P_1 + P_2 + P_3$

3.9. $V_1 = 3$ V, $V_2 = 5$ V, $V_3 = 7$ V, $V_{ab} = 12$ V

3.11. $I = -0.2$ A, $V_a = 6$ V, $V_b = 9.6$ V, $V_c = -2.4$ V, $V_1 = -3.6$ V The current is now flowing in the opposite direction. The polarities of V_1 are reversed.

3.13. $R_4 = 240\,\Omega$, $E = 60$ V, $I_T = 3.75$ A

3.15. $R_T = 75\,\Omega$, $I_T = 0.4$ A, $I_1 = 0.3$ A, $I_2 = 0.1$ A

3.17. $R_2 = 30\,\Omega$, $R_T = 20\,\Omega$, $I_1 = 0.15$ A, $I_2 = 0.3$ A, $I_T = 0.45$ A

3.19. $I_1 = 0.6$ A, $I_2 = 0.4$ A, $I_3 = 0.24$ A, $I_4 = 0.16$ A

3.21. $I_1 = 0.0625$ A, $I_2 = 0.05$ A, $I_3 = 0.0125$ A, $E = 8.125$ V, $V_1 = 3.125$ V

3.23. $E = 13.6$ V, $I_2 = 0.12$ A, $I_3 = 0.08$ A, $V_2 = 9.6$ V

4 PRACTICAL SOURCES

OBJECTIVES

- Gaining knowledge and appreciation of the fundamental differences between ideal and practical voltage and current sources, and the circuit parameters used to represent these source models in electric circuits.

- Ability to calculate and measure the relevant practical voltage and current source parameters.

- Understanding the applications of the load-line graphical analysis technique in determining the voltage and current in electrical circuits, particularly those that include nonlinear elements.

- Comprehension of the fact that a practical source can be treated and modelled as either a practical voltage source or a practical current source. As a consequence, the reader should be able to understand the relationships between the source models parameters and be able to convert a voltage source model to a current source model and vice versa, in order to facilitate certain circuit analysis requirements.

- Understanding the concepts and applications of the maximum power transfer theorem, the matching condition, and the power transfer efficiency.

- Ability to design various types of practical voltage dividers.

4.1 IDEAL AND PRACTICAL VOLTAGE SOURCES

As might have been observed from many of the previous examples, a 12 V dc voltage source drives 2 A in a 6 Ω resistor, 0.5 A in a 24 Ω resistor, or 0.1 A in a 120 Ω resistor. In all such cases,

$$V = IR = E$$

This type of dc voltage source is said to be an ideal voltage source.

Definition. An *ideal voltage source* is a device that provides a constant voltage across its terminals ($V = E$) no matter what current is drawn from it (i.e., independent of the value of the load resistance connected to it).

The model of an ideal dc voltage source of value E is shown in Fig. 4.1 inside the dashed box representing such a source.

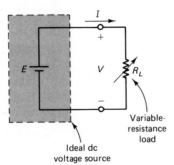

Variable-resistance load

Ideal dc voltage source

Figure 4.1

Graphically, one can represent the terminal V–I characteristics of an ideal dc voltage source as a horizontal line, parallel to the current axis and intercepting the voltage axis at the value E. This is shown in Fig. 4.2. Whether the current is small (with a high load resistance) or large (with a low load resistance), the terminal voltage of the ideal voltage source is the constant value E.

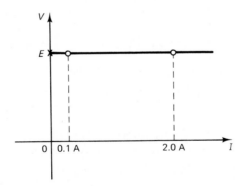

Figure 4.2

In practice, however, dc voltage sources do not exhibit such behavior. One usually observes that as the current I drawn from the source is increased, by connecting lower load resistance across the source terminals, the terminal voltage across the source (and load), V, decreases. One can intuitively conclude that such a drop in the source voltage V from its ideal value E must have occurred across some element inside the source.

Not only that, but as the current increases, the source voltage decreases, meaning that an increased voltage drop is now exhibited across this internal element. This element must then be a resistance, called the *internal resistance* of the source, R_{int}, since V_{int} is proportional to the current I. V_{int} is called the *internal voltage drop*.

The model that can appropriately explain the terminal $V-I$ characteristics of a practical voltage source is thus as shown inside the dashed box (source) in Fig. 4.3. A circuit model is only a tool used in circuit analysis to formulate and explain the specific behavior of the device modeled. In many cases such models are approximate and are correct only within a finite operating range. It should be noted, then, that R_{int} is not a real element and its terminals are inaccessible.

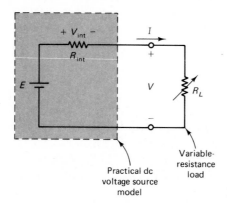

Practical dc
voltage source
model

Variable-
resistance
load

Figure 4.3

The circuit in Fig. 4.3 can be described using KVL. Here

$$E = V_{\text{int}} + V$$
$$= IR_{\text{int}} + V \tag{4.1}$$

or

$$\boxed{V = E - R_{\text{int}}I} \tag{4.2}$$

Also,

$$V = IR_L \quad \text{(Ohm's law for the load resistance)} \tag{4.3}$$

From Eqs. (4.3) and (4.1) it can easily be concluded that

$$I = \frac{E}{R_{\text{int}} + R_L} \tag{4.4}$$

Equation (4.2) describes the terminal $V-I$ characteristics of the practical voltage source. It is similar to the equation of a straight line:

$$y = a - mx$$

where y is the vertical axis variable (V), x the horizontal axis variable (I), a the intercept of the line with the y axis (E), and $-m$ is the slope of the line ($-R_{\text{int}}$). This $V-I$ characteristics is shown in Fig. 4.4. It is clear that by varying R_L, hence I, and measuring V, we can obtain such characteristics experimentally. As I increases, V decreases since $V_{\text{int}}(=I \times R_{\text{int}})$ is increased, exactly as explained in examining the behavior of practical voltage sources.

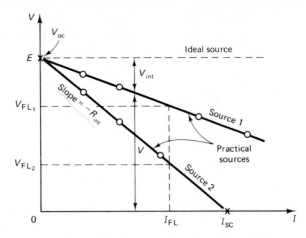

Figure 4.4 *V–I* characteristics of practical voltage sources.

The characteristics of two sources are shown in Fig. 4.4. Since the slope $(-m)$ is equal to $-R_{int}$, it is clear that source 1 has a lower internal resistance than that associated with source 2. The dashed line corresponds to the case of the ideal voltage source, with a zero slope; that is, the ideal voltage source has a zero internal resistance (as the model of Fig. 4.1 depicts).

Two pieces of information are needed to determine completely the characteristics of a given source. These are called the *source parameters*. They are:

1. *E*, the open-circuit voltage of the source. The reason for such a name is that when $R_L = \infty$ (open-circuit load, or terminals), according to Eq. (4.4), the current *I* from the source is zero. Substituting in Eq. (4.2) gives us

$$V \text{ (at } R_L = \infty) = E - R_{int} \times 0$$

that is,

$$V_{oc} = E \qquad (4.5)$$

V_{oc} (or *E*) can easily be measured using an ideal voltmeter (whose resistance R_V approaches infinity)—of course, while R_L is disconnected (no load or open circuit).

2. R_{int}, the value of the internal resistance in the source's circuit model.

This is similar to requiring two points on a line to determine the line completely. It should be noted that at a certain setting of the source voltage (if it is a variable dc source), the source parameters *E* and R_{int} are constants.

Based on these two parameters one can determine another interrelated source parameter. This is called the short-circuit current, I_{sc}. A short circuit means that R_L is zero (just a wire, or an ideal ammeter whose resistance R_A approaches zero). From Eq. (4.4),

$$I \text{ (at } R_L = 0) = \frac{E}{R_{int} + 0}$$

Therefore,

$$I_{sc} = \frac{E}{R_{int}} \qquad (4.6)$$

Substituting from Eq. (4.6) in Ohm's law for R_L [Eq. (4.3)], the terminal voltage across the source, V, is zero under short-circuit conditions. This is obvious since a zero load resistance has zero voltage across it, provided that the current flow in it is finite, as it is here. Figure 4.4 clearly shows that the intercepts of the V–I characteristics of the source with the voltage and current axes determine V_{oc} and I_{sc}, respectively.

In many practical situations, it is quite important to determine the source parameters, for example the model of an active transducer such as a photocell. One of the following experimental techniques can be used:

1. Connect a variable resistance as a load across the terminals of the source. Vary R_L and measure V and I. Plot such point coordinates on a graph paper. Ideally, only two points are required, but practically a sufficient number of measurements should be obtained to be able to draw a best-fitting line. The slop of the line is $-R_{int}$, while the intercept with the voltage axis (at $I = 0$) is the value of E.

2. Measure E, the open-circuit voltage of the source, using a high-resistance voltmeter. With the voltmeter still connected across the source terminals, also connect a variable-resistance load (a decade resistance box, for example) across the terminals of the source, starting with R_L very high. Reduce R_L while monitoring V until V is reduced to half its open-circuit reading (i.e., $0.5E$). At this setting, the value of R_L is equivalent to the internal resistance of the source model. Note that here

$$V_{int} = E - V = E - 0.5E = 0.5E = V$$

Therefore,

$$IR_{int} = IR_L$$
$$R_{int} = R_L$$

3. Use a voltmeter to measure $V_{oc}(= E)$. Then connect an ammeter across the terminals of the source to measure I_{sc}. From Eqs. (4.5) and (4.6) it is clear that

$$R_{int} = \frac{E}{I_{sc}} = \frac{V_{oc}}{I_{sc}} = \frac{\text{voltmeter reading (at open-circuit conditions)}}{\text{ammeter reading (at short-circuit conditions)}}$$

Even though the third method above looks to be the simplest, one should be warned that it is rarely used. This is because under short-circuit conditions, I_{sc} could be quite large. Allowing the source to drive such a current in a zero-resistance load, even momentarily, could permanently ruin some of the components inside the source. This method should be avoided unless one is sure beforehand that R_{int} is going to be quite large (i.e., I_{sc} is a very small and safe current). The V–I characteristics of the source are also called the source's *regulation curve* or *load line*, for reasons that will become apparent shortly.

Most sources are specified to operate up to a maximum load current, known as the full-load current, I_{FL}. This, in fact, corresponds to the minimum recommended load resistance. As can be observed from Fig. 4.4, this value of the source current determines what is known as the full-load voltage, V_{FL}. The open-circuit voltage is also called the no-load voltage, V_{NL}.

The measure of the variations in the source voltage, as the load is varied, is called the *voltage regulation* of the source. It is defined as

$$\text{V.R. (\%)} = \text{voltage regulation (\%)} = \frac{V_{NL} - V_{FL}}{V_{FL}} \times 100$$

The lower the voltage regulation of a source, the better the source would be, as this indicates a smaller internal voltage drop and smaller internal resistance. The ideal voltage source would correspond to zero voltage regulation.

Remark. In practice, many approximations are incorporated in circuit analysis. A good approximation usually simplifies the analysis, thus reducing the time required to obtain a numerical value of a circuit variable, called the *response*. If the value of the response thus obtained is within 1% of its measured (or exact) value, we can be confident about the suitability of such an approximation.

Electric circuits are easier to analyze if the sources are considered to be ideal, since in such cases the internal resistances are removed from the circuit, that is,

$$R_{int} \cong 0 \, \Omega$$

This was the approximation used in analyzing the circuits in previous chapters, since all sources are in general "practical," containing a finite internal resistance in their circuit models.

However, for an approximation to be "good," one has to identify the range of its suitability in practice. Considering Eqs. (4.4) and (4.2), and substituting I from Eq. (4.4) into Eq. (4.2), we obtain

$$V = E - \frac{R_{int}E}{R_{int} + R_L} = E\frac{R_L}{R_{int} + R_L} = E\frac{1}{1 + R_{int}/R_L} \tag{4.7}$$

For $R_L \gg R_{int}$, say that R_L is 100 times (or more) larger than R_{int}; then

$$V = \frac{1}{1 + 0.01} E = 0.99E \cong E$$

Thus, in practice, with a load resistance more than 100 times larger than the source resistance, the source can be considered "approximately" ideal. In such cases the internal resistance of the source can be neglected.

EXAMPLE 4.1

A practical voltage source delivers 0.5 A to a load consisting of three resistors, 30 Ω, 100 Ω, and 150 Ω, in series. When the three resistors are in parallel, the source's current is 3.75 A. Find
(a) The circuit model of the source and its short-circuit current.
(b) The source's terminal voltage and current if the load is the 30 Ω resistor in series with the parallel connection of the 100 Ω and 150 Ω resistors.

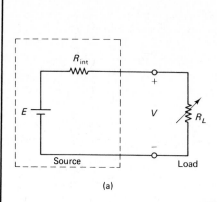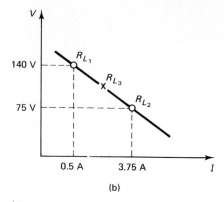

(a)   (b)

Figure 4.5

Solution   (a) E and R_{int}, the source model parameters, are unknown and have to be determined [see Fig. 4.5(a)]. Two conditions [i.e., two point coordinates for V and I as shown in Fig. 4.5(b)] are needed to define the source's characteristics completely. These conditions are provided in the problem, corresponding to the two values of the overall load resistance R_L.

$$R_{L_1} = \text{total resistance of the series connected elements}$$
$$= 30 + 100 + 150 = 280 \ \Omega$$

Here $I = 0.5 \ \text{A}$ and $V = IR_{L_1} = 0.5 \times 280 = 140 \ \text{V}$. Thus from the V–I characteristics of the source [Eq. (4.2)], we can write

$$140 = E - 0.5R_{\text{int}} \tag{4.8}$$

$R_{L_2} = \text{total resistance of the parallel-connected elements} = \dfrac{1}{\frac{1}{30} + \frac{1}{100} + \frac{1}{150}}$

$$= 20 \ \Omega$$

Here $I = 3.75 \ \text{A}$ and $V = IR_{L_2} = 3.75 \times 20 = 75 \ \text{V}$. Thus for this second point on the line, Eq. (4.2) gives

$$75 = E - 3.75R_{\text{int}} \tag{4.9}$$

It is now a simple matter to solve Eqs. (4.8) and (4.9) simultaneously. Subtracting Eq. (4.9) from Eq. (4.8) yields

$$140 - 75 = (E - 0.5R_{\text{int}}) - (E - 3.75R_{\text{int}})$$
$$65 = 3.25R_{\text{int}}$$

Therefore,

$$R_{\text{int}} = 20 \ \Omega$$

Substituting in Eq. (4.8) or (4.9) gives us

$$E = 140 + 10$$
$$= 75 + 75 = 150 \ \text{V}$$

Then, using Eq. (4.6), we have

$$I_{sc} = \frac{E}{R_{int}}$$

$$= \frac{150}{20} = 7.5 \text{ A}$$

(b) Now that the source parameters are known, all we need to solve this simple series circuit is the new value of the load resistance:

$$R_{L_3} = 30 + \frac{100 \times 150}{100 + 150} = 90 \ \Omega$$

Thus

$$I = \frac{E}{R_{int} + R_{L_3}} = \frac{150}{20 + 90} = 1.36 \text{ A}$$

and

$$V = IR_{L_3} = 122.73 \text{ V}$$

EXAMPLE 4.2

A certain power supply delivers 64 W if either a 4 Ω or a 9 Ω load resistance is connected to it. What should be the value of the load resistance so that the source's current becomes 2.5 A? What is the terminal voltage of the source in this case?

Solution Here also we have to start by determining the source parameters. The conditions required are given in terms of power in the load resistance R_L. It is easy to convert such information into the V–I form, using $P_L = I^2 R_L$. For $R_{L_1} = 4 \ \Omega$ and $P_L = 64 = I^2 \times 4$,

$$I = 4 \text{ A} \quad \text{and} \quad V = IR_{L_1} = 4 \times 4 = 16 \text{ V}$$

For $R_{L_2} = 9 \ \Omega$ and $P_L = 64 = I^2 \times 9$,

$$I = 2.667 \text{ A} \quad \text{and} \quad V = IR_{L_2} = \frac{8}{3} \times 9 = 24 \text{ V}$$

Using these V–I coordinates of each of the two points, the source characteristics [Eq. (4.2)] give

$$16 = E - 4R_{int} \qquad \text{(corresponding to the case of } R_{L_1}\text{)}$$

and

$$24 = E - 2.667R_{int} \qquad \text{(corresponding to the case of } R_{L_2}\text{)}$$

As in Example 4.1 the two equations above can be solved by subtraction. Here

$$8 = 1.333R_{int}$$

Therefore,

$$R_{int} = 6 \ \Omega$$

and

$$E = 16 + 4 \times 6 = 40 \text{ V}$$

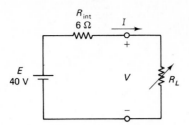

Figure 4.6

Now the problem simplifies to solving a simple series circuit, to find the value of the unknown load resistance R_L that draws 2.5 A from the source, as shown in Fig. 4.6. Alternatively, since the source characteristics are now completely defined,

$$V = E - R_{int}I \qquad \text{that is, } V = 40 - 6I$$

it is now easy to find V corresponding to $I = 2.5$ A. Here

$$V = 40 - 6 \times 2.5 = 25 \text{ V}$$

and

$$R_L = \frac{V}{I} = \frac{25}{2.5} = 10 \, \Omega$$

4.2 LOAD-LINE ANALYSIS

Any network that can be simplified through an equivalent circuit to a form similar to the practical voltage source model can be solved using the graphical technique called *load-line analysis*. This is quite a useful method. In particular, if the load (or the source) has a nonlinear V–I characteristic, for example diodes or lamps, this is the simplest method of analysis. The primary limitation of this technique is accuracy, due to its graphical nature. The basic idea behind this method is that the values of V and I at the terminals of the source are exactly the same as the voltage across and the current into the load R_L, as indicated in Fig. 4.7.

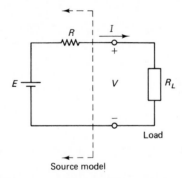

Source model

Figure 4.7

The method is quite simple. On the graph representing the I–V characteristics of the load, we also draw the I–V characteristics of the source. Notice that the axes have been interchanged compared with the source characteristics examined in Section 4.1. It is easier to draw the source characteristics using the two axes intercepts E and I_{sc}. This is shown in both Figs. 4.8 and 4.9.

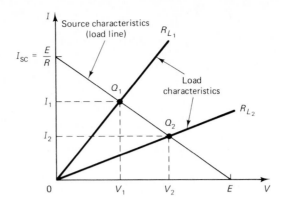

Figure 4.8 Variable-resistance linear load.

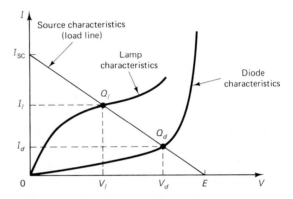

Figure 4.9 Nonlinear load.

The point of intersection of the load and the source characteristics represents the only condition where V and I are the same for both devices. Thus this point determines the coordinates of the current and voltage representing the solution of the circuit. This point is called the Q (for "quiescent") or *operating point*. A given source has specific fixed characteristics since E and R (or R_{int}) are constant parameters. Different loads, or a variable-resistance linear load, can be connected to this source. All the Q points of these various loads must fall on the source characteristics, as depicted by the points Q_1 and Q_2 in Fig. 4.8 and by the points Q_1 and Q_d in Fig. 4.9. For this reason the source characteristics are referred to as the *load line,* since the operating points of the loads connected to this source are all points on this line.

Linear resistance loads, obeying Ohm's law, have straight-line I–V characteristics. Certainly, in such cases circuit solution is very simple analytically, as the circuit is a simple series circuit. However, when the load has nonlinear I–V characteristics, analytical algebraic solution could be time consuming. Here the simplicity of the load-line technique is advantageous.

■ EXAMPLE 4.3

Find the voltage across and the current through the 5 kΩ load resistance in the circuit shown in Fig. 4.10.

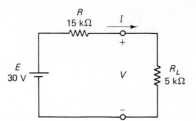

Figure 4.10

Solution   Using the load-line method, one starts first by determining the intercepts of the load line.

$$E = 30 \text{ V}$$

$$I_{sc} = \frac{E}{R} = \frac{30}{15 \text{ k}\Omega} = 2 \text{ mA}$$

The I–V source characteristics can then easily be drawn on graph paper, as shown in Fig. 4.11. Since the load R_L obeys Ohm's law,

$$V = I \times 5 \text{ k}\Omega$$

the I–V load characteristics is a straight line passing through the origin. Only one point is then needed to draw this line. If V is, say, 10 V, then I is

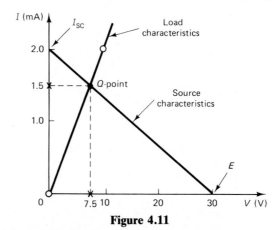

Figure 4.11

$10/5 \text{ k}\Omega = 2 \text{ mA}$. These are the coordinates of a point on the load characteristics, which can then be easily drawn. The point of intersection (the Q point) determines the solution of the circuit. Here

$$I = 1.5 \text{ mA} \qquad \text{and} \qquad V = 7.5 \text{ V}$$

Analytically, one can easily solve this simple series circuit. Using the voltage-division principle yields

$$V = E \frac{R_L}{R + R_L} = 30 \times \frac{5 \text{ k}\Omega}{15 \text{ k}\Omega + 5 \text{ k}\Omega} = 7.5 \text{ V}$$

Then

$$I = \frac{E}{R_L} = \frac{7.5}{5 \text{ k}\Omega} = 1.5 \text{ mA}$$

confirming the load-line results.

EXAMPLE 4.4

The source in Example 4.3 is connected to a lamp whose I–V characteristics are shown in Fig. 4.12. Using the load-line techniques, find the voltage across and the current through the lamp.

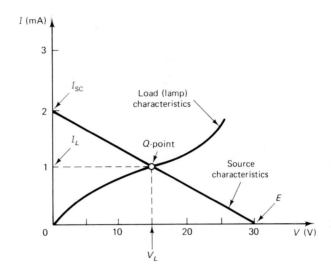

Figure 4.12

Solution The source characteristics (load line) is drawn on the same graph (Fig. 4.12) using the axes intercepts

$$E = 30 \text{ V}$$

and

$$I_{sc} = \frac{E}{R} = \frac{30}{15 \text{ k}\Omega} = 2 \text{ mA}$$

The coordinates of the point of intersection, the Q point, determine the solution of the circuit (i.e., the voltage and current of the load). Here, reading directly from the graph in Fig. 4.12, we obtain

$$V_L = 15 \text{ V}$$

and

$$I_L = 1 \text{ mA}$$

EXAMPLE 4.5

A photoconductor cell, whose I–V characteristics are shown in Fig. 4.13(a), is connected to a source as shown in Fig. 4.13(b). Determine the readings of an ideal voltmeter (V_{ph}) connected across the cell, when the light intensities are 0.1, 0.2, and 0.4 lumen. If the power dissipation capability of the photoconductor cell is 8 mW, show that this circuit operates safely at all times (no damage to the cell).

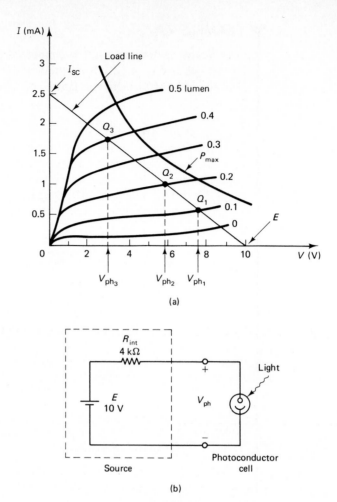

(a)

(b)

Figure 4.13

Solution The characteristics of the photoconductor cell indicate that this device is a nonlinear resistor whose $I-V$ characteristics change according to the intensity of the light falling on the cell. For a fixed voltage across the cell, the higher the intensity of the light, the more conductive the cell becomes, resulting in higher current flow, as shown in Fig. 4.13(a).

On the same graph, the load line is drawn using the axes intercepts as determined from the source parameters. Here

$$E = 10 \text{ V}$$

$$I_{sc} = \frac{E}{R_{int}} = \frac{10}{4 \text{ k}\Omega} = 2.5 \text{ mA}$$

The point of intersection of the load line with the different characteristics curves of the load (depending on the light intensity) determines the operating or Q points. Thus reading directly from the graph, we have:

At 0.1 lumen (Q_1):

$$V_{ph_1} = 7.6 \text{ V}$$

At 0.2 lumen (Q_2):

$$V_{\mathrm{ph}_2} = 5.8\,\mathrm{V}$$

At 0.4 lumen (Q_3):

$$V_{\mathrm{ph}_3} = 3.0\,\mathrm{V}$$

These voltages are the readings of an ideal voltmeter connected across the cell.

Since $P_{\mathrm{max}} = 8\,\mathrm{mW}$ is the maximum power dissipation capability of the cell, the maximum power dissipation hyperbola could be drawn on the same graph by determining the I–V coordinates of points, such that $VI = P_{\mathrm{max}}$. As the table below indicates, V is chosen arbitrarily and then I is calculated to satisfy this power relationship.

$$P_{\mathrm{max}} = 8\,\mathrm{mW}$$

V(V)	2	3	4	5	6	8	10
I(mA)	4	2.67	2	1.6	1.33	1.0	0.8

This P_{max} hyperbola is also shown in Fig. 4.13(a). Since the load line and the operating points are always below (or to the left) of the P_{max} hyperbola, this indicates that the power dissipation of the cell will always be below P_{max}. Thus the cell in this circuit operates safely at all times.

4.3 | IDEAL AND PRACTICAL CURRENT SOURCES

Definition. An ideal current source is a device that provides a constant current to any load resistance connected across it (i.e., independent of the voltage across its terminals).

The model of the ideal current source is shown in Fig. 4.14. Here

$$I = I_S \quad \text{(for any value of } R_L \text{ or } V\text{)}$$

Also,

$$V = IR_L = I_S R_L$$

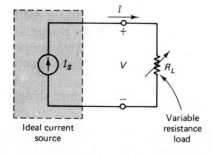

Ideal current source

Variable resistance load

Figure 4.14

As I_S is a constant, then whether R_L is small (V is small) or R_L is large (V is large), the current flowing from the source is always I_S. The V–I characteristics of such a source are then represented by the vertical dashed line in Fig. 4.15.

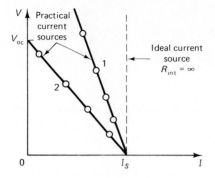

Figure 4.15 V–I characteristics of current sources.

In practice, however, one observes that the output current from the source is reduced as the voltage across the source's terminals is increased, by increasing the value of the load resistance R_L. To account for such a drop in the output current I, the model of the practical current source includes an internal resistance R_{int} connected in parallel with the ideal current source I_S as shown in Fig. 4.16.

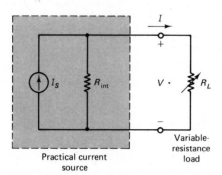

Figure 4.16 Model of a practical current source.

Applying KCL at the top node of the current source in Fig. 4.16 yields

$$I = I_S - \frac{V}{R_{int}} \tag{4.10}$$

Here I_S and R_{int} are the constant source parameters. Equation (4.10) shows that as V is increased, I decreases in accordance with the practical observation. If $R_{int} = \infty$, Eq. (4.10) reduces to

$$I = I_S \quad \text{(ideal current source case)}$$

Thus the ideal current source is a practical current source with infinite internal resistance. Notice that a practical current source becomes closer to being considered ideal as its internal resistance approaches ∞, in contrast with the practical voltage source, which approaches the ideal case as its resistance approaches zero.

When both sides of Eq. (4.10) are multiplied by R_{int} and the V and I terms are transposed, Eq. (4.10) can be rewritten in the form

$$V = I_S R_{int} - R_{int} I \tag{4.10a}$$

In Eq. (4.10a), when I is zero through open-circuiting R_L, the terminal voltage is

then

$$V = V_{oc} \quad \text{(open-circuit voltage of the current source)}$$
$$= I_S R_{int} \tag{4.11}$$

This is clear from Fig. 4.16, as the removal of R_L allows all of I_S to flow in R_{int}, creating V_{oc} across the source's terminals. V_{oc} is a constant for a given current source with finite source parameters I_S and R_{int}. Equation (4.10a) can then be rewritten as

$$V = V_{oc} - R_{int}I \tag{4.12}$$

This equation also constitutes what is referred to as the $V–I$ characteristics (or load line) of the current source. It describes a straight line with a slope of "$-R_{int}$." Its voltage axis intercept (at $I = 0$) is

$$V = V_{oc}$$

and its current axis intercept (at $V = 0$) is

$$I = \frac{V_{oc}}{R_{int}} = I_S \quad \text{[from Eq. (4.11)]}.$$

Such characteristics are shown in Fig. 4.15. Source 1 has a larger internal resistance (higher slope) than source 2. Thus source 1 is closer to the ideal case.

4.4 SOURCE CONVERSIONS

Examination of the $V–I$ characteristics for the practical voltage source [Eq. (4.2)] and the $V–I$ characteristics of the practical current source [Eq. (4.10a)], rewritten below:

$$V = E - IR_{int} \quad \text{(voltage source)}$$
$$V = I_S R_{int} - IR_{int} \quad \text{(current source)}$$

or

$$V = V_{oc} - IR_{int}$$

reveals that both equations would represent the same load line if

$$R_{int} \quad \text{(voltage source)} = R_{int} \quad \text{(current source)} \tag{4.13}$$

that is, both characteristics represent the same slope, and

$$E = V_{oc} = I_S R_{int} \tag{4.14}$$

that is, both characteristics represent the same voltage axis intercept.

Thus, as far as the load R_L in Fig. 4.17 is concerned, it does not really matter which model one chooses for the electrical source, as long as either model produces the same terminal $V–I$ characteristics (load line). In other words, if the source parameter equivalencies defined by Eqs. (4.13) and (4.14) are valid, both the voltage source model and the current source model produce the same voltage across the load R_L and the same current through that load. From the terminal-behavior point of view, a voltage source model is equivalent to a current source model, and vice versa.

These remarks allow us to convert a given voltage source with terminals a and b into an equivalent current source with the same terminal points as shown in

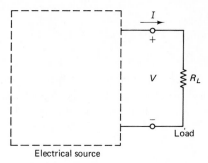

Electrical source   **Figure 4.17**

Fig. 4.18(a). The reverse conversion for a current source whose terminals are 1 and 2 into a voltage source is also possible, as indicated in Fig. 4.18(b). One should always be careful with the polarities. Both models produce the same terminal polarities when the current arrow in the current source model points toward the positive terminal in the voltage source model.

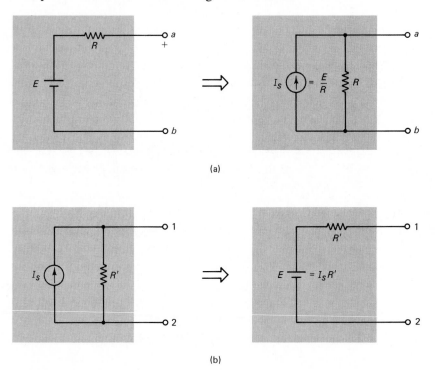

(a)

(b)

Figure 4.18 (a) Voltage source-to-current conversion; (b) current source-to-voltage source conversion.

This conversion process is useful in a variety of complicated networks because it can result in simplifying the networks, hence the analysis required, as will be shown in Chapter 5. Many examples involving this conversion process will be examined there.

Remark. Going back to the current source characteristics [eq. (4.10a)] and substituting for V by its equivalent, IR_L, we obtain

$$IR_L = I_S R_{\text{int}} - IR_{\text{int}}$$

Therefore,

$$I = I_S \frac{R_{int}}{R_{int} + R_L} = I \frac{1}{1 + (R_L/R_{int})} \tag{4.15}$$

This equation is, in fact, the current-division principle applied to the circuit of Fig. 4.16. If $R_L \ll R_{int}$ (say, R_L is less than $0.01 R_{int}$),

$$I = \frac{1}{1 + 0.01} I_S = 0.99 I_S \cong I_S$$

Then the current in the load is approximately equal to the current provided by the ideal current source (i.e., the effect of R_{int} can be neglected).

Based on this remark and the remark at the end of Section 4.1, we reach the following conclusions:

1. If $R_L \gg R_{int}$, the electrical source is a voltage source (approximately ideal).
2. If $R_L \ll R_{int}$, the electrical source is a current source (approximately ideal).

In between these two limits, one model is as good as the other.

EXAMPLE 4.6

Find V and I in the circuit shown in Fig. 4.19. Convert the source into an equivalent voltage source and recalculate V and I.

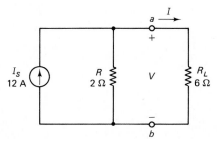

Figure 4.19

Solution Using the current-division principle for the circuit shown in Fig. 4.19,

$$I = I_S \frac{R}{R + R_L} = 12 \times \frac{2}{2 + 6} = 3 \text{ A}$$

Then

$$V = IR_L = 3 \times 6 = 18 \text{ V}$$

As shown in Fig. 4.20, the current source can be converted to the equivalent voltage source indicated between the same two load terminals, a and b. In this circuit

$$I = \frac{E}{R + R_L} = \frac{24}{2 + 6} = 3 \text{ A}$$

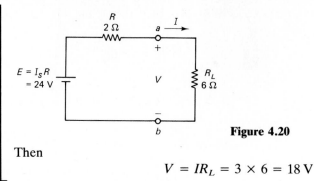

Figure 4.20

Then

$$V = IR_L = 3 \times 6 = 18 \text{ V}$$

■ Thus both types of the source models produce the same results.

4.5 MAXIMUM POWER TRANSFER THEOREM AND POWER TRANSFER EFFICIENCY

The circuit shown in Fig. 4.21 depicts the situation when a variable-resistance load, R_L, is connected to a network whose equivalent circuit has the form of a practical voltage source model. In the following discussion we examine the behavior of the circuit variables V, I, and P_L (the power dissipated in the load) as the value of R_L is varied.

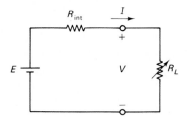

Figure 4.21

1. *Load current I.* In this simple series circuit,

$$I = \frac{E}{R_{\text{int}} + R_L} \tag{4.16}$$

when $R_L = 0$ (short circuit), $I = I_{\text{sc}} = E/R_{\text{int}}$. As R_L is increased, I decreases, until when $R_L = \infty$ (open circuit), $I = 0$. Also, when $R_L = R_{\text{int}}$,

$$I = \frac{E}{2R_{\text{int}}} = 0.5I_{\text{sc}}$$

Figure 4.22 shows this variation of I as a function of R_L.

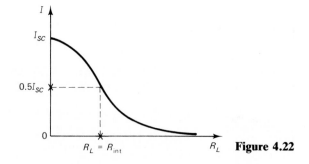

Figure 4.22

2. *Load voltage V.* Using the voltage-division principle for the circuit of Fig. 4.21, we obtain

$$V = E \frac{R_L}{R_L + R_{int}}$$

$$= \frac{E}{1 + (R_{int}/R_L)} \quad \text{[as in Eq. (4.7)]}$$

When $R_L = 0$ (short circuit), $V = 0$. As R_L increases, the denominator of the equation above decreases and V increases, until $R_L = \infty$ (open circuit),

$$\frac{R_{int}}{R_L} = 0 \quad \text{and} \quad V = V_{oc} = E$$

Also, when $R_L = R_{int}$, $V = 0.5E$. This behavior of V as a function of R_L is shown in Fig. 4.23.

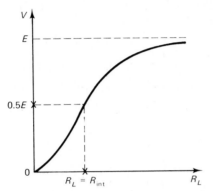

Figure 4.23

3. *Load power P_L.* The expression for the load power is obtained as follows:

$$P_L = VI = I^2 R_L = \left(\frac{E}{R_{int} + R_L}\right)^2 R_L \tag{4.17}$$

The load power is zero when R_L is zero, as V is zero in this short-circuit condition. The load power is also zero when R_L is ∞, as I is zero and V is finite ($= E$) in this open-circuit condition.

As R_L varies between zero and ∞, P_L, being a positive quantity, will increase from a zero value to reach a peak and then decreases toward a zero value again. If R_L is varied in relation to R_{int} and Eq. (4.17) is used to calculate P_L numerically, the graph of such a variation will be as shown in Fig. 4.24. From

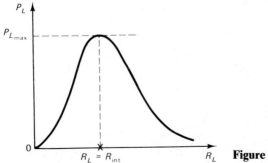

Figure 4.24

this graph it can be concluded that *maximum load power is obtained when the load resistance R_L is equal to the source resistance R_{int}.*

The condition

$$\boxed{R_L = R_{int}} \tag{4.18}$$

is called the *matching condition* or *maximum power transfer condition.* The resulting maximum load power is obtained from Eq. (4.17):

$$\boxed{P_{L\,max} = \left(\frac{E}{R_{int} + R_{int}}\right)^2 R_{int} = \frac{E^2}{4R_{int}} \text{ W}} \tag{4.19}$$

This result could also be obtained experimentally or by using calculus, as shown in Appendix A4.

4. *Power transfer efficiency η.* In the circuit of Fig. 4.21, the ideal voltage source E in the model provides power to both R_L and R_{int}:

$$P_T = P_L + P_{int}$$
$$EI = I^2 R_L + I^2 R_{int}$$

The power transfer efficiency is defined as

$$\eta = \frac{P_L}{P_T} = \frac{I^2 R_L}{I^2 R_L + I^2 R_{int}}$$

$$= \frac{R_L}{R_L + R_{int}} = \frac{1}{1 + (R_{int}/R_L)} \tag{4.20}$$

It is clear that the variation of η with R_L is similar to the variation of V with R_L, except that the maximum value of η is unity and at $R_L = R_{int}$, η is 0.5 (i.e., within a scale factor of E). Figure 4.25 shows this variation of η with R_L.

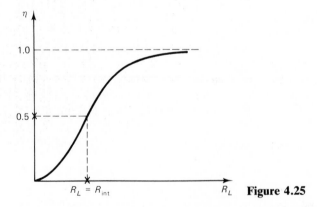

Figure 4.25

It is important to note that when the load power is maximum, $R_L = R_{int}$ and

$$I^2 R_L = I^2 R_{int}$$

This means that the source itself dissipates as much power in its internal resistance as the power absorbed by the load. That is why the efficiency is only 50% at this condition.

The maximum power transfer condition is very important in many communications aplications where the highest power in the load is desired, especially from a very small received signal. On the other hand, in most power systems (utilities) applications, a 50% efficiency is not desirable because of the wasted energy. In such situations a compromise might have to be made between the load power and the power transfer efficiency. For example, if $R_L = 2R_{int}$, then

$$P_L = 0.222 \frac{E^2}{R_{int}} \quad \text{[from Eq. (4.17)]}$$

and

$$\eta = 0.667 \quad \text{[from Eq. (4.20)]}$$

that is, the load power is only 11% less than its maximum possible value, while the efficiency of power transfer has improved by 33% (from 0.5 to 0.667).

■ EXAMPLE 4.7

In the circuit shown in Fig. 4.26, calculate
(a) The short-circuit current of the source.
(b) The power transfer efficiency.

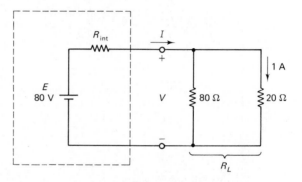

Figure 4.26

Solution (a) To find the short-circuit current, R_{int} has to be obtained. The circuit is a series parallel one, with

$$R_L = 80 \, \Omega \parallel 20 \, \Omega = \frac{80 \times 20}{100} = 16 \, \Omega$$

The terminal voltage of the source is $V = 1 \times 20 = 20 \, V$. Applying KCL to the upper node of the load yields

$$I = \frac{20}{80} + 1 = 1.25 \, A$$

Since $E = V + IR_{int}$,

$$80 = 20 + 1.25 \times R_{int}$$

Then

$$R_{int} = \frac{60}{1.25} = 48 \, \Omega$$

$$I_{sc} = \frac{E}{R_{int}} = \frac{80}{48} = 1.67 \text{ A}$$

(b) The power transfer efficiency is

$$\eta = \frac{R_L}{R_L + R_{int}} = \frac{16}{16 + 48} = 0.25 \text{ (or 25\%)}$$

EXAMPLE 4.8

A certain load is connected to a source with $20\ \Omega$ internal resistance. The source delivers 1 A and operates at 60% efficiency. Find the open-circuit voltage of this source.

Solution As the efficiency η and R_{int} are given, one can then calculate the value of R_L:

$$\eta = \frac{R_L}{R_L + R_{int}}$$

$$0.6 = \frac{R_L}{R_L + 20} \quad \text{or} \quad R_L = 0.6R_L + 12$$

Therefore,

$$R_L = \frac{12}{0.4} = 30\ \Omega$$

Refer to Fig. 4.21. All the resistances and the current in the circuit are known; thus

$$E = \text{open-circuit voltage of the source}$$
$$= I(R_L + R_{int}) = 1 \times 50 = 50 \text{ V}$$

EXAMPLE 4.9

When a $4\ \Omega$ load is connected to a given generator, its terminal voltage was found to be 160 V. The source's efficiency is 90% when a $9\ \Omega$ load is connected to it. Find the maximum power this generator can deliver.

Solution From the expression for the efficiency [Eq. (4.20)], one can find the value of the unknown internal resistance; noting that $\eta = 0.9$ when $R_L = 9\ \Omega$, we obtain

$$0.9 = \frac{9}{R_{int} + 9} \quad \text{or} \quad 0.9R_{int} + 8.1 = 9$$

Therefore,

$$R_{int} = \frac{0.9}{0.9} = 1\ \Omega$$

To find the maximum power available from this generator, the value of E has to be determined. Again referring to Fig. 4.21, since the output voltage is given at $R_L = 4\,\Omega$, one can use the voltage-divider principle [Eq. (4.7)]

$$V = E\frac{R_L}{R_L + R_{int}}$$

$$160 = E\frac{4}{4 + 1}$$

Therefore,

$$E = 200\text{ V}$$

Thus when $R_L = R_{int} = 1\,\Omega$,

$$P_{L\,max} = \frac{E^2}{4R_{int}} = \frac{(200)^2}{4 \times 1} = 10\text{ kW}$$

EXAMPLE 4.10

A source delivers 5 A when the load connected to it is 5 Ω and 2 A when the load is increased to 20 Ω. Find
(a) The maximum power available from this source.
(b) The source's efficiency with the 20 Ω load.
(c) The source's efficiency when it is delivering 45 W.

Solution (a) Here we are given two conditions from which the source parameters E and R_{int} can be found. When $R_L = 5\,\Omega$, $I = 5\text{ A}$, and $V = IR_L = 25\text{ V}$, then:

$$25 = E - 5R_{int}$$

When $R_L = 20\,\Omega$, $I = 2\text{ A}$, and $V = IR_L = 40\text{ V}$, then:

$$40 = E - 2R_{int}$$

Solving these two equations, first by subtraction, gives us

$$15 = 3R_{int}$$

Therefore,

$$R_{int} = 5\,\Omega \quad \text{and} \quad E = 25 + 5 \times 5 = 50\text{ V}$$

Thus when $R_L = R_{int} = 5\,\Omega$,

$$P_{Lmax} = \frac{E^2}{4R_{int}} = \frac{50 \times 50}{4 \times 5} = 125\text{ W}$$

(b) When $R_L = 20\,\Omega$, the power transfer efficiency is

$$\eta = \frac{R_L}{R_L + R_{int}} = \frac{20}{25} = 0.8 \quad \text{or} \quad 80\%$$

(c) To find the efficiency when the source is delivering 45 W, the value of the load resistance connected to the source must be found first. Using Eq. (4.17), which relates P_L to R_L,

$$45 = \frac{2500R_L}{(R_L + 5)^2}$$

$$R_L^2 + 10R_L + 25 = 55.556R_L$$

$$R_L^2 - 45.556R_L + 25 = 0$$

This second-order equation has two roots,

$$R_{L_1} = 45\ \Omega \qquad \text{and} \qquad R_{L_2} = 0.556\ \Omega$$

The graph of Fig. 4.24 is sketched in Fig. 4.27 for this numerical example. It is clear that for a given value of $P_L < P_{L\max}$, two values of the load resistance would dissipate this same amount of power.

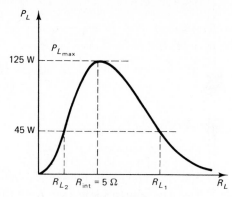

Figure 4.27

Since there are two solutions for this power condition, one would also expect two values of the efficiency, corresponding to R_{L_1} and R_{L_2}, respectively.

$$\eta_1 = \frac{R_{L_1}}{R_{L_1} + R_{\text{int}}} = \frac{45}{50} = 0.9 \quad \text{or} \quad 90\%$$

$$\eta_2 = \frac{R_{L_2}}{R_{L_2} + R_{\text{int}}} = \frac{0.556}{5.556} = 0.1 \quad \text{or} \quad 10\%$$

EXAMPLE 4.11 (Generalization of Example 4.10(c))

Two load resistances, R_1 and R_2, dissipate the same power when connected to a source whose internal resistance is R_{int}. Prove that
(a) $R_{\text{int}}^2 = R_1 R_2$.
(b) $\eta_1 + \eta_2 = 1$.

Solution   (a) As both load resistances dissipate the same power, then from Eq. (4.17),

$$P_L = \frac{E^2 R_1}{(R_1 + R_{\text{int}})^2} = \frac{E^2 R_2}{(R_2 + R_{\text{int}})^2}$$

Cancelling E^2 and cross-multiplying yields

$$R_1 R_2^2 + 2R_1 R_2 R_{\text{int}} + R_1 R_{\text{int}}^2 = R_2 R_1^2 + 2R_1 R_2 R_{\text{int}} + R_2 R_{\text{int}}^2$$

Therefore,

$$R_{\text{int}}^2(R_1 - R_2) = R_1 R_2 (R_1 - R_2)$$

Thus

$$R_{\text{int}}^2 = R_1 R_2 \tag{4.21}$$

as required.

(b) η_1 and η_2 are the efficiencies corresponding to the load resistances R_1 and R_2. Then

$$\eta_1 + \eta_2 = \frac{R_1}{R_1 + R_{\text{int}}} + \frac{R_2}{R_2 + R_{\text{int}}} = \frac{R_1 R_2 + R_1 R_{\text{int}} + R_1 R_2 + R_2 R_{\text{int}}}{(R_1 + R_{\text{int}})(R_2 + R_{\text{int}})}$$

$$= \frac{2R_1 R_2 + R_{\text{int}}(R_1 + R_2)}{R_1 R_2 + R_{\text{int}}^2 + R_{\text{int}}(R_1 + R_2)}$$

Substituting for $R_1 R_2$ by R_{int}^2, as obtained in Eq. (4.21), we get

$$\eta_1 + \eta_2 = \frac{2R_{\text{int}}^2 + R_{\text{int}}(R_1 + R_2)}{2R_{\text{int}}^2 + R_{\text{int}}(R_1 + R_2)} = 1 \tag{4.22}$$

as required.

EXAMPLE 4.12

Two sources are connected in parallel as indicated in the circuit diagram shown in Fig. 4.28. The meters were connected as shown, with the polarities indicated corresponding to upscale deflections. When the switch S was open, A_1 read 2 A and V read 10 V. When the switch S was closed, A_1 read 12 A and A_2 read 15 A. Find the values of E_1, E_2, R_1, and R_2.

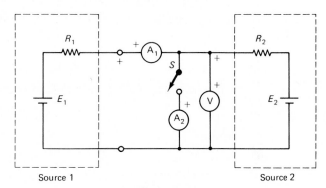

Figure 4.28

Solution The circuit is redrawn in Fig. 4.29(a) with the switch S open, to show the measured variables. By applying KVL to each half of the circuit, we get

$$10 = E_1 - 2R_1 = E_2 + 2R_2 \tag{4.23}$$

The circuit is redrawn in Fig. 4.29(b) to show the new measured variables when the switch S is closed. Note that here both sources are short circuited. Thus

$$I_1 = \frac{E_1}{R_1} = 12 \qquad \text{that is, } E_1 = 12R_1 \tag{4.24}$$

and through applying KCL to the top node of the short circuit:

$$15 = I_1 + \frac{E_2}{R_2} = 12 + \frac{E_2}{R_2} \qquad \text{that is, } E_2 = 3R_2 \tag{4.25}$$

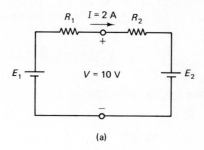

(a)

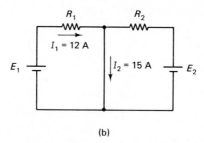

(b)

Figure 4.29

Using Eq. (4.24) and the first equality in Eq. (4.23) gives us

$$10 = 12R_1 - 2R_1 = 10R_1$$

Therefore,

$$R_1 = 1\,\Omega \qquad \text{and} \qquad E_1 = 12\,\text{V}$$

Again, using Eq. (4.25) and the second equality in Eq. (4.23), we have

$$10 = 3R_2 + 2R_2 = 5R_2$$

Therefore,

$$R_2 = 2\,\Omega \qquad \text{and} \qquad E_2 = 6\,\text{V}$$

4.6 DESIGN OF PRACTICAL VOLTAGE DIVIDERS

In practice, an electrical device (or load), or a complex electronic circuit requires specific finite voltages or currents to operate properly. Such voltages and currents are usually referred to as the *nominal voltages* or *nominal currents*. A fixed-value voltage source (e.g., a battery) can provide only one value of voltage, which might not be the required nominal voltage of the load. Even a variable-value voltage source can provide only one value of voltage at a certain node at a specific setting, not a number of voltages at different nodes simultaneously as required by a complex electronic circuit.

The solution for such practical requirements is provided through the application of the voltage-division principle. As discussed earlier, a string of resistors connected in series with a voltage source E as in Fig. 4.30 provide a number of voltages across each (or combination) of them. It should be important to remember that all these voltages are less than the source voltage E. Here

$$I = \frac{E}{R_1 + R_2 + R_3}$$

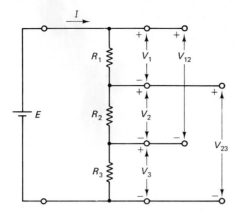

Figure 4.30

$$V_1 = IR_1 = E\frac{R_1}{R_1 + R_2 + R_3} \qquad V_2 = E\frac{R_2}{R_1 + R_2 + R_3} \cdots \text{etc.}$$

and

$$V_{12} = E\frac{R_1 + R_2}{R_1 + R_2 + R_3} \qquad V_{23} = E\frac{R_2 + R_3}{R_1 + R_2 + R_3}$$

To obtain a specific nominal voltage, we first choose a source whose voltage is larger than this nominal voltage. Then by proper choice of the values of the series-connected resistors, the required nominal value can be obtained. The ground connection (reference 0 V node) can be incorporated in the divider circuit to obtain positive and/or negative voltages (with respect to the ground) at the different output nodes of the circuit. Some examples are shown in Fig. 4.31.

Design of a voltage divider includes the determination of the values of the different resistive elements, their power ratings, and their interconnections, which are required to build a specific divider. All such voltage-divider circuits are in fact of the series–parallel type. The design process is an exercise involving the simultaneous application of KVL, KCL, and Ohm's law.

Consider the practical voltage-divider circuit shown in Fig. 4.32. It has a single load that requires a nominal operating voltage V_L, or equivalently, its nominal current is specified to be I_L, where

$$V_L = I_L R_L$$

To drop the source voltage E to the required load voltage V_L, a resistor R_{sd} is connected in series with the source. This resistor is called the *series dropping resistance* and it carries a current equal to the source current I_S. Applying KVL gives us

$$V_L = E - I_S R_{sd} \tag{4.27}$$

The voltage divider also contains a resistor R_b connected in parallel with the load. It is called the *bleeder resistance,* for reasons that will become apparent shortly. The equivalent resistance of the parallel combination of R_b and R_L is

$$R_{eq} = R_b \parallel R_L = \frac{R_b R_L}{R_b + R_L} \tag{4.28}$$

When the equivalent simple series circuit is drawn, with R_b and R_L replaced

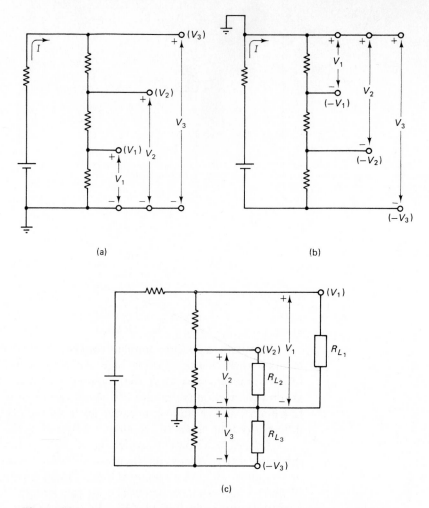

Figure 4.31 Use of ground to obtain different output voltage polarities: (a) all positive voltages; (b) all negative voltages; (c) V_1 and V_2 are positive voltages, V_3 is a negative voltage.

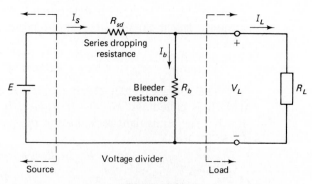

Figure 4.32

by R_{eq}, the application of the voltage-division principle gives

$$V_L = E \frac{R_{eq}}{R_{sd} + R_{eq}} \qquad (4.29)$$

Note that applying KCL at the top node of the load provides

$$I_S = I_b + I_L \qquad (4.30)$$

In many situations in practice, the load resistance R_L could vary from its nominal value and could increase to become ∞ (open circuit or no load). Even with such load variations, it is usually required that the load voltage V_L be relatively constant. To achieve this, the bleeder resistance R_b is usually chosen much lower in value than R_L. As discussed previously concerning the properties of parallel resistors, the equivalent resistance is almost equal to the smaller of the two, that is,

$$R_{eq} = \frac{R_b R_L}{R_b + R_L} \approx R_b \quad (\text{when } R_b \ll R_L)$$

Thus even when R_L varies over a wide range, R_{eq} is fairly constant. From Eq. (4.29), this results in V_L also fairly constant, as required. The smaller R_b is, the more constant V_L would be.

However, when R_b is reduced, the bleeder current I_b increases (note that the voltage across the bleeder is V_L, which is fairly constant). This results in an increased source current I_S, as is clear from Eq. (4.30). This increased source current means higher power dissipated by the circuit and more energy supplied by the source. This is called the *bleeding action* of the energy of the source and results in a shorter source lifetime.

The voltage regulation (V. R.) of the voltage divider is a measure of the relative variation of the load voltage. It is defined as

$$\text{V. R.}(\%) = \frac{V_L\big|_{oc} - V_L}{V_L} \times 100 \qquad (4.31)$$

where V_L is the nominal load voltage and

$$V_L\big|_{oc} = E \frac{R_b}{R_b + R_{sd}} \qquad (4.32)$$

= open-circuit voltage of the divider when R_L is ∞ (no-load condition)

In this case, as in many other design problems, the choice of the element values, in particular R_b, is usually a compromise between a desirable reduction in the voltage regulation (less variation in the load voltage) and an undesirable bleeding action of the source (more power dissipation).

The discussion above shows that KVL, KCL, and Ohm's law, used to analyze series-parallel circuits, are the tools required to design voltage-divider circuits.

■ EXAMPLE 4.13

Design a voltage divider to supply a 50 V load with a 25 mA current using a 250 V source. The bleeder resistance dissipates 5 W under nominal conditions. Find the voltage regulation of this divider.

Solution Since there is only one load to be supplied, the voltage divider required is similar to that of Fig. 4.32. The design involves finding the values and power ratings of R_{sd} and R_b in the voltage-divider circuit shown in Fig. 4.33.

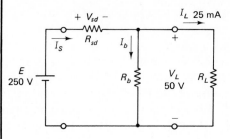

Figure 4.33

The bleeder power is known ($P_b = 5$ W) and the voltage across it is also known ($V_L = 50$ V). As $P_b = I_b V_L$, then

$$I_b = \frac{P_b}{V_L} = \frac{5}{50} = 0.1\,\text{A}$$

Therefore,

$$R_b = \frac{V_L}{I_b} = \frac{50}{0.1} = 500\,\Omega$$

Using KCL:

$$I_S = I_b + I_L = 100\,\text{m} + 25\,\text{m} = 125\,\text{mA}$$

Using KVL:

$$V_{\text{sd}} = E - V_L = 250 - 50 = 200\,\text{V}$$

Therefore,

$$R_{\text{sd}} = \frac{V_{\text{sd}}}{I_S} = \frac{200}{125\,\text{mA}} = 1.6\,\text{k}\Omega$$

The power rating of this series dropping resistance is

$$P_{\text{sd}} = I_S V_{\text{sd}} = 0.125 \times 200 = 25\,\text{W}$$

To find the voltage regulation of this divider, we first have to find $V_L|_{\text{oc}}$. With the load removed (open circuit or $R_L = \infty$), the voltage-division principle [Eq. (4.32)] gives

$$V_L|_{\text{oc}} = E\,\frac{R_b}{R_b + R_{\text{sd}}} = 250 \times \frac{500}{500 + 1600} = 59.52\,\text{V}$$

Thus, from Eq. (4.31),

$$\text{V. R.}(\%) = \frac{59.52 - 50}{50} \times 100 = 19\%$$

EXAMPLE 4.14

Using a 200 V power supply with negligible internal resistance, design a voltage divider to feed a 1 MΩ load whose nominal voltage is 125 V. The source power is limited to 260 mW. Find the voltage regulation of this divider.

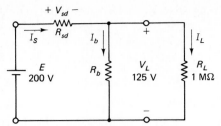

Figure 4.34

Solution A schematic of the voltage-divider circuit required is shown in Fig. 4.34. Since the power delivered by the source is given, the source current I_S can be determined:

$$I_S = \frac{P_S}{E} = \frac{260\,\text{mW}}{200} = 1.3\,\text{mA}$$

but

$$I_L = \frac{V_L}{R_L} = \frac{125}{1\,\text{M}\Omega} = 0.125\,\text{mA}$$

Thus, from KCL,

$$I_b = I_S - I_L = 1.3 - 0.125 = 1.175\,\text{mA}$$

Also, from KVL,

$$V_{sd} = E - V_L = 200 - 125 = 75\,\text{V}$$

Now the voltages across and the current through all the elements are obtained; then

$$R_{sd} = \frac{75}{1.3\,\text{mA}} = 57.7\,\text{k}\Omega \qquad P_{sd} = 75 \times 1.3\,\text{mA} = 97.5\,\text{mW}$$

and

$$R_b = \frac{125}{1.175\,\text{mA}} = 106.4\,\text{k}\Omega \qquad P_b = 125 \times 1.175\,\text{mA} = 146.9\,\text{mW}$$

Note that $P_L = 125 \times 0.125\,\text{mA} = 15.6\,\text{mW}$; hence

$$P_L + P_{sd} + P_b = 260\,\text{mW} = P_S \quad \text{(as given)}$$

$$V_L|_{oc} \text{ (with } R_L \text{ removed)} = E\frac{R_b}{R_b + R_{sd}} = 200 \times \frac{106.4\,\text{k}\Omega}{164.1\,\text{k}\Omega}$$

$$= 129.7\,\text{V}$$

This means that as R_L is increased from its nominal value to ∞, V_L increases from 125 V to 129.7 V. Using Eq. (4.31) yields

$$\text{V. R.} = \frac{129.7 - 125}{125} \times 100 = 3.76\%$$

EXAMPLE 4.15

Given a 15 V source (approximately ideal), it is required to design a voltage divider to supply a variable resistance load. V_L should vary between 11 and 12 V as R_L varies between 20 and 30 kΩ. What is the highest power delivered by the source in this circuit? (Refer to Fig. 4.32.)

Solution   Assuming that V_L is fairly constant, which is the objective in the voltage-divider design, then as R_L increases, I_L decreases. Thus I_S $(= I_b + I_L)$ is also decreased. This corresponds to a smaller voltage drop across R_{sd} and hence $V_L = E - V_{sd}$ is increased. In other words,

$$R_{L_1} = 30\,\text{k}\Omega \qquad \text{corresponds to} \qquad V_{L_1} = 12\,\text{V}$$

and

$$R_{L_2} = 20\,\text{k}\Omega \qquad \text{corresponds to} \qquad V_{L_2} = 11\,\text{V}$$

As the circuit (refer to Fig. 4.32) has two unknown elements, R_{sd} and R_b, two conditions are always needed to allow the determination of the two unknowns. Here the two conditions are given in the form above. What is needed now is to translate these conditions into two suitable equations.

Let us use KCL (we could also start by using KVL):

$$I_S = I_b + I_L$$

Each of these currents can be expressed in terms of the unknown elements,

$$I_S = \frac{V_{sd}}{R_{sd}} = \frac{E - V_L}{R_{sd}} \qquad I_b = \frac{V_L}{R_b} \qquad I_L = \frac{V_L}{R_L}$$

Hence

$$\frac{E - V_L}{R_{sd}} = \frac{V_L}{R_b} + \frac{V_L}{R_L}$$

Corresponding to the first load value, with $E = 15\,\text{V}$,

$$\frac{3}{R_{sd}} = \frac{12}{R_b} + \frac{12}{30 \times 10^3} = \frac{12}{R_b} + 0.4 \times 10^{-3} \qquad (4.33)$$

Corresponding to the second load value,

$$\frac{4}{R_{sd}} = \frac{11}{R_b} + \frac{11}{20 \times 10^3} = \frac{11}{R_b} + 0.55 \times 10^{-3} \qquad (4.34)$$

These two equations can then be solved simultaneously. Multiply Eq. (4.33) all through by 4 and multiply Eq. (4.34) all through by 3; then subtracting yields

$$0 = \frac{48}{R_b} - \frac{33}{R_b} + 1.6 \times 10^{-3} - 1.65 \times 10^{-3}$$

Therefore,

$$R_b = \frac{15}{0.05 \times 10^{-3}} = 300\,\text{k}\Omega$$

Substituting in Eq. (4.33) gives us

$$\frac{3}{R_{sd}} = 0.04 \times 10^{-3} + 0.4 \times 10^{-3} = 0.44 \times 10^{-3}$$

Therefore,

$$R_{sd} = \frac{3}{0.44 \times 10^{-3}} = 6.818\,\text{k}\Omega$$

The highest power delivered by the source is when the source current I_S is largest. This corresponds to V_L being the smaller of the two values, that is,

$$I_S \text{ (highest)} = \frac{E - V_{L_2}}{R_{sd}} = \frac{4}{6.818 \text{ k}\Omega} = 0.587 \text{ mA}$$

Therefore,

$$P_S \text{ (highest)} = 15 \times 0.587 \text{ mA} = 8.8 \text{ mW}$$

EXAMPLE 4.16

Design a voltage divider that supplies 30 mA at -3 V to a load. The source is a practical voltage source whose terminal voltage under loaded conditions is 20 V. The source dissipates 1 W, while the load and the voltage divider circuit dissipate 5 W. Find the elements of the voltage-divider circuit, then determine E and R_{int} of the source.

Solution Here also the voltage divider feeds a single load. The load voltage is negative; thus the positive terminal of the source should be connected to ground. The series dropping resistance is still needed to drop the voltage from 20 V to 3 V. The circuit is thus as shown in Fig. 4.35. Here

$$V_{sd} + 3 = 20 \qquad \text{that is, } V_{sd} = 17 \text{ V}$$

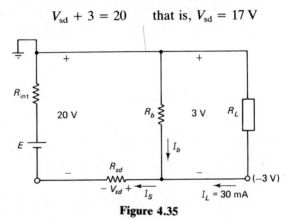

Figure 4.35

To find the elements of the voltage divider, let us use the power condition given:

$$5 = P_{sd} + P_b + P_L = 17I_S + 3I_b + 3 \times 0.03$$

Then

$$17I_S + 3I_b = 4.91$$

But from KCL, $I_S = I_b + 0.03$. Substituting in the previous equation gives us

$$17I_b + 0.51 + 3I_b = 4.91$$

Thus

$$I_b = \frac{4.4}{20} = 0.22 \text{ A} = 220 \text{ mA}$$

and

$$I_S = 220 \text{ mA} + 30 \text{ mA} = 250 \text{ mA}$$

Therefore,

$$R_b = \frac{3}{0.22} = 13.64 \text{ }\Omega \qquad P_b = 3 \times 0.22 = 0.66 \text{ W}$$

and

$$R_{sd} = \frac{17}{0.25} = 68\ \Omega \qquad P_{sd} = 17 \times 0.25 = 4.25\ \text{W}$$

To find the source parameters, first the source's power dissipation gives

$$1 = I_S^2 \times R_{int} \qquad \text{that is, } R_{int} = \frac{1}{(0.25)^2} = 16\ \Omega$$

Also, from KVL,

$$20 = E - I_S R_{int} = E - 0.25 \times 16$$

Therefore,

$$E = 24\ \text{V}$$

EXAMPLE 4.17

A 5 kΩ potentiometer with adjustable tap is used as a voltage divider. The tap is adjusted to feed a 40 V, 200 mW load from a 100 V source. What is the resistance of the bleeder portion of this divider?

Solution Figure 4.36 shows the circuit connection of the potentiometer, between the source and the load. Since the load power and voltage are known,

$$I_L = \frac{P_L}{V_L} = \frac{200\ \text{mW}}{40} = 5\ \text{mA}$$

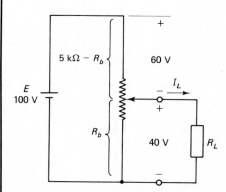

Figure 4.36

If we denote the bleeder portion of the potentiometer R_b, the series dropping resistance portion is $5\ \text{k}\Omega - R_b$. According to KVL, the voltage across this series dropping resistance is

$$V_{sd} = 100 - 40 = 60\ \text{V}$$

Now applying KCL at the potentiometer tap yields

$$\frac{60}{5000 - R_b} = \frac{40}{R_b} + 5 \times 10^{-3}$$

This equation has only R_b as unknown. Its solution will provide the value of R_b. Thus

$$60R_b = 2 \times 10^5 - 40R_b + 25R_b - 5 \times 10^{-3}R_b^2$$
$$R_b^2 + 1.5 \times 10^4 R_b - 4 \times 10^7 = 0$$

Therefore,
$$R_b = 2.31 \text{ k}\Omega$$

(*Note:* The other root is negative and is thus rejected.)

EXAMPLE 4.18

Using a 60 V source with negligible internal resistance, design a voltage divider to supply 40 mA at 30 V and 60 mA at −20 V, to two loads. The source power is limited to 6 W.

Solution   Resistors R_1 and R_2 are the two bleeders for the two loads. Ground should be connected as shown in Fig. 4.37 to provide both the positive and negative voltage nodes. A series dropping resistance is needed to drop the 60 V from the source to the total of 50 V required by the two loads.

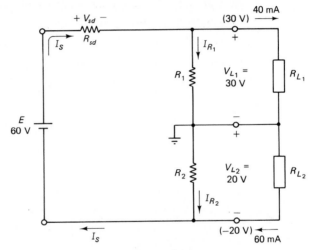

Figure 4.37

First the source current is obtained from

$$I_S = \frac{P_S}{E} = \frac{6}{60} = 0.1 \text{ A} = 100 \text{ mA}$$

and from KVL,

$$E = V_{\text{sd}} + V_{L_1} + V_{L_2}$$
$$60 = V_{\text{sd}} + 30 + 20 \qquad \text{that is, } V_{\text{sd}} = 10 \text{ V}$$

Thus

$$R_{\text{sd}} = \frac{10}{0.1} = 100 \ \Omega \qquad \text{and} \qquad P_{\text{sd}} = 10 \times 0.1 = 1 \text{ W}$$

Applying KCL at the top node gives us

$$100 \text{ mA} = I_{R_1} + 40 \text{ mA} \qquad \text{that is, } I_{R_1} = 60 \text{ mA}$$

Therefore,

$$R_1 = \frac{30}{60 \text{ mA}} = 500 \ \Omega \qquad \text{and} \qquad P_{R_1} = 30 \times 60 \text{ mA} = 1.8 \text{ W}$$

Also applying KCL at the bottom node, we have

$$100 \text{ mA} = I_{R_2} + 60 \text{ mA} \qquad \text{that is, } I_{R_2} = 40 \text{ mA}$$

and

$$R_2 = \frac{20}{40 \text{ mA}} = 500 \, \Omega \qquad \text{and} \qquad P_{R_2} = 20 \times 40 \text{ mA} = 0.8 \text{ W}$$

Note that it is not a good practice to apply KCL at the ground node because the ground wire can carry such a current, usually unknown, as to balance the current conditions for the circuit.

EXAMPLE 4.19

Design a voltage divider to deliver 1 W at 100 V, 2 W at -50 V, and 1.5 W at -75 V. The source has a 250 Ω internal resistance and supplies 100 mA of current. Use the minimum number of components. What is the open-circuit voltage of this source?

Solution From the specifications of the loads, the load currents can easily be found:

$$I_{L_1} = \frac{1}{100} = 10 \text{ mA}$$

$$I_{L_2} = \frac{2}{50} = 40 \text{ mA}$$

$$I_{L_3} = \frac{1.5}{75} = 20 \text{ mA}$$

Since we have to use the minimum number of components, the internal resistance of the source can be considered as a replacement for the series dropping resistance. The circuit and ground connections are as shown in Fig. 4.38. From KVL,

$$E = 75 + 100 + 0.1 \times 250 = 200 \text{ V}$$

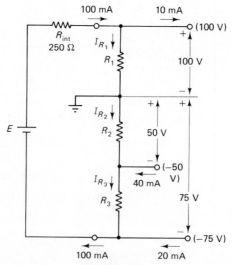

Figure 4.38

The currents in each of the bleeder resistors, R_1, R_2, and R_3 can be found by applying KCL to each of the load nodes. Thus

$$I_{R_1} = 100\,\text{mA} - 10\,\text{mA} = 90\,\text{mA}$$

Therefore,

$$R_1 = \frac{100}{90\,\text{mA}} = 1.11\,\text{k}\Omega \quad \text{and} \quad P_{R_1} = 100 \times 90\,\text{mA} = 9\,\text{W}$$

$$I_{R_3} = 100\,\text{mA} - 20\,\text{mA} = 80\,\text{mA}$$

Note that the voltage across R_3 is $-50 - (-75) = 25\,\text{V}$. Then

$$R_3 = \frac{25}{80\,\text{mA}} = 312.5\,\Omega \quad \text{and} \quad P_{R_3} = 25 \times 80\,\text{mA} = 2\,\text{W}$$

Now

$$I_{R_2} + 40\,\text{mA} = I_{R_3} = 80\,\text{mA} \quad \text{that is, } I_{R_2} = 40\,\text{mA}$$

Therefore,

$$R_2 = \frac{50}{40\,\text{mA}} = 1.25\,\text{k}\Omega \quad \text{and} \quad P_{R_2} = 50 \times 40\,\text{mA} = 2\,\text{W}$$

EXAMPLE 4.20

Using a voltage source whose open-circuit voltage is 100 V and whose internal resistance is 50 Ω, design a voltage-divider circuit to supply power to two loads. Load 1 requires 50 V and 1 W of power and load 2 requires -20 V and 1.2 W of power. The source current is limited to 100 mA. Also find
(a) The terminal voltage of the source under these nominal operating conditions.
(b) The voltage of load 1 if load 2 is disconnected (i.e., replaced by an open circuit).

Solution The voltage-divider circuit is connected as shown in Fig. 4.39 assuming that a series dropping resistance R_{sd} is needed. The validity of this assumption will be proved or disproved through the application of KVL. R_1 and R_2 are the two bleeder resistances required for the two loads.

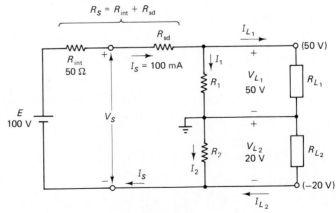

Figure 4.39

The ground connection is as shown to provide the positive and negative voltages required. From KVL,

$$E = I_S R_S + V_{L_1} + V_{L_2} = I_S(R_{int} + R_{sd}) + V_{L_1} + V_{L_2}$$

$$100 = 0.1 \times 50 + 0.1 \times R_{sd} + 50 + 20$$

$$= 75 + 0.1 R_{sd}$$

Therefore,

$$R_{sd} = \frac{25}{0.1} = 250\,\Omega \qquad \text{and} \qquad P_{sd} = 25 \times 0.1 = 2.5\,\text{W}$$

From the load powers, the load currents can be determined:

$$I_{L_1} = \frac{P_{L_1}}{V_{L_1}} = \frac{1}{50} = 20\,\text{mA} \qquad \text{with} \qquad R_{L_1} = \frac{50}{20\,\text{mA}} = 2.5\,\text{k}\Omega$$

and

$$I_{L_2} = \frac{P_{L_2}}{V_{L_2}} = \frac{1.2}{20} = 60\,\text{mA} \qquad \text{with} \qquad R_{L_2} = \frac{20}{60\,\text{mA}} = 333.3\,\Omega$$

Now applying KCL to the top and the bottom nodes of the two loads yields

$$I_S = I_1 + I_{L_1}$$

$$100\,\text{mA} = I_1 + 20\,\text{mA}$$

Therefore,

$$I_1 = 80\,\text{mA}$$

and

$$I_S = I_2 + I_{L_2}$$

$$100\,\text{mA} = I_2 + 60\,\text{mA}$$

Thus

$$I_2 = 40\,\text{mA}$$

Therefore,

$$R_1 = \frac{V_{L_1}}{I_1} = \frac{50}{80\,\text{mA}} = 625\,\Omega \qquad \text{and} \qquad P_{R_1} = V_{L_1} \times I_1$$

$$= 50 \times 80\,\text{mA} = 4\,\text{W}$$

and

$$R_2 = \frac{V_{L_2}}{I_2} = \frac{20}{40\,\text{mA}} = 500\,\Omega \qquad \text{and} \qquad P_{R_2} = V_{L_2} \times I_2$$

$$= 20 \times 40\,\text{mA} = 0.8\,\text{W}$$

The design of the voltage divider is thus complete.

(a) Under these nominal operating conditions, using KVL (see Fig. 4.39), we have

$$V_S = E - I_S R_{int}$$

$$= 100 - 0.1 \times 50 = 95\,\text{V}$$

(b) When R_{L_2} is disconnected (i.e., replaced by an open circuit), the source E sees R_S in series with ($R_1 \parallel R_{L_1}$) and R_2. Then the total resistance of the circuit in this case is

$$R_T = 50 + 250 + \frac{625 \times 2500}{625 + 2500} + 500$$

$$= 1300 \, \Omega$$

Therefore,

$$I'_S = \frac{E}{R_T} = \frac{100}{1300} = 76.92 \, \text{mA}$$

Now applying Ohm's law gives us

$$V'_{L_1} = I'_S(R_1 \parallel R_{L_1})$$

$$= 76.92 \times 10^{-3} \times \frac{625 \times 2500}{625 + 2500} = 38.46 \, \text{V}$$

that is, by removing the second load, the voltage across the first load is reduced from 50 V to 38.46 V.

Note that the voltage across R_S is also reduced because the source current is reduced. However, the voltage across the second load, V_{L_2}, will increase because the total of the voltage drops is constant ($= E$). Here $V'_{L_2} = I'_S \times R_2 = 76.92 \times 10^{-3} \times 500 = 38.46 \, \text{V}$ (increased from 20 V). (Check: Find V'_S and check the validity of KVL in this case.)

4.7 | CONCLUDING REMARKS

The material presented in this chapter showed the interrelationship between practical voltage sources and practical current sources. Limits on the ratio R_L/R_{int} indicate when we can consider a given source as approximately ideal voltage source, approximately ideal current source, or in general equivalent to one of the practical models. Model parameters are defined and practical methods used to measure them are discussed. The graphical interpretations of the source parameters are also examined, in relation to the regulation curve or load line. The simple graphical technique of circuit analysis using the load line is covered. It should again be stressed that this technique is very useful, especially in the cases when the load is a nonlinear resistor.

The theory of maximum power transfer and the variation of the load voltage, current, and power transfer efficiency are examined in relation to the load resistance variations.

Design of practical voltage dividers is covered and many examples are discussed. It is also shown that such circuits are in fact of the series-parallel type. Their design and analysis involves nothing more than the applications of KVL, KCL, and Ohm's law.

DRILL PROBLEMS

Section 4.1

4.1. Obtain the open-circuit voltage and the internal resistance of the practical voltage source that has a terminal voltage of 10 V when the load connected to it is 100 Ω, and a terminal voltage of 8 V when the load connected to it is reduced to 64 Ω.

4.2. Obtain the parameters of a practical voltage source that delivers 4 W to either a 100 Ω load or a 400 Ω load.

4.3. A source whose short circuit current is 0.5 A delivers 0.3 A to a 200 Ω load connected to it. What are the parameters of the practical voltage source model of this electrical energy source?

Section 4.2

4.4. Use the load-line technique to determine the load current and the load voltage when the following loads are connected to a source whose open-circuit voltage is 20 V and whose internal resistance is 50 Ω:
 (a) $R_L = 450$ Ω.
 (b) $R_L = 200$ Ω.
 (c) $R_L = 50$ Ω.
 (d) $R_L = 30$ Ω.

Section 4.4

4.5. Find V and I in the circuit shown in Fig. 4.40. Convert this source into its equivalent voltage source model and recaluate V and I.

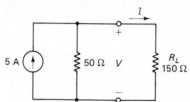

Figure 4.40

4.6. Repeat Drill Problem 4.5 if the load resistance, R_L, is changed to 200 Ω.

4.7. Find V and I in the circuit shown in Fig. 4.41. Convert this source into its equivalent current source model and recalculate V and I.

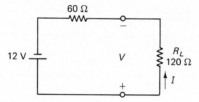

Figure 4.41

4.8. Repeat Drill Problem 4.7 if the load resistance R_L is changed to 240 Ω.

Section 4.5

4.9. Calculate the short-circuit current and the maximum power available from the source examined in Drill Problem 4.1. What are the power transfer efficiencies corresponding to the two load values given in that problem?

4.10. Calculate the short-circuit current and the maximum power available from the source examined in Drill Problem 4.2. Obtain the power transfer efficiencies corresponding to the two load resistances given in that problem.

Section 4.6

4.11. Design a voltage divider similar to that shown in Fig. 4.32 to deliver 20 mA to a 5 kΩ load from an ideal source whose open circuit voltage is 150 V. The source current is limited to 100 mA. Calculate the voltage regulation of this divider.

4.12. Redesign the voltage divider specified above under the condition that the bleeder current is to be 10 times the load current. What are the source current and the voltage regulation in this case?

4.13. Analyze the voltage-divider circuit shown in Fig. 4.42 to obtain the voltages at the two taps (V_1 and V_2), the two bleeder currents, I_1 and I_2, and the power ratings of the two bleeder resistances.

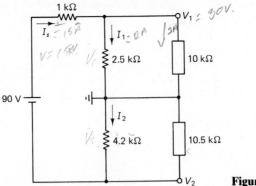

Figure 4.42

4.14. If the 10 kΩ load in the voltage-divider circuit examined above is disconnected, recalculate the values of V_1, V_2, I_1, I_2 and the power dissipated in the two bleeder resistances in this case.

PROBLEMS

Section 4.1

(i) **4.1.** A practical power supply provides 2 A to a load with a terminal voltage of 60 V. The terminal voltage drops to 40 V when the source delivers 4 A. Find
  (a) The open-circuit voltage and internal resistance of this source.
  (b) The current drawn from this supply when a 70 Ω load is connected to it.

(i) **4.2.** Connecting either a 4 Ω load or a 16 Ω load to a generator draws 4 W of power. Find the short-circuit current of this generator.

(i) **4.3.** A source delivers 0.3 A to a 100 Ω load resistance. When a 200 Ω load is connected to this source, it dissipates 8 W. What are the parameters of this source? What is the terminal voltage of this source when a 500 Ω load is connected to it?

(i) **4.4.** The graph in Fig. 4.43(a) shows the V–I characteristics of a certain source.

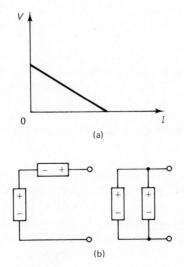

(a)

(b)

Figure 4.43

Superimpose on this graph the characteristics of

(a) Two such sources connected in series aiding.

(b) Two such sources connected in parallel aiding.

These source connections are shown in Fig. 4.43(b).

(ii) **4.5.** A source whose short-circuit current is 1.0 A delivers 0.4 A to a 250 Ω load. What is the voltage across, and the value of, the load resistance when this source is delivering 0.2 A?

Section 4.2

(i) **4.6.** Using the load-line analysis techniques, find the terminal voltage V and the load current I for the circuit shown in Fig. 4.44 when

(a) $R_L = 50\,\Omega$.

(b) $R_L = 150\,\Omega$.

(c) $R_L = 300\,\Omega$.

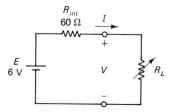

Figure 4.44

(i) **4.7.** A diode whose I–V characteristics is given in Fig. 4.45 is connected to a source whose open-circuit voltage is 3 V and whose internal resistance is 2 kΩ. Using the load-line technique, find the current in and the voltage across this diode.

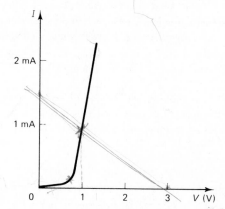

Figure 4.45

(i) **4.8.** The photoconductor cell shown in Fig. 4.46(a) has the I–V characteristics shown in Fig. 4.46(b). Its maximum power dissipation capability is 5 mW. What is the minimum allowable value of R that makes the cell operate safely at all times? (*Hint:* Draw the load line just touching the power dissipation hyperbola.) Find the voltage across the cell at light intensities of 0.1, 0.2, and 0.3 lumen.

Section 4.4

(i) **4.9.** Convert the voltage source in Problem 4.1 into an equivalent current source and then recalculate part (b) using this current source model.

(i) **4.10.** Recalculate the voltage across the source terminals in Problem 4.3 if an 800 Ω load resistance is connected to it. Use the equivalent-current source model.

(i) **4.11.** Solve Problem 4.5 using an equivalent current source model instead of the voltage source model.

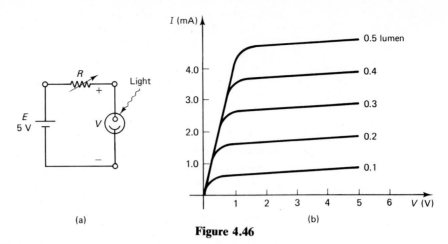

(a)

(b)

Figure 4.46

Section 4.5

(ii) **4.12.** The terminal voltage of a certain generator is 50 V when a 10 Ω load is connected to it. The short-circuit current of this source is 10 A. Find
    (a) The efficiency when a 90 Ω load is connected to this source.
    (b) The maximum power that this generator can deliver.
    (c) The efficiency of this generator if it is delivering 160 W.

(ii) **4.13.** When a 1000 Ω load is connected to a practical voltage source, it dissipates 10 W at an efficiency of 80%. What is the maximum power this source can then deliver and its efficiency?

(i) **4.14.** If a 225 Ω load is connected to a voltage source, it operates at 75% efficiency while the load dissipates 2.25 W. Find the open-circuit voltage of this supply.

(ii) **4.15.** A certain power supply delivers 108 W if a load of either 3 Ω or 12 Ω is connected to it. Find
    (a) The terminal voltage if a load of 6 Ω is connected to the source.
    (b) The maximum power this source can deliver.
    (c) The power delivered by the supply when it operates at 80% efficiency.
    (d) The efficiency if the source delivers 108 W.

(i) **4.16.** A source delivers 0.5 A to a 60 Ω load resistance. Its terminal voltage is 20 V when the load resistance is 20 Ω. Find
    (a) The efficiency in each case.
    (b) The maximum power this source can deliver.
    (c) The power delivered to a load operating at 75% efficiency.

(i) **4.17.** With the load shown in Fig. 4.47 connected to the practical voltage source, the source is operating at a 75% efficiency. Find
    (a) The source parameters.
    (b) The power delivered by the source in this case.
    (c) The maximum power this source can deliver.

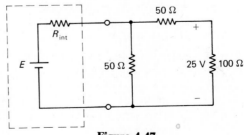

Figure 4.47

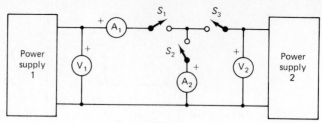

Figure 4.48

(ii) **4.18.** The following table shows some of the results obtained while testing the two power supplies connected as shown in Fig. 4.48. (c = closed, o = open). Find the open-circuit voltage and the internal resistance of each of these two sources and then fill in the blanks in the accompanying table.

Set	Switch positions			Meter readings			
	S_1	S_2	S_3	$A_1(A)$	$A_2(A)$	$V_1(V)$	$V_2(V)$
(i)	c	o	c	1.0		7.0	
(ii)	c	c	c	4.5	10.5		
(iii)	c	c	o				
(iv)	o	c	c				

Section 4.6

(i) **4.19.** Design a voltage divider to supply a 15 V, 0.3 W load from a 24 V source which could be considered ideal. The bleeder resistance is one-fifth of the load resistance. Find the voltage regulation of this divider and the power supplied by the source.

(i) **4.20.** Repeat Problem 4.19 but with the source being a practical voltage source with a 50 Ω internal resistance and the load voltage required is −15 V instead of +15 V.

(i) **4.21.** A 20 V ideal source whose power dissipation is limited to 0.5 W is used to design a voltage divider to supply a 10 V load with 0.1 W. Find the elements of the required voltage divider and its voltage regulation (V. R.).

(i) **4.22.** Design a voltage divider to supply a 12 V load with 10 mA, given a practical 18 V source whose internal resistance is 20 Ω. The bleeder resistance dissipates 0.75 W under nominal conditions. Find the power supplied by the source and the voltage regulation of this divider.

(ii) **4.23.** It is required to design a voltage divider to supply a variable resistance load. The load voltage should be in the range 10 to 10.5 V as the load resistance varies between 10 kΩ and 20 kΩ. The source available is an ideal 15 V source. Find the elements of the required voltage divider.

(ii) **4.24.** Using an ideal 24 V source, design a voltage divider to supply a 50 kΩ load with 0.2 mA. The voltage regulation of this divider is specified to be 5%.

(i) **4.25.** A 10 kΩ potentiometer with an adjustable tap is to be used to design a voltage divider to supply a 6 V, 20 mA load from a 12 V ideal source. What is the resistance of each of the bleeder and series dropping portions of this potentiometer?

(i) **4.26.** Repeat Problem 4.25 but with the load requiring 1 mA of current.

(i) **4.27.** Design a voltage divider to supply the following loads

$$100 \text{ mA at } +250 \text{ V}$$

$$40 \text{ mA at } +60 \text{ V}$$

$$0 \text{ mA at } -10 \text{ V}$$

The source used has an internal resistance of 200 Ω and supplies 200 mA. Use the minimum number of components. Find the open-circuit voltage of the source.

(i) **4.28.** Using an ideal source, design a voltage divider that has to deliver 20 mA at +20 V, 10 mA at +10 V, and 40 mA at −5 V simultaneously. The source supplies 120 mA of current. Find the minimum value of the open-circuit voltage of this source (i.e., no series dropping resistance is required).

(ii) **4.29.** Design a voltage divider to supply the following loads:

$$2 \text{ W at } +40 \text{ V} \qquad \text{and} \qquad 1 \text{ W at } -10 \text{ V}$$

The open-circuit voltage and the short-circuit current of the voltage source to be used are 125 V and 0.5 A, respectively. Find the resistances and their ratings required to design this voltage divider without using a series dropping resistance. What is the power supplied by this source?

(ii) **4.30.** Redesign the voltage divider in Problem 4.29 but with the source current limited to 150 mA. Find the load voltages if both loads are disconnected simultaneously (i.e., replaced by an open circuit). (*Hint:* A series dropping resistance is required in this case.)

(ii) **4.31.** **(a)** Design a voltage divider to supply the following loads:

$$25 \text{ W at } +250 \text{ V} \quad \text{(designated tap } A\text{)}$$
$$2.5 \text{ W at } +50 \text{ V} \quad \text{(designated tap } B\text{)}$$
$$1 \text{ W at } -10 \text{ V} \quad \text{(designated tap } C\text{)}$$

The total current drain from the power supply is 200 mA. The supply has a 50 Ω internal resistance. The supply voltage is adjustable; choose and specify the minimum required open-circuit voltage of the source.

(b) If the 25 W load is disconnected, find the tap voltage V_A, V_B, and V_C.

APPENDIX A4: PROOF OF THE MAXIMUM POWER TRANSFER THEOREM

Power in a variable resistance load was obtained in Eq. (4.17):

$$P_L = \frac{E^2 R_L}{(R_L + R_{int})^2}$$

This equation shows the variation of P_L as a function of R_L. The condition for a maximum in P_L corresponds to the case when the slope of the P_L curve becomes zero, that is,

$$\frac{dP_L}{dR_L} = 0$$

Thus differentiating P_L and equating to zero yields

$$E^2 \frac{(R_L + R_{int})^2 - 2R_L(R_L + R_{int})}{(R_L + R_{int})^4} = 0$$

That is,

$$(R_L + R_{int})^2 = 2R_L(R_L + R_{int})$$

or

$$R_L + R_{int} = 2R_L$$

Therefore,

$$R_L = R_{int}$$

is the condition that results in P_L being a maximum. Substituting this value of R_L into Eq. (4.17) results in

$$P_{L\text{max}} = \frac{E^2}{4R_{\text{int}}}$$

as obtained in Eq. (4.19). It should be noted that

$$\left.\frac{d^2 P_L}{dR_L^2}\right|_{R_L = R_{\text{int}}}$$

is a negative quantity, clarifying that this condition actually results in a maximum, not a minimum.

GLOSSARY

Bleeder Resistance: A resistive element in the voltage divider, connected in parallel with the load. Its function is to reduce the amount of voltage regulation of the divider.

Full-Load Current: The maximum allowable amount of current available from a power supply when the resistance of the load connected to the power supply is at its minimum allowable value.

Full-Load Voltage: The voltage across the terminals of a power supply when the resistance of the load connected to the power supply is at its minimum allowable value.

Ideal Current Source: An electrical energy source that delivers a constant amount of current to any load connected to it, regardless of the voltage across the terminals of the source.

Ideal Voltage Source: An electrical energy source that delivers a constant amount of voltage to any load connected to it, regardless of the amount of current drawn from this source.

Internal Resistance: An equivalent resistive component of the practical model of a voltage or a current source.

Internal Voltage Drop: The voltage drop across the internal resistance of the practical voltage source model.

Load-Line Analysis: A graphical technique used to obtain the voltage across and the current through a given load when that load is connected to a given electrical energy source. This technique uses the graphical I–V characteristics of the load and the source.

Maximum Power Transfer Condition (or Matching Condition): The relationship between the load resistance (or impedance) and the internal resistance (or impedance) of a source that results in the maximum possible power in the load.

Nominal Current: The normal value of the current required to properly operate a given electrical device (or load).

Nominal Voltage: The normal value of the voltage required to properly operate a given electrical device (or load).

Open-Circuit Voltage (or No-Load Voltage): The value of the voltage across the terminals of an electrical energy source when no load is connected to this source (i.e., when the terminals of the source are left open-circuited).

Power Transfer Efficiency: The ratio of the load power to the total power provided by the ideal voltage source.

Practical Current Source: An actual source that delivers a decreasing amount of current as the resistance (or impedance) of the load connected to it is increased. The model used for such a source includes an internal resistance in parallel with an ideal current source.

Practical Voltage Source: An actual voltage source that delivers a decreasing amount of voltage as the resistance (or impedance) of the load connected to it is decreased. The model used for such a source includes an internal resistance in series with an ideal voltage source.

Q (Quiescent Point or Operating Point): The point of intersection of the load and the source $I–V$ characteristics. Its coordinates determine the load current and voltage, which are also the terminal current and voltage of the source.

Regulation Curve or Load Line: The $V–I$ (or $I–V$) characteristics of a given source, or of a voltage divider, or in general of any network that can be represented by a practical voltage source model.

Series Dropping Resistance: A resistive component of the voltage-divider circuit. Its function is to provide a drop in voltage from that available from a source to the required value of the load voltage.

Short-Circuit Current: The value of the current provided by a source when its terminals are short circuited [i.e., when the resistance (or impedance) of the load is zero].

Source Transformation: The technique through which a practical voltage source model can be converted into its mathematically equivalent current source model, or vice versa. The equivalency is valid only as far as the terminal $V–I$ behavior of the two models are concerned.

Voltage Divider: The circuit used to provide the required nominal voltages and currents to different loads from a single source whose voltage is higher than the highest load operating voltage.

Voltage Regulation: A measure of the variation of a source (or a voltage-divider) voltage, as the load resistance is varied. It is expressed as a ratio (or percentage) of the voltage variation between the no-load (o.c.) condition and the full-load (nominal) condition, with respect to the full-load (nominal) voltage.

ANSWERS TO DRILL PROBLEMS

4.1. $R_{int} = 80\,\Omega$, $E = 18$ V

4.3. $R_{int} = 300\,\Omega$, $E = 150$ V

4.5. $V = 187.5$ V, $I = 1.25$ A; equivalent source: $E = 250$ V, $R_{int} = 50\,\Omega$

4.7. $V = 8$ V, $I = 0.0667$ A; equivalent source: $I_S = 0.2$ A, $R_{int} = 60\,\Omega$

4.9. $I_{sc} = 0.225$ A, $P_{max} = 1.0125$ W, $\eta_1 = 0.5556$, $\eta_2 = 0.4444$

4.11. $R_{sd} = 500\,\Omega$, $P_{sd} = 5$ W, $R_b = 1250\,\Omega$, $P_b = 8$ W, V. R. $= 7.14\%$

4.13. $V_1 = 30$ V, $V_2 = -45$ V, $I_1 = 12$ mA, $P_1 = 0.36$ W, $I_2 = 10.714$ mA, $P_2 = 0.482$ W

5 GENERAL RESISTIVE NETWORKS

OBJECTIVES

- Understanding the method of development of the loop analysis technique based on the systemic application of KVL.
- Understanding the method of development of the node analysis technique based on the systematic application of KCL.
- Ability to determine the most suitable method of network analysis in order to simplify the procedure of determining a certain unknown network variable (voltage and/or current).
- Proficiency of applying the general methods of loop-, and node-analysis techniques to any resistive electrical network.

The series equivalent and the parallel equivalent resistance models that were introduced in Chapter 3 and used to simplify some circuit analysis procedures are not applicable in all cases, for example, the general bridge circuit shown in Fig. 3.50. Other networks containing a number of voltage and current sources in a number of loops cannot be simplified using the techniques discussed in Chapter 3. Those techniques are also very difficult to apply to nonplanar networks (those circuits that when drawn in a single plane contain crossed wires, one can think of circuit elements geometrically arranged in the shape of a pyramid or a cube, etc.).

The new circuit analysis techniques introduced here are general and can be applied to any type of network. The loop or mesh analysis method is based on the systematic application of KVL, and the node analysis method is based on the systematic application of KCL. With both methods we end up with a number of linear simultaneous algebraic equations which can be solved to obtain the value(s) of the required unknown(s).

The most important criterion in determining which method is easier or more suitable to use is the number of simultaneous equations. It is better to use the method that results in the least number of simulatneous equations, because this will save the computation time required to solve these equations. Even though nonplanar networks will not be discussed here, it is important to mention that these types of networks are best solved using the node analysis technique.

5.1 LOOP (OR MESH) ANALYSIS METHOD

5.1.1 Development of the Technique

Any closed path in a circuit is a loop. A mesh is a special loop that does not contain any branches inside it. Considering the circuit shown in Fig. 5.1, there are two meshes, marked ① and ②. Each is also a loop. The outside path, however, including E_1, R_1, R_2, and E_2, is a loop but not a mesh, because the branch containing the element R_3 is inside this loop. Meshes resemble window panes (or partitions).

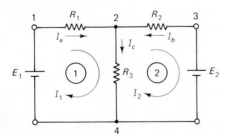

Figure 5.1

For any given circuit, the number of linear independent simultaneous equations required to solve the circuit using the loop (or mesh) analysis method is

$$n = B - (N - 1) = B - N + 1 \qquad (5.1)$$

where B is the number of branches and N is the number of nodes.

Note. Here all the sources should be voltage sources. Further explanation of this remark follows later.

In Fig. 5.1 the circuit shown has four nodes and five branches. Thus according to Eq. (5.1), two independent equations are required. This is also the same as the number of meshes. This last observation is general and applies to all planar networks. It is important to note that an independent equation is one that cannot be obtained from the other simultaneous equations through algebraic manipulations.

In general, any n loops (or meshes) may be considered to develop the required equations in n unknowns. Meshes are, however, easier to identify. Also, if one is not skillful enough in this technique, some of the loops chosen may result in dependent (i.e., redundant) equations. It is therefore recommended that while this important skill is being developed, meshes be the preferred choice.

Once the meshes are chosen, the technique itself is simply the application of KVL to each mesh (or loop) to obtain the required equations. The objective here is to develop a systematic procedure with which to write these equations in a straight-forward and simple manner.

To develop this procedure, let us first consider the branch currents, I_a, I_b, and I_c in the circuit shown in Fig. 5.1. Clearly, the unknown current I_c is dependent on I_a and I_b since from KCL,

$$I_c = I_a + I_b \tag{5.2}$$

that is, I_c can be easily obtained once I_a and I_b are determined.

Writing KVL for each of meshes ① and ② in the clockwise direction of the arrows indicated in Fig. 5.1, we have

$$-E_1 + I_a R_1 + I_c R_3 = 0 \tag{5.3}$$

and

$$E_2 - I_c R_3 - I_b R_2 = 0 \tag{5.4}$$

These two equations can be rewritten as follows, substituting for I_c from Eq. (5.2):

$$R_1 I_a + R_3(I_a + I_b) = E_1$$

and

$$R_2 I_b + R_3(I_a + I_b) = E_2$$

or

$$(R_1 + R_3)I_a + R_3 I_b = E_1 \tag{5.3a}$$

and

$$R_3 I_a + (R_2 + R_3)I_b = E_2 \tag{5.4a}$$

Clearly, Eqs. (5.3a) and (5.4a), for the two meshes, respectively, provide the required two equations in the two unknowns, I_a and I_b. They can then be solved to determine these two branch currents.

To produce a more systematic procedure, each mesh is assumed to allow a fictitious mesh current, I_1 and I_2 for meshes ① and ② in the circuit of Fig. 5.1, to flow in this mesh. The systematic procedure also requires that the direction of these mesh currents be the same, preferably clockwise, even though the counterclockwise direction is just as good. These mesh, or in general loop currents, can easily be related to the real branch currents.

In the case of the circuit of Fig. 5.1, we can easily deduce that

$$I_1 = I_a \tag{5.5}$$

and

$$I_2 = -I_b \tag{5.6}$$

by comparing the assumed mesh current flow to the branch current in each of the elements R_1 and R_2.

Equations (5.3a) and (5.4a) can now be written with I_a and I_b replaced by their equivalent mesh currents from Eqs. (5.5) and (5.6), as follows:

$$(R_1 + R_3)I_1 - R_3I_2 = E_1 \tag{5.7}$$

and

$$R_3I_1 - (R_2 + R_3)I_2 = E_2$$

or if we multiply both sides of this last equation by -1,

$$-R_3I_1 + (R_2 + R_3)I_2 = -E_2 \tag{5.8}$$

Equations (5.7) and (5.8) are called the *mesh equations* (or in general the *loop equations*). Solution of these equations provides the required value of the unknown mesh currents, I_1 and I_2 in this case. The different branch currents can then easily be deduced.

The current in an element common between two meshes is the net algebraic sum of the two mesh (or loop) currents. In Fig. 5.2 the current I_{n-1} in mesh number $(n-1)$ is flowing downward and the current I_n in mesh number n is flowing upward. Thus

current in the common element $= I_{n-1} - I_n$ (flowing downward)

or

current in the common element $= I_n - I_{n-1}$ (flowing upward)   (5.9)

Applying this result to the element R_3, common between meshes ① and ② in the circuit of Fig. 5.1, yields

$$I_c = I_1 - I_2 \quad \text{(flowing downward)}$$

This agrees with Eq. (5.2), with $I_a = I_1$ and $I_b = -I_2$ as in Eqs. (5.5) and (5.6).

With the loop (or mesh) current determined and the branch currents identified, the voltage across and the power in any component can easily be calculated. If the solution results in a negative value for either a mesh or a branch current, this means that the real current direction is opposite to the assumed one.

Remark. When KVL is applied to the outside loop of the circuit in Fig. 5.1, in the clockwise direction and in terms of the branch currents, then

$$-E_1 + I_aR_1 - I_bR_2 + E_2 = 0 \tag{5.10}$$

This loop equation is not independent because it is, in fact, obtainable from Eqs. (5.3) and (5.4) by adding the left-hand-side terms and equating them to the sum of the right-hand-side terms.

The point raised earlier about the possibility that some loop equations may be dependent (i.e., redundant) can now be appreciated. For an n-mesh network there are only n independent mesh (or loop) equations.

5.1.2 Generalized Systematic Procedure

The systematic procedure outlined above, resulting in the mesh equations (5.7) and (5.8), can be generalized by noting that:

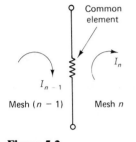

Common element

I_{n-1}

I_n

Mesh $(n-1)$ Mesh n

Figure 5.2

R_{11} = self-or total resistance of mesh 1
    = $R_1 + R_3$ (in Fig. 5.1)

$R_{12} = R_{21}$ = mutual or common resistance between meshes 1 and 2
    = R_3 (in Fig. 5.1)

R_{22} = self-or total resistance of mesh 2
    = $R_2 + R_3$ (in Fig. 5.1)

E_{11} = algebraic sum of all the source voltages in mesh 1, driving the mesh current, I_1, in the clockwise direction
    = E_1 (in Fig. 5.1)

E_{22} = algebraic sum of all the source voltages in mesh 2, driving the mesh current I_2 in the clockwise direction
    = $-E_2$ (in Fig. 5.1)

Therefore, the general form of the mesh equations for a two-mesh network [with reference to Eqs. (5.7) and (5.8)] can be written as follows:

$$R_{11}I_1 - R_{12}I_2 = E_{11} \qquad (5.11)$$

$$-R_{21}I_1 + R_{22}I_2 = E_{22} \qquad (5.12)$$

Similarly, one can extend such a systematic procedure to any network with any number of meshes. For example, for a three-mesh network, the mesh equations are in the form

$$R_{11}I_1 - R_{12}I_2 - R_{13}I_3 = E_{11} \qquad (5.13)$$

$$-R_{21}I_1 + R_{22}I_2 - R_{23}I_3 = E_{22} \qquad (5.14)$$

$$-R_{31}I_1 - R_{32}I_2 + R_{33}I_3 = E_{33} \qquad (5.15)$$

where R_{11}, R_{22}, and R_{33} are the self-(or total) resistance of meshes 1, 2, and 3, respectively. E_{11}, E_{22} and E_{33} are also the algebraic sums of the voltage sources in meshes 1, 2, and 3 respectively, driving the corresponding mesh currents in the clockwise direction. The mutual or common resistance between each pair of meshes are clearly

$$R_{12} = R_{21} \quad \text{(mutual resistance between meshes 1 and 2)}$$

$$R_{23} = R_{32} \quad \text{(mutual resistance between meshes 2 and 3)}$$

$$R_{31} = R_{13} \quad \text{(mutual resistance between meshes 3 and 1)}$$

These general mesh equations are also referred to as the *shortcut procedure* because they can be obtained in a straightforward manner, merely by inspection of the network.

■ EXAMPLE 5.1

(a) For the three-mesh network shown in Fig. 5.3, write KVL for each mesh in terms of the branch currents I_a, I_b and I_c.

(b) Rewrite the same equations in terms of the mesh currents I_1, I_2, and I_3 by comparing the equivalency of the corresponding branch and mesh currents.

(c) Show that the resulting mesh equations do, in fact, agree with the general form in Eqs. (5.13), (5.14), and (5.15) and can thus be obtained by inspection.

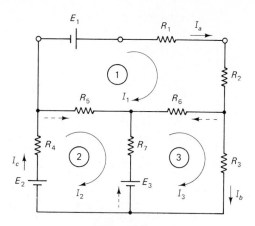

Figure 5.3

Solution   (a) The currents in the rest of the branches (shown as dashed arrows in Fig. 5.3) can easily be obtained from KCL. They are

$$I_{R_6} = I_a - I_b$$
$$I_{R_7} = I_b - I_c$$
$$I_{R_5} = I_c - I_a$$

Now applying KVL to mesh (or loop) 1 in the clockwise direction yields

$$- E_1 + R_1 I_a + R_2 I_a + R_6(I_a - I_b) - R_5(I_c - I_a) = 0 \qquad (5.16a)$$

Collecting terms and rearranging gives us

$$(R_1 + R_2 + R_6 + R_5)I_a - R_5 I_c - R_6 I_b = E_1 \qquad (5.16b)$$

For mesh 2, KVL in the clockwise direction provides

$$- E_2 + R_4 I_c + R_5(I_c - I_a) - R_7(I_b - I_c) + E_3 = 0 \qquad (5.17a)$$

Collecting terms and rearranging, we have

$$- R_5 I_a + (R_4 + R_5 + R_7)I_c - R_7 I_b = E_2 - E_3 \qquad (5.17b)$$

KVL in the clockwise direction for mesh 3 provides

$$- E_3 + R_7(I_b - I_c) - R_6(I_a - I_b) + R_3 I_b = 0 \qquad (5.18a)$$

Again collecting terms and rearranging yields

$$- R_6 I_a - R_7 I_c + (R_7 + R_6 + R_3)I_b = E_3 \qquad (5.18b)$$

(b) Now comparing the branch and the mesh currents, it is clear that

$$I_1 = I_a$$
$$I_2 = I_c$$

and

$$I_3 = I_b$$

Substituting the corresponding mesh currents in Eqs. (5.16b), (5.17b), and

(5.18b), we have

$$(R_1 + R_2 + R_5 + R_6)I_1 - R_5I_2 - R_6I_3 = E_1 \qquad (5.16c)$$

$$- R_5I_1 + (R_4 + R_5 + R_7)I_2 - R_7I_3 = E_2 - E_3 \qquad (5.17c)$$

$$- R_6I_1 - R_7I_2 + (R_3 + R_6 + R_7)I_3 = E_3 \qquad (5.18c)$$

These are the required mesh equations.

(c) Applying the general form equations (5.13), (5.14), and (5.15) to the circuit in Fig. 5.3 gives us the following:

$$R_{11} = \text{self-resistance of mesh 1} = R_1 + R_2 + R_6 + R_5$$
$$R_{22} = \text{self-resistance of mesh 2} = R_4 + R_5 + R_7$$
$$R_{33} = \text{self-resistance of mesh 3} = R_7 + R_6 + R_3$$
$$R_{12} = R_{21} = \text{mutual resistance between meshes 1 and 2} = R_5$$
$$R_{23} = R_{32} = \text{mutual resistance between meshes 2 and 3} = R_7$$
$$R_{31} = R_{13} = \text{mutual resistance between meshes 3 and 1} = R_6$$
$$E_{11} = \text{algebraic sum of source voltages in mesh 1 driving } I_1$$
$$\text{clockwise} = E_1$$
$$E_{22} = \text{algebraic sum of source voltages in mesh 2 driving } I_2$$
$$\text{clockwise} = E_2 - E_3$$
$$E_{33} = \text{algebraic sum of source voltages in mesh 3 driving } I_3$$
$$\text{clockwise} = E_3$$

Clearly, when these terms are inserted in the general form equations, the results are exactly as obtained in Eqs. (5.16c), (5.17c), and (5.18c). It is thus easier to apply the shortcut procedure (by inspection) to obtain the mesh equations directly.

5.1.3 Source Conversions

As explained in Chapter 4, it is always possible to convert a practical current source model into an equivalent voltage source model, and vice versa, as summarized in Fig. 5.4. This source equivalency is, however, valid as far as the load R_L or the rest of the circuit to the right of the source terminals a and b are concerned. It is important to note that inside the source model itself the equivalency is not valid.

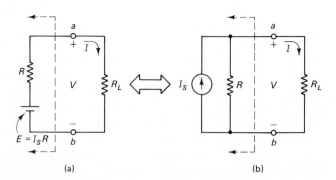

Figure 5.4 (a) Voltage source model; (b) current source model.

For the voltage source model in Fig. 5.4,

$$I \text{ (load current)} = \frac{E}{R + R_L} = \frac{I_S R}{R + R_L} \tag{5.19}$$

$$V \text{ (load voltage)} = IR_L = I_S \frac{RR_L}{R + R_L} \tag{5.20}$$

Here the current in the source's internal resistance R is the same as the load current I. The voltage across this internal resistance is

$$V_R = IR = I_S \frac{R^2}{R + R_L} \quad (\neq V) \tag{5.21}$$

For the current source model in Fig. 5.4,

$$I \text{ (load current using current-division principle)} = I_S \frac{R}{R + R_L} \tag{5.22}$$

$$V \text{ (load voltage)} = IR_L = I_S \frac{RR_L}{R + R_L} \tag{5.23}$$

However, in this case, the current in the source's internal resistance R is determined from the current-division principle (or Ohm's law) as

$$I_R = I_S \frac{R_L}{R + R_L} \quad (\neq I) \tag{5.24}$$

and the voltage across this internal resistance is

$$V_R = I_R R = I_S \frac{RR_L}{R + R_L} = V$$

(as is obvious from Fig. 5.4). Thus, as Eqs. (5.19) and (5.20) are exactly the same as Eqs. (5.22) and (5.23), the equivalency is valid as far as R_L (or the rest of the circuit) is concerned. However, for the internal resistance of the source, each model results in a different value of the voltage across it and the current flow in it, as examined in Eqs. (5.21) and (5.24). This proves the fact that this source equivalency is unfortunately not valid for the elements used in the transformation inside the models. This point is very important because it indicates that if the current or voltage for an element used in source conversions is required, the original circuit connections must be considered, not the equivalent (transformation) model.

In mesh (or loop) analysis, the procedure involves the application of KVL. But the voltage across an ideal current source is undetermined. However, in conjunction with a resistance in parallel with such an ideal current source, the terminal $V-I$ characteristics (or equation) can be defined as shown in Chapter 4. This is the same as saying that a practical current source can be converted to an equivalent practical voltage source. Particularly, to apply the generalized shortcut procedure, it is important that current sources be converted to voltage sources (i.e., all the sources in the network should preferably be of the voltage source model type). This is the recommended procedure in mesh analysis: *Before starting the analysis, always convert practical current sources to their equivalent voltage source models.*

This is not to say that without performing such source conversions mesh analysis cannot be applied. It still can. This involves choosing the loops (or meshes) carefully, with only one mesh current allowed to flow through an ideal current source (here $I_{mesh} = I_{source}$). Here writing the loop (or mesh) equations requires independent, careful application of KVL to each loop (or mesh). This makes the procedure of mesh analysis longer and more tedious than the generalized shortcut procedure.

In summary, the following steps are to be followed to apply systematically the shortcut generalized procedure of mesh analysis:

1. Convert all practical current sources that may exist in the network into equivalent voltage source models.
2. Identify meshes in the network and assign mesh currents to them all in the same direction, preferably clockwise.
3. Determine the self-(or total) resistance of each mesh.
4. Determine the mutual (or common) resistance(s) between each pair of meshes.
5. Determine the total algebraic sum of the source voltages in each mesh driving the corresponding mesh currents in the assumed (clockwise) direction. Source voltages are taken with a positive sign when driving the mesh current in the clockwise direction, and with a negative sign when opposing the flow of current in the assumed directions.
6. Substitute the required terms, as determined above, in the generalized mesh equations, as described in Eqs. (5.11) and (5.12) for a two-mesh network; Eqs. (5.13), (5.14) and (5.15) for a three-mesh network; and so on.
7. Solve the resulting mesh equations to determine the unknown mesh currents. Relate the mesh currents to any required branch (or element) current in the original circuit to obtain the desired result of the network analysis.

It should be noted that some important characteristics of this generalized (shortcut) mesh analysis procedure can be deduced by observing, for example, the three mesh equations (5.13), (5.14), and (5.15):

I. All the principal diagonal coefficients, R_{11}, R_{22}, R_{33}, and so on, have positive signs.
II. All the other off-diagonal coefficients, representing the mutual resistances, have negative signs. Each such coefficients appears twice ($R_{12} = R_{21}$, $R_{13} = R_{31}$, etc.). They are symmetrically distributed about the principal diagonal.

5.2 DETERMINANTS AND CRAMER'S RULE

In loop (or mesh) analysis or node analysis techniques used to analyze electrical networks, a system of linear simultaneous equations results and has to be solved. The study of algebra provides many techniques to solve such a system of linear simultaneous equations, among which is the Gaussian elimination method and

Cramer's rule, using determinants. Here a brief discussion of Cramer's rule and determinants is presented. As mentioned, this is a useful solution tool because of its systematic nature; however, the reader is encouraged to consult mathematical texts for more details on this and other methods of solutions.

Consider, for example, the following two linear simultaneous equations in two unknowns, x_1 and x_2:

$$a_1x_1 + b_1x_2 = k_1 \tag{5.25}$$

$$a_2x_1 + b_2x_2 = k_2 \tag{5.26}$$

The coefficients a_1, a_2, b_1, b_2, k_1, and k_2 are all known constants. The coefficient determinant, Δ, is written and its value calculated as follows:

$$\Delta = \begin{vmatrix} a_1 & b_1 \\ a_2 & b_2 \end{vmatrix} = a_1b_2 - b_1a_2 \tag{5.27}$$

This system determinant is a 2×2 array because the system consists of two equations. In the determination of the value of Δ as shown in Eq. (5.27), the product of the terms along the diagonal arrow pointing to the right (a_1b_2) is taken with a positive sign, while the product of the terms along the diagonal arrow pointing to the left (b_1a_2) is taken with a negative sign. The products themselves could be either positive, negative, or zero, depending on the value and sign of the coefficients. This is called the *diagonal rule*. As a numerical example, consider

$$\Delta = \begin{vmatrix} 4 & -3 \\ 2 & -5 \end{vmatrix} = [4(-5)] - [(-3)2] = -20 + 6 = -14$$

Cramer's rule states that the values of the unknowns x_1 and x_2 can be determined from

$$x_1 = \frac{\Delta_1}{\Delta} \tag{5.28}$$

and

$$x_2 = \frac{\Delta_2}{\Delta} \tag{5.29}$$

where the determinant Δ_1 is obtained by replacing the first column (i.e., the coefficients of x_1: a_1 and a_2) by the right-hand constants of the equations: k_1 and k_2. Thus

$$\Delta_1 = \begin{vmatrix} k_1 & b_1 \\ k_2 & b_2 \end{vmatrix} = k_1b_2 - b_1k_2 \tag{5.30}$$

Similarly, the determinant Δ_2 is obtained by replacing the second column (i.e., the coefficients of x_2: b_1 and b_2) by the right-hand constants of the equations: k_1 and k_2. Thus

$$\Delta_2 = \begin{vmatrix} a_1 & k_1 \\ a_2 & k_2 \end{vmatrix} = a_1k_2 - k_1a_2 \tag{5.31}$$

■ EXAMPLE 5.2

Determine the values of x and y in the following two simultaneous equations:

$$x - 3y = 6$$

$$2x + 4y = -5$$

Solution   Here the system determinant Δ is

$$\Delta = \begin{vmatrix} 1 & -3 \\ 2 & 4 \end{vmatrix} = 1 \times 4 - (-3) \times (2) = 4 + 6 = 10$$

while

$$\Delta_1 = \begin{vmatrix} 6 & -3 \\ -5 & 4 \end{vmatrix} = 6 \times 4 - (-3) \times (-5) = 24 - 15 = 9$$

and

$$\Delta_2 = \begin{vmatrix} 1 & 6 \\ 2 & -5 \end{vmatrix} = 1 \times (-5) - 6 \times 2 = -5 - 12 = -17$$

Therefore, from Cramer's rule,

$$x = \frac{\Delta_1}{\Delta} = \frac{9}{10} = 0.9$$

and

$$y = \frac{\Delta_2}{\Delta} = \frac{-17}{10} = -1.7$$

It is easy to check that these results are correct by substituting back into the two equations given and confirming that they are both satisfied.

The procedure outlined above can be extended to the case of three equations in three unknowns. For example, consider

$$a_1 x_1 + b_1 x_2 + c_1 x_3 = k_1 \tag{5.32}$$
$$a_2 x_1 + b_2 x_2 + c_2 x_3 = k_2 \tag{5.33}$$
$$a_3 x_1 + b_3 x_2 + c_3 x_3 = k_3 \tag{5.34}$$

Here the coefficient (or the system) determinant Δ is

$$\Delta = \begin{vmatrix} a_1 & b_1 & c_1 \\ a_2 & b_2 & c_2 \\ a_3 & b_3 & c_3 \end{vmatrix} \tag{5.35}$$

There are many ways of calculating the value of such a 3×3 array determinant (because we have a system of three equations). One such method also employs the diagonal rule, with the first two columns repeated to the right, as follows:

$$= a_1 b_2 c_3 + b_1 c_2 a_3 + c_1 a_2 b_3 - c_1 b_2 a_3 - a_1 c_2 b_3 - b_1 a_2 c_3 \tag{5.35a}$$

As before, the products of the terms along the diagonal arrows pointing to the right are taken with a positive sign, while the products of the terms along the diagonal arrows pointing to the left are taken with a negative sign (notice that the products themselves could be algebraically positive, negative, or zero numbers).

In this case, Cramer's rule to determine the value of the unknowns x_1, x_2, and x_3, is stated as follows:

$$x_1 = \frac{\Delta_1}{\Delta} \tag{5.36}$$

$$x_2 = \frac{\Delta_2}{\Delta} \tag{5.37}$$

and

$$x_3 = \frac{\Delta_3}{\Delta} \tag{5.38}$$

The determinant Δ_1 is the same as Δ with the first column representing the coefficients of x_1, $(a_1, a_2,$ and $a_3)$ replaced by the right-hand-side constants of the equations $(k_1, k_2,$ and $k_3)$, as was done in the case of the two equations. Similarly, for Δ_2 and Δ_3 the second and the third columns are, respectively, replaced by the right-hand-side constants, k_1, k_2 and k_3. Thus

$$\Delta_1 = \begin{vmatrix} k_1 & b_1 & c_1 \\ k_2 & b_2 & c_2 \\ k_3 & b_3 & c_3 \end{vmatrix}$$

$$\Delta_2 = \begin{vmatrix} a_1 & k_1 & c_1 \\ a_2 & k_2 & c_2 \\ a_3 & k_3 & c_3 \end{vmatrix}$$

and

$$\Delta_3 = \begin{vmatrix} a_1 & b_1 & k_1 \\ a_2 & b_2 & k_2 \\ a_3 & b_3 & k_3 \end{vmatrix}$$

Calculations of the values of these determinants are performed in a manner similar to that used for Δ, as shown in Eq. (5.35a).

The following examples will show typical numerical computations. We shall seldom encounter circuit analysis problems, particularly in this chapter, that result in more than three equations in three unknowns. However, in many practical situations, the analysis of electrical or electronic circuits may require solving a set of four, five, or more equations in the same number of unknowns. Solutions using determinants then become quite tedious, long, and prone to error, especially when done by hand. In such situations computer-aided analysis techniques become quite useful, if not in fact necessary. This topic is treated in Chapter 19.

5.2.1 Examples of Mesh (or Loop) Analysis

EXAMPLE 5.3

Find the mesh currents in the circuit shown in Fig. 5.5. What is the current flowing in the 12V battery?

Solution All the sources in this network are voltage sources. The two meshes are clearly identified and the corresponding mesh currents are assumed to flow

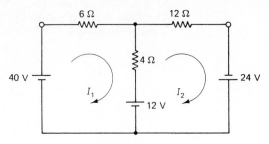

Figure 5.5

in a clockwise direction as shown. Two equations can be written for this two-mesh (or loop) network.

The self-resistance of mesh 1 is $(6 + 4) = 10\,\Omega$, while the self-resistance of mesh 2 is $(4 + 12) = 16\,\Omega$. The mutual resistance is the $4\,\Omega$ element. The net voltage of the sources in mesh 1 driving I_1 in the clockwise direction is $(40 - 12) = 28\,\text{V}$. Similarly, for mesh 2, the net voltage of the sources is $(12 + 24) = 36\,\text{V}$. Thus using the shortcut generalized procedure, we have

$$(6 + 4)I_1 - 4I_2 = (40 - 12) \Rightarrow 10I_1 - 4I_2 = 28$$

and

$$-4I_1 + (4 + 12)I_2 = (12 + 24) \Rightarrow -4I_1 + 16I_2 = 36$$

To use Cramer's rule to find the solutions for I_1 and I_2 from these two equations, the following determinants have to be evaluated:

$$\Delta = \begin{vmatrix} 10 & -4 \\ -4 & 16 \end{vmatrix} = 10 \times 16 - (-4) \times (-4) = 160 - 16 = 144$$

$$\Delta_1 = \begin{vmatrix} 28 & -4 \\ 36 & 16 \end{vmatrix} = 28 \times 16 - (-4) \times 36 = 448 + 144 = 592$$

and

$$\Delta_2 = \begin{vmatrix} 10 & 28 \\ -4 & 36 \end{vmatrix} = 10 \times 36 - 28 \times (-4) = 360 + 112 = 472$$

Therefore,

$$I_1 = \frac{\Delta_1}{\Delta} = \frac{592}{144} = 4.11\,\text{A}$$

and

$$I_2 = \frac{\Delta_2}{\Delta} = \frac{472}{144} = 3.28\,\text{A}$$

As I_1 is larger than I_2, the net current flowing in the 12 V battery will flow downward. Its value is

$$I_1 - I_2 = 4.11 - 3.28 = 0.83\,\text{A}$$

EXAMPLE 5.4

Find the current in each of the voltage sources and in the $6\,\Omega$ resistor in the network shown in Fig. 5.6.

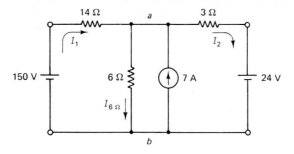

Figure 5.6

Solution The practical current source connected between nodes a and b must first be converted into an equivalent voltage source between the same two nodes, as shown in Fig. 5.7. The equivalent voltage source has the value $E = 6 \times 7 = 42\,\text{V}$ (notice that the positive terminal is upward, based on the current source direction).

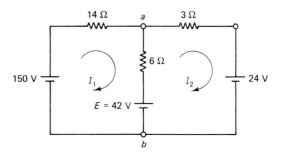

Figure 5.7

The two meshes and the corresponding mesh currents are identified as shown in Fig. 5.7. Now writing the mesh equations directly (this network is similar to the one in Example 5.3), we have

$$(14 + 6)I_1 - 6I_2 = 150 - 42 \Rightarrow 20I_1 - 6I_2 = 108$$
$$-6I_1 + (6 + 3)I_2 = 42 - 24 \Rightarrow -6I_1 + 9I_2 = 18$$

Here

$$\Delta = \begin{vmatrix} 20 & -6 \\ -6 & 9 \end{vmatrix} = 180 - (-6) \times (-6) = 180 - 36 = 144$$

$$\Delta_1 = \begin{vmatrix} 108 & -6 \\ 18 & 9 \end{vmatrix} = 108 \times 9 - (-6) \times 18 = 972 + 108 = 1080$$

and

$$\Delta_2 = \begin{vmatrix} 20 & 108 \\ -6 & 18 \end{vmatrix} = 20 \times 18 - 108 \times (-6) = 360 + 648 = 1008$$

Thus from Cramer's rule,

$$I_1 = \frac{\Delta_1}{\Delta} = \frac{1080}{144} = 7.5\,\text{A}$$

and

$$I_2 = \frac{\Delta_2}{\Delta} = \frac{1008}{144} = 7.0\,\text{A}$$

The mesh currents I_1 and I_2 in Fig. 5.7 are exactly the same as the branch currents I_1 and I_2 flowing in the two voltage sources in Fig. 5.6. However, the current in the $6\,\Omega$ resistor cannot be found from the equivalent circuit of Fig. 5.7, that is,

$$I_{6\Omega} \neq I_1 - I_2 \neq 7.5 - 7 \neq 0.5\,\text{A}$$

because the $6\,\Omega$ resistor was considered as an internal resistance in the source transformation (see the discussion in Section 5.1.3). From the original circuit of Fig. 5.6, applying KCL at node a, we have

$$I_1 + 7 = I_2 + I_{6\Omega}$$

Therefore,

$$I_{6\Omega} = I_1 + 7 - I_2 = 7.5 + 7 - 7 = 7.5\,\text{A}$$

EXAMPLE 5.5

Find the current in the $4\,\Omega$ resistor in the circuit shown in Fig. 5.8.

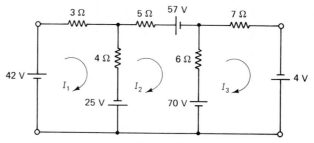

Figure 5.8

Solution No source conversions are needed here since all the sources are voltage sources. The three meshes are easily identified and the corresponding mesh currents are assumed to flow in a clockwise direction as shown.

Following the procedure described in conjunction with the shortcut method of Eqs. (5.13), (5.14), and (5.15), the three mesh equations can be written directly as indicated below. Note that there is no common (or mutual) resistance between meshes 1 and 3 (i.e., $R_{13} = R_{31} = 0$). Thus

$$(3 + 4)I_1 - 4I_2 - 0I_3 = (42 + 25)$$
$$\Rightarrow 7I_1 - 4I_2 - 0I_3 = 67$$
$$-4I_1 + (4 + 5 + 6)I_2 - 6I_3 = (-25 - 57 - 70)$$
$$\Rightarrow -4I_1 + 15I_2 - 6I_3 = -152$$

(All the voltage sources in the second mesh oppose the direction of flow of I_2.)

$$-0I_1 - 6I_2 + (6 + 7)I_3 = (70 + 4)$$
$$\Rightarrow -0I_1 - 6I_2 + 13I_3 = 74$$

(Note the symmetrical distribution of the equal negative terms, mutual resistance, about the principal diagonal in the three mesh equations above.)

Since the current in the 4 Ω resistor is the only unknown required, being

$$I_{4\Omega} = I_1 - I_2 \quad \text{(flowing downward)}$$

only I_1 and I_2 need to be determined. According to Cramer's rule, only the following determinants have to be evaluated:

$$\Delta = \begin{vmatrix} 7 & -4 & 0 \\ -4 & 15 & -6 \\ 0 & -6 & 13 \end{vmatrix}$$

$$= 7 \times 15 \times 13 + (-4) \times (-6) \times 0 + 0 \times (-4) \times (-6) - 0 \times 15 \times 0$$
$$- 7 \times (-6) \times (-6) - (-4) \times (-4) \times (13)$$
$$= 1365 - 252 - 208 = 905$$

$$\Delta_1 = \begin{vmatrix} 67 & -4 & 0 \\ -152 & 15 & -6 \\ 74 & -6 & 13 \end{vmatrix}$$

$$= 67 \times 15 \times 13 + (-4) \times (-6) \times 74 + 0 - 0 - 67 \times (-6) \times (-6)$$
$$- (-4) \times (-152) \times (13)$$
$$= 13,065 + 1776 - 2412 - 7904 = 4525$$

$$\Delta_2 = \begin{vmatrix} 7 & 67 & 0 \\ -4 & -152 & -6 \\ 0 & 74 & 13 \end{vmatrix}$$

$$= 7 \times (-152) \times 13 + 0 + 0 - 0 - 7 \times (-6) \times 74 - 67 \times (-4) \times 13$$
$$= -13,832 + 3108 + 3484 = -7240$$

Therefore,

$$I_1 = \frac{\Delta_1}{\Delta} = \frac{4525}{905} = 5.0 \text{ A}$$

and

$$I_2 = \frac{\Delta_2}{\Delta} = \frac{-7240}{905} = -8.0 \text{ A}$$

This means that the current in the second mesh is actually flowing in the counter-clockwise direction. Now

$$I_{4\Omega} = I_1 - I_2 = 5 - (-8) = 13 \text{ A} \quad \text{(flowing downward)}$$

EXAMPLE 5.6

Find the current in the 6 Ω resistor in the network shown in Fig. 5.9.

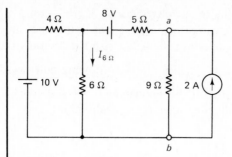

Figure 5.9

Solution The current source between nodes a and b is first converted into a voltage source between the same two nodes as shown in Fig. 5.10. The 9 Ω resistor is considered as the source's internal resistance, while

$$E = 2 \times 9 = 18 \text{ V}$$

The positive polarity of E is upward, based on the direction of the 2 A current source. The two meshes and the corresponding mesh currents are assigned as shown in Fig. 5.10.

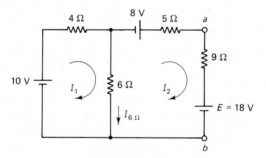

Figure 5.10

From the general procedure (shortcut method), the two mesh equations can be written directly as follows:

$$(4 + 6)I_1 - 6I_2 = 10 \Rightarrow 10I_1 - 6I_2 = 10$$
$$- 6I_1 + (6 + 5 + 9)I_2 = (8 - 18) \Rightarrow -6I_1 + 20I_2 = -10$$

To determine I_1 and I_2, the following determinants need to be evaluated:

$$\Delta = \begin{vmatrix} 10 & -6 \\ -6 & 20 \end{vmatrix} = 10 \times 20 - (-6) \times (-6) = 200 - 36 = 164$$

$$\Delta_1 = \begin{vmatrix} 10 & -6 \\ -10 & 20 \end{vmatrix} = 10 \times 20 - (-6) \times (-10) = 200 - 60 = 140$$

and

$$\Delta_2 = \begin{vmatrix} 10 & 10 \\ -6 & -10 \end{vmatrix} = 10 \times (-10) - 10 \times (-6) = -100 + 60 = -40$$

From Cramer's rule,

$$I_1 = \frac{\Delta_1}{\Delta} = \frac{140}{164} = 0.854 \text{ A}$$

and

$$I_2 = \frac{\Delta_2}{\Delta} = \frac{-40}{164} = -0.244 \text{ A}$$

(I_2 is actually flowing in the counterclockwise direction.) Since the 6 Ω resistor is not involved in source conversions, its current can be determined directly from Fig. 5.10. Here

$$I_{6\Omega} = I_1 - I_2 = 0.854 - (-0.244) = 1.098 \text{ A} \quad \text{(flowing downward)}$$

5.3 NODE VOLTAGE ANALYSIS METHOD

5.3.1 Development of the Technique

This technique of analyzing electric circuits is quite general, applicable to any kind of planar or nonplanar network. It is based on the application of KCL at the various nodes in a given circuit. The objective of solving the network by this method is to determine the values of the voltages at the different nodes. That is why this technique is usually referred to as the *node voltage analysis method*. Voltage is, by definition, the potential difference between two points (or nodes). The determination of any node voltage must therefore be with respect to a reference point.

   If a given network has N nodes, the reference node is usually chosen to be the one with the largest number of element terminals connected to it. This choice simplifies the solution because it avoids writing KCL at this reference node. All other node voltages in the network (now $N - 1$ unknowns) will be determined with respect to this chosen common reference node. For convenience, the voltage of this reference node is taken to be zero; in other words, this node could be assigned as the network ground (with a notation G or symbol $\perp$), either temporarily or permanently if this node is, in fact, connected to the real ground (earth). The nodes are assigned either numbers or symbols and the single-subscripted notations will be used to indicate the node voltages, all with respect to the chosen common reference ground point. It is then required to write $(N - 1)$ KCL equations at the different nodes, in terms of the $(N - 1)$ unknown node voltages; and solve the resulting system of linear simultaneous equations to determine the values of the unknowns.

   Based on the discussion of single-subscripted voltages in Chapter 3, consider an element R connected between the two nodes a and b, with node voltages V_a and V_b, as shown in Fig. 5.11 (this is just an isolated portion of a general large

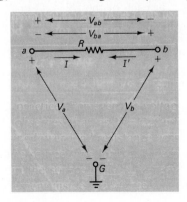

Figure 5.11

network). One can then express the voltage across and the current through this element as

$$V_{ab} = V_a - V_b \qquad (5.39)$$

or

$$V_{ba} = V_b - V_a = -V_{ab} \qquad (5.40)$$

The element current flowing out of node a is

$$I = \frac{V_{ab}}{R} = \frac{V_a - V_b}{R} \qquad (5.41a)$$

$$= G(V_a - V_b) \quad \text{where} \quad G = \frac{1}{R} \qquad (5.41b)$$

This current, I, is flowing into node b. Because the current in one direction is equal to the reverse of the current flowing in the opposite direction, that is,

$$I' = -I \qquad (5.42)$$

we can equally describe the situation in Fig. 5.11 by expressing the current flowing out of node b and into node a, the current I', as

$$I' = \frac{V_{ba}}{R} = \frac{V_b - V_a}{R} \qquad (5.43a)$$

$$= G(V_b - V_a) \qquad (5.43b)$$

Thus the element current expressions in Eqs. (5.41) and (5.43) can equally well be used to describe the same current because they satisfy Eq. (5.42). This also follows directly from the voltage expressions in Eq. (5.40).

To develop a systematic generalized procedure for node voltage analysis as was done for the mesh (or loop) analysis, we have to

1. Convert practical voltage sources into their equivalent current source models. Current in an ideal voltage source is undetermined; however, with a series-connected resistance (considered as an internal resistance), the V–I characteristics of the source is determined. Current source models are easier to deal with and more suitable for application of KCL at the different nodes.

2. Apply KCL at each of the $(N - 1)$ nodes in the form: sum of currents leaving the node due to the voltage drops between this node voltage and the other nodes in the circuit = algebraic sum of current sources entering this node. If a current source is in a direction that is actually leaving the node concerned, its value must be taken with a negative sign in the right-hand side of KCL as expressed above.

Consider, for example, the circuit shown in Fig. 5.12. There are three nodes in this circuit. Node ③ has the most element terminals connected to it (four). It is therefore assigned as the reference node, G, with a ground potential of zero volts. Nodes ① and ② are chosen as indicated, with the node voltages V_1 and V_2 being the voltages at these two nodes, respectively, with respect to the ground

than the voltage at the reference node, which is taken to be the 0 V ground potential in this analysis technique.

In summary, the recommended procedure for conducting the generalized systematic approach of the node voltage analysis technique (also called the shortcut procedure) is as follows:

1. Convert all the practical voltage sources into their equivalent current source models.
2. Identify the reference node and denote it as the ground node (0 V point). Assign numbers and the corresponding node voltages (e.g., V_1, V_2, etc.) to the other $N - 1$ nodes in the network.
3. Determine the self-(or total) conductance of each node.
4. Determine the mutual (or common) conductances between each pair of nodes.
5. Determine the total algebraic sum of all the current sources entering each of the circuit nodes.
6. Substitute the required terms in the appropriate forms of the generalized node equations, either (5.48) and (5.49) for $n = N - 1 = 3 - 1 = 2$ or (5.50), (5.51), and (5.52) for $n = N - 1 = 4 - 1 = 3$, and so on.
7. Solve the resulting node equations to determine the values of the unknown node voltages.

By observing the general forms of the node equations derived above, the following important characteristics can be deduced:

I. All the principal diagonal coefficients, G_{11}, G_{22}, G_{33}, and so on, have positive signs.
II. All the other off-diagonal coefficients, representing the mutual conductance terms, have negative signs. Each such coefficient appears twice ($G_{12} = G_{21}$, $G_{13} = G_{31}$, $G_{23} = G_{32}$, etc.). They are symmetrically distributed about the principal diagonal.

5.3.3 Source Conversions

As discussed in Section 5.1.3, the determination of the voltage across or the current through an element involved in a practical voltage-to-current source conversion must be calculated from the original circuit, not from the equivalent model. Even though the conversion of a practical voltage source to its equivalent current source model simplifies the node analysis procedure and permits direct use of the generalized shortcut method, this does not imply that the circuit cannot be analyzed by the node voltage technique, without such conversions. In fact, the network can still be analyzed but the process is more involved and requires special care.

To show some of the ideas that must be taken into account, consider the case shown in Fig. 5.13 (a). Here the voltage source E is ideal and cannot be converted into a current source model. One does not need to write KCL at this node because V_1, the voltage at node 1, is already known:

$$V_1 = E$$

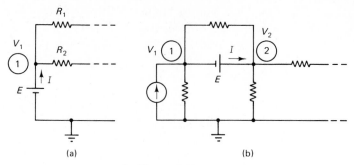

(a)   (b)

Figure 5.13

The current I is still unknown: its value can be found once the currents in the elements R_1 and R_2 are calculated from the rest of the network.

In the case of the circuit shown in Fig. 5.13 (b), E is an ideal voltage source, but

$$V_2 = V_1 + E \quad \text{(from KVL)} \tag{5.53}$$

that is, if one node voltage (either V_1 or V_2) is obtained, the other can be directly found. When writing KCL at nodes 1 and 2, the current I in the source E remains as an unknown quantity in both these equations. Once these two node equations are written, the unknown quantity I can be eliminated by the addition or subtraction of the two equations. The resulting equation can then be expressed in terms of either of the unknown node voltages, V_1 or V_2, using Eq. (5.53).

5.3.4 Examples of Node Voltage Analysis

EXAMPLE 5.7

Use nodal analysis to calculate the current in each of the three resistors in the circuit shown in Fig. 5.14.

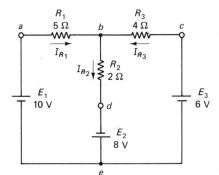

Figure 5.14

Solution   Similar two-mesh circuits have been analyzed previously using the mesh analysis technique. It looks as if this circuit has five nodes (i.e., four node equations would be required to solve it). However, this is deceiving since three node voltages are already known. Considering the reference node to be point e, denoting it the ground node G, we have

$$V_a = E_1 = 10\,V$$
$$V_c = E_3 = 6\,V$$
$$V_d = -E_2 = -8\,V$$

Thus only the voltage at node b is to be calculated, requiring only one node voltage equation. This is clear when each of the three practical voltage sources (E_1 and R_1, E_2 and R_2, E_3 and R_3) is converted into its equivalent current source model, as shown in Fig. 5.15. All the sources have the two common terminals: b and e (or G). Here

$$I_{S_1} = \frac{E_1}{R_1} = \frac{10}{5} = 2\,A$$

$$I_{S_2} = \frac{E_2}{R_2} = \frac{8}{2} = 4\,A$$

$$I_{S_3} = \frac{E_3}{R_3} = \frac{6}{4} = 1.5\,A$$

The directions of the current sources are as indicated in Fig. 5.15, based on the polarities of the corresponding voltage sources. This network now has two nodes, requiring only $N - 1 = 2 - 1 = 1$ node equation, as expected.

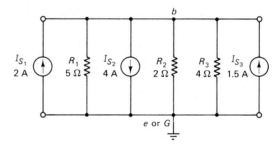

Figure 5.15

Assuming that the voltage at node b is V_b and applying KCL at this node yields

$$V_b\left(\frac{1}{R_1} + \frac{1}{R_2} + \frac{1}{R_3}\right) = I_{S_1} - I_{S_2} + I_{S_3}$$

This is of the form

$$V_b G_{11} = I_{11}$$

(only one self-conductance, but no mutual conductances). Substituting for the elements by their numerical values, we have

$$V_b(\tfrac{1}{5} + \tfrac{1}{2} + \tfrac{1}{4}) = 2 - 4 + 1.5$$
$$0.95V_b = -0.5$$

Therefore,

$$V_b = -0.526\,V$$

that is, the voltage at node b is less than the voltage at node e (or G) by 0.526 V. The currents in the three resistors cannot be determined directly from the circuit of Fig. 5.15 because each of these resistors was involved in the corresponding voltage-to-current source conversions.

Referring back to the original circuit, with all the node voltages now determined, it is then easy to use Ohm's law to calculate the currents in the three resistors. Thus

$$I_{R_1} = \frac{V_a - V_b}{R_1} = \frac{10 - (-0.526)}{5} = 2.105 \text{ A} \quad \text{(flowing from } a \text{ to } b\text{)}$$

$$I_{R_2} = \frac{V_b - V_d}{R_2} = \frac{-0.526 - (-8)}{2} = 3.737 \text{ A} \quad \text{(flowing from } b \text{ to } d\text{)}$$

$$I_{R_3} = \frac{V_c - V_b}{R_3} = \frac{6 - (-0.526)}{4} = 1.632 \text{ A} \quad \text{(flowing from } c \text{ to } b\text{)}$$

As a check,

$$I_{R_1} + I_{R_3} = 2.105 + 1.632 = 3.737 = I_{R_2}$$

satisfying KCL applied to node b in the original circuit.

EXAMPLE 5.8

Find the node voltages in the network shown in Fig. 5.16. What is the current I in the 10 V source?

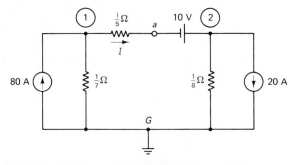

Figure 5.16

Solution Here the nodes are identified and the ground connection fixed. Assume that the node voltages are V_1 and V_2 with respect to the ground point. The practical voltage source can be converted to the equivalent current source model, between nodes 1 and 2 as shown in Fig. 5.17, with

$$I_S = \frac{10}{\frac{1}{5}} = 50 \text{ A}$$

Now

G_{11} = self-(or total) conductance of node ① = $7 + 5 = 12$ S

G_{22} = self-(or total) conductance of node ② = $8 + 5 = 13$ S

$G_{12} = G_{21}$ = mutual conductance between nodes ① and ② = 5 S

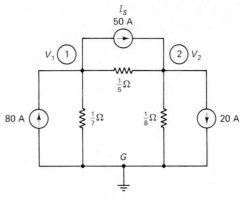

Figure 5.17

Therefore, based on the general form of the nodal equations (5.48) and (5.49), the two required node voltage equations are

$$12V_1 - 5V_2 = 80 - 50 = 30$$
$$-5V_1 + 13V_2 = 50 - 20 = 30$$

To obtain the solution for V_1 and V_2, the following determinants have to be evaluated:

$$\Delta = \begin{vmatrix} 12 & -5 \\ -5 & 13 \end{vmatrix} = 12 \times 13 - (-5) \times (-5) = 156 - 25 = 131$$

$$\Delta_1 = \begin{vmatrix} 30 & -5 \\ 30 & 13 \end{vmatrix} = 30 \times 13 - (-5) \times 30 = 390 + 150 = 540$$

and

$$\Delta_2 = \begin{vmatrix} 12 & 30 \\ -5 & 30 \end{vmatrix} = 12 \times 30 - 30 \times (-5) = 360 + 150 = 510$$

Now applying Cramer's rule, we obtain

$$V_1 = \frac{\Delta_1}{\Delta} = \frac{540}{131} = 4.122 \text{ V}$$

and

$$V_2 = \frac{\Delta_2}{\Delta} = \frac{510}{131} = 3.893 \text{ V}$$

To find the current in the 10 V source, the original circuit in Fig. 5.16 is reexamined, with the node voltages now determined. Noting that

$$V_a = V_2 - 10 = 3.893 - 10 = -6.107 \text{ V}$$

then

$$I = \frac{V_1 - V_a}{\frac{1}{5}} = 5 \times [4.122 - (-6.107)] = 51.145 \text{ A}$$

As a check, apply KCL at node 1 of the original circuit:

$$80 = \frac{V_1}{\frac{1}{7}} + I = \frac{4.122}{\frac{1}{7}} + 51.145 = 80$$

■ (i.e., KCL is satisfied).

EXAMPLE 5.9

Find the currents in the 5 Ω, 2 Ω, and 1 Ω resistors in the circuit shown in Fig. 5.18, using the node voltage technique.

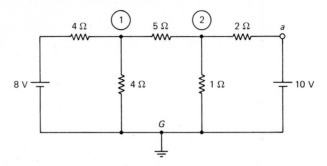

Figure 5.18

Solution   The given circuit has three meshes, requiring three loop (or mesh) equations to analyze it. However, when the two practical voltage sources are converted into their equivalent current source models, as shown in Fig. 5.19, the circuit has

$$N = \text{number of nodes} = 3$$

requiring only $n = N - 1 = 3 - 1 = 2$ node equations to analyze it. The ground (or reference) node G is chosen as shown in Fig. 5.19 and nodes ① and ② are identified. The same nodal notations are then indicated in the original circuit diagram in Fig. 5.18.

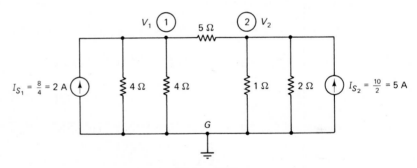

Figure 5.19

One can now proceed to write the nodal equations directly, using the shortcut procedure. Here

$$(\tfrac{1}{4} + \tfrac{1}{4} + \tfrac{1}{5})V_1 - (\tfrac{1}{5})V_2 = 2 \Rightarrow 0.7V_1 - 0.2V_2 = 2$$
$$-(\tfrac{1}{5})V_1 + (\tfrac{1}{5} + 1 + \tfrac{1}{2})V_2 = 5 \Rightarrow -0.2V_1 + 1.7V_2 = 5$$

The following determinants should then be evaluated:

$$\Delta = \begin{vmatrix} 0.7 & -0.2 \\ -0.2 & 1.7 \end{vmatrix} = 0.7 \times 1.7 - (-0.2) \times (-0.2)$$

$$= 1.19 - 0.04 = 1.15$$

$$\Delta_1 = \begin{vmatrix} 2 & -0.2 \\ 5 & 1.7 \end{vmatrix} = 2 \times 1.7 - (-0.2) \times 5 = 3.4 + 1.0 = 4.4$$

$$\Delta_2 = \begin{vmatrix} 0.7 & 2 \\ -0.2 & 5 \end{vmatrix} = 0.7 \times 5 - 2 \times (-0.2) = 3.5 + 0.4 = 3.9$$

Therefore, applying Cramer's rule gives us

$$V_1 = \frac{\Delta_1}{\Delta} = \frac{4.4}{1.15} = 3.826 \text{ V}$$

and

$$V_2 = \frac{\Delta_2}{\Delta} = \frac{3.9}{1.15} = 3.391 \text{ V}$$

Now to calculate the currents in the 5 Ω and 1 Ω resistors, the circuit in Fig. 5.19 can be used directly because these two elements are not involved in source transformations. Thus

$$I_{5\Omega} = \frac{V_1 - V_2}{5} = \frac{3.826 - 3.391}{5} = 0.087 \text{ A} \quad \text{(flowing from node 1 to node 2)}$$

and

$$I_{1\Omega} = \frac{V_2}{1} = 3.391 \text{ A}$$

The current in the 2 Ω resistor has to be determined from the original circuit, as this resistor was involved in the corresponding source transformation. Here

$$V_a = 10 \text{ V}$$

and

$$I_{2\Omega} = \frac{V_a - V_2}{2} = \frac{10 - 3.391}{2} = 3.304 \text{ A} \quad \text{(flowing from node } a \text{ to node 2)}$$

(Note that if the circuit in Fig. 5.19 is used,

$$I_{2\Omega} = \frac{V_2}{2} = \frac{3.391}{2} = 1.696 \text{ A}$$

which, as will be shown below, cannot be correct!)

To check the correctness of the results obtained above, apply KCL at node 2 of the original circuit. Here

$$I_{5\Omega} + I_{2\Omega} \qquad \text{(both flowing into node 2)}$$

$$= I_{1\Omega} \qquad \text{(flowing out of node 2)}$$

$$0.087 + 3.304 = 3.391$$

(i.e., KCL is satisfied). Clearly, the value of $I_{2\Omega}$ calculated in parentheses above cannot satisfy KCL at node 2 in the original circuit. (It does, however, satisfy KCL at node 2 in the equivalent modeled circuit of Fig. 5.19!)

EXAMPLE 5.10

Find V_1 and the currents in the two resistors in the circuit shown in Fig. 5.20.

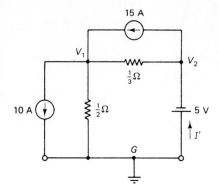

Figure 5.20

Solution Here V_2 is known, being

$$V_2 = -5\,\text{V}$$

as the ideal 5 V voltage source is connected between node 2 and the ground point. Thus KCL need not be written for node 2. For node 1, KCL provides

$$(3 + 2)V_1 - 3V_2 = 15 - 10$$

or

$$5V_1 - 3V_2 = 5$$

Substituting for the numerical value of V_2 in the equation above yields

$$5V_1 = 5 + 3V_2 = 5 + 3 \times (-5) = -10$$

Therefore,

$$V_1 = -2\,\text{V}$$

Thus

$$I_{(\frac{1}{2})\Omega} = \frac{0 - V_1}{\frac{1}{2}} = \frac{0 - (-2)}{\frac{1}{2}} = 4\,\text{A} \qquad \text{(flowing from } G \text{ to node 1)}$$

and

$$I_{(\frac{1}{3})\Omega} = \frac{V_1 - V_2}{\frac{1}{3}} = \frac{-2 - (-5)}{\frac{1}{3}} = 9\,\text{A} \qquad \text{(flowing from node 1 to node 2)}$$

As a check, consider again KCL at node 1:

$$I_{(\frac{1}{2})\Omega} + 15 \qquad \qquad \text{(currents into node 1)}$$
$$= 10 + I_{(\frac{1}{3})\Omega} \quad \text{(currents leaving node 1)}$$
$$4 + 15 = 10 + 9$$

(i.e., KCL is satisfied). If the current in the ideal 5 V voltage source, I', is required, then by applying KCL at node 2, we obtain

$$I' + I_{(\frac{1}{3})\Omega} = 15$$

Therefore,

$$I' = 15 - 9 = 6\,\text{A}$$

It is clear from the examples examined above that either the mesh or the node analysis technique can be applied to any planar circuit to obtain the values of the required unknown circuit quantities (whether voltages or currents). The question of which technique should be used depends on which of the two circuit analysis techniques requires fewer equations to solve. The mesh analysis technique requires $B - N + 1$ mesh equations and the nodal analysis technique requires $N - 1$ node voltage equations. The method that requires fewer equations should be the one chosen. It is preferable to check the number of equations required after the sources have been converted to the form suitable for the particular analysis method. If both techniques require the solution of the same number of equations, the method that involves fewer source conversions should be the one chosen.

DRILL PROBLEMS

Section 5.1

5.1. Write the mesh current equations for the network shown in Fig. 5.21.

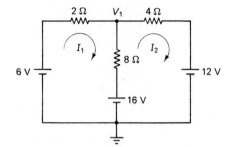

Figure 5.21

5.2. Write the mesh current equations for the network shown in Fig. 5.22.

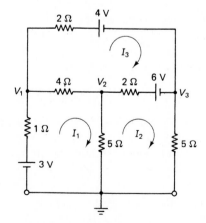

Figure 5.22

5.3. Convert the practical current sources in the network shown in Fig. 5.23 into their equivalent voltage source models and then write the appropriate mesh current equations.

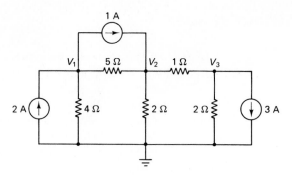

Figure 5.23

Section 5.3

5.4. Write the node voltage equations for the network shown in Fig. 5.23.

5.5. Convert the practical voltage sources in the network shown in Fig. 5.21 into their equivalent current source models and then write the appropriate node voltage equation(s).

5.6. Convert the practical voltage sources in the network shown in Fig. 5.22 into their equivalent current source models and then write the appropriate node voltage equations.

Sections 5.1 to 5.3

5.7. Solve, using Cramer's rule, the mesh equations obtained in Drill Problem 5.1 to determine the values of I_1 and I_2.

5.8. Solve, using Cramer's rule, the mesh equations obtained in Drill Problem 5.2 to determine the values of I_1, I_2, and I_3.

5.9. Solve the mesh equations obtained in Drill Problem 5.3 in order to determine the appropriate mesh current and hence the value of V_2.

5.10. Solve, using Cramer's rule, the node voltage equations obtained in Drill Problem 5.4 in order to determine the voltages V_1, V_2, and V_3. Does the value of V_2 obtained here agree with that obtained in Drill Problem 5.9?

5.11. Using the results obtained in Drill Problem 5.5, determine the value of V_1. Check the correctness of this value through the results obtained in Drill Problem 5.7.

5.12. Using Cramer's rule, solve the node voltage equations obtained in Drill Problem 5.6 in order to determine the voltages V_1, V_2, and V_3.

5.13. Obtain the value of V_2 in the circuit shown in Fig. 5.22 using the mesh currents determined in Drill Problem 5.8. Compare this value of V_2 with that obtained in Drill Problem 5.12.

PROBLEMS

Note: The problems below that are marked with asterisk (*) can be used as exercises in computer-aided network analysis or as programming exercises.

Section 5.2

(i) **5.1.** Evaluate each of the following determinants.

(a) $\begin{vmatrix} 3 & -2 \\ -4 & 6 \end{vmatrix}$ (b) $\begin{vmatrix} 7 & 0 & 3 \\ 0 & 1 & -1 \\ -2 & -1 & 4 \end{vmatrix}$ (c) $\begin{vmatrix} 5 & -1 & 0 \\ -1 & 6 & -2 \\ 0 & -2 & 7 \end{vmatrix}$

(i)   **5.2.** Use Cramer's rule to solve for the unknowns in the following set of equations.

$$2x - 8y = -20$$
$$5x + 2y = 10$$

(i)   **5.3.** Use Cramer's rule to solve for the unknowns in the following set of equations.

$$x_1 - 2x_2 + 3x_3 = 4$$
$$2x_1 + 6x_2 - 5x_3 = 0$$
$$-x_1 + x_2 - 2x_3 = -2$$

Sections 5.1 to 5.3

(i)   **5.4.** Using mesh analysis, find the current in the 5 Ω resistor in the network shown in Fig. 5.24.

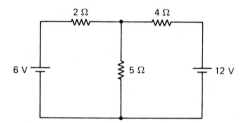

Figure 5.24

(i)   ***5.5.** Find the currents in each of the 5 Ω and 1 Ω resistors and the voltage across the 3 Ω resistor in the network of Fig. 5.25 using the mesh analysis technique.

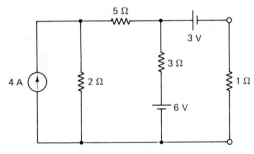

Figure 5.25

(i)   **5.6.** Solve Problem 5.4 using the node voltage analysis method.

(i)   **5.7.** Solve for the same requirements in Problem 5.5 using the node voltage analysis technique.

(i)   **5.8.** Using mesh analysis, find the current in the 3 Ω and 1 Ω resistors in the circuit shown in Fig. 5.26.

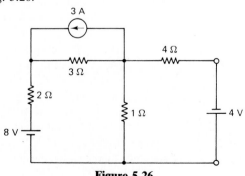

Figure 5.26

(i) *5.9. Applying the node voltage analysis technique to the circuit of Fig. 5.26, find the currents in the 3 Ω and 4 Ω resistors and the voltage across the 1 Ω resistor.

(i) *5.10. Using the nodal analysis technique, find the node voltages V_1 and V_2, then calculate the current in the 2 Ω resistor in the circuit shown in Fig. 5.27.

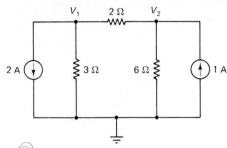

Figure 5.27

(i) 5.11. With source conversion, show that the mesh analysis technique can directly provide the value of the current in the 2 Ω resistor in the circuit of Fig. 5.27. From your results, recalculate the value of the voltages across the 3 Ω and 6 Ω resistors.

(ii) *5.12. Use mesh analysis to calculate the currents in the 3 V and 4 V sources in the network shown in Fig. 5.28.

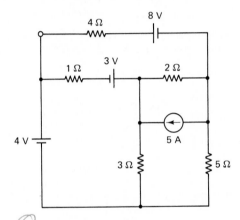

Figure 5.28

(ii) *5.13. Using nodal analysis, find the voltages across the 3 Ω and 5 Ω resistors in the network shown in Fig. 5.28.

(i) 5.14. Calculate the current in the 4 Ω resistor in Fig. 5.28, using
   (a) The results of Problem 5.12.
   (b) The results of Problem 5.13.

(ii) *5.15. Apply the mesh analysis technique to calculate the currents in the 2 Ω and 3 Ω resistors in the circuit of Fig. 5.29.

(ii) 5.16. Calculate the voltage across the 3 Ω and 5 Ω resistors in the circuit shown in Fig. 5.29 using the nodal analysis technique.

(i) 5.17. Using the results obtain in Problem 5.16, calculate
   (a) The current in the 2 Ω resistor.
   (b) The current in the 6 V source.
   (c) The current in the 1 Ω resistor.

(ii) *5.18. Using the mesh analysis technique, find the current in the 3 V source in the circuit shown in Fig. 5.30.

(ii) *5.19. Using the nodal analysis technique, find the node voltages V_1 and V_2 in the circuit shown in Fig. 5.30.

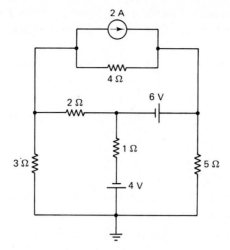

Figure 5.29

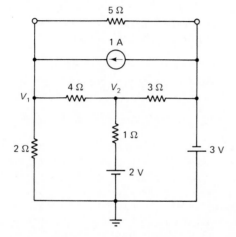

Figure 5.30

(i) **5.20.** Using the results obtained in Problem 5.19, recalculate the current in the 3 V source to check the result of Problem 5.18.

GLOSSARY

Cramer's Rule: An algebraic procedure to obtain the solution (the values of the unknowns) of a set of simultaneous equations using determinants.

Diagonal Rule: A method of expressing and obtaining the value of a determinant.

Mesh (Loop) Analysis: A technique for determining the mesh (loop) currents in a network. This method is based on the systematic application of KVL to each of the meshes (loops) in the network.

Mesh (Loop) Currents: A set of fictitious currents assigned to flow in each of the closed meshes (loops) of the network.

Mesh (Loop) Equations: A set of simultaneous equations in terms of the mesh (loop) currents. Each equation represents the KVL corresponding to one of the meshes of the given network.

Mutual (or Common) Conductance Between Two Nodes: The sum of the conductances of the common branch(es) connected between two nodes in a given network.

Mutual (or Common) Resistance Between Two Meshes (Loops): The sum of the resistances of the common branch(es) between two meshes (loops) in a given network.

Nodal Analysis: A technique for determining the node voltages in a network with respect to a chosen reference (ground) node.

Node Equations: A set of simultaneous equations in terms of the node voltages. Each equation represents the KCL corresponding to one of the nodes of the given network.

Node Voltages: The voltages between each of the nodes in a given network and the chosen reference (ground) node.

Self (or Total) Conductance of a Node: The sum of the conductances of all the branches connected directly between a specific node and all the other nodes of a given network.

Self (or Total) Resistance of a Mesh (Loop): The sum of all the resistances of all the branches constituting a specific closed mesh (loop) in a given network.

ANSWERS TO DRILL PROBLEMS

5.1. $10I_1 - 8I_2 = -10$, $-8I_1 + 12I_2 = 4$

5.3. $11I_1 - 2I_2 = 13$, $-2I_1 + 5I_2 = 6$

5.5. $0.875V_1 = 8$

5.7. $I_1 = -1.571$ A, $I_2 = -0.714$ A
Both these mesh currents are actually flowing in the opposite directions to those assumed.

5.9. $I_1 = 1.5098$ A, $I_2 = 1.8039$ A and $V_2 = -0.588$ V

5.11. $V_1 = 9.143$ V
From the results of Drill Problem 5.7 (Fig. 5.21), $V_1 = 16 + 8(I_1 - I_2) = 9.144$ V, as obtained above.

5.13. From the results of Drill Problem 5.8 and referring to Fig. 5.22, $V_2 = 5(I_1 - I_2) = 2.945$ V, as obtained in Drill Problem 5.12.

6 | NETWORK THEOREMS

OBJECTIVES

- Familiarity with, and ability to apply special network theorem tools in order to simplify and/or shorten the required network analysis process. In particular, the tools introduced and applied in this chapter are Thevenin's theorem, Norton's theorem, Superposition theorem, Maximum power transfer theorem, and Millman's theorem.

The general methods of circuit analysis presented in Chapter 5 are sufficient to solve any network in order to obtain the required response (e.g., an element's voltage, current, or power) due to the given excitation(s) (voltage and/or current sources). Network theorems are important tools for circuit analysis which when applied properly can shorten and simplify the required network analysis process.

In this chapter the most important of these theories are introduced and applied. Many other theories also exist but they are primarily of academic importance and are not discussed here. The theories of concern to us here apply to linear networks (or portions of them). Linear networks contain linear elements, that is, elements whose voltage–current relationship is representable as a straight-line equation,

$$v = ki$$

like a linear resistor (obeying Ohm's law). Later, other linear elements (capacitors and inductors) are introduced. Linear networks also contain voltage and current sources.

Network theorems can be thought of as techniques that reduce a complicated circuit to a simpler equivalent one, or replace parts of a network by equivalent models, resulting in a quicker and simpler analysis process. This is similar to the idea behind the equivalent resistance of a series or a parallel combination of elements.

Further discussion of these theories as they apply to ac circuits is presented in Chapter 16. In particular, the $Y - \Delta$ network conversion is examined in Chapter 16, because its main application is in three-phase ac circuits. Also, examination of circuits containing dependent sources is deferred until Chapter 16.

6.1 THE SUPERPOSITION THEOREM

This theorem states that *in a linear network containing many sources, the response (voltage or current) in an element is the algebraic sum (superposition) of the partial responses produced in this element due to each of the sources acting alone in the network.*

The network is analyzed with one source acting in the network at a time, while all the other sources in the network are temporarily "killed." The required response in the element concerned is calculated with proper polarities (or sign). This is referred to as the *partial response* due to the source considered. The process is repeated for each of the sources in the original network. The required response is the overall algebraic summation (i.e., the superposition) of these partial responses, including their polarities as calculated.

A dead or killed source does not produce an output. Thus to kill a voltage source V_S, we have to make its output

$$V_S = 0 \text{ V} \quad \text{(the equation of a short circuit)} \tag{6.1}$$

An element with zero volts across it is a *short circuit*. Thus an ideal voltage source is killed by replacing it with a short circuit or by connecting a short circuit wire across it (only in the circuit diagram!). Similarly, to kill a current source, I_S, we have to make its output:

$$I_S = 0 \text{ A} \quad \text{(the equation of an open circuit)} \tag{6.2}$$

An element with a zero current flowing in it is an *open circuit*. Thus an ideal current source is killed by replacing it with an open circuit.

It is important to note that the internal resistance of a practical voltage or current source is not affected when the source is killed. These internal resistances remain in the circuit even after the source concerned is killed because the killing operation affects only the ideal source in the practical source model.

If the required response is the power in a certain element, this theory cannot be applied directly because the power is a quadratic quantity (I^2R or V^2/R). However, we find either the overall current I or voltage V in the specific element as the required response and then use this quantity to calculate the power in the element. Therefore, never attempt to find the partial powers and then add these partial responses.

Sometimes it may seem that applying this theorem makes the solution quite long because the network should be analyzed as many times as there are sources. However, when the other sources are dead (i.e., when the network has only one source), it is usually in a much simpler form (because of the short-circuit and open-circuit replacements of voltage and current sources, respectively) and its solution is straightforward. If the process involved in finding the partial responses is more than applying Ohm's law or voltage and current division, the theorem is not really of greater advantage than, say, mesh or nodal analysis.

■ EXAMPLE 6.1

Use superposition to find the current I and the power dissipated in the 10 Ω resistor in the circuit shown in Fig. 6.1.

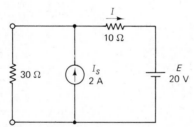

Figure 6.1

Solution Consider first the current source, with the voltage source killed, as shown in Fig. 6.2(a). The partial response I' in the 10 Ω resistor can be obtained in this circuit using the current-division principle. Here

$$I' = I_S \frac{30}{10 + 30} = 2 \times \frac{30}{40} = 1.5 \text{ A}$$

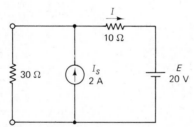

(a)   (b)

Figure 6.2

Now, when the voltage source acts alone while the current source is killed, as shown in Fig. 6.2(b), the partial response I'' in the $10\,\Omega$ resistor can easily be obtained from Ohm's law. Here

$$I'' = \frac{20}{10 + 30} = 0.5\,\text{A}$$

Notice that I' is flowing in the required reference direction of I, while I'' is flowing in the opposite direction. Thus by superposition,

$$I = I' - I'' = 1.5 - 0.5 = 1.0\,\text{A}$$

and

$$P_{10\Omega} = I^2 R = (1)^2 \times 10 = 10\,\text{W}$$

To show that the superposition of power gives erroneous results, consider

$$P_{10\Omega} = P'_{10\Omega} + P''_{10\Omega} = (I')^2 \times 10 + (I'')^2 \times 10$$
$$= (1.5)^2 \times 10 + (0.5)^2 \times 10 = 25\,\text{W}$$

This is obviously not the answer. Even if the polarities of the voltage source are reversed so that I'' is in the same direction as I', resulting in I being $2\,\text{A}$,

$$P_{10\Omega} = I^2 \times 10 = (2)^2 \times 10 = 40\,\text{W}$$

Again, 25 W is not the right answer!

EXAMPLE 6.2

Find the voltage V across the $5\,\Omega$ resistor in the circuit shown in Fig. 6.3 using superposition.

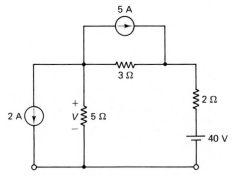

Figure 6.3

Solution The circuit is redrawn first with the voltage source acting alone while the two current sources are killed (i.e., each is replaced by an open circuit), as shown in Fig. 6.4(a). Here, using the voltage-division principle, we obtain

$$V' = 40 \times \frac{5}{2 + 3 + 5}$$

$$= 40 \times \frac{5}{10} = 20\,\text{V}$$

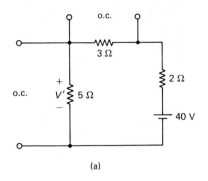

(a)

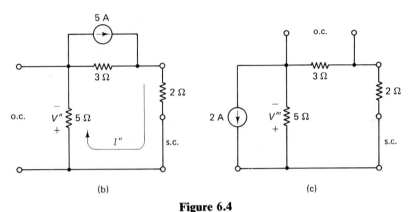

(b)

(c)

Figure 6.4

Figure 6.4(b) shows the same circuit with the 5 A source acting alone while the two other sources are killed. Here

$$V'' = 5 \times I'' = 5 \times 5 \times \frac{3}{3 + 2 + 5}$$

$$= 25 \times \frac{3}{10} = 7.5 \text{ V}$$

I'' is obtained from the current division principle.

When the 2 A source is acting alone while the other sources are dead, the circuit becomes as shown in Fig. 6.4(c). Here

$$V''' = 2 \times \frac{5(2 + 3)}{5 + 2 + 3}$$

$$= 2 \times \frac{5 \times 5}{10} = 5 \text{ V}$$

V''' is the voltage across the two 5 Ω parallel branches due to the 2 A source. Thus using superposition and accounting for the polarities as indicated yields

$$V = V' - V'' - V''' = 20 - 7.5 - 5 = 7.5 \text{ V}$$

EXAMPLE 6.3

Find the voltage V_{ab} in the circuit shown in Fig. 6.5 using the superposition theorem.

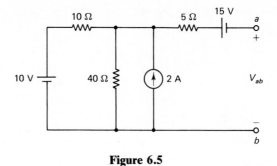

Figure 6.5

Solution   The circuit is redrawn in Fig. 6.6 with each source acting alone while the other sources are killed. Note that no current flows in the 5 Ω resistor due

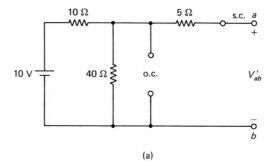

(a)

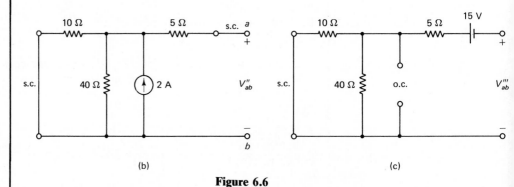

(b)

(c)

Figure 6.6

to the open circuit between terminals a and b. Thus using the voltage-division principle

$$V'_{ab} = 10 \times \frac{40}{10 + 40} = 8 \text{ V}$$

Using Ohm's law, we have

$$V''_{ab} = 2 \times \frac{10 \times 40}{10 + 40} = 16 \text{ V}$$

While

$$V'''_{ab} = -15 \text{ V}$$

because Figure 6.6(c) resembles a practical source with a

$$5 + \frac{10 \times 40}{10 + 40} = 13 \ \Omega$$

internal resistance, which is open circuited. Thus

$$V_{ab} = V'_{ab} + V''_{ab} + V'''_{ab}$$
$$= 8 + 16 - 15 = 9 \ \text{V}$$

6.2 THÉVENIN'S THEOREM

Thévenin's theorem is one of the most important theorems in network analysis. Its application allows us to replace an entire network between two terminals (say, a and b or 1 and 2, etc.) with an equivalent simple series circuit model containing one voltage source, V_{th}, in series with one resistance, R_{th}, resembling the practical voltage source model.

Figure 6.7 shows the Thévenin equivalent-circuit (model) for any general network, for example network A. As far as network B, or load R_L, connected between terminals a and b is concerned, both the original network A and the Thévenin equivalent circuit are exactly the same. Both produce the same voltage V between nodes a and b, and both result in the same current flow I.

The simplification provided through the application of Thevenin's theory allows for a much quicker calculation of V and I, instead of analyzing the original and more complicated network. Usually, the resulting equivalent circuit is nothing more than a simple series circuit (e.g., when network B is just a load R_L). Notice that the equivalency between the original network and Thévenin's circuit model does not imply equivalent voltages and currents inside network A and Thévenin's equivalent circuit.

To evaluate the elements of Thévenin's equivalent-circuit model, network B or load R_L is first removed, and terminals a and b are left open as shown in Fig.

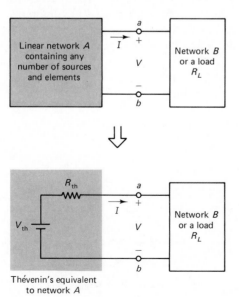

Thévenin's equivalent to network A

Figure 6.7

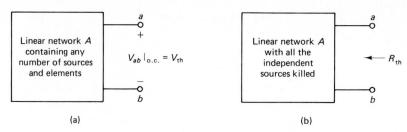

Figure 6.8

6.8(a). The open-circuit voltage between terminals a and b is then calculated and

$$\boxed{V_{\text{th}} = V_{ab}|_{\text{o.c.}}} \tag{6.3}$$

Next, again with network B (or load R_L) disconnected, all the independent sources in network A are killed, as discussed previously. Each of the ideal voltage sources is replaced by a short circuit and each of the ideal current sources is replaced by an open circuit. Then

$$\boxed{\begin{array}{c} R_{\text{th}} = \text{input resistance of the dead network } A, \\ \text{looking between terminals } a \text{ and } b \end{array}} \tag{6.4}$$

The polarity of V_{th} is the same as the polarity of the voltage measured or calculated between terminals a and b, $V_{ab}|_{\text{o.c.}}$.

Corollary I: Maximum power transfer theory. To determine the value of the load resistance R_L that will result in the maximum possible power developing in it, P_L, when connected to any two terminals (a and b) of a given network A, Thévenin's equivalent-circuit model for network A is first determined. This is shown in Fig. 6.9.

The network in this figure resembles that of a practical voltage source with the load R_L connected to the terminals of the source. The condition for maximum power transfer in this circuit was discussed in Chapter 4. It was shown there that when R_L is matched to the internal resistance of the source, maximum power will be developed in the load. Therefore, here

$$R_L = R_{\text{th}} \text{ (of network } A) \tag{6.5}$$

is the condition for maximum power transfer to the load. This maximum amount

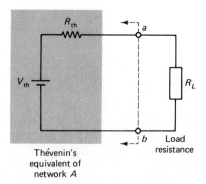

Thévenin's
equivalent of
network A

Figure 6.9

of load power is given by

$$P_{L \, \text{max}} = \frac{V_{\text{th}}^2}{4R_{\text{th}}}$$

(6.6)

as compared to the results obtained in Chapter 4.

Corollary II. When a nonlinear load R_L is connected between two terminals, say a and b, of a given newtork A, and it is required to determine the voltage across and the current in this nonlinear load element, Thévenin's theorem is of prime importance.

Network A is replaced by its Thévenin equivalent-circuit model and the non-linear load R_L is connected between the two terminals a and b. The circuit is then similar to that shown in Fig. 6.9. Again, Thévenin's equivalent circuit is similar to the practical voltage source model, and therefore the graphical load line analysis technique discussed in Chapter 4 can be applied to this circuit. The load-line intercepts are V_{th} on the voltage axis and

$$I_{\text{sc}} = \frac{V_{\text{th}}}{R_{\text{th}}}$$

(6.7)

on the current axis. The load voltage and current can then easily be determined.

Corollary III. Thévenin's equivalent-circuit model can be found for different parts of a given network. Thévenin's theorem can be applied successively to smaller and then larger parts of a given network, to simplify obtaining the Thévenin equivalent circuit for the entire network.

EXAMPLE 6.4

Find the Thévenin equivalent for the circuit shown in Fig. 6.10 between terminals a and b and then calculate the current in and the voltage across the 15 Ω load resistance. If R_L is a variable-load resistance, what should be its value for maximum power to be developed in R_L? What is the value of this maximum power?

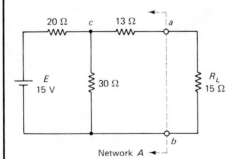

Figure 6.10

Solution To find the Thévenin equivalent, the load R_L is first removed (or disconnected) from the circuit, as shown in Fig. 6.11(a). Noting that no current now flows in the 13 Ω resistor, then

$$V_{ca} = V_{ac} = 0 \, \text{V}$$

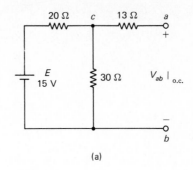

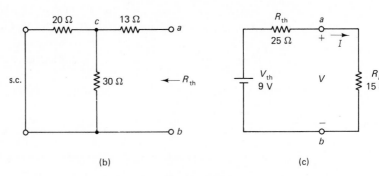

Figure 6.11

and

$$V_{\text{th}} = V_{ab} \big|_{\text{o.c.}} = V_{ac} + V_{cb} = V_{cb}$$

$$= 15 \times \frac{30}{20 + 30} = 9 \text{ V}$$

The voltage-division principle is used to obtain V_{cb} [across the 30 Ω resistor].

The circuit in Fig. 6.11(a) (i.e., network A) is redrawn in Fig. 6.11(b) with the voltage source killed (i.e., replaced by a short circuit). The resistance looking into the dead network between terminals a and b is then R_{th}:

$$R_{\text{th}} = 13 + \frac{20 \times 30}{20 + 30} = 13 + 12 = 25 \text{ Ω}$$

noting that the 20 Ω and 30 Ω resistors are in parallel in Fig. 6.11(b).

The Thévenin equivalent model for network A between terminals a and b is then drawn as in Fig. 6.11(c) and the load R_L is reconnected. From this simple series circuit,

$$I = \frac{V_{\text{th}}}{R_{\text{th}} + R_L} = \frac{9}{15 + 25} = 0.225 \text{ A}$$

and

$$V = IR_L = V_{\text{th}} \frac{R_L}{R_{\text{th}} + R_L} = 9 \times \frac{15}{25 + 15} = 3.375 \text{ V}$$

If R_L is variable, then for maximum power transfer in R_L,

$$R_L = R_{\text{th}} \quad \text{(matching condition)}$$

$$= 25 \text{ Ω}$$

and

$$P_{L\,max} = \text{maximum value of the load power}$$

$$= \frac{V_{th}^2}{4R_{th}} = \frac{(9)^2}{4 \times 25} = 0.81 \text{ W}$$

EXAMPLE 6.5

Find the Thévenin equivalent circuit between terminals a and b for the network shown in Fig. 6.12 and then calculate the value of the current I.

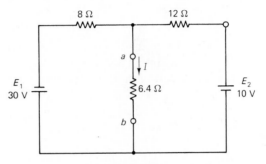

Figure 6.12

Solution   To obtain the Thévenin equivalent circuit for the given network, the 6.4 Ω load is first disconnected from terminals a and b, as shown in Fig. 6.13(a). Here

$$I' = \frac{E_2 + E_1}{8 + 12} = \frac{10 + 30}{20} = 2 \text{ A}$$

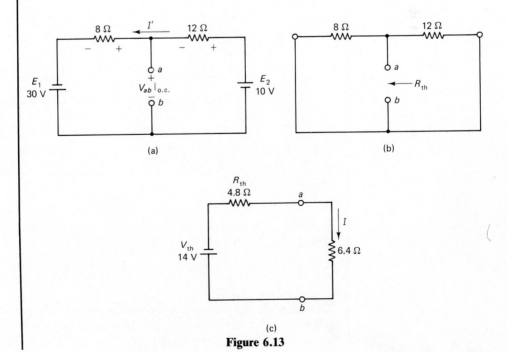

(a)   (b)

(c)

Figure 6.13

Therefore,

$$V_{th} = V_{ab}|_{o.c.} = I' \times 8 - E_1 = 2 \times 8 - 30 = -14\,\text{V}$$

(i.e., point a is lower in potential than point b). Also,

$$V_{th} = E_2 - I' \times 12 = 10 - 2 \times 12 = -14\,\text{V}$$

When the two voltage sources are killed, as shown in Fig. 6.13(b), the resistance between points a and b is then the parallel equivalent of the 8 Ω and 12 Ω resistors, that is,

$$R_{th} = \frac{8 \times 12}{8 + 12} = 4.8\,\Omega$$

The 6.4 Ω load resistor is reconnected to the Thévenin equivalent circuit between terminals a and b, as shown in Fig. 6.13(c). Notice the polarities of V_{th}, to conform with the calculation of $V_{ab}|_{o.c.}$ above, showing that point b is higher in potential than point a. Thus

$$I = -\frac{14}{4.8 + 6.4} = -1.25\,\text{A}$$

The negative sign of the current indicates that the actual current is flowing from point b to point a.

EXAMPLE 6.6

Find the value of R_L in the circuit shown in Fig. 6.14 that results in maximum power to be produced in R_L. What is this value of $P_{L\,\text{max}}$?

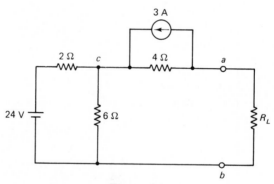

Figure 6.14

Solution The load is first disconnected from terminals a and b and the dead circuit is redrawn as in Fig. 6.15(a). Thus

$$R_{th} = 4 + \frac{2 \times 6}{2 + 6} = 5.5\,\Omega$$

To obtain V_{th}, the network is redrawn as in Fig. 6.15(b), with terminals a and b left open. The current source between terminals a and c is converted into the equivalent voltage source model. Here no current is flowing in the 4 Ω resistor;

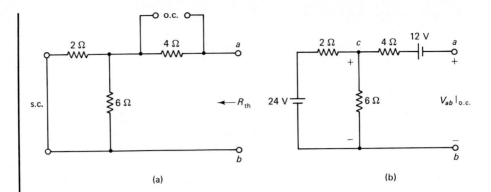

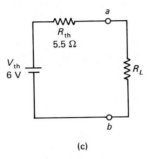

(c)

Figure 6.15

therefore,

$$V_{th} = V_{ab}|_{o.c.} = V_{ac} + V_{cb}$$

$$= -12 + 24 \times \frac{6}{6 + 2}$$

$$= -12 + 18 = 6 \text{ V}$$

Note that

$$V_{ac} = -V_{ca} = -12 \text{ V}$$

which is the value of the open-circuit voltage of the converted current source ($= 3 \text{ A} \times 4 \Omega$).

The load resistance R_L is reconnected to the Thévenin equivalent model of the given circuit, as shown in Fig. 6.15(c). Maximum power transfer condition requires that

$$R_L = R_{th} = 5.5 \Omega$$

Here

$$V_L = \tfrac{1}{2}V_{th} = \tfrac{1}{2} \times 6 = 3 \text{ V}$$

and

$$P_{L\max} = \frac{V_L^2}{R_L} = \frac{V_{th}^2}{4R_{th}} = \frac{(3)^2}{5.5} = 1.636 \text{ W}$$

6.3 NORTON'S THEOREM

This theorem is very similar to Thévenin's theorem. Its application allows one to replace an entire network between two available terminals (say a and b) with an equivalent simple circuit containing one current source, I_N, in parallel with one resistance, R_N, resembling the practical current source model. This model is called the *Norton equivalent circuit*.

Figure 6.16 shows the Norton equivalent-circuit model for any general network, as, for example, network A. Here also, as far as network B (or load R_L) is concerned, both the original network A and Norton's equivalent circuit produce the same voltage, V, across terminals a and b, and also the same current flow, I. It is clear that the network analysis process, to obtain V or I, is considerably simplified by using Norton's equivalent compared to analyzing the original more complicated network.

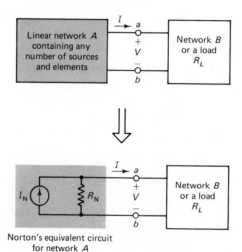

Figure 6.16

To evaluate the elements of the Norton equivalent-circuit model, network B or load R_L is first disconnected from terminals a and b. Then all the sources in the original network are killed; that is, with network A dead, the input resistance between terminals a and b is R_N. Thus as in Fig. 6.8,

$$R_N = \text{input resistance of the dead network } A,$$
$$\text{looking between terminals } a \text{ and } b = R_{th} \qquad (6.8)$$

If a short circuit is connected between terminals a and b of the original network A, as shown in Fig. 6.17, and $I_{ab}|_{s.c.}$ is calculated, then

$$I_N = I_{ab}|_{s.c.} \qquad (6.9)$$

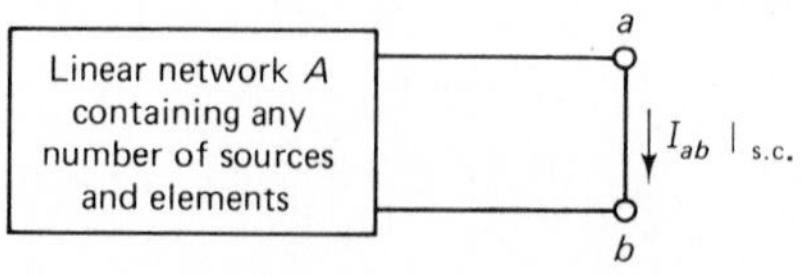

Figure 6.17

The load R_L, or network B, is then reconnected between terminals a and b of the Norton equivalent circuit, to calculate the required voltage V, or current I in Fig. 6.16. Notice that the direction of I_N is upward when $I_{ab}|_{\text{s.c.}}$ is downward, and vice versa. This is clearly the proper situation, because if R_L is replaced by a short circuit in the Norton equivalent circuit (see Fig. 6.16), $I_{ab}|_{\text{s.c.}}$ will flow downward.

Corollary I: Source conversions. According to Thévenin's and Norton's theorems, a given general linear network (the original network A), its Thévenin equivalent-circuit model, or its Norton equivalent-circuit model produce the same results for the voltage V across the current I in the load R_L, connected between terminals a and b.

This equivalency is represented as shown in Fig. 6.18. As a result, Thévenin's and Norton's equivalent-circuit models must also be equivalent to each other. This equivalency is valid for any value of R_L. As shown in Eq. (6.8),

$$R_{\text{th}} = R_N$$

Also, when $R_L = 0$, then according to Eqs. (6.7) and (6.9) (also from Fig. 6.18),

$$I_{ab}|_{\text{s.c.}} = \frac{V_{\text{th}}}{R_{\text{th}}} = I_N \tag{6.10}$$

Then

$$\boxed{V_{\text{th}} = I_N R_{\text{th}} = I_N R_N} \tag{6.11}$$

This last result can also be confirmed if R_L is ∞ (i.e., an open circuit) in Fig. 6.18.

$$V_{ab}|_{\text{o.c.}} = V_{\text{th}} \quad \text{[from Thévenin's equivalent circuit, Fig. 6.18(b)]}$$
$$= I_N R_n \quad \text{[from Norton's equivalent circuit, Fig. 6.18(c)]}$$

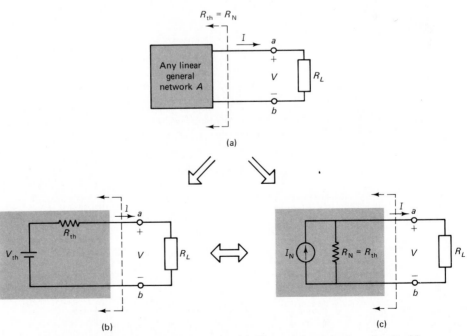

Figure 6.18 (a) Original network; (b) Thévenin's equivalent circuit; (c) Norton's equivalent circuit.

This is, in fact, the equivalency between the practical voltage source model and the practical current source model, discussed in Chapter 4. The conversion formulas obtained in Eqs. (6.8), (6.10), and (6.11) are exactly the same as those obtained in Chapter 4 and used extensively in Chapter 5. The only difference is in the notations designating the source parameters (i.e., E instead of V_{th}, I_S instead of I_N, and R_{int} instead of R_{th} or R_N).

Corollary II: Millman's theorem. This theorem follows directly from the source conversions discussed above. It applies to situations requiring the reduction of a number of parallel-connected voltage sources into a single voltage source, as shown in Fig. 6.19. Here the voltage sources are first converted into their equivalent current source models, which are easily added up because of their parallel connection. Thus

$$R_T = \frac{1}{G_T} = \frac{1}{G_1 + G_2 + G_3 + \cdots + G_n}$$

$$= \frac{1}{1/R_1 + 1/R_2 + 1/R_3 + \cdots + 1/R_n} \tag{6.12}$$

and

$$I_T = I_1 + I_2 + I_3 + \cdots + I_n = \frac{E_1}{R_1} + \frac{E_2}{R_2} + \frac{E_3}{R_3} + \cdots + \frac{E_n}{R_n}$$

$$= E_1 G_1 + E_2 G_2 + E_3 G_3 + \cdots + E_n G_n \tag{6.13}$$

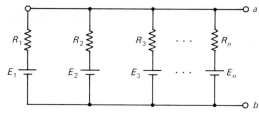

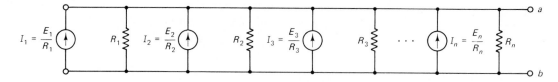

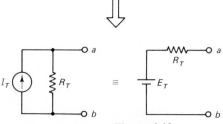

Figure 6.19

The summation in Eq. (6.13) is algebraic, based on the original polarity directions of the parallel connected voltage sources. Finally,

$$E_T = I_T R_T = \frac{E_1 G_1 + E_2 G_2 + E_3 G_3 + \cdots + E_n G_n}{G_1 + G_2 + G_3 + \cdots + G_n} \quad (6.14)$$

EXAMPLE 6.7

Solve Example 6.4 using the Norton equivalent-circuit model.

Solution Referring to Fig. 6.11, R_{th} was obtained as indicated in Fig. 6.11(b). Here

$$R_N = R_{th} = 25\ \Omega$$

Network A is redrawn in Fig. 6.20(a) with a short circuit connected between terminals a and b to calculate I_N. Here

$$I_N = I_{ab}\big|_{s.c.} = I' \times \frac{30}{30 + 13}$$

using the current-division principle. I' can be found from Ohm's law of the whole circuit. Thus

$$I_N = \frac{15}{20 + (30 \times 13)/(30 + 13)} \times \frac{30}{43} = \frac{450}{860 + 390} = 0.36\ \text{A}$$

Norton's equivalent circuit is shown in Fig. 6.20(b) with the load ($R_L = 15\ \Omega$) reconnected between terminals a and b. Applying the current-division pricniple to this simple parallel circuit yields

$$I = I_N \times \frac{25}{25 + 15} = 0.36 \times \frac{25}{40} = 0.225\ \text{A}$$

and

$$V = I \times R_L = 0.225 \times 15 = 3.375\ \text{V}$$

Also note that Eq. (6.11) (source conversion) is confirmed since

$$I_N R_N = 0.36 \times 25 = 9\ \text{V} = V_{th}$$

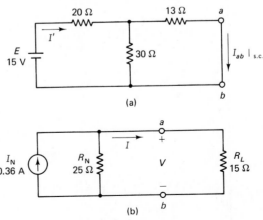

(a)

(b)

Figure 6.20

The matching condition (for maximum power transfer) is still

$$R_L = R_N = R_{th} = 25 \ \Omega$$

Using Norton's equivalent-circuit model, I_L in this case will be $\frac{1}{2}I_N$. Thus

$$P_{L \, max} = (I_L)^2 \times R_L = \frac{I_N^2}{4} \times R_N = \frac{(0.36)^2}{4} \times 25 = 0.81 \text{ W}$$

The results obtained here are exactly the same as those obtained in Example 6.4.

EXAMPLE 6.8

Solve Example 6.6 using the Norton equivalent-circuit model.

Solution R_N is obtained in exactly the same way as was used to find R_{th}, as shown in Fig. 6.15(a). Thus

$$R_N = R_{th} = 5.5 \ \Omega$$

The original network is redrawn in Fig. 6.21(a) with a short circuit connected between terminals a and b in order to calculate I_N. Here the superposition theorem will be applied to obtain $I_{ab}|_{s.c.}$.

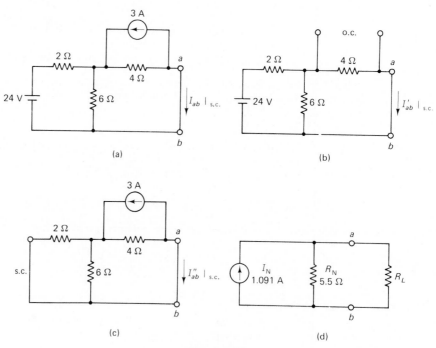

Figure 6.21

Figure 6.21(b) shows the circuit with the 24 V source acting alone (the current source is killed). Thus

$$I'_{ab}|_{s.c.} = \frac{24}{2 + (6 \times 4)/(6 + 4)} \times \frac{6}{6 + 4} = 3.273 \text{ A}$$

Figure 6.21(c) shows the circuit with the 3 A source acting alone (the voltage source is killed). Here the $2\,\Omega$ and $6\,\Omega$ resistors in parallel are equivalent to

$$R' = \frac{2 \times 6}{2 + 6} = 1.5\,\Omega$$

$I''_{ab}|_{\text{s.c.}}$ flows in this R'. Thus using the current-division principle, we have

$$I''_{ab}|_{\text{s.c.}} = -3 \times \frac{4}{4 + 1.5} = -2.182\,\text{A}$$

and

$$I_N = I_{ab}|_{\text{s.c.}} = I'_{ab}|_{\text{s.c}} + I''_{ab}|_{\text{s.c.}}$$
$$= 3.273 - 2.182 = 1.091\,\text{A}$$

[Check: $I_N R_N = 1.091 \times 5.5 = 6\,\text{V} = V_{\text{th}}$, satisfying the source conversion relationship, Eq. (6.11).]

Norton's equivalent-circuit model, with the load R_L reconnected between terminals a and b, is shown in Fig. 6.21(d). Compare this with the circuit in Fig. 6.15(c).

For maximum power transfer (matching condition),

$$R_L = R_N = 5.5\,\Omega$$

Thus

$$I_L = \tfrac{1}{2}I_N = 0.546\,\text{A}$$

and

$$P_{L\max} = (I_L)^2 \times R_L = (0.546)^2 \times 5.5 = 1.637\,\text{W}$$

(as obtained in Example 6.6).

EXAMPLE 6.9

Use Millman's theorem to simplify the network to the left of terminals a and b in the circuit of Fig. 6.22. Find the load current I_L and the load voltage V_L.

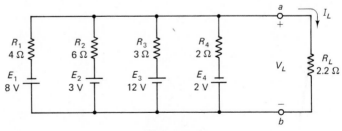

Figure 6.22

Solution Figure 6.23 shows the simplified equivalent circuit using Millman's theorem. From Eq. (6.12),

$$R_T = \frac{1}{G_1 + G_2 + G_3 + G_4}$$

$$= \frac{1}{\frac{1}{4} + \frac{1}{6} + \frac{1}{3} + \frac{1}{2}} = 0.8\,\Omega$$

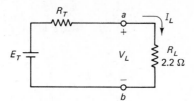

Figure 6.23

Using Eq. (6.13) to obtain I_T, and noting the appropriate signs corresponding to the different voltage source polarities, gives us

$$I_T = -\frac{E_1}{R_1} + \frac{E_2}{R_2} + \frac{E_3}{R_3} - \frac{E_4}{R_4}$$

$$= -\frac{8}{4} + \frac{3}{6} + \frac{12}{3} - \frac{2}{2} = 1.5 \text{ A}$$

Thus from Eq. (6.14), or source conversion,

$$E_T = I_T R_T = 1.5 \times 0.8 = 1.2 \text{ V}$$

Now using the circuit in Fig. 6.23, we have

$$I_L = \frac{E_T}{R_T + R_L} = \frac{1.2}{0.8 + 2.2} = 0.4 \text{ A}$$

and

$$V_L = I_L \times R_L = 0.4 \times 2.2 = 0.88 \text{ V}$$

EXAMPLE 6.10

Find the Thévenin equivalent for the network shown in Fig. 6.24. Use source conversion to obtain the Norton equivalent model. What is the value of the load resistance R_L to be connected between the a and b terminals for maximum power transfer? Calculate $P_{L\max}$.

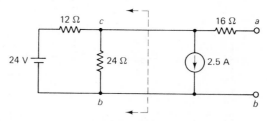

Figure 6.24

Solution The circuit can be partitioned as shown by the dashed lines in Fig. 6.24. As a start, the Thévenin equivalent for the part to the left of the current source (between nodes c and b) can easily be obtained:

$$R'_{\text{th}} = \frac{24 \times 12}{24 + 12} = 8 \,\Omega$$

and

$$V'_{\text{th}} = 24 \times \frac{24}{12 + 24} = 16 \text{ V}$$

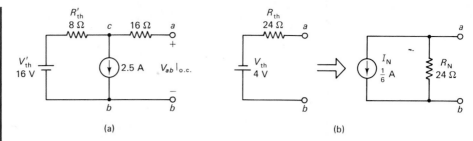

Figure 6.25

The original circuit then simplifies to the equivalent form shown in Fig. 6.25(a). From this circuit, with the sources killed (i.e., V'_{th} is shorted and the 2.5 A source is opened),

$$R_{th} = 16 + 8 = 24\,\Omega$$

Also,

$$V_{th} = V_{ab}|_{o.c.} = V_{cb}$$

(because no current can flow in the 16 Ω resistor). But from KCL at node c,

$$\frac{V'_{th} - V_{cb}}{R'_{th}} = 2.5 \qquad \text{that is,} \qquad \frac{16 - V_{cb}}{8} = 2.5$$

Therefore,

$$V_{th} = V_{cb} = 16 - 8 \times 2.5 = -4\,\text{V}$$

Because this result indicates that point b is higher in potential than point a, the Thévenin equivalent-circuit model would be as shown in Fig. 6.25(b) (notice the polarity reversal of V_{th}).

It is now a straightforward process to obtain Norton's equivalent model, through source conversion, as shown in Fig. 6.25(b). Here

$$I_N = \frac{V_{th}}{R_{th}} = \frac{4}{24} = \frac{1}{6}\,\text{A} \qquad \text{and} \qquad R_N = R_{th}$$

For maximum power transfer, the load connected between terminals a and b must be matched to R_{th} ($= R_N$), that is,

$$R_L = R_{th} = 24\,\Omega$$

Thus

$$P_{L\,max} = \frac{V_{th}^2}{4R_{th}} = \frac{(4)^2}{4 \times 24} = 0.1667\,\text{W}$$

EXAMPLE 6.11

Find the Thévenin equivalent-circuit model for the network to the left of terminals a and b in Fig. 6.26. Use this result to calculate I, the current in the 20 Ω load resistance.

Solution   The dead circuit is shown in Fig. 6.27(a). Thus

$$R_{th} = 2.5 + \frac{3(5 + 10)}{3 + (5 + 10)} = 5\,\Omega$$

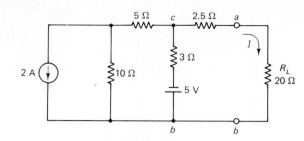

Figure 6.26

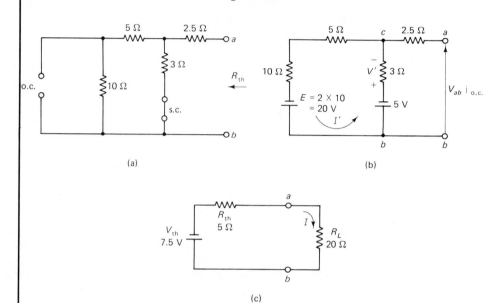

(a)

(b)

(c)

Figure 6.27

The current source (2 A in parallel with 10 Ω) is converted into an equivalent voltage source (notice the polarities) as shown in the original network, redrawn in Fig. 6.27(b), with terminals a and b left open. In this circuit

$$I' = \frac{20 - 5}{10 + 5 + 3} = 0.833 \text{ A}$$

Since no current is flowing in the 2.5 Ω resistor,

$$V_{th} = V_{ab}|_{o.c.} = V_{cb} = -5 - V'$$
$$= -5 - 0.833 \times 3 = -7.5 \text{ V}$$

The Thévenin equivalent circuit is shown in Fig. 6.27(c) , with R_L reconnected to terminals a and b. From this circuit,

$$I = -\frac{V_{th}}{R_{th} + R_L}$$

$$= -\frac{7.5}{5 + 20} = -0.3 \text{ A}$$

The negative sign in the result above indicates that the actual current flowing in R_L is 0.3 A from b to a (i.e., its direction is opposite to that assumed initially).

EXAMPLE 6.12

Find Norton's equivalent for the network to the left of terminals a and b in Fig. 6.28, then calculate the voltage V across R_L.

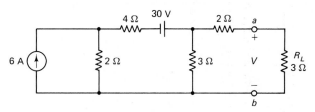

Figure 6.28

Solution   From the dead network shown in Fig. 6.29(a),

$$R_N = R_{th} = 2 + \frac{3 \times (4 + 2)}{3 + (4 + 2)} = 4\,\Omega$$

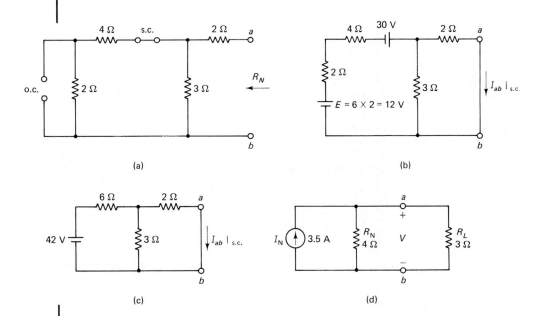

Figure 6.29

The original circuit is redrawn in Fig. 6.29(b) with the 6 A current source converted into a voltage source and a short circuit connected between terminals a and b, in order to calculate I_N. When the two series-connected practical voltage sources in the closed loop are combined, the simpler form of the circuit shown in Fig. 6.29(c) is obtained. Thus

$$I_N = I_{ab}\big|_{s.c.}$$

$$= \frac{42}{6 + (3 \times 2)/(3 + 2)} \times \frac{3}{3 + 2} = 3.5 \text{ A}$$

The calculation above involves obtaining the 42 V source current from Ohm's law and the use of the current-division principle.

The Norton equivalent circuit is shown in Fig. 6.29(d) with the 3 Ω load resistor reconnected between terminals *a* and *b*. From this circuit

$$V = I_N \frac{R_N R_L}{R_N + R_L}$$

$$= 3.5 \times \frac{4 \times 3}{4 + 3} = 6 \text{ V}$$

EXAMPLE 6.13

Solve Example 6.11 using only source conversions.

Solution Many techniques can be used to solve any given network, including Thévenin's or Norton's equivalents, superposition, source conversions, and loop or node analysis methods. This example shows just another way of analyzing the network in Fig. 6.26 to obtain the value of *I*.

The 2 A current source was first converted into a voltage source as shown in Fig. 6.27(b). When the series-connected resistors (the 5 Ω and the 10 Ω) are combined, the circuit in Fig. 6.30(a) is obtained.

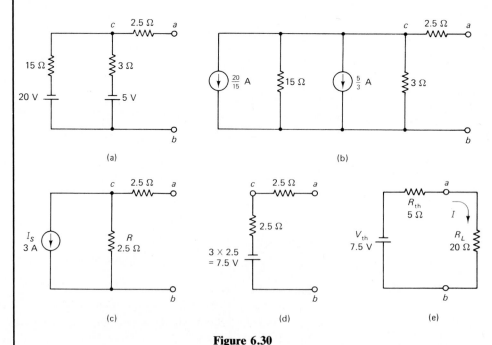

Figure 6.30

The two voltage sources can be replaced by their equivalent current sources as shown in Fig. 6.30(b). The two current sources can then be combined to produce

$$I_S = \frac{20}{15} + \frac{5}{3} = 3 \text{ A}$$

and

$$R = \frac{3 \times 15}{3 + 15} = 2.5 \ \Omega$$

This is shown in Fig. 6.30(c). Converting this current source into its equivalent voltage source produces the circuit shown in Fig. 6.30(d).

Finally, combining the two series resistors and reconnecting the 20 Ω load resistance R_L between terminals a and b, the Thévenin equivalent circuit model in Fig. 6.30(e) is obtained. This is exactly the same result as that obtained in Fig. 6.27(c). Thus

$$I = -\frac{7.5}{5 + 20} = -0.3 \text{ A}$$

as obtained before.

DRILL PROBLEMS

Section 6.1

6.1. Apply the superposition theorem to obtain the value of V in the network shown in Fig. 6.31.

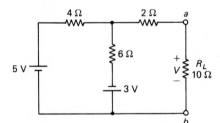

Figure 6.31

6.2. Use the superposition theorem to calculate the value of the current I in the circuit shown in Fig. 6.32.

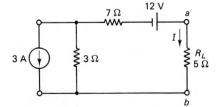

Figure 6.32

6.3. Use the superposition theorem to find the value of the voltage V in the circuit shown in Fig. 6.33.

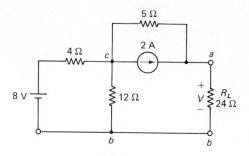

Figure 6.33

Section 6.2

6.4. Find the Thévenin's equivalent circuit for the network shown in Fig. 6.31 between terminals a and b. Use this model to obtain the value of V.

6.5. Obtain the Thévenin's equivalent circuit for the network shown in Fig. 6.32 between terminals a and b. Use this model to calculate the value of the current I.

6.6. What are the parameters of the Thévenin's equivalent model for the circuit shown in Fig. 6.33 between terminals a and b? Use this model to find the value of V.

Section 6.3

6.7. Find the Norton's equivalent circuit for the network shown in Fig. 6.31 between terminals a and b. Use this model to obtain the value of V.

6.8. Obtain the Norton's equivalent circuit for the network shown in Fig. 6.32 between terminals a and b. Use this model to calculate the value of I.

6.9. What are the parameters of the Norton's equivalent model for the circuit shown in Fig. 6.33 between terminals a and b? Use this model to find the value of V.

6.10. If the resistance of the load R_L in the networks shown in Figs. 6.31, 6.32, and 6.33 is variable, what should be the value of R_L in each case for maximum power transfer to the load? What is the value of the maximum power available from each of these networks?

PROBLEMS

Note: The problems below that are marked with an asterisk (*) can be used as exercises in computer-aided network analysis or as programming exercises.

Section 6.1

(i) **6.1.** Use the superposition theorem to find the value of V in the circuit shown in Fig. 6.34.

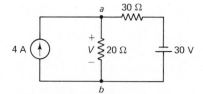

Figure 6.34

(i) ***6.2.** Use the superposition theorem to find the value of I in the circuit shown in Fig. 6.35.

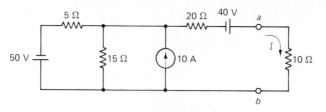

Figure 6.35

(ii)   ***6.3.**   Find the voltage across the 25 Ω load resistance in Fig. 6.36 using the superposition theorem.

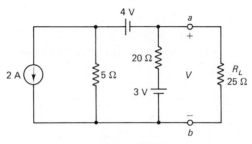

Figure 6.36

(i)   ***6.4.**   Obtain the value of the current I in Fig. 6.37 using the superposition theorem.

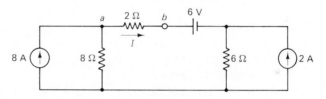

Figure 6.37

(i)   **6.5.**   Solve Example 6.5 using the superposition theorem.

(i)   **6.6.**   Apply the superposition theorem to solve Example 6.12.

Section 6.2

(i)   **6.7.**   With the 20 Ω resistor between terminals a and b in Fig. 6.34 removed, find the Thévenin equivalent circuit. Reconnect the 20 Ω resistor and calculate the value of V using the Thévenin equivalent model.

(ii)   **6.8.**   Find the Thévenin equivalent circuit for the network shown in Fig. 6.35, between terminals a and b. Use this result to obtain the value of the current I in the 10 Ω resistor. What should be the value of this resistor (if it was variable) to obtain maximum power transfer? What is the value of this maximum power?

(i)   **6.9.**   Obtain the Thévenin equivalent circuit between terminals a and b for the network shown in Fig. 6.36. Calculate V using the Thévenin circuit model. Replace R_L by the resistor required for maximum power transfer and find the value of this maximum power.

(i)   **6.10.**   Obtain the Thévenin equivalent between terminals a and b for the network shown in Fig. 6.37. Use this circuit model to calculate the value of I, the current in the 2 Ω resistor.

Problems

(i) **6.11.** Find the Norton equivalent circuit between terminals *a* and *b* for the network shown in Fig. 6.34. Use this circuit model to calculate the value of the voltage *V* across the 20 Ω resistor.

(i) **6.12.** Replace the network shown in Fig. 6.35 between terminals *a* and *b* by its Norton equivalent-circuit model. Calculate the value of the current *I* in the 10 Ω resistor.

(ii) **6.13.** Obtain the Norton equivalent circuit for the network shown in Fig. 6.36 between terminals *a* and *b* and then calculate the value of the voltage *V*. Show that the Norton equivalent circuit is the voltage-to-current source conversion of the Thévenin equivalent circuit obtained in Problem 6.9.

(i) **6.14.** Replace the network shown in Fig. 6.37 between terminals *a* and *b* by its Norton equivalent model and use your result to calculate the value of *I*. Confirm the source model equivalency by comparing your answers with those obtained in Problem 6.10.

(i) ***6.15.** Use Millman's theorem to simplify the circuit shown in Fig. 6.38. Calculate the values of the current *I* and the voltage *V*.

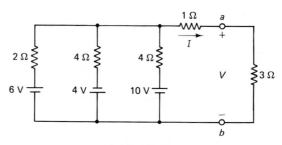

Figure 6.38

Sections 6.2 and 6.3

(ii) ***6.16.** Find the Thévenin equivalent for the network shown in Fig. 6.39 to the left of points *a* and *b*. Use this result to calculate *V* and *I*.

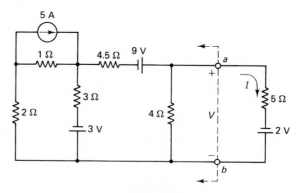

Figure 6.39

(ii) **6.17.** Solve Problem 6.16 using Norton's equivalent-circuit model.

(i) **6.18.** Through using source conversions only, prove the correctness of the Thévenin and Norton equivalent circuits obtained in Problems 6.16 and 6.17.

(ii) **6.19.** Obtain the Norton equivalent-circuit model between terminals *a* and *b* for the network shown in Fig. 6.40.

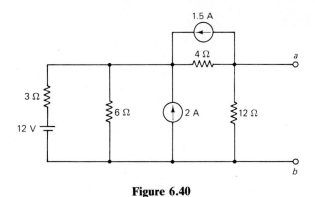

Figure 6.40

(ii) **6.20.** Find the Thévenin equivalent circuit for the network shown in Fig. 6.40. What should be the value of the load resistor to be connected between terminals *a* and *b* to obtain maximum power transfer to the load? What is the value of this maximum power?

GLOSSARY

Dead (or Killed) Source: A source whose output is zero. For a dead voltage source, $V_S = 0$ V, indicating that it should be replaced by a short circuit. For a dead current source, $I_S = 0$ A, indicating that it should be replaced by an open circuit.

Maximum Power Transfer (or Matching) Condition: The condition necessary to ensure that maximum power is developed in the load from a given network or source.

Millman's Theorem: A method based on source conversions, used to replace a number of parallel connected sources with an equivalent single voltage or current source.

Norton's Theorem: A theorem that permits the replacement of any two-terminal linear network with a simple equivalent practical current source model.

Source Equivalency: The conditions necessary to replace a practical voltage source by its equivalent practical current source model, or vice versa, such that the external response of either of the two source models is the same.

Superposition Theorem: A theorem that permits the evaluation of the required response (voltage or current) as an algebraic summation of the partial responses of the network; each partial response is obtained due to the influence of one source at a time while all the other sources in the networks are temporarily killed.

Thévenin's Theorem: A theorem that permits the replacement of any two-terminal linear network with a simple equivalent practical voltage source model.

ANSWERS TO DRILL PROBLEMS

6.1. $V' = 2.0833$ V, $V'' = -0.8333$ V, $V = 1.25$ V

6.3. $V' = 4.5$ V, $V'' = 7.5$ V, $V = 12$ V

6.5. $R_{th} = 10\,\Omega$, $V_{th} = 3$ V, $I = 0.2$ A

6.7. $R_N = 4.4\,\Omega$, $I_N = 0.4091$ A, $V = 1.25$ V

6.9. $R_N = 8\,\Omega$, $I_N = 2$ A, $V = 12$ V

7 | BASIC ELECTRICAL MEASUREMENT INSTRUMENTS

OBJECTIVES

- ■ Understanding the principles of operation of the basic current-sensing type meter movement and its design parameters.
- ■ Ability to design and calculate the loading effects of ammeters.
- ■ Ability to design and calculate the loading effects of voltmeters.
- ■ Ability to design ohmmeters based on D'Arsonval meter movement.
- ■ Understanding the principles of operation and the function of the main building blocks of the oscilloscope.

Once an electrical circuit is designed, the next step is to build it (i.e., implementation) and then test its performance. This last step is obviously of prime importance, to make sure that the circuit is operating according to the required objectives. In many other cases measurements have to be performed to check that a given network is functioning properly or to find out what component(s) may have caused a possible malfunction.

Basically, measurements involve the determination of current, voltage, power, and possibly the component values (i.e., resistances). Many instruments used in modern electronics laboratories perform measurements with accuracy, reliability, and automation that far exceed the standards of a few years ago. The level of sophistication of modern measuring instruments was achieved as a result of the enormous advances in computer engineering and electronics. Examination of the principles of operation of such instruments is beyond the scope of this introductory text. However, as sure as one can be of encountering these modern instruments and using them, many other types of still versatile and useful instruments of few years back will also undoubtedly be encountered and used.

The measurements of the fundamental electrical quantities, I, V, P, and R, are based on either one of the following:

1. *Current sensing.* The instruments are mostly of the electromagnetic meter movement type.
2. *Voltage sensing.* The instruments are mostly electronic in nature, using amplifiers and semiconductor devices (diodes and transistors).

In this chapter the discussion concentrates on current-sensing meters, as for example the VOM (Volt-Ohm-Milliammeter) meter in Fig. 7.1(a). The principles

(a)

Figure 7.1 (a) Current-sensing universal volt-ohm-milliammeter.

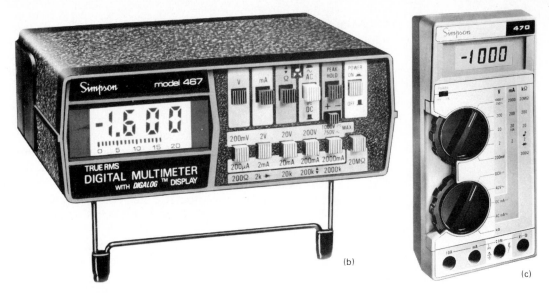

(b)

(c)

Figure 7.1 (b), (c) voltage-sensing digital multimeters. (Courtesy of Simpson Electric Company.)

of operation of one very important instrument, the oscilloscope, of the voltage-sensing type, are also discussed. Other voltage-sensing instruments [see Fig. 7.1(b) and (c)] can be studied after some electronics background has been gained. Measurement of current, voltage, and resistance is examined here; power measurement is dealt with in Chapter 13.

The instruments examined are mainly dc meters. Meters used in ac circuits usually require a rectifier circuit to produce a unidirectional current (or voltage). The response of the ac meter is proportional to the average value of the electrical quantity, but the scale could be calibrated to read any required value [e.g., the root-mean-square (RMS) value].

In addition to the construction, design parameters, and applications of these meters, the emphasis in this chapter is on loading-effect errors, errors introduced in the measurement process itself due to the nonideal nature of the measuring instruments used.

The central component of current-sensing meters is the meter movement. D'Arsonval meter movement is one such device.

7.1 D'ARSONVAL METER MOVEMENT

The D'Arsonval meter movement is also known as the permanent-magnet moving-coil movement. It is the most commonly used type of movement. The basic construction of such a meter movement is shown schematically in Fig. 7.2(a). An actual current-sensing meter movement is shown in Fig. 7.2(b).

When the meter current I_m flows in the wire coil in the direction indicated in Fig. 7.2(a), a magnetic field is produced in the coil. This electrically induced magnetic field interacts with the magnetic field of the horseshoe-type permanent magnet. The result of such an interaction is a force causing a mechanical torque to be exerted on the coil. Since the coil is wound and permanently fixed on a rotating cylindrical drum as shown, the torque produced will cause the rotation of

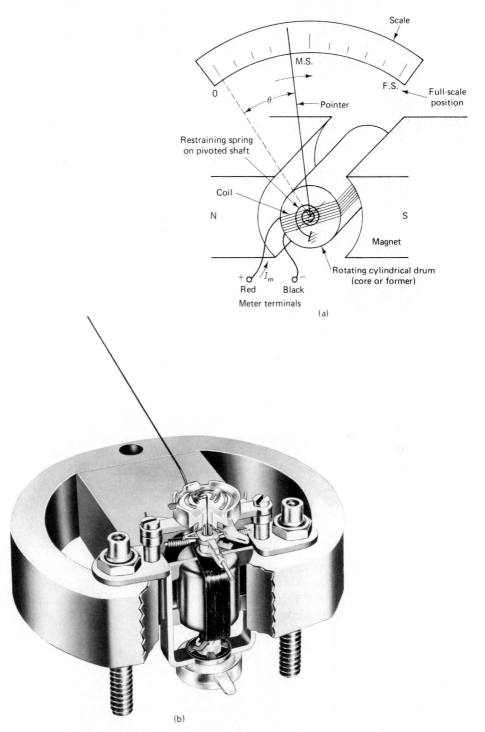

Figure 7.2 (a) Meter movement schematic; (b) currents-sensing meter movement. [(b) Courtesy of Simpson Electric Company.]

the drum (core) around its pivoted shaft. This process, called the *electromagnetic phenomenon,* is the principle of operation of electric motors (and generators) and is examined in more detail in Chapter 9.

A pointer is fixed to the rotating shaft and will deflect upscale against a fixed scale as shown in Fig. 7.2(a). At this point it should be noted that if the direction of flow of the meter current I_m is reversed, the direction of the rotational torque induced will also be reversed and the pointer's deflection will be in the opposite direction.

When the drum rotates, two restraining springs, one mounted in the front onto the shaft and the other mounted onto the back part of the shaft, will exhibit a countertorque opposing the rotation and restraining the motion of the drum. This spring-produced countertorque depends on the angle of deflection of the drum, θ, or the pointer. At a certain position (or deflection angle), the two torques are in equilibrium, or balanced, that is,

$$\text{electromagnetic-induced torque} = \text{spring countertorque}$$

$$k_1 I_m = k_2 \theta$$

or

$$I_m = k\theta \qquad \left(k = \frac{k_2}{k_1}\right) \tag{7.1}$$

The movement will then become stationary and the deflection of the pointer, θ, is proportional to the value of the meter current I_m, causing the electromagnetic torque. This linear proportionality is indicated in Eq. (7.1). The position of the pointer against the fixed scale can thus be calibrated to indicate directly the current magnitude.

It should be emphasized that this type of meter movement is polarized, that is, its terminals are polarity marked; $+$ is coded red and $-$ is coded black. The current must enter the meter movement from the $+$ terminal in order that the resulting torque be in the direction that causes an upscale deflection of the pointer. At the zero-deflection ($\theta = 0°$) position of the pointer, there is a stop pin. Thus permanent damage can be caused if the current is allowed to flow in the opposite direction for an extended period. Thus, as soon as a downscale deflection is observed, the circuit must be disconnected and the wires connected to the meter terminals reversed.

The values of the constants k_1, k_2, and k in Eq. (7.1) depend only on the construction parameters of the meter movement: for example, the physical dimensions, number of turns of the coil, and strength of the permanent magnet. Also of importance is the value of k_2, which depends on the strength of the restraining springs.

Each meter movement is characterized by two electrical quantities:

1. R_m: the meter resistance, which is due to the wire used to construct the coil.
2. $I_{F.S.}$: the meter current, which causes the pointer to deflect all the way up to the full-scale position on the fixed scale (this is marked F. S. in Fig. 7.2(a)). This value of the meter current is always referred to as the *full-scale current* of the meter movement.

These two meter movement parameters are functions of the construction parameters of the movement. Each movement is thus characterized by these two quantities. In particular, the value of k (which depends on k_1 and k_2) and the angle corresponding to the full-scale deflection, $\theta_{F.S.}$, determine the value of the full-scale current, since from Eq. (7.1),

$$I_{F.S.} = k\theta_{F.S.}$$
$$= \text{a fixed number for a given construction} \qquad (7.2)$$

For practical meter movements, $I_{F.S.}$ ranges from as small as $10\ \mu A$ (very sensitive movement) to as large as $30\ mA$ (very rugged movement).

7.1.1 Meter Movement Sensitivity or Ohm/Volt Rating

A movement is more sensitive if its full-scale current is very small. In other words, the meter sensitivity S is the inverse of the full-scale current,

$$\boxed{S = \frac{1}{I_{F.S.}}\ \Omega/V} \qquad (7.3)$$

The units of the sensitivity S are the ohms/volt since from Ohm's law,

$$I = \frac{V\ (V)}{R\ (\Omega)}\ A$$

or

$$1/I \text{ is in } A^{-1} \text{ or } \Omega/V.$$

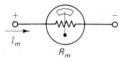

I_m

R_m

Figure 7.3

S can also be referred to as the ohm/volt rating of the meter movement. Either $I_{F.S.}$ or S (fixed value) is defined for a given meter movement and is usually noted on its casing or in the manufacturer's data sheets. Figure 7.3 indicates the electrical circuit symbol of the meter movement that will be used in this chapter.

■ EXAMPLE 7.1

Find the voltage across a $500\ \Omega$, $1\ k\Omega/V$ meter movement if
(a) The pointer is deflected to the full-scale (F. S.) position.
(b) The pointer is deflected to the midscale (M. S.) position.

Solution As specified in this example, this movement is characterized by

$$R_m = 500\ \Omega$$

and

$$S = \frac{1}{I_{F.S.}} = 1\ k\Omega/V = 1000\ \Omega/V$$

Thus

$$I_{F.S.} = \frac{1}{S} = 1\ mA$$

Figure 7.3 indicates that any meter movement can be treated in an electrical circuit as if it were a resistor of value R_m.
    (a) When the pointer is at the F. S. position,

$$I_m = I_{F.S.} = 1\ mA$$

and

$$V_m = V_{F.S.} = I_{F.S.}R_m = 1 \times 10^{-3} \times 500 = 0.5\ V$$

(b) When the pointer is at the M. S. (midscale) position,

$$I_m = \tfrac{1}{2}I_{\text{F.S.}} = 0.5\,\text{mA}$$

and

$$V_m = I_m R_m = 0.5 \times 10^{-3} \times 500 = 0.25\,\text{V}$$

7.2 CONSTRUCTION AND DESIGN OF AMMETERS

As pointed out above, the deflection of the pointer in the D'Arsonval meter movement is propotional to the meter current I_m. Therefore, this instrument can be used to measure current (i.e., it is an ammeter). However, the meter movement by itself is of limited use and capability. It cannot be used to measure any current above its full-scale current value $I_{\text{F.S.}}$, which corresponds to the full-scale deflection of the pointer. If the current allowed to flow in the movement, I_m, exceeds $I_{\text{F.S.}}$, permanent damage can result, in particular to the restraining springs. This may seem to suggest that one must start with a large number of different meter movements in order to measure different currents in different ranges. This is obviously impractical. Besides, we do not usually know the value of the current to be measured before the measurement is done.

To be able to measure currents higher in value than $I_{\text{F.S.}}$ of a given meter movement, a more practical solution must be found. The solution required involves the application of the current-division principle. If the circuit current to be measured, I, is multiplied by a known fraction before flowing in the meter movement, this corresponds to the meter movement current I_m, being a small known portion of I. The movement scale can then be recalibrated to account for such a linear fractional multiplication.

Figure 7.4 shows the construction of an ammeter. The implementation of the actual ammeter circuit takes into account the current-division principle. Here

$$I_m = I\frac{R_{\text{sh}}}{R_m + R_{\text{sh}}} \tag{7.4}$$

where I_m is the movement current, I the circuit current being measured, R_m the resistance of the movement, and R_{sh} the "shunt" resistance connected in parallel with the movement to provide a path for the portion of the circuit's current not allowed to flow in the meter movement. Equation (7.4) is called the *calibration equation* of the ammeter. The value of R_{sh} can be chosen as required to adjust the

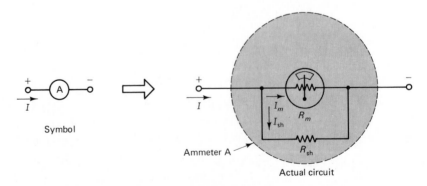

Figure 7.4

fractional multiplier $[R_{sh}/(R_m + R_{sh})]$ for proper operation of the ammeter, as will be specified below. Notice that

$$I = I_m + I_{sh} \tag{7.5}$$

and

$$R_m I_m = R_{sh} I_{sh} \tag{7.6}$$

R_{sh} is usually much smaller than R_m. Therefore, the shunt resistance carries the larger portion of the circuit's current.

When it is required to measure a circuit's current up to a maximum value of I_{max} using such a meter movement, then as

$$I = I_{max} \tag{7.7}$$

the meter movement current should correspond to the full-scale deflection of the pointer, that is,

$$I_m = I_{F.S.} \tag{7.8}$$

The value I_{max} is then the range setting of the ammeter.

Substituting the values of I and I_m from Eqs. (7.7) and (7.8) in the ammeter's calibration equation, Eq. (7.4), we obtain

$$\boxed{I_{F.S.} = I_{max} \frac{R_{sh}}{R_m + R_{sh}}} \tag{7.9}$$

This is the required design equation of the ammeter. For a given meter movement, $I_{F.S.}$ and R_m are known. Therefore, Eq. (7.9) provides the necessary value of R_{sh} for the required range setting, I_{max}. Each range setting of the ammeter's scale requires a different value of R_{sh}.

Thus a given meter movement can be used to build a multirange ammeter. Each range requires a different value of the shunt resistance. A multiposition make-before-break switch can be used to connect the proper shunt resistance corresponding to the range selected by the position of the switch. A three-range ammeter, requiring three different shunt resistors, is shown in Fig. 7.5.

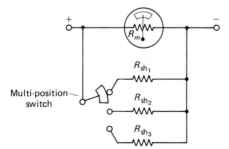

Figure 7.5

As can be deduced from Fig. 7.4, the ammeter actually corresponds to two parallel-connected resistors. The resistance of the ammeter R_A is thus

$$R_A = \frac{R_m R_{sh}}{R_m + R_{sh}} \tag{7.10}$$

Using Eq. (7.9), we obtain

$$R_A = R_m \frac{I_{F.S.}}{I_{max}} \tag{7.11}$$

7.2   Construction and Design of Ammeters   **239**

Equation (7.11) is actually Ohm's law for two parallel-connected resistors as shown in Fig. 7.4 because it can be written as follows:

voltage across the ammeter when the pointer is at the F. S. position

$$= I_{F.S.}R_m = I_{max}R_A \tag{7.11a}$$

Also, because usually

$$R_{sh} \ll R_m$$

then according to the properties of the parallel circuits,

$$R_A \cong R_{sh} \tag{7.12}$$

that is, the equivalent resistance (R_A) is approximately equal to the smaller of the two parallel-connected resistors (R_{sh}). Equation (7.12) indicates that as the range of a multirange ammeter is changed, R_{sh} will be different and the ammeter's resistance R_A will be different for the different ranges.

EXAMPLE 7.2

Design a two-range ammeter using a $100\,\Omega$, $1\,k\Omega/V$ meter movement. One range is to read up to $20\,mA$ and the second range is to read up to $100\,mA$. Find the required values of the shunt resistances and the resistance of the ammeter in each of the two ranges.

Solution The circuit diagram of such an ammeter is similar to that shown in Fig. 7.5, except that only two shunt resistors are required, corresponding to the two ranges. The switch required is only a two-position switch. The meter movement has the two parameters

$$R_m = 100\,\Omega$$

and

$$I_{F.S.} = \frac{1}{S} = \frac{1}{1000\,\Omega/V} = 1\,mA$$

For the 20 mA range, $I_{max} = 20\,mA$. Then from Eq. (7.9),

$$1\,mA = 20\,mA \frac{R_{sh_1}}{100 + R_{sh_1}}$$

$$100 + R_{sh_1} = 20R_{sh_1}$$

Therefore,

$$R_{sh_1} = \frac{100}{19} = 5.263\,\Omega$$

Here

$$R_{A_1} = \frac{R_m R_{sh_1}}{R_m + R_{sh_1}} = \frac{100 \times 5.263}{105.263} = 5.0\,\Omega$$

or

$$R_{A_1} = R_m \frac{I_{F.S.}}{I_{max}} = 100 \times \frac{1\,mA}{20\,mA} = 5.0\,\Omega$$

For the 100 mA range, $I_{max} = 100$ mA. Again using Eq. (7.9), we have

$$1\,\text{mA} = 100\,\text{mA}\,\frac{R_{sh_2}}{100 + R_{sh_2}}$$

$$100 + R_{sh_2} = 100 R_{sh_2}$$

Therefore,

$$R_{sh_2} = \frac{100}{99} = 1.01\,\Omega$$

Here

$$R_{A_2} = R_m \frac{I_{\text{F.S.}}}{I_{max}} = 100 \times \frac{1\,\text{mA}}{100\,\text{mA}} = 1.0\,\Omega$$

Notice that $R_A \cong R_{sh}$ and that the ammeter's resistance changes as the range used in the measurement is changed.

EXAMPLE 7.3

A given meter movement is used to build a 25 mA range ammeter. The resistance of the ammeter is $10\,\Omega$ and the shunt resistance used has a value $10.204\,\Omega$. Find the parameters of this movement.

Solution It is required here to determine the value of each of R_m and $I_{\text{F.S.}}$ (or S) of this meter movement. Since R_A and R_{sh} are given, we can use the relationship

$$R_A = \frac{R_m R_{sh}}{R_m + R_{sh}}$$

to determine R_m. Here

$$10 = \frac{10.204 R_m}{R_m + 10.204}$$

$$10 R_m + 102.04 = 10.204 R_m$$

Therefore,

$$R_m = \frac{102.04}{0.204} = 500\,\Omega$$

Now using the design equation [Eq. (7.9)] with $I_{max} = 25$ mA yields

$$I_{\text{F.S.}} = I_{max} \frac{R_{sh}}{R_m + R_{sh}} = 25\,\text{mA} \times \frac{10.204}{510.204} = 0.5\,\text{mA}$$

One could have also used Eq. (7.11a):

$$R_m I_{\text{F.S.}} = R_A I_{max}$$

Then

$$I_{\text{F.S.}} = 10 \times \frac{25\,\text{mA}}{500} = 0.5\,\text{mA}$$

as above. Therefore,

$$S = \text{sensitivity of the movement}$$

$$= \frac{1}{I_{\text{F.S.}}} = \frac{1}{0.5\,\text{mA}} = 2000\,\Omega/\text{V}$$

EXAMPLE 7.4 LOADING EFFECT OF THE AMMETER

Determine the ideal value of the currents I' and I in the circuit shown in Fig. 7.6. If the ammeter designed in Example 7.2 is used to measure the current I on its 20 mA range, what will be the reading of the ammeter?

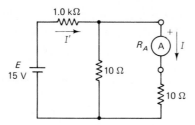

Figure 7.6

Solution   The series-parallel circuit shown in Fig. 7.6 can easily be analyzed. The resistance seen by the source is

$$R_T = 1000 + \frac{10 \times 10}{10 + 10} = 1005 \, \Omega$$

Therefore,

$$I' = \frac{E}{R_T} = \frac{15}{1005} = 14.925 \, \text{mA}$$

This current divides between the two $10 \, \Omega$ branches equally. Hence

$$I = \tfrac{1}{2}I' = 7.463 \, \text{mA}$$

These are the ideal values of the currents that one would expect to read if the ammeter used for measuring the currents is an ideal instrument.

The ammeter must be connected in series with the element through which the current is to be measured. The circuit must be disconnected to insert the ammeter. In order for this measurement process not to disturb the value of the current being measured, the resistance of the ammeter (which is series connected) should ideally be zero so that the resistance of the circuit remains unchanged. As shown in the discussion above, a practical ammeter has a finite but small resistance, R_A. If the ammeter is connected in series with a large resistance, R_A can easily be neglected in comparison, but if the ammeter is connected in series with a small comparable resistance, R_A cannot be neglected. In fact, it will cause large errors in the measurements (readings). This is what is called the meter's *loading effect*.

When the ammeter is connected in series with the $10 \, \Omega$ resistance to measure this branch's current I, as shown in Fig. 7.6, this branch will now have a resistance of $10 + R_{A_1} = 10 + 5 = 15 \, \Omega$. $R_{A_1} \, (= 5 \, \Omega)$ is used here because it is the resistance of the ammeter in Example 7.2 on its 20 mA range.

The circuit can now be reanalyzed as follows. Here

$$R_T = 1000 + \frac{10 \times 15}{10 + 15} = 1006 \, \Omega$$

Therefore,

$$I' = \frac{E}{R_T} = \frac{15}{1006} = 14.91 \, \text{mA}$$

Using current division to obtain the actual value of I in this case yields

$$I = I' \times \frac{10}{10 + 15} = 14.91 \text{ mA} \times \frac{10}{25}$$

$$= 5.964 \text{ mA (actual reading of the ammeter)}$$

This reading compares quite unfavorably with the ideal value of I, being 7.463 mA, as obtained above. It corresponds to a 20% error. This loading effect is over and above any inaccuracy in the meter itself.

However, if the same ammeter is used to measure the value of I', by connecting it in series with the source or the 1.0 kΩ, the new value of the total resistance of the circuit is

$$R_T = 1000 + R_{A_1} + \frac{10 \times 10}{10 + 10} = 1000 + 5.0 + 5.0 = 1010 \text{ Ω}$$

and the actual value of I' being measured (i.e., the reading of the ammeter) is

$$I' = \frac{E}{R_T} = \frac{15}{1010} = 14.851 \text{ mA}$$

This reading compares quite favorably with the ideal value of I', being 14.925 mA, only a 0.5% error.

The analysis presented here explains the procedure to be followed in examining the loading effect of the ammeter. The percentage error is calculated from:

$$\text{Error } (\%) = \frac{\text{ideal value} - \text{measured value}}{\text{ideal value}} \times 100$$

7.3 CONSTRUCTION AND DESIGN OF VOLTMETERS

The meter movement is modeled as a resistance of value R_m, as shown in Fig. 7.3. Therefore, Ohm's law applied to this movement provides

$$V_m = R_m I_m \qquad (7.13)$$

where V_m is the voltage across the meter movement when the current flowing in it is I_m. When the current in the movement is $I_{\text{F.S.}}$,

$$V_{\text{F.S.}} = R_m I_{\text{F.S.}} \qquad (7.14)$$

As V_m and I_m are linearly related, the scale can easily be recalibrated to indicate voltage instead of current. The voltage reading then corresponds to the voltage V_m across the movement terminals. The full-scale voltage determined by Eq. (7.14) is, however, quite small. As shown in Example 7.1, $V_{\text{F.S.}}$ is typically of the order of 0.5 V, which is an impractically small value.

To increase the full-scale voltage range of the movement when functioning as a voltmeter, the meter movement current I_m has to be lowered. This can easily be achieved by inserting a large resistance, called the multiplier resistance (R_{mult}), in series with the movement, as shown in Fig. 7.7. This figure represents the

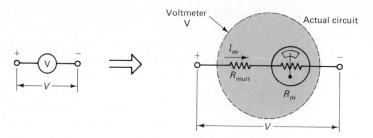

Figure 7.7

actual circuit of the voltmeter. Here

$$I_m = \frac{V}{R_m + R_{\text{mult}}}$$

or

$$V = I_m(R_m + R_{\text{mult}}) \tag{7.15}$$

Equation (7.15) is called the *calibration equation* of the voltmeter. For a specific value of R_{mult} (and R_m), the current scale I_m can be recalibrated into a voltage scale V using this equation. If the value of the multiplier resistance R_{mult} is increased, the same meter movement current I_m would flow as a result of a larger terminal voltage V. This indicates that the multiplier resistance can properly be chosen to provide a specific full-scale range for the required voltmeter.

If it is required to design a voltmeter to measure voltages up to a specific value V_{max}, this voltage when applied to the terminals of the voltmeter should result in $I_{\text{F.S.}}$ flowing in the meter movement (i.e., the pointer deflection is then up to the full-scale position). Thus, from Eq. (7.15),

$$\boxed{V_{\text{max}} = I_{\text{F.S.}}(R_m + R_{\text{mult}})} \tag{7.16}$$

V_{max} is then referred to as the scale range of the voltmeter. Equation (7.16) is called the voltmeter's *design equation*. For a given meter movement, with R_m and $I_{\text{F.S.}}$ specified, this equation can be used to determine the value of the multiplier resistance, R_{mult}, required to provide a voltage range V_{max} for the voltmeter.

Note that the resistance of the voltmeter, R_V, is the series combination of R_m and R_{mult}, as can be seen from Fig. 7.7. Therefore,

$$\boxed{\begin{aligned} R_V &= R_{\text{mult}} + R_m = \frac{V_{\text{max}}}{I_{\text{F.S.}}} \\[2mm] &= V_{\text{max}} \times \frac{1}{I_{\text{F.S.}}} \\[2mm] &= \text{range of voltmeter} \times S \end{aligned}} \tag{7.17a} \tag{7.17b}$$

as S is $1/I_{\text{F.S.}}$ according to Eq. (7.3). S is a fixed parameter of the movement and of the voltmeter.

Equation (7.17b) is quite important. It indicates that the higher the range of the voltmeter, the larger would be the internal resistance of the voltmeter, R_V. This is clear since a higher voltage range requires a larger multiplier resistance. Also, Eq. (7.17b) indicates that a more sensitive meter movement (higher S or lower $I_{\text{F.S.}}$) also results in a larger voltmeter resistance. These remarks are very useful for the discussion of the loading effects of the voltmeter.

Using the same meter movement, a multirange voltmeter can be designed, requiring a multiposition switch and a number of multiplier resistors, similar to the number of ranges required. A three-range voltmeter is shown schematically in Fig. 7.8. For each voltage range (i.e. different value of V_{max}) a different multiplier resistance R_{mult} is required. The proper multiplier resistance is then connected in series with the meter movement according to the position of the switch, providing a voltage range as indicated opposite to the switch position.

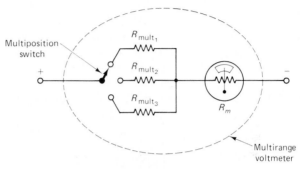

Figure 7.8

A voltmeter always draws current I_m in order to cause the deflection of the pointer. In a circuit, a voltmeter can be treated as a resistive element of value R_V. It always reads the real value of the voltage across its terminals.

7.3.1   Loading Effect of a Voltmeter

When the voltmeter is connected across the two terminals of an element in order to measure the voltage V across the element, it will draw a specific current I_m from the circuit, as shown in Fig. 7.9. This process could disturb the circuit conditions and result in reading errors called *loading-effect errors*.

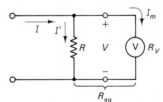

Figure 7.9

If the voltmeter were ideal, it would draw no current from the circuit. This means that an ideal voltmeter is like an open circuit ($R_V \cong \infty$). Electronic voltmeters (EVMs), which are voltage-sensing meters, draw a very small amount of current from the circuit and hence approach the ideal requirements (their resistances R_V are of the order of $1\,\text{M}\Omega\text{--}10\,\text{M}\Omega$).

Ideally (with no voltmeter connected), the voltage across the resistance R in Fig. 7.9 is

$$V_i = IR \tag{7.18}$$

When a voltmeter with a finite resistance R_V is connected across R to measure the voltage drop, the current I splits up such that

$$I = I' + I_m \tag{7.19}$$

and the voltage measured would now become

$$V = I'R \tag{7.20a}$$

$$= IR_{eq} = I\frac{RR_V}{R + R_V} \tag{7.20b}$$

Of course, the value of I itself may be changed, depending on how R_{eq} (instead of R) might affect the rest of the circuit. Equations (7.19) and (7.20), however, explain the loading-effect process. I' is less than I (by the value of I_m) or equivalently, R_{eq} replaces R in the circuit; R_{eq}, being the parallel equivalent of R and R_V, is less than R itself (see the properties of parallel circuits).

It is clear from the discussion above that to reduce the loading effect, I_m must be very small or R_V must be made as large as possible. Examining Eqs. (7.17) reveals that R_V is increased by using a higher-sensitivity (or lower $I_{F.S.}$) meter movement or by using a higher range (V_{max}) to read the voltage. The latter choice may not always be acceptable because of the increased possibility of parallax (reading) errors, especially near the lower end of the scale. The following examples illustrate more clearly the loading effects of voltmeters.

EXAMPLE 7.5

Design a two-range voltmeter; one range to read up to 10 V and the other range to read up to 25 V, using a 1 kΩ meter movement whose full-scale current is 0.25 mA.

Solution The voltmeter is constructed by connecting either one of two multiplier resistances in series with the meter movement, using a two-position switch as shown in Fig. 7.10. Here

$$R_m = 1\,\text{k}\Omega \qquad \text{and} \qquad I_{F.S.} = 0.25\,\text{mA}$$

or

$$S = \frac{1}{I_{F.S.}} = \frac{1}{0.25\,\text{mA}} = 4\,\text{k}\Omega/\text{V}$$

For the 10 V range ($V_{max} = 10$ V), using Eq. (7.16), we have

$$10 = 0.25 \times 10^{-3}(1000 + R_{mult_1})$$

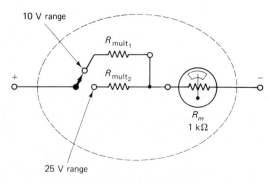

10 V range

25 V range

Figure 7.10

Therefore,

$$R_{\text{mult}_1} = \frac{10}{0.25 \times 10^{-3}} - 1000 = 40,000 - 100 = 39,000\ \Omega = 39\ \text{k}\Omega$$

or equivalently,

$$R_{V_1} = R_m + R_{\text{mult}_1} = \text{voltage range} \times S \quad \text{[from Eq. (7.17b)]}$$
$$1\ \text{k}\Omega + R_{\text{mult}_1} = 10\ \text{V} \times 4\ \text{k}\Omega/\text{V} = 40\ \text{k}\Omega$$

or

$$R_{\text{mult}_1} = 40\ \text{k}\Omega - 1\ \text{k}\Omega = 39\ \text{k}\Omega \quad \text{(as above)}$$

For the 25 V range ($V_{\text{max}} = 25$ V), using Eq. (7.17b), we have

$$R_{V_2} = R_m + R_{\text{mult}_2} = \text{voltage range} \times S$$
$$1\ \text{k}\Omega + R_{\text{mult}_2} = 25\ \text{V} \times 4\ \text{k}\Omega/\text{V} = 100\ \text{k}\Omega$$

Therefore,

$$R_{\text{mult}_2} = 100\ \text{k}\Omega - 1\ \text{k}\Omega = 99\ \text{k}\Omega$$

Using Eq. (7.16) will produce the same result.

EXAMPLE 7.6

If the voltmeter designed in Example 7.5 is used to measure the voltage across the 10 kΩ resistor in the circuit shown in Fig. 7.11, what will be its reading on each of the two ranges? Assume no parallax (reading) error. Compare these voltmeter readings with the ideal value of the voltage across the 10 kΩ resistor.

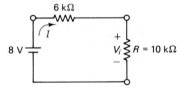

Figure 7.11

Solution For the simple series circuit shown in Fig. 7.11,

$$I = \frac{8}{6\ \text{k}\Omega + 10\ \text{k}\Omega}$$
$$= \frac{8}{16\ \text{k}\Omega} = 0.5\ \text{mA}$$

and

$$V_i = IR$$
$$= 0.5\ \text{mA} \times 10\ \text{k}\Omega = 5.0\ \text{V}$$

When the voltmeter is connected across R to measure the voltage drop, as shown in Fig. 7.12, its reading will be the exact value of the voltage V.

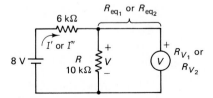

Figure 7.12

However, this circuit is now a series–parallel circuit since the voltmeter is equivalent to a resistance (R_{V_1} or R_{V_2}, depending on the range), and thus V will be different from V_i. With the voltmeter on the 10 V range, $R_{V_1} = 40\,k\Omega$ and

$$R_{eq_1} = \frac{RR_{V_1}}{R + R_{V_1}} = \frac{10\,k\Omega \times 40\,k\Omega}{50\,k\Omega} = 8\,k\Omega$$

Therefore, the source current is:

$$I' = \frac{8}{6\,k\Omega + 8\,k\Omega} = 0.571\,mA$$

and

$$V \text{ (voltmeter reading)} = I'R_{eq_1} = 0.571\,mA \times 8\,k\Omega = 4.57\,V$$

Thus the loading-effect reading error is

$$\text{error (\%)} = \frac{V_i - V}{V_i} \times 100 = \frac{5.0 - 4.57}{5.0} \times 100 = 8.57\%$$

With the voltmeter on the 25 V range, $R_{V_2} = 100\,k\Omega$ and

$$R_{eq_2} = \frac{RR_{V_2}}{R + R_{V_2}} = \frac{10\,k\Omega \times 100\,k\Omega}{110\,k\Omega} = 9.091\,k\Omega$$

Here the source current is:

$$I'' = \frac{8}{6\,k\Omega + 9.091\,k\Omega} = 0.53\,mA$$

and

$$V \text{ (voltmeter reading)} = I''R_{eq_2}$$
$$= 0.53\,mA \times 9.091\,k\Omega = 4.819\,V$$

Therefore, the loading-effect reading error in this case is

$$\text{error (\%)} = \frac{V_i - V}{V_i} \times 100 = \frac{5.0 - 4.819}{5.0} \times 100 = 3.62\%$$

It is thus clear that the voltmeter's reading approaches the ideal value (i.e., causing less loading-effect error) when the reading is taken on the higher voltage range (assuming no parallax error).

EXAMPLE 7.7

A Simpson voltmeter ($S = 1\,k\Omega/V$ and 5.0 V range) and an AVO meter ($S = 20\,k\Omega/V$ and 10 V range) are used to measure the open-circuit voltage of the practical voltage source shown in Fig. 7.13. What will be the reading of each of the voltmeters?

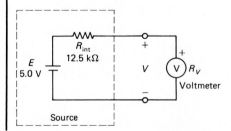

Figure 7.13

Solution The open-circuit voltage of a source (here it is ideally $E = 5.0\,\text{V}$) is measured by connecting a voltmeter across the terminals of the source without any load connected to the source's terminals. This example will show (drastically!) the loading effects of nonideal voltmeters.

For the Simpson voltmeter,

$$R_V = \text{voltage range} \times \text{sensitivity} = 5.0\,\text{V} \times 1\,\text{k}\Omega/\text{V} = 5.0\,\text{k}\Omega$$

The voltmeter will read the actual (exact) value of V, which by the voltage-division principle is:

$$V = E\frac{R_V}{R_V + R_{\text{int}}} = 5.0 \times \frac{5.0\,\text{k}\Omega}{5.0\,\text{k}\Omega + 12.5\,\text{k}\Omega} = 1.43\,\text{V}!$$

(quite different from the expected 5.0 V reading).

For the AVO meter, using its 10 V range, we have

$$R_V = \text{voltage range} \times S = 10 \times 20\,\text{k}\Omega/\text{V} = 200\,\text{k}\Omega$$

Here, also using the voltage-division principle, the voltmeter's reading is

$$V = E\frac{R_V}{R_V + R_{\text{int}}} = 5.0 \times \frac{200\,\text{k}\Omega}{200\,\text{k}\Omega + 12.5\,\text{k}\Omega} = 4.71\,\text{V}$$

(closer to the ideal value of 5.0 V).

More accurate readings are thus obtained with voltmeters having higher values of R_V. Here the higher value of R_V for the AVO meter compared to the Simpson voltmeter is due mainly to its higher sensitivity.

EXAMPLE 7.8

In the circuit shown in Fig. 7.14, the voltmeter used reads 18 V on its 25 V range. Find the sensitivity of this voltmeter. What will this voltmeter read on its 100 V range?

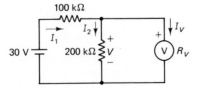

Figure 7.14

Solution The reading of the voltmeter is $V = 18\,\text{V}$ on the 25 V range. For this series-parallel circuit,

$$I_2 = \frac{V}{200\,\text{k}\Omega}$$

$$= \frac{18}{200\,\text{k}\Omega} = 0.09\,\text{mA}$$

The voltage across the 100 kΩ resistor $= 30 - 18 = 12\,\text{V}$; therefore,

$$I_1 = \frac{12}{100\,\text{k}\Omega} = 0.12\,\text{mA}$$

But from KCL,

$$I_1 = I_2 + I_V$$

Therefore,

$$I_V = I_1 - I_2 = 0.12\,\text{mA} - 0.09\,\text{mA} = 0.03\,\text{mA}$$

Hence, using Ohm's law,

$$R_V = \frac{V}{I_V} = \frac{18}{0.03\,\text{mA}} = 600\,\text{k}\Omega$$

$$= \text{voltage range} \times \text{sensitivity}\ (S)$$

Therefore,

$$S\ (\text{sensitivity of the voltmeter}) = \frac{600\,\text{k}\Omega}{25} = 24\,\text{k}\Omega/\text{V}$$

Now, on the 100 V range, the resistance of the voltmeter becomes

$$R_V' = \text{voltage range} \times S = 100\,\text{V} \times 24\,\text{k}\Omega/\text{V} = 2400\,\text{k}\Omega$$

$$= 2.4\,\text{M}\Omega$$

Here the equivalent resistance of the 200 kΩ resistor and the voltmeter (R_V') is

$$R_{\text{eq}} = \frac{200\,\text{k}\Omega \times 2400\,\text{k}\Omega}{200\,\text{k}\Omega + 2400\,\text{k}\Omega} = 184.615\,\text{k}\Omega$$

Using the voltage-division principle, the voltmeter's reading now becomes

$$V = 30 \times \frac{184.615\,\text{k}\Omega}{100\,\text{k}\Omega + 184.615\,\text{k}\Omega} = 19.46\,\text{V}$$

Notice that the ideal voltage should be

$$V_i = 30 \times \frac{200\,\text{k}\Omega}{100\,\text{k}\Omega + 200\,\text{k}\Omega} = 20\,\text{V}$$

that is, the voltmeter's reading on its higher range (100 V range) is much closer to the ideal value than its reading on the lower voltage range (25 V range), as explained previously.

7.4 DESIGN AND CONSTRUCTION OF OHMMETERS

If the meter movement's current I_m is somehow made to be proportional to the value of an unknown resistance to be measured, the meter's scale can be calibrated to read resistance directly. Thus the meter movement could also function as an ohmmeter. Here, however, a voltage source (e.g., a battery) must be added to the meter's circuit to drive the current necessary for the deflection of the pointer. The construction of the ohmmeter involves a circuit which is similar to a combination of an ammeter and a voltmeter. A typical ohmmeter circuit is shown in Fig. 7.15. This is a series–parallel circuit which can easily be analyzed to determine I_m as a function of R_x, the resistance being measured.

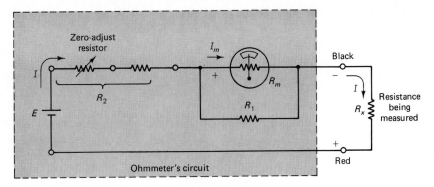

Figure 7.15

The total resistance seen by the voltage source E is

$$R_T = R_x + R_2 + \frac{R_1 R_m}{R_1 R_m} \tag{7.21a}$$

$$= R_x + R_{\text{M.S.}} \tag{7.21b}$$

where the quantity

$$R_2 + \frac{R_1 R_m}{R_1 + R_m}$$

is denoted $R_{\text{M.S.}}$, for reasons to become apparent shortly. The current I flowing in the source and R_x is

$$I = \frac{E}{R_T} = \frac{E}{R_x + R_2 + R_1 R_m/(R_1 + R_m)} = \frac{E}{R_x + R_{\text{M.S.}}} \tag{7.22}$$

The meter movement current I_m can be determined using the current-division principle,

$$I_m = I \frac{R_1}{R_1 + R_m}$$

$$= \frac{E}{R_x + R_{\text{M.S.}}} \times \frac{R_1}{R_1 + R_m} \tag{7.23}$$

Thus

$$I_m \propto \frac{1}{R_x}$$

that is, the meter current is inversely (hyperbolically) proportional to the value of the unknown resistance.

When $R_x = \infty$ (i.e., an open circuit), $I_m = 0$ but when $R_x = 0$ (i.e., a short circuit), I_m should be its maximum value, as the total resistance of the circuit is at its minimum value [see Eqs. (7.22) and (7.23)]. Therefore, $R_x = 0\,\Omega$ should correspond to I_m becoming $I_{\text{F.S.}}$, the full-scale current of the meter movement. Thus from Eq. (7.23),

$$\boxed{I_{\text{F.S.}} = \frac{E}{R_{\text{M.S.}}} \times \frac{R_1}{R_1 + R_m} = \frac{E}{R_2 + R_1 R_m/(R_1 + R_m)} \times \frac{R_1}{R_1 + R_m}} \tag{7.24}$$

The movement scale can thus be calibrated directly in terms of resistance, according to Eq. (7.23), which is called the *calibration equation* of the ohmmeter. Based on the discussion above, the zero-current position corresponds to the infinite-resistance value ($R_x = \infty \, \Omega$), while the full-scale current position corresponds to the zero-resistance value ($R_x = 0 \, \Omega$). This is shown in Fig. 7.16. Thus the ohmmeter (resistance) scale is inverted with respect to the current (or voltage) scale. Also, the inverse proportionality of Eq. (7.23) indicates that the resistance scale is nonlinear, sparse near the $R_x = 0$ position and crowded near the $R_x = \infty$ position.

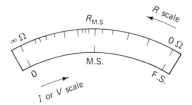

Figure 7.16 Ohmmeter's scale calibration.

If we substitute

$$R_x = R_{\text{M.S.}}$$

in Eq. (7.23), then

$$I_m = \frac{E}{2R_{\text{M.S.}}} \times \frac{R_1}{R_1 + R_m} = \frac{1}{2} I_{\text{F.S.}} \quad \text{[from Eq. (7.24)]}$$

This indicates that the meter movement current will be half its full-scale value when the value of the unknown resistance is $R_{\text{M.S.}}$ In other words, the midscale pointer deflection position corresponds to $R_x = R_{\text{M.S.}}$, as shown in Fig. 7.16. This specific value of resistance, $R_{\text{M.S.}}$, is called the midscale resistance calibration value and is given by

$$R_{\text{M.S.}} = R_2 + \frac{R_1 R_m}{R_1 + R_m} \tag{7.25}$$

as discussed above in conjunction with Eqs. (7.21) and (7.22).

To design an ohmmeter given a specific source E and a meter movement with specified parameters ($I_{\text{F.S.}}$ and R_m), the values of the required resistances, R_1 and R_2, in the ohmmeter circuit of Fig. 7.15 have to be determined. Equation (7.24) by itself is not sufficient because it has two unknowns, R_1 and R_2. Therefore, an extra parameter has to be specified. This is usually the midscale resistance value $R_{\text{M.S.}}$. Thus Eqs. (7.24) and (7.25) are the required design equations, providing two equations in two unknowns (R_1 and R_2), which can be solved.

Referring to Fig. 7.15, one should note that

1. The battery is connected so that it drives the meter movement current I_m into the positive (red) terminal of the movement, in order to cause an upscale deflection of the pointer. Therefore, the current in the external resistance flows from the negative (black) terminal toward the positive (red) terminal, as shown, due to the battery polarity connections.

2. R_2 is a series connection of two resistors: one of them fixed and the other one (about 20% of R_2) variable. The variable resistor is called the *zero-adjust resistor*. When R_x is zero (i.e., a short circuit is connected across the ohmmeter's terminals), the pointer should deflect all the way up to the full-scale position [see Eq. (7.24)]. Because the battery voltage E does decay (change) with time, readjustment of the value of R_2, through the zero-adjust resistor, is necessary to compensate for this battery voltage change. This is usually an initial checking procedure to ensure that the ohmmeter would function properly; otherwise, the source should be replaced by a new battery.

■ EXAMPLE 7.9

Given a meter movement with a sensitivity of $10 \text{ k}\Omega/\text{V}$ and an internal resistance of $1 \text{ k}\Omega$, design an ohmmeter with a midscale reading of $1000 \, \Omega$ using a 10 V battery. Calibrate the scale of the ohmmeter, showing at least five resistance readings. What is the highest possible midscale reading of such an ohmmeter?

Solution   Here the parameters of the meter movement are

$$R_m = 1000 \, \Omega$$

$$I_{\text{F.S.}} = \frac{1}{S} = \frac{1}{10 \text{ k}\Omega} = 0.1 \text{ mA}$$

Using Eq. (7.24), with $E = 10 \text{ V}$ and $R_{\text{M.S.}} = 1000 \, \Omega$, yields

$$0.1 \times 10^{-3} = \frac{10}{1000} \times \frac{R_1}{R_1 + 1000}$$

$$R_1 + 1000 = 100R_1$$

Therefore,

$$R_1 = \frac{1000}{99} = 10.1 \, \Omega$$

From Eq. (7.25),

$$1000 = R_2 + \frac{10.1 \times 1000}{1010.1} = R_2 + 10$$

Therefore,

$$R_2 = 990 \, \Omega$$

The two resistors required for the ohmmeter circuit of Fig. 7.15 are thus determined. Referring to the calibration equation of the ohmmeter [Eq. (7.23)], we have

$$I_m = \frac{10}{R_x + 1000} \times \frac{10.1}{1010.1} = \frac{0.1}{R_x + 1000}$$

The following calibration table can thus be calculated:

R_x (Ω)	0	250	500	1000	2000	4000	∞
I_m (mA)	0.1(F.S.)	0.08	0.067	0.05	0.033	0.02	0

Figure 7.17 shows a schematic of this ohmmeter's scale.

Figure 7.17

Equation (7.24) can be rewritten as

$$R_{M.S.} = \frac{E}{I_{F.S.}} \times \frac{R_1}{R_1 + R_m}$$

In this equation, E, $I_{F.S.}$, and R_m are all fixed quantities. The design parameter available, which can be changed, is the value of R_1. $R_{M.S.}$ would be its highest possible value if R_1 is made to be ∞ (i.e., an open circuit, or R_1 removed from the circuit; see Fig. 7.15), because then

$$\frac{R_1}{R_1 + R_m} = \frac{1}{1 + R_m/R_1} = 1$$

and

$$R_{M.S.}(\text{max}) = \frac{E}{I_{F.S.}} = \frac{10}{0.1 \times 10^{-3}} = 100 \text{ k}\Omega$$

Here, using Eq. (7.25), we have

$$100 \text{ k}\Omega = R_2 + \frac{R_1 R_m}{R_1 + R_m} = R_2 + R_m = R_2 + 1 \text{ k}\Omega$$

Therefore,

$$R_2 = 99 \text{ k}\Omega$$

EXAMPLE 7.10

What is the voltage of the battery needed to build an ohmmeter with a midscale reading of $800 \, \Omega$ using a meter movement whose resistance is $400 \, \Omega$ and full-scale current is $1 \, \text{mA}$? The resistance in series with the battery is $600 \, \Omega$. Using this battery and meter movement, redesign the ohmmeter to have a midscale reading of $80 \, \Omega$ only.

Solution   Referring to Fig. 7.15, R_1 is unknown and R_2 is $600 \, \Omega$. Since R_m is $400 \, \Omega$ and $R_{M.S.}$ is required to be $800 \, \Omega$, then from Eq. (7.25),

$$R_{M.S.} = R_2 + \frac{R_1 R_m}{R_1 + R_m}$$

$$800 = 600 + \frac{400 R_1}{400 + R_1}$$

$$200 = \frac{400 R_1}{400 + R_1}$$

$$400 + R_1 = 2R_1$$

Therefore,

$$R_1 = 400 \, \Omega$$

Now using Eq. (7.24), we obtain

$$I_{\text{F.S.}} = \frac{E}{R_{\text{M.S.}}} \times \frac{R_1}{R_1 + R_m}$$

$$1 \times 10^{-3} = \frac{E}{800} \times \frac{400}{800}$$

Then

$$E = 1 \times 10^{-3} \times 800 \times 2 = 1.6 \text{ V}$$

that is, the needed battery should have a voltage of 1.6 V.

To design the ohmmeter such that the midscale reading is only $80 \, \Omega$, using the same meter movement and battery, Eq. (7.24) is used first to determine R_1. Here

$$1 \times 10^{-3} = \frac{1.6}{80} \times \frac{R_1}{R_1 + 400}$$

$$R_1 + 400 = \frac{1.6 \times 10^3}{80} R_1 = 20R_1$$

Therefore,

$$R_1 = \frac{400}{19} = 21.05 \, \Omega$$

Equation (7.25) can now be used to determine the required value of R_2:

$$80 = R_2 + \frac{21.05 \times 400}{421.05} = R_2 + 20$$

Therefore,

$$R_2 = 60 \, \Omega$$

In studying and testing devices, networks, and systems, signals that vary with time (usually periodic in nature) are often encountered. The generation of ac sinusoidal waveforms is examined in Chapter 11 and the following chapters discuss the analysis of networks with such time-varying excitation sources. The oscilloscope has become a universal instrument, an integral part of electronics and physical sciences laboratories, because of its usefulness in the examination and the analysis of time-varying signals.

The oscilloscope is, in fact, a voltage-sensing voltmeter. It can be used to measure dc or ac current or voltage. Instead of the mechanical deflection of a metal pointer, the oscilloscope relies on the deflection of an electron beam against a fluorescent screen to indicate the magnitude of the quantity being measured. Because the plane of the screen is two-dimensional (x-y grid), the oscilloscope provides an extra degree of freedom in the measurement system. Thus the x axis and the y axis can be used to study the variation of one voltage as a function of another, or if the x axis is converted into time, as discussed below, the oscilloscope can then display voltage magnitude (y) as a function of time (x). Some typical standard laboratory-type scopes are shown in Fig. 7.18.

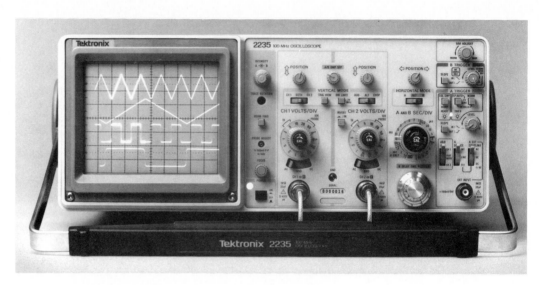

(a)

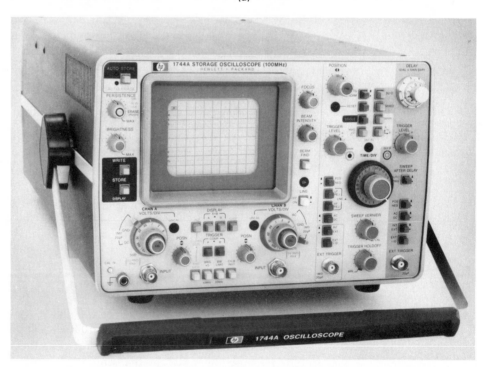

(b)

Figure 7.18 (a) Dual-channel 100-MHz oscilloscope; (b) dual-channel 100-MHz storage scope. [(a) Courtesy of Tektronix Canada Inc.; (b) courtesy of Hewlett-Packard (Canada) Ltd.]

The cathode ray tube (CRT) is the heart of the scope. A schematic diagram showing its structure and main components is shown in Fig. 7.19. The face plane of the CRT screen, showing the axis grid, is illustrated schematically in Fig. 7.20.

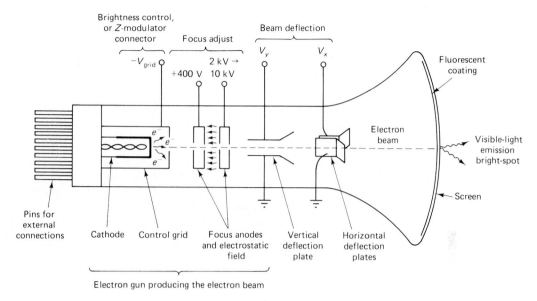

Figure 7.19

A brief discussion of the operating principles and use of a typical standard laboratory-type oscilloscope to display time-varying waveforms is presented here. The two main functions of the cathode ray tube are examined below.

1. *The electron gun*. This provides a sharply focused electron beam directed toward the fluorescent-coated screen. The thermally heated cathode emits electrons in many directions. The control grid provides an axial direction for the electron beam and controls the number and speed of electrons in the beam.

The momentum of the electrons (their number × their speed) determines the intensity, or brightness, of the light emitted from the fluorescent coating due to the electron bombardment. The light emitted is usually of the green

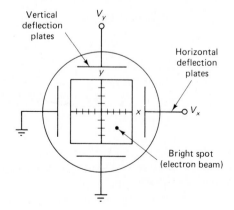

Figure 7.20

wavelength. Because electrons are negatively charged, a repulsion force is created by applying a negative voltage to the control grid, to adjust their number and speed. This negative control voltage can be made variable; a more negative voltage results in less number of electrons in the beam and hence decreased brightness of the beam spot. Most oscilloscopes provide an external connection, usually in the back of the scope, called the z-modulation terminal. This terminal is connected to the control grid. Thus the brightness of the trace generated by the electron beam can be externally controlled by connecting such an external control voltage to the z-modulation terminal.

Since the electron beam consists of many electrons, the beam tends to diverge. This is because the similar (negative) charges on the electrons repulse each other. To compensate for such repulsion forces, an adjustable electrostatic field is created between two cylindrical anodes, called the focusing anodes. The variable positive voltage on the second anode cylinder is therefore used to adjust the focus or sharpness of the bright beam spot.

2. *The deflection system.* This consists of two pairs of parallel plates, referred to as the vertical and horizontal deflection plates. One of the plates in each set is permanently connected to ground (0 V), while the other plate of each set is available through an external temination, called the Y input or the X input. Usually, an adjustable-gain amplifier stage (inside the scope) connects the external termination to the appropriate deflection plate (see Fig. 7.22).

As shown in Fig. 7.19, the electron beam passes through these plates. In reference to the schematic diagram in Fig. 7.20, a positive voltage applied to the Y-input terminal (V_y) causes the electron beam to deflect vertically upward, due to the attraction forces, while a negative voltage applied to the Y-input terminal will cause the electron beam to deflect vertically downward, due to the repulsion forces. Similarly, a positive voltage applied to X-input terminal (V_x) will cause the electron beam to deflect horizontally toward the right, while a negative voltage applied to the X-input terminal will cause the electron beam to deflect horizontally toward the left of the screen. The amount of vertical or horizontal deflection is directly proportional to the correspondingly applied voltage.

The face of the screen can be considered as an x-y plane. The (x, y) position of the bright spot is thus directly influenced by the horizontal and the vertical voltages applied to the deflection plates, V_x and V_y, respectively:

$$x = k_x V_x \quad \text{cm (or divisions)} \tag{7.26a}$$

and

$$y = k_y V_y \quad \text{cm (or divisions)} \tag{7.27a}$$

or

$$V_x = \frac{1}{k_x} x \tag{7.26b}$$

and

$$V_y = \frac{1}{k_y} y \tag{7.27b}$$

The quantities $1/k_x$ and $1/k_y$ in V/div (or V/cm) are called the *horizontal* and *vertical sensitivities,* respectively. Their values are selectable (and adjustable) through multipositional switches that control the gain of the corresponding amplifier stage. Some scopes have extra multipliers, $\times 10$ for y and $\times 5$ for x, that can be used to magnify the display by such factors (i.e., the display is 10 times or 5 times the real deflections, respectively). The bright spot of the electron beam can thus trace (or plot) the x-y relationship between the two voltages: V_x and V_y.

7.5.1 Display of Variations in Amplitude with Respect to Time

In numerous applications it will be required to display the voltage (or the current) as a function of time. By applying such a voltage to the Y input, the vertical deflection of the beam spot will be proportional to the magnitude of this voltage. It is then necessary to convert the x axis (horizontal deflection) into a time axis. A special unit inside the oscilloscope, called the *sweep generator,* provides a periodic voltage waveform that varies linearly with time, as shown in Fig. 7.21(a). This type of voltage is called the *sawtooth waveform.*

When such a voltage is connected to the horizontal deflection plates inside the scope, then during the trace time T_r,

$$V_x = mt \tag{7.28a}$$

where

$$m = \text{slope of the line} = \frac{V_s}{T_r} \tag{7.28b}$$

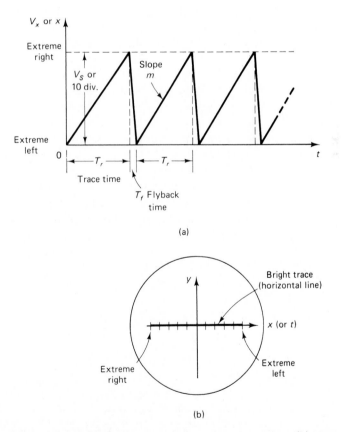

Figure 7.21 (a) Sawtooth waveform from sweep generator; (b) trace of bright spot corresponding to V_x.

Therefore, the corresponding horizontal deflection, x, from Eq. (7.26a), is

$$x = k_x V_x = k_x mt = \frac{k_x V_s}{T_r} t \tag{7.29a}$$

$$= ct \tag{7.29b}$$

where

$$c = \frac{k_x V_s}{T_r} = \frac{10 \text{ div}}{T_r} \tag{7.29c}$$

or

$$t = \frac{1}{c} x = \frac{T_r}{10 \text{ div}} x = (\text{time/div}) \times x \tag{7.30}$$

When V_x is zero volts, x is zero and the bright spot is at the extreme left-hand position, corresponding to the time $t = 0$. At the end of the trace time, $t = T_r$, V_x is V_s and the bright spot is at the extreme right-hand position, corresponding to a deflection of 10 horizontal divisions. The constant $1/c$ is called the time/div (or time/cm) calibration. It can be changed by changing the slope m, or equivalently the trace time T_r, as indicated in Eq. (7.29c). A multipositional switch available on the control panel of the scope (called the time base) can be used to select the appropriate time/div (i.e., the trace time of the sawtooth waveform).

During the flyback time T_f [Fig. 7.21(a)], which is usually very short compared to T_r, a high negative voltage pulse is applied to the control grid of the electron gun to blank the bright spot of the beam during this time. This prevents any reverse retrace (or shadow) as the beam is going back to the extreme left-hand position to start a new trace cycle (the next trace-time period).

Thus for a selected trace time T_r, the bright spot moves horizontally across the face of the screen, along the x-axis from left to right, with a constant speed, restarts again from the left, and repeats such traces. Depending on the speed of the bright spot and the persistence of vision, the trace produced by the spot will look like a horizontal straight line, as shown in Fig. 7.21(b). According to Eq. (7.30), the horizontal axis is now converted into a time axis.

The basic system units of an oscilloscope are shown in the block diagram of Fig. 7.22. Standard laboratory scopes have an input resistance (corresponding to R_V in the case of a voltmeter) of $1 \text{ M}\Omega$. The vertical amplifier provides a uniform adjustable gain for ac signals with frequencies (see Chapter 11) in the range 0 (or dc) to 10 MHz (or sometimes up to 50 MHz).

When a periodically variable voltage [e.g., the ac sinusoidal waveform $v(t)$ shown in Fig. 7.23] is applied to the Y-input terminal of the scope, the bright spot of the beam will move vertically up and down with a displacement y that depends on the V/div setting of the vertical amplifier gain control and the peak-to-peak voltage of this waveform. Here

$$V_{\text{pp}} = 2V_m = y \text{ (no. of vertical divisions)} \times \text{V/div setting of } Y \tag{7.31}$$

To display this voltage as a time function on the screen of the oscilloscope as shown in Fig. 7.23, the horizontal deflection x is made to be proportional to the time t using the sawtooth waveform of the sweep generator, as explained above. For the sinusoidal waveform shown in Fig. 7.23, and according to Eq. (7.30),

$$\text{period } T = x \text{ (no. of horizontal divisions for one complete cycle)}$$

$$\times \text{ time/div setting of the scope} \tag{7.32}$$

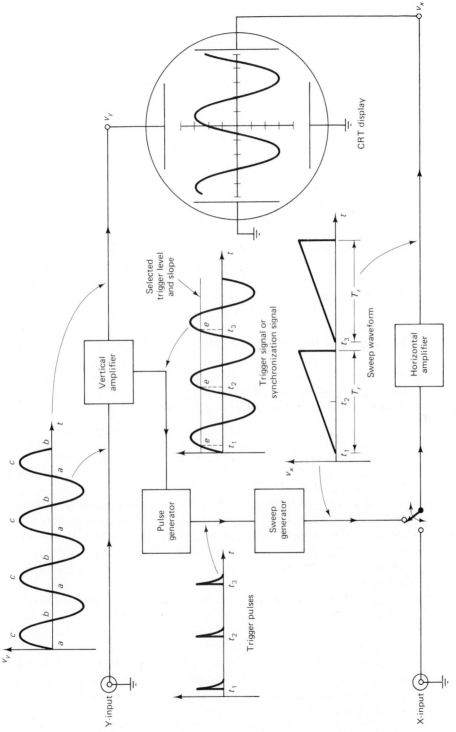

Figure 7.22 Oscilloscope block diagram and typical waveforms.

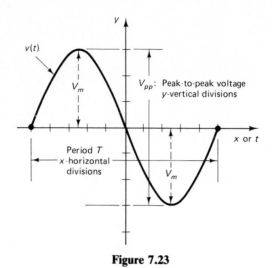

Figure 7.23

The two parameters of a sinusoidal voltage waveform, V_{pp} and T, can thus be measured using the scope as explained here.

7.5.2 Triggering and Synchronization of the Oscilloscope Display

In order that the appropriate measurements of the parameters of a certain waveform can be made, such a waveform must appear as a stationary display on the screen of the oscilloscope. For this to occur, an integral number n $(1, 2, 3, \ldots$ etc.) of the complete cycles of the waveform should be displayed during each trace period T_r. In other words,

$$T_r = nT \quad (T \text{ is the period of the waveform}) \qquad (7.33)$$

This situation implies that when the next trace period occurs, the bright spot retraces the next number of cycles, n, right on top of the previous display—hence the stationary appearance of the display.

The slope and the position of the "similar" starting points of the screen display are chosen by appropriate control knobs, called the $+$, 0, $-$, trigger switch, and level control. As shown in Fig. 7.22, these similar starting points are like the points denoted a (positive slope), b (negative slope), or c (zero slope). The function of the level-control knob, which controls the voltage of the starting point, will be examined below.

The condition specified in Eq. (7.33) is not always achievable. In such situations, the display appears as a moving waveform and is usually inde-cipherable. To obtain stable and stationary waveform displays always, the sweep generator should produce the trace periods T_r only in synchronism with the waveform being displayed. This synchronization process can be done either manually or automatically. In the "auto" mode, a signal (called the trigger signal) proportional to that being displayed, v_y, is applied to the pulse generator as shown in Fig. 7.22.

Depending on the selected level and slope of the starting point of the display (e.g., point e in Fig. 7.22), the output of the pulse generator will consist of narrow trigger pulses at the times t_1, t_2, t_3, and so on, separated from each other by one period T. Each time $(t_1, t_2, t_3,$ etc.) this triggering or synchronization signal crosses a preselected level (and a preselected slope), the pulse generator emits one narrow trigger pulse. The emitted pulse triggers the sweep generator to begin producing one cycle of the sweep waveform; its duration is the trace period T_r, which is controlled by the time/div setting. At the end of each sweep cycle, the sweep generator stops its output and awaits the arrival of the next trigger pulse before producing a new sweep cycle.

Notice that if the sweep generator receives a trigger pulse during its sweep cycle (i.e., during the trace period T_r), it will simply ignore the pulse and continue with the completion of its sweep cycle. The trigger pulse received after the completion of the trace period will initiate the new sweep cycle. This allows the scope to display more than one cycle, of period T, of the signal connected to its vertical deflection plates. In Fig. 7.22 the trace period starts at t_1, but because the trigger pulse at t_2 occurs while the sweep cycle (T_r) is being generated, this pulse is ignored. The new trace period is then generated at t_3, which occurs after the completion of T_r. This allows approximately two signal cycles to be displayed. When the new trace period is started at t_3, the signal level is now at the next starting point, e, and the new trace is right on top of the previous trace. Thus synchronization and stable displays are achieved.

■ EXAMPLE 7.11

The two waveforms shown in Fig. 7.24 were displayed by a two-beam scope (two vertical inputs: Y_A and Y_B, but one time base) on the CRT. The controls were adjusted as follows:

$$\text{time/div} = 0.5 \text{ ms} \quad (\times 5 \text{ mag.})$$

$$\text{V/div } Y_A = 2 \text{ V} \quad (\times 10 \text{ mag.})$$

$$\text{V/div } Y_B = 1 \text{ V} \quad (\times 1 \text{ mag.})$$

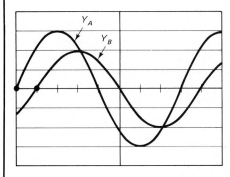

Figure 7.24

Find
(a) The period of these waveforms.
(b) The peak-to-peak voltage of each waveform.
(c) The phase shift in degrees between the two waveforms.

Solution (a) According to Eq. (7.32), the period T can be calculated, but we have to note that the x distance (i.e., no of divisions) is actually magnified $\times 5$.

Therefore,

$$T = x \text{ (no. of divisions of one complete cycle)} \times \text{(time/div)} \times \tfrac{1}{5}$$
$$= 8 \times 0.5 \times 10^{-3} \times \tfrac{1}{5}$$
$$= 0.8 \times 10^{-3} \, \text{s}$$
$$= 0.8 \, \text{ms}$$

(b) Again noting the magnification of the particular vertical distance (or no. of divisions) and using Eq. (7.31), we have

$$V_{pp}(Y_A) = 6 \times 2 \times \tfrac{1}{10} = 1.2 \, \text{V}$$

and

$$V_{pp}(Y_B) = 4 \times 1 = 4.0 \, \text{V}$$

(c) One period or one complete cycle is 360°. In this example, one complete cycle occupies eight divisions (x no. of divisions horizontally), as seen from Fig. 7.24. Therefore,

$$1 \, \text{div} = \frac{360°}{8} = 45°$$

Since the displacement between the two starting points of the two waveforms, which are circle marked in Fig. 7.24, is 1 div,

the phase shift between the two waveforms $= 1 \times 45° = 45°$.

DRILL PROBLEMS

Section 7.1
The following two meter movements will be used in the following drill exercises.

Meter movement A: $I_{\text{F.S.}} = 0.2 \, \text{mA}$, $R_m = 500 \, \Omega$
Meter movement B: $I_{\text{F.S.}} = 50 \, \mu\text{A}$, $R_m = 1000 \, \Omega$

7.1. Determine the sensitivities of the two meter movements given above.

7.2. What is the voltage across each of these two meter movements when the pointers deflect to the full-scale position?

Section 7.2
7.3. Design an ammeter to read up to 10 mA using the meter movement A. What is the resistance of this ammeter?

7.4. Design an ammeter to read up to 2 mA using the meter movement B. What is the resistance of this ammeter?

7.5. If the ammeter designed in Drill Problem 7.3 is used to measure the current in the circuit shown in Fig. 7.25, what will be its reading? Calculate the loading effect error in this case.

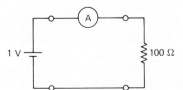

Figure 7.25

Section 7.3

7.6. Design a voltmeter to read up to 10 V using the meter movement A. What is the resistance of this voltmeter?

7.7. If the voltmeter designed in Drill Problem 7.6 is used to measure the voltage V in the circuit shown in Fig. 7.13, what will be its reading? Calculate the loading effect error in this case.

7.8. Design a voltmeter to read up to 20 V using the meter movement B. What is the resistance of this voltmeter?

7.9. If the voltmeter designed in Drill Problem 7.8 is used to measure the voltage V in the circuit shown in Fig. 7.14, what will be its reading? Calculate the loading effect error in this case.

Section 7.4

7.10. Given a 5 V battery and the meter movement A, design an ohmmeter with a midscale reading of 1000 Ω.

7.11. Repeat Drill Problem 7.10 but using the meter movement B.

7.12. Given a 10 V battery and the meter movement B, design an ohmmeter with a midscale reading of 5000 Ω.

PROBLEMS

Section 7.2

(i) **7.1.** Design a milliammeter with a 0.5 mA range using a meter movement with a sensitivity of 10 kΩ/V and an internal resistance of 1000 Ω. What is the resistance of this ammeter?

(i) **7.2.** A 200 Ω, 2 mA full-scale current movement is to be used to build an ammeter capable of measuring currents up to 100 mA on one range and 250 mA on the other range. Find the values of the required shunt resistances and the total resistance of the ammeter in each of these two ranges.

(ii) **7.3.** A 5 kΩ/V movement with a 100 Ω internal resistance is used to construct an ammeter to read currents up to 10 mA.
What is the value of the required shunt resistance?
What is the total resistance of this ammeter?
When the pointer is deflected to midscale value, determine the current in
(a) The movement.
(b) The shunt resistance.
(c) The circuit in which this ammeter is connected.

(ii) **7.4.** A certain ammeter shows half-scale deflection when connected in a circuit which has a 2.5 mA current flow. The resistance of this ammeter is 120 Ω, and the resistance of the meter movement used in its construction is 300 Ω. What is the sensitivity of this meter movement? What is the value of the shunt resistance?

(ii) **7.5.** The meter movement used in constructing the ammeter shown in Fig. 7.25 has a sensitivity of 1 kΩ/V and an internal resistance of 54 Ω. The resistance of the shunt element used is 6 Ω. What is the resistance of the ammeter? What is the maximum current that can be read by this ammeter? What will be the reading of this ammeter in the circuit shown in Fig. 7.25? Find the percentage error compared with the ideal current in this circuit due to the loading effect of this ammeter.

(ii) **7.6.** Find the readings of an ideal ammeter when connected in the circuit shown in Fig. 7.26 to measure I_S and I. If the ammeter designed in Problem 7.1 is used to measure these two currents, what will be its readings? Evaluate the loading-effect error of this ammeter.

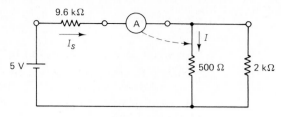

Figure 7.26

Section 7.3

(i) **7.7.** Find the value of the multiplier resistance required to construct a voltmeter using a $2\,k\Omega/V$, $500\,\Omega$ meter movement if the full-scale (maximum) voltage is
   (a) $5\,V$.
   (b) $10\,V$.
   (c) $50\,V$.

(i) **7.8.** Design a two-range voltmeter to read up to a maximum voltage of $25\,V$ on one range and $100\,V$ on the other range, using a $1000\,\Omega$ meter movement whose full-scale current is $2\,mA$.

(ii) **7.9.** A voltmeter with a sensitivity of $2\,k\Omega/V$ and a range of $100\,V$ is used to measure the voltages across each of R_1 and R_2 in the circuit shown in Fig. 7.27. What will be the two readings of this voltmeter? Compare these readings with the ideal voltages across R_1 and R_2.

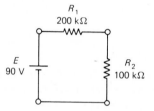

Figure 7.27

(ii) **7.10.** Design a $10\,V$-range voltmeter using a meter movement whose full-scale current is $1\,mA$ and whose resistance is $2.0\,k\Omega$. What would this voltmeter read when connected across the $10\,k\Omega$ resistor in the circuit shown in Fig. 7.28? What is the ideal value of the voltage across the $10\,k\Omega$ resistor? Evaluate the percentage error of the loading effect of this voltmeter.

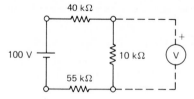

Figure 7.28

(i) **7.11.** Given a meter movement with the following parameters

$$I_{\text{F.S.}} = 500\,\mu A$$

$$V_{\text{F.S.}} = 50\,mV$$

design
   (a) A voltmeter with a $10\,V$ range.
   (b) An ammeter with a $10\,mA$ range.
   (c) An ohmmeter with a midscale reading of $1500\,\Omega$ using a $1.5\,V$ battery.

(i) **7.12.** The voltmeter and ammeter designed in Problem 7.11 are to be used to measure a nominal 2 kΩ resistance by the VA method (i.e., connected to measure the voltage across and the current through this resistance when connected to source E, and then using Ohm's law). Find the appropriate circuit connections you would make to minimize the errors of the measurements.

(ii) **7.13.** A 25 V-range voltmeter reads 15 V when connected to measure the voltage across the 250 kΩ resistor in the circuit shown in Fig. 7.29. What is the sensitivity of this voltmeter?

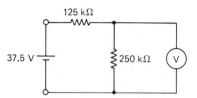

Figure 7.29

(i) **7.14.** **(a)** What is the ideal voltage across the 400 kΩ resistor in the circuit shown in Fig. 7.30?

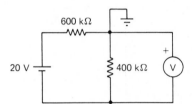

Figure 7.30

(b) Find the measured voltage across the 400 kΩ resistor if an AVO meter is used having a sensitivity of 20 kΩ/V on its dc ranges when the range selector is set at
  (i) 10 V.
  (ii) 100 V.
(c) Find the equivalent resistance of the AVO meter, R_V, in the dc range
  (i) 10 V.
  (ii) 100 V.

(ii) **7.15.** The voltmeter in the circuit shown in Fig. 7.31 has a sensitivity of 1 kΩ/V. It reads 7.5 V on its 10 V range. What is the value of E?

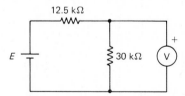

Figure 7.31

(i) **7.16.** A meter movement has a 10 kΩ/V sensititity and a 1 kΩ internal resistance.
  (a) Design an ammeter with a 5 mA range.
  (b) What is the percentage error of this meter if a 15 Ω shunt resistor is used?
  (c) Design a voltmeter with a 5 V range.
  (d) What is the percentage error of this meter if a 40 kΩ multiplier resistance is used?

(i) **7.17.** Using the meter movement in Problem 7.16 and a 6 V battery:

    (a) Design an ohmmeter with a midscale reading of 1 kΩ.

    (b) Design an ohmmeter with a midscale reading of 10 kΩ.

    (c) What is the deflection of each of the ohmmeters designed in parts (a) and (b) in percent of the full scale, if used to measure a 5 kΩ resistor?

Section 7.5

(i) **7.18.** The two waveforms shown in Fig. 7.32 were displayed on the CRT of a two-beam oscilloscope when the controls were adjusted as follows:

$$\text{time/div} = 0.2\,\text{ms} \quad (\times 1\,\text{mag.})$$
$$\text{V/div } Y_A = 5\,\text{V} \quad (\times 10\,\text{mag.})$$
$$\text{V/div } Y_B = 0.2\,\text{V} \quad (\times 1\,\text{mag.})$$

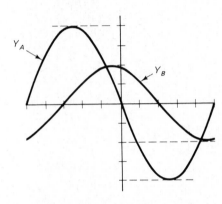

Figure 7.32

Find

(a) The period of the waveforms.

(b) The peak-to-peak voltage of each waveform.

(c) The phase shift in degrees between the two waveforms.

(i) **7.19.** The oscilloscope display shown in Fig. 7.33 was obtained when the controls were adjusted to

$$\text{time/div} = 0.5\,\text{ms} \quad (\times 5\,\text{mag.})$$
$$\text{V/div } Y_A = 10\,\text{V} \quad (\times 1\,\text{mag.})$$
$$\text{V/div } Y_B = 2\,\text{V} \quad (\times 1\,\text{mag.})$$

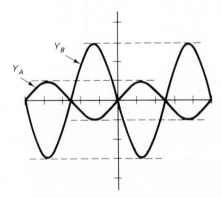

Figure 7.33

Find

(a) The period of the waveforms, T.

(b) The trace period T_r.

(c) V_{pp} for Y_A and Y_B.

(d) The phase shift in degrees between the two waveforms.

(i) **7.20.** For the display on the CRT of a scope, as shown in Fig. 7.34, the controls were adjusted to

$$\text{V/div } Y_A = 0.5 \text{ V} \quad (\times 1 \text{ mag.})$$

$$\text{V/div } Y_B = 0.1 \text{ V} \quad (\times 10 \text{ mag.})$$

$$\text{time/div} = 0.1 \text{ ms} \quad (\times 1 \text{ mag.})$$

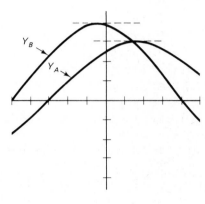

Figure 7.34

Find

(a) The period of the waveforms.

(b) The peak-to-peak voltage of each waveform.

(c) The phase shift in degrees between the two waveforms.

GLOSSARY

Ammeter: The instrument used to measure the value of the electric current flowing in any element (branch) of an electric network. The ammeter is always connected in series with the element through which the current is to be measured.

Calibration Equation: The relationship between the response of a meter and the electrical variable being measured.

Cathode Ray Tube (CRT): This is the heart of an oscilloscope. It is a vacuum tube in which the electron beam is generated, deflected, controlled, focused, and eventually displayed as a bright spot on the screen face of the tube.

D'Arsonval Meter Movement: The basic building block of the current-sensing type of electrical measurement instruments. Its operation is based on the electromagnetic phenomenon.

Deflection Plates: The two sets of horizontal and vertical plates, to which external voltages are applied in order to cause the deflection of the electron beam in a CRT.

Electromagnetic Phenomenon: The physical phenomenon relating to the inter-action of electricity and magnetism. It indicates that an electric current flow produces a magnetic field, and the motion of a conductor in a magnetic field produces an electrical current (and voltage).

Electron Gun: The portion of the CRT that produces a sharp and focused electron beam.

Full-Scale Current: The amount of electric current that results in the full-scale deflection of the pointer in a current-sensing type of meter movement.

Loading Effect: The error in measurement caused by the disturbances in the value of the currents and voltages in a circuit, due to the nonideal nature of the measurement device, particularly its internal resistance.

Meter Movement Resistance: The resistance of the wire-wound moving coil of the current-sensing meter movement.

Multiplier Resistance: A resistor connected in series with the meter movement in the construction of a voltmeter. Its value determines the maximum voltage that can be measured by the voltmeter (i.e., the scale range setting).

Ohmmeter: The instrument used to measure the value of the resistance of an electric device.

Oscilloscope: The oscilloscope is basically a voltage-sensing electronic voltmeter. It is used to display, study, and measure the parameters of any time-varying signal waveform.

Shunt Resistance: A resistance connected in parallel with the meter movement in the construction of an ammeter. Its value determines the maximum current that can be measured by the ammeter (i.e., the scale range setting).

Sweep Generator: One of the electronic subsystems of the oscilloscope. This unit (stage) generates a periodic voltage waveform that is linearly proportional to time, called the sawtooth waveform.

Voltmeter: The instrument used to measure the value of the voltage across any element (between two nodes) in an electric network. The voltmeter is always connected in parallel with the element across which the voltage is to be measured.

ANSWERS TO DRILL PROBLEMS

7.1. $S_A = 5 \, \text{kV}/\Omega$, $S_B = 20 \, \text{kV}/\Omega$

7.3. $R_{sh} = 10.204 \, \Omega$, $R_A = 10 \, \Omega$

7.5. I (measured) = 9.091 mA, loading effect error = 9.09%

7.7. $V = 4 \, \text{V}$, loading effect error = 20%

7.9. $V = 17.143 \, \text{V}$, loading effect error = 14.29%

7.11. $R_1 = 10.101 \, \Omega$, $R_2 = 990 \, \Omega$

8 | CAPACITANCE AND TRANSIENTS IN *RC* CIRCUITS

OBJECTIVES

- Familiarity with the physical parameters affecting the value of a capacitor, the various types of practical capacitors and the amount of charge and energy that can be stored in capacitors.
- Ability to calculate the equivalent value of capacitance and the charge distribution in series, parallel, and series-parallel connections of capacitors.
- Understanding the meaning and application of Ohm's law for a capacitor: current proportionality to the rate of change of voltage.
- Understanding the physical process of the charging and discharging cycles in RC dc circuits, and the concept of the time constant associated with this physical phenomena.
- Ability to calculate, analytically and graphically, the voltages, currents, charge, and the rate of change of these circuit variables in various examples of charging and discharging in RC dc circuits.
- Ability to derive the initial and steady state conditions in RC dc circuits and thus use the analytical expression for the generalized form of the transient voltage across a capacitor in such circuits.

As has been examined previously, linear resistors obey Ohm's law,

$$V = RI$$

They are therefore characterized by the fact that the voltage across each of them is directly proportional to the magnitude of the current flowing in each of them.

In this chapter another linear passive circuit element is examined. It is called the capacitor. As shown here, the current in this element is not determined by the voltage across the element, but by *the rate of change* of this voltage with respect to time. This element is also different from a resistor in that it can *store energy* (i.e., has a memory), as opposed to a resistor, which spends energy or dissipates power. The electronic flash circuit is an example of the practical application of a capacitor. Here energy is stored in the capacitor until needed later to power the flashlight. This process is called the charging of a capacitor from a source, and then discharging it in a load. It is studied in detail in this chapter.

8.1 CAPACITORS AND CAPACITANCE

The simplest form of a capacitor consists of two conducting (metal) plates separated by an insulating material such as air, paper, mica, or ceramic, as shown in Fig. 8.1. The electric circuit symbol of an ideal capacitor is shown in Fig. 8.2. The resemblance between the physical capacitor and its symbol is apparent.

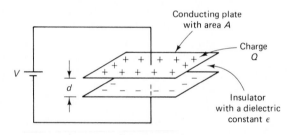

Figure 8.1

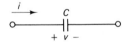

Figure 8.2

The most important property of such a device is its ability to store electric charge (Q) and consequently electric energy. When a potential energy source, for example V in Fig. 8.1, is connected to the terminals of a capacitor, some of the electrons from the top metal plate will be attracted by the positive terminal of the source, which is connected to this plate. The same number of electrons (equivalent to the electric charge) will be repelled by the negative terminal of the source and will thus accumulate on the bottom metal plate. The result of this process is that the top plate will be deficient of electrons and hence positively charged, while the bottom plate will acquire excess electrons (the same number that left the top plate) and will then become negatively charged. This motion of electric charge takes place through the conducting wires and the source (resembling an energy pump) but not through the insulating material separating the plate. In fact, the capacitor's plates resemble an open switch.

When additional electrons try to go through the same process, the already positively charged top plate will try to attract them, whereas the already negatively charged bottom plate will try to repulse them. The potential energy source (battery V) will have to spend energy in this process of charge separation. Eventually, a balance will be established between the potential energy supplied by the source and the attraction and repulsion forces exhibited by the steady-state amount of charge, Q, accumulating on each of the two plates. Evidently, this charging process takes time and as a result, "capacitors cannot be charged instantly." This is a very important remark that must never be forgotten.

It is also clear from this discussion that an increase in the magnitude of the energy source, V, will allow more work to be done in separating the charges on the metal plates of the capacitor, resulting in an increase in the amount of charge, Q (of equal magnitude and opposite polarity), accumulating on the plates. The capacitor is said to "store" the electric charge Q, when the voltage source V is applied to it. One can also say that the capacitor is "charged" to a voltage V. The amount of charge stored, Q, is directly proportional to the capacitor's voltage V. Thus

$$Q \propto V \tag{8.1}$$

or

$$\boxed{Q = CV} \tag{8.2a}$$

This basic charge-voltage relationship of a capacitor can also be written as

$$C = \frac{Q}{V} \tag{8.2b}$$

The constant of proportionality in this equation, C, is a measure of the ability of the device to store charge per unit voltage applied. C is therefore called the *capacitance* of the capacitor. The larger the value of the capacitance, the higher would be the amount of charge stored for the same magnitude of the applied voltage. The formal definition of capacitance is thus: *the amount of charge stored on each plate of a capacitor per unit voltage applied between the plates.*

Since capital letter symbols (e.g., Q and V) refer to quantities that are constant with respect to time (i.e., dc values) and because the dependence on time is clearly apparent (capacitors do take time to charge), Eq. (8.2) can be generalized, using lowercase letter symbols, that is,

$$q = Cv \tag{8.3a}$$

or

$$C = \frac{q}{v} \tag{8.3b}$$

where q and v are in general variable time functions.

As the equations above indicate, the voltage v and the charge q for a capacitor are linearly related. Therefore, "the voltage across the capacitor cannot be changed instantly" because it takes time for the charges on the capacitor's plates to reach their steady-state value of Q. This is due to the action of opposition to the separation and accumulation of charges, as explained above. Capacitance can therefore also be defined as *the property of an electric circuit element that opposes the change of the voltage across its terminals.*

According to Eqs. (8.2) and (8.3), the unit of capacitance is coulomb/volt, which is termed the farad, abbreviated F, in the SI system of units, in honor of the British physicist Michael Faraday. However, 1 F is quite a large unit since most of the charges dealt with in electrical circuits are of the order of milli- or microcoulombs (or even lower). The more practical units of capacitance are therefore

$$1\,\mu\text{F} = 10^{-6}\,\text{F} \qquad 1\,\text{nF} = 10^{-9}\,\text{F} \qquad 1\,\text{pF} = 10^{-12}\,\text{F}$$

The value of a capacitor is usually stamped on the body of the physical capacitor. Sometimes color codes (in the form of dots or bands), such as those used for physical resistors, are used to indicate the value of a capacitor in picofarads.

Capacitance, whether wanted (as in physical capacitor elements) or not (in which case it is termed *stray* or *parasitic* capacitance), is always exhibited in any electrical circuit as long as there are two conductors (e.g., wires) separated by an insulating material (e.g., air). It is therefore important to examine how the value of a capacitance depends on physical parameters and dimensions. When charges accumulate on two plates separated by an insulating (or dielectric) material, an electric field is established in the space between the plates (the dielectric material). In this field, forces of attraction or repulsion are exhibited, acting on an isolated electrical charge placed in this space. The different dielectric materials affect the establishment of such electrostatic fields to different degrees. The quantity used to quantify such differences is called the *permittivity* (or *dielectric constant*) ϵ of the material, where

$$\epsilon = \epsilon_r \epsilon_0 \tag{8.4}$$

where ϵ_0 is the permittivity of free space (=8.85 pF/m) and ϵ_r is the relative permittivity of the dielectric material. For mica, ϵ_r, is 5.0, for glass ϵ_r is 7.5, and for ceramic ϵ_r is 7500. The existence of this electric field influences the forces of attraction and repulsion that are exhibited during the charge separation process explained above, and hence affects the amount of overall charge stored in the capacitor, Q. Therefore, the value of the capacitance is directly influenced by the permittivity ϵ of the dielectric material.

Also, the larger the area of the metallic plates, A, constituting the capacitor (see Fig. 8.1), the more would be the amount of charge stored on them, resulting in a higher value of the capacitance. On the other hand, when the distance separating the two plates, d, is reduced, the charges already accumulating on the plates have an increased influence on the charge separation process, actually counteracting the attraction and repulsion forces explained above. This again allows more charge to accumulate on the plates and thus increases the value of the capacitance. In terms of these parameters, the value of the capacitance can be expressed as

$$\boxed{C = \epsilon \frac{A}{d} \quad \text{F}} \tag{8.5}$$

This relationship indicates that any required capacitance value, whether fixed or variable, can be obtained by manipulating the basic physical dimensions A and/or d and by choosing the appropriate dielectric.

It is also important to note that any dielectric material breaks down and loses its insulation property if the voltage applied to it is sufficiently increased. This breakdown voltage depends on the type of dielectric material and the separation distance d between the two plates. A maximum allowable operating (or working) voltage is specified for each type of capacitors and is usually printed on the body of practical capacitors. This phenomenon of insulator breakdown may have been observed in practice, for example in the case of sparking through the air that may occur while opening a switch in certain electric circuits.

When a certain amount of charge is stored in a capacitor, an electrical field is present in the dielectric material of the capacitor. That field represents the storage of a finite amount of electrical energy. Actually, this is the amount of energy w spent by the voltage source during the process of charge separation. Through calculus, it can be shown that the electrical energy stored in the electrostatic field of a capacitor is given by

$$w = \tfrac{1}{2}Cv^2 \quad \text{J} \tag{8.6}$$

For those interested, a proof of this energy formula is given in Appendix A8.1.

■ EXAMPLE 8.1

Calculate the value of the capacitance for each of the following physical configurations.
(a) The plate area is $8\,\text{cm}^2$, the plate separation is $0.2\,\text{cm}$, and the insulator is air.
(b) The plate area is $16\,\text{cm}^2$, the plate separation is $0.1\,\text{cm}$, and the dielectric is ceramic.

Solution In the expression for capacitance [Eq. (8.5)], the units used should be consistent with the SI system of units.

(a) $C = \epsilon_0 \dfrac{A}{d} = 8.85\,\text{pF/m} \times \dfrac{8 \times 10^{-4}\,\text{m}^2}{0.2 \times 10^{-2}\,\text{m}} = 3.54\,\text{pF}$

(b) For the ceramic dielectric, $\epsilon = \epsilon_r \epsilon_0$, where ϵ_r is 7500. Thus here

$$C = \epsilon_r \epsilon_0 \dfrac{A}{d} = 7500 \times 8.85\,\text{pF/m} \times \dfrac{16 \times 10^{-4}\,\text{m}^2}{0.1 \times 10^{-2}\,\text{m}} = 0.106\,\mu\text{F}$$

Notice that even though the physical dimensions in both cases above are of the same order of magnitude, the nature of the dielectric can increase the value of the capacitance by many orders of magnitude.

■ EXAMPLE 8.2

A voltage is applied to each of the capacitors in Example 8.1. The value of this voltage is 100 V. Find
(a) The charge stored in each of the capacitors.
(b) The energy stored in each of the capacitors.

Solution (a) Since $Q = CV$, then for each of the capacitors in Example 8.1,

$$Q_1 = C_1 V = 3.54 \times 10^{-12} \times 100 = 354\,\text{pC}$$
$$Q_2 = C_2 V = 0.106 \times 10^{-6} \times 100 = 10.6\,\mu\text{C}$$

(b) The energy stored in a capacitor is

$$W = \tfrac{1}{2}CV^2$$

Therefore, for each of the capacitors in Example 8.1,

$$W_1 = \tfrac{1}{2}C_1 V^2 = \tfrac{1}{2} \times 3.54 \times 10^{-12} \times (100)^2$$
$$= 1.77 \times 10^{-8} = 17.7\,\text{nJ}$$

and

$$W_2 = \tfrac{1}{2}C_2V^2 = \tfrac{1}{2} \times 0.106 \times 10^{-6} \times (100)^2$$
$$= 0.53 \times 10^{-3} = 0.53 \, \text{mJ}$$

8.2 TYPES OF PRACTICAL CAPACITORS

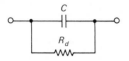

Figure 8.3

Ideally, the dielectric insulating material corresponds to an infinite resistance (i.e., resembling an open switch), but practically, the dielectric material has a finite but high resistance, called the leakage resistance, R_d. A more accurate model of a fixed practical capacitor includes this leakage resistance in parallel with an ideal capacitor, as shown in Fig. 8.3. Manufacturers specify either R_d or the product $R_d C$. The higher this product is, the better the quality of the capacitor. This leakage (or dissipation) resistor is the reason a charged practical capacitor cannot store the charge on it indefinitely. The charge is said to *leak off* through this leakage resistance. In most circuit analysis situations, R_d is very large (about 1000 MΩ) compared to the other resistors in the circuilt and can therefore be neglected without any appreciable error.

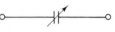

Figure 8.4

Capacitors are classified in general according to whether they are fixed or variable. The electric circuit symbol for a *variable capacitor* is shown in Fig. 8.4. Examples of variable air capacitors are shown in Fig. 8.5. The capacitor consists of a number of metal plates which can be rotated with respect to another set of plates. The rotation (through the capacitor's shaft) changes the overlap area of the metal plates and thus the value of the capacitor. The plates are ganged, resembling a number of equal capacitors connected in parallel. Variable air capacitors are sometimes referred to as *trimmers*. They are available in many sizes.

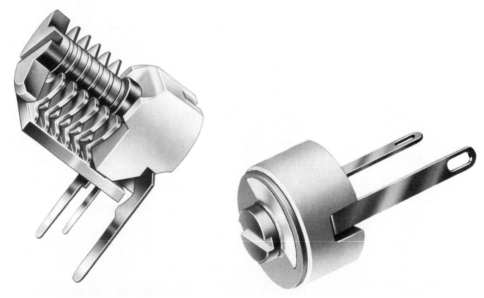

Figure 8.5 Examples of various shapes and sizes of variable capacitors (trimmers or turners). (Courtesy of E. F. Johnson Company, Components Division.)

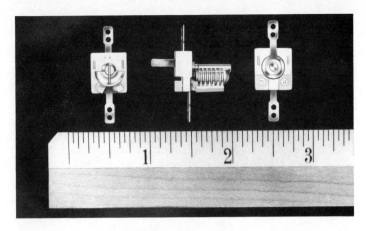

Figure 8.5 (*continued*)

Fixed capacitors are produced in many sizes and shapes, usually as disks, cans, tubes, or microminiature chips and also integrated-circuit construction (using a reverse-biased semiconductor device). Some typical examples of fixed capacitors are shown in Figs. 8.6 and 8.7. Fixed capacitors are classified according to the type of dielectric material used in their construction, such as mica, paper, polystyrene, Mylar, ceramic, and tantalum. The simple tubular-type capacitors have concentric rolls of metal (usually aluminum) foils, separated by the dielectric material (e.g., waxed paper). Conductors are attached to each metal-foil sheet to constitute the terminals of the capacitor.

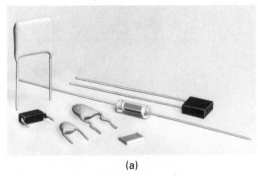

(a)

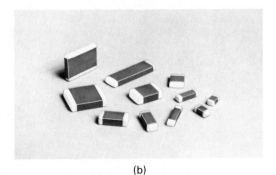

(b)

Figure 8.6 (a) Various types of fixed capacitors; (b) chip capacitors. (Courtesy of Centralab Inc., A North America Philips Co.)

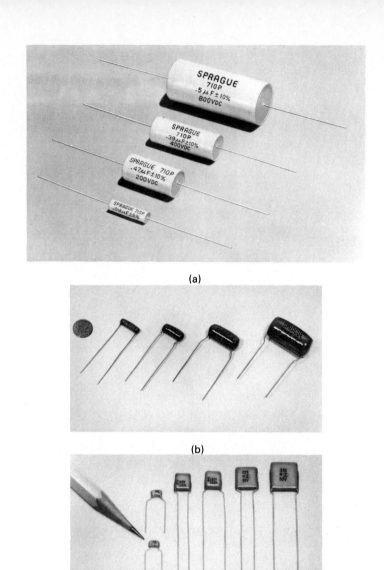

(a)

(b)

(c)

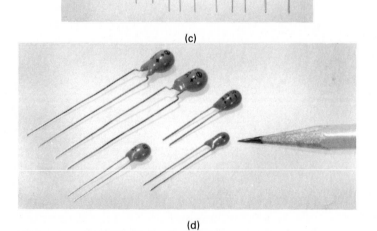

(d)

Figure 8.7 (a)–(d) Various types of fixed capacitors; (e) electrolytic capacitors. (Courtesy of Sprague Electric Co.)

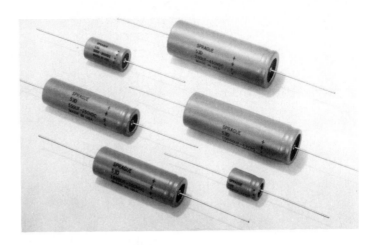

Figure 8.7 *(continued)*

Electrolytic capacitors usually possess the largest capacitance values. A typical construction of such capacitors is made from an aluminum sheet as one conductor and electrolytic paste as the other. The insulating material is aluminum oxide, formed when the applied voltage has the proper polarities that polarizes the metal plate. The capacitance attains relatively high values due to the thinness of this aluminum oxide insulation. The proper voltage polarities are usually marked on the cylindrical body of such capacitors. If the applied voltage has the incorrect polarities, the oxide will be reduced and conduction will occur between the terminals of the device.

Two important characteristics are usually indicated on the outside bodies of capacitors. These are the value of the capacitor, with the relevant tolerance, and the working voltage of the device. These numerical values are either stamped or color coded in the form of dots or bands. The color coding is similar to that used to identify resistors' values.

8.3 SERIES AND PARALLEL CONNECTIONS OF CAPACITORS

When a number of capacitors are connected in series as shown in Fig. 8.8 (three in this case), the source will provide work to separate a charge Q such that the top plate of C_1 has a charge of $+Q$ deposited on it while the bottom plate of C_3 has a charge $-Q$ deposited on it.

Since the charge on each plate of a given capacitor must be equal, but with opposite polarities, this implies that the charge on each of the series-connected capacitors (C_1, C_2, and C_3 in this case) must be the same, Q, as shown in Fig. 8.8.

To replace such a string of series-connected capacitors by a single equivalent capacitor C_{eq}, as indicated in Fig. 8.8, the same voltage source V must do the same amount of work and separate the same amount of charge, Q, as in the original circuit. Then, according to the voltage-charge relationship for a capacitor

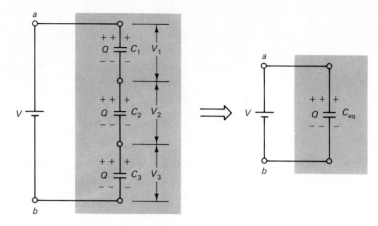

Figure 8.8

[Eq. (8.2)], the following expressions for the capacitors in Fig. 8.8 can be written:

$$V_1 = \frac{Q}{C_1} \tag{8.7a}$$

$$V_2 = \frac{Q}{C_2} \tag{8.7b}$$

$$V_3 = \frac{Q}{C_3} \tag{8.7c}$$

and

$$V = \frac{Q}{C_{eq}} \tag{8.7d}$$

As KVL applies to the original circuit in Fig. 8.8,

$$V = V_1 + V_2 + V_3 \tag{8.8}$$

Substituting from Eqs. (8.7) into Eq. (8.8) yields

$$\frac{Q}{C_{eq}} = \frac{Q}{C_1} + \frac{Q}{C_2} + \frac{Q}{C_3}$$

Dividing both sides of the equation above by Q, we obtain

$$\frac{1}{C_{eq}} = \frac{1}{C_1} + \frac{1}{C_2} + \frac{1}{C_3} \tag{8.9a}$$

For n capacitors in series, it is easy to conclude that

$$\boxed{\frac{1}{C_{eq}} = \frac{1}{C_1} + \frac{1}{C_2} + \frac{1}{C_3} + \cdots + \frac{1}{C_n}} \tag{8.9b}$$

while for two capacitors in series

$$\frac{1}{C_{eq}} = \frac{1}{C_1} + \frac{1}{C_2} = \frac{C_1 + C_2}{C_1 C_2} \tag{8.10a}$$

280   Chapter 8   Capacitance and Transients in *RC* Circuits

or

$$C_{eq} = \frac{C_1 C_2}{C_1 + C_2}$$ (8.10b)

The results obtained in Eqs. (8.9) and (8.10) are similar to those obtained for the equivalent resistance of parallel-connected resistors. Here C_{eq} is smaller than the smallest of the series-connected capacitors.

When a number of capacitors are connected in parallel as shown in Fig. 8.9 (three in this case), the voltage across each of them will be the same, equal to the value of the voltage source V in this case, since all the parallel capacitors are connected between the same two nodes, a and b. Each of the capacitors will, however, be charged by a different amount of charge, such that [according to Eq. (8.2)]

$$Q_1 = C_1 V$$ (8.11a)

$$Q_2 = C_2 V$$ (8.11b)

and

$$Q_3 = C_3 V$$ (8.11c)

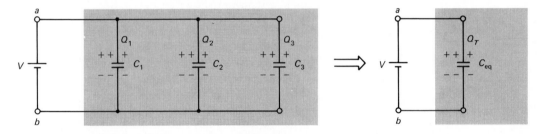

Figure 8.9

The energy source provided the total work necessary to separate the total amount of charge, Q_T, which is

$$Q_T = Q_1 + Q_2 + Q_3,$$ (8.12)

To replace these parallel-connected capacitors by a single equivalent capacitor C_{eq} between the same two nodes, a and b, as shown in Fig. 8.9, the same voltage source, V, should do the same amount of work to separate the same total amount of charge Q_T as in the original network. Applying Eq. (8.2) to C_{eq} in Fig. 8.9 provides

$$Q_T = C_{eq} V$$ (8.13)

Now substituting from Eqs. (8.11) and (8.13) into Eq. (8.12) results in

$$C_{eq} V = C_1 V + C_2 V + C_3 V$$

Dividing both sides of the equation above by V, we obtain

$$C_{eq} = C_1 + C_2 + C_3$$ (8.14a)

For the case of n capacitors in parallel, it is easy to extend the result above such that

$$C_{eq} = C_1 + C_2 + C_3 + \cdots + C_n \qquad (8.14b)$$

The result obtained in Eqs. (8.14) is similar to that obtained for the equivalent resistance of series-connected resistors. Here C_{eq} is the sum of all the parallel-connected capacitors and is therefore larger than the largest of these capacitors.

Series–parallel connections of capacitors may also be encountered. These forms of connections can be dealt with by successively applying the appropriate expressions derived above, as shown in the following examples.

■ EXAMPLE 8.3

A 0.01 μF and a 0.04 μF capacitor are connected in parallel to a 500 V supply. Find
(a) The equivalent capacitance.
(b) The charge on each of the capacitors and the total charge separated by the source.
(c) The energy stored in each of the capacitors.

Solution The original parallel connection of the two capacitors and the equivalent circuit are shown in Fig. 8.10.

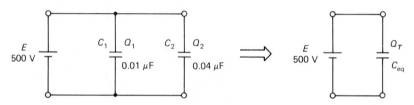

Figure 8.10

(a) The equivalent capacitance is therefore

$$C_{eq} = C_1 + C_2 = 0.01 \ \mu\text{F} + 0.04 \ \mu\text{F} = 0.05 \ \mu\text{F}$$

(b) The voltage across each of C_1, C_2, and C_{eq} is the same, being the source voltage, 500 V. Applying Eq. (8.2) to each of the capacitors, we have

$$Q_1 = C_1 E = 0.01 \ \mu\text{F} \times 500 = 5 \ \mu\text{C}$$
$$Q_2 = C_2 E = 0.04 \ \mu\text{F} \times 500 = 20 \ \mu\text{C}$$

and

$$Q_T = C_{eq} E = 0.05 \ \mu\text{F} \times 500 = 25 \ \mu\text{C}$$

Note that $Q_T = Q_1 + Q_2$, as expected. The total charge separated by the source is Q_T.

(c) The energy stored in each capacitor can be calculated using Eq. (8.6). Here

$$W_1 = \tfrac{1}{2}C_1 E^2 = \tfrac{1}{2} \times 0.01 \ \mu\text{F} \times (500)^2 = 1.25 \ \text{mJ}$$
$$W_2 = \tfrac{1}{2}C_2 E^2 = \tfrac{1}{2} \times 0.04 \ \mu\text{F} \times (500)^2 = 5.0 \ \text{mJ}$$

and

$$W_T = \tfrac{1}{2}C_{eq} E^2 = \tfrac{1}{2} \times 0.05 \ \mu\text{F} \times (500)^2 = 6.25 \ \text{mJ}$$

Clearly, $W_T = W_1 + W_2$, which confirms the fact that the work done by the source in separating the charge (being equivalent to the energy stored) is the same in both the original and the equivalent circuits.

EXAMPLE 8.4

The two capacitors in Example 8.3 are connected in series to the same voltage source. Find
(a) The equivalent capacitance.
(b) The charge on and the voltage across each of the capacitors.
(c) The energy stored in each capacitor in this case.

Solution   The series-connected original circuit and the equivalent circuit are shown in Fig. 8.11.

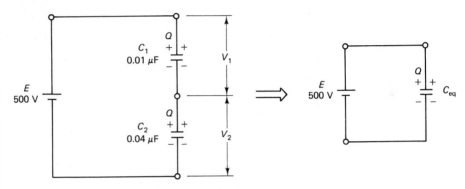

Figure 8.11

(a) From Eq. (8.10b),

$$C_{eq} = \frac{C_1 C_2}{C_1 + C_2} = \frac{0.01 \ \mu F \times 0.04 \ \mu F}{0.01 \ \mu F + 0.04 \ \mu F} = 0.008 \ \mu F$$

(b) In this case of series connection of capacitors, the charge separated by the source Q is the same amount of charge on each of C_1, C_2, and C_{eq}. Applying Eq. (8.2) for C_{eq}, the voltage across which is E (= 500 V), we obtain

$$Q = C_{eq}E = 0.008 \ \mu F \times 500 = 4 \ \mu C$$

Also using Eq. (8.2), the voltage across each of C_1 and C_2 can easily be calculated. Here

$$V_1 = \frac{Q}{C_1} = \frac{4 \ \mu C}{0.01 \ \mu F} = 400 \ V$$

and

$$V_2 = \frac{Q}{C_2} = \frac{4 \ \mu C}{0.04 \ \mu F} = 100 \ V$$

Clearly, KVL for the original circuit is satisfied since

$$E = V_1 + V_2$$

(c) Using Eq. (8.6), the energy stored in each capacitor can be calculated.

$$W_1 = \tfrac{1}{2}C_1V_1^2 = \tfrac{1}{2} \times 0.01 \ \mu F \times (400)^2 = 0.8 \ mJ$$
$$W_2 = \tfrac{1}{2}C_2V_2^2 = \tfrac{1}{2} \times 0.04 \mu F \times (100)^2 = 0.2 \ mJ$$
$$W_T = \tfrac{1}{2}C_{eq}E^2 = \tfrac{1}{2} \times 0.008 \ \mu F \times (500)^2 = 1.0 \ mJ$$

Here also, $W_T = W_1 + W_2$, confirming the remark made in Example 8.3.

EXAMPLE 8.5

Find the equivalent capacitance of the series-parallel capacitive circuit shown in Fig. 8.12, as seen by the source. Calculate the voltage across and the amount of charge on each of the capacitors.

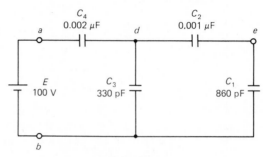

Figure 8.12

Solution   Capacitors C_1 and C_2 are in series. They can be replaced by a single equivalent capacitor, C_{12}, between points d and b, as shown in Fig. 8.13(a),

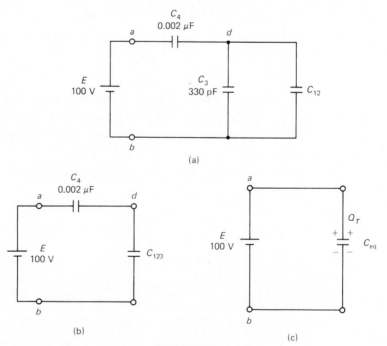

Figure 8.13

where

$$C_{12} = \frac{C_1 C_2}{C_1 + C_2}$$

$$= \frac{860 \text{ pF} \times 1000 \text{ pF}}{1860 \text{ pF}} = 462.4 \text{ pF}$$

Note that $C_2 = 0.001 \text{ } \mu\text{F} = 10^{-9} \text{ F} = 1000 \text{ pF}$.

In the equivalent circuit of Fig. 8.13(a), C_3 and C_{12} are in parallel. They can be replaced by a single equivalent capacitor, C_{123}, between points d and b, as shown in Fig. 8.13(b), where

$$C_{123} = C_{12} + C_3$$

$$= 462.4 \text{ pF} + 330 \text{ pF} = 792.4 \text{ pF}$$

The source then sees C_4 in series with C_{123}, as indicated in the equivalent circuit of Fig. 8.13(b). Therefore, the overall equivalent capacitance is

$$C_{eq} = \frac{C_4 \times C_{123}}{C_4 + C_{123}}$$

$$= \frac{2000 \text{ pF} \times 792.4 \text{ pF}}{2792.4 \text{ pF}} = 567.5 \text{ pF}$$

The simplified circuit is shown in Fig. 8.13(c). Since C_{eq} is across E, then from Eq. (8.2),

$$Q_T = C_{eq}E = 567.5 \text{ pF} \times 100 = 56.75 \text{ nC}$$

This is also the charge on the series-connected capacitors C_4 and C_{123} in Fig. 8.13(b). Therefore, from Eq. (8.2),

$$V_4 = V_{ad} = \text{voltage across } C_4 = \frac{Q_T}{C_4} = \frac{56.75 \text{ nC}}{2000 \text{ pF}} = 28.4 \text{ V}$$

and

$$V_{123} = V_{db} = \text{voltage across } C_3 \text{ and } C_{12} \qquad \text{[see Fig. 8.13(a)]}$$

$$= \frac{Q_T}{C_{123}} = \frac{56.75 \text{ nC}}{792.4 \text{ pF}} = 71.6 \text{ V}$$

Notice that $V_4 + V_{123} = E$ (i.e., KVL), as expected.

Again applying Eq. (8.2) for each of C_3 and C_{12} (the voltage across each is $V_{db} = 71.6 \text{ V}$),

$$Q_3 = \text{charge on } C_3 = C_3 V_{db} = 330 \text{ pF} \times 71.6 = 23.63 \text{ nC}$$

and

$$Q_{12} = \text{charge on } C_{12} = \text{charge on } C_1 \text{ and } C_2 \text{ in Fig. 8.12}$$
$$\text{because of the series connection}$$

$$= C_{12} V_{db} = 462.4 \text{ pF} \times 71.6 = 33.11 \text{ nC}$$

Note that Q_T (on C_{123}) $= Q_3 + Q_{12}$, in accordance with the properties of the parallel connection of capacitors, in Fig. 8.13(a).

Since the charge on each of C_1 and C_2 is now known (Q_{12} above), then from Eq. (8.2),

$$V_1 = \text{voltage across } C_1 = \frac{Q_{12}}{C_1} = \frac{33.11 \text{ nC}}{860 \text{ pF}} = 38.5 \text{ V}$$

and

$$V_2 = \text{voltage across } C_2 = \frac{Q_{12}}{C_2} = \frac{33.11 \text{ nC}}{1000 \text{ pF}} = 33.1 \text{ V}$$

As a final check,

$$V_1 + V_2 = 38.5 + 33.1 = 71.6 \text{ V} = V_{db}$$

satisfying KVL, as expected.

8.4 CURRENT IN A CAPACITOR: OHM'S LAW FOR THE CAPACITOR

Electric current in any circuit element is defined as the rate of change of charge with respect to time, or in other words, the rate of flow of charge. If a charge Δq flows into one terminal of an element in a time Δt (where Δ refers to small or incremental variations), then

$$i = \frac{\Delta q}{\Delta t} \qquad (8.15)$$

In the dc circuits considered in previous chapters, the rate of flow of charge is the same at every instant of time. This clearly implies that

$$\frac{\Delta q}{\Delta t} = \text{current} = \text{constant} = I \qquad (8.16)$$

As $\Delta q / \Delta t$ is the slope of the graphical relationship between q and t, a constant slope corresponds to a straight-line relationship between q and t, as shown in Fig. 8.14(a). This situation therefore corresponds to a constant current at every instant of time.

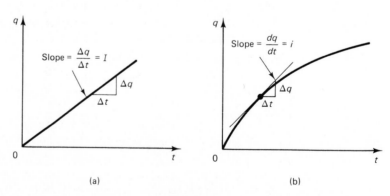

(a)   (b)

Figure 8.14

In circuits containing capacitors, the charge accumulating on the capacitor does not build up linearly or instantaneously with time, as explained earlier, due to the repulsion and attraction forces. Therefore, q is not a linear function of time but a curve, as shown for example in Fig. 8.14(b). The slope of the curve is different at different instants of time, implying that the current in the capacitor is in general a function of time t. Also, the slope of the curve at any instant of time (at a point), which is the tangent line, coincides with the curve itself when Δt tends to zero.

Therefore, extending the definition of current to this situation, we can say that

$$i = \lim_{\Delta t \to 0} \frac{\Delta q}{\Delta t} = \frac{dq}{dt} \tag{8.17}$$

which is the differential of q with respect to t, or the derivative of the function $q(t)$ with respect to t, according to the methods of calculus.

For the capacitor

$$\boxed{q = Cv} \tag{8.18}$$

when both the charge q and the voltage v are functions of time, with C being a constant. Therefore,

$$\Delta q = C \, \Delta v$$

or

$$i = \frac{\Delta q}{\Delta t} = C \frac{\Delta v}{\Delta t} = C \times \text{rate of change of voltage} \tag{8.19}$$

Hence the capacitor's current is

$$\boxed{i = \lim_{\Delta t \to 0} \frac{\Delta q}{\Delta t} = \lim_{\Delta t \to 0} C \frac{\Delta v}{\Delta t} = C \frac{dv}{dt}} \tag{8.20}$$

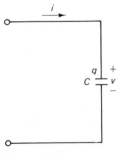

Figure 8.15

where dv/dt is the derivative of $v(t)$ with respect to t according to the methods of calculus, which also represents the rate of change of the voltage with respect to time.

Equation (8.20) is therefore Ohm's law for a capacitor, representing the voltage–current relationship for this element. This relationship governs the behavior of i with v for the capacitor, with Fig. 8.15 showing the circuit representation and polarity convention of these electrical circuit variables. One can then conclude that the current in a capacitor is proportional to the rate of change of the voltage across its terminals. The constant of proportionality is C.

When the voltage across the capacitance is constant with respect to time, or in other words, a steady-state situation is reached, v does not vary with t, or dv/dt, the rate of change of voltage, is zero. Then

$$i = C \frac{dv}{dt} = C \times 0 = 0 \, \text{A} \tag{8.21}$$

In dc circuits, this steady-state situation is reached within a certain time called the *transient time*. In other words, when the steady state is reached in dc circuits, the voltage across the capacitor (and also the charge stored in it) will be constant, while the current flow in it is zero. In this case a capacitor resembles an open circuit (or open switch) and blocks the flow of dc current.

The rest of this chapter deals with the study and analysis of transients in circuits containing capacitors, dc sources, and resistors. An important aspect of this study concerns the behavior of the voltage across the capacitor [and the charge on it since they are linearly related according to Eq. (8.18)] as a function of time.

If the instant at which the circuit conditions are changed, for example through opening or closing a switch, is arbitrarily defined to correspond to $t = 0$ s, it must be important to differentiate between the conditions right before the change (0^- s) and right after the change (0^+ s). This is shown in the graphs of Fig. 8.16. Notice that the notation 0^- s (right before) and 0^+ s (right after) indicate instants of time infinitesimally close to the instant of actual change. The notations concerning the values of the electrical quantities at these instants are

$$v(0^-) = \text{voltage across the capacitor, say, right before the change}$$

$$v(0^+) = \text{voltage across the capacitor, say, right after the change}$$

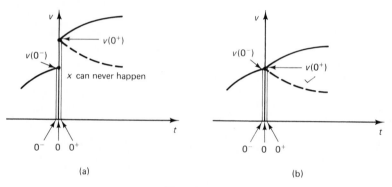

Figure 8.16

Based on the discussion in Section 8.1, it has been indicated that the voltage across the capacitor cannot change instantly (i.e., at a certain instant of time, say $t = 0$); the capacitor's voltage must have one and only one unique value and cannot jump to a different value. This means that the situation represented in Fig. 8.16(a) can never happen because

$$v(0^-) \neq v(0^+) \tag{8.22}$$

While the situation represented in Fig. 8.16(b) is the corrrect one because

$$v(0^-) = v(0^+) \tag{8.23}$$

The verification for the conclusions above can easily be arrived at using the definition of the capacitor's current in Eq. (8.19). In Fig. 8.16(a) and Eq. (8.22), Δv is a finite quantity (not equal to zero) at the instant $t = 0$, while in Fig. 8.16(b) and Eq. (8.23), Δv is zero at the instant $t = 0$. In both cases Δt tends to zero at this instant (the two instants 0^- and 0^+ converge). Therefore, the case corresponding to Fig. 8.16(a) results in

$$i = C\frac{\Delta v}{\Delta t} = \infty$$

while the case corresponding to Fig. 8.16(b) results in

$$i = C\frac{\Delta v}{\Delta t} = C\frac{0}{0}$$

which can be finite.

Since there is no physical source or physical element that can generate or stand an infinite amount of current, even instantaneously, one can conclude that the situation corresponding to Fig. 8.16(a) and Eq. (8.22) is practically and physically impossible. This proves that *the voltage across and the charge on a capacitor cannot experience an instantaneous jump in value.*

8.5 TRANSIENTS IN *RC* DC CIRCUITS

8.5.1 Charging Cycle

Figure 8.17 illustrates an example of an *RC* circuit that can be used to charge and discharge a capacitor.

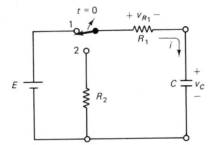

Figure 8.17

When the switch is in position 1, the voltage source E is connected through the resistance R_1 to the capacitor C. This is the practical version of the idealized situation shown in Fig. 8.1, which includes no resistors. Any practical source has a finite internal resistance and connecting wires also have finite nonzero (but very small) resistance. The resistance R_1 can account for those parasitic resistors, or it could be an actual series-connected discrete element. The value of R_1 (even though it could be very small) cannot be neglected in such circuits; otherwise, an infinite value of the current i may result [see Eq. (8.31)], which is an impossible situation.

The source E will provide energy to allow the charge to move in the circuit, accumulating on the plates of the capacitor C and thus charging it. As the charge q on the capacitor plates increases, so will the voltage v_C across it, since according to Eq. (8.18),

$$v_C = \frac{q}{C}$$

This is called the *charging cycle*. It continues until the charge accumulating on the capacitor, and the voltage across it, reach steady-state values, provided that the switch remains in position 1 long enough.

Steady-state conditions. At steady state the capacitor's charge and the voltage across it are constant and do not change with time. According to Eq. (8.20),

$$i = C\frac{dv_C}{dt} = C \times 0 = 0\,\text{A}$$

as obtained in Eq. (8.21). The capacitor resembles an open circuit. The steady-state values of the electric circuit variables can thus be easily obtained when the circuit is redrawn with C replaced by an open circuit, as shown in Fig. 8.18. Thus

$$i(\infty) = 0 \tag{8.24}$$

$$v_C(\infty) = E \tag{8.25}$$

and

$$q(\infty) = Cv_C(\infty) = CE = Q \tag{8.26}$$

The notation $i(\infty)$, and so on, indicates the value of the variable after a very long time ($t = \infty$), that is, the steady state value of the variable.

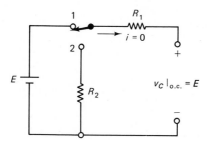

Figure 8.18 Steady-state conditions.

Here the capacitor is said to be fully charged. This condition is achieved when the work provided by the source is balanced by the effects of the repulsion and attraction forces, as described in Section 8.1, in conjunction with Fig. 8.1 and Eq. (8.2).

As discussed earlier, the charging process does take time since the voltage across, and the charge on, the capacitor cannot change (or jump) instantaneously. The time span during which q, i, and v_C are changing (i.e., are functions of time) is the transient time; v_C is the transient voltage and i is the transient current during this time span. How long this transient time is, how we quantify it, and how the value of any of the circuit variables can be determined at a given instant of time are the subjects of the following analysis.

At every instant of time, Ohm's law for each of the components applies:

$$i = C\frac{dv_C}{dt} \tag{8.20}$$

$$v_{R_1} = iR_1 \tag{8.27a}$$

$$= R_1C\frac{dv_C}{dt} \quad \text{[using Eq. (8.20)]} \tag{8.27b}$$

and referring to Fig. 8.17, the application of KVL produces

$$E = v_{R_1} + v_C \tag{8.28a}$$

$$= iR_1 + v_C \tag{8.28b}$$

$$= R_1 C \frac{dv_C}{dt} + v_C \tag{8.28c}$$

Equation (8.28c), describing the variation of v_C, is called a *first-order linear differential equation*. Its solution can be obtained using calculus. For those interested, such a solution is discussed in Appendix A8.2. A quantitative examination of this equation is, however, of prime importance. Before such a solution for v_C (as a function of time) can be found, it is essential that the value of v_C at $t = 0^+$ (right after closing the switch at, say, $t = 0\,\mathrm{s}$) be known. This is referred to as the *initial condition*, expressed as $v_C(0^+)$. Other initial conditions may also be important. They are denoted similarly [e.g., $i(0^+)$, $dv_C(0^+)/dt$, etc.].

Assuming that the voltage across the capacitor before the instant of switching, $v_C(0^-)$, is known to be V_0 volts, and since the voltage across the capacitor cannot change instantly [Eq. (8.23)],

$$v(0^+) = v(0^-) = V_0 \tag{8.29}$$
$$= \text{initial voltage across the capacitor}$$

For example, if there was no charge, $q(0^-)$, initial stored on the capacitor, then

$$v(0^+) = v(0^-) = V_0 = \frac{q(0^-)}{C} = 0\,\mathrm{V} \tag{8.30}$$

As KVL [Eq. (8.28a)] applies at every instant of time,

$$v_{R_1}(0^+) = R_1 i(0^+) = E - v_C(0^+)$$
$$= E \quad [\text{using condition (8.30)}]$$

Therefore,

$$i(0^+) = I_0 = \text{initial value of the capacitor's charging current}$$

$$= \frac{E}{R_1} \tag{8.31}$$

This is the highest value that the capacitor's charging current can assume. As the current is the rate of flow of charge, the charge will rush with the highest speed to charge the capacitor. Also from Ohm's law for the capacitor [Eq. (8.20)],

$$\frac{dv_C}{dt}(0^+) = \frac{i(0^+)}{C} = \frac{E}{R_1 C} \tag{8.32}$$

Equation (8.32) means that the voltage across the capacitor will increase with the highest rate (highest slope) at this moment of time ($t = 0^+$).

As the charge accumulates on the capacitor, q increases and v_C also increases. Since according to KVL,

$$v_{R_1} = R_1 i = E - v_C$$

then v_{R_1} and i decrease. But i is the rate of flow of charge. Therefore, the rate of accumulation of charge decreases (i.e., the charging process slows down). This is also evident from

$$\frac{dv_C}{dt} = \frac{1}{C} i$$

which means that the rate of change of the capacitor's voltage decreases with time as i decreases with the increase in v_C.

This process can be graphed as shown in Fig. 8.19, indicating the variation of $v_C (= q/C)$, $v_{R_1} (= E - v_C)$ and i as functions of time. The charging process ends when the steady-state conditions are reached [Eqs. (8.24) and (8.25)], indicating the completion of the charging cycle when the capacitor's voltage reaches the value of E and the charging current is reduced to zero.

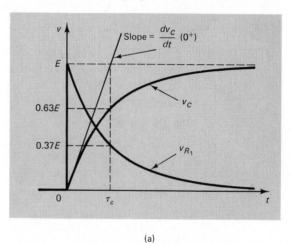

(a)

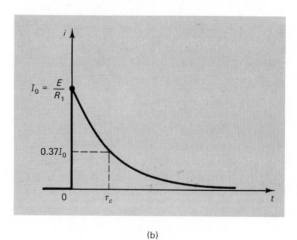

(b)

Figure 8.19

Notice that the capacitor's charging current does jump from

$$i(0^-) = 0 = \text{current right before closing the switch}$$

to

$$i(0^+) = I_0 = \frac{E}{R_1} = \text{current right after closing the switch}$$

Also, $v_{R_1}(=iR_1)$ does experience a jump in value at $t = 0$, the moment of closing the switch at position 1.

The curves shown in Fig. 8.19 are called *exponential curves*, exponential rise for v_C and exponential decay for v_{R_1} and i. The tangent to v_C at $t = 0^+$ has a slope given by Eq. (8.32):

$$\frac{dv_C}{dt}(0^+) = \frac{E}{R_1C} = \frac{\text{rise}}{\text{run}} = \frac{E}{\tau_C} \quad \text{(from Fig. 8.19)}$$

Therefore,

$$\boxed{\begin{aligned} \tau_C &= R_1C \quad \text{s} \\ &= \text{total series resistance} \times \text{capacitance value} \end{aligned}} \quad (8.33)$$

τ_C is called the charging time constant. Its units are seconds since

$$R_1 \times C = \frac{\text{volts}}{\text{ampere}} \times \frac{\text{coulombs}}{\text{volt}} = \frac{\text{coulombs}}{\text{coulomb/second}} = \text{seconds}$$

At the instant of time when $t = \tau_C$, v_C reaches 63.2% of its steady-state value (assuming that V_0 is zero). In fact, after $5\tau_C$ seconds of time has elapsed, v_C reaches 99.3% of its steady-state value. Even though theoretically v_C does not reach the exact value E except after a very long time ($t = \infty$), for all practical purposes the time span $5\tau_C$ seconds is considered to indicate the end of the transient cycle. One then says that the capacitor is now fully charged and steady state has been reached.

To obtain numerical values of the electrical circuit variables ($v_C, i, \ldots$ etc.) during this charging cycle, the analytical solution of Eq. (8.28c) should be known. As is proved in Appendix A8.2, with τ_C as defined in Eq. (8.33):

1. With the initial condition of the capacitor's voltage being in general,

$$v_C(0^+) = v_C(0^-) = V_0$$

the capacitor's voltage at any instant of time, $v_C(t)$, is obtained from

$$\boxed{v_C(t) - V_0 = (E - V_0)(1 - e^{-t/\tau_C})} \quad (8.34)$$

in words,

$$\begin{aligned} \text{capacitor voltage at time } t &- \text{ initial voltage} \\ = (\text{steady-state voltage} &- \text{ initial voltage})(1 - e^{-t/\tau_C}) \end{aligned}$$

e is 2.7183, the base of the natural logarithm.

2. In the case when the initial capacitor's voltage is

$$v_C(0^+) = v_C(0^-) = V_0 = 0$$

as in condition (8.30), then

$$\boxed{v_C(t) = E(1 - e^{-t/\tau_C})} \quad (8.35)$$

Either Eq. (8.34) or (8.35) allows one to determine the capacitor's voltage at any instant of time, or to determine the time at which the capacitor's voltage has a certain given value. Calculators can easily be used to carry out such numerical computations.

Since every charging (or discharging) circuit may have different component values and different sources and initial conditions, Eqs. (8.34) and (8.35) can be normalized by letting

$$x = \frac{t}{\tau_C} \tag{8.36}$$

that is, the time axis is normalized with respect to τ_C, and the voltage axis becomes

$$y = \frac{v_C(t)}{E} \quad \text{[in Eq. (8.35)]} \tag{8.37a}$$

$$= \frac{v_C(t) - V_0}{E - V_0} \text{[in Eq. (8.34)]} \tag{8.37b}$$

$$= \frac{\text{change in capacitor voltage}}{\text{maximum change in capacitor voltage}}$$

Then both equations can be expressed as

$$y = 1 - e^{-x} \tag{8.38}$$

This is the equation of the universal exponential rise curve, shown in Fig. 8.20. The horizontal axis, x, is in units of time constant, while the vertical axis, y, varies from 0 to 1 or 0 to 100%. A graphical solution to evaluate the circuit variables during the transient time is thus possible using these normalized exponential curves, as will be shown in the following examples.

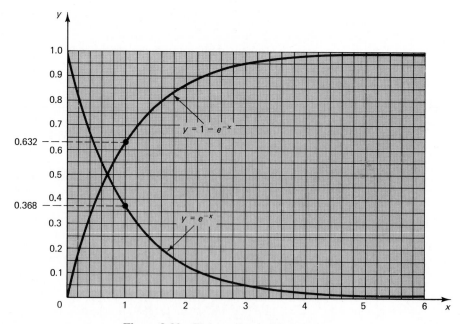

Figure 8.20   Universal exponential curves.

Once the capacitor's voltage, $v_C(t)$, is known, it is a simple matter to use Ohm's law and KVL to evaluate the rest of the circuit variables. For example, to obtain the charging current i, using Eqs. (8.27a) and (8.28a),

$$i = \frac{v_{R_1}}{R_1} = \frac{E - v_C}{R_1} = \frac{E}{R_1}\left(1 - \frac{v_C}{E}\right)$$

Substituting from Eq. (8.35) for v_C gives us

$$i = \frac{E}{R_1}[1 - (1 - e^{-t/\tau_C})] = \frac{E}{R_1} e^{-t/\tau_C} = I_0 e^{-t/\tau_C} \qquad (8.39)$$

where

$$I_0 = \frac{E}{R_1} \qquad (8.40)$$

Equation (8.39) can also be normalized in the following way:

$$y = \frac{i}{I_0} = e^{-x} \qquad \left(x = \frac{t}{\tau_C}\right) \qquad (8.41)$$

which is the universal exponential decay curve, also shown in Fig. 8.20.

The charge on the capacitor at any time is

$$q = C v_C$$

while the rate of change of the capacitor's voltage, dv_C/dt, can be determined from Eq. (8.20), using the value of i calculated above, that is,

$$\frac{dv_C}{dt} = \frac{i}{C}$$

Since from KVL,

$$v_{R_1} + v_C = E \quad \text{(always constant)}$$

then a change in $v_{R_1}(\Delta v_{R_1})$ must be exactly equal and opposite to the change in $v_C(\Delta v_C)$, since the sum of the two variables is always the constant source voltage E. In other words,

$$\Delta v_{R_1} + \Delta v_C = 0 \qquad \text{or} \qquad \Delta v_{R_1} = -\Delta v_C$$

Dividing both sides of the equation above by Δt and taking the limit as Δt tends to zero, we obtain the rate of change of the resistor's voltage as

$$\frac{dv_{R_1}}{dt} = -\frac{dv_C}{dt} = -\frac{i}{C} \qquad (8.42)$$

But since $v_{R_1} = R_1 i$, then

$$R_1 \frac{di}{dt} = -\frac{i}{C}$$

or

$$\text{rate of change of the charging current } \frac{di}{dt} = -\frac{i}{R_1 C} = -\frac{i}{\tau_C} \qquad (8.43)$$

The negative rate of change (slope) indicates a decrease in the value of i.

Since practically,

$$\text{duration of the transient time} = 5\tau_C = 5R_1C \tag{8.44}$$

the smaller τ_C is (by decreasing C or R_1), the faster the charging cycle would be because it would be completed in a shorter time span.

8.5.2 Discharging Cycle: Source-Free RC Circuits

When the switch in the circuit shown in Fig. 8.17 is placed in position 2, the circuit becomes as shown in Fig. 8.21. It contains no energy sources and is therefore referred to as a *source-free RC circuit*. The current i and the voltage polarities are left in the same directions as those used in the discussions of Section 8.5.1. The capacitor is assumed to have accumulated a certain charge Q and reached a certain voltage V_0 during the previous charging cycle. To study this discharging circuit, it will be assumed that the time reference has been shifted such that the switch is placed in position 2 at $t = 0$. In other words, the discharging cycle will also start at $t = 0$, for the sake of simplifying the analysis.

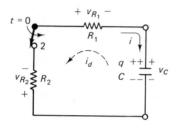

Figure 8.21

The positive and negative charges stored on the two plates of the capacitor will now try to neutralize each other. This means that the energy stored in the capacitor during the charging cycle will be used to move the charges in the opposite direction during this neutralization process. In so doing, the capacitor acts as a temporary energy source, delivering energy (equal to the same amount stored in it) to the rest of the components in the discharging circuit and driving the discharge current i_d, shown by the dashed arrow in Fig. 8.21. This means that the expected solution for i will result in a sign change (negative in this case) indicating the reversal in direction explained above.

Following the direction of the current i, KVL for this circuit can be expressed as follows:

$$v_{R_2} + v_{R_1} + v_C = 0 \tag{8.45}$$

Therefore, using Ohm's law for the two resistors, we have

$$i(R_1 + R_2) + v_C = 0$$

or

$$i = -\frac{1}{(R_1 + R_2)} v_C \tag{8.46}$$

Also, since

$$i = C\frac{dv_C}{dt} \tag{8.20}$$

then

$$C \frac{dv_C}{dt} = -\frac{1}{R_1 + R_2} v_C$$

or

$$\boxed{\frac{dv_C}{dt} = -\frac{1}{(R_1 + R_2)C} v_C} \qquad (8.47)$$

Equation (8.47) is a first-order linear differential equation, describing the variation of v_C in this discharge circuit. Its solution also involves calculus and is given in Appendix A8.2. The initial condition here is the voltage across the capacitor at the moment of switching to position 2. As indicated above (and since the voltage across the capacitor cannot change instantly),

$$v_C(0^+) = V_0 \qquad (8.48)$$

Since the current i, is directly related to v_C as Eq. (8.46) indicates, the initial value of the current right after the switching moment is given by

$$i(0^+) = -\frac{1}{R_1 + R_2} v_C(0^+)$$

or

$$I_{d0} = i(0^+) = -\frac{V_0}{R_1 + R_2} \qquad (8.49)$$

This initial value of the discharge current is negative, indicating a reversal in the direction of flow compared to i, as anticipated above.

When the discharge current starts to flow, the charge on the capacitor's plates, q, decreases due to the neutralization of the positive and negative charges. Therefore, $v_C(=q/c)$ also decreases and consequently the discharge current i decreases, as Eq. (8.46), relating i and v_C, indicates. The decrease in the rate of flow of charge (i.e., the current) means that the discharging process is slowing down. This is confirmed by the decrease in the rate of change of the capacitor's voltage, as can be deduced from Eq. (8.47).

At the end of this discharging cycle, the complete neutralization of the charge stored on the capacitor's plates is achieved. When q becomes zero, so will v_C and i. In other words, the steady state reached at the conclusion of the discharging cycle corresponds to

$$i(\infty) = 0 \qquad (8.50a)$$

and

$$v_C(\infty) = 0 \qquad q(\infty) = 0 \qquad \frac{dv_C}{dt}(\infty) = 0 \qquad (8.50b)$$

This is easily seen from Fig. 8.21 because at the steady state the capacitor resembles an open circuit and the current flow becomes zero.

The behavior of the circuit's variables described above is represented graphically as shown in Fig. 8.22. The inital slope of the $v_C(t)$ curve at $t = 0^+$ is

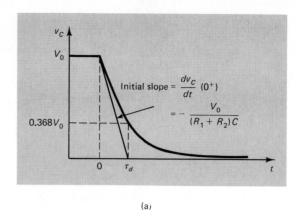

(a)

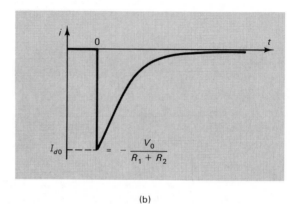

(b)

Figure 8.22

obtained using Eq. (8.47):

$$\frac{dv_C}{dt}(0^+) = -\frac{1}{(R_1 + R_2)C}v_C(0^+) = -\frac{V_0}{(R_1 + R_2)C} \tag{8.51a}$$

$$= -\frac{\text{rise}}{\text{run}} \quad \text{[from Fig. 8.22(a)]}$$

$$= -\frac{V_0}{\tau_d} \tag{8.51b}$$

where

$$\tau_d = (R_1 + R_2)C \quad s$$
$$= \text{total series resistance} \times \text{capacitance value} \tag{8.52}$$

τ_d is called the discharge time constant. In general, it is different in value from τ_C, the charging time constant. If R_2 were made to be zero (i.e., a short circuit), τ_d and τ_C would be equal.

At the moment when $t = \tau_d$, the capacitor's voltage v_C drops to 36.8% of its initial value. After an elapsed time of $5\tau_d$ the capacitor's voltage reaches 0.7% of its initial value. Even though theoretically v_C reaches a zero value (steady-state

Chapter 8   Capacitance and Transients in *RC* Circuits

condition) only after a very long time ($t = \infty$), for all practical purposes the discharge cycle is considered to end after a time span of $5\tau_d$. At such time the steady-state conditions are said to be reached.

Both of the variations of v_C and i as functions of time are exponential decay curves. The capacitor voltage $v_C(t)$ does not experience a sudden jump in value at $t = 0^+$ (right after the switching instant), but the discharge current does experience such a sudden jump in value at $t = 0^+$, as indicated in Fig. 8.22. As mentioned before, the discharge current i_d is opposite in direction to the charge current i. This explains the negative values for the current curve shown in Fig. 8.22(b).

To calculate numerical values for the circuit variables during the discharge cycle, the analytical solution of Eq. (8.47) should be known. As proved in Appendix A8.2, the variation of $v_C(t)$ as a function of time during the discharge cycle is given by

$$\boxed{v_C(t) = V_0 e^{-t/\tau_d}} \tag{8.53}$$

where V_0 is the initial capacitor's voltage $v_C(0^+)$ and τ_d is the discharge time constant as defined in Eq. (8.52).

Equation (8.53) can also be normalized by letting

$$\frac{t}{\tau_d} = x \quad \text{and} \quad \frac{v_C}{V_0} = y \tag{8.54a}$$

Then

$$y = e^{-x} \tag{8.54b}$$

which is the universal exponential decay curve shown in Fig. 8.20. x varies in units of τ_d and y varies from 1 to 0 or 100% to 0%. The use of this curve allows us to obtain the required circuit solutions using graphical techniques, as will be shown in the following examples.

Once v_C is determined, Ohm's law or KVL [Eq. (8.46) or (8.47)] can be used to obtain the values of the other circuit variables. Here

$$q = Cv_C$$

and

$$i = -\frac{v_C}{R_1 + R_2} = \frac{-V_0}{R_1 + R_2} e^{-t/\tau_d}$$

$$= I_{d0} e^{-t/\tau_d} \tag{8.55}$$

where I_{d0} is as defined in Eq. (8.49). This current expression can also be normalized using

$$x = \frac{t}{\tau_d} \quad \text{and} \quad y = \frac{i}{I_{d0}}$$

Then

$$y = e^{-x}$$

which is clearly the same universal exponential decay curve in Fig. 8.20.

The rate of change of the capacitor's voltage can be obtained from Eq. (8.20):

$$\frac{dv_C}{dt} = \frac{i}{C}$$

using the current value determined above, or from Eq. (8.47) in terms of v_C:

$$\frac{dv_C}{dt} = -\frac{1}{(R_1 + R_2)C} v_C = -\frac{v_C}{\tau_d} \qquad (8.56)$$

This is always a negative slope indicating a decrease in v_C toward its steady-state value of zero.

The rate of change of the discharge current can be obtained by differentiating both sides of Eq. (8.46), that is,

$$\frac{di}{dt} = -\frac{1}{R_1 + R_2}\frac{dv_C}{dt} = -\frac{1}{R_1 + R_2}\frac{i}{C} = -\frac{i}{\tau_d} \qquad (8.57)$$

As the transient time of the discharge cycle is

$$5\tau_d = 5(R_1 + R_2)C \qquad (8.58)$$

it is clear that a faster discharge can be obtained by reducing τ_d [i.e., either C and/or $(R_1 + R_2)$].

8.5.3 Generalized Form for the Transient Voltage Solution

All the circuits considered in this chapter contain only one capacitor. However, some of the charging and discharging circuits to be encountered could contain more than one source and more than one resistor, with arbitrary configuration. When the capacitor terminals, say a and b, are pulled as the load terminals of the network, Thévenin's theorem can be applied to the rest of the network, to the left-hand side of the load terminals a and b.

As shown in Chapter 6, the Thévenin's equivalent circuit contains one voltage source (V_{th}) in series with one resistance (R_{th}). This makes the equivalent circuit plus the capactive load resemble the general charging circuit shown in Fig. 8.17. Hence the solution can be obtained as described above, noting that E is replaced by V_{th} and R_1 is replaced by R_{th}. In a source-free circuit all resistors can be combined into one equivalent resistance connected to the capacitor, which can then easily be analyzed as was shown for the discharge circuit of Fig. 8.21.

The time constant is an important parameter in the transient analysis. Since any circuit can be made to resemble a simple RC circuit, driven or source free, the charging (τ_C) or discharging (τ_d) time constant can be found easily as the product of the capacitance times its series-connected resistance.

During the charging cycle, the general expression for the capacitor's voltage is as given by Eq. (8.34):

$$\begin{aligned} v_C(t) &= V_0 + (E - V_0)(1 - e^{-t/\tau_C}) \\ &= V_0 + E - V_0 - (E - V_0)e^{-t/\tau_C} \\ &= E - (E - V_0)e^{-t/\tau_C} = v_{Css} + v_{Ctr} \end{aligned} \qquad (8.59)$$

During the discharge cycle, v_C is obtained from Eq. (8.53):

$$v_C(t) = V_0 e^{-t/\tau_d} = v_{Css} + v_{Ctr} \qquad (8.60)$$

Both of these expressions can be written as a sum of two terms, v_{Css}, the steady-state capacitor's voltage, and v_{Ctr}, the transient part of the capacitor's voltage. The transient solution is the term that contains the exponential decay function $e^{-t/\tau}$ (τ being the circuit's time constant). It is in general in the form

$$v_{Ctr} = ke^{-t/\tau} \quad (k \text{ is a constant}) \tag{8.61}$$

This transient solution always decays to zero, usually in a very short time. The steady-state solution v_{Css} is the constant term in the voltage expression, either E or 0, etc.

It is very easy to obtain the steady-state part of the solution by analyzing the circuit with the capacitor replaced by an open circuit (as explained above) and determining the voltage across these open-circuited terminals of the capacitor. This constant voltage is the steady-state solution, v_{Css}. Once this is obtained and the circuit's time constant τ is determined, the total solution can be written as

$$\boxed{v_C(t) = v_{Css} + ke^{-t/\tau}} \tag{8.62}$$

What is now left to determine the capacitor's voltage completely is to calculate the value of the constant k. The initial condition given or determined before the instant of switching,

$$v_C(0^+) = v_C(0^-) = V_0$$

can be substituted into Eq. (8.62), obviously at $t = 0$. The resulting equation can easily be solved to determine k. Hence the complete solution for the capacitor's voltage is known as a function of time, as required.

■ EXAMPLE 8.6

The capacitor in the circuit shown in Fig. 8.23 was initially charged to a voltage of 10 V before the circuit was connected. Find the capacitor's voltage $v_C(t)$ for all $t \geq 0$. Sketch this capacitor voltage variation with time.

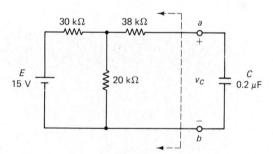

Figure 8.23

Solution   The circuit on the left-hand side of the terminals a and b of the capacitor can be replaced by its Thévenin equivalent as shown in Fig. 8.24(a). Here

$$V_{th} = 15 \times \frac{20\,k\Omega}{30\,k\Omega + 20\,k\Omega} = 6\,V$$

and

$$R_{th} = 38\,k\Omega + \frac{30\,k\Omega \times 20\,k\Omega}{30\,k\Omega + 20\,k\Omega} = 50\,k\Omega$$

$$\tau = R_{th} \times C = 50\,k\Omega \times 0.2\,\mu F = 10\,ms$$

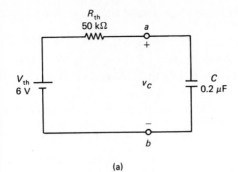

(a)

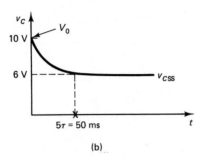

(b)

Figure 8.24

When the capacitor is replaced by an open circuit, the steady-state voltage can be determined:

$$V_{ab}|_{o.c.} = v_{Css} = 6\,V$$

Therefore,

$$v_C(t) = v_{Css} + ke^{-t/\tau}$$
$$= 6 + ke^{-t/(10\times10^{-3})}$$
$$= 6 + ke^{-100t}$$

To determine k, the initial condition given can be used:

$$v_C(0) = 10\,V$$

Substituting in the expression above at $t = 0$ yields

$$10 = 6 + ke^{-0} = 6 + k$$

and

$$k = 4$$

Therefore,

$$v_C(t) = 6 + 4e^{-100t} \qquad V$$

A sketch of this voltage variation as a function of time is shown in Fig. 8.24(b). Note that the transient ends after 5τ (50 ms).

EXAMPLE 8.7

Determine the capacitor's voltage $v_C(t)$ in the circuit shown in Fig. 8.25 for all $t \geq 0$. The switch shown was opened for a very long time and then closed at $t = 0$.

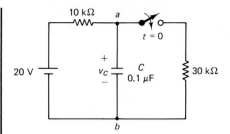

Figure 8.25

Solution The notion of very long time implies that steady state has been reached. Thus before closing the switch, the circuit can be represented as shown in Fig. 8.26(a), with the capacitor replaced by an open circuit. Thus

$$v_C(0^-) = v_C(0^+) = 20 \text{ V}$$

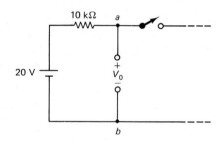

(a)

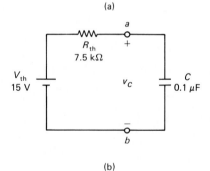

(b)   Figure 8.26

With the switch closed at $t = 0$, the capacitor terminals a and b are pulled to the right and the rest of the circuit is replaced by its Thévenin equivalent as shown in Fig. 8.26(b). Here

$$R_{th} = \frac{10 \text{ k}\Omega \times 30 \text{ k}\Omega}{10 \text{ k}\Omega + 30 \text{ k}\Omega} = 7.5 \text{ k}\Omega$$

and

$$V_{th} = 20 \times \frac{30 \text{ k}\Omega}{10 \text{ k}\Omega + 30 \text{ k}\Omega} = 15 \text{ V}$$

Thus

$$\tau = R_{th} \times C = 7.5 \times 10^3 \times 0.1 \times 10^{-6} = 0.75 \text{ ms}$$

The steady state voltage can easily be determined in the equivalent circuit of Fig. 8.26(b), again with C replaced by an open circuit. Thus

$$v_{Css} = V_{th} = 15 \text{ V}$$

Hence

$$v_C(t) = 15 + ke^{-t/0.75ms} = 15 + ke^{-1333.3t} \quad \text{V}$$

But at $t = 0$, $v_C(0^+) = 20\,\text{V}$ (as determined above). Therefore, substituting in the expression above gives us

$$20 = 15 + ke^{-0} = 15 + k$$

and

$$k = 5$$

Thus

$$v_C(t) = 15 + 5e^{-1333.3t} \quad \text{V}$$

8.6 GENERAL EXAMPLES OF TRANSIENTS IN RC CIRCUITS

EXAMPLE 8.8

The switch in the RC circuit shown in Fig. 8.27 was closed at $t = 0$. There was no charge initially stored on the capacitor. Find
(a) v_R, the voltage across the resistor at $t = 60\,\mu s$.
(b) dv_C/dt, the rate of change of the capacitor's voltage at $t = 100\,\mu s$.
(c) The time at which v_R is 2 V.
(d) The time at which dv_C/dt is $10\,\text{kV/s}$.
(e) The time at which dv_R/dt is $-15\,\text{kV/s}$.

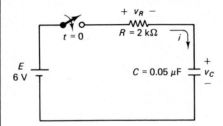

Figure 8.27

Solution   Since the initial charge on the capacitor is

$$q(0^-) = q(0^+) = 0\,\text{C}$$

then

$$V_0 = v_C(0^+) = 0\,\text{V}$$

When the switch is closed, the current magnitude jumps from 0 A to I_0, where

$$I_0 = \frac{E}{R} = \frac{6}{2\,\text{k}\Omega} = 3\,\text{mA}$$

The capacitor's voltage rises towards $E = 6\,\text{V}$, while the current drops toward zero, with a time constant

$$\tau = RC = 2 \times 10^3 \times 0.05 \times 10^{-6} = 0.1\,\text{ms} = 100\,\mu s$$

A sketch of the variations of v_C and i with time is shown in Fig. 8.28.
   (a) Using the graphical technique for this example, the capacitor's voltage rises exponentially:

$$y = 1 - e^{-x} \quad \text{where} \quad y = \frac{v_C}{E} \quad \text{and} \quad x = \frac{t}{\tau}$$

[see Eqs. (8.36) and (8.37a)].

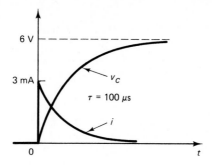

Figure 8.28

At $t = 60\ \mu s$, $x = t/\tau = 60\ \mu s/100\ \mu s = 0.6$. Therefore, from Fig. 8.20, this value of x corresponds to:

$$y = 0.45 = \frac{v_C}{6}$$

and

$$v_C = 0.45 \times 6 = 2.7\ \text{V}$$

Since $E = v_R + v_C$ (KVL always applies),

$$v_R = E - v_C = 6 - 2.7 = 3.3\ \text{V}$$

(b) From Eq. (8.20),

$$\frac{dv_C}{dt} = \frac{i}{C}$$

The current i at $t = 100\ \mu s$ needs to be determined. The current follows the exponential decay curve $y = e^{-x}$, where $y = i/I_0$ and $x = t/\tau$ [see Eq. (8.41)]. At $t = 100\ \mu s$, $x = t/\tau = 100\ \mu s/100\ \mu s = 1$. Using the universal exponential curves in Fig. 8.20, this value of x corresponds to

$$y = 0.368 = \frac{i}{3\ \text{mA}}$$

and

$$i = 0.368 \times 3\ \text{mA} = 1.104\ \text{mA}$$

Therefore,

$$\frac{dv_C}{dt} = \frac{i}{C} = \frac{1.104\ \text{mA}}{0.05\ \mu\text{F}} = 22.08\ \text{kV/s}$$

(c) When $v_R = 2\ \text{V}$

$$i = \frac{v_R}{R} = \frac{2}{2\ \text{k}\Omega} = 1\ \text{mA}$$

Then

$$y = \frac{i}{I_0} = \frac{1\ \text{mA}}{3\ \text{mA}} = 0.333$$

This value of y corresponds to $x = 1.1$, using the universal exponential decay curve. As

$$x = 1.1 = \frac{t}{\tau} = \frac{t}{100\ \mu s}$$

then

$$t = 110 \, \mu s$$

(d) $i = C(dv_C/dt) = 0.05 \times 10^{-6} \times 10 \times 10^3 = 0.5 \, \text{mA}$. It is then required to find the time at which i is $0.5 \, \text{mA}$. As was done in part (c),

$$y = \frac{i}{I_0} = \frac{0.5 \, \text{mA}}{3 \, \text{mA}} = 0.167$$

Then

$$x = 1.79 = \frac{t}{\tau} = \frac{t}{100 \, \mu s}$$

and

$$t = 179 \, \mu s$$

(e) This value of dv_R/dt must first be related to i or v_C. From Eq. (8.42),

$$\frac{dv_R}{dt} = -\frac{i}{C}$$

and

$$i = -0.05 \times 10^{-6} \times (-15 \times 10^3) = 0.75 \, \text{mA}$$

Therefore,

$$y = \frac{i}{I_0} = \frac{0.75 \, \text{mA}}{3 \, \text{mA}} = 0.25$$

Using the universal exponential decay curve, this value of y corresponds to

$$x = 1.39 = \frac{t}{\tau} = \frac{t}{100 \, \mu s}$$

and

$$t = 139 \, \mu s$$

EXAMPLE 8.9

Initially, there was no charge on the capacitor in the circuit shown in Fig. 8.29 and the switch was in position 2. At $t = 0$, the switch was moved to position 1.
(a) What is the voltage across the $4 \, k\Omega$ resistor at $t = 2 \, s$?
(b) Find the time at which the charge on the capacitor is $4 \, \text{mC}$.
(c) At $t = 3 \, s$, the switch was placed back into position 2 and remained there. Find the voltage V_{12} across the switch contacts at $t = 5 \, s$.

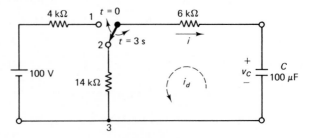

Figure 8.29

Solution The voltage across the capacitor, v_C, rises from $V_0 = 0\,$V toward a maximum of 100 V during the charging cycle when the switch is in position 1. The charging time constant is

$$\tau_C = (4\,\text{k}\Omega + 6\,\text{k}\Omega) \times 100\,\mu\text{F} = 10 \times 10^3 \times 100 \times 10^{-6} = 1\,\text{s}$$

At $t = 3\,$s, the charging cycle is interrupted and the capacitor starts to discharge. With the switch in position 2, the discharge time constant is

$$\tau_d = (14\,\text{k}\Omega + 6\,\text{k}\Omega) \times 100\,\mu\text{F} = 2\,\text{s}$$

A sketch of the variation of v_C with time is shown in Fig. 8.30. The analytical method will be used here.

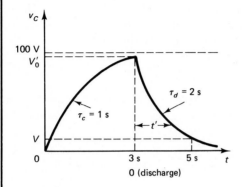

Figure 8.30

(a) From Eq. (8.35),

$$v_C = E(1 - e^{-t/\tau_C}) = 100(1 - e^{-t}) \quad \text{V}$$

At $t = 2\,$s,

$$v_C = 100(1 - e^{-2}) = 86.47\,\text{V}$$

Using KVL,

$$E = v_C + v_{4k\Omega} + v_{6k\Omega}$$

$$100 = 86.47 + i(10 \times 10^3)$$

and

$$i = \frac{13.53}{10 \times 10^3} = 1.353\,\text{mA}$$

Thus

$$v_{4k\Omega} = 1.353\,\text{mA} \times 4\,\text{k}\Omega = 5.41\,\text{V}$$

(b) When $q = 4 \times 10^{-3}\,$C,

$$v_C = \frac{q}{C} = \frac{4 \times 10^{-3}}{100 \times 10^{-6}} = 40\,\text{V}$$

Substituting in the expression above for v_C yields

$$40 = 100(1 - e^{-t}) = 100 - 100e^{-t}$$

$$e^{-t} = \frac{100 - 40}{100} = 0.6$$

Therefore,

$$t = -\ln 0.6 = 0.511 \, \text{s}$$

(c) The discharge cycle starts at $t = 3 \, \text{s}$. At this time:

$$v_C = V_0' = 100(1 - e^{-3}) = 95.02 \, \text{V}$$

Shifting the time reference to this moment of time (discharge starts here), the capacitor's voltage is determined from Eq. (8.53)

$$v_C = V_0' e^{-t'/\tau_d} = 95.02 e^{-0.5t'} \qquad \text{V}$$

where t' is the time measured from the onset of the discharge cycle. At $t = 5 \, \text{s}$, this moment corresponds to $t' = 2 \, \text{s}$ into the discharge cycle. Thus substituting $t' = 2 \, \text{s}$ into the expression above yields

$$v_C = 95.02 e^{-0.5 \times 2} = 95.02 e^{-1} = 34.96 \, \text{V}$$

Using Ohm's law for the discharge circuit [Eq. (8.46)], we find that the discharge current i_d (see Fig. 8.29) at this instant of time is

$$i_d = \frac{34.96}{14 \, \text{k}\Omega + 6 \, \text{k}\Omega} = 1.748 \, \text{mA}$$

Therefore,

$$V_{23} = 1.748 \, \text{mA} \times 14 \, \text{k}\Omega = 24.47 \, \text{V}$$

Since the switch is now in position 2, then

$$V_{13} = \text{E} = 100 \, \text{V}$$

Thus

$$V_{12} = V_{13} - V_{23} = 100 - 24.47 = 75.53 \, \text{V}$$

EXAMPLE 8.10

The switch in the circuit shown in Fig. 8.31 has been in position 1 for a very long time. At $t = 0$ it was placed in position 2 for 8 ms and then replaced into position 1. Find
(a) $dv_C/dt(0^+)$, the initial rate of change of v_C.
(b) v_C at $t = 8 \, \text{ms}$.
(c) The last time at which v_C is 10 V.
(d) The time at which $i = 20 \, \text{mA}$.
(e) The time at which $i = -2 \, \text{mA}$.

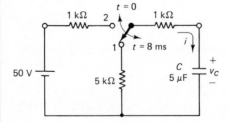

Figure 8.31

Solution This circuit is very similar to that analyzed in Example 8.9. It has the same initial condition of

$$V_0 = 0 \text{ V}$$

since the capacitor would be completely discharged when the switch is in position 1 for a very long time. Here

$$\tau_C(\text{switch in position 2}) = (1 \text{ k}\Omega + 1 \text{ k}\Omega) \times 5 \text{ }\mu\text{F}$$
$$= 2 \times 10^3 \times 5 \times 10^{-6} = 10 \text{ ms}$$
$$\tau_d(\text{switch in position 1}) = (5 \text{ k}\Omega + 1 \text{ k}\Omega) \times 5 \text{ }\mu\text{F}$$
$$= 6 \times 10^3 \times 5 \times 10^{-6} = 30 \text{ ms}$$

The capacitor's voltage, v_C, rises during the charging cycle toward a maximum of 50 V ($= E$), but the charging cycle is interrupted at $t = 8$ ms, to start the discharge cycle. A sketch of the variation of v_C with time is shown in Fig. 8.32. The graphical technique will be used to analyze this circuit.

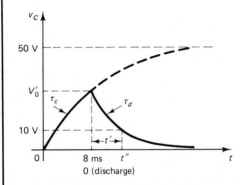

Figure 8.32

(a) The initial rate of change of the capacitor's voltage was shown to be

$$\frac{dv_C}{dt}(0^+) = \frac{i(0^+)}{C} = \frac{E}{R_1 C} \quad [\text{according to Eq. (8.32)}]$$

where here R_1 is $1 \text{ k}\Omega + 1 \text{ k}\Omega = 2 \text{ k}\Omega$, with the switch in position 2. Thus

$$\frac{dv_C}{dt}(0^+) = \frac{50}{2 \times 10^3 \times 5 \times 10^{-6}} = 5 \text{ kV/s}$$

(b) v_C follows the universal exponential rise curve $y = 1 - e^{-x}$, where $y = v_C/E$. At $t = 8$ ms,

$$x = \frac{t}{\tau_C} = \frac{8 \text{ ms}}{10 \text{ ms}} = 0.8$$

Thus from Fig. 8.20,

$$y = 0.55 = \frac{V_0'}{E} = \frac{V_0'}{50}$$

Therefore,

$$V_0' = 0.55 \times 50 = 27.5 \text{ V}$$

(c) v_C is 10 V for the last time during the discharge cycle. Here v_C follows the universal exponential decay curve $y = e^{-x}$, where $y = v_C/V_0'$.

Thus at $y = 10/27.5 = 0.364$, Fig. 8.20 gives

$$x = 1 = \frac{t'}{\tau_d} = \frac{t'}{30\,\text{ms}}$$

and

$$t' = 30\,\text{ms}$$

Therefore, the time required (see Fig. 8.32) is

$$t'' = 8\,\text{ms} + 30\,\text{ms} = 38\,\text{ms}$$

(d) When $i = 20\,\text{mA}$, this occurs during the charging cycle (switch in position 2). Here one can use KVL to obtain v_C:

$$v_C = E - i(1\,\text{k}\Omega + 1\,\text{k}\Omega) = 50 - 2 \times 10^3 \times 20 \times 10^{-3} = 10\,\text{V}$$

Thus

$$y = \frac{v_C}{E} = \frac{10}{50} = 0.2$$

From the universal exponential rise curve in Fig. 8.20,

$$x = 0.223 = \frac{t}{\tau_C} = \frac{t}{10\,\text{ms}}$$

Therefore,

$$t = 2.23\,\text{ms}$$

(e) When $i = -2\,\text{mA}$, this occurs during the discharge cycle (with the switch in position 1). Here from Ohm's law (see Fig. 8.31),

$$v_C = -i(1\,\text{k}\Omega + 5\,\text{k}\Omega) = 2 \times 10^{-3} \times 6 \times 10^3 = 12\,\text{V}$$

Thus

$$y = \frac{v_C}{V_0'} = \frac{12}{27.5} = 0.436$$

From the universal exponential decay curve in Fig. 8.20,

$$x = 0.83 = \frac{t'''}{\tau_d} = \frac{t'''}{30\,\text{ms}}$$

Therefore,

$$t = 8\,\text{ms} + t''' = 8\,\text{ms} + 0.83 \times 30\,\text{ms} = 32.9\,\text{ms}$$

EXAMPLE 8.11

The circuit shown in Fig. 8.33 is at steady-state conditions. At $t = 0$ the switch was placed in position 1, then at $t = 2\,\text{s}$ it was placed back in position 2 and remains there. Find
(a) The capacitor's voltage at $t = 2\,\text{s}$.
(b) The times at which the capacitor's voltage is zero.

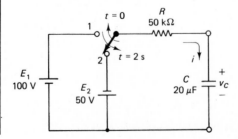

Figure 8.33

Solution   When the switch is in position 2 and steady state has been reached, the capacitor is equivalent to an open circuit. Noting the polarities and the reference direction given in Fig. 8.33, the initial voltage on the capacitor is

$$v_C(0^+) = V_0 = -50 \, \text{V}$$

As the switch is placed in position 1, the capacitor will charge toward 100 V, with a time constant τ:

$$\tau = RC = 50 \times 10^3 \times 20 \times 10^{-6}$$
$$= 1 \, \text{s}$$

Here according to Eq. (8.34),

$$v_C = V_0 + (E_1 - V_0)(1 - e^{-t/\tau})$$
$$= -50 + (100 - (-50))(1 - e^{-t}) = -50 + 150 - 150e^{-t}$$
$$= 100 - 150e^{-t} \quad \text{V} \quad (0 \le t \le 2 \, \text{s})$$

(a) At $t = 2 \, \text{s}$,

$$v_C = V_0' = 100 - 150e^{-2} = 79.7 \, \text{V}$$

(b) As the switch is then placed back in position 2, the capacitor will charge toward $-50 \, \text{V}$ $(-E_2)$, with the same time constant, τ, as above. Figure 8.34 shows a sketch of the variation of v_C with time. Also using Eq. (8.34) in this case and noting that here the time reference is displaced to $t = 2 \, \text{s}$, we obtain

$$v_C = V_0' + (-E_2 - V_0')(1 - e^{-t'/\tau})$$
$$= 79.7 + (-50 - 79.7)(1 - e^{-t'})$$
$$= 79.7 - 129.7 + 129.7e^{-t'}$$
$$= -50 + 129.7e^{-t'}$$

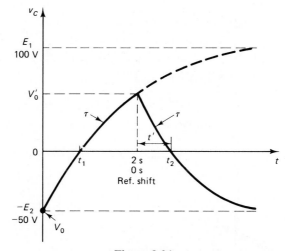

Figure 8.34

where t' is the time measured from the onset of placing the switch in position 2. With the switch in position 1, substituting $v_C = 0$ in the corresponding expression for v_C yields

$$0 = 100 - 150e^{-t_1}$$

$$e^{-t_1} = \frac{100}{150} = 0.667$$

Therefore,

$$t_1 = -\ln 0.667 = 0.405 \text{ s}$$

With the switch in position 2, substituting $v_C = 0$ in the corresponding expression for v_C gives us

$$0 = -50 + 129.7e^{-t'}$$

$$e^{-t'} = \frac{50}{129.7} = 0.3855$$

Therefore,

$$t' = -\ln 0.3855 = 0.953 \text{ s}$$

$$t_2 = 2 + 0.953 = 2.953 \text{ s}$$

EXAMPLE 8.12

In the circuit analyzed in Example 8.11, find
(a) The capacitor's current at $t = 3$ s.
(b) The times at which the current is 2.25 mA.

Solution Since the capacitor's voltage have been determined in Example 8.11 for the switching cycle described, the current can be obtained from the application of KVL. When the switch is in position 1,

$$E_1 = iR + v_C$$

Therefore,

$$i = \frac{1}{R}(E_1 - v_C)$$

$$= \frac{1}{50 \times 10^3} \times [100 - (100 - 150e^{-t})]$$

$$= 3 \times 10^{-3}e^{-t} \quad \text{A} \quad (0 \le t \le 2 \text{ s})$$

When the switch is in position 2,

$$iR + v_C + E_2 = 0$$

Therefore,

$$i = \frac{-1}{R}(E_2 + v_C)$$

$$= -\frac{1}{50 \times 10^3}[50 + (-50 + 129.7e^{-t'})]$$

$$= -2.594 \times 10^{-3}e^{-t'} \quad \text{A}$$

Figure 8.35 shows a sketch of the variation of i with time. Notice the shift in

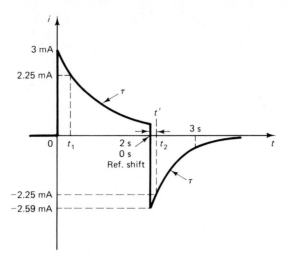

Figure 8.35

the time reference to the moment $t = 2\,\text{s}$, used to describe the current variation when the switch is in position 2.

(a) At $t = 3\,\text{s}$ this moment corresponds to a time of 1 s into the switching cycle when the switch is in position 2. Thus substituting $t' = 1\,\text{s}$ in the corresponding current equation, we have

$$i = -2.594 \times 10^{-3}e^{-1} = -0.954\,\text{mA}$$

(b) The current is $2.25\,\text{mA}$ first when $t = t_1$ (see Fig. 8.35), with the switch in position 1. Thus here

$$2.25 \times 10^{-3} = 3 \times 10^{-3}e^{-t_1}$$

$$e^{-t_1} = 0.75$$

Therefore,

$$t_1 = -\ln 0.75 = 0.288\,\text{s}$$

The next time the current magnitude is $2.25\,\text{mA}$ occurs at $t = t_2$, with the switch in position 2. Thus here

$$-2.25 \times 10^{-3} = -2.594 \times 10^{-3}e^{-t'}$$

$$e^{-t'} = 0.867$$

and

$$t' = -\ln 0.867 = 0.142\,\text{s}$$

Thus

$$t_2 = 2 + 0.142 = 2.142\,\text{s}$$

EXAMPLE 8.13

The pulse voltage waveform e_S in Fig. 8.36(a) resembles an intermittent dc source, producing either a maximum of $10\,\text{V}$ or a minimum of $0\,\text{V}$. Its waveform is shown in Fig. 8.36(b). At $t = 0$ there was no voltage on the capacitor.

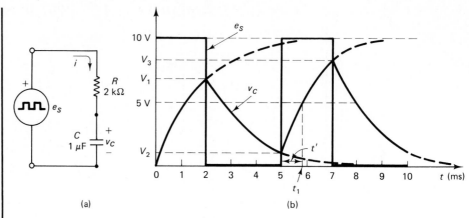

(a)   (b)

Figure 8.36

(a) Superimpose the capacitor's voltage on the source's waveform in Fig. 8.36(b).

(b) When is $v_C = 5$ V for the third time?

(c) Find i, dv_C/dt, and di/dt at the moment of time determined in part (b).

Solution   The source in this circuit acts like a switched dc source, continuously charging and discharging the capacitor, with a time constant τ:

$$\tau = RC = 2 \times 10^3 \times 1 \times 10^{-6} = 2\,\text{ms}$$

(a) The capacitor's voltage rises toward 10 V during each charging cycle and decays towards zero during each discharge cycle. However, neither cycle would be completed due to the source's waveform. The voltages reached at the end of each of the first three cycles are denoted V_1, V_2, and V_3, respectively. The capacitor's waveform is as sketched in Fig. 8.36(b)

(b) The graphical technique will be used to analyze this circuit. The first charging cycle follows the universal exponential rise curve:

$$y = 1 - e^{-x} \qquad \text{where} \qquad y = \frac{v_C}{10} \qquad \text{and} \qquad x = \frac{t}{\tau} = \frac{t}{2\,\text{ms}}$$

At the end of the first charging cycle, $t = 2$ ms,

$$x = \frac{2\,\text{ms}}{2\,\text{ms}} = 1$$

The corresponding value of y from the universal exponential rise curve in Fig. 8.20 is

$$y = 0.632 = \frac{V_1}{10}$$

Therefore,

$$V_1 = 6.32\,\text{V}$$

During the following discharge cycle, v_C follows the universal exponential decay curve:

$$y = e^{-x} \qquad \text{where} \qquad y = \frac{v_C}{V_1}$$

At the end of this discharge cycle ($t = 5$ ms), the discharge time is 3 ms. Thus

$$x = \frac{3\text{ ms}}{2\text{ ms}} = 1.5$$

The corresponding value of y from the universal exponential decay curve in Fig. 8.20 is

$$y = 0.223 = \frac{V_2}{V_1}$$

Therefore,

$$V_2 = 0.223 \times 6.32 = 1.41 \text{ V}$$

The following recharge cycle starts from V_2 toward 10 V (i.e., the maximum change in v_C in this case is $10 - V_2 = 10 - 1.41 = 8.59$ V). Here again v_C follows the universal exponential rise curve:

$$y = 1 - e^{-x} \quad \text{where} \quad y = \frac{v_C - 1.41}{8.59} \quad \text{[see Eq. (8.37b)]}$$

At the end of this recharge cycle ($t = 7$ ms), the recharge time is 2 ms. Thus

$$x = \frac{2\text{ ms}}{2\text{ ms}} = 1$$

The corresponding value of y from the universal exponential rise curve in Fig. 8.20 is

$$y = 0.632 = \frac{V_3 - 1.41}{8.59}$$

Therefore,

$$V_3 = 0.632 \times 8.59 + 1.41 = 6.84 \text{ V}$$

v_C is 5 V for the third time during this recharge cycle, at t_1. Here

$$y = \frac{v_C - 1.41}{8.59} = \frac{5 - 1.41}{8.59} = 0.418$$

The corresponding value of x from the universal exponential rise curve in Fig. 8.20 is

$$x = 0.54 = \frac{t'}{\tau} = \frac{t'}{2\text{ ms}}$$

Therefore,

$$t' = 1.08 \text{ ms}$$

and

$$t_1 = 5 + 1.08 = 6.08 \text{ ms}$$

(c) During the recharge cycle, KVL gives

$$e_S = v_C + iR$$

then

$$i = \frac{1}{R}(e_S - v_C)$$

At the moment v_C is 5 V,

$$i = \frac{1}{2 \times 10^3}(10 - 5) = 2.5\,\text{mA}$$

As $i = C\,dv_C/dt$ [Ohm's law for the capacitor, Eq. (8.20)],

$$\frac{dv_C}{dt} = \frac{i}{C} = \frac{2.5 \times 10^{-3}}{1 \times 10^{-6}} = 2.5\,\text{kV/s}$$

From Eq. (8.57)

$$\frac{di}{dt} = -\frac{i}{\tau} = -\frac{2.5 \times 10^{-3}}{2 \times 10^{-3}} = -1.25\,\text{A/s}$$

EXAMPLE 8.14

The switch in the circuit shown in Fig. 8.37 has been in position 1 for a very long time. It was placed in position 2 for 2 ms and then returned back to position 1.

(a) Superimpose on a common time axis the waveforms of v_{in}, v_C, and v_R for a total of 12 ms after the initial switching action.

(b) What is the minimum value of the capacitor's voltage, $v_{C\text{min}}$?

(c) How long does v_C remain below 5 V?

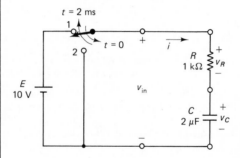

Figure 8.37

Solution   (a) For the first 2 ms, with the switch in position 2, v_{in} is zero due to the short circuit. When the switch is replaced back into position 1, v_{in} is E (10 V). Before the switching action, the switch was in position 1 for a very long time. This means that with the steady state achieved, the capacitor resembles an open circuit and its initial voltage is then E (10 V), that is,

$$v_C(0^+) = V_0 = 10\,\text{V}$$

The capacitor's voltage thus undergoes a discharging cycle for the first 2 ms and then a recharging cycle, with the switch back in position 1. Its voltage will rise toward the steady state voltage of E (10 V). During the first 2 ms, KVL provides

$$v_R + v_C = 0 \qquad \text{that is, } v_R = -v_C$$

After the switch is replaced back into position 1, v_R can be obtained from KVL for the circuit. In this case

$$v_R = E - v_C$$

316

The sketch of the variations described above is shown in Fig. 8.38, indicating the waveforms of v_{in}, v_C, and v_R.

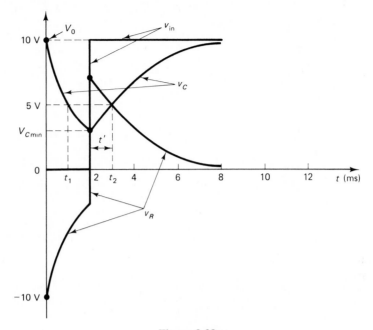

Figure 8.38

(b) During both the discharge and charge cycles, the time constant τ is the same:

$$\tau = RC = 1 \times 10^3 \times 2 \times 10^{-6} = 2 \text{ ms}$$

The capacitor's voltage initially decays exponentially according to

$$v_C = V_0 e^{-t/\tau} = 10 e^{-t/2 \text{ ms}}$$

The minimum value of the capacitor's voltage is reached at $t = 2$ ms. Therefore,

$$V_{C\text{min}} = 10 e^{-2 \text{ ms}/2 \text{ ms}} = 10 e^{-1} = 3.68 \text{ V}$$

(c) v_C is 5 V for the first time during the discharge cycle, at $t = t_1$. Thus here

$$5 = 10 e^{-t_1/2 \text{ ms}}$$
$$e^{-t_1/2 \text{ ms}} = 0.5$$

Therefore,

$$t_1 = -2 \text{ ms} \ln 0.5 = 1.386 \text{ ms}$$

During the recharge cycle the capacitor's voltage rises from its initial value of $V_{C\text{min}}$ ($=3.68$ V) toward a maximum of 10 V. Using Eq. (8.34), we have

$$v_C = V_{C\text{min}} + (E - V_{C\text{min}})(1 - e^{-t/\tau})$$
$$= 3.68 + 6.32 - 6.32 e^{-t/2 \text{ ms}} = 10 - 6.32 e^{-t/2 \text{ ms}} \quad \text{V}$$

Notice that the time here is being accounted for starting from the moment $t = 2$ ms. Thus when v_C reaches 5 V for the second time, at $t = t_2$ (see Fig. 8.38), corresponding to t' seconds during the recharge cycle,

$$5 = 10 - 6.32e^{-t'/2 \text{ ms}}$$

$$e^{-t'/2 \text{ ms}} = \frac{10 - 5}{6.32} = 0.791$$

Therefore,

$$t' = -2 \text{ ms} \ln 0.791 = 0.469 \text{ ms}$$

and

$$t_2 = 2 \text{ ms} + 0.469 \text{ ms} = 2.469 \text{ ms}$$

Thus the time during which v_C remains below 5 V is

$$t_2 - t_1 = 2.469 \text{ ms} - 1.386 \text{ ms} = 1.083 \text{ ms}$$

DRILL PROBLEMS

Section 8.1

8.1. Find the capacitance of the parallel-plate capacitor given that the area of each plate is 16 cm^2 and the distance between the plates is 0.1 mm. The dielectric is
(a) Air.
(b) Mica.
(c) Ceramic.

8.2. If a 20 V source is connected to each of the capacitors in Drill Problem 8.1, calculate the amount of charge and the amount of energy stored in each.

8.3. What is the area of the parallel plates of a 0.1 nF capacitor if the dielectric is a sheet of mica whose thickness is 0.2 mm?

Section 8.3

8.4. Obtain the total (equivalent) capacitance of the capacitors connected as shown in Fig. 8.39(a) and those connected as shown in Fig. 8.39(b).

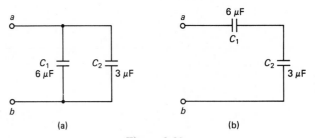

(a)   (b)

Figure 8.39

8.5. If a 12 V source is connected to points a and b of the capacitor connections shown in Fig. 8.39(a) and (b), find the charge stored in each capacitor and the stored charge in the equivalent capacitors obtained in Drill Problem 8.4.

8.6. Calculate the energy stored in each of the capacitors considered in Drill Problem 8.5.

8.7. Obtain the equivalent capacitance of the capacitors connected as shown in Fig. 8.40(a) and those connected as shown in Fig. 8.40(b).

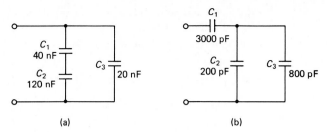

Figure 8.40

Section 8.4

8.8. Obtain the current in a $0.2\,\mu\mathrm{F}$ capacitor if the voltage across this capacitor has the waveform shown in Fig. 8.41.

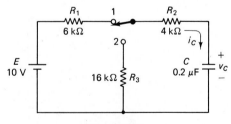

Figure 8.41

Section 8.5

8.9. The capacitor in the circuit shown in Fig. 8.42 was initially uncharged. Obtain the analytical expression for the voltage $v_C(t)$ if the switch was placed in position 1 at $t = 0\,\mathrm{s}$. Determine the value of v_C at $t = 1$, 2, and 5 ms.

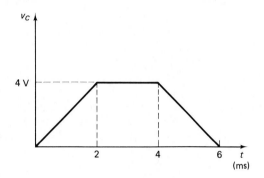

Figure 8.42

8.10. Obtain the analytical expression for the capacitor's current $i_C(t)$, in the network shown in Fig. 8.42 and analyzed in Drill Problem 8.9. Calculate the value of i_C at $t = 0\,\mathrm{s}$, 2 ms, and 5 ms.

8.11. What is the value of v_C, i_C, and v_{R_T} in the network shown in Fig. 8.42 at $t = 3\tau$ and at $t = 6\tau$, where τ is the charging cycle time constant?

8.12. At $t = 6\tau$, the switch in the network shown in Fig. 8.42 was placed in position 2. Considering this moment as the new time reference and using the initial capacitor's voltage as obtained in Drill Problem 8.11, obtain the new analytical expression for $v_C(t)$ during this discharge cycle.

8.13. Obtain the expression for $i_C(t)$ during the discharge cycle considered in Drill Problem 8.12.

8.14. Calculate the values of v_C and i_C at $t = 0$ s, 5 ms, and 10 ms during the discharge cycle examined in Drill Problems 8.12 and 8.13.

8.15. Obtain the values of v_{12} (the voltage across the switch terminals), dv_C/dt (the rate of change of the capacitor's voltage), and di_C/dt (the rate of change of the capacitor's current) at the three instants of time considered in Drill Problem 8.14.

PROBLEMS

Note: The problems below that are marked with an asterisk (*) can be used as exercises in computer-aided network analysis or as programming exercises.

Section 8.1

(i) **8.1.** Find the capacitance of a parallel-plate capacitor if the area of each plate is 50 cm², the distance between the plates is 0.5 mm, and the dielectric is
   (a) Air.
   (b) Mica.

(i) **8.2.** Find the separation between the plates of the following capacitors, each having an area of 20 cm².
   (a) A 1 μF capacitor; the dielectric is ceramic.
   (b) A 0.22 nF capacitor; the dielectric is glass.
   (c) A 5.1 pF capacitor; the dielectric is mica.

(i) **8.3.** Calculate the charge and energy stored in each of the following capacitors:
   (a) A 2 μF capacitor with a voltage of 25 V.
   (b) A 3.3 nF capacitor with a voltage of 10 V.

Section 8.3

(i) **8.4.** Find the equivalent capacitance for the capacitive circuit shown in Fig. 8.43. Calculate the voltage across, the charge, and the energy stored in each of these capacitors.

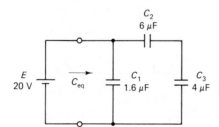

Figure 8.43

(i) **8.5.** Repeat Problem 8.4 for the circuit shown in Fig. 8.44.

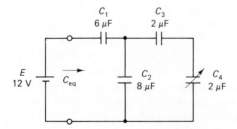

Figure 8.44

Chapter 8   Capacitance and Transients in *RC* Circuits

(i) **8.6.** If C_4 in Problem 8.5 (Fig. 8.44) is made variable, find its value so that
  (a) C_{eq} is 3.5 μF.
  (b) C_{eq} is 3.7 μF.

Section 8.5

(i) **8.7.** In the source-free circuit shown in Fig. 8.45, $v_C(0^+)$ is 20 V. Find $v_C(t)$ and $i(t)$ for all $t \geq 0$. Use these expressions to obtain v_C and i at
  (a) $t = 2$ ms.
  (b) $t = 4$ ms.

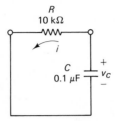

Figure 8.45

(i) **8.8.** The switch in the circuit shown in Fig. 8.46 remained in position 1 for a very long time. At $t = 0$ the switch was placed in position 2. Determine $v_C(0^+)$ and express v_C and i as functions of time for all $t \geq 0$. Find v_C and i at $t = 0.3$ s. At this instant determine v_{12}, the voltage across the switch terminals.

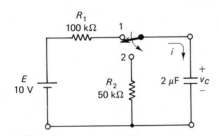

Figure 8.46

(i) ***8.9.** In the circuit shown in Fig. 8.46, the switch remained in position 2 for a very long time. It was placed in position 1 at $t = 0$. Determine v_C and i as functions of time for all $t \geq 0$. Also find the values of v_C and i at $t = 0.3$ s.

(ii) **8.10.** Steady state has been reached in the circuit shown in Fig. 8.47 before the switch was opened at $t = 0$. Find $v_C(0^+)$, then determine v_C and i as functions of time for all $t \geq 0$. What will be the voltage across the switch contacts at $t = 10$ ms?

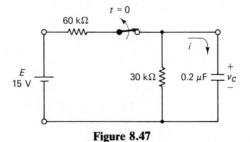

Figure 8.47

(i) ***8.11.** Use the generalized solution form to obtain v_C and i_s in the circuit shown in Fig. 8.48 if $v_C(0^+)$ is 10 V. What will be the values of v_C and i_s after a very long time?

Problems

321

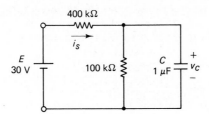

Figure 8.48

(i) *8.12. With the switch in the circuit shown in Fig. 8.49 being opened for a very long time, determine the initial voltage across the capacitor. The switch was then closed at $t = 0$. Determine $v_C(t)$ using the generalized solution technique.

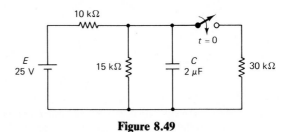

Figure 8.49

(i) *8.13. Repeat Problem 8.12 but with the switch being closed for a very long time and then opened at $t = 0$.

(i) 8.14. As the switch in the circuit shown in Fig. 8.50(a) is being closed, which of the waveforms shown represents the voltage across the resistor?

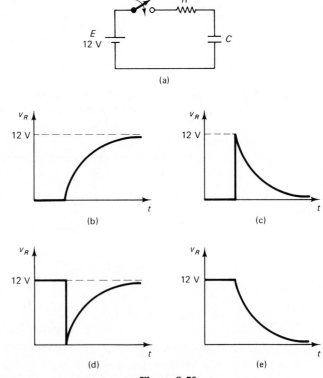

Figure 8.50

(ii) **8.15.** There is no charge initially stored on the capacitor in the circuit shown in Fig. 8.51. At $t = 0$, the switch was placed in position 1. It was then moved to position 2 at $t = 1.5$ ms, and it remained there. Find

    (a) v_R at $t = 1$ ms and at $t = 2$ ms.
    (b) dv_C/dt at $t = 1$ ms and at $t = 2$ ms.
    (c) During this switching procedure, how long does the voltage across the capacitor remain above 5 V?

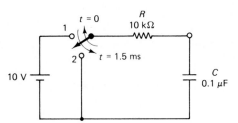

Figure 8.51

(ii) **8.16.** The switch in the circuit shown in Fig. 8.52 remained in position 2 for a very long time. At $t = 0$ it was placed in position 1 and then at $t = 2$ s it was moved back to position 2, where it remained. Find v_{AB} at $t = 1$ s and at $t = 3$ s.

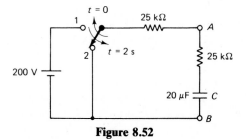

Figure 8.52

(ii) **8.17.** In the circuit of Fig. 8.53, the switch has been open for a very long time. It was closed at $t = 0$ and reopened again at $t = 1.6$ s, remaining open thereafter. Find

    (a) The source current at $t = 1$ s.
    (b) dv_C/dt at $t = 4.1$ s.
    (c) The voltage across the switch's terminals at $t = 4.1$ s.

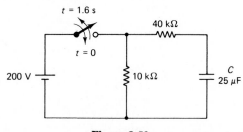

Figure 8.53

(ii) **8.18.** The switch has been in position 2 and steady state was reached in the circuit of Fig. 8.54. The switch was placed in position 1 at $t = 0$ and then replaced back into position 2 at $t = 2$ s. Find the voltage across the 8 kΩ resistor at $t = 3$ s.

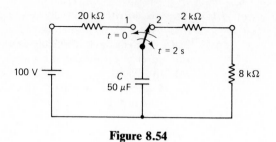

Figure 8.54

(ii) **8.19.** The switch in the circuit shown in Fig. 8.55 has been in position 1 until steady state is reached. At $t = 0$ the switch was moved to position 2. At $t = 1$ ms, the switch was returned to position 1 and left there. Superimpose on a common time axis between $t = -1$ ms and $t = 9$ ms plots of v_A, v_B, and v_{AB}, indicating the values of these voltages at $t = 1$ ms.

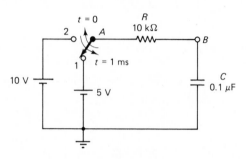

Figure 8.55

(ii) **8.20.** In the circuit shown in Fig. 8.56, there was initially no charge on the capacitor. The switch is moved from position A to position B for 10 ms and then returned to position A. Superimpose on a common time axis, for t varying from 0 to 20 ms, the voltages v_{in}, v_R, and v_C, indicating the values of these voltages at $t = 10$ ms.

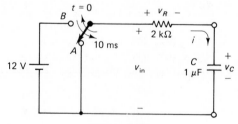

Figure 8.56

(ii) **8.21.** In Problem 8.20, calculate
  (a) v_C and i at $t = 4$ ms.
  (b) v_C and i at $t = 14$ ms.
  (c) di/dt at $t = 12$ ms.

(iii) **8.22.** Steady state was reached with the switch in position 1 in the circuit shown in Fig. 8.57. At $t = 0$ the switch was moved to position 2 for 80 ms and then returned to position 1.
Find
  (a) v_C and dv_C/dt at $t = 80^+$ ms (right after the switching action).
  (b) v_{12}, the voltage across the switch's contacts at $t = 80^+$ ms.

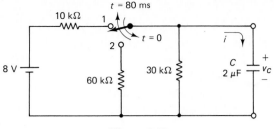

Figure 8.57

(ii) **8.23.** In the circuit shown in Fig. 8.57, the switch remained in position 2 for a very long time. It was then moved to position 1 for 45 ms and then returned back to position 2. Find

(a) v_C and dv_C/dt at $t = 45^+$ ms (right after the switching action).

(b) v_{12}, the voltage across the switch's contacts at $t = 45^+$ ms.

APPENDIX A8.1: ENERGY STORED IN A CAPACITOR

For the general time-varying electrical quantities, the electrical current i is the rate of change of charge q with respect to time, that is,

$$i = \lim_{\Delta t \to 0} \frac{\Delta q}{\Delta t} = \frac{dq}{dt} \quad \text{A} \tag{8A.1.1}$$

while the power p and energy w are defined as

$$p = vi = \frac{dw}{dt} \tag{8A.1.2}$$

Notice that the electrical power is also defined as the rate of change of energy with respect to time. Also, for a capacitor,

$$q = Cv \tag{8A.1.3}$$

Using the calculus relationship between integrals and differentials of variable quantities, Eq. (8A.1.2) can be expressed as

$$w = \int p \, dt = \int vi \, dt \tag{8A.1.4}$$

But the capacitor's current is

$$i = \frac{dq}{dt} = \frac{d(Cv)}{dt} = C\frac{dv}{dt} \tag{8A.1.5}$$

Now substituting from Eq. (8A.1.5) into Eq. (8A.1.4) for the capacitor's current i, we have

$$w = \text{energy stored in a capacitor}$$

$$= \int vC\frac{dv}{dt} \, dt = C\int v \, dv = \tfrac{1}{2}Cv^2 \quad \text{J} \tag{8A.1.6}$$

APPENDIX A8.2: DETERMINATION OF THE CAPACITOR'S VOLTAGE $v_c(t)$ DURING TRANSIENTS [SOLUTION OF EQS. (8.28c) AND (8.47)]

The first order linear differential equation describing the variation of v_C in an RC circuit driven by a dc source E [Eq. (8.28c), which is KVL for the circuit] is rewritten

$$R_1 C \frac{dv_C}{dt} + v_C = E \qquad (8A.2.1)$$

with the general condition that the initial voltage across the capacitor is

$$v_C(0^+) = v_C(0^-) = V_0 \qquad \text{V} \qquad (8A.2.2)$$

Denoting

$$R_1 C = \tau_C = \text{charging time constant} \qquad (8A.2.3)$$

and dividing all terms in Eq. (8A.2.1) by the charging time constant give us

$$\frac{dv_C}{dt} + \frac{v_C}{\tau_C} = \frac{E}{\tau_C}$$

or

$$\frac{dv_C}{dt} = \frac{1}{\tau_C}(E - v_C)$$

Separating the variables v_C and t yields

$$\frac{dv_C}{E - v_C} = \frac{1}{\tau_C} dt$$

Now, integrating both sides of the equation above, we obtain

$$\int \frac{1}{E - v_C} dv_C = \int \frac{1}{\tau_C} dt + \text{constant } (k)$$

$$-\ln(E - v_C) = \frac{t}{\tau_C} + k$$

Then

$$E - v_C = e^{-t/\tau_C} e^{-k} = k' e^{-t/\tau_C}$$

where the arbitrary constant $k'(=e^{-k})$ is yet to be determined. Therefore,

$$v_C(t) = E - k' e^{-t/\tau_C} \qquad (8A.2.4)$$

This expression for $v_C(t)$ is valid at every instant of time t. Therefore, using the initial condition [Eq. (8A.2.2)], the arbitrary constant k' can be determined as follows:

$$v_C(0^+) = V_0 = E - k' e^{-0} = E - k'$$

Thus

$$k' = E - V_0$$

Substituting this value of k' into Eq. (8A.2.4) gives

$$v_C(t) = E - (E - V_0)e^{-t/\tau_C} \qquad (8A.2.5)$$

Subtracting the term V_0 from both sides of the equation above yields

$$v_C(t) - V_0 = E - V_0 - (E - V_0)e^{-t/\tau_C}$$

Thus

$$v_C(t) - V_0 = (E - V_0)(1 - e^{-t/\tau_C}) \qquad (8A.2.6)$$

This is the required general expression for $v_C(t)$ as a function of time t.

In the case when the initial voltage on the capacitor is zero (capacitor initially uncharged), that is,

$$v_C(0^+) = v_C(0^-) = V_0 = 0$$

then substituting $V_0 = 0$ in Eq. (8A.2.6) will simplify it, resulting in

$$v_C(t) = E(1 - e^{-t/\tau_C}) \qquad \text{V} \qquad (8A.2.7)$$

For the source-free circuit shown in Fig. 8.21, the first-order linear differential equation describing the variation of v_C during the discharge cycle is given by Eq. (8.47) (which is KVL for this circuit). Here

$$\frac{dv_C}{dt} = -\frac{1}{(R_1 + R_2)C}v_C = -\frac{1}{\tau_d}v_C \qquad (8A.2.8)$$

where

$$\tau_d = \text{discharge time constant} = (R_1 + R_2)C \qquad (8A.2.9)$$

The initial voltage across the capacitor is

$$v_C(0^+) = V_0 \qquad (8A.2.10)$$

Through the separation of variables, Eq. (8A.2.8) can be rewritten as

$$\frac{dv_C}{v_C} = -\frac{dt}{\tau_d}$$

Integrating both sides of the equation above gives us

$$\int \frac{1}{v_C}\, dv_C = -\int \frac{1}{\tau_d}\, dt + \text{constant } (k)$$

$$\ln v_C = -\frac{t}{\tau_d} + k$$

Thus

$$v_C(t) = e^{-t/\tau_d}e^k = K'e^{-t/\tau_d} \qquad (8A.2.11)$$

The arbitrary constant $K'(=e^k)$ can be determined, using the initial condition [Eq. (8A.2.10)]. As Eq. (8A.2.11) is valid at every instant of time, then substituting $t = 0^+$,

$$v_C(0^+) = V_0 = K'e^{-0} = K'$$

Therefore,

$$v_C(t) = V_0 e^{-t'/\tau_d} \qquad (8A.2.12)$$

This is the expression required, describing $v_C(t)$ as a function of time during the discharge cycle.

GLOSSARY

Breakdown Voltage: The voltage required to produce conduction in an insulating material.

Capacitance: The measure of the capacitor's ability to store charge and energy. It is the amount of charge stored in the capacitor per unit voltage applied to it.

Capacitor: The electrical circuit element that exhibits opposition to the change in voltage across its two terminals. It has the ability of storing charge on its plates and of storing energy in its electrostatic field.

Dielectric (or Insulator): Material that does not normally allow the flow of electric current. It exhibits extremely large resistance value (approaching infinity) (e.g., air, mica, ceramic, polystyrene, etc.).

Electric Field: The physical space in which the electrical forces of attraction or repulsion are exhibited, affecting a unit positive charge placed in this space.

Permittivity: The measure of the ability of the dielectric material to permit the establishment of an electric field within the dielectric.

Relative Permittivity: The permittivity of a dielectric material relative to that of free space (air).

Steady-State Conditions: The values of the voltages and currents in an electric circuit that are attained at the end of the transient phase.

Stray Capacitance: Capacitance that exists not through design, but due to two conducting surfaces separated by a very thin dielectric material.

Time Constant (τ): The product of the value of the equivalent resistance in series with the capacitor, times the value of the capacitance in RC circuits. It is the time after which the voltage across the capacitor rises to 63.2% of its steady-state value during the charging cycle; and it is also the time after which the voltage across the capacitor drops to 36.8% of its initial value during the discharge cycle.

Transients: The variations in the voltage and current waveforms, occurring within a finite span of time (transient time), due to a switching action in an electric circuit in order to charge or discharge energy storage elements in this circuit. The transients usually occur before the steady-state conditions in the circuit have been achieved.

ANSWERS TO DRILL PROBLEMS

8.1. (a) 141.6 pF, (b) 708 pF, (c) 1.062 μF

8.3. $A = 4.52$ cm^2

8.5. (a) $Q_1 = 72\,\mu$C, $\quad Q_2 = 36\,\mu$C, $Q_T = 108\,\mu$C (b) $\quad Q_1 = Q_2 = Q_T = 24\,\mu$C, $V_1 = 4$ V, $V_2 = 8$ V

8.7. (a) $C_{eq} = 50$ nF, (b) $C_{eq} = 750$ pF

8.9. $v_C(t) = 10(1 - e^{-500t})$ V, v_C (1 ms) = 3.935 V, v_C (2 ms) = 6.321 V, v_C (5 ms) = 9.179 V

8.11. $v_C(3\tau) = 9.502$ V, $\quad v_C(6\tau) = 9.975$ V, $\quad i_C(3\tau) = 0.0498$ mA, $i_C(6\tau) = 2.479\,\mu$A, $\quad v_{R_T}(3\tau) = 0.498$ V, $v_{R_T}(6\tau) = 0.025$ V

8.13. $i_C(t) = -0.4988e^{-250t}$ mA

8.15. $v_{12}(0) = 2.019$ V, $\quad \dfrac{dv_C}{dt}(0) = -2494$ V/s, $\quad v_{12}(5\,\text{ms}) = 7.714$ V, $\dfrac{dv_C}{dt}(5\,\text{ms}) = -714.5$ V/s, v_{12} (10 ms) = 9.346 V, $\dfrac{dv_C}{dt}(10\,\text{ms}) = -204.5$ V/s, $\dfrac{di_C}{dt}(0) = 0.1247$ A/s, $\dfrac{di_C}{dt}(5\,\text{ms}) = 0.03573$ A/s, $\dfrac{di_C}{dt}(10\,\text{ms}) = 0.01023$ A/s

9 | INTRODUCTION TO MAGNETISM

OBJECTIVES

- Familiarity with the concept of the magnetic field and its properties.
- Understanding the physical interrelationship between electricity and magnetism as exhibited through the electromagnetic phenomena.
- Familiarity with Maxwell's right-hand rule and its application.
- Ability to calculate the parameters of simple magnetic circuits, including reluctance, magnetomotive force, magnetic field intensity, flux and flux density.
- Ability to use Ohm's law for magnetic circuits and Ampere's Circuital Law in simple applications.
- Familiarity with magnetic permeability and the hysteresis phenomena.
- Understanding the concepts of Electromagnetic Induction, Faraday's and Lenz's laws and their use in practical electromagnetic applications, particularly the generator and motor principles.

Many physical phenomena, such as gravity, electrical forces of attraction and repulsion between charged bodies, and magnetism, are classified as field phenomena because the forces they exhibit act on bodies placed in their space of influence without physical contact. The space in which these forces of attraction and repulsion are exhibited is called the *field* of the physical phenomena.

A magnet is made of a material such as iron or steel. It has two different types of polarizations at its ends, as shown in Fig. 9.1. These *poles* are referred to as the north (N) and south (S) poles. These polarizations are produced due to the orientation of the resultant of atomic magnetic moments produced by electron spins (around their axis) while rotating in their orbits around their parent atoms.

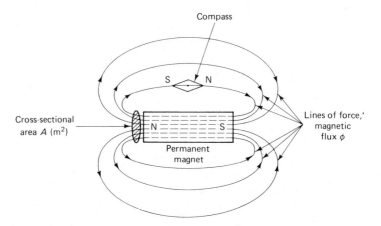

Figure 9.1   Magnetic field of a permanent magnet.

The space around the magnet in which other magnets (or magnetic materials, e.g., a compass needle as in Fig. 9.1) are affected by the invisible field forces is called the *magnetic field* of the magnet. In the figure, the compass needle will deflect such that its N pole is attracted toward the S pole of the permanent magnet (i.e., unlike poles attract each other and like poles repulse each other). Thus the magnetic poles behave in a manner analogous to the positively and negatively charged bodies in electrostatic fields. In fact, as shown in this chapter, electricity and magnetism are so closely interrelated that they can be considered as one physical phenomenon, called the *electromagnetic field phenomenon*.

As shown in Fig. 9.1, the magnetic field is represented by invisible *lines of force*, called the *magnetic flux* (ϕ). Lines of force always form a closed path with no specific origin. They possess direction, which is conventionally taken to be the direction in which a hypothetically isolated north (N) pole would move when placed in the magnetic field. Obviously, this N pole would be repulsed by the N pole of the permanent magnet and attracted toward the S pole of this magnet. Thus, outside the magnet (i.e., in the magnetic field), the lines of force are always in the direction from north to south, but inside the magnet the direction is from south to north (i.e., a continuous closed path). The concept of lines of force was formulated through observing the way in which sprinkled iron dust would be shaped if placed on a sheet of paper on top of the magnet (i.e., in the magnetic field). A stronger magnetic field has a larger field space associated with it (i.e., it must be represented by more lines of force). The overall number of these lines of force is representative of the magnetic quantity referred to above as the magnetic flux ϕ.

In the MKS (SI) system of units, the unit of ϕ is the *weber,* abbreviated Wb. As will be shown later, this unit is related to the electrical units. In fact,

$$1 \text{ weber} = 1 \text{ volt} \cdot \text{second}.$$

Of even more importance in quantifying the magnetic field is the density (or concentration) of these lines of force per unit cross-sectional area perpendicular to the direction of the lines of force. This quantity is called the flux density B and is defined at any given position as

$$B = \frac{\phi}{A} \quad \text{Wb/m}^2 \text{ or tesla (T)} \tag{9.1}$$

where A is the cross-sectional area (m²), as indicated in Fig. 9.1.

The magnetic lines of force (flux lines) possess some basic characteristics which are very useful in understanding the effects of the magnetic fields:

1. They possess direction (N to S) and always form closed loops.
2. They never intersect.
3. They tend to repulse each other and hence diverge as one moves away from poles.
4. They exhibit tension along their length and tend to be as short as possible, like an expanded elastic band.
5. They tend to take up the path of least resistance; that is, they are easier to set up (more lines of force) in a soft-iron medium (i.e., a magnetic material) than they are in a nonmagnetic material such as air, copper, or glass.

A magnetic material is an element that can be polarized (N and S poles).

It is easy to see that some of the properties of magnetic flux mentioned above are also common to electric current. Thus the analogy: *Magnetic flux in a magnetic circuit resembles electric current in electric circuits.* The main contrast in this analogy is that there is no real flow attached to magnetic flux, versus the flow of electric charge carriers constituting the electric current.

In Fig. 9.2(a), based on properties 1, 2, and 4, the two magnets, when placed as shown, will experience a force of attraction. On the other hand, when

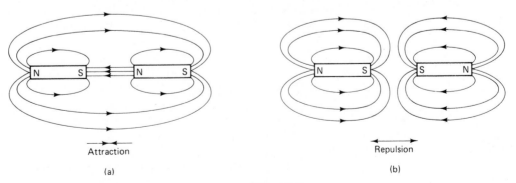

Attraction

(a)

Repulsion

(b)

Figure 9.2

the two magnets are placed as shown in Fig. 9.2(b), properties 1, 2, and 3 explain why the force between the two magnets is now a force of repulsion.

In Fig. 9.3(a), property 5 is clearly exhibited. The copper rod placed in the magnetic field hardly changes the orientation of the lines of force, whereas the soft-iron rod is an easy path for the lines of force to take versus the air, hence the change of direction shown. This property also explains why soft-iron magnetic shields are placed around sensitive electronic devices, circuits, or measuring instruments to redirect the magnetic field and create a magnetic field-free space, thus minimizing or preventing the interference of the magnetic field with the function of the electronic device or circuit, as shown in Fig. 9.3(b).

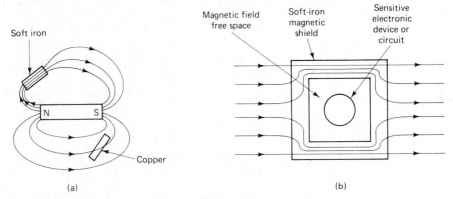

Figure 9.3

9.2 ELECTROMAGNETIC PHENOMENA

Oersted and Ampère observed that a conductor carrying an electric current I exhibits a magnetic field in the space surrounding it. Figure 9.4 indicates that the current-induced magnetic flux lines are in the form of concentric circles. This can be confirmed by sprinkling iron dust on a sheet of paper placed perpendicular to the conductor. These magnetic flux lines also possess direction, as indicated by the deflection of a compass needle placed on this sheet of paper.

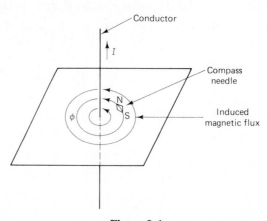

Figure 9.4

This observation means that electric current produces magnetic flux. This is called *electromagnetism*. The flux ϕ is increased if the current is increased. An easy method is devised to determine the direction of this electrically induced magnetic flux lines. This is called *Maxwell's right-hand rule* (RHR). It states that *if the conductor is held with the right hand such that the thumb points in the direction of the current flow I, the fingers circling the conductor point in the direction of the magnetic flux lines.*

This experimental observation clearly indicates that electricity and magnetism are, in fact, interrelated phenomena. Also, as observed above and in the rest of this chapter, every magnetic quantity has an analogous electric quantity. This interrelationship also indicates that electrical units can be used to express magnetic quantities (i.e., electrical and magnetic units are interrelated).

If the current-carrying conductor is formed in the shape of a loop as shown in Fig. 9.5, the induced magnetic flux lines will pass through the inside of the loop in the same direction, into the plane of the loop. The "×" symbol indicates the end of the flux arrow. Thus the field is strengthened by concentrating it in a smaller area. Further concentration of the magnetic field is achieved if the conductor is formed into a coil (solenoid) of n turns or loops as shown in Fig. 9.6. Also, if a magnetic material (e.g., iron, steel, or cobalt) is placed in the center of the coil (called the former), further concentration of the magnetic field is achieved. Evidently, this produces an electromagnet with north and south poles. The direction of the flux lines are still in accordance with RHR. Note that every loop produces magnetic flux lines from right to left, in the same direction, inside the coil, which clearly continue from left to right (N and S poles) outside the coil. The magnetic field thus produced can be controlled in strength by varying the current magnitude, using the variable resistance R in Fig. 9.6.

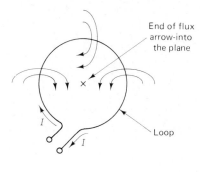

Figure 9.5

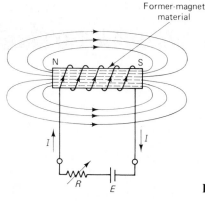

Figure 9.6

Two parallel conductors carrying current in the same or opposite directions, as shown in Fig. 9.7, will induce magnetic fields in the form of concentric circles as indicated. Note that "×" indicates the end of the current arrow (i.e., current in) and "·" indicates the tip of the current arrow (i.e., current out). The RHR is applied to obtain the directions of the magnetic lines of force. As shown in Fig. 9.7(a), when the conductors' currents are in the same direction, the magnetic lines of force between the conductors cancel out in the middle and the lines of force on the outside will encircle both conductors, being in the same direction. According to the fourth property of the lines of force, they exhibit tension along their length and thus result in a force of attraction between the two conductors. On the other hand, Fig. 9.7(b) shows the lines of field mapping of the magnetic flux when the two conductors carry current in the opposite direction. In the middle of the field space all the lines of force are in the same direction (crowding). According to the third property of the lines of force, they will tend to repulse each other, resulting in a repulsion force between the two conductors.

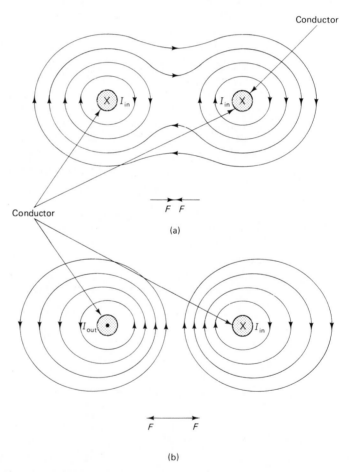

Figure 9.7 (a) Conductor currents in the same direction result in attraction force; (b) conductor currents in the opposite direction result in repulsion force.

9.3 SIMPLE MAGNETIC CIRCUIT: THE MAGNETIC FIELD INTENSITY

Based on the electromagnetic phenomena introduced above, if a current-carrying conductor is formed into a coil with n turns around a doughnut-shaped iron (magnetic material) core as shown in Fig. 9.8, the electrically induced magnetic flux lines will be concentrated, flowing in closed circular loops inside the iron core (in accordance with the properties of the magnetic lines of force). The result is an example of a simple magnetic circuit.

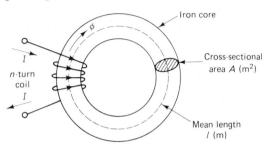

Figure 9.8   Simple magnetic circuit.

The flux ϕ in this magnetic circuit is analogous to the electric current in an electric circuit, even though no real flow exists here. Other analogies between a magnetic circuit and an electric circuit can also be observed. In the preceding section it was concluded that the cause of the induced magnetic flux is the flow of electric current. In Fig. 9.8 there are n turns, each having a current flow of I. Thus the cause of the induced flux ϕ is the current flow nI. The analogy in electric circuits is the fact that a voltage potential energy difference is the cause of the flow of current carriers [i.e., V (or E) causes I].

The quantity nI [in ampere-turns (At)] is called the *magnetomotive force* (MMF). It is the driving force for the existence of the magnetic flux ϕ. Thus

$$F_m = \text{magnetomotive force (MMF)} = NI \quad \text{At} \tag{9.2}$$

If the same magnetomotive force is applied to similar iron cores, each with a different mean length l (in meters), one would expect that stronger magnetic fields will result in iron cores with shorter l. Another magnetic quantity is thus defined, being a reflection of the magnetic field intensity, or the *magnetizing force*:

$$H = \text{magnetic field intensity} = \frac{F_m}{l} = \frac{nI}{l} \quad \text{At/m} \tag{9.3}$$

9.4 OHM'S LAW FOR A MAGNETIC CIRCUIT: RELUCTANCE AND PERMEABILITY

9.4.1 Reluctance

When the current I or the number of of turns n is increased in the simple magnetic circuit of Fig. 9.8, the magnetomotive force F_m is increased, resulting in a higher flux ϕ in the magnetic core, in accordance with the electromagnetic phenomena. Thus

$$F_m \propto \phi$$

Therefore,

$$F_m = k\phi \tag{9.4}$$

Based on the analogies established previously, E or V is analogous to F_m ($= nI$) and I is analogous to ϕ, Eq. (9.4) is then similar to Ohm's law for electric circuits: $V = RI$.

Thus the constant of proportionality k in Eq. (9.4) is actually a measure of opposition to the establishment of magnetic flux. This quantity is called the reluctance $\mathcal{R}$ of the magnetic circuit, and it is analogous to the resistance R of an electric circuit. Hence Ohm's law for a magnetic circuit can be expressed as

$$F_m = \mathcal{R}\phi \tag{9.5}$$

$$\boxed{\mathcal{R} = \frac{F_m}{\phi} \quad \text{At/Wb}} \tag{9.5a}$$

[notice the units of reluctance in Eq. (9.5a)]. Clearly, the higher the reluctance, the smaller would be the resulting magnetic flux ϕ for a certain fixed magnetomotive force F_m, and vice versa.

9.4.2 Permeability

It is easier to establish or set up the magnetic flux lines in some materials (e.g., iron) than it is in other materials (e.g., air). The magnetic lines of force, like electric current, always try to follow the path of least resistance. *Permeability is the property of materials that measures its ability to permit the establishment of magnetic lines of force.* It is analogous to conductivity in electric circuits. Air (or vacuum) is taken as the reference material. Its permeability is called μ_0. The permeability μ of any other material is

$$\mu = \mu_r\mu_0 \tag{9.6}$$

where μ_r is called the relative permeability; it is a dimensionless quantity. Non-magnetic materials (e.g., air, glass, copper, and aluminum) are characterized by their μ_r, which is approximately unity. Materials that have lower permeability than air (μ_r is a fraction) are called *diamagnetic materials*. Those materials that have slightly higher permeability than air (i.e., μ_r is 1 to 10) are called *paramagnetic materials*. On the other hand, magnetic materials such as iron, steel, cobalt, nickel, and alloys of such materials are called *ferromagnetic materials*, being characterized by their high values of μ_r (from 100 to 100,000).

From the definitions of reluctance $\mathcal{R}$ and permeability μ, it is clear that one refers to the opposite of the other (i.e., $\mathcal{R}$ is inversely proportional to μ). It is also easy to see that when the length of the magnetic circuit, l, is increased, the longer path of the magnetic lines of force implies higher reluctance. On the other hand, the higher cross-sectional area of the magnetic circuit means that more lines of force can be established (i.e., smaller reluctance). Based on these observations,

$$\boxed{R = \frac{l}{\mu A} \quad \text{At/Wb}} \tag{9.7}$$

From this equation it is easy to deduce the units of μ (or μ_0), and also a very important relationship between B, the flux density, and H, the magnetic field intensity causing it:

$$\mu = \frac{1}{\mathcal{R}}\frac{l}{A}$$

$$= \frac{\phi}{F_m}\frac{l}{A} \quad \text{[using Eq. (9.5)]}$$

$$= \frac{\phi}{A}\frac{1}{F_m/l} = \frac{B}{H} \quad \text{[using Eqs. (9.1) and (9.3)]} \quad (9.8)$$

or

$$\boxed{B = \mu H} \quad (9.8a)$$

Thus the unit of μ is

$$\frac{\text{webers/m}^2}{\text{ampere-turns/m}} = \text{webers/A} \cdot \text{m}$$

Note that the turns have no units. For air

$$\mu_0 = 4\pi \times 10^{-7}\,\text{Wb/A} \cdot \text{m}$$

As derived in Eq. (9.8), μ is the ratio of the flux density B and the corresponding field intensity H, producing the magnetic flux. For a given magnetic material, μ is a constant only in a fairly narrow range of H, as shown in Fig. 9.9. In general, then, μ is a function of H, resulting in a nonlinear relationship between B and H, as the curve in Fig. 9.9 indicates. When H is increased beyond a certain value, the resulting flux density B is minimally affected, and eventually B does not increase any further. This is called the *saturation region*. In this region, B is approximately constant at the value B_{sat}.

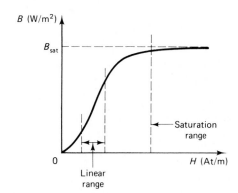

Figure 9.9 Typical *B–H* curve for a magnetic material.

EXAMPLE 9.1

The simple magnetic circuit of Fig. 9.8 has a cross-sectional area of 50 cm^2 and a mean length of 2 m. The relative permeability of the magnetic material of the core is 80. Find the reluctance of this magnetic circuit. If the current coil has 150 turns and the resulting flux is 80 μWb, what is the value of the current flowing in the coil?

Solution Since $\mu_r = 80$,

$$\mu = \mu_r \mu_0 = 80 \times 4\pi \times 10^{-7} = 1.005 \times 10^{-4} \, \text{Wb/A} \cdot \text{m}$$

Now as μ, l, and A are known, it is easy to find $\mathcal{R}$ from Eq. (9.7):

$$\mathcal{R} = \frac{l}{\mu A} = \frac{2}{1.005 \times 10^{-4} \times 0.005} = 3.98 \times 10^6 \, \text{At/Wb}$$

(note that $50 \, \text{cm}^2 = 0.005 \, \text{m}^2$). Using Ohm's law for this magnetic circuit gives us

$$\mathcal{R} = \frac{F_m}{\phi}$$

Therefore,

$$F_m = \mathcal{R} \times \phi = 3.98 \times 10^6 \times 80 \times 10^{-6} = 318.4$$
$$= nI = 150I$$
$$I = \frac{318.4}{150} = 2.123 \, \text{A}$$

EXAMPLE 9.2

The simple magnetic circuit in Fig. 9.8 has a magnetomotive force of 200 At. l is 0.4 m and μ is $6 \times 10^{-4} \, \text{Wb/A} \cdot \text{m}$. Find the flux density B in the iron core.

Solution Here

$$H = \frac{F_m}{l} = \frac{200}{0.4} = 500 \, \text{At/m}$$

Since from Eq. (9.8), $\mu = B/H$,

$$B = \mu H = 6 \times 10^{-4} \times 500 = 0.3 \, \text{Wb/m}^2$$

9.5 HYSTERESIS

When a specimen of an initially unmagnetized magnetic material is subjected to a magnetomotive force (ni), as in the setup of Fig. 9.10(a), an induced magnetic field will be established (ϕ). If the average length and cross-sectional area of the magnetic circuit are known, it is possible to plot the variation of H ($= ni/l$), the field intensity, versus the resulting B ($= \phi/A$), the flux density, as i is varied in magnitude and direction.

As i is increased from zero, in the assumed positive direction indicated in Fig. 9.10(a), H will increase, resulting in an increase in the flux density B, following the normal B–H curve (as in Fig. 9.9) of this magnetic material. This corresponds to the curve oa in Fig. 9.10(b). At point a the material is saturated and no further increase in B will result even when the current is increased. When the current i is then reduced back to zero, the B–H curve will follow the portion ab in Fig. 9.10(b). At point b, H is zero but B is finite at the value B_r, called the residual flux density, indicating that the iron core has now become a magnet, retaining a finite magnetic flux. This lagging of B behind H is the phenomenon called *hysteresis*.

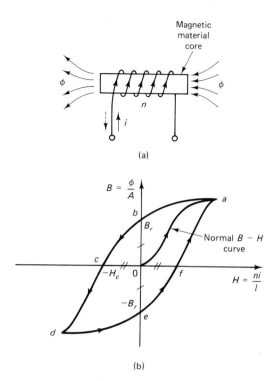

Figure 9.10 (b) Typical hysteresis curve for iron.

To demagnetize the magnet thus formed, the direction of the current i must be reversed, as indicated by the dashed arrow in Fig. 9.10(a). Once i is reversed, H becomes negative. As i is then increased in this reverse direction, H increases in the negative direction and the flux density B will first drop until it reaches zero at point c (i.e., the magnet is now completely demagnetized) and then the magnetic flux density increases in the opposite direction (i.e., magnet polarity reversal), with continued increase in i until saturation is again reached in this reversed polarity condition, at point d. In other words, the $B-H$ relationship is now following the curve bcd, as indicated in Fig. 9.10(b). Now, if i is dropped back again to zero, resulting in H becoming zero, we would notice that the iron will retain the same amount of flux density, $-B_r$, resulting in the iron core now becoming a magnet with a polarity reversal. The $B-H$ relationship traces the curve de in this condition.

When the current is then increased from zero, in the same positive direction as before, first the reverse-polarity magnet will be completely demagnetized at a value of i corresponding to point f. With continued increase in the magnitude of i, the magnet will again reach saturation at the same point a as in in the initial magnetization process. The $B-H$ relationship is now tracing the curve efa, completing the hysteresis cycle.

The *abcdefa* hysteresis loop can be retraced again and again if the current magnitude is allowed to vary following the cycle described above. Energy from the magnetizing source (i) is lost during this magnetizing–demagnetizing–magnetizing process. This lost energy, referred to as the *hysteresis loss,* is proportional to the area inside the hysteresis loop.

For some materials, such as the ferrite switching cores used in the early computers' memory banks, the hysteresis loop is nearly rectangular, as shown in Fig. 9.11. It has only two stable states (corresponding to storage of two bits of information). The switching from the B_r polarization state to the $-B_r$ polarization state occurs only if the magnetizing current is dropped to slightly below the value $-I'$. Polarization remains at this $-B_r$ level until the magnetizing current is increased to slightly above the value I', at which condition the polarization is switched back to the stable state, B_r.

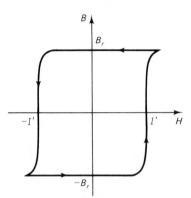

Figure 9.11 $B-H$ hysteresis loop for ferrite switching cores.

9.6 SIMPLE SERIES AND PARALLEL MAGNETIC CIRCUITS

As established previously, the following is a summary of the magnetic quantities encountered in the magnetic circuits and their corresponding analogous electrical quantities in electric circuits:

Electrical quantity	Magnetic quantity	Defining equation
E V	F_m At	$F_m = nI, \quad H = \dfrac{F_m}{l}$
I A	ϕ Wb	$\phi = BA$
R Ω	$\mathcal{R}$ At/Wb	$\mathcal{R} = \dfrac{F_m}{\phi} = \dfrac{1}{\mu}\dfrac{l}{A}$
σ $\mho \cdot m$	μ Wb/A $\cdot$ m	$\mu = \mu_r\mu_0 = \dfrac{B}{H}$
V V	F_m At	$F_m = \phi\mathcal{R} = Hl$

Similar to KVL and KCL, used to analyze series and parallel electric circuits, Ampère's circuital law and the conservation of flux law are used to analyze simple series and parallel magnetic circuits.

9.6.1 Ampère's Circuital Law

Figure 9.12 shows an example of a simple series magnetic circuit, made up of three different materials, including the air gap. Since there is only one path for the magnetic flux lines ϕ, it must be the same in all parts of this series magnetic

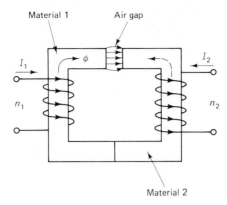

Figure 9.12   Simple series magnetic circuit.

circuit. This is similar to the case of a series electric circuit, where the current is the same in all series-connected components.

As E and V are analogous to F_m ($= nI$) and $\phi\mathcal{R}$ or Hl, a law similar to KVL (algebraic sum of voltage rise and drop in a closed loop = 0) also applies to the closed-loop series magnetic circuit. It is Ampère's circuital law. It states that *the sum of the magnetomotive force (MMF or F_m) rises equals the sum of the MMF drops around any closed path of a magnetic circuit*. In general, Ampère's circuital law can be stated as follows:

$$
\begin{aligned}
&\text{algebraic sum of the applied MMFs} \\
&\qquad = \phi\mathcal{R}_1 + \phi\mathcal{R}_2 + \cdots = \phi\mathcal{R}_T \qquad\qquad (9.9)\\
&\qquad = H_1l_1 + H_2l_2 + H_3l_3 + \cdots \qquad\qquad (9.10)
\end{aligned}
$$

where ϕ is the same amount of magnetic flux in the series magnetic circuit, $\mathcal{R}_1$ is the reluctance of part 1 of the circuit (similarly for $\mathcal{R}_2, \mathcal{R}_3, \ldots$), H_1 is the magnetic field intensity of part 1 of the circuit, and l_1 is the length of this part (and similarly for $H_2, l_2, H_3, l_3, \ldots$). From Eq. (9.9) one can easily observe that in a series magnetic circuit,

$$
\mathcal{R}_T = \text{total series reluctance} = \mathcal{R}_1 + \mathcal{R}_2 + \mathcal{R}_3 + \cdots \qquad (9.11)
$$

Applying Ampère's circuital law to the magnetic circuit of Fig. (9.12) gives us

$$
n_1I_1 - n_2I_2 = \phi(\mathcal{R}_1 + \mathcal{R}_2 + \mathcal{R}_3) = \phi\mathcal{R}_T
$$

that is, the net algebraic sum of the applied MMFs in the assumed positive direction of ϕ is: $\phi \times$ the sum of the reluctances of each part of the series circuit, consisting of materials 1 and 2 and an air gap (3). Similarly,

$$
n_1I_1 - n_2I_2 = H_1l_1 + H_2l_2 + H_3l_3
$$

The negative sign associated with the term n_2I_2 is due to the fact that this MMF is trying to induce a flux in the opposite direction of ϕ, as indicated in Fig. 9.12. Form (9.9) of Ampère's circuital law is used if the dimensions and the permeabilities of each portion of the circuit are known, so that the reluctances can be calculated. On the other hand, form (9.10) of Ampère's circuital law is more useful if μ is unknown but the $B-H$ curves of the magnetic materials are given.

9.6.2 Conservation of Flux Law

In the magnetic circuit shown in Fig. 9.13, the flux ϕ induced by the MMF (nI) can split up into two parts, since there are two different paths for the magnetic flux lines, in either branch a (ϕ_a) or branch b (ϕ_b). The sum of the two magnetic fluxes must be the same as the total flux ϕ. Here

$$\phi = \phi_a + \phi_b \qquad (9.12)$$

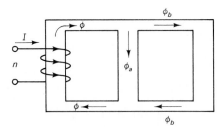

Figure 9.13

This is the law of conservation of flux. It is similar to KCL at a node of an electric circuit. Clearly, it can be expanded to express the different portions of the magnetic fluxes if more than two magnetic paths exist. If the reluctance of branch a is $\mathcal{R}_a$ and the reluctance of branch b is $\mathcal{R}_b$, then in a manner similar to the equivalent resistance of parallel branches in electrical circuits, the equivalent reluctance of branches a and b is

$$\mathcal{R}_{eq} = \frac{\mathcal{R}_a \mathcal{R}_b}{\mathcal{R}_a + \mathcal{R}_b} \qquad (9.13)$$

EXAMPLE 9.3

The magnetic circuit for a relay is shown in Fig. 9.14. If the average length of the iron core is 40 cm and the length of the air gap is 0.2 cm, while the average cross-sectional area is 2.5 cm², find the current required to produce a flux density of 0.1 Wb/m² in the air gap. The number of turns of the coil is 50 and μ_r (for iron) is 200.

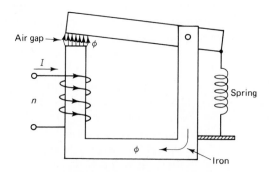

Figure 9.14 Magnetic relay.

Solution Since

$$\mu_0 = 4\pi \times 10^{-7} = 1.257 \times 10^{-6} \, \text{Wb/A} \cdot \text{m}$$

and

$$\mu(\text{iron}) = \mu_r\mu_0 = 2.513 \times 10^{-4} \, \text{Wb/A} \cdot \text{m}$$

and the dimensions of the magnetic circuit are known, it is easy to find the reluctance of each portion of this series magnetic circuit. Here

$$\mathcal{R}_{\text{iron}} = \frac{1}{\mu}\frac{l_{\text{iron}}}{A} = \frac{0.4}{2.513 \times 10^{-4} \times 2.5 \times 10^{-4}} = 6.366 \times 10^6 \, \text{At/Wb}$$

and

$$\mathcal{R}_{\text{air gap}} = \frac{1}{\mu_0}\frac{l_{\text{air gap}}}{A} = \frac{0.002}{1.257 \times 10^{-6} \times 2.5 \times 10^{-4}} = 6.364 \times 10^6 \, \text{At/Wb}$$

Notice that even though the air gap has a very small length compared to the length of the iron, it contributes a reluctance of almost the same magnitude as that of the iron part of the magnetic circuit. In other cases it might even be much larger! Therefore,

$$\mathcal{R}_T(\text{of the series magnetic circuit}) = \mathcal{R}_{\text{air gap}} + \mathcal{R}_{\text{iron}} = 1.273 \times 10^7 \, \text{At/Wb}$$

To find the driving MMF $(= nI)$, Ampère's circuital law [Eq. (9.9)] can be used:

$$F_m = \phi\mathcal{R}_T = BA\mathcal{R}_T \quad (\text{where } B = \phi/A)$$
$$= 0.1 \times 2.5 \times 10^{-4} \times 1.273 \times 10^7 = 318.25 \, \text{At}$$
$$= nI = 50I$$

Therefore,

$$I = \text{required magnetizing current} = 6.365 \, \text{A}$$

EXAMPLE 9.4

The $B\text{--}H$ curves of the iron and steel materials used in the magnetic circuit shown in Fig. 9.15(a) are shown in Fig. 9.15(b). The average length of each part of the circuit (both are equal) is 0.5 m, while the cross-sectional area is 4 cm². In Fig. 9.15(a) it is required that the magnetic flux ϕ be 2×10^{-4} Wb, while $n_1 = 100$ turns, $n_2 = 200$ turns, and $I_1 = 6$ A. Find I_2.

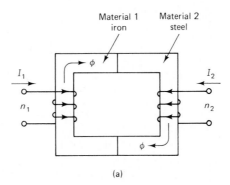

(a)

Figure 9.15

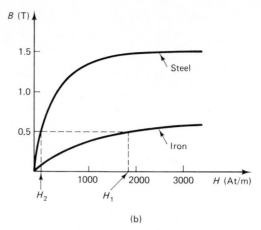

(b)

Figure 9.15 *(continued)*

Solution   The flux ϕ and the flux density B are the same in both parts of this series magnetic circuit, where

$$B = \frac{\phi}{A} = \frac{2 \times 10^{-4}}{4 \times 10^{-4}} = 0.5 \text{ Wb/m}^2 \text{ (or tesla)}$$

Since μ is unknown for each material, but the B–H curves are given, the magnetizing forces for each part of the magnetic circuit can be obtained from the B–H curves, as indicated, corresponding to the required value of B above. Here

$$H_1 \text{ (iron)} = 1800 \text{ At/m}$$

$$H_2 \text{ (steel)} = 200 \text{ At/m}$$

Both of the applied MMFs produce flux in the same direction (from the RHR; i.e., they aid each other). Thus the net applied MMF is

$$F_m = n_1 I_1 + n_2 I_2$$

Using the second form of Ampère's circuital law [Eq. (9.10)]:

$$F_m = H_1 l_1 + H_2 l_2 = 1800 \times 0.5 + 200 \times 0.5 = 1000 \text{ At}$$

$$= n_1 I_1 + n_2 I_2 = 100 \times 6 + 200 \times I_2$$

$$I_2 = \frac{1000 - 600}{200} = 2 \text{ A}$$

9.7 | ELECTROMAGNETIC INDUCTION: FARADAY'S AND LENZ'S LAWS

In the discussion of Section 9.2 it was shown that electric current flow in a wire conductor produces a magnetic field. Can a magnetic field result in current flow or induced voltage; that is, is the electromagnetic induction process reversible? The answer is yes.

Faraday and Henry observed that when the magnetic lines of force (i.e., flux ϕ) linking a conductor are changed, voltage will be induced across the terminals of the conductor. The magnetic flux linkage can be changed by either moving the conductor or the magnetic field itself in such a way that the conductor cuts across the magnetic lines of force. The induced voltage and the resulting induced current are produced only if the cutting action is exhibited. If there is no motion or if the relative motion of the conductor is parallel to the magnetic lines of force so that the magnetic flux lines are not cut, the induced voltage and the resulting current are zero.

In Fig. 9.16, ϕ_{orig} is the original magnetic flux of the stationary horseshoe magnet. If the conductor is moved perpendicular to the magnetic lines of force, as shown, the galvanometer's pointer deflects, indicating an induced current flow (i_{ind}) resulting from an induced voltage (v_{ind}). When the motion is upward as indicated, Fig. 9.16 shows the direction of flow of i_{ind} and the polarities of v_{ind}; more about this later. If the motion is downward, the polarities of v_{ind} and the direction of i_{ind} reverse. In both cases, cutting of the magnetic flux linkage is exhibited. If the conductor is moved from left to right (or vice versa), its motion is parallel to the flux lines and no cutting occurs. Here v_{ind} and i_{ind} are both zeros.

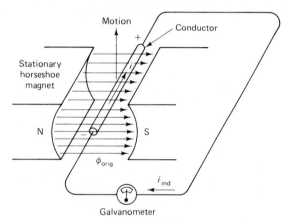

Figure 9.16

In Fig. 9.17 the original flux, ϕ_{orig}, of the permanent magnet will link with the coil conductor if the magnet is moved toward the coil as shown, or if the magnet was close to the coil (or even inside it) and is moved away from the coil.

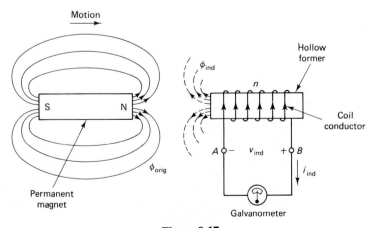

Figure 9.17

In both cases, the coil conductor cuts across the magnetic flux lines, generating the induced voltage v_{ind} and the induced current i_{ind}, whose polarities and direction are as indicated when the motion is from left to right, as Fig. 9.17 shows. If the motion is reversed, again the galvanometer's pointer deflection will be reversed, indicating reversed polarities of v_{ind} and reversed current flow, i_{ind}. Also indicated in Fig. 9.17 is the magnetic flux ϕ_{ind} (the induced flux) resulting from the flow of i_{ind}, obtained through the application of the RHR.

Faraday also observed that the faster the relative motion of the conductor (or coil) with respect to the magnetic field, the higher will be the deflection of the galvanometer's pointer (i.e., the higher is the induced voltage v_{ind}). This faster motion (with respect to time) indicates a higher rate of change of the magnetic flux linkage. The induced voltage can also be increased by increasing the magnitude of the flux linkage with the coil (see Fig. 9.17); i.e., if the coil has a larger number of turns n.

Faraday's Law. *The induced voltage in a circuit, due to electromagnetic induction, is proportional to the rate of change (with respect to time) of the magnetic flux linkage with the circuit.* In mathematical terms,

$$v_{ind} = \text{EMF} = n\frac{\Delta\phi}{\Delta t} \tag{9.14}$$

At the limit, when Δt tends to zero, $\Delta\phi/\Delta t$ becomes $d\phi/dt$, that is,

$$v_{ind} = \text{EMF} = n\frac{d\phi}{dt} \tag{9.14a}$$

The units in Eq. (9.14) indicate that volts = webers/second, which is the observation noted in Section 9.1. The induced voltage is, in fact, an EMF causing the flow of the induced current, i_{ind}.

Lenz's Law. As was observed above referring to Figs. 9.16 and 9.17, the reversal of the motion resulted in the reversal of the polarities of v_{ind} and hence the reversal of the direction of current flow, i_{ind}. These polarities and the direction of current flow can be determined through the application of Lenz's law. This law is based on Newton's third law of mechanics, which states that *for every action, there is an equal and opposite reaction.*

Lenz's law states that *the induced effect (e.g., voltage) will have polarities such that the resulting current flow should produce a flux ϕ_{ind} (due to the flow of i_{ind}) opposing the change in the cause (e.g., the original flux ϕ_{orig}).* It is very important to keep in mind that the induced flux ϕ_{ind} does not necessarily oppose ϕ_{orig}, but opposes the change in ϕ_{orig} due to the motion.

In the example of Fig. 9.17, when the magnet is moved approaching the coil from its left-hand side, the original flux at this location is increasing from left to right. By applying Lenz's law, the induced flux ϕ_{ind} should oppose the increase of the original flux, and thus it is in the direction from right to left, as indicated in the figure by the dashed plot of ϕ_{ind}. This flux, ϕ_{ind}, is due to the flow of the current i_{ind}. Through the application of the RHR, i_{ind} should flow up from terminal A and down from terminal B of the coil. The induced voltage in the coil, v_{ind}, is in effect an EMF (i.e., a source). Considering the network external to the coil as a load, consisting here of the galvanometer's resistance, the current i_{ind} will flow in it from B and A. According to the v–i polarity convention for a load, point B should be the higher potential point (positive terminal) with respect to point A. Hence the polarities of v_{ind} are determined as shown.

A cross section of the example of Fig. 9.16 is shown in Fig. 9.18. This example demonstrates the electric *generator principle*. When the conductor is forced to move upward (i.e., external force is used to move the conductor), then according to Lenz's law, the induced flux ϕ_{ind} should be set up in such a direction to oppose this motion (the cause). According to the third and fourth properties of the magnetic lines of force, the induced flux should be set up so that more lines of force are exhibited above the conductor, aiding the original magnetic field and thus contributing to higher opposition to the cause (the force causing the motion). As indicated in Fig. 9.18, this induced field must have been produced by the induced current flowing in through the cross-sectional plane (the "×" indicates the end of the current arrow flowing in), according to Maxwell's RHR. A modification of the RHR is suggested to help in remembering the directions associated with the generator principle (see Fig. 9.19): *Let the flux lines with the arrow direction fall in the palm of the open right hand, the thumb perpendicular to the fingers points in the direction of force (or motion), while the extended fingers point in the direction of current flow.* This modified RHR is very easy to apply and check with the example of Fig. 9.18 (or Fig. 9.16). Referring to Fig. 9.16, the induced voltage in the conductor (which is equivalent to an EMF) should have its positive polarity where the current leaves this simple generator into the load (i.e., at the back terminal of the conductor, as shown).

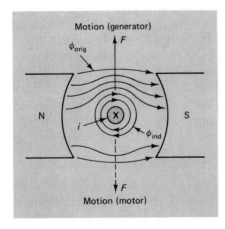

Figure 9.18

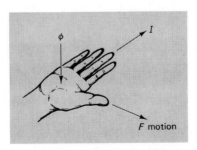

Figure 9.19   Right-hand rule.

As can be concluded from this discussion, motion of the conductor relative to the magnetic field, resulting in the cutting of the magnetic lines of force, will produce electric voltage (EMF = v_{ind}). This is the principle used to generate electric voltage and energy in hydrostations, using the force generated from steam turbines or waterfalls to rotate the conductors in the magnetic field. This principle will be used in Chapter 11 to demonstrate the generation of ac voltage.

The electromagnetic phenomenon can also be used in the reverse process. This is called the *motor principle* or action. Here the current is allowed to flow in a conductor placed in a magnetic field. The field induced by the current flow interacts with the original magnetic field, resulting in a mechanical force causing the conductor to move (i.e., creating force). In the example of Fig. 9.18, the conductor was originally placed in the magnetic field. Current i is forced to flow in the conductor in the direction indicated, through connecting an external source to the conductor. As indicated in the figure, the induced flux ϕ_{ind} due to the current flow is in accordance with Maxwell's RHR. Thus more magnetic lines of force appear above the conductor (crowding) and less lines of force appear below the conductor due to cancellation. According to the properties of the magnetic lines of force, this crowding causes a force, pushing the conductor downward, as indicated in Fig. 9.18. This is, in fact, the principle of operation of the electric motors, and also the meter movement. To remember the directions associated with the motor action (i.e., current flow causing force or motion), the left hand is used to replace the right hand in the modified RHR stated above for the generator. The rule is then referred to as the LHR. It is easy and straightforward to check its application directly in the example of Fig. 9.18.

EXAMPLE 9.5

Find the direction in which the magnet was moved to cause the induced current to flow as shown in the single-turn coil of Fig. 9.20.

Solution Using Maxwell's RHR, the induced flux ϕ_{ind}, due to the flow of current i_{ind}, can be mapped as indicated by the dashed lines of force in Fig. 9.20. According to Lenz's law, the action that caused the induced flux reaction shown must have been the increase in the original flux linkage with the coil from right to left. Obviously, this is the case if the magnet was moved toward the coil.

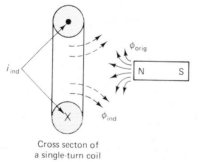

Cross secton of
a single-turn coil

Figure 9.20

EXAMPLE 9.6

The magnet shown in Fig. 9.21 is moved away from the coil, resulting in an induced current flow in the single-turn coil as indicated. Find the polarities of the magnet.

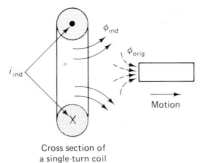

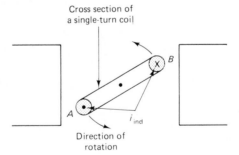

Figure 9.21

Solution As the magnet is being moved away, the number of flux lines linking with the single-turn coil is reduced. The induced flux ϕ_{ind} can be easily mapped using Maxwell's RHR, as shown. According to Lenz's law, ϕ_{ind} is in such a direction as to oppose the change in the original flux. The change in this example is a reduction, or decay, in the number of lines of force. Hence ϕ_{ind} is actually set up in the direction of ϕ_{orig} to help the decaying flux. Thus ϕ_{orig} is mapped in the same direction as ϕ_{ind}, from left to right, entering the pole closest to the single-turn coil. This is shown by the dashed lines in Fig. 9.21. Thus the pole closest to the coil is the south (S) pole.

EXAMPLE 9.7

When the coil in Fig. 9.22 is rotated counterclockwise, the current induced has the direction shown. What are the polarities of the permanent magnet?

Figure 9.22

Solution Since the coil is being rotated, producing the induced current i_{ind}, this is an example of a generator. Thus applying the modified RHR to either conductor (either side of the coil), noting that side A is moving down and side B is moving up, the unknown direction of the original flux will be falling into the open palm of the right hand from the left to right. The left side of the magnet is then the north (N) pole.

EXAMPLE 9.8

If the current is forced to flow in the coil in Fig. 9.23 in the direction shown, what is the direction of rotation of the coil?

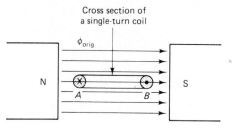

Cross section of
a single-turn coil

Figure 9.23

Solution   This is an example of the motor action, as current flow is the cause of the forces of rotation (actually torque). Using the LHR in this case, side A is found to be moving down, while side B is found to be moving up. Thus the coil will rotate in a counterclockwise direction.

EXAMPLE 9.9

Find the polarities of the secondary winding;
(a) When the switch S in the primary circuit is being closed [Fig. 9.24(a)].
(b) When the switch S in the primary circuit is being opened [Fig. 9.24(b)].

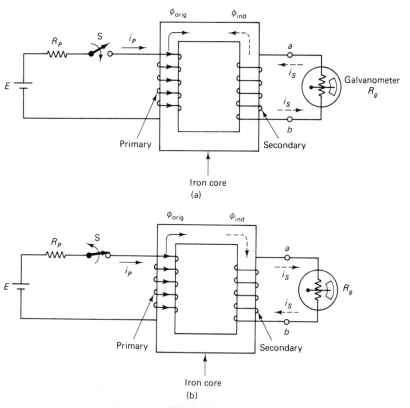

Figure 9.24

Solution   (a) As the switch S in Fig. 9.24(a) is being closed, the current i_p flowing in the primary winding will create the flux ϕ_{orig} in the iron core. The number of the magnetic flux lines in the iron core is thus increasing, since the magnetic flux used to be zero when S was open and i_p was zero. The direction of ϕ_{orig} is obtained from Maxwell's RHR. According to Lenz's law, the induced flux ϕ_{ind} should be in such a direction as to oppose the change in ϕ_{orig}. Since ϕ_{orig} is increasing, ϕ_{ind} will be opposing it, as indicated by the dashed line in Fig. 9.24(a). Now, using Maxwell's RHR on any turn of the secondary winding, it can easily be concluded that i_s is flowing from b and a in the external load R_g. This is indicated by the dashed i_s in Fig. 9.24(a). As the secondary winding is now acting as a source with a voltage $v_{s\,ind}$, its b terminal is the higher potential point (i.e., b is $+$ and a is $-$).

(b) Here ϕ_{orig} due to i_p is in the same direction as before. However, as the switch S is being opened, the number of the magnetic lines of force, ϕ_{orig}, is decreasing as the flux decays to zero when S is open (i_p will also be zero). According to Lenz's law, the induced flux ϕ_{ind} should be set up in such a way as to oppose the change in ϕ_{orig}. Thus here ϕ_{ind} is in the same direction as ϕ_{orig} to help this decaying flux and oppose the decrease in the number of the magnetic flux lines. The dashed line of ϕ_{ind} in Fig. 9.24(b) shows that the secondary flux linkage is reversed. Thus from Maxwell's RHR, i_s is reversed, as shown by the dashed line of i_s in Fig. 9.24(b). The polarities of the voltage now induced in the secondary winding will also be reversed. Here a is $+$ and b is $-$.

Note that under steady-state conditions (i.e., when $i_p = E/R_p$ while S is closed) ϕ_{orig} reaches a constant steady-state value and does not change with respect to time. Since there is no change in ϕ_{orig}, ϕ_{ind} is zero and i_s is zero. Thus the galvanometer's pointer deflects only while S is being closed or opened and shows no deflection (i.e., i_s and $v_{s\,ind} = 0$) at any other time.

EXAMPLE 9.10

In the circuit of Fig. 9.25 it was observed that with the switch in position A, as the magnet was moved, the galvanometer's pointer deflection is opposite to that obtained when the switch is in position B. What is the direction of motion of the permanent magnet?

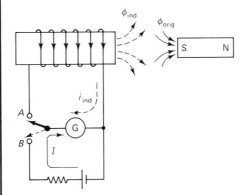

Figure 9.25

Solution Since the galvanometer's deflection in position B is due to the current I shown, then with the switch in position A, the induced current i_{ind} must flow in the opposite direction compared to I, to cause an opposite galvanometer's deflection. The current i_{ind} is shown dashed in Fig. 9.25. This induced current produces the induced flux ϕ_{ind}, which is also shown dashed in the figure, in accordance with Maxwell's RHR.

By examining the directions of the induced flux lines ϕ_{ind}, in comparison with the original flux lines, ϕ_{orig} in Fig. 9.25, it is clear that ϕ_{ind} is in the same direction as ϕ_{orig}, helping it. According to Lenz's law, since ϕ_{ind} opposes the change in ϕ_{orig}, this change must have been a decrease in the number of the flux lines linking the coil and the magnet. The decrease, or decay in ϕ_{orig}, must then have been caused by moving the magnet away from the coil. Thus the direction of motion of the magnet is from left to right.

PROBLEMS

Section 9.4

For Problems 9.1–9.4, refer to Fig. 9.8.

(i) **9.1.** For a simple magnetic circuit whose cross-sectional area is $8\,\text{cm}^2$ and whose mean length is $40\,\text{cm}$, the relative permeability of the core material is 200. If the applied current to the coil of 6 A results in a core magnetic flux of 3 m Wb, find
    (a) The reluctance of the magnetic circuit.
    (b) The flux density in the core.
    (c) The field intensity (magnetizing force) H.
    (d) The number of turns of the coil.

(i) **9.2** The applied MMF to a simple magnetic circuit is 350 At. It was found that the resulting magnetic field density is $0.7\,\text{Wb/m}^2$. The average length of this magnetic circuit is $0.5\,\text{m}$ and its cross-sectional area is $4\,\text{cm}^2$. Find
    (a) The relative permeability of the magnetic material of the core.
    (b) The reluctance of the magnetic circuit.
    (c) The amount of the magnetic flux ϕ.

(i) **9.3.** The relative permeability of a simple magnetic circuit is 100. If the applied MMF of 200 At results in a magnetic flux of $4 \times 10^{-4}\,\text{Wb}$, find the reluctance of this magnetic circuit. If the mean length is $75\,\text{cm}$, find
    (a) The cross-sectional area.
    (b) The magnetizing force H.
    (c) The flux density B.

(i) **9.4.** Find the flux density B in a simple magnetic circuit whose average length is $1.5\,\text{m}$ if the current flowing in a 150-turn coil is 5 A and the relative permeability of the core material is 400.

Section 9.6

(i) **9.5.** In the series magnetic circuit of Fig. 9.26, the mean length of the part made of material 1 is $0.5\,\text{m}$, with a relative permeability of 800, and the mean length of the part made of material 2 is $0.8\,\text{m}$, with a relative permeability of 1200. The cross-sectional area of both parts is $5\,\text{cm}^2$. If the flux density B in the circuit is $1.0\,\text{Wb/m}^2$, with $n = 180$ turns, find the value of the applied current I.

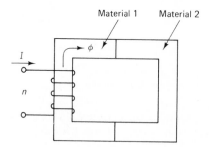

Figure 9.26

(ii) **9.6.** In the series magnetic circuit of Problem 9.5 (same dimensions), at a flux density of $1.2 \, \text{Wb/m}^2$, it was found from the $B-H$ curves of materials 1 and 2 that the corresponding H_1 is 800 At/m and H_2 is 1400 At/m. Find the applied current in this case. What is the overall circuit's reluctance in this situation?

(ii) **9.7.** For the series magnetic circuit of Fig. 9.12, the part made of material 1 has an average length of 0.4 m and the part made of material 2 has an average length of 0.6 m, while the air gap is 0.5 cm. If μ_{r_1} is 150 and μ_{r_2} is 240 and the cross-sectional area is $4 \, \text{cm}^2$, find the overall reluctance of this circuit. If I_1 is 10 A, n_1 is 100 turns, and I_2 is 2 A, find n_2 so that the flux density in the air gap is $0.1 \, \text{Wb/m}^2$, in the direction indicated in Fig. 9.12.

(i) **9.8.** In Problem 9.7, find the value of n_2 that will cause the flux density in the air gap to be $0.1 \, \text{Wb/m}^2$, but in the opposite direction to that indicated in Fig. 9.12.

(i) **9.9.** In the magnetic relay circuit of Fig. 9.14, the average cross-sectional area of the core is $6 \, \text{cm}^2$, while its mean length is 1 m. The mean length of the air gap is 0.5 cm. The relative permeability of the iron core is 75. If I is 5 A and n is 200 turns, what is the flux density in the air gap?

Section 9.7

(i) **9.10.** In which direction will the coil shown in Fig. 9.27 rotate if current is allowed to flow in the single-turn coil in the direction indicated?

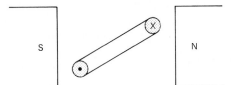

Figure 9.27

(i) **9.11.** The magnet in Fig. 9.28 is moved toward the coil from right to left as indicated. What is the polarity of terminal a of the coil?

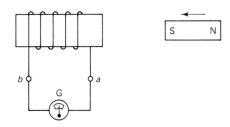

Figure 9.28

(i) **9.12.** If R is decreased in the circuit shown in Fig. 9.29, what will be the polarity of terminal a?

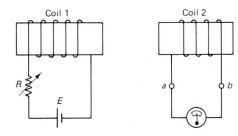

Figure 9.29

(i) **9.13.** In Fig. 9.29, R is fixed. What will be the polarity of terminal a if coil 2 is moved away from coil 1?

(ii) **9.14.** Which of the following conditions caused induced current to flow in the conductor ring shown in Fig. 9.30?
   (a) The switch has been closed for a long time.
   (b) The switch has been opened for a long time.
   (c) The switch is just being closed.
   (d) The switch is just being opened.

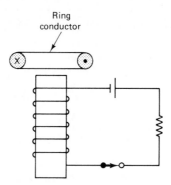

Ring conductor

Figure 9.30

(ii) **9.15.** Current was allowed to flow in the single-turn coil shown in Fig. 9.31. It resulted in a repulsion force causing the permanent magnet to move as indicated. What is the direction of the current flow in the single-turn coil?

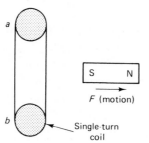

a

S   N

F (motion)

b

Single-turn coil

Figure 9.31

(ii) **9.16.** If the resistance R in the circuit of Fig. 9.32 is increased, what is the polarity of terminal a of coil 1?

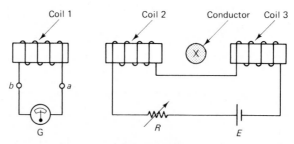

Coil 1   Coil 2   Conductor   Coil 3

b   a

G   R   E

Figure 9.32

(ii) **9.17.** A conductor carrying current (as indicated) is placed between coils 2 and 3 in the circuit shown in Fig. 9.32. If R is increased or decreased, what will be the direction of motion of this conductor?

GLOSSARY

Ampère's Circuital Law: The law establishing that the algebraic sum of the rises and drops in the magnetomotive forces around a closed loop of a magnetic circuit is equal to zero.

Diamagnetic Materials: Materials that have permeabilities slightly less than that of free space (e.g., glass, copper, and aluminum).

Electromagnetic Induction: The generation of electric current and voltage through the relative change in the magnetic field around (or linking with) an electrical conductor. This is also called the generator principle.

Faraday's Law: The law that establishes the relationship between the amount of induced voltage in an electric circuit through electromagnetic induction, and the rate of change of the magnetic field linking with this circuit.

Ferromagnetic Materials: Materials whose permeabilities are much larger than that of free space (e.g., iron, steel, cobalt, and nickel).

Flux Density (B): A measure of the amount of the magnetic flux per unit cross-sectional area perpendicular to the direction of the flux lines. It is measured in Wb/m^2 or tesla (T).

Hysteresis: The lagging effect between the flux density of a magnetic material and the magnetizing force applied to this material.

Left-Hand Rule (LHR): The rule used to determine the direction of motion resulting from the interaction between two magnetic fields, one of them induced through the flow of electrical current.

Lenz's Law: The law establishing the direction of the induced effect (current, voltage, and magnetic flux) due to electromagnetic induction. It is a restatement of Newton's third law as it applies to the electromagnetic phenomena.

Magnetic Field: The space in which the forces of attraction and repulsion between magnetic poles are exhibited. It is represented (mapped) by magnetic flux lines.

Magnetic Field Intensity (or Magnetizing Force; H): The amount of magnetomotive force per unit length of the magnetic path, required to establish the magnetic field (flux). It is measured in Ampere $\cdot$ turns/meter (At/m).

Magnetic Flux (ϕ): A measure of the strength of the magnetic field. The lines representing the magnetic flux are continuous and possess direction. It is measured in webers (Wb).

Magnetic Materials: Materials that are affected by noncontact forces when placed in a magnetic field.

Magnetomotive Force (MMF): The driving force required to establish a magnetic field in a magnetic material. It is measured in ampere $\cdot$ turns (At).

Motor Principle: The principle that governs the creation of force (or torque) when two magnetic fields interact. Either or both of the magnetic fields could be electrically induced.

Paramagnetic Materials: Materials that have permeabilities slightly higher than that of free space.

Permanent Magnet: A ferromagnetic material that remains magnetized for a long period of time without the aid of an external magnizing force. It retains a finite amount of magnetic flux.

Permeability (μ): A measure of the ability of a material to allow the establishment of magnetic flux (ϕ). It is measured in Wb/A · m.

Relative Permeability (μ,): The ratio of the permeability of a material to that of free space.

Reluctance (ℜ): A quantity determined by the physical dimensions and properties of a material. It provides the measure that indicates the reluctance of the material to the establishment of magnetic flux lines within it. It is measured in At/Wb.

Right-Hand Rule (RHR): A rule establishing the directional relationship between the magnetic flux, the force acting on a conductor placed in the magnetic field, and the current resulting from the cutting of the magnetic flux lines due to the action of the force. Another version of the RHR is used to determine the direction of the magnetic field induced due to the flow of current in an electrical conductor.

10 INDUCTANCE AND TRANSIENTS IN *RL* CIRCUITS

OBJECTIVES

- Familiarity with the physical phenomena underlying the behavior of an inductor, the parameters that control the value of the inductance, and the ability of an inductor to store energy.

- Understanding the meaning and application of Ohm's law for an inductor: voltage proportionality to the rate of change of current.

- Ability to calculate the equivalent value of inductance for series, parallel, and series–parallel connections of inductors.

- Understanding the physical process of current rise and decay in *RL* dc circuits and the concept of the time constant associated with this physical phenomena.

- Ability to calculate, analytically and graphically, the current(s) and voltages and the rate of change of these circuit variables in various examples involving current rise and decay in *RL* dc circuits.

- Ability to derive the initial and steady state conditions in *RL* dc circuits and thus use the analytical expression for the generalized form of the transient current flowing in an inductor in such circuits.

An inductor is a two-terminal element made of a conducting wire wound in the form of a coil as shown in Fig. 10.1. The coil could have any number of turns n. It could be wound on a core with effective length l and cross-sectional area A, made of air, nonmagnetic material, or any magnetic material (e.g., iron, steel, or ferrite) whose permeability is μ.

Consider the coil shown in Fig. 10.1. As the current i starts to increase from zero, passing first in turn A, a magnetic flux, ϕ_{orig}, is produced in this turn, linking with turns B and C due to their proximity to turn A. According to Lenz's law, an induced flux ϕ_{ind} is produced in both turns B and C, trying to oppose the change (increase) in ϕ_{orig}. The direction of ϕ_{ind} is opposite to ϕ_{orig}, as shown in the figure. This induced flux is due to an induced current i_{ind} trying to flow in the direction opposite to i. This indicates that the coil exhibits the property of opposition to the change in current flowing in it. The induced current is due to the generation of a counter EMF in both turns B and C, with a polarity such as to oppose the flow of current i. Both counter EMFs are in series, aiding each other. The same effect will be exhibited when the current i flows next in turn B. Here the induced flux, induced current, and counter EMF of both turns A and C will be in the same direction as discussed above, again in opposition to the change in ϕ_{orig} and i.

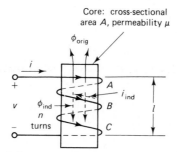

Core: cross-sectional area A, permeability μ

Figure 10.1

The overall counter EMF thus induced in the coil will have the polarities indicated in Fig. 10.1, and denoted by the voltage drop v, that is,

$$v = \text{counter EMF} \quad \text{V}$$

The direction of i and the associated polarities of v obey the v–i conventions of a load. This is why the counter EMF is referred to as a voltage drop. This is in contrast to the situations discussed in Chapter 9, where the coil was acting as a source due to its interaction with magnetic fields external to it. This phenomenon of electromagnetic induction in the coil itself is what is called the *self-inductance L*, or simply *inductance*.

The most important property of the coil is its *opposition to the change in current* flowing in it, but not the current itself.

Definition. The *inductance L* is the property of the electric circuit element that exhibits opposition to the change of current flowing in it.

For this reason, current in an inductor cannot change instantaneously. This observation should be compared with the voltage across a capacitor, which also cannot change instantaneously.

Ohm's law for an inductor. From Faraday's law (Chapter 9),

$$v = \text{counter EMF} = n\frac{\Delta\phi}{\Delta t} = n\frac{d\phi}{dt} \tag{10.1}$$

Since

$$\phi = \frac{F_m}{\mathcal{R}} = \frac{ni}{\mathcal{R}}$$

(Ohm's law for the magnetic circuit of the coil, i.e., the path of flux ϕ_{orig}), then

$$v = n\frac{d}{dt}\frac{ni}{\mathcal{R}} = \frac{n^2}{\mathcal{R}}\frac{di}{dt}$$

Note that n and $\mathcal{R}$ are constants, independent of time; only i can be dependent on time. The constant in the equation above is the quantity referred to as the inductance L, that is,

$$\boxed{v = L\frac{di}{dt}} \tag{10.2}$$

where

$$L = \frac{n^2}{\mathcal{R}} \tag{10.3}$$

Equation (10.2) is Ohm's law for the inductor L. The units of L can easily be derived from this equation. The unit of L is called the *henry*, abbreviated H, where

$$\text{volts} = \text{henrys(ampere/second)}$$

Therefore,

$$1\text{ henry} = (1\text{ volt/ampere}) \cdot \text{second} = 1\text{ ohm} \cdot \text{second}$$

The circuit symbol and the associated v–i polarities of an ideal inductor are shown in Fig. 10.2. The symbol is actually derived from the shape of the coil.

Since the coil is made of a conducting wire, it has a small but finite resistance (R_L). The equivalent circuit model of the inductor indicated in Fig. 10.3 accounts for this small resistance. Unless otherwise specified, the value of R_L is usually very small and can be easily neglected in most cases. Also, since the conductive turns of the coil are separated by an insulator (or air), a finite (but small) stray capacitance can be accounted for in the equivalent-circuit model of a practical inductor as shown in Fig. 10.3. This capacitance is usually neglected unless the coil is used in very high frequency applications.

Ohm's law for an inductor [Eq. (10.2)], which is in fact derived from Lenz's and Faraday's laws, clearly shows the most important physical property of this

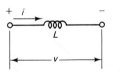

Figure 10.2 Circuit symbol of an ideal inductor.

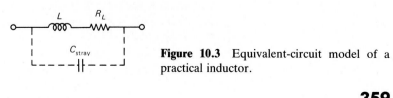

Figure 10.3 Equivalent-circuit model of a practical inductor.

element. The induced voltage v across the inductor (i.e., the counter EMF) is proportional not to the current but to the rate of change of the current with respect to time, di/dt. Thus if the current does not change with time (dc current), di/dt is zero and the voltage across the inductor is zero. This is because when i is constant, the magnetic flux ϕ is constant and no flux line variation (i.e., no cutting of lines of force) occurs. Hence no induced voltage will be developed. This observation means that the inductor is equivalent to a short circuit under steady-state dc conditions. Obviously, this is true since the inductor (i.e., coil) is nothing more than a conducting wire in this case. On the other hand, if i changes very rapidly with time (e.g., high-frequency current), its rate of change with respect to time, di/dt, is very high, meaning that the magnetic flux lines ϕ are varying and cutting the conductor wire at a high rate, resulting in a large induced voltage drop v, in opposition to such changes in current. The inductor (coil) is sometimes referred to as a "choke" because of its choking (opposition) reaction to the high rate of change of current flowing in it. In Chapter 11 this point will be discussed further to quantify the magnitude of such opposition.

The value of the inductance L as obtained in Eq. (10.3) depends only on the physical properties and construction parameters (dimensions) of the coil. The reluctance $\mathcal{R}$ is the overall reluctance of the magnetic circuit in which the flux lines ϕ are established. If the core of the coil is air or any other nonmagnetic material whose μ is approximately μ_0, the reluctance of the magnetic circuit is essentially that of the core since the effective cross-sectional area of the path outside the core is much higher than that of the core. Referring to Fig. 10.1, we see that

$$\mathcal{R} = \frac{l}{\mu A} = \frac{l}{\mu_0 A} \tag{10.4}$$

substituting in Eq. (10.3) yields

$$\boxed{L = \frac{n^2}{\mathcal{R}} = \frac{n^2 \mu_0 A}{l} \quad \text{H}} \tag{10.5}$$

Examining Eq. (10.5), it is clear that L depends on the construction dimensions of the coil, its number of turns, and the permeability of the core.

Also, since

$$F_m = ni = \phi \mathcal{R} = \phi \frac{n^2}{L} \quad \text{[using Eq. (10.3)]}$$

then

$$\boxed{Li = n\phi} \tag{10.6}$$

or

$$\phi = \frac{L}{n} i \tag{10.6a}$$

An inductor whose ϕ versus i relationship [Eq. (10.6a)] is linear is called a *linear inductor*. This is true only if L is not dependent on the value of i. By examining Eqs. (10.4) and (10.5), L is independent of i, if μ is constant, as is the case when the core is air or any other nonmagnetic material. However, some inductors are constructed with the core made of a magnetic material (e.g., ferrite). This is typically the case of adjustable (or variable) inductors, where the value of L depends on the extent of insertion of a ferrite rod into the core of the coil. Since μ of such magnetic material core is not always constant (see Fig. 9.9 for a typical $B-H$ curve of a magnetic material) but is a function of i, the inductor could be nonlinear.

In practice, inductors can be constructed with an inductance L as low as a fraction of a μH (as in high-frequency communications applications) or as high as 100 H (as in high-voltage power applications). Figure 10.4 shows some examples of practical inductors.

Ideal inductors (Fig. 10.2), like ideal capacitors, do not dissipate energy. As discussed in Chapter 11, their power dissipation is zero. Inductors, like capacitors, are energy-storage elements. In inductors, energy is stored in the

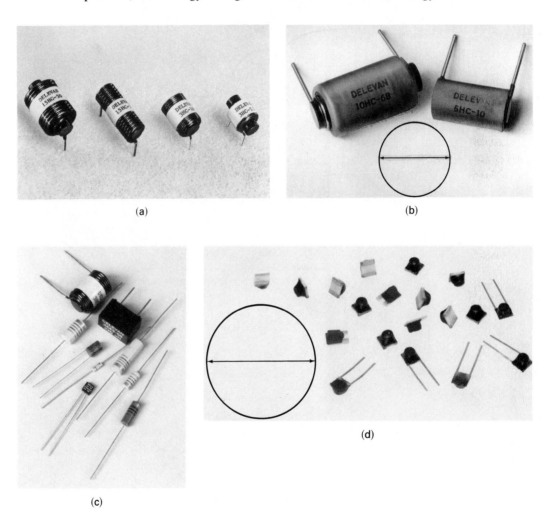

(a)

(b)

(c)

(d)

Figure 10.4 (a) Open-coil high-current filter chokes; (b) coated high-current filter chokes; (c) different sizes of fixed inductors; (d) micro-current chip inductors; (e) surface-mounted micro-current chip inductors; (f) micro-current inductor in a hybrid circuit. (Courtesy of Delevan, Division of American Precision Industries Inc.)

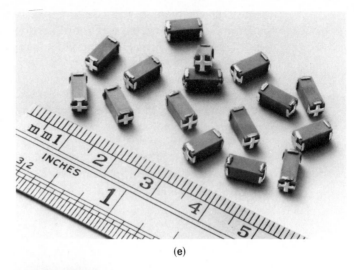

(e)

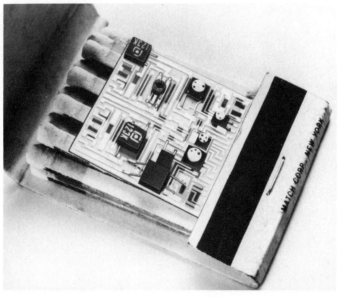

(f)

Figure 10.4 (*continued*)

magnetic field. The amount of energy stored in an inductor is

$$w = \tfrac{1}{2}Li^2 \quad \text{J} \tag{10.7}$$

where i is the instantaneous value of the current flowing in the inductor L. This relation can be proved through calculus, as follows:

$$w = \int_0^t p(t)\, dt = \int_0^t iv\, dt = \int_0^t iL\frac{di}{dt}\, dt$$

$$= \int_0^i Li\, di$$

$$= \tfrac{1}{2}Li^2 \quad \text{J (as above)}$$

EXAMPLE 10.1

If the current i flowing in a 0.1 H inductor varies with time as shown in Fig. 10.5, what will be the inductor's voltage as a function of time?

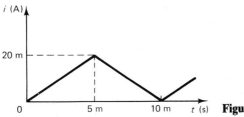

Figure 10.5

Solution   Since

$$v_L = L\frac{di}{dt} = 0.1\frac{di}{dt}$$

The slope di/dt is constant $= 20\,\text{mA}/5\,\text{ms} = 4\,\text{A/s}$ for $0 < t < 5\,\text{ms}$, and the slope di/dt is constant $= -20\,\text{mA}/5\,\text{ms} = -4\,\text{A/s}$ for $5\,\text{ms} < t < 10\,\text{ms}$. The voltage v_L can then easily be plotted, as shown in Fig. 10.6.

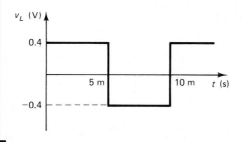

Figure 10.6

10.2   SERIES AND PARALLEL CONNECTIONS OF INDUCTORS

Inductors can be connected in series or in parallel or in general in a series-parallel configuration as in the case for resistors and capacitors. Figure 10.7 shows a series connection of a number of inductors. To find the single equivalent inductor that can replace these series-connected inductors, the terminal v–i relationships for the original and the equivalent circuits must be the same. Since the current i is the same in all series-connected elements and $v_1, v_2, v_3, \ldots$ are the voltages across

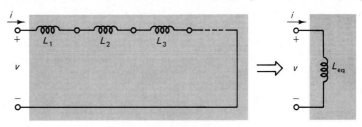

Figure 10.7

each of inductors $L_1, L_2, L_3, \ldots$, then from KVL,

$$v = v_1 + v_2 + v_3 + \cdots$$

$$= L_1 \frac{di}{dt} + L_2 \frac{di}{dt} + L_3 \frac{di}{dt} + \cdots$$

$$= (L_1 + L_2 + L_3 + \cdots) \frac{di}{dt}$$

$$= L_{eq} \frac{di}{dt} \quad \text{(from the equivalent circuit)}$$

Hence

$$\boxed{L_{eq} = L_1 + L_2 + L_3 + \cdots} \tag{10.8}$$

(equivalent of series-connected inductors)

Figure 10.8 shows a parallel connection of a number of inductors. Also here, to find the single equivalent inductor that can replace these parallel-connected inductors, the terminal v–i relationships for the original and the equivalent circuits must be the same. Here the voltage v is the same across each inductor in parallel, while the currents $i_1, i_2, i_3, \ldots, i_n$ are the currents in each of the parallel-connected inductors. From KCL,

$$i = i_1 + i_2 + i_3 + \cdots + i_n \quad \text{(differentiating both sides)}$$

$$\frac{di}{dt} = \frac{di_1}{dt} + \frac{di_2}{dt} + \frac{di_3}{dt} + \cdots + \frac{di_n}{dt}$$

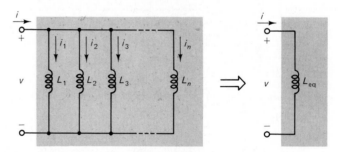

Figure 10.8

Since from Ohm's law for each inductor,

$$v = L_1 \frac{di_1}{dt} = L_2 \frac{di_2}{dt} = L_3 \frac{di_3}{dt} = L_n \frac{di_n}{dt}$$

$$= L_{eq} \frac{di}{dt} \quad \text{(from the equivalent circuit)}$$

then

$$\frac{v}{L_{eq}} = \frac{v}{L_1} + \frac{v}{L_2} + \frac{v}{L_3} + \cdots + \frac{v}{L_n} \quad \text{(dividing both sides by } v\text{)}$$

$$\frac{1}{L_{eq}} = \frac{1}{L_1} + \frac{1}{L_2} + \frac{1}{L_3} + \cdots + \frac{1}{L_n} \tag{10.9}$$

or

$$L_{eq} = \frac{1}{1/L_1 + 1/L_2 + 1/L_3 + \cdots + 1/L_n}$$ (10.9a)

(equivalent of parallel connected inductors)

For only two inductors in parallel (e.g., L_1 and L_2),

$$L_{eq} = \frac{1}{1/L_1 + 1/L_2} = \frac{L_1 L_2}{L_1 + L_2}$$ (10.10)

It is clear from the discussion above that inductors in series or in parallel are treated in exactly the same manner as resistors in series or in parallel, respectively.

■ EXAMPLE 10.2

Find the value of the equivalent inductor that can replace the series-parallel connection of inductors shown in Fig. 10.9.

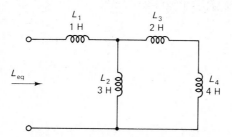

Figure 10.9

Solution   As L_3 and L_4 are in series, their equivalent is

$$L_{34} = L_3 + L_4 = 2 + 4 = 6\,H$$

Now L_{34} is in parallel with L_2; their equivalent is

$$L_{234} = \frac{L_2 L_{34}}{L_2 + L_{34}} = \frac{3 \times 6}{3 + 6} = 2\,H$$

The circuit is thus equivalent to L_1 in series with L_{234}. Thus the total equivalent inductor is

$$L_{eq} = L_1 + L_{234} = 1 + 2 = 3\,H$$

10.3 TRANSIENTS IN RL BRANCHES: CURRENT RISE

To understand the difference between the way an ideal resistor and an ideal inductor behave when a switch is closed in a dc circuit, the circuits in Fig. 10.10 will be examined. The switch was open for a long time and no initial current flows in either circuit. This is expressed mathematically as

$$i(0^-) = 0\,A \qquad t = 0^- \text{ means right before the switching instant}$$

1.   In a purely resistive circuit [Fig. 10.10(a)], at $t = 0^+$ (right after the

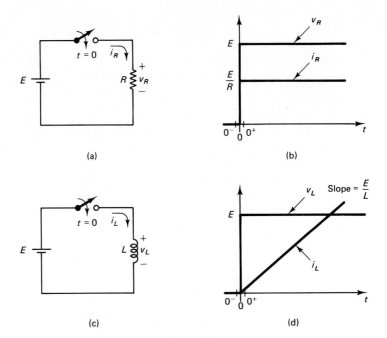

Figure 10.10

switch is closed), Ohm's law and KVL apply, that is,

$$E = v_R = Ri_R$$

Therefore,

$$i_R = \frac{E}{R}$$

Thus the current in a resistor can jump instantaneously to its steady-state value. Figure 10.10(b) illustrates this behavior.

2. In a purely inductive circuit [Fig. 10.10(c)] at $t = 0^+$ (right after the switch is closed), again Ohm's law and KVL apply:

$$E = v_L = L\frac{di_L}{dt}$$

Therefore,

$$\frac{di_L}{dt} = \frac{E}{L} = \text{constant slope } (m) \tag{10.11}$$

Thus even though the voltage across the inductor v_L can instantaneously change from 0 to E, the current in the inductor i_L will rise with a constant slope as derived in Eq. (10.11). This constant slope indicates that i_L versus t is a straight line:

$$i_L = mt = \frac{E}{L}t \tag{10.12}$$

The inductor's current then rises in a straight-line behavior with respect to time, as shown in Fig. 10.10(d). The line passes by the origin ($t = 0$, $i_L = 0$) because the current in the inductor cannot change instan-

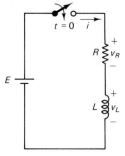

Figure 10.11

taneously, as explained in Section 10.1, due to Lenz's law and the induced counter EMF opposing the change in current.

Since the slope in Eqs. (10.11) and (10.12) (i.e., the rate of change of i_L) is inversely proportional to L, the higher the inductance is, the lower would be the rate of change of current, confirming the definition of L as the property exhibiting the opposition to change in current.

Now, consider the series connection of R and L as in the series RL branch in Fig. 10.11, again with no initial current flow in the circuit. Notice that now the current in R and L should be the same due to the series connection. Right after the switch is closed at $t = 0^+$, KVL applies, that is,

$$v_R + v_L = E \qquad (10.13)$$

Therefore,

$$iR + L\frac{di}{dt} = E \quad \text{[using Eq. (10.2) for } v_L\text{]}$$

or

$$\frac{di}{dt} + \frac{R}{L}i = \frac{E}{L} \qquad (10.14)$$

This equation is a first-order linear differential equation. The solution of this equation provides the functional relationship between i and t, $i(t)$. Similar to the discussion of the capacitor's voltage in RC circuits, this solution cannot be fully determined unless the initial condition is specified. Here this condition is usually in the form

$$i(0^-) = i(0^+) = I_0 \qquad \text{(some constant value)} \qquad (10.15)$$

The complete solution of Eq. (10.14) is provided in Appendix A.10.

Consider first the simplest case, where no initial current is flowing in the RL branch, that is,

$$i(0^-) = i(0^+) = 0$$

Right after the switch is closed, the substitution in KVL [Eq. (10.14)] provides

$$\frac{di}{dt}(0^+) = \frac{E}{L} = \text{initial slope}$$

$$= \text{initial rate of change of current} \qquad (10.16a)$$

Also,

$$v_R(0^+) = Ri(0^+) = 0 \qquad (10.16b)$$

and

$$v_L(0^+) = E - v_R = E \quad \text{[from Eq. (10.13)]} \qquad (10.16c)$$

Thus the initial rate of change of the branch current is the same as in the case of an ideal purely inductive branch [see Fig. 10.10(c) and (d) and Eq. (10.11)]. Equations (10.16b) and (10.16c) indicate that at the moment the switch is closed, the voltage across R is zero and the voltage across L jumps instaneously from 0 to E volts.

As the rate of change of current is a positive quantity, the current starts to increase. With i increasing, v_R also increases, but v_L decreases since their sum is always E, as stated in Eq. (10.13). But from Eq. (10.2),

$$\frac{di}{dt} = \frac{v_L}{L}$$

Therefore, the rate of change of current (i.e., the slope of the curve) will decrease. Thus the current i will rise fast initially, but its rate of increase will become smaller with time. This results in the exponential rise curve shown in Fig. 10.12.

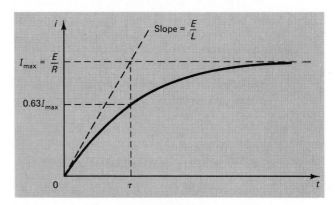

Figure 10.12

Eventually, when the rate of change of the current drops to zero (i.e., when the current becomes constant),

$$\frac{di}{dt} = 0 \tag{10.17a}$$

$$v_L = L\frac{di}{dt} = 0 \tag{10.17b}$$

$$v_R = E - v_L = E = I_{max}R \tag{10.17c}$$

Therefore,

$$I_{max} = \frac{E}{R} \tag{10.17d}$$

All the source voltage is now across the resistance as v_L is zero. Equations (10.17) define the steady-state conditions for i, v_R, and v_L. After that there is no more change with time.

As discussed in Section 10.1, the inductor is equivalent to a short circuit ($v_L = 0$) under steady-state dc conditions, confirming the results obtained above. It is then very easy to calculate the maximum value that the current in such RL branches would reach, as in Eq. (10.17d), being limited only by the value of R. Figure 10.13 shows the variation with time of the resistor's voltage $v_R(= Ri)$ and the inductor's voltage $v_L(= E - v_R)$.

The time span during which current is changing toward its steady-state value is called the transient time. In Figure 10.12 the point on the time axis marked τ

Chapter 10   Inductance and Transients in *RL* Circuits

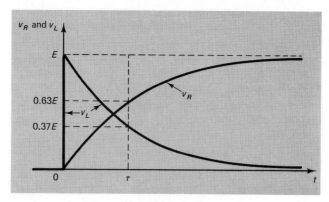

Figure 10.13

has a special practical importance. From the graph

$$\text{initial slope of } i(t) = \frac{di}{dt}(0^+) = \frac{E}{L}$$

$$= \frac{\text{rise}}{\text{run}} = \frac{I_{max}}{\tau} = \frac{E}{R\tau}$$

Therefore,

$$\boxed{\tau = \frac{L}{R} = \frac{\text{total inductance connected in series with } R \text{ and } E}{\text{total resistance connected in series with } L \text{ and } E}} \quad (10.18)$$

Since L has the units of henry ($=$ ohm $\cdot$ seconds) and R has the units of ohm, then τ does in fact have the units of time (i.e., seconds). τ is called the time constant of the RL branch. The current in the RL branch rises to 63% of its maximum value after a time of 1τ. In fact, after a time of 5τ the current reaches more than 99% of its final value. For most practical cases one says that the steady state has been reached after 5τ. Obviously, higher inductance in a branch corresponds to a larger value of τ and the current reaches steady state after a longer period of time. Higher resistance in a branch has the opposite effect.

The complete solution to KVL [Eq. (10.14)] is needed in order that numerical values of i, v_R, and v_L can be obtained at different values of time. This solution is obtained using calculus as shown in Appendix A.10. In general, with the initial condition $i(0^+) = I_0$, the solution according to Eq. (10A.2) is

$$\boxed{i - I_0 = (I_{max} - I_0)(1 - e^{-t/\tau})} \quad (10.19)$$

In the case when I_0 is zero (no initial current flow), then

$$\boxed{i = I_{max}(1 - e^{-t/\tau})} \quad (10.20)$$

The general solution [Eq. (10.19)] can also be expressed in the form

$$i = I_0 + (I_{max} - I_0) - (I_{max} - I_0)e^{-t/\tau}$$

$$= I_{max} - (I_{max} - I_0)e^{-t/\tau}$$

Hence

$$\boxed{i = i_{ss} \text{ (steady-state current)} + i_{tr} \text{ (transient current)}} \qquad (10.21)$$

where

$$i_{tr} = ke^{-t/\tau} \text{ (k is an arbitrary constant)}$$

$$i_{ss} = \text{current in the inductor at steady-state conditions}$$
$$\text{(i.e., with L replaced by a short circuit)}$$

Thus, in general, τ is easily obtained from the RL branch and i_{ss} is calculated from the circuit as the current in the RL branch with L replaced by a short circuit. The only unknown in Eq. (10.21) is the value of k. This is determined by substituting in Eq. (10.21) using the initial condition (i.e., at $t = 0$, $i = I_0$). Hence the complete solution is determined.

Graphical techniques using the normalized universal exponential curves in Fig. 10.14 can also be used to compute the values of the unknown variables in an RL circuit, as was done in the RC circuit. Here

$$x = \frac{t}{\tau} \quad \text{(normalized time axis)}$$

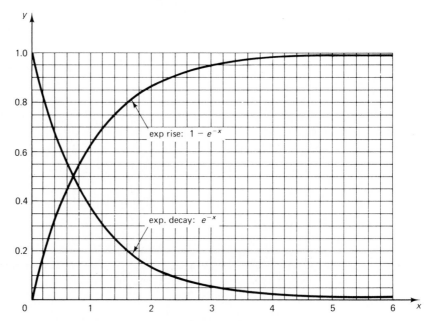

Figure 10.14

and from Eq. (10.19),

$$y = \frac{i - I_0}{I_{max} - I_0} = \frac{\text{change in current from the initial value}}{\text{maximum change in current from the initial value}}$$

$$= 1 - e^{-x} \quad \text{(normalized exponential rise curve)}$$

If $I_0 = 0$, then

$$y = \frac{i}{I_{max}} = 1 - e^{-x} \quad \text{(as above)}$$

Once i is known, then

$$v_R = Ri$$

If, for example, Eq. (10.20) is used, with $I_{max} = E/R$, then

$$v_R = E(1 - e^{-t/\tau}) \tag{10.22}$$

which can also be easily normalized as follows:

$$y = \frac{v_R}{E} = 1 - e^{-x} \quad \text{(also normalized exponential rise curve)}$$

and

$$v_L = E - v_R$$

If Eq. (10.22) is substituted in the above,

$$v_L = E - E(1 - e^{-t/\tau})$$
$$= Ee^{-t/\tau} \tag{10.23}$$

This also can be normalized as

$$y = \frac{v_L}{E} = e^{-x} \quad \text{(normalized exponential decay curve)}$$

With v_L known, then from Eq. (10.2),

$$\frac{di}{dt} = \frac{v_L}{L} \tag{10.24}$$

and

$$\frac{dv_R}{dt} = \frac{d}{dt}(Ri) = R\frac{di}{dt} = \frac{R}{L}v_L = \frac{v_L}{\tau} \tag{10.25}$$

Since

$$v_R + v_L = E \quad \text{(constant)}$$

then

$$\Delta v_R \text{ (change in } v_R) + \Delta v_L \text{ (change in } v_L) = 0$$

Therefore,

$$\Delta v_L = -\Delta v_R$$

and

$$\frac{dv_L}{dt} = -\frac{dv_R}{dt} = -\frac{v_L}{\tau} \tag{10.26}$$

■ EXAMPLE 10.3

In the circuit shown in Fig. 10.15, find
(a) i at $t = 0.4\,\text{s}$
(b) t at which $v_L = 70\,\text{V}$.
(c) t at which $v_R = v_L$.
There is no current initially flowing in L.

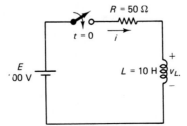

Figure 10.15

Solution Here

$$\tau = \frac{L}{R} = \frac{10}{50} = 0.2 \, \text{s}$$

and at steady state (L is equivalent to a short circuit)

$$I_{\text{max}} = \frac{E}{R} = \frac{100}{50} = 2 \, \text{A}$$

A sketch of i versus t in this case is shown in Fig. 10.16. Using the graphical technique (the universal exponential curves of Fig. 10.14), we obtain:

(a) At $t = 0.4 \, \text{s}$,

$$x = \frac{t}{\tau} = \frac{0.4}{0.2} = 2$$

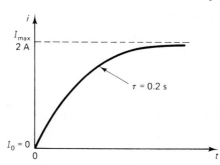

Figure 10.16

From the exponential rise curve ($y = 1 - e^{-x}$),

$$y = \frac{i}{I_{\text{max}}} = \frac{i}{2} = 0.865$$

Therefore,

$$i = 1.73 \, \text{A}$$

(b) When $v_L = 70 \, \text{V}$ and $v_R = E - v_L = 100 - 70 = 30 \, \text{V}$,

$$i = \frac{v_R}{R} = \frac{30}{50} = 0.6 \, \text{A}$$

Hence,

$$y = \frac{i}{I_{\text{max}}} = \frac{0.6}{2} = 0.3$$

Again from the exponential rise curve, this value of y corresponds to

$$x = \frac{t}{\tau} = \frac{t}{0.2} = 0.357$$

Therefore,

$$t = 0.071 \, \text{s}$$

(c) When $v_R = v_L$, each will be equal to $\dfrac{E}{2} = 50 \, \text{V}$,

$$i = \frac{v_R}{R} = \frac{50}{50} = 1 \, \text{A}$$

Thus,

$$y = \frac{i}{I_{max}} = \frac{1}{2} = 0.5$$

From the exponential rise curve, this value of y corresponds to

$$x = \frac{t}{\tau} = \frac{t}{0.2} = 0.693$$

Therefore,

$$t = 0.139\,s$$

EXAMPLE 10.4

In Fig. 10.17 there was no current initially flowing in L. The switch S_2 was closed 1 s after the switch S_1 is closed. Find i, v_L, and di/dt 1 s after S_2 is closed.

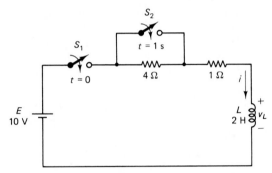

Figure 10.17

Solution At $t = 0$, S_1 is closed but S_2 is open. Here

$$\tau_1 = \frac{L}{4 + 1} = \frac{2}{5} = 0.4\,s$$

and

$$I_{m1} = \frac{10}{4 + 1} = 2\,A$$

At $t = 1\,s$,

$$x = \frac{t}{\tau_1} = \frac{1}{0.4} = 2.5$$

Using the exponential rise curve ($y = 1 - e^{-x}$) yields

$$y = \frac{I_0}{I_{m1}} = \frac{I_0}{2} = 0.918$$

$$I_0 \text{ (current at } t = 1\,s) = 2 \times 0.918 = 1.836\,A$$

The inductor's current remains unchanged at the moment of switching, at $i = I_0$. For $t \geq 1\,s$, S_2 will be closed, short circuiting the 4 Ω resistor. Thus here

$$\tau_2 = \frac{2}{1} = 2\,s$$

and with L equivalent to a short circuit (i.e., after the transient has died away)

$$I_{m2} = \frac{10}{1} = 10 \text{ A}$$

The current i will then rise from I_0 toward its new maximum I_{m2}, starting at $t = 1\,\text{s}$. The variation of i versus t is sketched in Fig. 10.18. If the time axis is now shifted to the point $t = 1\,\text{s}$ (i.e., we start counting time from this instant), then at $t = 2\,\text{s}$ (1 s from the new origin)

$$x = \frac{t}{\tau_2} = \frac{1}{2} = 0.5$$

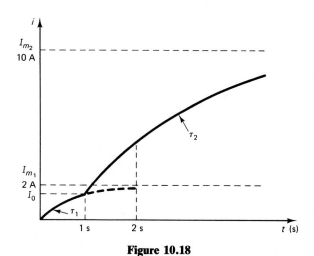

Figure 10.18

From the exponential rise curve, this value of x corresponds to

$$y = 1 - e^{-x} = 0.393$$

Notice that this situation corresponds to the general case [Eq. (14.19)] with I_0 finite (and not zero). Thus here

$$y = 0.393 = \frac{i - I_0}{I_{m2} - I_0} = \frac{i - 1.836}{10 - 1.836}$$

Therefore,

$$i = 1.836 + 8.164 \times 0.393 = 5.04 \text{ A}$$

The circuit is now the series connection of E, L, and the $1\,\Omega$ resistor. From KVL,

$$v_L = 10 - 5.04 \times 1 = 4.96 \text{ V}$$

$$= L\frac{di}{dt}$$

Therefore,

$$\frac{di}{dt} = \frac{v_L}{L} = \frac{4.96}{2} = 2.48 \text{ A/s}$$

EXAMPLE 10.5

Find $i(t)$ in the circuit of Fig. 10.19, if $i(0^+) = 6$ A. What is the value of v_L at $t = 1$ s?

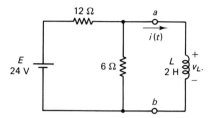

Figure 10.19

Solution Using Thévenin's theorem, the network to the left of points a and b can first be converted to a single-resistance single-source equivalent form as shown in Fig. 10.20. This equivalent circuit can then be analyzed as discussed above. Here

$$R_{th} = \frac{6 \times 12}{6 + 12} = \frac{72}{18} = 4 \, \Omega$$

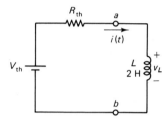

Figure 10.20

and

$$V_{th} = 24 \times \frac{6}{6 + 12} = \frac{144}{18} = 8 \, V$$

As the algebraic form of $i(t)$ is required, one can use the method of Eq. (10.21). τ (from the equivalent circuit) is

$$\frac{L}{R_{th}} = \frac{2}{4} = 0.5 \, s$$

i_{ss} (steady-state current in L when it is replaced by a short circuit), is

$$\frac{V_{th}}{R_{th}} = \frac{8}{4} = 2 \, A$$

Therefore,

$$i(t) = i_{ss} + i_{tr} = 2 + ke^{-t/0.5}$$
$$= 2 + ke^{-2t}$$

To determine k, the initial condition can be substituted in the equation above. Thus

$$6 = 2 + ke^{-0} = 2 + k$$

Therefore,

$$k = 4$$

and
$$i(t) = 2 + 4e^{-2t} \quad A$$

At $t = 1$ s,
$$i = 2 + 4e^{-2} = 2.54 \quad A$$

From the equivalent circuit, using KVL,
$$v_L = V_{th} - iR_{th} = 8 - 2.54 \times 4 = -2.16 \, V$$

10.4 SOURCE-FREE RL BRANCHES: CURRENT DECAY

Almost always, a discharge resistance R_2 is connected across (in parallel with) any *RL* branch, as shown in Fig. 10.21. The importance of such a resistance will be apparent from the following discussion.

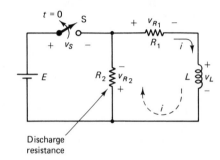

Figure 10.21

With the switch S closed for some time, the current flow in the inductor will be finite, say I_0. At the moment the switch is reopened ($t = 0$), the inductor's current remains momentarily at the value I_0 since it cannot change instantaneously. If R_2 is not connected (i.e., $R_2 = \infty$, open circuit), all this current would be flowing in the leakage resistance of the switch R_{sw}, which could be quite high (about $1 \, M\Omega$). Thus

$$v_S = \text{voltage across the switch terminals} = I_0 R_{sw}$$

v_S could reach such a high level that may easily break down the insulation in the switch (air) and sparking would be observed. In fact, the ignition in automobiles, generated by the spark plugs, follows the same principle discussed here. Also, since KVL is always applicable, still with R_2 disconnected,

$$E = v_S + v_{R_1} + v_L$$

(note that $v_{R_1} = E$ at the end of the current rise), v_L would also be of the same magnitude as v_S (with negative sign). The inductor L will then sustain permanent damage (as well as the switch) due to such tremendously high voltages. These problems can be overcome and controlled by connecting the appropriate value of the discharge resistance R_2, in parallel with the *RL* branch.

There is no external source connected in the discharge circuit, consisting of the series connection of L, R_2, and R_1. This situation is referred to as *source-free circuit*. Because there is no source to sustain the current flow in L, the magnetic field in L starts to collapse. According to Lenz's law, a counter EMF will be induced in the inductor L trying to aid the magnetic flux and oppose its decay. The polarities of this induced voltage across L will then be reversed compared to the situation that existed during the buildup of current and flux. In other words, while the current in the inductor will still flow in the same direction, but with a magnitude decaying toward zero, the voltage v_L appears as if it was a source trying to assist the current flow. Applying KVL to the discharge circuit in Fig. 10.21, one obtains

$$v_L + v_{R_2} + v_{R_1} = 0$$

The dashed loop current indicates that current is still flowing in the same direction as previously. KVL can also be rewritten in the following forms:

$$v_L = -i(R_1 + R_2) = -iR_T \tag{10.27}$$

where $R_T = R_1 + R_2$ = total resistance in series with L, and by using Eq. (10.2) for v_L,

$$L\frac{di}{dt} + i(R_1 + R_2) = 0$$

Therefore,

$$\boxed{\frac{di}{dt} + \frac{R_T}{L}i = 0} \tag{10.28}$$

Equation (10.28) is a first-order differential equation relating $i(t)$ and t. The initial condition in this case is considered to be the value of the inductor's current at the moment the switch is reopened (at $t = 0$), that is,

$$i(0^+) = I_0$$

The initial value of the inductor's voltage at the moment S is reopened is obtained from Eq. (10.27):

$$V_{L_0} = -I_0 R_T = -I_0(R_1 + R_2) \tag{10.29}$$

Notice the negative sign indicating the reversal in polarities, as discussed above. The initial voltage across the terminals of the switch S as it is reopened can easily be obtained by applying KVL to the loop containing E and the switch S. From Fig. 10.21,

$$E + v_{R_2} - v_S = 0 \qquad \text{(valid for any value of } t\text{)}$$

Therefore,

$$V_{S_0} = v_S(0^+) = E + v_{R_2}(0^+)$$
$$= E + I_0 R_2 \tag{10.30}$$

Through examining Eqs. (10.29) and (10.30), it is obvious that the voltages across the inductor and across the switch are much smaller than they could have been if R_2 is removed ($R_2 = \infty$). In fact, the values of these voltages are controlled by the choice of the value of R_2 and depend on the value of the inductor's current I_0 at the moment the switch was reopened.

Using Eq. (10.2), the initial slope of the inductor's current is

$$\frac{di}{dt}(0^+) = \frac{v_L(0^+)}{L} = \frac{V_{L_0}}{L}$$
$$= -I_0\frac{R_T}{L} \tag{10.31}$$

Consider now the circuit in Fig. 10.21 while the switch S is closed; here the branch containing R_2 and the branch containing R_1 and L in series are both in parallel with the external source E. Due to the properties of parallel branches, the currents in each of the branches are independent. Thus the current in the R_1–L branch would rise exponentially from its initial value (say, $0A$) toward a maximum of E/R_1 amperes, with a time constant of $\tau_1(=L/R_1)$, as discussed in Section 10.3. Also, the voltage across L initially jumps to the value of E and decays exponentially toward zero volts, with the same time constant, τ_1. The moment at which the switch S is reopened is now to be considered as the new origin of the time axis, $t = 0$ (i.e., an axis shift). At this moment the current in the inductor (initial current) is I_0. According to Eq. (10.31), the current in the inductor will now change with an initial rate which is a negative quantity. This means a decay, as shown in Fig. 10.22(a). As the current i reduces in magnitude, v_L also reduces in magnitude, according to Eq. (10.27). Since

$$\frac{di}{dt} = \frac{v_L}{L}$$

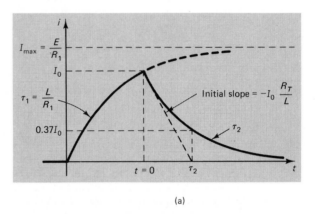

(a)

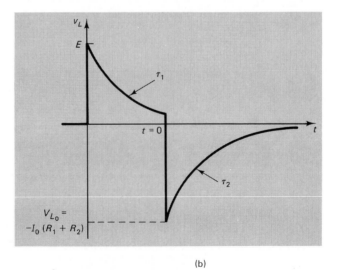

(b)

Figure 10.22

the rate of change of current is also reduced (i.e., the current reduction becomes slower). This action results in the exponential decay curves for i and v_L shown in Fig. 10.22(a) and (b). From the graph for $i(t)$,

$$\text{initial slope} = \frac{di}{dt}(0^+) = -I_0\frac{R_1 + R_2}{L}$$

$$= -\frac{\text{rise}}{\text{run}} = -\frac{I_0}{\tau_2}$$

Therefore,

$$\boxed{\begin{aligned}\tau_2 &= \text{current decay time constant} = \frac{L}{R_T} = \frac{L}{R_1 + R_2}\\[2mm] &= \frac{\text{total series inductance}}{\text{total resistance in series with } L \text{ in the source-free circuit}}\end{aligned}} \quad (10.32)$$

It should be noted here that the exponential rise and exponential decay time constants, τ_1 and τ_2, respectively, are in general unequal. Here τ_2 is less than τ_1. With a reconnection of the switch S and replacing R_2 by a short circuit (wire), both time constants could be made equal, as will be shown in the examples.

At $t = \tau_2$ the current is reduced to 37% of its initial value (i.e., $0.37I_0$). After a time span of $5\tau_2$ (transient time), the current is reduced to less than 1% of its initial value. For all practical purposes, complete decay is then reached. The transient has ended and steady state has been reached. At the steady-state conditions of such source-free circuits,

$$i_{ss} = 0 \quad \text{and} \quad v_{Lss} = 0$$

In order to obtain numerical values of the circuit's variables, the functional behavior of i versus t [i.e., $i(t)$] must be known. This is the solution of Eq. (10.28). As derived in Appendix A.10, Eq. (10A.4) gives

$$\boxed{i(t) = I_0 e^{-t/\tau_2}} \quad (10.33)$$

Once $i(t)$ is known, then from Eq. (10.27),

$$v_L(t) = -i(t)R_T$$

and

$$\frac{di}{dt}(t) = \frac{v_L(t)}{L} = -i(t)\frac{R_T}{L} = -\frac{i(t)}{\tau_2} \quad (10.34)$$

Also, from differentiating both sides of Eq. (10.27),

$$\frac{dv_L}{dt}(t) = -R_T\frac{di(t)}{dt} = -\frac{R_T v_L(t)}{L} = -\frac{v_L(t)}{\tau_2} \quad (10.35)$$

The equations above describe the complete algebraic solution of the source-free RL circuits. Graphical solution techniques, using the universal exponential decay curve of Fig. 10.14, can also be used, with Eq. (10.33) normalized as follows:

$$y = \frac{i(t)}{I_0} = e^{-x} \quad \left(\text{with } x = \frac{t}{\tau_2}\right)$$

EXAMPLE 10.6

The switch in the circuit shown in Fig. 10.23 remained in position 1 for a very long time. At $t = 0$ it was switched to position 2. Find $i(t)$ and $v(t)$ for $t \geq 0$.

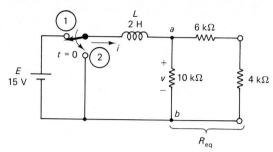

Figure 10.23

Solution   Between nodes a and b the three resistors can be replaced by

$$R_{eq} = \frac{10\,k\Omega(6\,k\Omega + 4\,k\Omega)}{10\,k\Omega + 6\,k\Omega + 4\,k\Omega} = \frac{100\,k\Omega}{20} = 5\,k\Omega$$

The statement "for a very long time" means that steady state has been reached. Thus with the switch in position 1, i_{ss} is obtained with L equivalent to a short circuit:

$$i_{ss} = I_0 = \frac{E}{R_{eq}} = \frac{15}{5\,k\Omega} = 3\,mA$$

Notice that here the current i flows in the same elements: L and R_{eq}, whether it is rising or decaying, that is,

$$\tau_1 = \tau_2 = \frac{L}{R_{eq}} = \frac{2}{5000} = 0.4\,ms$$

As the functional behavior of $i(t)$ is required, then from Eq. (10.33),

$$i(t) = I_0 e^{-t/\tau_2} = 3e^{-2500t} \quad mA$$

Therefore,

$$v(t) = -R_{eq}i(t) = -15e^{-2500t} \quad V$$

In this circuit the inductor's current will be interrupted during the switching instant; hence sparking will be observed.

EXAMPLE 10.7

In the circuit shown in Fig. 10.24, there was no current initially flowing in L. The switch S remained closed for 0.24 s before being reopened again. Find
(a) i, v_L, and v_S at the moment the switch is reopened ($t = 0.24^+$ s).
(b) i, v_L, and di/dt at $t = 0.34$ s.

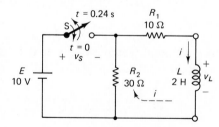

Figure 10.24

Solution   (a) With the switch closed ($0 \le t \le 0.24$ s), the current i in the inductive branch (R_1 and L) would rise toward a maximum value of

$$I_{max} = \frac{E}{R_1} = \frac{10}{10} = 1 \text{ A}$$

The time constant of this exponential rise behavior is

$$\tau_1 = \frac{L}{R_1} = \frac{2}{10} = 0.2 \text{ s}$$

A sketch of $i(t)$ versus t is shown in Fig. 10.25. At $t = 0.24$ s,

$$x = \frac{t}{\tau_1} = \frac{0.24}{0.2} = 1.2$$

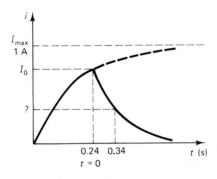

Figure 10.25

From the universal exponential rise curve, $y = 1 - e^{-x}$, corresponding to $x = 1.2$,

$$y = 0.7 = \frac{I_0}{I_{max}}$$

Therefore,

$$I_0 = 0.7 \text{ A}$$

At this moment the switch is reopened. Applying KVL to the loop containing E and S, as was done to obtain Eq. (10.30):

$$E + I_0 R_2 - V_{S_0} = 0$$

Therefore,

$$V_{S_0} \text{ (at } t = 0.24 \text{ s)} = E + I_0 R_2$$
$$= 10 + 0.7 \times 30 = 31 \text{ V}$$

From the source-free circuit (L, R_1, and R_2), KVL provides [refer to Eq. (10.29)]

$$V_{L_0} = v_L(t = 0.24 \text{ s}) = -I_0(R_1 + R_2)$$
$$= -0.7 \times 40 = -28 \text{ V}$$

(b) In the source-free circuit, the current i will decay from I_0 toward zero with a time constant τ_2:

$$\tau_2 = \frac{L}{R_1 + R_2} = \frac{2}{40} = 0.05 \text{ s}$$

At $t = 0.34$ s [i.e., after 0.1 s from the moment $(t = 0)$ the switch was reopened],

$$x = \frac{0.1}{\tau_2} = \frac{0.1}{0.05} = 2$$

Using the universal exponential decay curve: $y = e^{-x}$, corresponding to $x = 2$,

$$y = \frac{i}{I_0} = 0.135$$

Therefore,

$$i(t = 0.34 \text{ s}) = 0.135 \times 0.7 = 0.095 \text{ A}$$

From KVL of the source-free circuit,

$$v_L = -i(R_1 + R_2) = -0.095 \times 40 = -3.8 \text{ V}$$

and

$$\frac{di}{dt} = \frac{v_L}{L} = -\frac{3.8}{2} = -1.9 \text{ A/s}$$

10.5 GENERAL EXAMPLES OF RL CIRCUIT TRANSIENTS

EXAMPLE 10.8

There was no current initially flowing in the inductor in the circuit shown in Fig. 10.26. At $t = 0$ the switch was closed and then reopened again 2 s later.

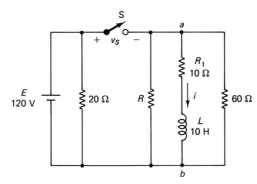

Figure 10.26

Find
(a) The time at which di/dt is 5 A/s.
(b) v_L at $t = 1$ s.
(c) The value of R if at $t = 2^+$ s, the voltage across the switch, v_S, is 200 V.

Solution (a) With S closed, all the branches are in parallel with the source E. The current in the inductor will rise exponentially toward a steady-state value, I_m (when L is equivalent to a short circuit):

$$I_m = \frac{E}{R_1} = \frac{120}{10} = 12 \text{ A}$$

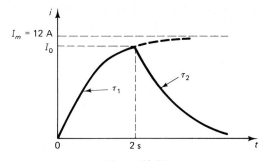

Figure 10.27

and

$$\tau_1 = \frac{L}{R_1} = \frac{10}{10} = 1\,\text{s}$$

A sketch of i versus t is shown in Fig. 10.27.

To find the time at which $di/dt = 5\,\text{A/s}$, we first observe that this rate of change of current is a positive quantity (i.e., current is rising). Thus the required time is somewhere between 0 and 2 s. The current at this instant can be found if v_L is calculated first.

$$v_L = L\frac{di}{dt} = 10 \times 5 = 50\,\text{V}$$

Therefore,

$$i = \frac{E - v_L}{R_1} = \frac{120 - 50}{10} = 7\,\text{A}$$

Now using the algebraic relationship for the current rise i versus t (with $I_0 = 0$) [Eq. (10.20)], we have

$$i = I_m(1 - e^{-t/\tau_1})$$
$$7 = 12(1 - e^{-t})$$
$$e^{-t} = 1 - 0.583 = 0.417$$

Therefore,

$$t = 0.875\,\text{s}$$

(b) At $t = 1\,\text{s}$, the rising current in the inductor is

$$i = 12(1 - e^{-1/1}) = 12(1 - 0.368) = 7.585\,\text{A}$$

Therefore,

$$v_L = E - iR_1 = 120 - 7.585 \times 10 = 44.15\,\text{V}$$

(c) At the moment the switch is reopened, $t = 2^+\,\text{s}$, the current in the inductor is

$$I_0 = 12(1 - e^{-2}) = 10.376\,\text{A}$$

From KVL applied to the loop containing the switch

$$-v_S + E + v_R = 0$$

(with the bottom terminal, b, assumed to be the positive polarity for v_R);

therefore,

$$v_R = v_S - E = 200 - 120 = 80 \text{ V}$$

The value of R can be determined if i_R (the current in it) can be found. But the 80 V is also across the 60 Ω resistor, then

$$i_{60\Omega} = \frac{80}{60} = 1.333 \text{ A}$$

From KCL applied at the bottom terminal, b, at $t = 2^+$ s,

$$i = I_0 = i_{60\Omega} + i_R$$

Therefore,

$$i_R = 10.376 - 1.333 = 9.043 \text{ A}$$

and

$$R = \frac{v_R}{i_R} = \frac{80}{9.043} = 8.85 \text{ Ω}$$

EXAMPLE 10.9

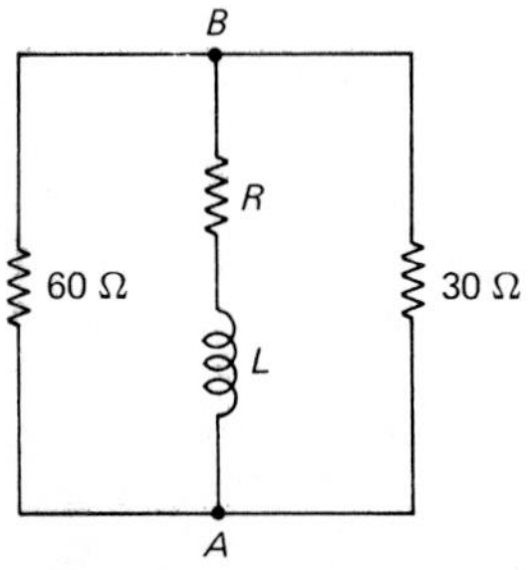

In the source-free circuit shown in Fig. 10.28, at $t = 0$ v_L was 150 V and v_{AB} was −120 V. Find

(a) R.

(b) L if at $t = 0.14$ s, v_L was 75 V.

Solution (a) If the 60 Ω and 30 Ω resistors are combined in parallel, the equivalent circuit of Fig. 10.29 results, with

$$v_{BA} = -v_{AB} = 120 \text{ V} \qquad (\text{at } t = 0)$$

and

$$R' = \frac{30 \times 60}{30 + 60} = 20 \text{ Ω}$$

Therefore,

$$I_0 = i_L \text{ (at } t = 0) = \frac{120}{20} = 6 \text{ A}$$

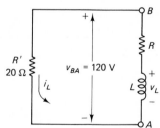

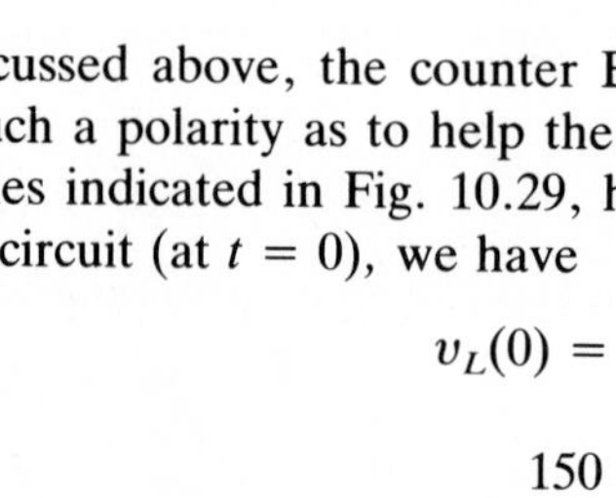

Figure 10.29

As discussed above, the counter EMF induced in the inductor v_L should be with such a polarity as to help the decaying current i_L. Thus v_L will have the polarities indicated in Fig. 10.29, helping the current i_L. Now, applying KVL to this circuit (at $t = 0$), we have

$$v_L(0) = i_L(0)R + v_{BA}(0)$$

or

$$150 = 6R + 120$$

Figure 10.28

60 Ω B R L 30 Ω A

Therefore,

$$R = \frac{30}{6} = 5\,\Omega$$

(b) The inductor's current decays according to Eq. (10.33) as

$$i_L = I_0 e^{-t/\tau}$$

At $t = 0.14$ s and with $v_L = 75$ V,

$$i_L = \frac{v_L}{R + R'} = \frac{75}{5 + 20} = 3\,\text{A}$$

Thus

$$3 = 6e^{-0.14/\tau}$$

$$e^{0.14/\tau} = 2$$

$$\frac{0.14}{\tau} = \ln 2 = 0.693$$

and

$$\tau = 0.202\,\text{s}$$

but here

$$\tau = \frac{L}{R + R'}$$

Therefore,

$$L = 0.202 \times 25 = 5.05\,\text{H}$$

EXAMPLE 10.10

The switch in the circuit shown in Fig. 10.30 has been closed for a very long time. At $t = 0$ it was opened. At this instant the direction of i_{R_2} was reversed, but its magnitude remained unchanged. At $t = 1$ s, i_{R_4} was 0.26 A. Find
(a) R_3 and L.
(b) i_{R_1}, v_L, and v_S at $t = 2$ s.
(c) The time at which di_{R_2}/dt is -4 A/s.

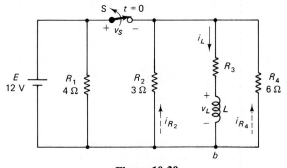

Figure 10.30

Solution (a) Since the switch has been closed for a very long time, steady-state conditions were achieved. Here

$$I_0 = i_L(0) = \frac{E}{R_3}$$

However, neither I_0 nor R_3 are known. But with the switch closed,

$$I_{R_2} = i_{R_2}(0) = \frac{E}{R_2} = \frac{12}{3} = 4\text{ A}$$

When the switch is opened (i.e., $t \geq 0$), the inductor's current i_L will then split at node b between R_2 and R_4, causing the reversal in the direction of the current in R_2. From the current-division principle,

$$i_{R_2} = i_L \frac{R_4}{R_2 + R_4} = \frac{6}{6+3} i_L = \frac{2}{3} i_L$$

At $t = 0$,

$$i_{R_2}(0) = 4 = \tfrac{2}{3} I_0$$

Therefore,

$$I_0 = 6\text{ A}$$

Thus

$$R_3 = \frac{E}{I_0} = \frac{12}{6} = 2\ \Omega$$

The current i_L will then decay according to Eq. (10.33):

$$i_L = I_0 e^{-t/\tau} = 6 e^{-t/\tau}$$

To find L, τ has to be determined [i.e., another value for $i_L(t)$ has to be known]. But at $t = 1$ s and from the current-division principle,

$$i_{R_4}(t = 1\text{ s}) = 0.26 = i_L(t = 1\text{ s}) \frac{R_2}{R_2 + R_4}$$

$$= \tfrac{1}{3} i_L(t = 1\text{ s})$$

Thus

$$i_L(t = 1\text{ s}) = 3 \times 0.26 = 0.78\text{ A}$$

and

$$0.78 = 6 e^{-1/\tau}$$

$$e^{1/\tau} = \frac{6}{0.78} = 7.692$$

Therefore,

$$\tau = \frac{1}{\ln(7.692)} = 0.49\text{ s}$$

With the two parallel resistors R_2 and R_4 replaced by their equivalent, $R'(= 2\ \Omega)$, the circuit is redrawn as shown in Fig. 10.31 (S is open). Thus here

$$\tau = 0.49 = \frac{L}{R_3 + R'} = \frac{L}{2 + 2}$$

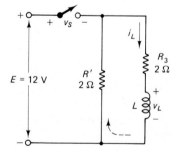

Figure 10.31

Therefore,
$$L = 1.96\,\text{H}$$

(b) R_1 is always in parallel with E, whether S is open or closed. Thus
$$i_{R_1} = \frac{E}{R_1} = \frac{12}{4} = 3\,\text{A}$$

for all values of t. Now, at $t = 2\,\text{s}$,
$$i_L = 6e^{-2/0.49} = 0.1\,\text{A}$$

From the equivalent circuit of Fig. 10.31, using KVL, we have
$$v_L = -i_L(R_3 + R') = -0.1 \times 4 = -0.4\,\text{V}$$

Also from KVL,
$$-v_S + E + i_L R' = 0$$

Therefore,
$$v_S = 12 + 0.1 \times 2 = 12.2\,\text{V}$$

(c) As derived in part (a),
$$i_{R_2} = \tfrac{2}{3} i_L$$

Thus
$$\frac{di_{R_2}}{dt} = \frac{2}{3}\frac{di_L}{dt}$$

$$\frac{di_L}{dt} = \frac{3}{2}(-4) = -6\,\text{A/s}$$

But
$$v_L = L\frac{di_L}{dt} = 1.96(-6) = -11.76\,\text{V}$$

and from the equivalent circuit of Fig. 10.31,
$$i_L = \frac{-v_L}{R_3 + R'} = \frac{11.76}{4} = 2.94\,\text{A}$$

To find the time corresponding to these values of v_L and i_L, either graphical or algebraic techniques can be used. From $i_L(t)$ above,
$$2.94 = 6e^{-t/0.49}$$

$$e^{t/0.49} = \frac{6}{2.94} = 2.041$$

Therefore,
$$t = 0.49 \ln(2.041) = 0.35\,\text{s}$$

EXAMPLE 10.11

In the circuit shown in Fig. 10.32, find the value of R so that the voltage across the switch never exceeds 500 V under any conditions of closing or opening of the switch. If the switch has been closed for 1 min and then reopened, how long will the voltage across the switch remain above 200 V with the value of R being as determined?

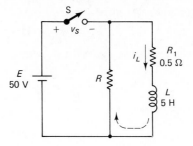

Figure 10.32

Solution Applying KVL to the loop containing the switch, the voltage across the switch is given by

$$v_S = E + i_L R \quad \text{(with S open)}$$

The highest value v_S can have occurs when S is closed for a very long time and i_L reaches its maximum value [i.e., at the moment S is reopened ($t = 0^+$)]:

$$i_L(0^+) = I_0 = I_{\text{max}} = \frac{E}{R_1} = \frac{50}{0.5} = 100 \text{ A}$$

Thus

$$500 = 50 + 100R$$

and

$$R = 4.5 \text{ } \Omega$$

A sketch of i_L versus t is shown in Fig. 10.33. With this value of R,

$$\tau_2 \text{ (decay time constant)} = \frac{L}{R_1 + R} = \frac{5}{5} = 1 \text{ s}$$

while

$$\tau_1 \text{ (rise time constant)} = \frac{L}{R_1} = \frac{5}{0.5} = 10 \text{ s}$$

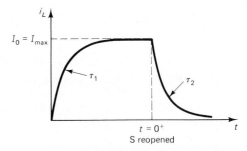

Figure 10.33

Since the switch has been closed for 1 min ($= 60$ s, i.e., $6\tau_1$), the inductor's current i_L can be considered to have reached its steady-state value I_{max} of 100 A [in fact, $i_L = I_{\text{max}}(1 - e^{-t/\tau_1}) = 100(1 - e^{-60/10}) = 99.75$ A]. As i_L decays exponentially according to

$$i_L = I_0 e^{-t/\tau_2} = 100e^{-t} \quad \text{A}$$

v_S also decays exponentially from its maximum value of 500 V. When v_S is

388

constant

$$\tau = \frac{5}{6 + 4} = 0.5 \, \text{s}$$

To determine v_S at $t = 1 \, \text{s}$, the value of i_L at this time must first be found. Using the expression for the exponentially decaying current, we have

$$i_L = I_0 e^{-t/\tau} = 2e^{-1/0.5} = 0.271 \, \text{A}$$

Applying KVL to the loop containing the switch and noting that no current is flowing in R_{th} while the switch is open yields

$$v_S = V_{\text{th}} + i_L \times 4 = 20.57 + 0.271 \times 4 = 21.65 \, \text{V}$$

EXAMPLE 10.14

With no current initially flowing in the inductor, the switch in the network shown in Fig. 10.37(a) was closed at $t = 0$ and reopened at $t = 8 \, \text{ms}$. A recording voltmeter connected across the switch provided the graph shown in Fig. 10.37(b). Find
(a) E.
(b) L.

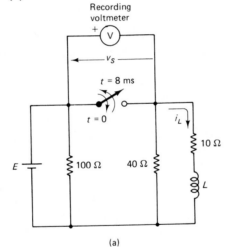

(a)

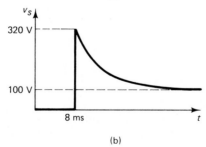

(b)

Figure 10.37

Solution (a) A sketch of the inductor's current i_L variation with time is shown in Fig. 10.38. However, neither τ_1, τ_2, I_{max}, or I_0 are known. But when the switch has been reopened for a very long time, i_L would have decayed to zero,

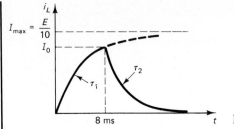

Figure 10.38

that is,

$$v_S = E + i_L \times 40 = E$$

From Fig. 10.37(b), this situation corresponds to

$$E = 100 \text{ V}$$

(b) At the moment the switch is reopened ($t = 8$ ms), i_L is momentarily at the value I_0 and the voltage across the switch is 320 V. Thus

$$v_S = E + I_0 \times 40$$

$$320 = 100 + I_0 \times 40$$

Therefore,

$$I_0 = \frac{220}{40} = 5.5 \text{ A}$$

Also,

$$I_{max} = \frac{E}{10} = \frac{100}{10} = 10 \text{ A}$$

Now the point on the exponential rise curve at $t = 8$ ms, $i_L = I_0 = 5.5$ A, is completely known and τ_1 can be determined using either algebraic or graphical techniques. From the universal exponential rise curve, with

$$y = \frac{i_L}{I_{max}} = \frac{I_0}{I_{max}} = \frac{5.5}{10} = 0.55$$

$$x = 0.8 = \frac{t}{\tau_1} = \frac{8 \text{ ms}}{\tau_1}$$

Therefore,

$$\tau_1 = 10 \text{ ms}$$

$$= \frac{L}{10}$$

and

$$L = 10 \text{ m} \times 10 = 100 \text{ mH} = 0.1 \text{ H}$$

DRILL PROBLEMS

Section 10.1

10.1. Find the value of the inductance whose coil has an effective length of 10 cm, an effective cross-sectional area of 2 cm², and which consists of 400 turns. The core is air.

10.2. What will be the number of turns of an iron core inductor ($\mu_r = 1000$) whose value is to be 0.2 H? The effective length of the core is 15 cm and its cross-sectional area is 1.2 cm².

10.3. What is the value of the flux linking the inductor examined in Drill Problem 10.2 if the current flowing in the coil is 0.1 A? How much energy is stored in this inductor in this case?

10.4. What are the rate of change of the current and the rate of change of the flux in the inductor examined in Drill Problem 10.2 if the voltage across the inductor is measured to be 3 V?

Section 10.2

10.5. Obtain the value of the equivalent inductance for the inductive circuit shown in Fig. 10.39.

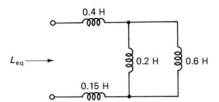

Figure 10.39

10.6. Obtain the value of the equivalent inductance for the inductive circuit shown in Fig. 10.40.

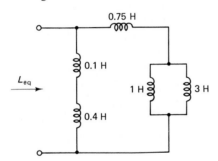

Figure 10.40

Sections 10.3 to 10.5

10.7. There was no current initially flowing in the inductor in the network shown in Fig. 10.41. At $t = 0$ s, the switch S was closed. Obtain the analytical expressions for $i(t)$ and $v_L(t)$.

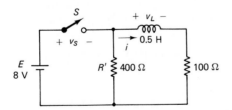

Figure 10.41

10.8. Determine the values of $i(t)$ and $v_L(t)$ obtained in Drill Problem 10.7 at
  (a) $t = 5$ ms.
  (b) $t = 10$ ms.
  (c) $t = 20$ ms.

10.9. At $t = 30$ ms, switch S in the network analyzed in Drill Problem 10.7 and shown

Drill Problems

in Fig. 10.41 was opened. Considering this moment to be the origin of a new time axis ($t' = 0$ s), determine the initial values: $i(0^+)$, $v_L(0^+)$, and $v_S(0^+)$. Note that $t' = 0^+$ s refers to the moment right after opening the switch.

10.10. Determine the analytical expressions for $i(t')$ and $v_L(t')$ after switch S in the network shown in Fig. 10.41 is opened. Use the initial conditions and the time axis as obtained and defined in Drill Problem 10.9.

10.11. Determine the values of i, v_L, and v_S in Drill Problem 10.10, with switch S open, at
- **(a)** $t' = 1$ ms.
- **(b)** $t' = 3$ ms.
- **(c)** $t' = 5$ ms.

10.12. What should be the value of R' in the circuit shown in Fig. 10.41 if the voltage v_S across switch S is never to exceed 24 V under any conditions of opening or closing the switch?

PROBLEMS

Note: The problems below that are marked with an asterisk (*) can be used as exercises in computer-aided network analysis or as programming exercises.

Section 10.1

(i) **10.1.** Find the value of the inductance of a 100-turn coil, with an air core, if its effective length is 20 cm and its cross-sectional area is 2×10^{-3} m^2.

(i) **10.2.** How many turns are required to obtain a 1 mH inductance from the air-core coil whose dimensions are given in Problem 10.1?

(i) **10.3.** The coil in Fig. 10.42 is made of 200 turns and its doughnut-shaped iron core has a relative permeability of 400, an average length of 0.5 m, and a cross-sectional area of 25 cm^2. What is the inductance of the coil?

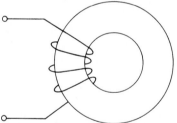

Figure 10.42

(i) **10.4.** A current of 2 A, changing at the rate of 50 A/s, is allowed to flow in the coil of Problem 10.3. At this instant of time, what is the voltage across the coil? What is the amount of magnetic flux in the core of the coil?

(i) **10.5.** The current whose waveform is shown in Fig. 10.43 is allowed to flow in a 20 mH inductor. Draw the expected waveform of the voltage across this inductor. What is the problem that this inductor may face?

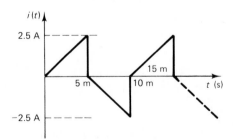

Figure 10.43

Section 10.2

(i)　**10.6.** What is the value of the single equivalent inductor L_{eq} that can replace the inductive ladder network shown in Fig. 10.44?

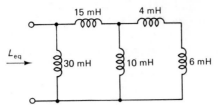

Figure 10.44

Sections 10.3, 10.4, and 10.5

(ii)　**10.7.** Initially there was no current flowing in the coil in the circuit shown in Fig. 10.45. At $t = 0$ the switch was closed and then reopened at $t = 0.75$ s. Find
   (a) The time at which the inductor's voltage is 250 V.
   (b) The voltage across the switch at $t = 1$ s.

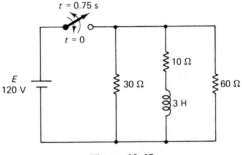

Figure 10.45

(ii)　**10.8.** In the source-free circuit of Fig. 10.46, at $t = 0$, V_{AB} is -60 V. At $t = 2$ s the voltage across the inductor was 15 V. Find the value of L.

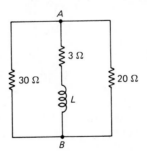

Figure 10.46

(i)　***10.9.** The circuit shown in Fig. 10.47 is in the steady-state conditions. At $t = 0$ the

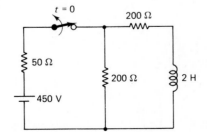

Figure 10.47

Problems

395

switch was opened. Find

(a) The time at which the current in the inductor is 1 A.

(b) The rate of change of the current in the inductor at $t = 15$ ms.

(i) **10.10.** The periodic current waveform $i(t)$, shown in Fig. 10.48, is flowing in the series *RL* branch indicated. Calculate and plot the variation of V_{AB} versus t for $0 \le t \le 12$ ms.

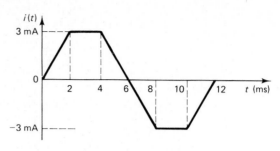

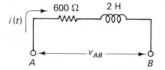

Figure 10.48

(i) **10.11.** Find the expression for $i_L(t)$ in the circuit shown in Fig. 10.49 if at $t = 0$, $i_L(0^+) = 40$ mA. What is the voltage across L, v_L, at $t = 0.7$ ms? [*Hint:* Use Thévenin's theorem for the network to the left of the inductor.]

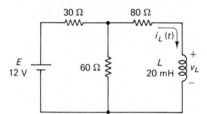

Figure 10.49

(i) **10.12.** The switch in the circuit shown in Fig. 10.50 was closed at $t = 0$ and reopened at $t = 2$ s. There was no current initially flowing in the inductor. Find

(a) i_L at $t = 0.5$ s.

(b) The time at which dv_L/dt is -250 V/s.

(c) The value of di_L/dt when the voltage across the resistance is 25 V.

(d) The current at 2.2 s and the physical condition of the switch at this time.

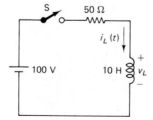

Figure 10.50

(ii) **10.13.** In the circuit shown in Fig. 10.51, the switch was initially open and no current was flowing in *L*. At $t = 0$ the switch was closed and then reopened at $t = 5$ ms.

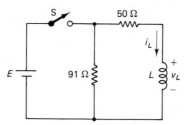

Figure 10.51

At $t = 0$, v_L was 100 V, and at $t = 4$ ms the currents in both resistors were equal. Find

(a) The values of E and L.

(b) The time at which the rate of change of the current i_L is -355 A/s.

(c) The voltage across the switch at the instant of time specified in part (b).

(ii) **10.14.** In the circuit shown in Fig. 10.52, the switch was initially open and no current was flowing in L. At $t = 0$ the switch was closed and then reopened at $t = 10$ ms. At $t = 0$, di_L/dt was 1000 A/s. Find

(a) The value of L.

(b) The peak voltage across the switch.

(c) The time at which di_L/dt is -200 A/s.

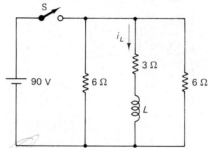

Figure 10.52

(i) **10.15.** Initially there was no current in the coil in the circuit shown in Fig. 10.53. At $t = 0$ the switch was closed and later reopened at $t = 14$ ms. At $t = 20$ ms find

(a) i_1 and i_2.

(b) v_L.

(c) v.

(d) di_1/dt.

(e) The voltage across the switch terminals.

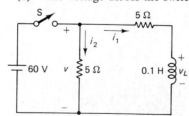

Figure 10.53

(ii) **10.16.** There was no current initially flowing in the coil in Fig. 10.54. At $t = 0$ the switch

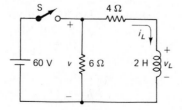

Figure 10.54

was closed and later reopened at $t = 1.25$ s. Find

(a) The source's current at $t = 1$ s.
(b) v_L at $t = 1$ s.
(c) di_L/dt at $t = 1$ s.
(d) v_L and di_L/dt at $t = 1.5$ s.

(i) *10.17. There was no current initially flowing in L in the circuit shown in Fig. 10.55. At $t = 0$ the switch was closed and then reopened at $t = 48$ ms. Find

(a) v_L at $t = 21$ ms.
(b) The time at which the magnitude of i_1 is 240 mA.
(c) The voltage across the switch v_S at $t = 63$ ms.

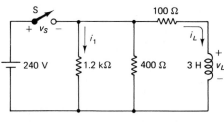

Figure 10.55

(ii) 10.18. In the circuit of Fig. 10.54, what is the peak voltage across the switch? How long should the switch be closed for this peak voltage to occur? If this switch voltage is required not to exceed 100 V at any time, no matter what switching conditions occur, what should be the value of the resistance to be added in parallel with the 6 Ω resistor to achieve this requirement?

APPENDIX A.10: TRANSIENT CURRENT IN AN RL BRANCH UNDER DC EXCITATION

With reference to Fig. 10.11, KVL for this circuit was obtained in Eq. (10.14):

$$\frac{di}{dt} + \frac{R}{L}i = \frac{E}{L}$$

or

$$\frac{di}{dt} + \frac{1}{\tau}i = \frac{E}{L}$$

(where $\tau = L/R =$ time constant of the RL branch). In general, the initial condition [Eq. (10.15)] is

$$i(0^+) = I_0$$

The solution of this first-order linear differential equation can be obtained through separation of variables and integration, that is,

$$\frac{di}{dt} = \frac{E}{L} - \frac{i}{\tau}$$

$$\int \frac{di}{E/L - i/\tau} = \int dt$$

398

then

$$-\tau \ln\left(\frac{E}{L} - \frac{i}{\tau}\right) = t + k'$$

$$\ln\left(\frac{E}{L} - \frac{i}{\tau}\right) = -\frac{t}{\tau} + k''$$

and

$$\frac{E}{L} - \frac{i}{\tau} = e^{-t/\tau}e^{k''} = k'''e^{-t/\tau}$$

Thus

$$i = \frac{E}{L}\tau - \tau k'''e^{-t/\tau} = \frac{E}{R} - ke^{-t/\tau} \qquad (10A.1)$$

This solution is valid for all values of $t \geq 0$. At $t = 0^+$, $i(0^+) = I_0$. Thus

$$I_0 = \frac{E}{R} - k$$

and

$$k = \frac{E}{R} - I_0$$

Substituting in Eq. (10A.1) gives

$$i = \frac{E}{R} - \left(\frac{E}{R} - I_0\right)e^{-t/\tau} + I_0 - I_0$$

Denoting $E/R = I_{max}$, and noting the addition and subtraction of the equal terms I_0 in the equation above, we have

$$i - I_0 = (I_{max} - I_0) - (I_{max} - I_0)e^{-t/\tau}$$

and

$$i - I_0 = (I_{max} - I_0)(1 - e^{-t/\tau}) \qquad (10A.2)$$

Equation (10A.2) is the required general solution. In the special case of no initial current flow [i.e., when $i(0^+) = I_0 = 0$],

$$i = I_{max}(1 - e^{-t/\tau}) \qquad (10A.3)$$

If $E = 0$, the circuit resembles a source-free RL branch, with R being the total series resistance of the circuit. Here

$$I_{max} = \frac{E}{R} = 0$$

Then Eq. (10A.2) becomes

$$i = I_0 - I_0(1 - e^{-t/\tau})$$
$$= I_0 e^{-t/\tau} \qquad (10A.4)$$

GLOSSARY

Choke: Another name for an inductor; the term arises from the ability of the inductor to exhibit very high opposition to the fast changes in the current flowing in it, (i.e., it "chokes" high-frequency current waveforms).

Inductor: A two-terminal linear circuit element that exhibits opposition to the change in the current flowing in it. It is usually constructed from a wire wound into a number of turns, forming a coil around a magnetic or nonmagnetic core.

Henry (H): The unit of inductance. One henry is $1\,\Omega \cdot s$.

Self-Inductance (L): The property of the electric circuit element that exhibits opposition to the change in current flowing in it. This property is due to the counter EMF induced in the coil as a result of the time-varying magnetic field linking it, in accordance with Faraday's and Lenz's laws.

ANSWERS TO DRILL PROBLEMS

10.1. $L = 0.402\,\text{mH}$

10.3. $\phi = 44.84\,\mu\text{Wb}$, $W = 1\,\text{mJ}$

10.5. $L_{eq} = 0.7\,\text{H}$

10.7. $i(t) = 80(1 - e^{-200t})\,\text{mA}$,
$v_L(t) = 8e^{-200t}\,\text{V}$

10.9. $i(0^+) = 79.8\,\text{mA}$,
$v_L(0^+) = -39.9\,\text{V}$,
$v_S(0^+) = 39.92\,\text{V}$

10.11. (a) $i(1\,\text{ms}) = 29.36\,\text{mA}$,
$v_L(1\,\text{ms}) = -14.678\,\text{V}$,
$v_S(1\,\text{ms}) = 19.744\,\text{V}$
(b) $i(3\,\text{ms}) = 3.97\,\text{mA}$,
$v_L(3\,\text{ms}) = -1.987\,\text{V}$,
$v_S(3\,\text{ms}) = 9.588\,\text{V}$
(c) $i(5\,\text{ms}) = 0.54\,\text{mA}$,
$v_L(5\,\text{ms}) = -0.269\,\text{V}$,
$v_S(5\,\text{ms}) = 8.216\,\text{V}$

11 | ALTERNATING CURRENT AND VOLTAGE

OBJECTIVES

- Understanding how ac voltage is generated based on the electromagnetic phenomenon, and the parameters required to fully define sinusoidal waveforms.

- Full appreciation of the lead/lag concepts: the phase angle of a waveform and the phase shift between waveforms.

- Appreciation of the meaning of the average and the effective values of a waveform, and ability to calculate these values for various types of waveforms.

- Ability to represent sinusoidal waveforms by phasor quantities, and to identify the lead/lag interrelationships between waveforms using the phasor diagram.

- Proficiency in complex numbers algebraic manipulations.

- Understanding the time domain interrelationships between the voltage, current, and power waveforms for the basic three circuit elements: R, L, and C, when connected to ac voltage sources.

- Ability to represent the voltage/current relationships for R, L, and C using phasor diagrams.

- Understanding the meaning of the inductive and the capacitive reactances, and their behaviour as functions of the frequency.

- Understanding the impedance and admittance concepts relating to R, L, and C.

- Ability to do simple manipulations involving the application of KVL and KCL in their phasor forms.

AC stands for alternating current. For historical reasons, this misnomer is still being used to designate current and voltage; however, its meaning refers to the fact that such voltages and currents continuously reverse their polarities and direction as time goes on. Also, they do not have constant magnitudes with time (as do dc voltage and current); their magnitudes continuously vary with time. The most common shape of such magnitude variations with time is the sinusoidal waveform.

Figure 11.1 shows the simplest form of ac voltage generator. It is constructed from an *n*-turn coil placed in the magnetic field of a permanent magnet. The terminals of the coil are soldered to each of two freely rotating slip rings (without contact with each other). This prevents twisting of the two sides of the coil as it is forced to rotate in the magnetic field. Terminals *a* and *b* of this generator are stationary and they are always connected to the respective conducting sides of the coil, sides *A* and *B,* through the two brushes (made of carbon) which are always kept in touch with the slip rings by spring forces (not shown).

This generator operates based on the electromagnetic phenomenon and Faraday's law, as explained in Chapter 9. When mechanical work is applied to the coil in the form of rotating torque, the two conducting sides of the coil move, cutting the magnetic lines of force and inducing voltage. Notice that the two sides of the coil are moving in opposite directions. By applying the generator RHR to each side of the coil, as in Fig. 11.1, it is clear that at this instant the current *i* flows in the coil in the direction indicated. The marked polarities indicate that voltages induced in each side of the coil are in such a direction as to help each

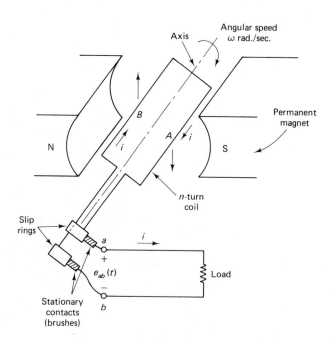

Figure 11.1

other (i.e., they are in series aiding). This results in the induced terminal voltage $e_{ab}(t)$ driving the current i in the external load.

To examine the shape of the generated voltage waveform $e_{ab}(t)$, consider first the cross-sectional diagrams in Fig. 11.2. Starting with the coil in position 1, perpendicular to the field, when the coil first starts to rotate, the two sides are moving parallel to the field, resulting in no cutting action, and the induced voltage $e_{ab}(t)$ is zero at this instant. As the coil continues to rotate to position 2, parallel to the field, the two sides of the coil will then be moving perpendicular to the field, achieving the maximum rate of cutting $(d\phi/dt)$ of the lines of force. Hence at this instant of time, according to Faraday's law, the induced voltage e_{ab} is maximum, E_m. In between these two positions, the rate of change of flux is increasing and the induced voltage increases from 0 V to E_m V. During the next 90° of rotation, the rate of change of flux is reducing and so would be the magnitude of the terminal voltage, e_{ab}, until sides A and B exchange positions. At this instant the angle of rotation, θ, is 180° and the induced voltage e_{ab} is again zero (no cutting action of the lines of force).

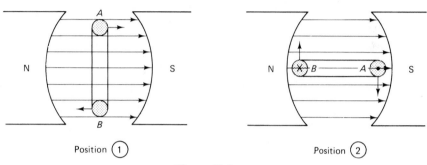

Figure 11.2

This process is repeated during the next half-cycle (180°), except that side A will now be moving upward and side B will now be moving downward (i.e., the direction of motion of the sides is opposite to that during the first half-cycle). Referring to Fig. 11.1, with the sides exchanged, we can easily see that the current in the external load will then exit the generator (source) from terminal b instead of a, resulting in a reversal in polarity of $e_{ab}(t)$. $e_{ab}(t)$ will then have a negative magnitude compared to the first half-cycle, as illustrated in the waveform shown in Fig. 11.3. After 270° of rotation (i.e., reversal of the sides compared to position 2), e_{ab} will be $-E_m$. When the cycle is finished, after 360° of rotation, sides A and B will be back in their original location (i.e., position 1) and e_{ab} will again be zero. The variation of $e_{ab}(t)$ as a function of time is referred to as the *waveform* of the generated voltage. The continuous rotation of the coil generates repeated cycles of the waveform, exact replicas of the first one. Because of the continuous reversal of the polarities of e_{ab} and the alternating flow of the current in the external load, such waveforms are referred to as ac waveforms.

The generated waveform can be expressed algebraically as

$$e_{ab}(t) = E_m \sin \theta \quad \text{V} \tag{11.1}$$

where θ is the angle of rotation of the coil. For more sophisticated generator

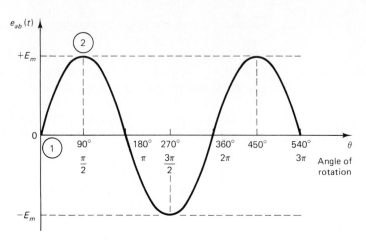

Figure 11.3   Sine waveform.

constructions, this angle is called the *electrical angle of rotation*, but for our simple example, it is also the same as the mechanical angle of rotation.

The quantity E_m, which is reached twice per cycle at $\theta = 90°$ and $\theta = 270°$, is called the *maximum* or *peak voltage* of the waveform. It is directly proportional to n (the number of turns of the coil), the magnetic flux ϕ, and the angular speed of rotation of the coil, ω.

The angular speed ω is defined in a manner similar to linear velocity:

$$\omega = \frac{\text{angle}}{\text{time}} = \frac{\theta}{t} \quad \text{rad/s} \tag{11.2}$$

or

$$\theta = \omega t \quad \text{rad} \tag{11.3}$$

ω is always expressed in radians/second, while θ is expressed in radians.

A radian is a unit of angle where

$$\pi \text{ (rad)} = 3.1416 \text{ rad} = 180° \quad \text{(refer to Fig. 11.3)}$$

Therefore,

$$1 \text{ rad} = \frac{180°}{3.1416} = 57.3° \tag{11.4}$$

that is,

$$\theta° = \theta \text{ (rad)} \times 57.3° \tag{11.5a}$$

or

$$\theta \text{ (rad)} = \frac{\theta°}{57.3°} \tag{11.5b}$$

Substituting for θ from Eq. (11.3) into Eq. (11.1) yields

$$e_{ab}(t) = e(t) = E_m \sin \omega t \quad \text{V} \tag{11.6}$$

Equation (11.6) thus expressed e, the instantaneous voltage, as a function of time t. Note that the quantity ωt is always in radians.

The circuit symbol for the ac voltage source is shown in Fig. 11.4. Even though the polarities of the source are continuously reversing every half cycle, the positive sign on the graph indicates that

$$e_{ab} = E_m \sin \omega t \qquad \text{not } -E_m \sin \omega t$$

Figure 11.4

404

There is a big difference between the two waveforms. This will become even more clear when phasors are introduced.

Sinusoidal ac voltage, which is the waveform of the voltage generated in hydro (or nuclear) stations and supplied to homes and factories, can also be generated electronically through conversion of dc into ac.

11.2 PERIOD AND FREQUENCY OF A SINUSOIDAL AC WAVEFORM

The quantity ω is referred to as the angular frequency, in rad/s. It is a constant for a given waveform. As Eq. (11.3) indicates, the horizontal axis θ can easily be expressed in terms of time t. The scaling here is done by dividing every value of θ by ω.

When one complete cycle is produced [i.e., θ is 2π rad (or 360°)], the corresponding elapsed time is t seconds. T is called the *period* of the sinusoidal waveform. From Eq. (11.2),

$$\omega = \frac{\theta}{t} = \frac{2\pi}{T} \quad \text{rad/s} \qquad \text{or} \qquad T = \frac{2\pi}{\omega} \quad \text{s} \tag{11.7}$$

This value of ω is used to convert the θ axis into a time axis, as shown in Fig. 11.5. Notice that T is the same whether it is measured between the two consecutive zero voltage, positive-going points, or between the two consecutive peaks or the two consecutive bottoms, or any other two consecutive symmetrically located points. It is always the time of one complete cycle. The frequency f of any periodic waveform (symmetrically repeated cycles every T seconds) is simply the number of cycles in 1 s. It is easy to deduce from the definitions of T and f that

$$f = \frac{1}{T} \quad \text{hertz (Hz; or cycles/second)} \tag{11.8}$$

For example, if the period t is 0.5 s, two complete cycles $(1/T)$ are produced in

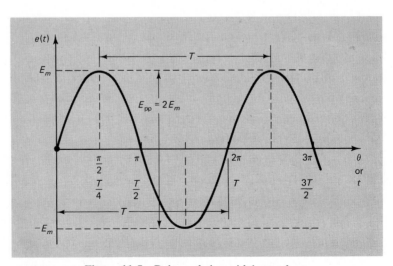

Figure 11.5 Balanced sinusoidal waveform.

1 s. If T is 0.1 s, then 10 cycles $(1/T)$ exist in 1 s. The commonly used unit for the frequency is the hertz (Hz).

From Eqs. (11.7) and (11.8),

$$\omega = 2\pi \frac{1}{T} = 2\pi f \quad \text{rad/s} \tag{11.9}$$

Thus the sinusoidal ac voltage waveform can be expressed as

$$e(t) = E_m \sin \omega t$$
$$= E_m \sin 2\pi f t$$

Again note that $\theta = 2\pi f t$ is in radians. If conversion to degrees is required, Eq. (11.5a) could be used.

In many cases of laboratory measurements using the oscilloscope, the sinusoidal waveform may not be balanced (i.e., it is not symmetrically located with respect to the central horizontal axis). In such cases, the quickest way of determining the parameters of the ac waveform is to measure T between two consecutive peaks or bottoms, and also to measure the peak-to-peak voltage (from top to bottom), E_{pp}. Referring to Fig. 11.5, we see that

$$E_m = \tfrac{1}{2}E_{pp} \tag{11.10}$$

Some useful trigonometric relationships are

$$\sin(2\pi \pm \theta) = \sin(2n\pi \pm \theta) = \sin(\pm\theta) = \pm\sin\theta \quad (n \text{ is an integer})$$
$$\cos(2\pi \pm \theta) = \cos(2n\pi \pm \theta) = \cos(\pm\theta) = \cos\theta$$

Clearly,

$$\sin(-\theta) = -\sin\theta \text{ and } \cos(-\theta) = \cos\theta$$

Also,

$$\sin(\pi \pm \theta) = \mp \sin\theta$$
$$\cos(\pi \pm \theta) = -\cos\theta$$
$$\sin\left(\frac{\pi}{2} \pm \theta\right) = \cos(\pm\theta) = \cos\theta$$
$$\cos\left(\frac{\pi}{2} \pm \theta\right) = \sin(\mp\theta) = \mp\sin\theta$$

11.3 PHASE SHIFT (OR PHASE ANGLE) OF AN AC WAVEFORM

The origin of the time (or angle) axis is arbitrary. For example, in the ac generator discussed in Section 11.1, the position of the coil at the instant of starting the observation of voltage $(t = 0)$ could, in fact, be anywhere. This means that the magnitude of $e(0)$, the initial value of the generator's voltage, could have been a positive, zero, or negative value, as indicated by the three different waveforms shown in Fig. 11.6. Throughout the rest of the text, unless specified otherwise, whenever the designation "ac waveform" is used, the

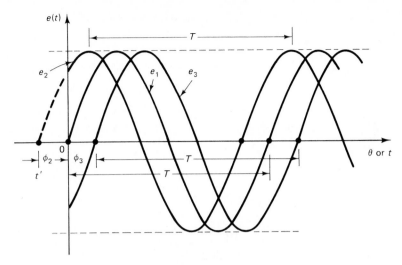

Figure 11.6

sinusoidal waveform is implied. Other waveforms will also be encountered in electrical engineering applications.

The general expression used to describe any ac waveform is written as

$$e(t) = E_m \sin(\omega t + \phi°) \quad \text{V}$$
$$\text{in rad} \quad \text{in deg}$$
$$\theta$$

(11.11)

where ϕ, usually expressed in degrees, is called the *phase angle* (or *phase shift*) or simply the *phase* of the waveform. It may be a positive or a negative number, or simply zero. One has to be careful in calculating the overall angle θ due to the mixed units used in the general expression of Eq. (11.11). The starting point of the waveform is conventionally considered to be the point in time at which the magnitude [e.g., $e(t)$] is zero and positive going (i.e., increasing in the positive direction). Such starting points are marked with circles in Fig. 11.6. Also notice that all the waveforms shown in Fig. 11.6 have the same period and hence the same values of the radian frequency ω. The starting point of the waveform therefore corresponds to the condition

$$\theta = 0° \quad (\text{not } \theta = 180°)$$

that is,

$$e(t') = E_m \sin(\omega t' + \phi) = 0$$

Therefore,

$$\omega t' + \phi = 0$$

and

$$t' = -\frac{\phi}{\omega}$$

(11.12)

In Eq. (11.12), ϕ must be expressed in radians to obtain the proper value of t'. Thus, if ϕ is positive, t' is negative, and vice versa. The three waveforms in Fig.

11.6 can then be expressed as

$$e_1(t) = E_m \sin \omega t \quad \text{V} \tag{11.13}$$

Here $\phi = 0$ and t' (the starting point) $= 0$, that is, the waveform starts at the origin.

$$e_2(t) = E_m \sin(\omega t + \phi_2) \quad \text{V} \tag{11.14}$$

Here $\phi = \phi_2$ and $t' = -\phi_2/\omega$ seconds (i.e., the waveform starts to the left of the origin).

$$e_3(t) = E_m \sin(\omega t - \phi_3) \quad \text{V} \tag{11.15}$$

Here $\phi = -\phi_3$ and $t' = \phi_3/\omega$ seconds (i.e., the waveform starts to the right of the origin).

As will be apparent in later discussions, an electric circuit driven by an ac source could have many voltage and current waveforms with different phases. A concept that will facilitate the analysis of ac circuits can easily be developed here, based on the comparison of the starting points of the waveforms in Fig. 11.6 when one imagines going along the increasing direction of the time axis.

A *leading* waveform, compared to another, is the one whose starting point occurs first in time, while a *lagging* waveform, compared to another, is the one whose starting point occurs later in time. Referring to Fig. 11.6, we would say

$$e_2 \text{ leads } e_1 \text{ by a phase } \phi_2 \text{ and } e_1 \text{ leads } e_3 \text{ by a phase } \phi_3$$

or

$$e_3 \text{ lags } e_1 \text{ by a phase } \phi_3 \text{ and } e_1 \text{ lags } e_2 \text{ by a phase } \phi_2$$

EXAMPLE 11.1

Draw a sketch for one cycle of the voltage waveform:

$$e(t) = 170 \sin(377t + 45°) \quad \text{V}$$

and determine E_m, ω, f, T, ϕ, and t' (the starting point) for this waveform. Also find the instantaneous value of e at $t = 1$ ms.

Solution Comparing the given waveform, $e(t)$, to the general expression of Eq. (11.11), one can easily conclude that

$$E_m = 170 \text{ V}$$

$$\omega = 377 \text{ rad/s} \quad \text{(coefficient of } t\text{)}$$

and

$$\phi = 45° = \frac{\pi}{4} = 0.785 \text{ rad}$$

From Eqs. (11.9) and (11.8)

$$f = \frac{\omega}{2\pi} = \frac{377}{6.283} = 60 \text{ Hz}$$

and

$$T = \frac{1}{f} = \frac{1}{60} = 16.67 \text{ ms}$$

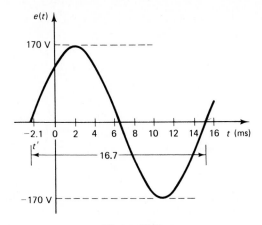

Figure 11.7

Using the definition of the starting point in Eq. (11.12) yields

$$t' = -\frac{\phi}{\omega} = -\frac{0.785}{377} = -2.08 \text{ ms}$$

At $t = 1$ ms,

$$
\begin{aligned}
e(t) &= 170 \sin (377 \times 1 \times 10^{-3} \text{ rad} + 45°) \\
&= 170 \sin (0.377 \text{ rad} + 45°) \\
&= 170 \sin (0.377 \times 57.3° + 45°) \\
&= 170 \sin (21.6° + 45°) = 156 \text{ V}
\end{aligned}
$$

The sketch of one cycle of this waveform is shown in Fig. 11.7.

11.4 VOLTAGE-CURRENT RELATIONSHIP FOR A RESISTOR IN AC CIRCUITS

When a sinusoidal ac voltage source, $e(t)$, is applied to a simple resistive circuit as in Fig. 11.8, Ohm's law and Kirchhoff's laws apply. Here

$$v(t) = e(t) = V_m \sin (\omega t + \phi) \quad \text{V} \tag{11.16}$$

and thus from Ohm's law,

$$i(t) = \frac{v(t)}{R} = \frac{V_m}{R} \sin (\omega t + \phi) \tag{11.17}$$

$$= I_m \sin (\omega t + \phi) \quad \text{A}$$

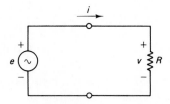

Figure 11.8

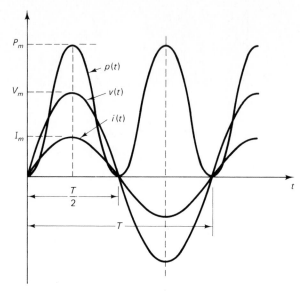

Figure 11.9

where

$$I_m = \frac{V_m}{R} \tag{11.17a}$$

$v(t)$ and $i(t)$ are displayed in Fig. 11.9 for the case when the phase angle ϕ is zero. From Fig. 11.9 and from Eqs. (11.16) and (11.17), it is clear that $i(t)$ is zero when $v(t)$ is zero. Also, $i(t)$ is maximum positive or maximum negative at the same time when $v(t)$ is of maximum positive or maximum negative magnitude, respectively. $i(t)$ has the same starting point in time as $v(t)$ and also the same period T (i.e., frequency f or ω).

Both waveforms thus follow each other in time. It can then be said that the voltage and the current waveforms for a resistive element are in phase. Thus except for the scale factor of $1/R$, both waveforms are similar.

Since power in any electric circuit element (here it is a resistive load) is obtained from

$$p = v \times i \quad \text{W} \tag{11.18}$$

power is then also a time function. In the case of the resistive element R discussed here,

$$
\begin{aligned}
p(t) &= V_m I_m \sin^2(\omega t + \phi) = P_m \sin^2(\omega t + \phi) \\
&= \tfrac{1}{2} V_m I_m [1 - \cos(2\omega t + 2\phi)] \quad \text{W}
\end{aligned} \tag{11.19}
$$

where the trigonometric identity $\cos 2\theta = 1 - 2\sin^2\theta$ has been used, and

$$P_m = V_m I_m = \frac{V_m^2}{R} = I_m^2 R \tag{11.19a}$$

$p(t)$ is plotted in Fig. 11.9 for the same zero phase angle condition. For any other phase angle, the whole graph is shifted to either the left or the right. As Eq. (11.19) and Fig. 11.9 indicate, the $p(t)$ waveform is always a positive quantity for the resistive element. Thus at any instant of time, the resistance always dissipates power. Notice that one cycle of the power waveform has a period of $T/2$ (i.e., half the period of the voltage or current waveforms); that is, $p(t)$ has double the frequency of the corresponding $v(t)$ or $i(t)$.

11.5 | AVERAGE AND EFFECTIVE (RMS) VALUES OF AC WAVEFORMS

Any periodic waveform, [e.g., $f(t)$ in Fig. 11.10] with a period T has an average value F_{av} calculated with respect to one cycle, but being the same throughout, as

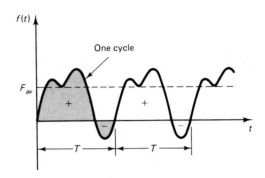

Figure 11.10

follows:

$$F_{av} = \frac{1}{T} \times \text{net area under the } f(t) \text{ curve over one period} \qquad (11.20)$$

or, using calculus,

$$F_{av} = \frac{1}{T} \int_0^T f(t)\, dt \qquad (11.21)$$

Positive and negative areas may be encountered as shown in Fig. 11.10. The net area is the algebraic sum of the component areas over one period of the waveform.

For any sinusoidal ac waveform, with any period (or frequency) and with any phase angle (sin or cos function), the net area over one complete period is always zero, due to the symmetry of both halves of the waveform, as shown in Fig. 11.11(a). Thus for ac voltage or current waveforms,

$$V_{av} = I_{av} = 0 \qquad (11.22)$$

and in general,

$$\int_0^T \sin(\omega t + \phi)\, dt = 0 \qquad (11.23)$$

Note that $\sin(\omega t + 90°) = \cos \omega t$.

If the negative half of the sine waveform is removed, as indicated by the solid curve of the graph in Fig. 11.11(b), the waveform will then be called a

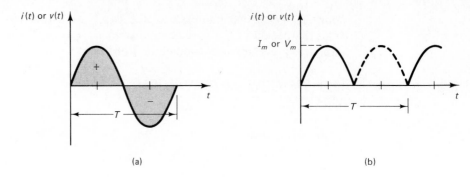

(a)   (b)

Figure 11.11

half-wave-rectified sine wave and

$$I_{av} = \frac{1}{\pi} I_m = 0.318 I_m \qquad (11.24)$$

or

$$V_{av} = \frac{1}{\pi} V_m = 0.318 V_m$$

When the negative half of the sine (or cosine) waveform is inverted, as shown by the dashed curve of the graph in Fig. 11.11(b), the waveform will then be called a *full-wave-rectified sine wave* and

$$I_{av} = \frac{2}{\pi} I_m = 0.637 I_m \qquad (11.25)$$

or

$$V_{av} = \frac{2}{\pi} V_m = 0.637 V_m$$

Equation (11.24) can easily be verified through calculus, while Eq. (11.25) follows directly from Eq. (11.24) since the area under the curve in one period is doubled.

Recalling the power dissipated in a resistance due to an ac voltage or current waveform [Eq. (11.19)], the term

$$\cos(2\omega t + 2\phi) = \sin(2\omega t + 2\phi + 90°)$$

has an average value of zero over one (or half a) period, as a result of Eq. (11.23). Thus the average power dissipated in a resistance R in an ac circuit is

$$P_{av} = \frac{1}{2} V_m I_m = \frac{1}{2} \frac{V_m^2}{R} = \frac{1}{2} I_m^2 R = \frac{1}{2} P_m \qquad (11.26)$$

where V_m and I_m are the maxima of the corresponding $v(t)$ and $i(t)$, respectively, and $V_m = I_m R$.

In general, the power function

$$p(t) = R i^2(t) = \frac{1}{R} v^2(t)$$

has an average value defined as in Eqs. (11.20) and (11.21), in the form

$$P_{av} = R\left[\frac{1}{T} \times \text{net area under the } i^2(t) \text{ waveform in one period}\right] \quad (11.27a)$$

$$= \frac{1}{R}\left[\frac{1}{T} \times \text{net area under the } v^2(t) \text{ waveform in one period}\right] \quad (11.27b)$$

or using calculus,

$$P_{av} = R\frac{1}{T}\int_0^T i^2(t)\, dt = \frac{1}{R}\frac{1}{T}\int_0^T v^2(t)\, dt \quad (11.28)$$

The quantity inside the brackets in Eqs. (11.27a) and (11.27b) is the average value of $i^2(t)$ and $v^2(t)$, respectively.

The *effective (or RMS) value* of any periodic waveform is defined as follows: *It is the constant (dc) equivalent of the waveform that will produce an amount of power in a given resistance R equal to the average power dissipated in this same resistance R due to the periodic waveform.* The effective value is usually denoted by I_{eff}, I_{RMS}, or simply I, and similarly, V_{eff}, V_{RMS}, or simply V, for the current waveform $i(t)$ and the voltage waveform $v(t)$, respectively.

Considering a sinusoidal ac current waveform, then according to the definition above and Eq. (11.26),

$$I^2 R = P_{av} = \tfrac{1}{2}I_m^2 R$$

Therefore,

$$\boxed{I = I_{eff} = I_{RMS} = \frac{I_m}{\sqrt{2}} = 0.707 I_m} \quad (11.29)$$

Similarly, for a sinusoidal ac voltage waveform,

$$\frac{V^2}{R} = P_{av} = \frac{1}{2}\frac{V_m^2}{R}$$

Therefore,

$$\boxed{V = V_{eff} = V_{RMS} = \frac{V_m}{\sqrt{2}} = 0.707 V_m} \quad (11.30)$$

Using the effective (i.e., the RMS) quantities of $i(t)$ and $v(t)$, the average power in a given resistance R can be expressed as

$$\boxed{P_{av} = VI = \frac{V^2}{R} = I^2 R} \quad (11.31)$$

(i.e., exactly the same as in the dc cases). The factor $\frac{1}{2} = (1/\sqrt{2}(1/\sqrt{2})$ is absorbed into V_m and I_m or I_m^2 or V_m^2 to produce the corresponding effective (or RMS) quantity.

AC measuring instruments (i.e., ammeters and voltmeters) are calibrated to read the RMS value of the sinusoidal waveform [$i(t)$ or $v(t)$] directly. However, these instruments will produce erroneous readings for other types of waveforms (e.g., square, triangular, pulse train, etc.). This is because the factors 0.637 (for the average of a full-wave-rectified ac waveform) and 0.707 (for the effective value of an ac waveform) apply only for sinusoidal waveforms.

For other types of waveforms, the definition of the effective (or RMS) value results in the following general mathematical expression:

$$I_{RMS}^2 R = P_{av} = R \times \text{av. vlaue of } i^2(t) \text{ over one period}$$

$$= R \frac{1}{T} \int_0^T i^2(t)\, dt$$

Therefore,

$$\boxed{I_{RMS} = \sqrt{\frac{1}{T} \int_0^T i^2(t)\, dt}} \qquad (11.32a)$$

Similarly,

$$\boxed{V_{RMS} = \sqrt{\frac{1}{T} \int_0^T v^2(t)\, dt}} \qquad (11.32b)$$

Notice that the letters RMS refer to the square *root* of the *mean* (or average) of the *s*quared variable.

The *form factor* of any waveform is defined as

$$F_f = \frac{\text{RMS value of the waveform}}{\text{average value of the full-wave rectified waveform}} \qquad (11.33)$$

For sinusoidal ac waveforms,

$$F_f = \frac{0.707 I_m}{0.637 I_m} = 1.11$$

EXAMPLE 11.2

The voltage across a certain circuit element is

$$v(t) = 800 \sin(628t + 30°) \quad V$$

and the current flowing in this element is

$$i(t) = 5 \sin(628t + 30°) \quad A$$

Find
(a) The nature and magnitude of this element and the power dissipated in it.
(b) f and T of these waveforms and the readings of the instruments used to measure them.

Solution   (a) Since $v(t)$ and $i(t)$ start from the same point in time (i.e., both waveforms have the same phase angle), they are in phase and the element is a resistor. Here

$$V_m = 800 \text{ V}$$

$$I_m = 5 \text{ A}$$

Therefore,

$$R = \frac{V_m}{I_m} = \frac{800}{5} = 160 \ \Omega$$

The average power dissipated in this resistor is

$$P_{av} = \frac{1}{2} V_m I_m = \frac{1}{2} I_m^2 R = \frac{1}{2} \frac{V_m^2}{R}$$

$$= \frac{1}{2} \times 800 \times 5 = 2000 \ W$$

(b) $\quad f = \dfrac{\omega}{2\pi} = \dfrac{628}{6.28} = 100 \ Hz$

$$t = \frac{1}{f} = \frac{1}{100} = 10 \ ms$$

The meters read the RMS values:

$$V_{RMS} = \frac{1}{\sqrt{2}} V_m = 0.707 \times 800 = 565.6 \ V$$

$$I_{RMS} = \frac{1}{\sqrt{2}} I_m = 0.707 \times 5 = 3.565 \ A$$

Note that

$$P_{av} = V_{RMS} I_{RMS} = 565.6 \times 3.565 = 2000 \ W$$

11.6 | PHASOR REPRESENTATION OF SINUSOIDAL AC WAVEFORMS

A sinusoidal waveform, whether a voltage or a current variable, for example,

$$v(t) = V_m \sin{(\omega t + \phi)} \quad V$$

has three parameters: $V \ (= 0.707V_m)$, the magnitude; ω (or f), the frequency; and ϕ, the phase. In examining ac resistive circuits, it was observed that $v(t)$ and $i(t)$ have the same frequency. This is also the case for ac inductive and capacitive circuits, as will be shown in the following sections. In any RLC circuit, under the influence of an ac source, all currents and voltages anywhere in the circuit will have the same frequency (i.e., ω is the same throughout), even though each voltage or current variable could have different magnitude and different phase.

The typical notation above for $v(t)$ is called the *time-domain* description of the variable. When Ohm's law, KCL, and KVL are applied to solve ac circuits, the addition, subtraction, multiplication, or division operations encountered will involve such time-domain sinusoidal quantities. To find the required unknown, a point-by-point computation at each value of time may have to be performed. Obviously, this is a time-consuming and tedious process.

To overcome this difficulty, a much simpler computational technique was devised by C. Steinmetz. It is based on the observation that since ω (the frequency) is the same throughout a given circuit, this parameter can be dropped from consideration temporarily. Thus all the voltage and current variables (sinusoidal quantities) would then need only two parameters to define each of them, the magnitude V (the RMS value $= 0.707V_m$), or I for the current, and the phase ϕ. A quantity defined by two such parameters is a vector (or a complex number), written as

$$\mathbf{V} = V\underline{/\phi} \quad V$$

To stress the point that such vectors do, in fact, represent functions of time, they are called *phasors*. The reason behind using the RMS value of the waveform to represent the magnitude of the phasor is that practical ac measuring instruments do indicate such magnitude, as mentioned previously. This process of conversion from time-domain representation to phasor representation is reversible and can be expressed as follows:

$$\boxed{v(t) = V_m \sin{(\omega t + \phi)} \overset{\omega}{\Leftrightarrow} \mathbf{V} = V\underline{/\phi}} \qquad (11.34)$$

with $V = $ RMS value of the waveform $= 0.707V_m$.

Using phasors, the solution of circuit problems involving KVL, KCL, and Ohm's law becomes a familiar algebraic manipulation process (addition, subtraction, multiplication, and division) of complex numbers. The mathematical proof that the algebraic summation of the time-domain sinusoidal waveforms provides the same result as the equivalent algebraic summation of the corresponding phasor quantities is given in Appendix A11. It should be kept in mind that phasors are only mathematical tools for manipulation purposes. Once the required circuit variable is obtained through the circuit's solution, it can easily be reconverted into its proper time-domain representation, reintroducing the frequency variable, as indicated in Eq. (11.34).

■ EXAMPLE 11.3

Find the phasors for the following voltage and current waveforms.
(a) $v_1(t) = 170 \sin{(377t + 30°)}$ volts.
(b) $v_2(t) = -141.4 \sin{(377t + 60°)}$ volts.
(c) $i(t) = 6 \cos{(377t - 30°)}$ amperes.

Solution   (a) All three waveforms have $\omega = $ constant $= 377 \text{ rad/s}$. To find the phasor representation for $v_1(t)$, Eq. (11.34) can be used directly:

$$\mathbf{V}_1 = 170 \times 0.707\underline{/30°} = 120\underline{/30°} \text{ V}$$

(b) The minus sign in the expression for $v_2(t)$ can be converted into an extra 180° phase shift, according to the trigonometric identities:

$$v_2(t) = -141.4 \sin{(377t + 60°)} \quad V$$
$$= 141.4 \sin{(377t + 60° + 180°)} \quad V$$

Thus, from Eq. (11.34),

$$\mathbf{V}_2 = 141.4 \times 0.707\underline{/240°} = 100\underline{/240°} \text{ V}$$
$$= 100\underline{/-120°} \text{ V}$$

(c) The cosine waveform must first be rewritten as a sine waveform.

Using the trigonometric identities

$$i(t) = 6\cos(377t - 30°) = 6\sin(377t - 30° + 90°)$$

Then from Eq. (11.34),

$$\mathbf{I} = 6 \times 0.707\underline{/60°} = 4.243\underline{/60°} \quad A$$

EXAMPLE 11.4

Convert $\mathbf{V} = 70.7\underline{/-45°}$ V into its time-domain representation if f is 50 Hz.

Solution Here $\omega = 2\pi f = 2 \times 3.14 \times 50 = 314\,\text{rad/s}$.

$$V_m = \sqrt{2}\,V = 1.414 \times 70.7 = 100\,\text{V}$$

Then, from Eq. (11.34),

$$V(t) = 100\sin(314t - 45°) \quad V$$

Lead and lag concepts in phasor diagrams. Graphically, a phasor is represented by a vector whose phase angle ϕ is the angle with the horizontal positive direction. Positive angles are measured in the counterclockwise direction and negative angles are measured in the clockwise direction with respect to the reference horizontal line. The length of the line representing the vector, measured from the origin point, is proportional to the magnitude of the phasor, depending on the scale chosen. Such a graphical representation of phasors is called a *phasor diagram*.

Consider again the three ac waveforms discussed in Section 11.3 [Eqs. (11.13), (11.14), and (11.15)]. Their phasor representations are

$$e_1(t) = E_m \sin \omega t \Leftrightarrow \mathbf{E}_1 = E\underline{/0°}\,V \qquad \text{where} \qquad E = \frac{1}{\sqrt{2}}E_m$$

$$e_2(t) = E_m \sin(\omega t + \phi_2) \Leftrightarrow \mathbf{E}_2 = E\underline{/\phi_2°}\,V$$

$$e_3(t) = E_m \sin(\omega t - \phi_3) \Leftrightarrow \mathbf{E}_3 = E\underline{/-\phi_3°}\,V$$

Graphical representation of these phasors (i.e., the phasor diagram) is shown in Fig. 11.12.

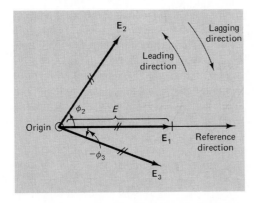

Figure 11.12

Based on the starting point of a waveform in the time domain, compared to other waveforms, it was observed that e_2 leads e_1 and e_1 leads e_3. The leading direction in the phasor diagram conforms with this concept. Going counterclockwise from one phasor (say, $\mathbf{E}_1$) to another (say, $\mathbf{E}_2$), one says that $\mathbf{E}_2$ leads $\mathbf{E}_1$ by an angle ϕ_2. Similarly, $\mathbf{E}_1$ leads $\mathbf{E}_3$ by an angle ϕ_3 and $\mathbf{E}_2$ leads $\mathbf{E}_3$ by an angle $(\phi_2 + \phi_3)$. The reverse direction (i.e., the clockwise direction) is the lagging direction. Going from one phasor (say, $\mathbf{E}_2$) to another (say, $\mathbf{E}_1$) in the clockwise direction, one says that $\mathbf{E}_1$ lags $\mathbf{E}_2$ by an angle ϕ_2 and similarly, $\mathbf{E}_3$ lags $\mathbf{E}_1$ by an angle ϕ_3.

11.7 | REVIEW OF COMPLEX-NUMBER ALGEBRAIC MANIPULATIONS

A complex number, representing a point or a vector (i.e., a phasor) in a two-dimensional complex plane, requires two pieces of information to define it. As shown in Fig. 11.13, the complex number $\mathbf{R}$ is completely defined by either of the following:

1. The x (real) and y (imaginary) coordinates from a reference origin point 0; this is called the Cartesian or rectangular form.
2. The length or magnitude R of the line drawn from the origin to the point and the phase angle ϕ that this line makes with the positive real axis direction; this is called the polar form.

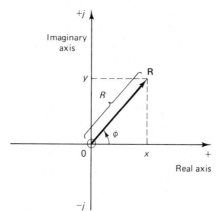

Figure 11.13

Thus the algebraic representation of a complex number or a vector (phasor) can take either of the following forms:

$$\mathbf{R} = R\underline{/\phi^\circ} \quad \text{(polar form)} \tag{11.35a}$$

or

$$\mathbf{R} = x + jy \quad \text{(rectangular form)} \tag{11.35b}$$

where

$$j = \sqrt{-1} \quad \text{or} \quad j^2 = -1 \tag{11.36a}$$

$$j \cdot j = -1 \quad \text{or} \quad \frac{1}{j} = -j \tag{11.36b}$$

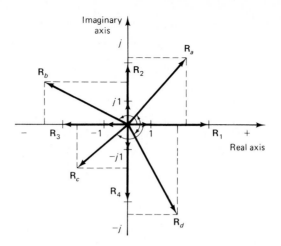

Figure 11.14

It is important to keep in mind a mental picture of the location of the complex number in one of the four quadrants, as shown in Fig. 11.14. This relates the angle ϕ to the signs of the rectangular coordinates x and y. When ϕ is between $0°$ and $90°$ (first quadrant), x and y are positive ($\mathbf{R}_a$). When ϕ is between $90°$ and $180°$ (second quadrant), x is negative and y is positive ($\mathbf{R}_b$). When ϕ is between $180°$ and $270°$, or $-90°$ and $-180°$ (third quadrant), both x and y are negative ($\mathbf{R}_c$). When ϕ is between $270°$ and $360°$, or $0°$ and $-90°$ (fourth quandrant), x is positive and y is negative ($\mathbf{R}_d$).

Continuous conversions between the two forms of complex numbers are frequently needed in the analysis of ac circuits because some algebraic manipulations of complex numbers are performed more easily in one form than the other, as will be shown shortly below. Conversions are performed based on the trigonometric relationships of the diagram in Fig. 11.13.

To convert from polar form to rectangular form,

$$x = R \cos \phi \qquad \text{and} \qquad y = R \sin \phi$$

Therefore,

$$\boxed{R\underline{/\phi°} = R \cos \phi + jR \sin \phi} \tag{11.37}$$

To convert from rectangular form to polar form,

$$R = \sqrt{x^2 + y^2} \qquad \text{and} \qquad \tan \phi = \frac{y}{x} \qquad \text{or} \qquad \phi = \arctan \frac{y}{x}$$

Therefore,

$$\boxed{x + jy = \sqrt{x^2 + y^2}\,\underline{/\tan^{-1} y/x}} \tag{11.38}$$

Many of the presently available hand-held calculators can perform these conversions easily and efficiently. In a direct manner, x and y are entered and R and ϕ are obtained by pressing the conversion button, or vice versa. Indirectly, the algebraic relationships in Eqs. (11.37) and (11.38) can be used.

It is very convenient to keep in mind the polar–rectangular forms of four special complex numbers, since they occur very frequently in circuit analysis.

Refer to Fig. 11.14:

$$\mathbf{R}_1 = R_1\underline{/0°} = R_1$$
($\phi = 0°$ corresponds to a purely real complex number;
imaginary part $= 0$)

$$\mathbf{R}_2 = R_2\underline{/90°} = jR_2$$
($\phi = 90°$ corresponds to a purely imaginary complex number;
real part $= 0$, i.e., $1\underline{/90°} = +j1$)

$$\mathbf{R}_3 = R_3\underline{/180°} = -R_3$$
($\phi = 180°$ corresponds to a purely negative real number,
i.e., $1\underline{/180°} = -1$)

$$\mathbf{R}_4 = R_4\underline{/-90°} = -jR_4$$
($\phi = -90°$ corresponds to a purely negative imaginary number,
i.e., $1\underline{/-90°} = -j1$)

Additions and subtractions of complex numbers (or phasors) are performed very easily using the rectangular form of these numbers. In general,

$$\mathbf{R}_1 \pm \mathbf{R}_2 = (x_1 + jy_1) \pm (x_2 + jy_2)$$
$$= (x_1 \pm x_2) + j(y_1 \pm y_2) \qquad (11.39)$$

Real parts are added or subtracted, separate from the addition or subtraction of the imaginary parts. The addition or subtraction is algebraic, including the signs.

Multiplication and division of complex numbers (or phasors) are performed very easily in the polar form. To appreciate such simplicity, the multiplication and division operations will be examined below, in a general manner, once in polar form and once in rectangular form.

$$\mathbf{R}_1 \cdot \mathbf{R}_2 = R_1\underline{/\phi_1°} \cdot R_2\underline{/\phi_2°} = R_1 R_2\underline{/\phi_1° + \phi_2°} \qquad (11.40)$$

(i.e., the magnitudes are multiplied while the phases are added algebraically). Or

$$\mathbf{R}_1 \cdot \mathbf{R}_2 = (x_1 + jy_1)(x_2 + jy_2) = x_1 x_2 + jy_1 x_2 + jy_2 x_1 - y_1 y_2$$
$$= (x_1 x_2 - y_1 y_2) + j(x_1 y_2 + x_2 y_1) \qquad (11.41)$$

Also,

$$\frac{\mathbf{R}_1}{\mathbf{R}_2} = \frac{R_1\underline{/\phi_1}}{R_2\underline{/\phi_2}} = \frac{R_1}{R_2}\underline{/\phi_1° - \phi_2°} \qquad (11.42)$$

(i.e., the magnitudes are divided while the phase of the denominator's complex number is algebraically subtracted from the phase of the numerator's complex number). Or

$$\frac{\mathbf{R}_1}{\mathbf{R}_2} = \frac{x_1 + jy_1}{x_2 + jy_2} = \frac{x_1 + jy_1}{x_2 + jy_2} \frac{x_2 - jy_2}{x_2 - jy_2}$$

(this step is called multiplication by the conjugate of the denominator)

$$\frac{\mathbf{R}_1}{\mathbf{R}_2} = \frac{x_1 x_2 + jy_1 x_2 - jy_2 x_1 + y_1 y_2}{x_2^2 + jy_2 x_2 - jy_2 x_2 + y_2^2}$$
$$= \frac{x_1 x_2 + y_1 y_2}{x_2^2 + y_2^2} + j\frac{y_1 x_2 - y_2 x_1}{x_2^2 + y_2^2} \qquad (11.43)$$

If $\mathbf{R} = R\underline{/\phi^\circ} = x + jy$ is a complex number, its complex conjugate is defined as

$$\mathbf{R}^* = R\underline{/-\phi^\circ} = x - jy$$

(i.e., the conjugate of a complex number is obtained from the original number by reversing the sign of its phase, or equivalently reversing the sign of its imaginary part). Note that

$$\mathbf{R} + \mathbf{R}^* = 2x$$

$$\mathbf{R} \cdot \mathbf{R}^* = R^2 = x^2 + y^2 \quad \text{(always a positive real number} = \text{magnitude}^2\text{)}$$

■ EXAMPLE 11.5

Find the resultant of each of the following algebraic operations.
(a) $5\underline{/30^\circ} - 3\underline{/-90^\circ} \times 2\underline{/30^\circ}$
(b) $(3 + j4)(2 - j1)(6 + j3)$
(c) $1 + j2)/(2 - j1)$

Solution (a) After performing the multiplication, the rectangular forms should be used. Thus

$$\mathbf{R} = 5\underline{/30^\circ} - 3\underline{/-90^\circ} \times 2\underline{/30^\circ}$$
$$= 5\underline{/30^\circ} - 6\underline{/-90^\circ + 30^\circ}$$
$$= 5\underline{/30^\circ} - 6\underline{/-60^\circ}$$
$$= (4.33 + j2.5) - (3 - j5.2)$$
$$= (4.33 - 3) + j[2.5 - (-5.2)]$$
$$= 1.33 + j7.7 = 7.81\underline{/80.2^\circ}$$

(b) This multiplication process can be performed more easily if each of the complex numbers is first converted into polar form. Thus

$$\mathbf{R} = (3 + j4)(2 - j1)(6 + j3)$$
$$= 5\underline{/53.1^\circ} \times 2.236\underline{/-26.6^\circ} \times 6.708\underline{/26.6^\circ}$$
$$= 75\underline{/53.1^\circ - 26.6^\circ + 26.6^\circ} = 75\underline{/53.1^\circ}$$
$$= 45 + j60$$

(c) Again in the division process, it is better if both the numerator and denominator are first expressed in polar form. Here

$$\mathbf{R} = \frac{1 + j2}{2 - j1} = \frac{2.236\underline{/63.4^\circ}}{2.236\underline{/-26.6^\circ}} = 1\underline{/63.4^\circ - (-26.6^\circ)}$$
$$= 1\underline{/90^\circ} = j1$$

Using the conjugate and recalculating this resultant in rectangular form yields

$$\mathbf{R} = \frac{1 + j2}{2 - j1} \frac{2 + j1}{2 + j1}$$
$$= \frac{2 - 2 + j4 + j1}{4 + 1}$$
$$= \frac{j5}{5} = j1 \quad \text{(as above)}$$

When an ac voltage source is applied to an inductor L, a current i will flow in it, obeying Ohm's law for the inductor at every instant of time. Thus applying Eq. (10.2) to the circuit in Fig. 11.15, we have

$$v_L(t) = e(t) = L\frac{di(t)}{dt} \tag{11.44}$$

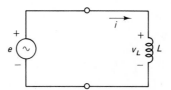

Figure 11.15

To examine the $v - i$ characteristics of the inductor in ac circuits, assume first that the current in the inductor is the sinusoidal waveform

$$i(t) = I_m \sin \omega t \quad \text{A} \tag{11.45a}$$

whose phasor representation is

$$\mathbf{I} = I\underline{/0°}\,\text{A} \quad (I = 0.707 I_m) \tag{11.45b}$$

Qualitatively, Eq. (11.44) states that the voltage across the inductor v_L is proportional to the rate of change of the current, $di(t)/dt$. Thus v_L would attain maximum positive or negative amplitude when $di(t)/dt$ is maximum positive or negative, as at $t = 0$ or $t = T/2$, respectively. Also, v_L would be zero at the instants of time at which $di(t)/dt$ is zero, as at $t = T/4, 3T/4, \dots$. As shown in Fig. 11.16, the waveform for $v_L(t)$ corresponding to $i(t)$ of Eq. (11.45a) is

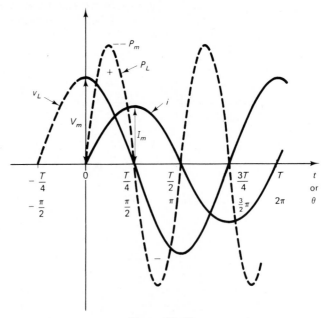

Figure 11.16

actually a cosine waveform, or in fact a sine waveform shifted to the left of the origin by $T/4$ (or 90°). The period (or the frequency f) of the voltage waveform is exactly the same as that of the current waveform. Thus

$$v_L(t) = V_m \cos \omega t = V_m \sin(\omega t + 90°) \quad \text{V} \tag{11.46a}$$

and the phasor representation of $v_L(t)$ is

$$\mathbf{V}_L = V\underline{/90°} \quad \text{V} \quad (V = 0.707V_m) \tag{11.46b}$$

Hence *the voltage across the inductor always leads the current flowing in this inductor by 90°*. The phasor diagram representing the phasors $\mathbf{I}$ and $\mathbf{V}$ [Eqs. (11.45b) and (11.46b)] in this case is shown in Fig. 11.17.

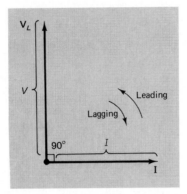

Figure 11.17

The power in the inductor is also a function of time, obtained from

$$P_L(t) = v_L(t)i(t) = V_m I_m \sin \omega t \cos \omega t$$
$$= \tfrac{1}{2}V_m I_m \sin 2\omega t \quad \text{W} \tag{11.47}$$

The power waveform is also shown in Fig. 11.16. As in the case of the resistive ac circuit, the power waveform has half the period (or double the frequency) of the corresponding voltage or current waveforms. However, an important difference can be observed here. The average power in the inductor is zero, since the average of any sinusoidal waveform over one period is zero. Thus

$$P_{Lav} = 0 \quad \text{W} \tag{11.48}$$

Inductors, therefore, do not dissipate power. In fact, inductors store the energy in their magnetic field every quarter of the cycle of the current waveform, and then return this energy back to the circuit during the next quarter of the cycle, without dissipating any amount of it.

The inductive reactance X_L is defined as the ratio

$$\boxed{X_L = \frac{V_m}{I_m} = \frac{\sqrt{2}\,V}{\sqrt{2}\,I} = \frac{V}{I} \quad \Omega} \tag{11.49}$$

It reflects the opposition to the current flow exhibited by the inductor. Since it is a ratio of volts to amperes, its unit is also the ohm. X_L is called the *reactance* to imply that there is a phase shift between the voltage and current waveforms and that it does not dissipate power. These properties are in exact contrast to the resistance R.

If the current waveform has a fixed magnitude I_m (or I), but its frequency ω is increased, this corresponds to a reduction in the period T ($T = 2\pi/\omega$). Thus $di(t)/dt$ (the slope) becomes larger. Referring to Eq. (11.44), one can conclude that the magnitude of the voltage waveform V_m (or V) is proportional to L, I_m, and ω, as discussed above. Hence

$$X_L \propto L\omega$$

The proportionality constant is unity, as will be shown below. Thus

$$\boxed{X_L = \omega L = 2\pi f L \quad \Omega}$$

(11.49a)

This result can be confirmed experimentally. The graphs in Fig. 11.18 show the typical variation of X_L as a function of L and f.

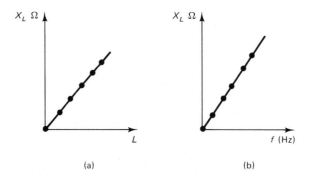

Figure 11.18   (a) f is fixed; (b) L is fixed.

Quantitatively, the conclusions reached above regarding the v–i relationship for the inductor and the magnitude of X_L can easily be confirmed through calculus. Substituting for $i(t)$ from Eq. (11.45a) into Ohm's law for the inductor [Eq. (11.44)], we have

$$v_L(t) = L\frac{di(t)}{dt} = L\frac{d}{dt}(I_m \sin \omega t)$$

$$= I_m L\omega \cos \omega t$$

$$= V_m \sin(\omega t + 90°) \quad \text{V}$$

where $V_m = I_m L\omega$. Therefore,

$$X_L = \frac{V_m}{I_m} = \frac{V}{I} = \omega L \quad \Omega$$

Another important difference between the inductive reactance X_L and the resistance R can be reached from the discussion above. While R is independent of the value of the frequency f of the current or voltage waveform, X_L is linearly proportional to f. Thus X_L is zero (i.e., a short circuit) when f is zero (dc), confirming the conclusions reached in Chapter 10. As f is increased, X_L also increases. At very high frequencies (i.e., when f tends to infinity) X_L also tends to infinity. The inductor then behaves like an open circuit, hence the choking action for the high-frequency currents.

When an ac source is applied to a simple capacitive circuit as shown in Fig. 11.19, the current flowing in the capacitor obeys Ohm's law for the capacitor:

$$i(t) = C\frac{dv_C(t)}{dt} \qquad (11.50)$$

where

$$v_C(t) = e(t) = V_m \sin \omega t \quad \text{V} \qquad (11.51a)$$

The phasor representation of this sinusoidal voltage is

$$\mathbf{V}_C = V\underline{/0°}\,\text{V} \quad (V = 0.707V_m) \qquad (11.51b)$$

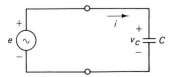

Figure 11.19

As the current $i(t)$ is proportional to the rate of change of the voltage across the capacitor, the magnitude of the current waveform will attain maximum positive or negative amplitude when $dv_C(t)/dt$ is maximum positive or negative, at $t = 0$ or $t = T/2, \ldots$. When $dv_C(t)/dt$ is zero, at $t = T/4, 3T/4, \ldots$, the magnitude of the current waveform will be zero. These waveforms are shown in Fig. 11.20. Therefore, the current waveform compared to the voltage waveform $v_C(t)$ is a

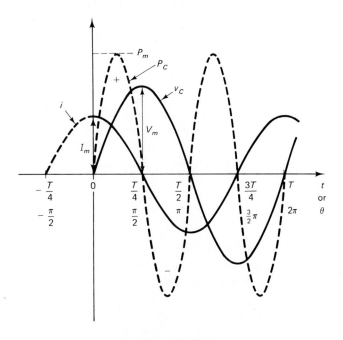

Figure 11.20

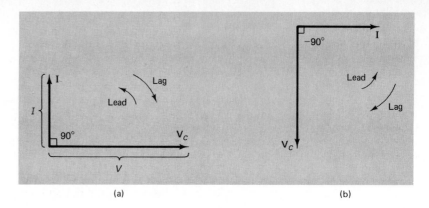

Figure 11.21

cosine function, or in other words, it is a sine-wave function shifted to the left by $T/4$ (or 90°). Both $i(t)$ and $v_C(t)$ have the same period T. Therefore,

$$i(t) = I_m \cos \omega t = I_m \sin (\omega t + 90°) \quad \text{A} \tag{11.52a}$$

The phasor representation for this current waveform is

$$\mathbf{I} = I\underline{/90°} \text{ A} \quad (I = 0.707I_m) \tag{11.52b}$$

The current in a capacitor, i, thus leads the voltage v_C across this capacitor by a phase angle of 90°, or in other words, *the capacitor's voltage lags the capacitor's current by 90°*. The phasor diagram for the case discussed here [i.e., Eqs. (11.51b) and (11.52b)] is shown in Fig. 11.21(a). Notice that it is possible to rotate the whole phasor diagram by 90° clockwise, with the same lead and lag relationships still applying, as shown in Fig. 11.21(b). In Fig. 11.21(a), $\mathbf{V}_C$ is said to be the reference phasor, while in Fig. 11.21(b), $\mathbf{I}$ is said to be the reference phasor.

The power in the capacitor is obtained from

$$p_C(t) = v_C(t)i(t) = V_m I_m \sin \omega t \cos \omega t$$
$$= \tfrac{1}{2} V_m I_m \sin 2\omega t \quad \text{W} \tag{11.53}$$

This waveform is also shown in Fig. 11.20. As was observed in the case of the inductor in Section 11.8, here also the power in the capacitor has half the period (or double the frequency) of the corresponding voltage and current waveforms. The average power dissipated in the capacitor is zero, either from observing the waveform in Fig. 11.20 or directly from Eq. (11.53), based on the result of Eq. (11.23). Thus

$$P_{Cav} = 0 \quad \text{W} \tag{11.54}$$

Capacitors therefore also do not dissipate any power, but they do store energy in their electric fields and then return back the stored energy to the circuit every one-quarter of the voltage (or current) waveform.

Capacitors also exhibit opposition to current flow. This property is called the capacitive reactance X_C, reflecting the two observations developed above: first, it does not dissipate any power, and second, it implies the existence of a phase shift between the voltage and current waveforms for the capacitor. The capacitive reactance is defined as

$$X_C = \frac{V_m}{I_m} = \frac{\sqrt{2}\,V}{\sqrt{2}\,I} = \frac{V}{I} \quad \Omega \tag{11.55}$$

Because it is a ratio of voltage to current magnitudes, its unit is also the ohm.

If the magnitude of $v_C(t)$, V_m in Eq. (11.51a), is kept constant, while ω (or f) is increased, this will produce a higher rate of change of the capacitor's voltage, $dv_C(t)/dt$, since the period T is reduced. According to Eq. (11.50), the magnitude of the current flowing in the capacitor will thus increase with the frequency. Hence

$$I_m \propto C\omega V_m$$

or

$$X_C = \frac{V_m}{I_m} \propto \frac{1}{\omega C}$$

This inverse proportionality of X_C with respect to C and ω (or f) can be confirmed experimentally. The graphs in Fig. 11.22 show the variation of X_C as a function of C and ω.

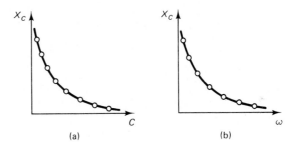

Figure 11.22   (a) ω is fixed; (b) C is fixed.

The proportionality constant in the relationship above is unity, as will be shown below. Thus

$$X_C = \frac{1}{\omega C} = \frac{1}{2\pi f C} \quad \Omega \tag{11.56}$$

Quantitatively, the conclusions reached above regarding the v–i relationship for a capacitor and the magnitude of X_C can easily be confirmed through calculus. Substituting for $v_C(t)$ from Eq. (11.51a) into Ohm's law for the capacitor [Eq. (11.50)] yields

$$i(t) = C\frac{dv_C(t)}{dt} = C\frac{d}{dt}(V_m \sin \omega t)$$

$$= CV_m\omega \cos \omega t$$

$$= I_m \sin (\omega t + 90°) \quad A$$

where $I_m = V_m C \omega$. Therefore

$$X_C = \frac{V_m}{I_m} = \frac{1}{\omega C} \quad \Omega$$

In contrast to the resistance R, which does not change with frequency, and the inductive reactance X_L, which increases linearly with frequency, the capacitive reactance X_C decreases with frequency. At $f = 0$ (i.e., dc), $X_C = \infty$; that is, the capacitor is then equivalent to an open circuit, confirming previous observations. As f increases, X_C decreases. When f tends to infinity, X_C is zero. This means that the capacitor would be equivalent to a short circuit at extremely high frequencies.

11.10 IMPEDANCE AND ADMITTANCE CONCEPTS

Based on the previous discussions, it can be concluded that the frequency of the current waveform is the same as the frequency of the corresponding voltage waveform for each of the three different electric circuit load elements, R, L, and C. Therefore, the time dependence can be dropped temporarily and circuit analysis can then be carried out using the phasor representations of the different currents and voltages in the given circuit.

To account completely for both the magnitude ratio and the phase shift that occurs between the voltage and current waveforms for any type of element, the impedance Z of each element is defined as the ratio of the voltage phasor $\mathbf{V}$ to the current phasor $\mathbf{I}$ of that element. Thus

$$\boxed{Z = \frac{\mathbf{V}}{\mathbf{I}} \quad \Omega \qquad \text{or} \qquad \mathbf{V} = Z\mathbf{I} \quad V} \qquad (11.57)$$

This equation represents Ohm's law in phasor form for any type of load element. The impedance Z also indicates opposition to current flow and it has the units of ohms. It completely defines the load element's behavior in an ac circuit. Z is in general a complex number, not a phasor, since it is not a function of time. In fact, the time dependence drops out, as Z is a ratio of two phasors (i.e., a ratio of two complex-number quantities).

The magnitude of Z reflects the amount of opposition to current flow exhibited by the element, while the phase of Z reflects the phase shift between the voltage and current waveforms caused by this element. When Z has a positive phase, this indicates that the voltage phasor $\mathbf{V}$ leads the corresponding current phasor $\mathbf{I}$, while a negative phase indicates that $\mathbf{V}$ lags $\mathbf{I}$ for the specific load element. The real time-domain representation of the load element and its associated voltage and current waveforms is shown in Fig. 11.23(a). The equivalent *frequency-domain representation* of the load element is shown in Fig. 11.23(b), indicating the impedance of the load element and the associated voltage and current phasors. The reason for referring to this equivalent circuit as the frequency-domain circuit is that Z is in general a function of the frequency, as discussed above for the cases of the inductive and capacitive loads.

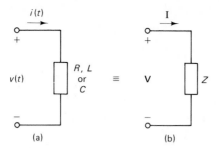

Figure 11.23 (a) Time-domain representation of voltage and current in a load element; (b) equivalent phasor representation of the load element (frequency domain).

The inverse of Z is called the admittance Y of the load element, having the units of siemens (or mhos or ℧). Hence Ohm's law can also be expressed as

$$Y = \frac{1}{Z} = \frac{\mathbf{I}}{\mathbf{V}} \quad \text{S (or ℧)}$$

or

$$\mathbf{I} = Y\mathbf{V} \quad \text{A} \tag{11.58}$$

Referring to the discussions of Sections 11.4, 11.8, and 11.9, let the voltage

$$v(t) = V_m \sin(\omega t + \phi°) \quad \text{V} \Leftrightarrow \mathbf{V} = V\underline{/\phi°} \quad \text{V} \tag{11.59}$$

be applied to each of the R, L, and C load elements in turn. $\phi°$ is any arbitrary phase. Then:

1. For a resistor R, $i(t)$ is in phase with $v(t)$,

$$i(t) = I_m \sin(\omega t + \phi°) \quad \text{A} \Leftrightarrow \mathbf{I} = I\underline{/\phi°} \quad \text{A} \tag{11.60}$$

where

$$I_m = \frac{V_m}{R} \quad \text{or} \quad \frac{V_m}{I_m} = \frac{V}{I} = R$$

Then

$$Z_R = \frac{\mathbf{V}}{\mathbf{I}} = \frac{V\underline{/\phi°}}{I\underline{/\phi°}} = \frac{V}{I}\underline{/0°} = R\underline{/0°} \quad \Omega \text{ (in polar form)}$$

$$= R + j0 \quad \Omega \text{ (in rectangular form)} \tag{11.61}$$

and

$$Y_R = \frac{1}{Z_R} = \frac{1}{R\underline{/0°}} = \frac{1}{R}\underline{/0°} = G\underline{/0°} \quad \text{S (in polar form)}$$

$$= G + j0 \quad \text{S (in rectangular form)} \tag{11.62}$$

where

$$G = \text{conductance} = \frac{1}{R} \quad \text{S} \tag{11.63}$$

2. For an inductor L, $i(t)$ lags $v(t)$ by $90°$,

$$i(t) = I_m \sin(\omega t + \phi° - 90°) \quad \text{A} \Leftrightarrow \mathbf{I} = I\underline{/\phi° - 90°} \quad \text{A} \tag{11.64}$$

where

$$I_m = \frac{V_m}{X_L} \quad \text{or} \quad \frac{V_m}{I_m} = \frac{V}{I} = X_L = \omega L \quad \Omega$$

Then

$$Z_L = \frac{\mathbf{V}}{\mathbf{I}} = \frac{V\underline{/\phi^\circ}}{I\underline{/\phi^\circ - 90^\circ}} = \frac{V}{I}\underline{/90^\circ} = X_L\underline{/90^\circ} \quad \Omega \text{ (in polar form)}$$

$$= jX_L \quad \Omega \text{ (in rectangular form)}$$

(11.65)

and

$$Y_L = \frac{1}{Z_L} = \frac{1}{X_L\underline{/90^\circ}} = \frac{1}{X_L}\underline{/-90^\circ} = B_L\underline{/-90^\circ} \quad \text{S (in polar form)}$$

$$= -jB_L \quad \text{S (in rectangular form)}$$

(11.66)

where

$$B_L = \text{inductive susceptance} = \frac{1}{X_L} = \frac{1}{\omega L} \quad \text{S} \qquad (11.67)$$

3. For a capacitor C, $i(t)$ leads $v(t)$ by 90°,

$$i(t) = I_m \sin(\omega t + \phi^\circ + 90^\circ) \quad \text{A} \Leftrightarrow \mathbf{I} = I\underline{/\phi^\circ + 90^\circ} \quad \text{A} \quad (11.68)$$

where

$$I_m = \frac{V_m}{X_C} \quad \text{or} \quad \frac{V_m}{I_m} = \frac{V}{I} = X_C = \frac{1}{\omega C} \quad \Omega$$

Then

$$Z_C = \frac{\mathbf{V}}{\mathbf{I}} = \frac{V\underline{/\phi^\circ}}{I\underline{/\phi^\circ + 90^\circ}} = \frac{V}{I}\underline{/-90^\circ} = X_C\underline{/-90^\circ} \quad \Omega \text{ (in polar form)}$$

$$= -jX_C \quad \Omega \text{ (in rectangular form)} \qquad (11.69)$$

and

$$Y_C = \frac{1}{Z_C} = \frac{1}{X_C\underline{/-90^\circ}} = \frac{1}{X_c}\underline{/90^\circ} = B_C\underline{/90^\circ} \quad \text{S (in polar form)}$$

$$= jB_C \quad \text{S (in rectangular form)} \quad (11.70)$$

where

$$B_C = \text{capacitive susceptance} = \frac{1}{X_C} = \omega C \quad \text{S} \qquad (11.71)$$

The impedance Z or admittance Y for any of the three different element types is independent of the initial phase of $v(t)$, ϕ. It is clear from the above that such an angle cancels out in the division process $(\mathbf{V}/\mathbf{I})$, due to the phase subtraction. Either of the polar or rectangular forms for Z (or Y) can be used to represent the elements in the frequency-domain equivalent circuits and in the phasor form of Ohm's law when applied to such ac circuits. The phasor diagrams for the three

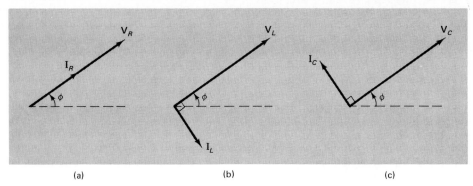

Figure 11.24 (a) Phasor diagram for R; (b) phasor diagram for L; (c) phasor diagram for C.

elements R, L, and C, with any arbitrary phase for **V**, are shown in Fig. 11.24, reflecting the **V**–**I** phase relationships discussed above.

It is also important to note that the nature of the load element (R, L, or C) can be found from its impedance Z (or admittance Y). Purely real impedance means that there is not phase shift between **V** and **I** and the element that satisfies this condition is the resistor. Purely positive imaginary impedance (i.e., the phase of the impedance is 90°) means that **V** leads **I** by 90° and the element that satisfies this condition is the inductor. Purely negative imaginary impedance (i.e., the phase of the impedance is −90°) means that **V** lags **I** by 90° and the element that satisfies this condition is the capacitor. Similar deductions can be reached based on the admittances.

Phasor forms for KVL and KCL. KVL and KCL always apply in any electric circuit. Figure 11.25 shows a general series–parallel ac circuit. Here, in the time domain,

$$\text{KVL:} \quad e = v_1 + v_2 \tag{11.72a}$$

and

$$\text{KCL:} \quad i_1 = i_2 + i_3 \tag{11.72b}$$

Since the elements in the ac circuit are R, L, or C (or combination thereof), all the waveforms for the currents and voltages anywhere in the circuit have the same frequency ω (or f) as that of the source. Thus the sinusoidal ac waveforms can be replaced by their corresponding phasors (see Appendix A11), as the time dependence can be temporarily dropped as discussed above. Hence, in phasor form,

$$\boxed{\text{KVL:} \quad \mathbf{E} = \mathbf{V}_1 + \mathbf{V}_2} \tag{11.73a}$$

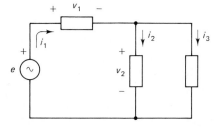

Figure 11.25

and

$$\boxed{\text{KCL:} \quad \mathbf{I}_1 = \mathbf{I}_2 + \mathbf{I}_3} \qquad\qquad (11.73\text{b})$$

Similarly, KVL and KCL can be written for any ac circuit with an arbitrary configuration or topology.

11.11 GENERAL EXAMPLES OF SIMPLE R, L, OR C AC CIRCUITS

■ **EXAMPLE 11.6**

The ac source applied to the simple inductive circuit shown in Fig. 11.26 is described by $e(t) = 100 \sin (628t + 90°)$ V. Find
(a) The source's frequency and the impedance of the load.
(b) The time-domain expression for $i(t)$.
(c) The phasor diagram for this circuit.

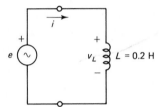

Figure 11.26

Solution   (a) Here ω (the coefficient of t) can easily be recognized from $e(t)$,

$$\omega = 628 \text{ rad/s}$$

Then

$$f = \frac{\omega}{2\pi} = \frac{628}{6.28} = 100 \text{ Hz}$$

and

$$X_L = \omega L = 628 \times 0.2 = 125.6 \text{ }\Omega$$

Thus

$$Z_L = X_L\underline{/90°} = 125.6\underline{/90°} = j125.6 \text{ }\Omega$$

(b) The phasor form for $e(t)$ can easily be deduced,

$$\mathbf{E} = 0.707 \times 100\underline{/90°} = 70.7\underline{/90°} \text{ V}$$

Applying Ohm's law to this inductive load gives us

$$\mathbf{E} = \mathbf{V}_L = Z_L\mathbf{I}$$

Therefore,

$$\mathbf{I} = \frac{\mathbf{E}}{Z_L} = \frac{70.7\underline{/90°}}{125.6\underline{/90°}} = 0.563\underline{/0°} \text{ A} \qquad \text{that is, } I_{\text{RMS}} = 0.563 \text{ A}$$

For this current phasor above, the corresponding time-domain expression

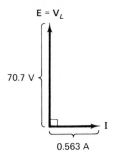

E = V_L

70.7 V

I

0.563 A

Figure 11.27

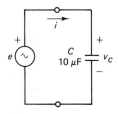

Figure 11.28

(same frequency ω) is

$$i(t) = 0.563 \times 1.414 \sin (628t + 0°) \text{ A}$$
$$= 0.796 \sin (628t) \text{ A}$$

(c) The phasor diagram is drawn in Fig. 11.27, showing $\mathbf{E}$ ($= \mathbf{V}_L$) and $\mathbf{I}$ as obtained above. It is clear that $\mathbf{V}_L$ leads $\mathbf{I}$ by 90°, as expected.

EXAMPLE 11.7

In the circuit shown in Fig. 11.28, the applied source* has a voltage of 35.4 V and a frequency of 159 Hz. Write the time-domain expressions for the voltage e and the current i if at $t = 0$, $i = 0$. Draw the phasor diagram for this circuit.

Solution   Here $\omega = 2\pi f = 6.28 \times 159 = 1000 \text{ rad/s}$.

$$Z_C = X_C\underline{/-90°} = \frac{1}{\omega C}\underline{/-90°} = \frac{1}{1000 \times 10 \times 10^{-6}}\underline{/-90°} = 100\underline{/-90°} \ \Omega$$
$$= -j100 \ \Omega$$

Since $i = 0$ at $t = 0$, this means that $i(t)$ is the reference waveform (or phasor) having a phase angle of zero (i.e., the starting point is at the origin):

$$i(t) = I_m \sin \omega t \Leftrightarrow \mathbf{I} = I\underline{/0°} \text{ A}$$

From Ohm's law applied to this circuit, we have

$$\mathbf{E} = E\underline{/\phi°} = \mathbf{I}Z_C$$
$$35.4\underline{/\phi°} = I\underline{/0°} \times 100\underline{/-90°} = 100I\underline{/-90°}$$

Comparing the magnitudes and phases of both sides of this equation, it is easy to conclude that

$$\phi = -90°$$

and

$$35.4 = 100I$$

Therefore,

$$I = 0.354 \text{ A}$$

Hence,

$$\mathbf{I} = 0.354\underline{/0°} \text{ A} \quad \text{and} \quad \mathbf{E} = 35.4\underline{/-90°} \text{ V}$$

The phasor diagram can then be drawn as shown in Fig. 11.29. The phasors above can easily be converted to time functions:

$$i(t) = 0.354 \times 1.414 \sin (1000t + 0°) = 0.5 \sin (1000t) \text{ A}$$

and

$$e(t) = 35.4 \times 1.414 \sin (1000t - 90°) = 50 \sin (1000t - 90°) \text{ V}$$

Notice that $\mathbf{V}_C$ ($= \mathbf{E}$) lags $\mathbf{I}$ by 90° as expected.

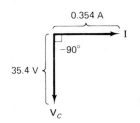

0.354 A

I

−90°

35.4 V

V_C

Figure 11.29

* When a voltage source is defined as in this example, its magnitude is the RMS value because it is the reading of an ac voltmeter connected across the source.

EXAMPLE 11.8

A source, whose voltage waveform is $e(t) = -50 \cos (2000t - 45°)$ V, is applied to a single-element load, resulting in a current flow, $i(t) = -5 \sin (2000t + 45°)$ A. Find

(a) The type and numerical value of the element.
(b) The readings of ideal ac voltmeter and ammeter used to measure the voltage and current in this circuit.
(c) The power dissipated in this element.
(d) The reading of the ammeter if the frequency of the source is doubled.

Solution (a) Rewriting the waveforms in terms of the positive sine-wave functional representation

$$e(t) = 50 \sin (2000t - 45° + 90° \pm 180°)$$

$$= 50 \sin (2000t - 135°) \text{ V} \Leftrightarrow \mathbf{E} = \frac{50}{\sqrt{2}} \underline{/-135°}$$

$$= 35.36 \underline{/-135°} \text{ V}$$

$$i(t) = 5 \sin (2000t + 45° \pm 180°) \text{ A}$$

$$= 5 \sin (2000t - 135°) \text{ A} \Leftrightarrow \mathbf{I} = \frac{5}{\sqrt{2}} \underline{/-135°}$$

$$= 3.536 \underline{/-135°}$$

with $\omega = 2000$ rad/s, therefore,

$$Z = \frac{\mathbf{E}}{\mathbf{I}} = \frac{35.36 \underline{/-135°}}{3.536 \underline{/-135°}} = 10 \underline{/0°} = 10 \, \Omega$$

Thus the element is a resistor of value 10 Ω. This is obvious since $v(t)$ and $i(t)$ (and the corresponding phasors) are in phase.

(b) The voltmeter reading = E[RMS value of $e(t)$] = 35.36 V;
the ammeter reading = I[RMS value of $i(t)$] = 3.536 A.

(c) P_{av} = average power dissipated in this resistor

$$= E \times I = I^2 R = \frac{E^2}{R} = 125 \text{ W}$$

(d) The value of R is independent of the source's frequency; Z thus remains the same. Since $\mathbf{E}$ is unchanged, $\mathbf{I}$ has the same value as above. Thus the ammeter's reading is unchanged from the value obtained in part (b).

EXAMPLE 11.9

In a single-element single-source circuit, the readings of an ideal voltmeter and an ideal ammeter were found to be 10 V and 4 A at the source's frequency setting of 1 kHz. It was also noted that $e(t)$ leads $i(t)$ and that at $t = 0$, the current's magnitude is half of its peak value and decreasing. Find

(a) $i(t)$ and $e(t)$.
(b) The nature and magnitude of the element.
(c) The power dissipated in the element.
(d) The ammeter's reading if the source frequency is doubled.
(e) The frequency of the source at which the ammeter's reading is 20 A.

Solution (a) Based on the information given, a sketch of $i(t)$ and $e(t)$ can be made as shown in Fig. 11.30. The angle by which $e(t)$ leads $i(t)$ must be 90° since the circuit contains only one element. The element is thus an inductor.

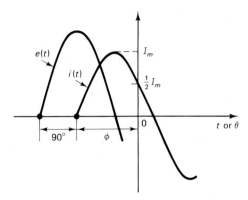

Figure 11.30

In general, $i(t)$ is expressed as

$$i(t) = I_m \sin (\omega t + \phi°)$$

at $t = 0$, $i(t) = \frac{1}{2}I_m$. Substituting in the expression above yields

$$0.5 = \sin \phi°$$
$$\phi = 30 \text{ or } 150°$$

The second answer is the correct one since $i(t)$ would be decreasing (as specified) with such a phase (see Fig. 11.30); therefore,

$$I_m = \sqrt{2}\,I = 1.414 \times 4 = 5.657 \text{ A}$$

and

$$E_m = \sqrt{2}\,E = 1.414 \times 10 = 14.14 \text{ V}$$

Also,

$$\omega = 2\pi f = 6.28 \times 1000 = 6280 \text{ rad/s}$$

Therefore,

$$i(t) = 5.657 \sin (6280t + 150°) \text{ A} \Leftrightarrow \mathbf{I} = 4\underline{/150°} \text{ A}$$

and with $e(t)$ leading $i(t)$ by 90°,

$$e(t) = 14.14 \sin (6280t + 150° + 190°)$$
$$= 14.14 \sin (6280t + 240°) \text{ V} \Leftrightarrow \mathbf{E} = 10\underline{/240°} \text{ V}$$

(b) As concluded in part (a), the element is an inductor. This can be confirmed by calculating the impedance of this element from Ohm's law:

$$Z = \frac{\mathbf{E}}{\mathbf{I}} = \frac{10\underline{/240°}}{4\underline{/150°}} = 2.5\underline{/90°} = X_L\underline{/90°} \text{ }\Omega$$

Therefore,

$$X_L = \omega L = 2.5 \text{ }\Omega$$

11.11   General Examples of Simple R, L, or C AC Circuits

Hence

$$L = \frac{2.5}{6280} = 0.398 \text{ mH}$$

(c) Inductors do not dissipate power. Thus

$$P_{av} = 0 \text{ W}$$

(d) If f (or ω) is doubled, $X_L(= \omega L)$ will be also double, becoming 5Ω; therefore,

$$I = \text{ammeter reading} = \frac{V_L}{X_L} = \frac{E}{X_L} = \frac{10}{5} = 2 \text{ A}$$

(i.e., the ammeter reading is half its original reading).

(e) When $I = 20 A$,

$$X_L = \frac{E}{I} = \frac{10}{20} = 0.5 \Omega$$

but $X_L = \omega L = 2\pi f L$; therefore,

$$f = \frac{X_L}{2\pi L} = \frac{0.5}{6.28 \times 0.398 \times 10^{-3}} = 200 \text{ Hz}$$

This result agrees with our expectation that since X_L is directly proportional to f, then as X_L is reduced by a factor of 5 (from 2.5Ω to 0.5Ω), f should be lower by the same factor of 5.

EXAMPLE 11.10

In the circuit shown in Fig. 11.31, $e_1(t) = 8 \sin(1000t + 15°)$ V, $e_2(t) = 8 \sin(1000t + 75°)$ V, and $i(t) = 2 \cos(1000t - 135°)$ A. Find
(a) The reading of an ideal voltmeter connected between points a and b.
(b) The nature and magnitude of the single-element load and the power dissipated in it.

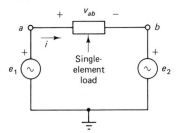

Figure 11.31

Solution   (a) Applying KVL to this circuit yields

$$v_{ab} + e_2 - e_1 = 0 \qquad \text{(in the time domain)}$$

Since all the waveforms have the same frequency ($\omega = 1000 \text{ rad/s}$), then in phasor form:

$$\mathbf{V}_{ab} = \mathbf{E}_1 - \mathbf{E}_2$$

But

$$\mathbf{E}_1 = 8 \times 0.707\underline{/15°} = 5.657\underline{/15°} \text{ V}$$

and

$$\mathbf{E}_2 = 8 \times 0.707\underline{/75°} = 5.657\underline{/75°} \text{ V}$$

Therefore,

$$\mathbf{V}_{ab} = 5.657\underline{/15°} - 5.657\underline{/75°}$$
$$= (5.464 + j1.464) - (1.464 + j5.464)$$
$$= 4 - j4 = 5.657\underline{/-45°} \text{ V}$$

The voltmeter's reading = V_{ab} (RMS value) = 5.657 V. Notice that due to the phasor algebraic subtraction process, all the magnitudes of the three voltages (e_1, e_2, and v_{ab}) in this circuit are the same!

(b) Rewriting the expression for $i(t)$ in terms of the positive sine-wave function, we have

$$i(t) = 2\cos(1000t - 135°) = 2\sin(1000t - 135° + 90°)$$
$$= 2\sin(1000t - 45°) \text{ A} \Leftrightarrow \mathbf{I} = 2 \times 0.707\underline{/-45°}$$
$$= 1.414\underline{/-45°} \text{ A}$$

Now applying Ohm's law for the single-element load,

$$Z = \frac{\mathbf{V}_{ab}}{\mathbf{I}} = \frac{5.657\underline{/-45°}}{1.414\underline{/-45°}} = 4\underline{/0°} = 4 \ \Omega$$

Thus the element is a resistor whose value is 4 Ω.

$$P_{av} = \text{average power dissipated in the resistor load element}$$
$$= V_{ab}I = I^2R = 5.657 \times 1.414 = 8 \text{ W}$$

EXAMPLE 11.11

A 50 Hz, 10 V source is connected across an ideal inductor. The ammeter in the circuit reads 0.5 A.
(a) Assuming that $e = 0$ at $t = 0$, write the time-domain expressions for $e(t)$ and $i(t)$. Draw the phasor diagram for this circuit.
(b) Find the value of L.
(c) If the frequency of the source is reduced to 10 Hz, what will be the reading of the ammeter?

Solution   (a) Since $e(t) = 0$ at $t = 0$, this indicates that $e(t)$ starts at the origin of time (i.e., its phase angle is zero). Also,

$$E_m = \sqrt{2}\,E = 1.414 \times 10 = 14.14 \text{ V}$$

and

$$\omega = 2\pi f = 6.28 \times 50 = 314 \text{ rad/s}$$

Therefore,

$$e(t) = 14.14\sin(314t) \text{ V} \Leftrightarrow \mathbf{E} = 10\underline{/0°} \text{ V}$$

For the inductor, the current waveform lags the voltage waveform by 90° (i.e., its starting point is shifted 90° to the right of the origin). Here

$$I_m = \sqrt{2}\,I = 1.414 \times 0.5 = 0.707 \text{ A}$$

Therefore,

$$i(t) = 0.707 \sin(314t - 90°) \text{ A} \Leftrightarrow \mathbf{I} = 0.5\underline{/-90°} \text{ A}$$

The phasor diagram is as shown in Fig. 11.32.

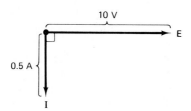

Figure 11.32

(b) Since $\mathbf{E}\,(=\mathbf{V}_L)$ and $\mathbf{I}$ are now known, the impedance of the element can be calculated.

$$Z_L = \frac{\mathbf{E}}{\mathbf{I}} = \frac{10\underline{/0°}}{0.5\underline{/-90°}} = 20\underline{/90°} \text{ } \Omega = j20 \text{ } \Omega = jX_L$$

Therefore,

$$X_L = 20 = \omega L$$

and

$$L = \frac{20}{314} = 63.7 \text{ mH}$$

(c) When $f = 10$ HZ,

$$X_L = \omega L = 2\pi f L = 6.28 \times 10 \times 63.7 \times 10^{-3} = 4 \text{ } \Omega$$

Therefore,

$$Z_L = X_L\underline{/90°} = 4\underline{/90°} \text{ } \Omega$$

For an ideal voltage source, the magnitude remains the same for all values of the source's frequency. Thus using Ohm's law gives us

$$\mathbf{I} = \frac{\mathbf{E}}{Z_L} = \frac{10\underline{/0°}}{4\underline{/90°}} = 2.5\underline{/-90°} \text{ A}$$

Thus the ammeter's reading = 2.5 A. Notice that the phase shift between $e(t)$ and $i(t)$ remains the same (90°) for all values of the source's frequency.

EXAMPLE 11.12

In the circuit shown in Fig. 11.33, $e_1(t) = 100\sqrt{2} \sin(2513t)$ V and $i(t) = 50\sqrt{2} \sin(2513t - 68.8°)$ A. Find
(a) The value of C.
(b) The voltage across the capacitor in polar form.
(c) $e_2(t)$.

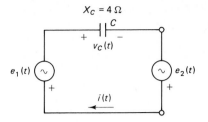

Figure 11.33

Solution   (a) Since

$$X_C = \frac{1}{\omega C}$$

then

$$C = \frac{1}{\omega X_C} = \frac{1}{2513 \times 4} = 99.5 \ \mu F$$

    (b) Converting the time-domain functions into their phasor form, we have

$$\mathbf{E}_1 = 100\underline{/0°} \ V$$

and

$$\mathbf{I} = 50\underline{/-68.8°} \ A$$

Also,

$$\mathbf{Z}_C = X_C\underline{/-90°} = 4\underline{/-90°} \ \Omega$$

Applying Ohm's law to the capacitor yields

$$\mathbf{V}_C = \mathbf{Z}_C\mathbf{I} = 4\underline{/-90°} \times 50\underline{/-68.8°} = 200\underline{/-158.8°} \ V$$

    (c) In the time domain, KVL for this circuit is

$$e_1 + v_C - e_2 = 0 \qquad \text{or} \qquad e_2 = e_1 + v_C$$

In phasor form,

$$\begin{aligned}
\mathbf{E}_2 = \mathbf{E}_1 + \mathbf{V}_C &= 100\underline{/0°} + 200\underline{/-158.8°} \\
&= 100 + (-186.5 - j72.3) \\
&= -86.5 - j72.3 \\
&= 112.7\underline{/-140°} \ V
\end{aligned}$$

The phasor $\mathbf{E}_2$ can then be easily converted into its corresponding time function,

$$\begin{aligned}
e_2(t) &= 112.7\sqrt{2} \sin(2513t - 140°) \\
&= 159.4 \sin(2513t - 140°) \ V
\end{aligned}$$

■ EXAMPLE 11.13

The voltage applied to a single-element load is $v(t) = 70.7 \sin(1000t + 45°)$ V. The resulting current flow is $i(t) = 14.14 \cos(1000t - 135°)$ A.

(a) Find the phasors **V** and **I** and draw the phasor diagram.

(b) Find the nature and magnitude of the element.

(c) If the element is replaced by a capacitor, resulting in an ammeter reading of 5 A, find the value of C and write the expression for $i(t)$ in this case.

Solution (a) Here $\omega = 1000\ \text{rad/s}$,

$$v(t) = 70.7 \sin (1000t + 45°)\ \text{V} \Leftrightarrow \mathbf{V} = 50\underline{/45°}\ \text{V}$$

and

$$i(t) = 14.14 \cos (1000t - 135°) = 14.14 \sin (1000t - 135° + 90°)$$

$$= 14.14 \sin (1000t - 45°)\ \text{A} \Leftrightarrow \mathbf{I} = 10\underline{/-45°}\ \text{A}$$

The phasor diagram is shown in Fig. 11.34.

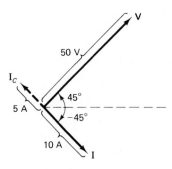

Figure 11.34

(b) From Ohm's law applied to this element,

$$Z = \frac{\mathbf{V}}{\mathbf{I}} = \frac{50\underline{/45°}}{10\underline{/-45°}} = 5\underline{/90°}\ \Omega = j5\ \Omega = jX_L$$

The element is thus an inductor (**I** lags **V** by 90°), with

$$X_L = 5 = \omega L$$

Therefore,

$$L = \frac{5}{1000} = 5\ \text{mH}$$

(c) When L is replaced by a capacitor C, the capacitor's current (5 A) would lead the voltage across it by 90°, as indicated in the phasor diagram of Fig. 11.34. Therefore,

$$\mathbf{I}_C = 5\underline{/135°}\ \text{A} \Leftrightarrow i_C(t) = 5\sqrt{2} \sin (1000t + 135°)\ \text{A}$$

Using Ohm's law, we have

$$Z_C = \frac{\mathbf{V}}{\mathbf{I}_C} = \frac{50\underline{/45°}}{5\underline{/135°}} = 10\underline{/-90°}\ \Omega = -j10\ \Omega = -jX_C$$

Therefore,

$$X_C = 10 = \frac{1}{\omega C}$$

and

$$C = \frac{1}{X_C \omega} = \frac{1}{10 \times 1000} = 100 \ \mu F$$

DRILL PROBLEMS

Section 11.1 to 11.3

11.1. convert the following angles from degrees to radians:
(a) 30°; (b) 60°; (c) 45°; (d) 150°; (e) 36.9°.

11.2. Convert the following angles from radians to degrees:
(a) 1.8 rad; (b) $\pi/6$; (c) $\pi/8$; (d) 1.5π; (e) 0.2π.

11.3. For the given angular frequencies (ω), determine the corresponding frequencies (f) and periods (T):
(a) 100 rad/s; (b) 2000 rad/s; (c) 3770 rad/s; (d) 31,400 rad/s.

11.4. Determine E_m (or I_m), ω, f, T, ϕ (phase shift angle), and t' (the starting point) for each of the following waveforms:
(a) $v_1(t) = 28.284 \sin(1000t + 45°)$ V.
(b) $v_2(t) = -14.14 \cos(1000t)$ V.
(c) $v_3(t) = 35.355 \cos(1000t - 75°)$ V.
(d) $i_1(t) = 0.7071 \sin(1000t - 120°)$ A.
(e) $i_2(t) = -1.414 \sin(1000t + 30°)$ A.
(f) $i_3(t) = 2 \cos(1000t - 60°)$ A.

11.5. Obtain the lead/lag phase relationship between the waveforms of each of the following sets. The waveforms designations refer to the waveforms defined in Drill Problem 11.4.
(a) v_1 and v_2.
(b) v_1 and v_3.
(c) v_2 and v_3.
(d) i_1 and i_2.
(e) i_2 and i_3.
(f) v_1 and i_3.
(g) v_2 and i_2.

Section 11.4 and 11.5

11.6. For each of the following currents flowing in turn in a 20 Ω resistor, find the resulting voltage, $v(t)$, the RMS values of the current and voltage, I and V, and the power dissipated in the resistance:
(a) $i(t) = 0.1414 \sin(400t + 30°)$ A.
(b) $i(t) = -1.5 \cos(1000t - 20°)$ A.
(c) $i(t) = 0.7071 \sin(3000t - 60°)$ A.

11.7. For each of the following voltages in turn across a 500 Ω resistor, determine the resulting current, $i(t)$, the RMS values of the current and voltage, I and V, and the power dissipated in the resistance:
(a) $v(t) = 100 \sin(200t - 45°)$ V.
(b) $v(t) = 50 \cos(10t + 30°)$ V.
(c) $v(t) = 70.71 \sin(500t + 10°)$ V.

Section 11.6

11.8. Obtain the phasors for the waveforms given in Drill Problem 11.4.

11.9. Using the phasors obtained in Drill Problem 11.8, determine the lead/lag phase relationships for each of the following combinations:

(a) $\mathbf{V}_1$, $\mathbf{I}_1$, $\mathbf{V}_2$.

(b) $\mathbf{V}_2$, $\mathbf{V}_3$, $\mathbf{I}_2$.

(c) $\mathbf{I}_3$, $\mathbf{V}_1$, $\mathbf{I}_1$.

11.10. Convert the following phasors into the corresponding time-domain waveform, given that the frequency of the source is 120 Hz:

(a) $\mathbf{V}_1 = 10\underline{/-37°}$ V.

(b) $\mathbf{V}_2 = 14.14\underline{/45°}$ V

(c) $\mathbf{I}_1 = 0.7071\underline{/210°}$ A.

(d) $\mathbf{I}_2 = 0.1\underline{/-150°}$ A.

Section 11.7

11.11. Convert each of the following complex numbers from rectangular form to polar form:

(a) $2 + j3$.

(b) $5 - j8$.

(c) $15 + j25$.

(d) $-20 - j40$.

(e) $-10 + j14$.

(f) $0.1 - j0.9$.

11.12. Convert each of the following complex numbers from polar form to rectangular form:

(a) $3\underline{/60°}$.

(b) $12\underline{/-53°}$.

(c) $100\underline{/45°}$.

(d) $0.5\underline{/-30°}$.

(e) $36\underline{/120°}$.

(f) $0.85\underline{/-150°}$.

11.13. Perform the following algebraic operations and express the answer in both polar and rectangular forms:

(a) $10\underline{/-30°} + (4 + j8)$.

(b) $14.14\underline{/45°} - (11 - j2)$.

(c) $(10 + j10) \times (3 - j4)$.

(d) $(0.2 - j0.4)/(1 + j2)$.

(e) $(1 - j2) \times (2 + j4) - (3 - j5)$.

(f) $(9 + j12) \times 4\underline{/-120°}/12.5\underline{/60°}$.

(g) $(20\underline{/37°} + 30\underline{/-60°}) \times (0.1 - j0.2)$.

Section 11.8

11.14. Determine the reactance (X_L) of the inductors in each of the following cases:

(a) $L = 0.1$ H, $f = 1$ kHz.

(b) $\omega = 2000$ rad/s, $L = 3.6$ mH.

(c) $f = 300$ Hz, $L = 12$ mH.

(d) $f = 1$ MHz, $L = 150$ μH.

11.15. Determine the value of the inductance (L) or the frequency of operation (f) in the following cases:

(a) $X_L = 50$ Ω, $f = 20$ kHz.

(b) $X_L = 11.6$ Ω, $L = 17$ mH.

(c) $f = 120$ Hz, $X_L = 0.75$ Ω.

(d) $L = 0.25$ H, $X_L = 25$ Ω.

11.16. Obtain the voltage, $v(t)$, across each of the following inductors and determine the voltmeter's reading in each case:
- **(a)** $i(t) = 1.2 \sin(400t + 20°)$ A, $L = 0.3$ H.
- **(b)** $i(t) = 0.7 \cos(1000t - 60°)$ A, $L = 0.15$ H.
- **(c)** $i(t) = 2.828 \sin(500t - 45°)$ A, $L = 200$ mH.
- **(d)** $i(t) = 0.3535 \cos(12{,}000t + 30°)$ A, $L = 60$ mH.

11.17. Obtain the current, $i(t)$, through each of the following inductors and determine the ammeter's reading in each case:
- **(a)** $v(t) = 14.14 \sin(5000t)$ V, $L = 70$ mH.
- **(b)** $v(t) = 200 \cos(377t - 25°)$ V, $L = 0.4$ H.
- **(c)** $v(t) = 35 \sin(800t + 50°)$ V, $L = 0.25$ H.
- **(d)** $v(t) = 9 \cos(1000t + 60°)$ V, $L = 40$ mH.

Section 11.9

11.18. Determine the reactance (X_C) of the capacitor in each of the following cases:
- **(a)** $C = 20$ nF, $\omega = 10{,}000$ rad/s.
- **(b)** $C = 1\,\mu F$, $f = 318$ Hz.
- **(c)** $C = 220$ nF, $f = 570$ Hz.
- **(d)** $C = 0.5\,\mu F$, $f = 1200$ Hz.

11.19. Determine the value of the capacitance (C) or the frequency of operation (f) in the following cases:
- **(a)** $X_C = 500\,\Omega$, $f = 800$ Hz.
- **(b)** $X_C = 20\,\Omega$, $C = 390$ pF.
- **(c)** $f = 700$ Hz, $X_C = 1\,k\Omega$.
- **(d)** $C = 2.2\,\mu F$, $X_C = 5\,k\Omega$.

11.20. Obtain the voltage, $v(t)$, across each of the following capacitors and determine the voltmeter's reading in each case:
- **(a)** $i(t) = 0.5 \cos(10{,}000t)$ A, $C = 1\,\mu F$.
- **(b)** $i(t) = 0.2 \sin(500t - 30°)$ A, $C = 20$ nF.
- **(c)** $i(t) = 0.707 \sin(4000t + 40°)$ A, $C = 0.2\,\mu F$.
- **(d)** $i(t) = 0.08 \cos(1000t + 20°)$ A, $C = 50$ nF.

11.21. Obtain the current, $i(t)$, through each of the following capacitors and determine the ammeter's reading in each case:
- **(a)** $v(t) = 141.4 \sin(2000t + 60°)$ V, $C = 500$ pF.
- **(b)** $v(t) = 10 \cos(10{,}000t - 30°)$ V, $C = 2\,\mu F$.
- **(c)** $v(t) = 27 \sin(300t + 75°)$ V, $C = 0.12\,\mu F$.
- **(d)** $v(t) = 420 \cos(500t + 23°)$ V, $C = 100$ nF.

Sections 11.10 and 11.11

11.22. Determine the nature, magnitude, and the amount of power dissipated in the element (R, L, or C) associated with each of the following cases involving a voltage–current pair of waveforms:
- **(a)** $v(t) = 100 \cos(250t)$ V
  $i(t) = 2 \sin(250t)$ A.
- **(b)** $v(t) = 141.4 \sin(1000t + 30°)$ V
  $i(t) = 0.707 \cos(1000t - 60°)$ A.
- **(c)** $v(t) = 150 \sin(4000t - 45°)$ V
  $i(t) = 0.3 \sin(4000t + 45°)$ A.
- **(d)** $v(t) = 28.28 \cos(500t - 60°)$ V
  $i(t) = 3.535 \sin(500t - 60°)$ A.

11.23. Draw the phasor diagram for each of the cases examined in Drill Problem 11.22 and determine the impedance of the corresponding element in each case.

Drill Problems

PROBLEMS

Sections 11.2 to 11.5

(i)   **11.1.** Express each of the following waveforms as positive sine-wave functions. Determine their peak amplitude, frequency, period, and phase.

   (a) $v(t) = 50 \cos 314t$ V.
   (b) $v(t) = -100 \sin (5000t - 45°)$ V.
   (c) $v(t) = -70.7 \cos (1000t + 30°)$ V.
   (d) $i(t) = -1.414 \sin (400t + 60°)$ A.
   (e) $i(t) = 0.5 \cos (377t + 45°)$ A.

(i)   **11.2.** Determine the RMS (effective) value of each of the waveforms in Problem 11.1. Write the corresponding phasor representation.

(i)   **11.3.** Determine the instantaneous values of each of the waveforms in Problem 11.1 at **(a)** $t = 0$ s, **(b)** $t = 0.5$ ms, **(c)** $t = 1$ ms. (Note the different angle units.)

(i)   **11.4.** If the waveforms in Problem 11.1 represent the waveforms of the sources applied to a 100 Ω resistor (either voltage or current sources), find the corresponding waveforms of the current or voltage responses in this resistor. For each case calculate the average power dissipated in this resistor.

(i)   **11.5.** Find the average value of the voltage waveforms shown in Fig. 11.35.

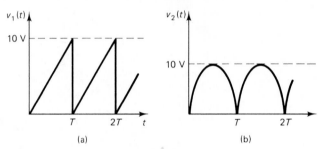

(a)   (b)

Figure 11.35

(i)   **11.6.** Find the average and effective values of the voltage waveform shown in Fig. 11.36.

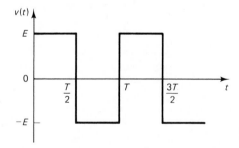

Figure 11.36

Section 11.6

(i)   **11.7.** Convert the following waveforms into phasors and draw the corresponding phasor diagram. Based on the time-domain waveforms, calculate the starting point of each waveform in time, then determine the lead and lag relationships between these waveforms from the time-domain and phasor diagram.

(a) $v_1(t) = 200 \cos{(400t)}$ V.

(b) $v_2(t) = -100 \sin{(400t - 45°)}$ V.

(c) $v_3(t) = 141.4 \sin{(400t - 30°)}$ V.

(i) **11.8.** Convert the following phasors into the time-domain waveform if f is 120 HZ. Find the time-domain starting point of each waveform and its instantaneous value at $t = 0$ s.

(a) $\mathbf{I}_1 = 2\underline{/-60°}$ A.   **(b)** $\mathbf{I}_2 = 7.07\underline{/45°}$ A.

(c) $\mathbf{I}_3 = 10\underline{/-90°}$ A.

Section 11.7

(i) **11.9.** Convert the following complex numbers into their polar form, and write the conjugate of each.

(a) $3 + j4$

(b) $-6 + j8$.

(c) $10 - j17.32$.

(d) $-20 - j20$.

(e) $-4 + j10$.

(i) **11.10.** Convert the following complex numbers into their rectangular forms and write the conjugate of each.

(a) $10\underline{/-45°}$.

(b) $6\underline{/60°}$.

(c) $14.14\underline{/-135°}$.

(d) $20\underline{/-30°}$.

(e) $16\underline{/150°}$.

(ii) **11.11.** Perform the following algebraic manipulations to find the resultant.

(a) $10\underline{/-45°} - 14.14\underline{/-135°}$.

(b) $20\underline{/-30°} + 16\underline{/150°} - 6\underline{/60°}$.

(c) $(3 + j4)(-6 + j8)$.

(d) $\dfrac{10 - j17.32}{-20 - j20}$.

(e) $(-4 + j10) \times 6\underline{/60°} - 20\underline{/-30°}$.

(ii) **11.12.** Find the resultant of each of the following algebraic manipulations.

(a) $\dfrac{j8}{2 - j2} + j4$.

(b) $(-3 + j4)(6 + j3) - 10\underline{/90°}$.

(c) $\dfrac{10\underline{/53.1°}}{2 + j4} \times (1 - j2)$.

(d) $(13 - j20) + \dfrac{10 - j10}{1 + j1}$.

(e) $\dfrac{14 + j20 - 10\underline{/90°}}{3 + j4} + 12\underline{/-60°}$.

Sections 11.8 to 11.11

(i) **11.13.** If the current in a 0.2 H inductor is $i(t) = 4 \sin{(100t + 30°)}$ A, find the voltage across this inductor in the time domain. Draw the phasor diagram for this circuit.

(i) **11.14.** If the voltage across a $0.1\,\mu$F capacitor is $v(t) = 70.7 \cos{(5000t)}$ V, find the current flowing in this capacitor in the time domain. Draw the phasor diagram for this circuit.

(i) **11.15.** Find the impedance of each of the following three single-element loads at $f = 200\,\text{Hz}$ and $f = 1.5\,\text{kHz}$.
   (a) A resistor of value $1\,\text{k}\Omega$.
   (b) A capacitor of value $0.2\,\mu\text{F}$.
   (c) An inductor of value $1.8\,\text{mH}$.

(ii) **11.16.** The voltage source: $e(t) = 50 \sin{(2000t + 30°)}\,\text{V}$ results in a current flow of: $i(t) = 0.5 \cos{(2000t)}\,\text{A}$. Find the phasors of these waveforms, hence obtain the impedance of this circuit in polar and rectangular forms.

(ii) **11.17.** The voltage and current in a single-element load are, respectively, $v(t) = 70.7 \sin{(1000t + 30°)}\,\text{V}$ and $i(t) = 7.07 \cos{(1000t + 30°)}\,\text{A}$. Find
   (a) The nature and magnitude of the element and draw the phasor diagram.
   (b) The power dissipated by this element.
   (c) The readings of ideal voltmeter and ammeter connected in this circuit.

(ii) **11.18.** The voltage and current in a single element load are, respectively, $v(t) = 800 \sin{(377t + 30°)}\,\text{V}$ and $i(t) = 5 \cos{(377t - 60°)}\,\text{A}$. Find
   (a) The nature and magnitude of the element and draw the phasor diagram.
   (b) The power dissipated in this element.
   (c) The frequency and period of these waveforms.

(i) **11.19.** Write the corresponding time-domain expression of each sinusoidal voltage waveform shown in Fig. 11.37. Draw the corresponding phasor diagram.

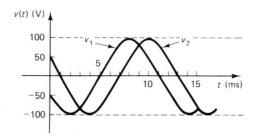

Figure 11.37

(ii) **11.20.** In the circuit shown in Fig. 11.38, $e_1(t) = 141.4 \sin{(2513t)}\,\text{V}$ and $i(t) = 50\sqrt{2} \sin{(2513t - 68.8°)}\,\text{A}$. Find
   (a) The voltage across the inductor in polar form.
   (b) The value of the inductor and the power dissipated in it.
   (c) $e_2(t)$, and draw the phasor diagram for this circuit.

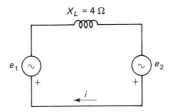

Figure 11.38

(ii) **11.21.** In the circuit shown in Fig. 11.39, $e_1(t) = 165 \sin{(314t)}\,\text{V}$ and $e_2(t) = 165 \sin{(314t - 90°)}\,\text{V}$.
   (a) Write the expression for $i(t)$.
   (b) Find the power dissipated in the resistor.
   (c) Repeat parts (a) and (b) for the case where the polarities of $e_2(t)$ are reversed.

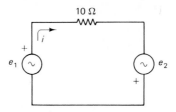

Figure 11.39

(ii) **11.22.** In the circuit shown in Fig. 11.40, $e_1(t) = 14.14 \sin(2513t)$ V, $e_2(t) = 28.28 \sin(2513t + 45°)$ V, and $e_3(t) = 42.43 \sin(2513t - 60°)$ V. Find

(a) The reading of an ideal voltmeter connected across the 5 Ω resistor.

(b) The time-domain expression for $i(t)$.

(c) The power dissipated in the 5 Ω resistor.

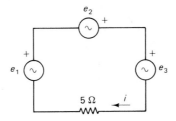

Figure 11.40

(ii) **11.23.** In the circuit shown in Fig. 11.41, $e_1(t) = 230\sqrt{2} \sin(120\pi t)$ V and $e_2(t) = 230\sqrt{2} \sin(120\pi t + 120°)$ V.

(a) Find the impedance and admittance of C.

(b) Write the time-domain expression for $i(t)$ and draw the phasor diagram for the circuit. What is the reading of the ammeter?

(c) Repeat parts (a) and (b) if the capacitance and frequency are doubled simultaneously.

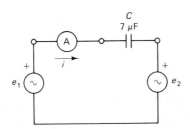

Figure 11.41

(ii) **11.24.** In the circuit shown in Fig. 11.42, $e_1(t) = 80 \cos(2000t)$ V, $e_2(t) = 80 \sin(2000t)$ V, and $i(t) = 10 \sin(2000t + 45°)$ A. Find

(a) The nature, magnitude, and power dissipated in the element.

(b) The readings of the voltmeter and ammeter. Draw the phasor diagram for this circuit.

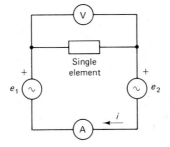

Figure 11.42

APPENDIX A11: PROOF OF EQUATIONS (11.73)

The two sinusoidal voltage signals

$$v_1(t) = \sqrt{2}\, V_1 \sin(\omega t + \phi_1)\ \text{V} \tag{11A.1}$$

and

$$v_2(t) = \sqrt{2}\, V_2 \sin(\omega t + \phi_2)\ \text{V} \tag{11A.2}$$

can be added algebraically in the time domain as shown below. It is important to note that both signals have the same frequency and that the phase shifts ϕ_1 and ϕ_2 could have any positive, negative, or zero value. V_1 and V_2 are the RMS values of the signals. Thus

$$
\begin{aligned}
v_T(t) &= v_1(t) + v_2(t) \\
&= \sqrt{2}\, V_1 \sin(\omega t + \phi_1) + \sqrt{2}\, V_2 \sin(\omega t + \phi_2) \\
&= \sqrt{2}\, V_1 (\sin \omega t \cos \phi_1 + \cos \omega t \sin \phi_1) \\
&\quad + \sqrt{2}\, V_2 (\sin \omega t \cos \phi_2 + \cos \omega t \sin \phi_2) \\
&= \sqrt{2}\, (V_1 \cos \phi_1 + V_2 \cos \phi_2) \sin \omega t \\
&\quad + \sqrt{2}\, (V_1 \sin \phi_1 + V_2 \sin \phi_2) \cos \omega t \\
&= \sqrt{2}\, V_T \cos \phi_T \sin \omega t + \sqrt{2}\, V_T \sin \phi_T \cos \omega t\ \text{V}
\end{aligned}
\tag{11A.4}
$$

(11A.3)

In the step above, the following substitutions were used:

$$V_T \cos \phi_T = V_1 \cos \phi_1 + V_2 \cos \phi_2 \tag{11A.5}$$

and

$$V_T \sin \phi_T = V_1 \sin \phi_1 + V_2 \sin \phi_2 \tag{11A.6}$$

The expression in Eq. (11A.4) can then be written in its final form:

$$v_T(t) = \sqrt{2}\, V_T \sin(\omega t + \phi_T)\ \text{V} \tag{11A.7}$$

where [from Eqs. (11A.5) and (11A.6)]

$$
\begin{aligned}
V_T^2 &= (V_1 \cos \phi_1 + V_2 \cos \phi_2)^2 + (V_1 \sin \phi_1 + V_2 \sin \phi_2)^2 \\
&= V_1^2 + V_2^2 + 2V_1 V_2 (\cos \phi_1 \cos \phi_2 + \sin \phi_1 \sin \phi_2)
\end{aligned}
$$

Hence

$$V_T = [V_1^2 + V_2^2 + 2V_1 V_2 \cos(\phi_1 - \phi_2)]^{1/2} \tag{11A.8}$$

and

$$\tan \phi_T = \frac{V_1 \sin \phi_1 + V_2 \sin \phi_2}{V_1 \cos \phi_1 + V_2 \cos \phi_2} \tag{11A.9}$$

Now, considering the phasor representations of each of the two time signals $v_1(t)$ and $v_2(t)$, as obtained from the definitions given in Eqs. (11.34) and (11.37),

$$\mathbf{V}_1 = V_1 \underline{/\phi_1} = V_1 \cos \phi_1 + jV_1 \sin \phi_1\quad \text{V} \tag{11A.10}$$

and

$$\mathbf{V}_2 = V_2 \underline{/\phi_2} = V_2 \cos \phi_2 + jV_2 \sin \phi_2\quad \text{V} \tag{11A.11}$$

The corresponding algebraic summation of the two signal phasors is therefore

$$\mathbf{V}_T = \mathbf{V}_1 + \mathbf{V}_2 \tag{11A.12}$$
$$= (V_1 \cos \phi_1 + V_2 \cos \phi_2) + j(V_1 \sin \phi_1 + V_2 \sin \phi_2)$$
$$= V_T \cos \phi_T + jV_T \sin \phi_T$$
$$= V_T \underline{/\phi_T} \tag{11.A13}$$

where

$$V_T \cos \phi_T = V_1 \cos \phi_1 + V_2 \cos \phi_2 \tag{11A.14}$$

and

$$V_T \sin \phi_T = V_1 \sin \phi_1 + V_2 \sin \phi_2 \tag{11A.15}$$

The two equations above are exactly the same as Eqs. (11A.5) and (11A.6), obtained previously through the algebraic summation in the time domain. Therefore, the values of V_T and ϕ_T will be exactly the same as those obtained in Eqs. (11A.8) and (11A.9), respectively.

The resultant phasor obtained in Eq. (11A.13) can then be converted into the corresponding time-domain signal, using Eq. (11.34), to provide

$$v_T(t) = \sqrt{2}\, V_T \sin(\omega t + \phi_T) \tag{11A.16}$$

The resultant signal obtained in Eq. (11A.16) through the algebraic phasor summation is exactly the same as the resultant signal obtained in Eq. (11A.7) through the direct algebraic summation of the time-domain signals. This proves the validity of using phasors to represent time-domain signals. The results obtained through the phasor analysis of electric circuits involving the use of Ohm's law, KVL, and KCL will therefore be exactly the same as those obtained through the direct time-domain manipulations.

GLOSSARY

Ac: Alternating current waveform. The "ac" notation is given to waveforms that exhibit a change in polarities/direction every prescribed amount of time. The magnitude of such a waveform varies continuously with time.

Admittance (Y): The inverse of impedance, normally expressed in a complex number format.

Angular Speed (ω): The velocity in angles/second with which a radius line rotates about the center of the circle it describes.

Average Value: The constant value parameter of a waveform that encloses an area bound by it and the horizontal axis, exactly equal to the total algebraic area enclosed by the complete original waveform and the horizontal axis over one period. Areas above the axis are positive and areas below the axis are negative.

Complex Conjugate: The conjugate of a complex number is obtained by changing the sign of the imaginary part of the rectangular form of the complex number, or equivalently, by changing the sign of the angle of the polar form of the complex number.

Complex Number: A number that represents a point (or a line vector) in a two-dimensional plane. It requires two distinct pieces of information to be completely defined. These are the (x, y) coordinates with respect to two specific axis, or the magnitude (R) and phase (θ) coordinates with respect to the origin of the axes and the positive x direction. In the first case the complex number is said to be expressed in rectangular form, and in the second case the complex number is said to be expressed in polar form.

Cycle: The portion of a waveform contained in one complete period of time.

Derivative: The rate of change of a function with respect to its independent variable.

Effective Value: The equivalent constant (dc) value of any time-varying waveform that produces the same power in a given resistance as the original waveform would produce in that resistance.

Form Factor (F_f):   The ratio of the RMS value of a waveform to the average value of the full-wave rectified waveform.

Frequency (f):   The number of cycles of a periodic waveform that occur in 1 second.

Impedance (Z):   The opposition to current flow exhibited by an element, or a connection of elements. The impedance is usually expressed in a complex number format and it normally implies that its magnitude and phase are in general functions of the frequency.

Instantaneous Value: The magnitude of a waveform at a specific instant of time. Instantaneous values are normally denoted by lowercase letters.

Leading and Lagging Phase Relationship: The phase shift angle in degrees between two waveforms. The waveform whose starting point occurs first in time is said to be leading the other waveform by the phase shift angle, while the waveform whose starting point occurs later in time is said to be lagging the other waveform by the phase shift angle.

Peak-to-Peak Value: The magnitude of the total swing of a signal from its highest point to its lowest point. The peak-to-peak value (p–p) is twice the maximum value of the signal waveform.

Period (T):   The time interval between two successive symmetrically located points of a periodic waveform.

Periodic Waveform: A waveform that continually repeats its basic form every specific amount of time.

Phasor: An equivalent mathematical representation of a waveform, with the time dependence dropped. Phasors are represented as vectors. The magnitude of the phasor is the RMS value of the waveform, while the phase angle of the phasor is the phase shift of the starting point of the waveform from the origin point of the time axis.

Phasor diagram: A diagram representing the phase relationships of the different quantities in an electric circuit in phasor format. Each of these quantities (voltages and currents) is represented as a phasor (vector) in the phasor diagram.

Radian: A unit of angle measurement that defines a finite segment of a circle. One radian is equal to approximately $57.3°$, or 2π radians (3.1416 rad) are equivalent to $360°$.

Reactance (X):   The opposition of an inductor or a capacitor to the flow of electric current. Reactances are usually dependent on the frequency of the signal waveform and it implies a phase shift between the voltage across the element and the current flow in the element.

RMS Value: The root-mean-square or effective value of a waveform.

Susceptance (B): The inverse of the reactance of an inductor or a capacitor.

ANSWERS TO DRILL PROBLEMS

11.1.
(a) 0.5236 rad,
(b) 1.0472 rad,
(c) 0.7854 rad,
(d) 2.618 rad,
(e) 0.644 rad

11.3.
(a) 15.915 Hz, 62.83 ms;
(b) 318.31 Hz, 3.1416 ms;
(c) 600 Hz, 1.667 ms;
(d) 4997.5 Hz, 0.2001 ms

11.5.
(a) $v_1(t)$ leads $v_2(t)$ by 135°
(b) $v_1(t)$ leads $v_3(t)$ by 30°
(c) $v_3(t)$ leads $v_2(t)$ by 105°
(d) $i_1(t)$ leads $i_2(t)$ by 30°
(e) $i_3(t)$ leads $i_2(t)$ by 180°
(f) $v_1(t)$ leads $i_3(t)$ by 15°
(g) $v_2(t)$ leads $i_2(t)$ by 60°

11.7.
(a) $i(t) = 0.2 \sin(200t - 45°)$ A,
$I = 0.14142$ A,
$V = 70.71$ V, $P = 10$ W
(b) $i(t) = 0.1 \sin(10t + 120°)$ A,
$I = 0.07071$ A,
$V = 35.355$ V, $P = 2.5$ W
(c) $i(t) = 0.14142 \sin(500t + 10°)$ A,
$I = 0.1$ A, $V = 50$ V,
$P = 5$ W

11.9.
(a) $\mathbf{V}_1$ leads $\mathbf{V}_2$ by 135°, while $\mathbf{V}_2$ leads $\mathbf{I}_1$ by 30°
(b) $\mathbf{V}_3$ leads $\mathbf{V}_2$ by 105°, while $\mathbf{V}_2$ leads $\mathbf{I}_2$ by 60°
(c) $\mathbf{V}_1$ leads $\mathbf{I}_3$ by 15°, while $\mathbf{I}_3$ leads $\mathbf{I}_1$ by 150°

11.11.
(a) $3.6056 \underline{/56.31°}$,
(b) $9.434 \underline{/-58°}$,
(c) $29.155 \underline{/59.04°}$,
(d) $44.721 \underline{/-116.57°}$,
(e) $17.205 \underline{/125.54°}$,
(f) $0.9055 \underline{/-83.66°}$

11.13.
(a) $12.66 + j3 = 13.01 \underline{/13.33°}$
(b) $-1 + j12 = 12.0416 \underline{/94.76°}$
(c) $70.71 \underline{/-8.13°} = 70 - j10$
(d) $0.2 \underline{/-126.86°} = -0.12 - j0.16$

(e) $7 + j5 = 8.602 \underline{/35.54°}$
(f) $4.8 \underline{/-126.87°} = -2.88 - j3.84$
(g) $7.595 \underline{/-87.67°} = 0.3088 - j7.5887$

11.15.
(a) $L = 0.3979$ mH
(b) $f = 108.6$ Hz
(c) $L = 0.9947$ mH
(d) $f = 15.915$ Hz

11.17.
(a) $i(t) = 0.0404 \sin(5000t - 90°)$ A,
$I = 28.57$ mA
(b) $i(t) = 1.326 \sin(377t - 25°)$ A, $I = 0.9378$ A
(c) $i(t) = 0.175 \sin(800t - 40°)$ A, $I = 0.1237$ A
(d) $i(t) = 0.225 \sin(1000t + 60°)$ A, $I = 0.1591$ A

11.19.
(a) $C = 397.9$ nF
(b) $f = 20.4$ MHz
(c) $C = 227.36$ nF
(d) $f = 14.47$ Hz

11.21.
(a) $i(t) = 0.1414 \sin(2000t + 150°)$ mA, $I = 0.1$ mA
(b) $i(t) = 0.2 \sin(10,000t + 150°)$ A, $I = 0.1414$ A
(c) $i(t) = 0.972 \sin(300t + 165°)$ mA, $I = 0.687$ mA
(d) $i(t) = 0.021 \sin(500t - 157°)$ A, $I = 14.85$ mA

11.23.
(a) $\mathbf{V} = 70.71 \underline{/90°}$ V,
$\mathbf{I} = 1.4142 \underline{/0°}$ A,
$\mathbf{Z}_L = 50 \underline{/90°} = j50$ Ω
(b) $\mathbf{V} = 100 \underline{/30°}$ V,
$\mathbf{I} = 0.5 \underline{/30°}$ A,
$\mathbf{Z}_R = 200 \underline{/0°} = 200$ Ω
(c) $\mathbf{V} = 106.07 \underline{/-45°}$ V,
$\mathbf{I} = 0.2121 \underline{/45°}$ A,
$\mathbf{Z}_C = 500 \underline{/-90°} = -j500$ Ω
(d) $\mathbf{V} = 20 \underline{/30°}$ V,
$\mathbf{I} = 2.5 \underline{/-60°}$ A,
$\mathbf{Z}_L = 8 \underline{/90°} = j8$ Ω

12 | STEADY-STATE ANALYSIS OF SIMPLE AC CIRCUITS

OBJECTIVES

- Ability to analyze ac series and parallel circuits containing $R, L,$ and C elements, using impedances, admittances, and phasors to obtain any required circuit quantity (eg., current, voltage, power, elements values, and frequency).

- Ability to use impedance and admittance triangles and draw phasor diagram in order to help understand the performance of a given RLC series or parallel circuit.

- Ability to correlate the frequency domain phasor manipulations to the actual time domain signals and their properties, and vice versa.

- Ability to calculate the power dissipated in any series, parallel, or series-parallel RLC ac circuit.

- Ability to convert a series connected ac load into an equivalent parallel connected load and vice versa.

- Ability to apply KVL and KCL to any series-parallel ac circuit with any general type branch impedance.

- Understanding the performance of ac impedance bridges used to measure unknown inductances and capacitances.

Starting from the simple series-connected *RL* and *RC* loads and progressing to the general type of series–parallel *RLC* loads, the topics discussed in this chapter develop the skills required to analyze such ac circuits in the phasor (frequency domain) form and correlate such analysis to the time domain. The average power dissipated in these ac loads is examined throughout.

12.1 SERIES *RL* LOADS

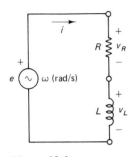

Figure 12.1

To emphasize the relationship between time-domain and frequency-domain (phasor-impedance) analysis, the *RL* series circuit will be examined in both domains. In any series circuit, the current is the same in all components and KVL is always applicable. The common quantity (in this case the current) is usually taken to be the reference waveform, with zero phase angle (i.e., the reference phasor), unless otherwise specified. This choice usually simplifies the analysis.

In the simple series *RL* ac circuit shown in Fig. 12.1, in the time domain, the current waveform $i(t)$ is assumed to be the reference quantity,

$$i(t) = I_m \sin \omega t \quad \text{A} \Leftrightarrow \mathbf{I} = I\underline{/0°}\ \text{A} \tag{12.1}$$

Therefore, since $v_R(t)$ is in phase with $i(t)$ and $v_L(t)$ leads $i(t)$ by 90°, the following expressions can be written based on the discussion in Chapter 11:

$$v_R(t) = V_{Rm} \sin \omega t = I_m R \sin \omega t \quad \text{V} \Leftrightarrow \mathbf{V}_R = V_R\underline{/0°} = IR\underline{/0°}\ \text{V} \tag{12.2}$$

and

$$v_L(t) = V_{Lm} \sin (\omega t + 90°)$$
$$= I_m X_L \sin (\omega t + 90°) \text{ V} \Leftrightarrow \mathbf{V}_L = V_L\underline{/90°} = IX_L\underline{/90°}\ \text{V} \tag{12.3}$$

The magnitudes of the phasors above are the RMS values of the corresponding waveforms, as discussed in Chapter 11.

Applying KVL to the circuit of Fig. 12.1, we have

$$e(t) = v_R(t) + v_L(t) \tag{12.4}$$

The time-domain representations of the waveforms above are shown in Fig. 12.2. The summation of $v_R(t)$ and $v_L(t)$ at each instant of time produces the total source voltage $e(t)$. As Fig. 12.2 indicates, $e(t)$ is also a sine waveform, with the same frequency (and period) as i, v_R, and v_L, but shifted from the time origin $(t = 0)$ by an angle $\phi°$. Thus in the *RL* series circuit, $e(t)$ leads $i(t)$ by the phase angle $\phi°$, that is,

$$e(t) = E_m \sin (\omega t + \phi°) \text{ V} \Leftrightarrow \mathbf{E} = E\underline{/\phi°}\ \text{V} \tag{12.5}$$

It is also important to note that the instant of time at which $e(t)$ reaches its maximum amplitude, E_m, does not coincide with either the maximum amplitude of $v_R(t)$, V_{Rm}, or the maximum amplitude of $v_L(t)$, V_{Lm}. Hence

$$E_m \neq V_{Rm} + V_{Lm} \tag{12.6}$$

Because of the tediousness and inefficiency of solving electric circuits using such time-domain operations, the phasor-impedance concepts (frequency-domain equivalent circuits) were introduced in Chapter 11. The use of the frequency-domain method in solving ac electric circuits is certainly quicker, more accurate, and hence more efficient than the time-domain operations.

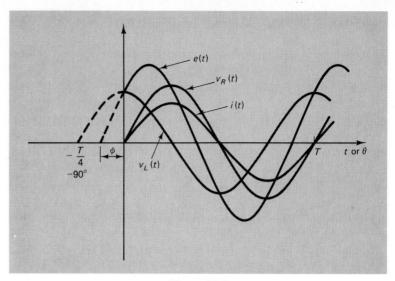

Figure 12.2

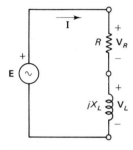

Figure 12.3

The *RL* series circuit is redrawn in the frequency domain as shown in Fig. 12.3. The voltages and current are represented by their phasors and the elements by their impedances. Ohm's law, KVL, KCL, and in fact all the other relationships and techniques developed for dc circuits also apply for the ac frequency-domain equivalent circuits. Here, however, phasors and impedances are represented by complex numbers instead of real numbers as in dc circuits. The analysis of ac circuits thus involves complex-number algebraic manipulations. Here, applying Ohm's law, we have

$$\mathbf{V}_R = \mathbf{I}Z_R \tag{12.7}$$

$$= I\underline{/0°}\, R\underline{/0°} = IR\underline{/0°}\ \mathrm{V} = V_R\underline{/0°}\ \mathrm{V} \tag{12.7a}$$

and

$$\mathbf{V}_L = \mathbf{I}Z_L \tag{12.8}$$

$$= I\underline{/0°}\, X_L\underline{/90°} = IX_L\underline{/90°} = V_L\underline{/90°}\ \mathrm{V} \tag{12.8a}$$

Equations (12.7) and (12.8) are the general form of Ohm's law as applied to the resistance and inductance, respectively. Equations (12.7a) and (12.8a) are the results when **I** is considered to be the reference phasor, as suggested above. Now applying KVL to the circuit in Fig. 12.3 gives us

$$\boxed{\mathbf{E} = \mathbf{V}_R + \mathbf{V}_L} \tag{12.9}$$

Substituting from Eqs. (12.7a) and (12.8a) for $\mathbf{V}_R$ and $\mathbf{V}_L$, respectively, we have

$$\mathbf{E} = V_R + jV_L = \sqrt{V_R^2 + V_L^2}\ \underline{\Big/ \tan^{-1}\frac{V_L}{V_R}} \tag{12.10a}$$

$$= IR + jIX_L = I\sqrt{R^2 + X_L^2}\ \underline{\Big/ \tan^{-1}\frac{X_L}{R}} \tag{12.10b}$$

$$= E\underline{/\phi°}\ \mathrm{V} \tag{12.10c}$$

where

$$E = \sqrt{V_R^2 + V_L^2} = I\sqrt{R^2 + X_L^2} = I\sqrt{R^2 + \omega^2 L^2} \quad \text{V} \qquad (12.11)$$

and

$$\phi = \tan^{-1}\frac{V_L}{V_R} = \tan^{-1}\frac{X_L}{R} = \tan^{-1}\frac{\omega L}{R} \qquad (12.12)$$

Once the phasor **E** is obtained, the corresponding time-domain waveform is

$$e(t) = \sqrt{2}\,E \sin{(\omega t + \phi^\circ)} \quad \text{V}$$

as arrived at in Eq. (12.5) based on the time-domain operations. Figure 12.4 shows the phasor diagram for this circuit. The vectorial sum of $\mathbf{V}_R$ and $\mathbf{V}_L$ [Eq. (12.9)] produces the resultant **E**. The phasor diagram is very helpful in the interpretation and understanding of the phase relationships between the different phasors (time variables) in a given circuit. If it is drawn to scale, approximate results can be obtained; however, the algebraic relations give more accurate results (using a hand-held calculator).

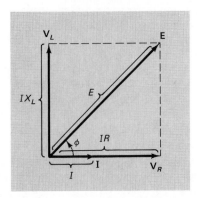

Figure 12.4

In general, any of the phasors in a given network could be taken as the reference phasor (i.e., when this phasor has a zero phase angle). Also, in many cases, no phasor may have a zero angle (e.g., $\mathbf{I} = I\underline{/\theta^\circ}$ A). However, the phase relationships between the phasors always hold, no matter which phasor is the reference. For example, a general phasor diagram for the *RL* series circuit (with $\mathbf{I} = I\underline{/\theta^\circ}$ A) is shown in Fig. 12.5. The voltage across R, $\mathbf{V}_R$, is always in phase with $\mathbf{I}$. The voltage across L, $\mathbf{V}_L$, is always leading $\mathbf{I}$ by 90°. The source voltage **E** $(=\mathbf{V}_R + \mathbf{V}_L)$ always leads $\mathbf{I}$ by an angle ϕ°, as derived in Eq. (12.12). In fact,

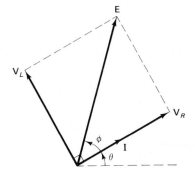

Figure 12.5

this phasor diagram is exactly the same as that shown in Fig. 12.4, but the entire diagram is rotated by the angle $\theta°$.

In the RL series load, the angle $\phi°$ [Eq. (12.12)] by which $\mathbf{E}$ leads $\mathbf{I}$, or $\mathbf{I}$ lags $\mathbf{E}$ is a function of ω (the frequency), L, and R. Its value varies within the range

(for purely resistive load, $0° \leq \phi \leq 90°$ (for purely inductive load, or when $f = 0$ Hz)   or when $f = \infty$ Hz)   (12.13)

The impedance triangle. From KVL [Eq. (12.9)],

$$\mathbf{E} = \mathbf{V}_R + \mathbf{V}_L$$

$$\boxed{\begin{aligned}\mathbf{E} &= \mathbf{I}Z_R + \mathbf{I}Z_L = \mathbf{I}(Z_R + Z_L)\\ &= \mathbf{I}Z_T\end{aligned}} \qquad (12.14)$$

Therefore, the total impedance of the series RL load is

$$Z_T = \frac{\mathbf{E}}{\mathbf{I}} = Z_R + Z_L$$

$$= R + jX_L = \sqrt{R^2 + X_L^2}\;\Big/\!\tan^{-1}\frac{X_L}{R}$$

$$= \sqrt{R^2 + \omega^2 L^2}\;\Big/\!\tan^{-1}\frac{\omega L}{R} \qquad (12.15)$$

Also,

$$Z_T = |Z_T|\underline{/\phi°}\ \Omega$$
$$= |Z_T|\cos\phi + j\,|Z_T|\sin\phi = R + jX_L \qquad (12.16)$$

that is,

$$|Z_T| = \sqrt{R^2 + X_L^2}$$
$$R = |Z_T|\cos\phi\ \Omega$$
$$X_L = |Z_T|\sin\phi\ \Omega \qquad (12.16a)$$

and

$$\phi = \tan^{-1}\frac{X_L}{R} = \tan^{-1}\frac{\omega L}{R} \qquad (12.16b)$$

The total impedance, being a vector or a complex number and not a phasor, is obtained by adding algebraically the impedances of the series-connected elements. It can be represented graphically by the impedance triangle shown in Fig. 12.6, in accordance with Eqs. (12.15) and (12.16). The real-axis component is the resistance of the circuit R, while the imaginary-axis component is the inductive reactance of the circuit, X_L. In the impedance triangle, the inductive reactance is always perpendicular to the resistance.

Since from Ohm's law for the entire circuit,

$$Z_T = \frac{\mathbf{E}}{\mathbf{I}} \qquad \text{[Eq. (12.14)]}$$

the phase angle of Z_T, $\phi°$, is exactly the same as the angle by which $\mathbf{E}$ leads $\mathbf{I}$, as obtained in Eq. (12.12). In fact, if the magnitude of each side of the impedance

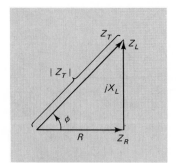

Figure 12.6

Chapter 12   Steady-State Analysis of Simple AC Circuits

triangle is multiplied by I (the current magnitude), the magnitudes of the corresponding phasors result, as can be observed from the phasor diagram in Fig. 12.4. Here also,

$$\mathbf{V}_R = \mathbf{I}Z_R$$

$$= \frac{\mathbf{E}}{Z_T}Z_R = \mathbf{E}\frac{Z_R}{Z_R + Z_L} \tag{12.17}$$

$$\mathbf{V}_L = \mathbf{I}Z_L$$

$$= \frac{\mathbf{E}}{Z_T}Z_L = \mathbf{E}\frac{Z_L}{Z_R + Z_L} \tag{12.18}$$

voltage-division principle

Power in the RL series load. As concluded above, the voltage $\mathbf{V}$ (or $\mathbf{E}$) across an RL series impedance load, leads the current $\mathbf{I}$ in this load by an angle $\phi°$. Thus if $i(t)$ is given by

$$i(t) = I_m \sin(\omega t + \theta°) \text{ A} \Leftrightarrow \mathbf{I} = I\underline{/\theta°} \text{ A}$$

then

$$v(t) = V_m \sin(\omega t + \theta° + \phi°) \text{ V} \Leftrightarrow \mathbf{V} = V\underline{/\theta° + \phi°} \text{ V}$$

The circuit and the corresponding phasor diagram are shown in Fig. 12.7(a) and (b), respectively. Hence, in the time domain the instantaneous power in the load is

$$p(t) = v(t)i(t) = V_m I_m \sin(\omega t + \theta°) \sin(\omega t + \theta° + \phi°)$$

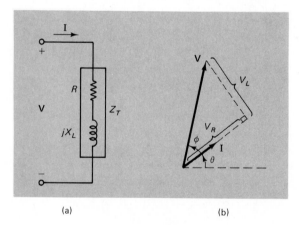

(a)   (b)

Figure 12.7

Using the trigonometric relationship

$$\sin A \sin B = \tfrac{1}{2}[\cos(A - B) - \cos(A + B)]$$

we obtain

$$p(t) = \tfrac{1}{2}V_m I_m[\cos \phi° - \cos(2\omega t + 2\theta° + \phi°)] \tag{12.19}$$

Since the average of any sinusoidal (or cos) function over any period is zero, the average of the second term in Eq. (12.19) is zero. Thus

$$\boxed{\begin{aligned} P_{av} &= \tfrac{1}{2}V_m I_m \cos\phi = \frac{V_m}{\sqrt{2}}\frac{I_m}{\sqrt{2}}\cos\phi \\ &= VI\cos\phi \quad \text{W} \end{aligned}} \qquad (12.20)$$

This expression for the average power in the load Z_T [Eq. (12.20)] is quite general, whether the angle ϕ is positive (as it is in the *RL* series load) or negative (as in some of the cases to be discussed later). The quantity

$$\cos\phi = \cos(-\phi) = \text{power factor of the load} = F_p \qquad (12.21)$$

Thus the real average power dissipated in a load in an ac circuit is the product of the RMS voltage across the load and the RMS current flowing in the load times the power factor of the load (i.e., the cosine of the angle between the two phasors **V** and **I**).

In the case of the *RL* impedance,

$$\boxed{\begin{aligned} P_{av} &= IV\cos\phi = IV_R \\ &= IIR = I^2R \quad \text{W} \end{aligned}} \quad \text{[refer to the geometry of Fig. 12.7(b)]} \qquad (12.22)$$

Hence only the resistive component is responsible for dissipating all of the average power, as would be expected since inductances do not dissipate power, but store energy.

EXAMPLE 12.1

A 120 V, 60 Hz source is applied to a solenoid (practical inductor) whose inductance is 0.5 H and whose internal resistance is 100 Ω. Assuming that $e(t)$ is the reference waveform, find $i(t)$. Draw the phasor diagram and calculate the power dissipated in this solenoid.

Solution The equivalent frequency-domain circuit is shown in Fig. 12.8(a).

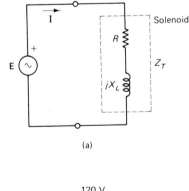

(a)

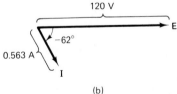

(b)

Figure 12.8

Here

$$\mathbf{E} = 120\underline{/0°}\text{ V} \quad \text{(reference waveform or phasor)}$$

$$\omega = 2\pi f = 6.28 \times 60 = 377\text{ rad/s}$$

and

$$Z_T = R + jX_L = R + j\omega L$$

$$= 100 + j377 \times 0.5 = 100 + j188.5 = 213\underline{/62°}\text{ }\Omega$$

Using Ohm's law for this circuit yields

$$\mathbf{I} = \frac{\mathbf{E}}{Z_T} = \frac{120\underline{/0°}}{213\underline{/62°}} = 0.563\underline{/-62°}\text{ A}$$

Therefore,

$$i(t) = 0.563\sqrt{2}\sin(377t - 62°)\text{ A}$$

$$= 0.797\sin(377t - 62°)\text{ A}$$

The corresponding phasor diagram is shown in Fig. 12.8(b). The power dissipated in this load is

$$P_{av} = EI\cos\phi = 120 \times 0.563\cos 62° = 31.7\text{ W}$$

$$= I^2 R = (0.563)^2 \times 100 = 31.7\text{ W}$$

EXAMPLE 12.2

In the circuit shown in Fig. 12.9, V_R and V_L were measured and found to be 10 V each. Assume that $i(t)$ is the reference waveform (with zero phase angle). Find

(a) The frequency f and the current $i(t)$.
(b) Z_T, the total circuit's impedance, and draw the impedance triangle.
(c) $e(t)$, and draw the phasor diagram.

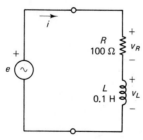

Figure 12.9

Solution (a) The voltage measurements indicate RMS magnitudes of the voltage across each of the two elements. Thus

$$V_R = IR$$

$$I = \frac{10}{100} = 0.1\text{ A}$$

and

$$V_L = IX_L$$

Further,

$$X_L = \frac{10}{0.1} = 100\ \Omega = \omega L$$

$$\omega = \frac{X_L}{L} = \frac{100}{0.1} = 1000\ \text{rad/s}$$

$$f = \frac{\omega}{2\pi} = \frac{1000}{6.28} = 159\ \text{Hz}$$

Therefore,

$$\mathbf{I} = 0.1\underline{/0^\circ}\ \text{A}$$

(as it is taken to be the reference waveform) and

$$i(t) = 0.1\sqrt{2}\sin(1000t) = 0.1414\sin(1000t)\ \text{A}$$

(b) The components of Z_T, the total circuit's impedance, are now both known. Thus

$$Z_T = R + jX_L$$
$$= 100 + j100 = 141.4\underline{/45^\circ}\ \Omega$$

The impedance triangle is shown in Fig. 12.10(a).

(c) Now applying Ohm's law for the overall circuit gives us

$$\mathbf{E} = \mathbf{I}Z_T = 0.1\underline{/0^\circ} \times 141.4\underline{/45^\circ}$$
$$= 14.14\underline{/45^\circ}\ \text{V}$$

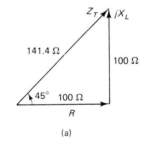

(a)

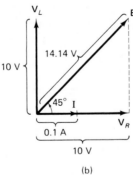

(b)

Figure 12.10

Also, since $\mathbf{I}$ is the reference phasor, as in the discussion above [Eq. (12.10a)],

$$\mathbf{E} = V_R + jV_L = 10 + j10$$
$$= 14.14\underline{/45^\circ}\ \text{V}$$

and

$$e(t) = 14.14\sqrt{2} \sin (1000t + 45°)$$
$$= 20 \sin (1000t + 45°) \text{ V}$$

The corresponding phasor diagram is shown in Fig. 12.10(b).

EXAMPLE 12.3

A series RL circuit has a current of 5 A when a 100 V, 159 Hz source is applied to it. The voltage drop across the resistance is 60 V. Find the value of L.

Solution Considering that $\mathbf{I}$ is the reference phasor, a sketch of the phasor diagram of the circuit is shown in Fig. 12.11. As is clear from the geometry relationships,

$$E^2 = V_R^2 + V_L^2$$
$$(100)^2 = (60)^2 + V_L^2$$

Therefore,

$$V_L = \sqrt{10,000 - 3600} = 80 \text{ V}$$
$$= IX_L$$
$$X_L = \frac{80}{5} = 16 \text{ }\Omega$$
$$= \omega L = 2\pi \times 159L$$

and

$$L = \frac{16}{6.28 \times 159} = 16 \text{ mH}$$

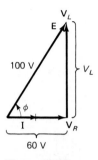

Figure 12.11

EXAMPLE 12.4

A selenoid (practical inductor), in series with an external resistor R, were connected to a 120 V, 60 Hz source. The voltages measured across the solenoid and the resistance R were both 80 V. The current was measured to be 0.5 A. Find the values of R, the inducance of the solenoid L, and the resistance of the solenoid r_L. The circuit is shown in Fig. 12.12.

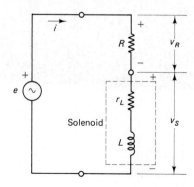

Figure 12.12

Solution The total impedance of this circuit, Z_T, is

$$Z_T = R + Z_S = (R + r_L) + jX_L$$

where Z_S is the impedance of the solenoid $= r_L + jX_L$. The impedance triangle of this circuit, showing Z_S and Z_T, is drawn in Fig. 12.13(a).

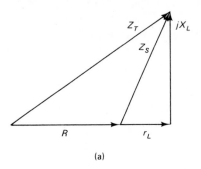

(a)

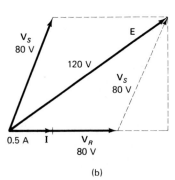

(b)

Figure 12.13

With **I** assumed to be the reference phasor, the phasor diagram can also be drawn as shown in Fig. 12.13(b), noting that the voltage phasors are the same as the corresponding impedances, multiplied by **I** (since the phase of **I** is zero).

Since no phase information is given in this example, the solution can be derived based on the phasor and impedance magnitudes only, using Ohm's law. Here

$$Z_R = \frac{\mathbf{V}_R}{\mathbf{I}} \qquad \text{that is, } R = \frac{V_R}{I} = \frac{80}{0.5} = 160 \ \Omega$$

$$Z_S = \frac{\mathbf{V}_S}{\mathbf{I}} \qquad \text{that is, } |Z_S| = \sqrt{r_L^2 + X_L^2} = \frac{V_S}{I} = \frac{80}{0.5} = 160 \ \Omega$$

Therefore,

$$r_L^2 + X_L^2 = (160)^2 = 25{,}600 \tag{12.23}$$

and

$$Z_T = \frac{\mathbf{E}}{\mathbf{I}} \qquad \text{that is,}$$

$$|Z_T| = \sqrt{(R + r_L)^2 + X_L^2} = \frac{E}{I} = \frac{120}{0.5} = 240 \ \Omega$$

and

$$(R + r_L)^2 + X_L^2 = (160 + r_L)^2 + X_L^2 = (240)^2 = 57{,}600 \tag{12.24}$$

Equations (12.23) and (12.24) express two equations in the two unknowns, r_L

and X_L. Through subtraction:

$$(160 + r_L)^2 - r_L^2 = 57{,}600 - 25{,}600$$

$$25{,}600 + 320r_L = 32{,}000$$

$$r_L = \frac{32{,}000 - 25{,}600}{320} = \frac{6400}{320} = 20\ \Omega$$

Now, substituting in Eq. (12.23) yields

$$X_L = \sqrt{25{,}600 - 400} = 158.75\ \Omega = \omega L = 2\pi \times 60\,L$$

Therefore,

$$L = \frac{158.75}{6.28 \times 60} = 0.421\ \text{H}$$

EXAMPLE 12.5

The source voltage and the resulting current in a two-element series circuit are given by $e(t) = 200 \sin (1000t - 30°)$ V and $i(t) = 4 \cos (1000t - 173°)$ A.
(a) Find the nature and magnitude of the two elements. What is the power dissipated in this circuit?
(b) What is the reading of an ideal voltmeter when connected across each element? Draw the phasor diagram.
(c) Express the voltage across each component in the time domain.

Solution (a) Here the phasors of **E** and **I** can be easily derived from the time-domain expressions

$$e(t) = 200 \sin (1000t - 30°)\ \text{V} \Leftrightarrow \mathbf{E} = 141.4\underline{/-30°}\ \text{V}$$

$$\omega = 1000\ \text{rad/s}$$

$$i(t) = 4 \cos (1000t - 173°)\ \text{A}$$

$$= 4 \sin (1000t - 83°)\ \text{A} \Leftrightarrow \mathbf{I} = 2.828\underline{/-83°}\ \text{A}$$

These two phasors can be drawn as in Fig. 12.14. Clearly, **E** leads **I**, indicating that the two elements are R and L. This can be confirmed and their values obtained by determining Z_T, using Ohm's law:

$$Z_T = \frac{\mathbf{E}}{\mathbf{I}} = \frac{141.4\underline{/-30°}}{2.828\underline{/-83°}} = 50\underline{/53°}\ \Omega$$

$$= 30 + j40\ \Omega = R + jX_L.$$

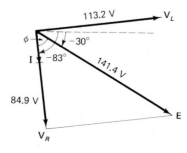

Figure 12.14

Therefore,
$$R = 30 \, \Omega$$
and
$$X_L = 40 \, \Omega = \omega L = 1000L$$
Therefore,
$$L = 40 \, \text{mH}$$

The average power dissipated is
$$EI \cos \phi = 141.4 \times 2.828 \cos 53°$$
$$= 240 \, \text{W}$$
Also,
$$EI \cos \phi = I^2 R = (2.828)^2 \times 30 = 240 \, \text{W}$$

(b) To find $\mathbf{V}_R$ and $\mathbf{V}_L$, Ohm's law is used for each element:
$$\mathbf{V}_R = \mathbf{I}Z_R = 2.828\underline{/-83°} \quad 30\underline{/0°} = 84.9\underline{/-83°} \, \text{V}$$
$$\mathbf{V}_L = \mathbf{I}Z_L = 2.828\underline{/-83°} \quad 40\underline{/90°} = 113.2\underline{/7°} \, \text{V}$$

Thus the voltmeter would read 84.9 V across R and 113.2 V across L.

These two phasors are also drawn in Fig. 12.14. Notice that $\mathbf{V}_R$ is in phase with $\mathbf{I}$, while $\mathbf{V}_L$ leads $\mathbf{I}$ by 90°, as they should. Also, the projection of $\mathbf{E}$ along the $\mathbf{I}$ phasor ($E \cos \phi$) results in V_R and the projection of $\mathbf{E}$ perpendicular to $\mathbf{I}$ ($E \sin \phi$) results in V_L.

(c) The $\mathbf{V}_R$ and $\mathbf{V}_L$ phasors can now be converted into the time domain ($\omega = 1000 \, \text{rad/s}$):
$$v_R = 84.9\sqrt{2} \sin (1000t - 83°) = 120 \sin (1000t - 83°) \, \text{V}$$
$$v_L = 113.2\sqrt{2} \sin (1000t + 7°) = 160 \sin (1000t + 7°) \, \text{V}$$

EXAMPLE 12.6

A given RL series circuit has an applied source voltage of $100\underline{/0°} \, \text{V}$ and the resulting current is $(10 - j10) \, \text{A}$ at a frequency of 1 kHz.
(a) Find the values of R and L.
(b) Find the reading of the ammeter in the circuit and the reading of a voltmeter connected across L.
(c) Write the expression for $e(t)$ if the time reference is shifted such that at $t = 0 \, \text{s}, i = 0 \, \text{A}$.

Solution (a) Since $\mathbf{E} = 100\underline{/0°} \, \text{V}$ and $\mathbf{I} = 10 - j10 = 14.14\underline{/-45°} \, \text{A}$, then from Ohm's law for the circuit,

$$Z_T = \frac{\mathbf{E}}{\mathbf{I}} = \frac{100\underline{/0°}}{14.13\underline{/-45°}} = 7.07\underline{/45°} \, \Omega$$
$$= 5 + j5 \, \Omega = R + jX_L$$

Thus
$$R = 5 \, \Omega$$
and
$$X_L = 2\pi \times 1000 \times L = 5 \, \Omega$$
Therefore,
$$L = 0.796 \, \text{mH}$$

(b) The ammeter reading is $I = 14.14$ A. The voltmeter connected across L will read the value of V_L, that is, $V_L = IX_L = 14.14 \times 5 = 70.7$ V.

(c) The phasor diagram in Fig. 12.15(a) shows $\mathbf{E}$ and $\mathbf{I}$ as given above. If $\mathbf{I}$ is rotated to be the reference phasor as required here, $\mathbf{E}$ will still lead $\mathbf{I}$ by $45°$, and the phasor diagram will be as shown in Fig. 12.15(b). Therefore,

$$\mathbf{E} = 100\underline{/45°} \text{ V}$$

and with $\omega = 2\pi f = 6280$ rad/s,

$$e(t) = 141.4 \sin{(6280t + 45°)} \text{ V}$$

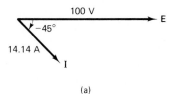

(a)

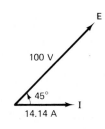

(b)

Figure 12.15

EXAMPLE 12.7

The oscilloscope inputs were connected to the circuit in Fig. 12.16(a) as indicated. The oscillogram obtained is shown in Fig. 12.16(b), with the scope controls adjusted as follows:

$$\text{time/div: } 0.1 \text{ ms, } \times \text{ mag.: } \times 1$$
$$\text{V/div } (A): 0.05, \text{ V/div } (B): 10$$

The black box contains two series-connected elements. Find the value of each of these elements.

Solution The Y_B waveform is $e(t)$; therefore,

$$E_m = 4 \times 10 = 40 \text{ V} \qquad \text{and} \qquad E = \frac{40}{\sqrt{2}} = 28.28 \text{ V}$$

The Y_A waveform shows the voltage $0.5 \times i$; thus

$$I_m = \frac{2 \times 0.05}{0.5} = 0.2 \text{ A} \qquad \text{and} \qquad I = 0.1414 \text{ A}$$

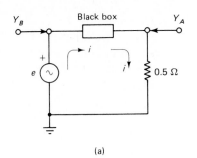

(a)

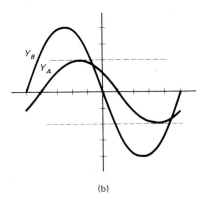

(b)

Figure 12.16

Also, $Y_B(=e)$ leads $Y_A(\propto i)$ by 1 division, that is,

$$\phi = \frac{360°}{10 \text{ div}} \times 1 \text{ div} = 36°$$

$$T(\text{the period}) = 10 \times 0.1 \text{ ms}$$
$$= 1 \text{ ms}$$

Therefore,

$$\omega = 2\pi f = \frac{2\pi}{T} = 6280 \text{ rad/s}$$

If the current phasor is taken as the reference, then

$$\mathbf{I} = 0.1414\underline{/0°} \text{ A}$$

and

$$\mathbf{E} = 28.28\underline{/36°} \text{ V}$$

The phasor diagram is thus as shown in Fig. 12.17. Hence

$$Z_T = \frac{\mathbf{E}}{\mathbf{I}} = \frac{28.28\underline{/36°}}{0.1414\underline{/0°}} = 200\underline{/36°} = 161.8 + j117.6 \, \Omega$$

$$= Z_{\text{box}} + 0.5$$

Therefore,

$$Z_{\text{box}} = 161.3 + j117.6 \, \Omega$$
$$= R + jX_L$$

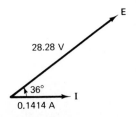

Figure 12.17

That is, the elements inside the box are a resistor and an inductor in series,

where

$$R = 161.3 \, \Omega$$
$$X_L = 117.6 \, \Omega = \omega L = 6280L$$
$$L = 18.7 \, \text{mH}$$

and

12.2 SERIES *RC* LOADS

The frequency-domain equivalent circuit for the *RC* series connected load is shown in Fig. 12.18, where

$$X_C = \frac{1}{\omega C} \, \Omega$$

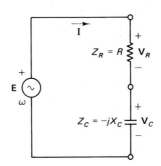

Figure 12.18

Here, as in any series-connected circuit, the current **I** is the same in all components. Ohm's law and KVL provide

$$\mathbf{V}_R = \mathbf{I}Z_R \tag{12.25}$$
$$\mathbf{V}_C = \mathbf{I}Z_C \tag{12.26}$$

and

$$\mathbf{E} = \mathbf{V}_R + \mathbf{V}_C \tag{12.27}$$

The equations above apply in general. If, for simplicity, the current waveform (and phasor) is taken to be the reference, that is,

$$i(t) = I_m \sin(\omega t) \, \text{A} \Leftrightarrow \mathbf{I} = I\underline{/0^\circ} \, \text{A} \tag{12.28}$$

then

$$\mathbf{V}_R = \mathbf{I}Z_R = I\underline{/0^\circ} \, R\underline{/0^\circ} = IR\underline{/0^\circ} \, \text{V} = V_R\underline{/0^\circ} \, \text{V} \tag{12.29}$$

and

$$\mathbf{V}_C = \mathbf{I}Z_C = I\underline{/0^\circ} \, X_C\underline{/-90^\circ} = IX_C\underline{/-90^\circ} = V_C\underline{/-90^\circ} \, \text{V} \tag{12.30}$$

Thus from KVL,

$$\mathbf{E} = \mathbf{V}_R + \mathbf{V}_C = V_R - jV_C = \sqrt{V_R^2 + V_C^2}\underline{\bigg/-\tan^{-1}\frac{V_C}{V_R}} \, \text{V}$$

$$= IR - jIX_C = I\sqrt{R^2 + X_C^2}\underline{\bigg/-\tan^{-1}\frac{X_C}{R}} \, \text{V}$$

$$= E\underline{/\phi^\circ} \, \text{V} \tag{12.31}$$

where

$$E = \sqrt{V_R^2 + V_C^2} = I\sqrt{R^2 + X_C^2} \qquad (12.31a)$$

and

$$\phi° = -\tan^{-1}\frac{V_C}{V_R} = -\tan^{-1}\frac{X_C}{R} = -\tan^{-1}\frac{1}{\omega CR} \qquad (12.31b)$$

Note that ϕ is a negative angle, being a function of the element values R and C, and also the frequency ω (or f). The corresponding phasor diagram is shown in Fig. 12.19.

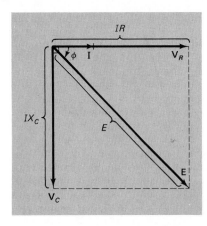

Figure 12.19

Once the phasor **E** is obtained as in Eq. (12.31), then

$$e(t) = \sqrt{2}\, E \sin(\omega t + \phi°) \quad \text{V}$$

As ϕ is a negative angle, $e(t)$ lags $i(t)$ in this type of circuit. This is also clear from the phasor diagram, being exactly the opposite to the RL series load case.

Examination of the phasor diagram reveals that the voltage across R, $\mathbf{V}_R$, is always in phase with the current **I**; the voltage across C, $\mathbf{V}_C$, always lags **I** by 90°, while the total source voltage **E** lags the current **I** by an angle ϕ. The general type of phasor diagram shown in Fig. 12.20 confirms these phase relationships. Depending on the elements values and the frequency, the negative angle ϕ varies

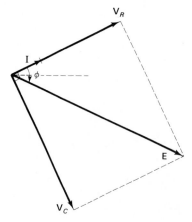

Figure 12.20

in the range

$$\text{(purely capacitive load} \leftarrow -90° \leq \phi \leq 0° \rightarrow \text{(purely resitive load} \quad (12.32)$$
$$\text{or } \omega = 0 \text{ rad/s)} \qquad\qquad\qquad \text{or } \omega = \infty \text{ rad/s)}$$

The impedance triangle. From KVL for the series RC circuit of Fig. 12.18,

$$\mathbf{E} = \mathbf{V}_R + \mathbf{V}_C$$

and substituting from the Ohm's law relationships of Eqs. (12.25) and (12.26), we have

$$\mathbf{E} = \mathbf{I}Z_R + \mathbf{I}Z_C = \mathbf{I}(Z_R + Z_C)$$

Therefore, the total impedance of the circuit is

$$Z_T = \frac{\mathbf{E}}{\mathbf{I}}$$
$$= Z_R + Z_C$$
$$= R - jX_C = \sqrt{R^2 + X_C^2}\,\bigg/\!-\tan^{-1}\frac{X_C}{R}\;\;\Omega \qquad (12.33a)$$

$$(12.33)$$

and

$$Z_T = |Z_T|\,\underline{/\phi°}\;\Omega$$
$$= |Z_T|\cos\phi + j\,|Z_T|\sin\phi \qquad (12.33b)$$

where

$$|Z_T| = \sqrt{R^2 + X_C^2} = \sqrt{R^2 + \left(\frac{1}{\omega C}\right)^2}\;\;\Omega \qquad (12.34a)$$

$$\phi° = -\tan^{-1}\frac{X_C}{R} = -\tan^{-1}\frac{1}{\omega CR} \qquad (12.34b)$$

$$R = |Z_T|\cos\phi\;\;\Omega \quad\text{and}\quad X_C = -|Z_T|\sin\phi\;\;\Omega \qquad (12.34c)$$

The total impedance Z_T is a vector that can be represented graphically by the impedance triangle shown in Fig. 12.21, in accordance with Eqs. (12.33) and (12.34). The real-axis component of Z_T is the resistance of the circuit, while the imaginary-axis component of Z_T is the capacitive reactance of the circuit. The phase angle of Z_T in the RC series circuit is the negative angle ϕ. It is the same angle by which $\mathbf{E}$ lags $\mathbf{I}$, that is, it is the phase shift between the two phasors $\mathbf{E}$ and $\mathbf{I}$, as Ohm's law for the circuit [Eq. (12.33)] indicates.

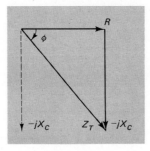

Figure 12.21

Using Eq. (12.33), the voltages across R and C can be written as

$$\left.\begin{array}{l} \mathbf{V}_R = \mathbf{I}Z_R = \dfrac{\mathbf{E}}{Z_T} Z_R = \mathbf{E}\dfrac{Z_R}{Z_R + Z_C} \\[4mm] \mathbf{V}_C = \mathbf{I}Z_C = \dfrac{\mathbf{E}}{Z_T} Z_C = \mathbf{E}\dfrac{Z_C}{Z_R + Z_C} \end{array}\right\} \text{voltage-division principle}$$

$$(12.35)$$

$$(12.36)$$

Power in RC loads. Following a procedure similar to that outlined for the *RL* loads, it can easily be shown that the average power dissipated in the *RC* loads, P_{av}, is also given by the same expression obtained in Eq. (12.20), that is,

$$P_{av} = VI \cos \phi \quad W$$

Even though ϕ is a negative angle in this case,

$$\cos(-\phi) = \cos\phi$$
$$= F_p = \text{power factor}$$

Here, from the phasor diagram of Fig. 12.22,

$$V \cos\phi = V_R = IR$$

Therefore,

$$P_{av} = I^2 R \quad W$$

This is similar to the result obtained in Eq. (12.22), indicating that while the capacitor stores energy, only the resistive component is responsible for dissipating all the power in these loads.

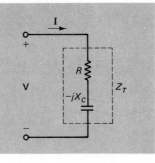

(a)

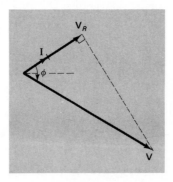

(b) **Figure 12.22**

EXAMPLE 12.8

The current in the series RC circuit shown in Fig. 12.23 is $i(t) = 2 \cos 5000t$ A. Find $e(t)$ and the power dissipated in this circuit. Write the phasors for $\mathbf{V}_R$ and $\mathbf{V}_C$ and draw the phasor diagram.

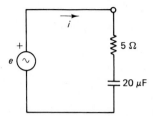

Figure 12.23

Solution Here

$$i(t) = 2 \cos 5000t$$
$$= 2 \sin (5000t + 90°) \text{ A}$$

Therefore,

$$\mathbf{I} = \frac{2}{\sqrt{2}} \underline{/90°} = 1.414 \underline{/90°} \text{ A}$$

and $\omega = 5000$ rad/s. The frequency-domain equivalent circuit is shown in Fig. 12.24(a), with

$$Z_C = -jX_C = -j\frac{1}{\omega C}$$

$$= -j\frac{1}{5000 \times 20 \times 10^{-6}} = -j10 \quad \Omega$$

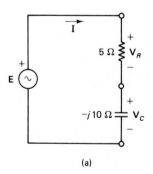

(a)

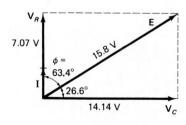

(b)

Figure 12.24

Thus the total impedance of the circuit is

$$Z_T = Z_R + Z_C = R - jX_C$$
$$= 5 - j10 = 11.2\underline{/-63.4°}\,\Omega$$

Now, applying Ohm's law to the circuit as a whole and to each component separately, we have

$$\mathbf{E} = \mathbf{I}Z_T = 1.414\underline{/90°} \times 11.2\underline{/-63.4°} = 15.8\underline{/26.6°}\,\mathbf{V}$$

Therefore,

$$e(t) = 15.8\sqrt{2}\sin(5000t + 26.6°)\,\mathrm{V}$$
$$= 22.4\sin(5000t + 26.6°)\,\mathrm{V}$$

and

$$V_R = \mathbf{I}Z_R = 1.414\underline{/90°} \times 5\underline{/0°} = 7.07\underline{/90°}\,\mathrm{V}$$
$$\mathbf{V}_C = \mathbf{I}Z_C = 1.414\underline{/90°} \times 10\underline{/-90°} = 14.14\underline{/0°}\,\mathrm{V}$$

The phasor diagram is shown in Fig. 12.24(b). Notice that **E** lags **I** (as it should) by the angle $\phi(=63.4°)$. Thus

$$P_{av} = EI\cos\phi = 15.8 \times 1.414\cos 63.4° = 10\,\mathrm{W}$$
$$= I^2R = (1.414)^2 \times 5 = 10\,\mathrm{W}\quad(\text{check!})$$

EXAMPLE 12.9

In a series RC circuit, the voltage across the resistor is 15 V and the voltage across the capacitor is 10 V. The current was measured to be 5 A. The frequency of the source was 10,000 rad/s. Find the values of R and C. Taking **I** as the reference phasor, find **E**.

Solution With **I** as the reference phasor, the phasor diagram is drawn as shown in Fig. 12.25. Here

$$V_R = IR$$

Then

$$R = \frac{V_R}{I} = \frac{15}{5} = 3\,\Omega$$

and similarly,

$$V_C = IX_C$$

Therefore,

$$X_C = \frac{V_C}{I} = \frac{10}{5} = 2\,\Omega = \frac{1}{\omega C} = \frac{1}{10,000C}$$

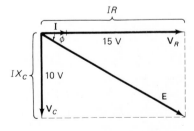

Figure 12.25

and

$$C = 50 \ \mu\text{F}$$

As

$$\mathbf{I} = 5\underline{/0°} \text{ A}$$

and

$$Z_T = R - jX_C = 3 - j2 = 3.6\underline{/-33.7°} \ \Omega$$

then

$$\mathbf{E} = \mathbf{I}Z_T = 5\underline{/0°} \times 3.6\underline{/-33.7°} = 18\underline{/-33.7°} \text{ V}$$

The result above can also be arrived at from the geometry of the phasor diagram in Fig. 12.25:

$$\mathbf{E} = \sqrt{V_R^2 + V_C^2}\underline{/-\tan^{-1}\frac{V_C}{V_R}} = \sqrt{(15)^2 + (10)^2}\underline{/-\tan^{-1}0.667}$$

$$= 18\underline{/-33.7°} \text{ V}$$

EXAMPLE 12.10

In the RC circuit shown in Fig. 12.26, $e(t) = 70.7 \sin 2000t$ V. Find the value of C that limits the power dissipation to 150 W. Write the expression for $i(t)$.

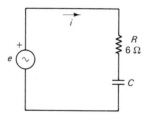

Figure 12.26

Solution   Since P_{av} and the value of R are given, then using $P_{av} = I^2R$,

$$I = \sqrt{\frac{P_{av}}{R}} = \sqrt{\frac{150}{6}} = 5 \text{ A} \qquad \text{and} \qquad \mathbf{E} = 50\underline{/0°} \text{ V}$$

To determine $\mathbf{I}$ completely, its phase shift ϕ with respect to the phasor $\mathbf{E}$ should be obtained. From the expression for the average power dissipated, we have

$$P_{av} = EI \cos \phi$$

Therefore,

$$\cos \phi = \frac{P_{av}}{EI} = \frac{150}{50 \times 5} = 0.6$$

$$\phi = 53.1°$$

In RC circuits $\mathbf{I}$ leads $\mathbf{E}$; thus

$$\mathbf{I} = 5\underline{/53.1°} \text{ A}$$

and

$$i(t) = 5\sqrt{2} \sin (2000t + 53.1°)$$

$$= 7.07 \sin (2000t + 53.1°) \text{ A}$$

Also, the total impedance of the circuit can be determined:

$$Z_T = \frac{\mathbf{E}}{\mathbf{I}} = \frac{50\underline{/0°}}{5\underline{/53.1°}} = 10\underline{/-53.1°} \; \Omega$$

$$= 6 - j8 \; \Omega = R - jX_C$$

Therefore,

$$X_C = 8 \; \Omega = \frac{1}{\omega C} = \frac{1}{2000C}$$

$$C = \frac{1}{8 \times 2000} = 62.5 \; \mu F$$

EXAMPLE 12.11

The oscilloscope inputs were connected to the network shown in Fig. 12.27, as indicated. Examination of the oscillogram thus obtained revealed that $T = 3.14$ ms, V_{pp} (beam A) = 60 V, and V_{pp} (beam B) = 40 V. Also, the Y_A waveform was lagging the Y_B waveform.
(a) What are the values of R and the single element X?
(b) Write the time-domain expression for the source voltage $e(t)$.

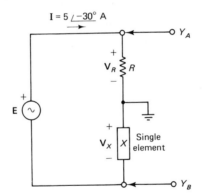

Figure 12.27

Solution

(a) $\omega = \dfrac{2\pi}{T} = \dfrac{6.28}{3.14 \times 10^{-3}} = 2000$ rad/s. Y_A represents v_R; therefore,

$$V_R = \frac{60}{2} \times 0.707 = 21.21 \; V$$

Y_B represents $-v_X$ (notice the polarities as indicated in the circuit diagram); therefore,

$$V_X = \frac{40}{2} \times 0.707 = 14.14 \; V$$

In the phasor diagram shown in Fig. 12.28 $\mathbf{V}_R$ (Y_A) is in phase with $\mathbf{I}$. The lag angle between $Y_A (= \mathbf{V}_R$, proportional to the current flowing in the element) and $Y_B (= -\mathbf{V}_X$, the voltage across the element) can only be 90° because X is a single element. Thus $-\mathbf{V}_X$ (Y_B) is then drawn leading $\mathbf{V}_R$ (Y_A) by 90°, and hence $\mathbf{V}_X$, drawn 180° shifted from $-\mathbf{V}_X$ (i.e., in the opposite direction), lags $\mathbf{I}$

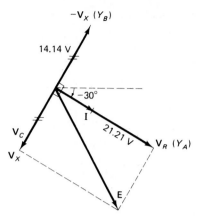

Figure 12.28

by 90°. Thus the element must be a capacitor because the voltage across it ($\mathbf{V}_C = \mathbf{V}_X$) lags the current flowing in it by 90°. Here

$$R = \frac{V_R}{I} = \frac{21.21}{5} = 4.24 \ \Omega$$

and

$$X_C = \frac{V_C}{I} = \frac{14.14}{5} = 2.828 \ \Omega = \frac{1}{\omega C} = \frac{1}{2000C}$$

Therefore,

$$C = 176.8 \ \mu\text{F}$$

(b) The total impedance of the circuit is

$$Z_T = R - jX_C = 4.24 - j2.828 = 5.1\underline{/-33.7°} \ \Omega$$

Using Ohm's law gives us

$$\mathbf{E} = \mathbf{I}Z_T = 5\underline{/-30°} \times 5.1\underline{/-33.7°} = 25.5\underline{/-63.7°} \ \text{V}$$

$\mathbf{E}$ lags $\mathbf{I}$ by 33.7° (i.e., the angle of Z_T); therefore,

$$e(t) = 25.5\sqrt{2} \sin(2000t - 63.7°)$$
$$= 36.04 \sin(2000t - 63.7°) \ \text{V}$$

EXAMPLE 12.12

The oscillogram in Fig. 12.29(a) was obtained during the analysis of the circuit shown in Fig. 12.29(b), with the oscilloscope controls adjusted as follows:

$$\text{time/div} = 0.1 \ \text{ms}, \quad \times \text{mag.:} \quad \times 5$$
$$\text{V/div (beam } A) = \text{unknown}$$
$$\text{V/div (beam } B) = 2 \ \text{V}$$

(a) With the help of the phasor diagram for this circuit, label the two signals with Y_A and Y_B, as appropriate.
(b) What is the phase angle between the applied voltage and the resulting current?

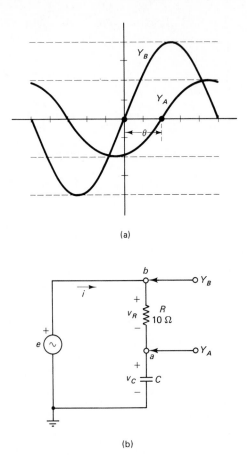

(a)

(b)

Figure 12.29

(c) What is the V/div setting of beam A?

(d) Find the values of V_R, I, and C.

Solution (a) From the circuit diagram connections, it can be concluded that Y_B displays the voltage from point b to ground [i.e., it displays $e(t)$] and Y_A displays the voltage from point a to ground [i.e., it displays $v_C(t)$]. Considering that $i(t)$ (or $\mathbf{I}$) is the reference waveform, a typical sketch for the phasor diagram of this circuit is shown in Fig. 12.30, with $\mathbf{E}$ ($= Y_B$) and $\mathbf{V}_C$ ($= Y_A$) indicated. From the phasor diagram it is clear that Y_B ($\mathbf{E}$) leads Y_A ($\mathbf{V}_C$) by the angle θ. The leading waveform (considering the circled starting points) is thus labeled Y_B in the oscillogram of Fig. 12.29(a).

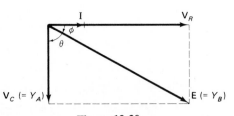

Figure 12.30

Chapter 12   Steady-State Analysis of Simple AC Circuits

(b) From the oscillogram, the phase shift between the two waveforms, θ, is

$$\theta \Rightarrow 2 \text{ div} = \frac{360°}{10} \times 2 = 72°$$

Therefore, the phase shift between **I** and **E**, from the phasor diagram in Fig. 12.30, is

$$\phi = 90° - 72° = 18° \quad \text{with } \mathbf{E} \text{ lagging } \mathbf{I}$$

(c) The voltage monitored by Y_B (i.e., e) can be determined from the oscillogram. Here

$$E = 0.707 \times 4 \text{ div} \times 2 \text{ V/div} = 5.656 \text{ V}$$

From the geometry of the phasor diagram,

$$V_R = E \cos \phi = 5.656 \cos 18° = 5.38 \text{ V}$$

and

$$V_C = E \sin \phi = 5.656 \sin 18° = 1.75 \text{ V}$$
$$V_{C\text{max}} = \sqrt{2} \, V_C = 2.47 \text{ V}$$

Since v_C is monitored by Y_A, $V_{C\text{max}}$ then corresponds to 2 divisions; thus

$$\text{V/div} \,(Y_A) = \frac{2.47}{2} = 1.235 \text{ V/div}$$

[i.e., 1 V/div (uncalibrated)].

(d) From part (c),

$$V_R = 5.38 \text{ V} = IR = 10I$$

Therefore,

$$I = 0.538 \text{ A}$$

Also from the oscillogram,

$$T = 10 \text{ div} \times \frac{0.1 \text{ m}}{5} = 0.2 \text{ ms}$$

The division by 5 (above) is necessary here because the time (or X) axis is magnified 5 times; therefore,

$$\omega = \frac{2\pi}{T} = \frac{6.28}{0.2 \times 10^{-3}} = 31.4 \text{ krad/s}$$

$$X_C = \frac{V_C}{I} = \frac{1.75}{0.538} = 3.25 \, \Omega = \frac{1}{\omega C}$$

and

$$C = \frac{1}{3.25 \times 31.4 \times 10^3} = 9.8 \, \mu\text{F}$$

EXAMPLE 12.13

The oscillogram in Fig. 12.31(a) is obtained while analyzing the circuit shown in Fig. 12.31(b). The black box contains two elements in series. The

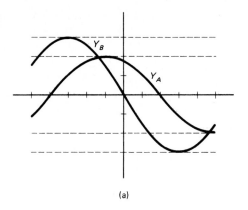

(a)

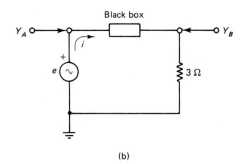

(b)

Figure 12.31

oscilloscope controls were adjusted as follows:

$$\text{time/div} = 5\,\text{ms}$$
$$\text{V/div}\,(Y_A) = 50\,\text{V}$$
$$\text{V/div}\,(Y_B) = 0.02\,\text{V}$$

(a) Find the values of the two elements.
(b) Express $e(t)$ if at $t = 0\,\text{s}$, $i = 0\,\text{A}$

Solution   (a) Y_B represents the voltage across the $3\,\Omega$ resistor as the circuit diagram connections indicate (i.e., Y_B corresponds to $3 \times i$ volts); therefore,

$$I = \tfrac{1}{3} \times 0.707 \times 3\,\text{div} \times 0.02\,\text{V/div} = 0.01414\,\text{A}$$

Y_A represents the source's voltage waveform, $e(t)$; therefore,

$$E = 0.707 \times 2\,\text{div} \times 50\,\text{V/div} = 70.7\,\text{V}$$

As $Y_A(=e)$ lags $Y_B(=3i$, directly proportional to the current) by the angle ϕ, with

$$\phi = 2\,\text{div} \times \frac{180°}{6\,\text{div}} = 60°$$

the circuit must then be an *RC* circuit. If I is taken to be the reference phasor,

$$\mathbf{I} = 0.01414\underline{/0°}\text{ A}$$

then

$$\mathbf{E} = 70.7\underline{/-60°}\text{ V}$$

The total impedance of the circuit is

$$Z_T = \frac{\mathbf{E}}{\mathbf{I}} = \frac{70.7\underline{/-60°}}{0.01414\underline{/0°}} = 5000\underline{/-60°}\text{ }\Omega$$

$$= 2500 - j4330\text{ }\Omega$$

$$= Z_{\text{box}} + 3\text{ }\Omega \quad \text{(from the circuit diagram)}$$

Thus

$$Z_{\text{box}} = 2497 - j4330 = R - jX_C$$

and

$$R = 2497\text{ }\Omega$$

$$X_C = 4330\text{ }\Omega = \frac{1}{\omega C}$$

But

$$\omega = \frac{2\pi}{T} = \frac{6.28}{2 \times 6\text{ div.} \times 5 \times 10^{-3}\text{ sec/div.}} = 104.7\text{ rad/s}$$

Therefore,

$$C = \frac{1}{4330 \times 104.7} = 2.2\text{ }\mu\text{F}$$

(b) Since $i = 0$ A at $t = 0$ s [i.e., $i(t)$ is the reference waveform], then

$$i(t) = 0.01414\sqrt{2}\sin(104.7t)$$

$$= 0.02\sin(104.7t)\text{ A}$$

E was obtained above, lagging **I** by 60°; then

$$e(t) = 70.7\sqrt{2}\sin(104.7t - 60°)$$

$$= 100\sin(104.7t - 60°)\text{ V}$$

12.3 SERIES *RLC* CIRCUITS

The frequency-domain equivalent circuit for the series *RLC* load is shown in Fig. 12.32. The current **I** is the same in all such series-connected elements. Applying Ohm's law to each component yields

$$\mathbf{V}_R = \mathbf{I}Z_R \tag{12.37}$$

$$\mathbf{V}_L = \mathbf{I}Z_L \tag{12.38}$$

and

$$\mathbf{V}_C = \mathbf{I}Z_C \tag{12.39}$$

Let $\mathbf{V}_X$ be the voltage across both *L* and *C* (i.e., across the total reactance of the circuit),

$$\mathbf{V}_X = \mathbf{V}_L + \mathbf{V}_C \tag{12.40}$$

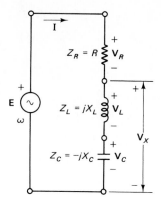

Figure 12.32

From KVL for this circuit,

$$\boxed{\mathbf{E} = \mathbf{V}_R + \mathbf{V}_L + \mathbf{V}_C = \mathbf{V}_R + \mathbf{V}_X} \tag{12.41}$$

The relations above apply in general. For simplicity, again let $\mathbf{I}$ be considered as the reference phasor, that is,

$$\mathbf{I} = I\underline{/0°}\ \text{A} \Leftrightarrow i(t) = \sqrt{2}\,I \sin(\omega t)\ \text{A} \tag{12.42}$$

Then

$$\mathbf{V}_R = I\underline{/0°} \times R\underline{/0°} = IR\underline{/0°}\ \text{V} = V_R\underline{/0°}\ \text{V}$$
$$\mathbf{V}_L = I\underline{/0°} \times X_L\underline{/90°} = IX_L\underline{/90°}\ \text{V} = V_L\underline{/90°}\ \text{V}$$
$$\mathbf{V}_C = I\underline{/0°} \times X_C\underline{/-90°} = IX_C\underline{/-90°}\ \text{V} = V_C\underline{/-90°}\ \text{V}$$

Therefore,

$$\begin{aligned}
\mathbf{E} &= V_R\underline{/0°} + V_L\underline{/90°} + V_C\underline{/-90°} \\
&= V_R + j(V_L - V_C) \\
&= V_R + jV_X \\
&= \sqrt{V_R^2 + V_X^2}\ \underline{/\tan^{-1}\dfrac{V_X}{V_R}} = E\underline{/\phi°}\ \text{V}
\end{aligned} \tag{12.43}$$

where

$$\begin{aligned}
\mathbf{V}_X &= \mathbf{V}_L + \mathbf{V}_C \quad \text{(vectorial sum)} \\
&= j(V_L - V_C) \\
&= jV_X
\end{aligned} \tag{12.44}$$

The phasors above are shown graphically in the phasor diagram for such circuits, as in Fig. 12.33.

In terms of the element values and the current, Eq. (12.43) can be written in the form

$$\begin{aligned}
\mathbf{E} &= IR + jI(X_L - X_C) \\
&= I\sqrt{R^2 + (X_L - X_C)^2}\ \underline{/\tan^{-1}\dfrac{X_L - X_C}{R}}\ \text{V} \\
&= E\underline{/\phi°}\ \text{V}
\end{aligned} \tag{12.45}$$

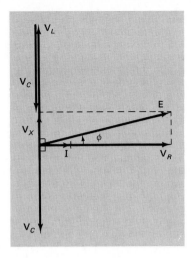

Figure 12.33

where

$$E = \sqrt{V_R^2 + V_X^2} = I\sqrt{R^2 + (X_L - X_C)^2}$$

$$= I\sqrt{R^2 + \left(\omega L - \frac{1}{\omega C}\right)^2} \, V \qquad (12.45a)$$

and

$$\phi = \tan^{-1}\frac{V_X}{V_R} = \tan^{-1}\frac{X_L - X_C}{R} = \tan^{-1}\frac{\omega L - 1/\omega C}{R} \qquad (12.45b)$$

Once **E** is determined, then

$$e(t) = 2\sqrt{E}\sin(\omega t + \phi°) \quad V \qquad (12.46)$$

As the phasor diagram indicates, the voltage across R, $\mathbf{V}_R$, is always in phase with **I**, while the voltage across L, $\mathbf{V}_L$, always leads **I** by 90° and the voltage across C, $\mathbf{V}_C$, always lags **I** by 90°. Thus the voltage $\mathbf{V}_X$ across the combined reactances of L and C, being the vectorial algebraic sum of $\mathbf{V}_L$ and $\mathbf{V}_C$, will always be leading or lagging **I** by 90°. Note that $\mathbf{V}_L$ and $\mathbf{V}_C$ are always opposite to each other, as they are 180° out of phase with each other.

If

$$X_L > X_C \qquad \text{that is, } \omega L > \frac{1}{\omega C}$$

then

$$IX_L > IX_C \qquad \text{that is, } V_L > V_C$$

Hence $\mathbf{V}_X$ leads **I** by 90°. Here **E** leads **I** by the angle ϕ (positive angle) and the circuit is said to be *predominantly inductive*.

If

$$X_L < X_C \qquad \text{that is, } \omega L < \frac{1}{\omega C}$$

then

$$IX_L < IX_C \qquad \text{that is, } V_L < V_C$$

Hence $\mathbf{V}_X$ lags **I** by 90°. Here **E** lags **I** by the angle ϕ (negative angle) and the circuit is said to be *predominantly capacitive*.

These phase relationships apply in general, with any phasor (or none) being considered as the reference. The phasor diagram can be thought of as being rotated (as a whole) to accommodate whichever phasor is used as the reference.

The case when

$$X_L = X_C \qquad \text{that is, } \omega L = \frac{1}{\omega C} \qquad\qquad (12.47)$$

corresponds to

$$IX_L = IX_C \qquad \text{that is, } V_L = V_C$$

For this case

$$\mathbf{V}_X = j(V_L - V_C) = 0 \qquad\qquad (12.48)$$

and

$$\mathbf{E} = \mathbf{V}_R \qquad\qquad (12.49)$$

that is, here $\mathbf{E}$ and $\mathbf{I}$ are in phase. This is called the *resonance condition* of the series *RLC* circuit. The frequency at which this condition [Eq. (12.47)] is satisfied is called the *resonance frequency*, ω_0, where

$$\boxed{\omega_0 = \frac{1}{\sqrt{LC}} \qquad \text{rad/s}} \qquad\qquad (12.47a)$$

or

$$\boxed{f_0 = \frac{\omega_0}{2\pi} = \frac{1}{2\pi\sqrt{LC}} \qquad \text{Hz}} \qquad\qquad (12.47b)$$

Resonance is treated in more detail in Chapter 14.

The impedance triangle. From KVL,

$$\mathbf{E} = \mathbf{I}(Z_R + Z_L + Z_C)$$

The total impedance of the series *RLC* circuit is

$$\boxed{\begin{aligned} Z_T &= \frac{\mathbf{E}}{\mathbf{I}} \\[4pt] &= Z_R + Z_L + Z_C \\[4pt] &= R + j(X_L - X_C) = R + j\left(\omega L - \frac{1}{\omega C}\right) \\[4pt] &= \sqrt{R^2 + (X_L - X_C)^2} \bigg/ \tan^{-1}\frac{X_L - X_C}{R} = |Z_T|\,\underline{/\phi^\circ}\ \Omega \end{aligned}} \qquad (12.50)$$

$$(12.51)$$

where

$$|Z_T| = \sqrt{R^2 + (X_L - X_C)^2} = \sqrt{R_2 + \left(\omega L - \frac{1}{\omega C}\right)^2}\ \Omega \qquad (12.51a)$$

and

$$\phi = \tan^{-1}\frac{X_L - X_C}{R} = \tan^{-1}\frac{\omega L - 1/\omega C}{R} \qquad\qquad (12.51b)$$

Both the magnitude and phase of Z_T are functions not only of the element values but also of the frequency. The total impedance Z_T is a vector and can be represented graphically by the impedance triangle shown in Fig. 12.34, satisfying Eqs. (12.51).

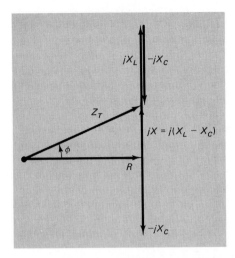

Figure 12.34

According to Ohm's law for the whole circuit [Eq. (12.50)], the phase angle of Z_T(i.e., ϕ) is the same phase shift between the phasors **E** and **I**. For predominantly inductive circuits ($X_L > X_C$) the angle ϕ is positive, while for predominantly capacitive circuits ($X_C > X_L$) the angle ϕ is negative and the total reactance $X(=X_L - X_C)$ is also negative. At resonance, $X_L = X_C$; that is, X is zero and Z_T will be a purely real number equal to R, the resistance of the circuit.

Substituting from Ohm's law for the entire circuit [Eq. (12.50)] into Ohm's law for each component [Eqs. (12.37), (12.38), and (12.39)], we obtain

$$\mathbf{V}_R = \mathbf{I}Z_R = \frac{\mathbf{E}}{Z_T} Z_R = \mathbf{E}\frac{Z_R}{Z_R + Z_L + Z_C} \tag{12.52}$$

$$\mathbf{V}_L = \mathbf{I}Z_L = \frac{\mathbf{E}}{Z_T} Z_L = \mathbf{E}\frac{Z_L}{Z_R + Z_L + Z_C} \quad \begin{matrix}\text{voltage-division}\\ \text{principle}\end{matrix} \tag{12.53}$$

$$\mathbf{V}_C = \mathbf{I}Z_C = \frac{\mathbf{E}}{Z_T} Z_C = \mathbf{E}\frac{Z_C}{Z_R + Z_L + Z_C} \tag{12.54}$$

Since the *RLC* circuit behaves as either predominantly inductive, capactive, or purely resistive, the power dissipated in this load (Z_T) is also obtained as before, using

$$P_{av} = EI \cos \phi \quad \text{W} \quad (\phi \text{ is the phase shift between } \mathbf{E} \text{ and } \mathbf{I})$$

From the phasor diagram in Fig. 12.33 it is clear that

$$E \cos \phi = V_R = IR$$

Therefore,

$$P_{av} = I^2R \quad \text{W}$$

Thus, as always, only the resistive element is responsible for dissipating all of the circuit's power.

Note. In any of the series circuits discussed above, *RL, RC,* or *RLC*, an extra reactive component (*L* or *C*) can be added to the circuit in series without affecting the magnitude of the total impedance $|Z_T|$, and hence the magnitude of the current $I(= E/|Z_T|)$ remains unchanged. This remark is true because the two complex-conjugate numbers $R + jX$ and $R - jX$ have the same magnitude,

$$|Z_T| = \sqrt{R^2 + X^2} \quad \Omega$$

Thus if

$$Z_T = R + jX \quad \Omega$$

then

$$Z_T' = Z_T + Z_{added} = R - jX$$

(i.e., both Z_T and Z_T' have the same magnitude); therefore,

$$Z_{added} = R - jX - Z_T = R - jX - (R + jX)$$
$$= -j2X \quad \Omega \tag{12.55}$$

A similar procedure applies if the initial total impedance Z_T is $R - jX$ ohms.

■ EXAMPLE 12.14

In the circuit shown in Fig. 12.35, $e(t) = 70.7 \sin (10{,}000t)$ V. Find
(a) The total circuit impedance Z_T.
(b) The expression for $i(t)$.
(c) The power dissipated in the circuit, and draw the complete phasor diagram.
(d) The value of the extra element that can be added to this circuit without changing the ammeter's reading. What is $i(t)$ in this case?

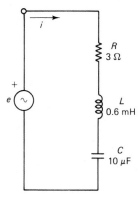

Figure 12.35

Solution   (a) Here

$$\mathbf{E} = 0.707 \times 70.7\underline{/0°} \text{ V} = 50\underline{/0°} \text{ V} \quad \text{and} \quad \omega = 10^4 \text{ rad/s}$$

Therefore,

$$Z_T = R + jX_L - jX_C = R + j\left(\omega L - \frac{1}{\omega C}\right)$$

$$= 3 + j\left(10^4 \times 0.6 \times 10^{-3} - \frac{1}{10^4 \times 10 \times 10^{-6}}\right)$$

$$= 3 + j(6 - 10) \quad \Omega \quad \text{(notice that } X_C > X_L\text{)}$$

$$= 3 - j4 = 5\underline{/-53.1°} \, \Omega$$

Thus the circuit is predominantly capacitive. It is then expected that the current **I** would lead the voltage **E**.

(b) From Ohm's law for the entire circuit,

$$\mathbf{I} = \frac{\mathbf{E}}{Z_T} = \frac{50/0°}{5/-53.1°} = 10/53.1° \text{ A}$$

Therefore,

$$i(t) = 10\sqrt{2} \sin (10{,}000t + 53.1°)$$
$$= 14.14 \sin (10{,}000t + 53.1°) \text{ A}$$

(c) The phase shift between **E** and **I**, ϕ, is 53.1°, the same as the angle of Z_T. Therefore,

$$P_{\text{av}} = EI \cos \phi = 50 \times 10 \times \cos 53.1° = 300 \text{ W}$$

and

$$P_{\text{av}} = I^2 R = (10)^2 \times 3 = 300 \text{ W}$$

To draw the complete phasor diagram, the voltage phasors (across the elements) must be determined. Here

$$\mathbf{V}_R = \mathbf{I}Z_R = 10/53.1° \times 3/0° = 30/53.1° \text{ V}$$
$$\mathbf{V}_L = \mathbf{I}Z_L = 10/53.1° \times 6/90° = 60/143.1° \text{ V}$$
$$\mathbf{V}_C = \mathbf{I}Z_C = 10/53.1° \times 10/-90° = 100/-36.9° \text{ V}$$

The phasor diagram is thus as shown in Fig. 12.36.

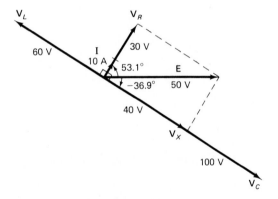

Figure 12.36

Note that in many ac circuits (e.g., this example) many of the element voltages might be larger in magnitude than the source voltage (E). Here both V_L and V_C are larger than E. This should not be a cause for alarm since voltage summation (i.e., KVL) is vectorial (including angles), not linear.

(d) Since here

$$Z_T = 3 - j4 \quad \Omega$$

the new total impedance of the circuit (with the added extra element) is

$$Z'_T = 3 + j4 = 5/53.1° = Z_T + Z_{\text{added}}$$

Note that both $|Z_T|$ and $|Z'_T|$, have the same magnitude (i.e., I remains unchanged). Thus

$$Z_{\text{added}} = 3 + j4 - (3 - j4) = j8 \ \Omega = jX_{L_{\text{added}}}$$

The added element is an inductor:

$$L_{\text{added}} = \frac{X_{L_{\text{added}}}}{\omega} = \frac{8}{10^4} = 0.8 \ \text{mH}$$

Here

$$\mathbf{I} = \frac{\mathbf{E}}{Z'_T} = \frac{50\underline{/0°}}{5\underline{/53.1°}} = 10\underline{/-53.1°} \ \text{A}$$

Therefore,

$$i(t) = 14.14 \sin (10{,}000t - 53.1°) \ \text{A}$$

The current, then, is lagging the source voltage because the circuit is now predominantly inductive.

EXAMPLE 12.15

A coil whose impedance is $10\underline{/59.5°} \ \Omega$ is connected in series with an unknown impedance Z to a 50 V, 20 krad/s source, as shown in Fig. 12.37. Find Z such that the resulting current is 5 A, leading the source voltage by 45°. What is the power dissipated in this circuit?

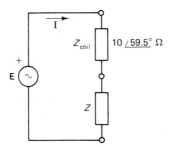

Figure 12.37

Solution   Let $\mathbf{I}$ be the reference phasor, that is,

$$\mathbf{I} = 5\underline{/0°} \ \text{A} \qquad \omega = 20 \times 10^3 \ \text{rad/s}$$

As $\mathbf{I}$ leads $\mathbf{E}$ by 45°, then

$$\mathbf{E} = 50\underline{/-45°} \ \text{V}$$

Therefore, from Ohm's law of the whole circuit and from the circuit diagram,

$$Z_T = \frac{\mathbf{E}}{\mathbf{I}} = \frac{50\underline{/-45°}}{5\underline{/0°}} = 10\underline{/-45°} = 7.07 - j7.07 \ \Omega$$

$$= Z_{\text{coil}} + Z = 10\underline{/59.5°} + Z = 5.07 + j8.62 + Z$$

and

$$Z = 7.07 - j7.07 - (5.07 + j8.62) = 2 - j15.69 \ \Omega$$

$$= R - jX_C$$

Thus the impedance Z is made of a series connection of a resistor R and a capacitor C, where

$$R = 2\,\Omega$$

$$X_C = 15.69\,\Omega = \frac{1}{\omega C}$$

and

$$C = \frac{1}{15.69 \times 20 \times 10^3} = 3.19\,\mu F$$

Here the average power dissipated is easier to calculate using the general expression, where ϕ is 45°,

$$P_{av} = EI\cos\phi = 50 \times 5 \times \cos 45° = 176.8\,\text{W}$$

EXAMPLE 12.16

If the frequency is doubled in the circuit shown in Fig. 12.38, find the current **I**.

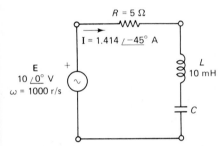

Figure 12.38

Solution The impedance is a function of the frequency. The unknown capacitor has to be determined first. At the given frequency, $\omega = 1{,}000\,\text{rad/s}$,

$$Z_T = R + j(X_L - X_C) = 5 + j(10 - X_C)\quad\Omega$$

Also from Ohm's law for the entire circuit,

$$Z_T = \frac{\mathbf{E}}{\mathbf{I}} = \frac{10\underline{/0°}}{1.414\underline{/-45°}} = 7.07\underline{/45°}\,\Omega = 5 + j5\,\Omega$$

Comparing the values of Z_T above (real to real and imaginary to imaginary), we have

$$10 - X_C = 5$$

Therefore,

$$X_C = 10 - 5 = 5\,\Omega = \frac{1}{1000\,C}$$

and

$$C = 200\,\mu F$$

Now, when ω is doubled ($\omega' = 2000\,\text{rad/s}$), the total impedance of the circuit becomes

$$Z_T' = R + j\left(\omega'L - \frac{1}{\omega'C}\right) = 5 + j(20 - 2.5)$$

$$= 5 + j17.5 = 18.2\underline{/74°}\,\Omega$$

Thus the current in the circuit is then

$$\mathbf{I}' = \frac{\mathbf{E}}{Z'_T} = \frac{10\underline{/0°}}{18.2\underline{/74°}} = 0.55\underline{/-74°} \text{ A}$$

EXAMPLE 12.17

In the circuit shown in Fig. 12.39, $V_{AD} = 200$ V, $V_{BF} = 346$ V, and the phase angle between these two voltages is 30°.

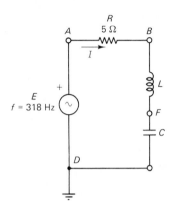

Figure 12.39

(a) Which of these two voltages is leading?
(b) Find the values of I, L, and C.
(c) If C is changed, what should be its new value so that the current magnitude remains unchanged?
(d) In part (c), what is the phase angle between **E** and **I**, and what is the phase angle between the oscillogram waveforms if Y_A and Y_B are connected to points A and B, respectively?

Solution (a) Taking **I** as the reference phasor, the phasor diagram can be drawn as shown in Fig. 12.40(a), noting that

$$V_{AD} = E = 200 \text{ V} \qquad \text{and} \qquad V_{BF} = V_L = 346 \text{ V}$$

From the phasor diagram, it is clear that $\mathbf{V}_L$ (V_{BF}) leads **E** (V_{AD}) by 30°, as stated in the problem.

(b) From the geometry of the phasor diagram,

$$V_R = E \sin 30° = 200 \times 0.5 = 100 \text{ V}$$

and

$$V_X = E \cos 30° = 200 \times 0.866 = 173.2 \text{ V}$$

Since

$$V_R = IR$$

$$I = \frac{V_R}{R} = \frac{100}{5} = 20 \text{ A}$$

and

$$V_L = 346 \text{ V} = IX_L$$

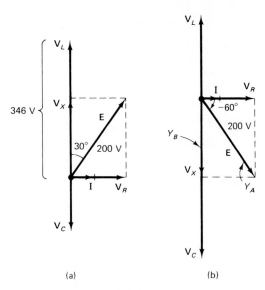

Figure 12.40

then

$$X_L = \frac{346}{20} = 17.3\,\Omega = \omega L = (2\pi \times 318)L$$

and

$$L = \frac{17.3}{1998} = 8.66\,\text{mH}$$

But

$$V_X = V_L - V_C$$

Therefore,

$$V_C = V_L - V_X = 346 - 173.2 = 172.8\,\text{V}$$
$$= IX_C$$

Then

$$X_C = \frac{V_C}{I} = \frac{172.8}{20} = 8.64\,\Omega = \frac{1}{\omega C}$$

and

$$C = \frac{1}{X_C\omega} = \frac{1}{8.64 \times 1998} = 57.9\,\mu\text{F}$$

(c) The total impedance of the circuit is

$$Z_T = R + j(X_L - X_C) = 5 + j(17.3 - 8.64)$$
$$= 5 + j8.66 = 10\underline{/60^\circ}\,\Omega$$

For the current magnitude to remain unchanged as C is changed to C', the new impedance of the circuit, Z'_T, should have the same magnitude as Z_T, that is,

$$|Z'_T| = |Z_T|$$

Then

$$Z'_T = 5 - j8.66\,\Omega = 10\underline{/-60^\circ} = R + j(X_L - X'_C)$$
$$= 5 + j(17.3 - X'_C)$$
$$-8.66 = 17.3 - X'_C$$

Therefore,

$$X'_C = 17.3 + 8.66 = 25.96 \ \Omega = \frac{1}{\omega C'}$$

and

$$C' = 19.4 \ \mu F$$

(d) The phase angle between **E** and **I** is the same as the phase angle of $Z'_T \ (= -60°)$. Thus, here, **E** lags **I** by 60°, as the circuit is now predominantly capacitive. With **I** still the reference phasor, the phasor diagram in this case is as shown in Fig. 12.40(b). The scope connections indicate that Y_A monitors e (i.e., **E**) and Y_B monitors v_X (i.e., $\mathbf{V}_X = \mathbf{V}_L + \mathbf{V}_C$). From the phasor diagram in Fig. 12.40(b), it can be concluded then that Y_A leads Y_B by 30°.

EXAMPLE 12.18

The oscillogram shown in Fig. 12.41(a) was obtained while analyzing the circuit in Fig. 12.41(b). The oscilloscope controls were adjusted as follows:

$$\text{time/div} = 5 \ \text{ms}$$
$$\text{V/div} \ (Y_A) = 2 \ V$$
$$\text{V/div} \ (Y_B) = \text{unknown}$$

Find the values of L and C and V/div (Y_B).

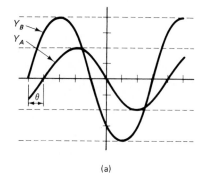

(a)

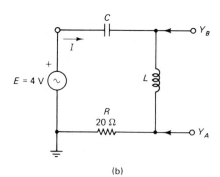

(b)

Figure 12.41

Solution Since the two beam displays are not labeled, they have to be identified first. A sketch of the phasor diagram for this circuit is shown in Fig. 12.42, with Y_A representing v_R (i.e., $\mathbf{V}_R$) and Y_B representing v_{RL} ($\mathbf{V}_{RL}$) according to the connections in Fig. 12.41(b). From the phasor diagram, it is

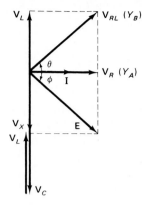

Figure 12.42

clear that $Y_B(\mathbf{V}_{RL})$ is the leading waveform. Thus the oscillogram display can be labeled.

From the oscillogram, Y_B leads Y_A by 1 div, that is,

$$\theta = 1 \text{ div} \times \frac{180°}{4 \text{ div}} = 45° \quad \text{and} \quad T = 8 \times 5 \text{ ms} = 40 \text{ ms}$$

and from beam A,

$$V_R = 0.707 \times 2 \text{ div} \times 2 \text{ V/div} = 2.828 \text{ V}$$
$$= IR$$

Therefore,

$$I = \frac{V_R}{R} = \frac{2.828}{20} = 0.1414 \text{ A}$$

Using the geometry of the phasor diagram (Fig. 12.42) gives us

$$V_R = V_{RL} \cos \theta$$

Therefore,

$$V_{RL} = \frac{2.828}{\cos 45°} = 4 \text{ V}$$

and

$$V_L = V_{RL} \sin \theta = 4 \times \sin 45° = 2.828 \text{ V}$$

As

$$V_L = IX_L$$

then

$$X_L = \frac{V_L}{I} = \frac{2.828}{0.1414} = 20 \text{ }\Omega$$
$$= \omega L$$

But

$$\omega = 2\pi f = \frac{2\pi}{T} = \frac{6.28}{40 \times 10^{-3}} = 157 \text{ rad/s}$$

Therefore,

$$L = \frac{X_L}{\omega} = \frac{20}{157} = 127.4 \text{ mH}$$

Also, as Y_B represents $\mathbf{V}_{RL}$, then

$$Y_B(\text{peak}) = 1.414 \times 4 = 5.657 \text{ V}$$

corresponding to four divisions; thus

$$V/div\ (Y_B) = \frac{5.657}{4} = 1.414\ V/div$$

(i.e., 1 V/div, uncalibrated).

To find C, X_C must be obtained, and to find X_C, V_C must first be calculated. The problem specifies that E is 4 V (i.e., $= V_{RL}$). This means that either V_C is zero (unacceptable since X_C is finite) or V_C is $2V_L$ in order that

$$\mathbf{V}_X = \mathbf{V}_L + \mathbf{V}_C = jV_L - j2V_L = -jV_L$$

resulting in

$$E = \sqrt{V_R^2 + V_X^2} = \sqrt{V_R^2 + V_L^2} = V_{RL}$$

This condition is indicated in the phasor diagram of Fig. 12.42. Here

$$V_C = 2V_L = 2 \times 2.828 = 5.656\ V$$
$$= IX_C$$

Therefore,

$$X_C = \frac{V_C}{I} = \frac{5.656}{0.1414} = 40\ \Omega$$

$$= \frac{1}{\omega C}$$

and

$$C = \frac{1}{157 \times 40} = 159\ \mu F$$

EXAMPLE 12.19

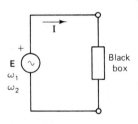

Figure 12.43

The black box in the circuit shown in Fig. 12.43 contains a number of series-connected elements. With the source voltage fixed, $\mathbf{E} = 10\underline{/0°}$ V, two tests were conducted. At $\omega_1 = 1000$ rad/s, $\mathbf{I}$ was found to be $2\underline{/60°}$ A, and at $\omega_2 = 2000$ rad/s, the current lags the voltage by 45°. Find the nature and the values of the elements inside the black box.

Solution   At $\omega = 1000$ rad/s, the circuit is capacitive since $\mathbf{I}$ leads $\mathbf{E}$. At $\omega = 2000$ rad/s, the circuit is inductive because then the current lags the source voltage. In both cases the phase shift between $\mathbf{E}$ and $\mathbf{I}$ is not $\pm 90°$ (i.e., the circuit must also contain a resistance). Thus

$$Z_T = R + j\left(\omega L - \frac{1}{\omega C}\right)$$

At $\omega = 1000$ rad/s,

$$Z_T = \frac{\mathbf{E}}{\mathbf{I}} = \frac{10\underline{/0°}}{2\underline{/60°}} = 5\underline{/-60°}\ \Omega = 2.5 - j4.33\ \Omega$$

$$= R + j\left(1000L - \frac{1}{1000C}\right)$$

Comparing these two equal forms of Z_T,

$$R = 2.5 \, \Omega$$

$$1000L - \frac{1}{1000C} = -4.33$$

At $\omega = 2000$ rad/s, the current lags the source voltage by 45°. Thus Z_T is inductive and its phase angle is 45°. Here

$$\tan 45° = 1 = \frac{X_L - X_C}{R} = \frac{2000L - 1/2000C}{2.5}$$

Then

$$2000L - \frac{1}{2000C} = 2.5$$

The two equations above, relating L and C, can now be solved to obtain the values of L and C. Multiplying the second equation by 2 and subtracting, we have

$$4000L - 1000L = 5 - (-4.33) = 9.33$$

Therefore,

$$L = \frac{9.33}{3000} = 3.11 \, \text{mH}$$

Substituting in the first equation yields

$$\frac{1}{1000C} = 3.11 + 4.33 = 7.44$$

Therefore,

$$C = \frac{1}{1000 \times 7.44} = 134.4 \, \mu\text{F}$$

EXAMPLE 12.20

In the circuit shown in Fig. 12.44(a), the voltmeter and the ammeter readings were 70.7 V and 4.24 A, respectively. With the scope inputs connected as shown, the oscillogram in Fig. 12.44(b) was obtained while V/div (Y_A) = 20 V and V/div (Y_B) = 50 V. Find the values of R, L, and C.

Solution   In this circuit,

$$E = 70.7 \, \text{V} \qquad \text{and} \qquad I = 4.24 \, \text{A}$$

Y_B represents v_C:

$$V_C = 0.707 \times 2 \, \text{div} \times 50 \, \text{V/div} = 70.7 \, \text{V}$$

Y_A represents $v_X (= v_L + v_C)$:

$$V_X = 0.707 \times 4 \, \text{div} \times 20 \, \text{V/div} = 56.57 \, \text{V}$$

Also, $\mathbf{V}_X$ is 180° out of phase with $\mathbf{V}_C$, as the oscillogram indicates.

Taking $\mathbf{I}$ as the reference phasor, the phasor diagram for this circuit can be drawn as shown in Fig. 12.45. Notice that $\mathbf{V}_X$, being in the opposite

(a)

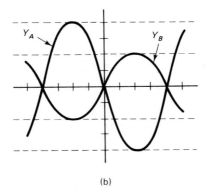

(b)

Figure 12.44

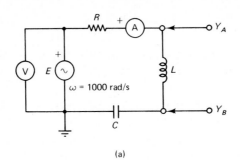

Figure 12.45

direction of $\mathbf{V}_C$, is in phase with $\mathbf{V}_L$ (i.e., V_L is larger in magnitude than V_c). As

$$V_X = V_L - V_C$$
$$V_L = V_X + V_C = 56.57 + 70.7 = 127.27 \text{ V}$$

Also,

$$E^2 = V_X^2 + V_R^2$$

Thus

$$V_R = \sqrt{(70.7)^2 - (56.57)^2} = 42.4 \text{ V}$$

Hence

$$R = \frac{V_R}{I} = \frac{42.4}{4.24} = 10 \ \Omega$$

$$X_L = \frac{V_L}{I} = \frac{127.27}{4.24} = 30 \ \Omega = \omega L = 1000L$$

Therefore,

$$L = 30 \text{ mH}$$

and

$$X_C = \frac{V_C}{I} = \frac{70.7}{4.24} = 16.67 \ \Omega = \frac{1}{\omega C} = \frac{1}{1000C}$$

Thus

$$C = 60 \ \mu\text{F}$$

As was the case in the dc parallel circuits, the ac parallel branches also have the same two important characteristics:

1. The voltage across each of the parallel branches (between two nodes, e.g., *a* and *b*) is the same.
2. KCL always applies: The algebraic (or vectorial) sum of currents entering a node (e.g., *a*) = the algebraic (or vectorial) sum of currents leaving this node.

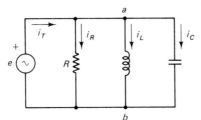

Figure 12.46

The parallel circuit in Fig. 12.46 has three load branches, each consisting of a single pure element. An ideal ac voltage source is connected across these three branches. Here the voltage across each branch is the same, equal to the source voltage *e(t)*. The currents in all the branches have the same frequency as the applied source. Thus instead of the time-domain circuit in Fig. 12.46, the equivalent frequency-domain circuit in Fig. 12.47 can be analyzed, using phasors and element impedances (or admittances).

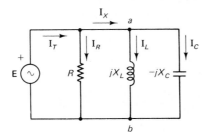

Figure 12.47

The common quantity in such parallel circuits is the source voltage. To simplify the analysis, this voltage will be considered as the reference waveform (i.e., the reference phasor),

$$e(t) = E_m \sin \omega t \text{ V} \Leftrightarrow \mathbf{E} = E\underline{/0°}\text{ V} \quad (\text{where } E = 0.707E_m) \quad (12.56)$$

As defined in Chapter 11, the admittance of an element, *Y* in siemens, is

$$Y = \frac{1}{Z} = \frac{\mathbf{I}}{\mathbf{V}} \text{ S} \tag{12.57a}$$

or

$$\mathbf{I} = \frac{\mathbf{V}}{Z} = Y\mathbf{V} \quad (\text{Ohm's law}) \tag{12.57b}$$

where $\mathbf{I}$ is the current in the element and $\mathbf{V}$ is the voltage across the element. Here

$$\mathbf{I}_R = \frac{\mathbf{E}}{\mathbf{Z}_R} = \frac{E\underline{/0°}}{R\underline{/0°}} = \frac{E}{R}\underline{/0°} = I_R\underline{/0°}\ \text{A}$$

Also,

$$\mathbf{I}_R = \mathbf{E}\mathbf{Y}_R = E\underline{/0°}\ G\underline{/0°}\ \text{A} = EG\underline{/0°}\ \text{A} = I_R\underline{/0°}\ \text{A} \qquad (12.58)$$

where G is the conductance $= 1/R$ S (or $\mho$).

$$\mathbf{I}_L = \frac{\mathbf{E}}{\mathbf{Z}_L} = \frac{E\underline{/0°}}{X_L\underline{/90°}} = \frac{E}{X_L}\underline{/-90°} = I_L\underline{/-90°}\ \text{A}$$

Also,

$$\mathbf{I}_L = \mathbf{E}\mathbf{Y}_L = E\underline{/0°}\ B_L\underline{/-90°} = EB_L\underline{/-90°} = I_L\underline{/-90°}\ \text{A} \qquad (12.59)$$

where B_L is the inductive susceptance $= 1/X_L = 1/\omega L$ siemens, and

$$\mathbf{I}_C = \frac{\mathbf{E}}{\mathbf{Z}_C} = \frac{E\underline{/0°}}{X_C\underline{/-90°}} = \frac{E}{X_C}\underline{/90°} = I_C\underline{/90°}\ \text{A}$$

Also,

$$\mathbf{I}_C = \mathbf{E}\mathbf{Y}_C = E\underline{/0°}\ B_C\underline{/90°} = EB_C\underline{/90°} = I_C\underline{/90°}\ \text{A} \qquad (12.60)$$

where B_C is the capacitive susceptance $= 1/X_C = \omega C$ siemens. The phasor diagram showing the voltage and the currents in this circuit is drawn as shown in Fig. 12.48.

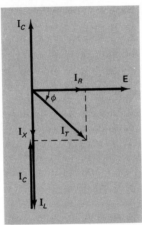

Figure 12.48

As in Fig. 12.47, we can denote

$$\mathbf{I}_X = \text{total reactive current} = \mathbf{I}_L + \mathbf{I}_C \qquad (12.61a)$$

$$= -jI_L + jI_C = -j(I_L - I_C) = -jI_X \qquad (12.61b)$$

Applying KCL to node a provides

$$\boxed{\begin{aligned} \mathbf{I}_T &= \mathbf{I}_R + \mathbf{I}_L + \mathbf{I}_C = \mathbf{I}_R + \mathbf{I}_X \\ &= I_R - j(I_L - I_C) = I_R - jI_X \\ &= \sqrt{I_R^2 + I_X^2}\ \Big/{-\tan^{-1}\frac{I_X}{I_R}} \\ &= I_T\underline{/\phi°}\ \text{A} \end{aligned}} \qquad (12.62)$$

where

$$I_T = \sqrt{I_R^2 + I_X^2} = \sqrt{I_R^2 + (I_L - I_C)^2} \qquad (12.62a)$$

and

$$\phi = -\tan^{-1}\frac{I_X}{I_R} = -\tan^{-1}\frac{I_L - I_C}{I_R} \qquad (12.62b)$$

The time-domain expression for $i(t)$ can then be written as

$$i(t) = \sqrt{2}\, I_T \sin\left(\omega t + \phi^\circ\right)$$

Phase Relationships. Through an examination of the phasor diagram in Fig. 12.48 it is clear that the current in the resistor $\mathbf{I}_R$ is in phase with the voltage $\mathbf{E}$ across it, while the inductor's current $\mathbf{I}_L$ lags the voltage across it, $\mathbf{E}$, by 90°, and the capacitor's current $\mathbf{I}_C$ leads the voltage across it, $\mathbf{E}$, by 90°. Thus $\mathbf{I}_L$ and $\mathbf{I}_C$ are always 180° out of phase with each other. Their vectorial sum, $\mathbf{I}_X$, is always perpendicular to $\mathbf{E}$ and results in a net subtraction of the two branch currents. Three possibilities can result:

1. If $X_L < X_C$ (i.e., $B_L > B_C$), then $I_L > I_C$ [according to Eqs. (12.59) and (12.60)]. In this case $\mathbf{I}_X$ is in the direction of $\mathbf{I}_L$ because the inductive current is larger than the capacitive current, as in the phasor diagram in Fig. 12.48. The total current $\mathbf{I}_T$ thus lags $\mathbf{E}$ by the phase angle ϕ°, and the circuit is said to be predominantly inductive.

2. If $X_L > X_C$ (i.e., $B_L < B_C$), then $I_L < I_C$. Here $\mathbf{I}_X$ is in the direction of $\mathbf{I}_C$ because the capacitive current is larger than the inductive current. This situation is represented in the general type of phasor diagrams for such circuits, as shown in Fig. 12.49, where no particular phasor is taken as the reference. Thus here the total current $\mathbf{I}_T$ leads $\mathbf{E}$ by the angle ϕ° and the circuit is said to be predominantly capacitive. Note that $\mathbf{I}_X$ is always perpendicular to $\mathbf{E}$ (or $\mathbf{I}_R$).

3. The special case when $X_L = X_C$ (i.e., $B_L = B_C$) corresponds to

$$\omega L = \frac{1}{\omega C} \qquad (12.63)$$

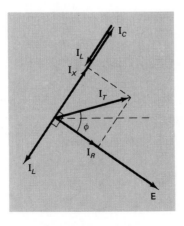

Figure 12.49

When ω is the variable, the solution of the equation above is

$$\omega_0 = \text{resonance frequency} = \frac{1}{\sqrt{LC}} \quad \text{rad/s} \qquad (12.63a)$$

that is,

$$f_0 = \frac{\omega_0}{2\pi} = \frac{1}{2\pi\sqrt{LC}} \quad \text{Hz} \qquad (12.63b)$$

Here $I_L = I_C$ and thus $\mathbf{I}_X = 0$. Therefore, $\mathbf{I}_T = \mathbf{I}_R$ and is in phase with $\mathbf{E}$ ($\phi = 0°$). The circuit is thus resistive in nature. This case is called the *parallel resonance condition*. Notice that the resonance frequency f_0 is the same as in the series resonance condition. This subject is discussed in more detail in Chapter 14.

The admittance triangle. The total impedance (and admittance) of the parallel branches in Fig. 12.47 is defined from Ohm's law for the whole circuit as the impedance (or admittance) "seen" by the source:

$$\boxed{Z_T = \frac{1}{Y_T} = \frac{\mathbf{E}}{\mathbf{I}_T}} \qquad (12.64a)$$

Therefore,

$$\boxed{\mathbf{I}_T = \frac{\mathbf{E}}{Z_T} = Y_T\mathbf{E}} \qquad (12.64b)$$

From KCL for the circuit in Fig. 12.47,

$$\mathbf{I}_T = \mathbf{I}_R + \mathbf{I}_L + \mathbf{I}_C$$

Substituting in the above equation using Ohm's law for each branch and for the entire circuit [i.e., Eqs. (12.58), (12.59), (12.60), and (12.64b)] gives us

$$\frac{\mathbf{E}}{Z_T} = \frac{\mathbf{E}}{Z_R} + \frac{\mathbf{E}}{Z_L} + \frac{\mathbf{E}}{Z_C}$$

Dividing both sides of this equation by $\mathbf{E}$ yields

$$\frac{1}{Z_T} = \frac{1}{Z_R} + \frac{1}{Z_L} + \frac{1}{Z_C} \qquad (12.65a)$$

Therefore,

$$\boxed{\begin{aligned} Y_T &= Y_R + Y_L + Y_C && (12.65b) \\ &= G\underline{/0°} + B_L\underline{/-90°} + B_C\underline{/90°} \\ &= G - jB_L + jB_C = G - j(B_L - B_C) \quad \text{S} && (12.66a) \\ &= G - jB \quad \text{S} \end{aligned}}$$

(where $B = B_L - B_C$ = net susceptance). Also,

$$Y_T = \frac{1}{R} - j\left(\frac{1}{\omega L} - \omega C\right) \quad \text{S} \qquad (12.66b)$$

$$= \sqrt{G^2 + B^2} \underline{\left/-\tan^{-1}\frac{B}{G}\right.} \quad \text{S}$$

$$= |Y_T|\underline{/\phi°} \quad \text{S} \qquad (12.66c)$$

where

$$|Y_T| = \sqrt{G^2 + (B_L - B_C)^2} = \sqrt{\frac{1}{R^2} + \left(\frac{1}{\omega L} - \omega C\right)^2} \qquad (12.67a)$$

and

$$\phi = -\tan^{-1}\frac{B_L - B_C}{G} = -\tan^{-1}\left[R\left(\frac{1}{\omega L} - \omega C\right)\right] \qquad (12.67b)$$

The magnitude and phase of the total admittance of the circuit, Y_T, is a function of not only the element values, but also of the frequency. It is independent of the initial choice of the reference phasor **E**, as this quantity drops out in the KCL equation above. The total admittance Y_T being a vector (i.e., a complex number), can be represented by an admittance triangle as shown in Fig. 12.50, in accordance with Eqs. (12.66) and (12.67). The phase angle of Y_T is the same as the phase shift between $\mathbf{I}_T$ and **E**, as Ohm's law for the entire circuit, Eq. (12.64b), indicates.

As discussed above, if $B_L > B_C$, Y_T will have a negative phase angle (i.e., $Z_T = 1/Y_T$ will have a positive phase angle), indicating that the circuit is predominantly inductive, and $\mathbf{I}_T$ will lag **E**. The reverse is true when $B_C > B_L$, with Y_T having a positive phase angle and the circuit becoming predominantly capacitive. In this case $\mathbf{I}_T$ will lead **E**. At resonance

$$B_L = B_C \qquad \text{and} \qquad Y_T = G \quad \text{(purely real)}$$

that is, the circuit is equivalent to the resistive component only, as if the L and C branches are equivalent to an open circuit. This is confirmed by the fact that at resonance $\mathbf{I}_X$ (in the LC branches) is zero.

Note. As discussed in the case of the RLC series circuit, here also an extra susceptance (either a pure L with susceptance B_L or a pure C with susceptance B_C) can be added in parallel to the original circuit without affecting the magnitude of the total circuit's current, $|I_T|$. For example, let

$$Y_T = G \pm jB \quad \text{S} \quad (+\text{indicates capacitive and} - \text{indicates inductive})$$

Then the new circuit admittance, Y'_T, would be

$$Y'_T = G \mp jB \quad \text{S} \quad \text{(i.e., of the opposite nature to } Y_T)$$

Here

$$|Y_T| = |Y'_T| = \sqrt{G^2 + B^2} \quad \text{S}$$

and

$$|I_T| = |Y_T|\,E = |Y'_T|\,E \quad \text{A}$$

Thus the total circuit current remains unchanged. The added branch admittance is

$$Y_{\text{added}} = Y'_T - Y_T = (G \mp jB) - (G \pm jB) = \mp j2B \quad \text{S} \qquad (12.68)$$

Power in the RLC parallel circuits. In a manner similar to the discussions concerning power in series circuits, if the voltage across the parallel load, Y_T, is

$$v(t) = V_m \sin\left(\omega t + \theta°\right) \text{V} \Leftrightarrow \mathbf{V} = V\underline{/\theta°} \text{ V}$$

resulting in the current flow

$$i_T(t) = I_{Tm} \sin\left(\omega t + \theta° + \phi°\right) \text{A} \Leftrightarrow \mathbf{I}_T = I_T\underline{/\theta° + \phi°} \text{ A}$$

where $\phi°$ may be a positive or negative phase angle (see Fig. 12.51), then it can

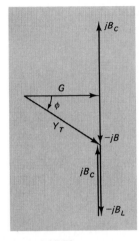

Figure 12.50

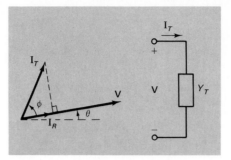

Figure 12.51

be shown that the average power dissipated in the *RLC* parallel circuit is

$$P_{av} = VI_T \cos \phi \quad W \tag{12.69}$$
$$= VI_R$$
$$= V\frac{V}{R} = \frac{V^2}{R} \tag{12.70a}$$
$$= RI_R I_R = RI_R^2 \tag{12.70b}$$

V and I_T are the RMS values of the corresponding voltage and current phasors, while $\phi°$ is the phase angle between the two phasors as shown in the phasor diagram of Fig. 12.51. Notice that I_R is the magnitude of the current in the resistive branch. Hence Eqs. (12.70) again indicate that only the resistive branch is responsible for the power dissipation in such parallel loads.

Equivalence between series and parallel *RLC* loads. When a voltage source **V** drives the same current **I** in two different networks (e.g., N_1 and N_2 in Fig. 12.52), the two networks are said to be *electrically equivalent* because their terminal voltage-current relationship is the same. Thus in Fig. 12.52,

$$\frac{\mathbf{V}}{\mathbf{I}} = Z = \frac{1}{Y} \tag{12.71}$$

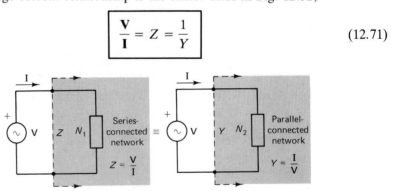

Figure 12.52

Z, the total impedance of N_1, is called the driving-point impedance, and Y, the total admittance of N_2, is called the driving-point admittance.

From the relationship in Eq. (12.71), any series form of impedance can be converted into a parallel form of admittance, or vice versa, through a process of complex-number inversion. If

$$Z = R + jX = |Z| \underline{/\theta°} \quad \Omega \tag{12.71a}$$

then

$$Y = \frac{1}{Z} = \frac{1}{|Z|\,\underline{/\phi^\circ}} = |Y|\,\underline{/-\phi^\circ} \quad S$$

$$= G - jB \quad S \tag{12.71b}$$

The nature of the network is preserved in such conversions; that is, an RL series circuit converts into an RL parallel circuit, while a series RC circuit converts into a parallel RC circuit. The reverse process of converting parallel circuits into equivalent series form is performed in a similar manner, also preserving the nature of the circuit.

Since X and B are functions of the frequency ω (or f), these equivalent circuits are correct only at the fixed specified frequency. If the frequency is changed, the elements values (both the resistive and the reactive components) of the equivalent circuit will also change.

■ EXAMPLE 12.21

The current i_T in the circuit shown in Fig. 12.53 is $i_T(t) = 7.07 \sin(2\pi \times 20{,}000t)$ A. Find
(a) The total admittance of the circuit and draw the equivalent series impedance.
(b) $e(t)$ and the average power dissipated P_{av}.
(c) The current in each branch and draw the phasor diagram for this circuit.

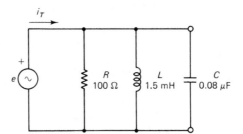

Figure 12.53

Solution Here $i_T(t)$ is the reference waveform, that is,

$$\mathbf{I}_T = 5\underline{/0^\circ}\,A \qquad \text{and} \qquad \omega = 2\pi \times 20{,}000 = 125.6\,krad/s$$

(a) $Y_R = G\underline{/0^\circ} = \dfrac{1}{R}\underline{/0^\circ} = 0.01\underline{/0^\circ} = 0.01\,S$

$$Y_L = B_L\underline{/-90^\circ} = \frac{1}{\omega L}\underline{/-90^\circ} = 0.0053\underline{/-90^\circ} = -j0.0053\,S$$

$$Y_C = B_C\underline{/90^\circ} = \omega C\underline{/90^\circ} = 0.01\underline{/90^\circ} = j0.01\,S$$

Therefore,

$$Y_T = Y_R + Y_L + Y_C$$
$$= 0.01 - j0.0053 + j0.01 = 0.01 + j0.0047$$
$$= 0.011\underline{/25.2^\circ}\,S$$

Notice that this circuit is predominantly capacitive because B_C is greater than B_L. One then expects the equivalent series circuit to be predominantly capacitive and that $e(t)$ would lag $i_T(t)$. Here

$$Z_{Teq} = \frac{1}{Y_T} = \frac{1}{0.011/25.2°} = 90.5/-25.2° \ \Omega$$

$$= 81.9 - j38.5$$

$$= R_{eq} - jX_{Ceq}$$

Therefore,

$$R_{eq} = 81.9 \ \Omega$$

and

$$C_{eq} = \frac{1}{\omega X_{Ceq}} = \frac{1}{125.6 \times 10^3 \times 38.5} = 0.207 \ \mu F$$

The equivalent series circuit is shown in Fig. 12.54(a). The elements and their values are the correct equivalent to the circuit in Fig. 12.53 only at the given frequency.

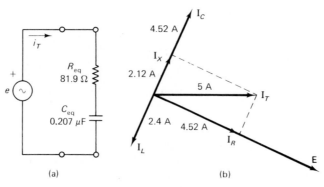

(a)

(b)

Figure 12.54

(b) Using Ohm's law for the entire circuit, we have

$$E = I_T Z_T = \frac{I_T}{Y_T} = \frac{5/0°}{0.011/25.2°} = 452.5/-25.2° \ V$$

Therefore,

$$e(t) = 452.5\sqrt{2} \sin (125.6 \times 10^3 t - 25.2°) \ V$$

$$= 640 \sin (125.6 \times 10^3 t - 25.2°) \ V$$

As expected, $e(t)$ lags $i_T(t)$ by $25.2°$.

$$P_{av} = EI_T \cos \phi_{E,I_T} = 454.5 \times 5 \times \cos 25.2° = 2047 \ W$$

Also,

$$P_{av} = \frac{E^2}{R} = \frac{(452.5)^2}{100} = 2047 \ W$$

(c) To find the current in each branch, Ohm's law is used for each branch:

$$\mathbf{I}_R = \frac{\mathbf{E}}{Z_R} = \mathbf{E}Y_R = 452.5\underline{/-25.2°} \times 0.01\underline{/0°}$$

$$= 4.525\underline{/-25.2°}\ \mathrm{A}$$

$$\mathbf{I}_L = \frac{\mathbf{E}}{Z_L} = \mathbf{E}Y_L = 452.5\underline{/-25.2°} \times 0.0053\underline{/-90°}$$

$$= 2.4\underline{/-115.2°}\ \mathrm{A}$$

and

$$\mathbf{I}_C = \frac{\mathbf{E}}{Z_C} = \mathbf{E}Y_C = 452.5\underline{/-25.2°} \times 0.01\underline{/90°}$$

$$= 4.525\underline{/64.8°}\ \mathrm{A}$$

The complete phasor diagram is shown in Fig. 12.54(b).

EXAMPLE 12.22

In the parallel circuit of Fig. 12.55, Y represents the admittance of a single element. What is the nature and magnitude of this element if $e(t) = 150\sin (3000t)$ V and $i_T(t) = 16.5\sin (3000t + 72.4°)$ A?

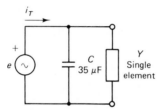

Figure 12.55

Find the value of the extra element that can be connected in parallel with this load without affecting the magnitude of i_T. What is the new current $i_T(t)$ in this case?

Solution The total admittance Y_T of the circuit shown in Fig. 12.55 is

$$Y_T = Y + jB_C = Y + j\omega C = Y + j0.105\quad \mathrm{S}$$

where ω is 3000 rad/s. Also, since

$$\mathbf{E} = 0.707 \times 150\underline{/0°} = 106.1\underline{/0°}\ \mathrm{V}$$

$$\mathbf{I}_T = 0.707 \times 16.5\underline{/72.4°} = 11.67\underline{/72.4°}\ \mathrm{A}$$

then from Ohm's law,

$$Y_T = \frac{\mathbf{I}_T}{\mathbf{E}} = \frac{11.67\underline{/72.4°}}{106.1\underline{/0°}} = 0.11\underline{/72.4°}\ \mathrm{S} = 0.0333 + j0.105\ \mathrm{S}$$

Comparing the two expressions above for Y_T, it is clear that

$$Y = 0.0333\ \mathrm{S} = \frac{1}{R},\qquad \text{that is, } R = 30\ \Omega$$

Thus Y is the admittance (conductance) of a 30 Ω resistor.

In order that the magnitude of i_T remain the same, with the added parallel element, the new total admittance, Y'_T, must be the conjugate of Y_T, that is,

$$Y'_T = 0.11\underline{/-72.4°}\,\text{S} = 0.0333 - j0.105\,\text{S}$$
$$= Y_T + Y_{\text{added}}$$

Therefore,

$$Y_{\text{added}} = Y'_T - Y_T = 0.0333 - j0.105 - (0.0333 + j0.105)$$

$$= -j0.21 = -jB_L = -j\frac{1}{\omega L}$$

Thus the added element is an inductor L,

$$L = \frac{1}{\omega B_L} = \frac{1}{3000 \times 0.21} = 1.59\,\text{mH}$$

Here

$$\mathbf{I}'_T = \mathbf{E}Y'_T = 106.1\underline{/0°} \times 0.11\underline{/-72.4°}$$
$$= 11.67\underline{/-72.4°}\,\text{A}$$

Therefore,

$$i'_T(t) = 16.5 \sin(3000t - 72.4°)\,\text{A}$$

Thus with the addition of this parallel inductor, the magnitude of the total current remains unchanged, but the nature of the circuit changes from predominantly capacitive (i_T leading e) to become predominantly inductive (i'_T lagging e).

EXAMPLE 12.23

The two branches in the circuit shown in Fig. 12.56 contain a single element each. The branch current $\mathbf{I}_1$ is in phase with $\mathbf{E}$ and its magnitude is 3 A. The source voltage $\mathbf{E}$ leads $\mathbf{I}_T$. Find the value of each element and the equivalent simple series circuit at the given frequency.

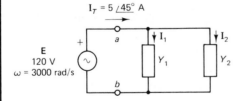

Figure 12.56

Solution Since the current $\mathbf{I}_1$ in Y_1 is in phase with $\mathbf{E}$, Y_1 must be a resistor. Also as $\mathbf{E}$ leads $\mathbf{I}_T$, the circuit is predominantly inductive [i.e., Y_2 (being a single element) must be an inductor]. The current $\mathbf{I}_2$ in Y_2 must then lag $\mathbf{E}$ by 90°. A sketch of the phasor diagram for this circuit can thus be drawn as shown in Fig. 12.57(a). As given above, I_1 is 3 A.

From the geometry of the phasor diagram,

$$\cos \phi = \frac{I_1}{I_T} = \frac{3}{5} = 0.6 \qquad \text{that is, } \phi = 53.1°$$

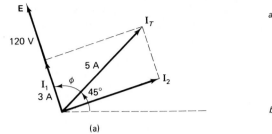
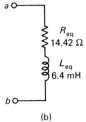

Figure 12.57

Also,

$$I_T^2 = I_1^2 + I_2^2$$

Therefore,

$$I_2 = \sqrt{(5)^2 - (3)^2} = 4 \text{ A}$$

Thus,

$$R \text{ (the component of } Y_1) = \frac{E}{I_1} = \frac{120}{3} = 40 \text{ } \Omega$$

and

$$X_L \text{ (the component of } Y_2) = \frac{E}{I_2} = \frac{120}{4} = 30 \text{ } \Omega = \omega L$$

Therefore,

$$L = \frac{30}{3000} = 10 \text{ mH}$$

Here

$$\mathbf{E} = 120\underline{/45° + 53.1°} = 120\underline{/98.1°} \text{ V}$$

To find the equivalent simple series circuit,

$$Z_{T\text{eq}} = \frac{\mathbf{E}}{\mathbf{I}_T} = \frac{120\underline{/98.1°}}{5\underline{/45°}} = 24\underline{/53.1°} \text{ } \Omega$$

$$= 14.42 + j19.2$$

$$= R_{\text{eq}} + jX_{L\text{eq}}$$

Therefore,

$$R_{\text{eq}} = 14.42 \text{ } \Omega$$

$$X_{L\text{eq}} = 19.2 = \omega L_{\text{eq}}$$

that is,

$$L_{\text{eq}} = \frac{19.2}{3000} = 6.4 \text{ mH}$$

The equivalent series circuit is shown in Fig. 12.57(b).

EXAMPLE 12.24

In the circuit shown in Fig. 12.58, $\mathbf{E}$ is $120\underline{/150°}$ V, I_T is $10A$ and $i'(t) = 8\sqrt{2}\sin(1000t + 60°)$ A. Find the value of R and the single element. Express the total current in phasor and in time-domain forms.

Solution The current in the single element is $\mathbf{I}' = 8\underline{/60°}$ A; it lags $\mathbf{E}$ by 90°. Thus the single element is an inductor L. $\mathbf{I}_R$ is in phase with $\mathbf{E}$. Thus $\mathbf{I}_T$ lags $\mathbf{E}$

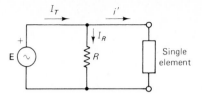

Figure 12.58

by an angle ϕ. A sketch of the corresponding phasor diagram is drawn in Fig. 12.59.

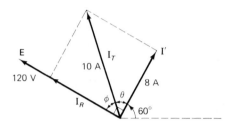

Figure 12.59

From the geometry of the phasor diagram,

$$I_T^2 = I_R^2 + I'^2$$

Therefore,

$$I_R = \sqrt{(10)^2 - (8)^2} = 6 \text{ A}$$

Also,

$$\sin \theta = \frac{I_R}{I_T} = \frac{6}{10} = 0.6, \qquad \text{that is, } \theta = 36.9°$$

Thus

$$\mathbf{I}_T = 10\underline{/60° + \theta°} = 10\underline{/96.9°} \text{ A}$$

and

$$i_T(t) = 10\sqrt{2} \sin (1000t + 96.9°) \text{ A}$$

As E, I_R, and I' are all now determined, then

$$R = \frac{E}{I_R} = \frac{120}{6} = 20 \ \Omega$$

$$X_L \text{ (single element)} = \frac{E}{I'} = \frac{120}{8} = 15 \ \Omega = \omega L$$

Therefore,

$$L = \frac{15}{1000} = 15 \text{ mH}$$

EXAMPLE 12.25

In the circuit of Fig. 12.60, I_R is 3 A, ω is 333.3 rad/s, and I_T is 5 A, $\mathbf{I}_T$ lags $\mathbf{I}_X$. Find
(a) The source voltage $\mathbf{E}$ in phasor form, taking $\mathbf{I}_T$ as the reference.
(b) The value of the unknown single element.
(c) The source current I_T when ω is reduced to zero.
(d) The source current when ω is changed such that phase shift between $\mathbf{I}_T$ and $\mathbf{I}_X$ is 30°.

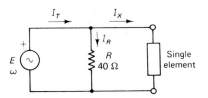

Figure 12.60

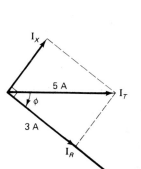

Figure 12.61

Solution (a) In Fig. 12.61, $\mathbf{I}_T$ is drawn as the reference phasor. $\mathbf{I}_X$ is drawn leading $\mathbf{I}_T$. Since

$$\mathbf{I}_T = \mathbf{I}_X + \mathbf{I}_R$$

then $\mathbf{I}_R$ (in phase with $\mathbf{E}$) must be lagging $\mathbf{I}_T$ by the phase angle ϕ, as indicated. Also, as the black box contains a single element, whose current $\mathbf{I}_X$ leads the voltage across it, $\mathbf{E}$, this element must be a capacitor. The angle between $\mathbf{E}$ and $\mathbf{I}_X$ is then 90°.

From the geometry of the phasor diagram,

$$\cos \phi = \frac{I_R}{I_T} = \frac{3}{5} = 0.6 \qquad \text{that is, } \phi = 53.1°$$

and

$$I_T^2 = I_R^2 + I_X^2$$

Therefore,

$$I_X = \sqrt{(5)^2 - (3)^2} = 4\ \text{A}$$

As

$$E = I_R R = 3 \times 40 = 120\ \text{V}$$

then

$$\mathbf{E} = 120\underline{/-53.1°}\ \text{V}$$

(b) The capacitor reactance X_C is thus

$$X_C = \frac{E}{I_X} = \frac{120}{4} = 30\ \Omega = \frac{1}{\omega C}$$

Therefore,

$$C = \frac{1}{\omega X_C} = \frac{1}{333.3 \times 30} = 100\ \mu\text{F}$$

(c) As ω tends to zero, X_C becomes ∞ (i.e., an open-circuit branch). Thus

$$\mathbf{I}_X = 0$$

and

$$\mathbf{I}_T = \mathbf{I}_R \qquad (I_R \text{ is constant independent of frequency})$$

Therefore,

$$I_T = I_R = 3\ A$$

(d) When ω is varied, I_X changes while I_R remains constant. Hence I_T and its phase shift with respect to $\mathbf{I}_R$ (or $\mathbf{E}$) also change. At the frequency setting of the source, corresponding to phase shift between $\mathbf{I}_X$ and $\mathbf{I}_T$ of 30°,

$$\phi = 90° - 30° = 60°$$

$$\cos \phi = \cos 60° = 0.5 = \frac{I_R}{I_T} = \frac{3}{I_T}$$

Therefore,

$$I_T = \frac{3}{0.5} = 6 \text{ A}$$

EXAMPLE 12.26

The black box in Fig. 12.62 contains a number of ideal parallel-connected elements. At $\omega_1 = 1000$ rad/s, the source current $\mathbf{I}$ was $2\sqrt{2}\underline{/-45°}$ A. At $\omega_2 = 2000$ rad/s, $\mathbf{I}$ was leading $\mathbf{E}$ by $60°$. Find the nature and magnitude of the least number of components inside the black box that satisfy these measurement results.

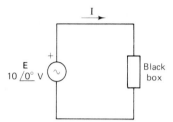

Figure 12.62

Solution   Since at ω_1 the current lags the voltage by a phase angle less than $90°$, the circuit is predominantly inductive (equivalent to parallel R and parallel L branches). But at ω_2 the current leads the voltage also by a phase angle less than $90°$; the circuit is then predominantly capacitive (equivalent to parallel R and parallel C branches). These changes in the $\mathbf{I}$–$\mathbf{E}$ characteristics of the circuit with frequency indicate that it must at least contain three ideal R, L, and C parallel branches. Hence

$$Y_T = G - jB_L + jB_C = \frac{1}{R} - j\frac{1}{\omega L} + j\omega C$$

In the first case, at $\omega = \omega_1 = 1000$ rad/s,

$$Y_T = \frac{\mathbf{I}}{\mathbf{E}} = \frac{2\sqrt{2}\underline{/-45°}}{10\underline{/0°}} = 0.2828\underline{/-45°} = 0.2 - j0.2 \text{ S}$$

Comparing this result with the general expression for Y_T above, we conclude that

$$G = \frac{1}{R} = 0.2 \text{ S} \qquad \text{that is, } R = \frac{1}{0.2} = 5 \text{ }\Omega$$

and

$$-\frac{1}{1000L} + 1000C = -0.2 \tag{12.72}$$

In the second case, at $\omega = \omega_2 = 2000$ rad/s, $\mathbf{I}$ leads $\mathbf{E}$ by $60°$ (i.e., the phase angle of Y_T is also $60°$). Here the circuit is predominantly capacitive, (i.e., B_C is greater than B_L). Thus from the relationship for the phase of Y_T,

$$\tan \phi = \tan 60° = \frac{\omega_2 C - 1/\omega_2 L}{1/R}$$

and

$$1.732 = 5\left(2000C - \frac{1}{2000L}\right)$$

$$1000C = \frac{1}{4000L} + 0.1732 \qquad (12.73)$$

Substituting from Eq. (12.73) into Eq. (12.72) yields

$$-\frac{1}{1000L} + \frac{1}{4000L} + 0.1732 = -0.2$$

Therefore,

$$-\frac{3}{4000L} = -0.3732$$

and

$$L = 2.01 \text{ mH}$$

Now, substituting into Eq. (12.73) gives us

$$C = 297.6 \,\mu\text{F}$$

EXAMPLE 12.27

The black box in the circuit shown in Fig. 12.63(a) contains two parallel-connected elements. The oscillogram obtained is shown in Fig. 12.63(b), with the oscilloscope controls adjusted to

$$\text{time/div} = 2\,\text{ms}$$
$$\text{V/div}\ (Y_A) = 0.02\,\text{V}$$
$$\text{V/div}\ (Y_B) = 20\,\text{V}$$

Find the nature and value of the two elements.

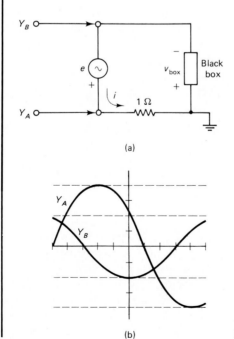

(a)

(b)

Figure 12.63

Solution Y_A displays the voltage across the 1 Ω sense resistor, (i.e., it displays $1 \times i$ volts); therefore,

$$I_{max} = 4 \times 0.02 = 0.08 \text{ A}$$

$$I = 0.707 I_{max} = 0.05657 \text{ A}$$

Y_B displays the inverted voltage across the black box (i.e., it displays $-v_{box}$). Thus the v_{box} waveform is Y_B inverted.

$$V_{box} = 0.707 \times 2 \times 20 = 28.28 \text{ V}$$

Also (with Y_B inverted), i leads v_{box} by an angle ϕ, where

$$\phi = 2 \text{ div} \times \frac{180°}{6 \text{ div}} = 60°$$

If **I** is taken to be the reference phasor,

$$\mathbf{I} = 0.05657\underline{/0°} \text{ A}$$

then

$$\mathbf{V}_{box} = 28.28\underline{/-60°} \text{ V} \quad (\text{as } \mathbf{V}_{box} \text{ lags } \mathbf{I} \text{ by } 60°)$$

Using Ohm's law gives us

$$Y_{box} = \frac{\mathbf{I}}{\mathbf{V}_{box}} = \frac{0.05657\underline{/0°}}{28.28\underline{/-60°}} = 0.002\underline{/60°} \text{ S}$$

$$= 0.001 + j0.001732 \text{ S} = G + jB_C = \frac{1}{R} + j\omega C$$

where

$$\omega = \frac{2\pi}{T} = \frac{2\pi}{12 \text{ div} \times 2 \times 10^{-3} \text{ s/div}} = 261.8 \text{ rad/s}$$

The admittance of the box thus indicates that the box contains a resistive branch R, in parallel with a capacitive branch C, where

$$R = \frac{1}{G} = \frac{1}{0.001} = 1000 \text{ Ω}$$

and

$$\omega C = B_C = 0.001732 \text{ S}$$

Therefore,

$$C = \frac{0.001732}{261.8} = 6.62 \text{ μF}$$

12.5 | GENERAL SERIES–PARALLEL AC CIRCUITS

In general, an impedance Z_n of branch n in a given ac circuit could be composed of a number of single R, L, and C elements connected in series or in parallel. In the steady-state ac analysis, the electric circuit is drawn in the frequency domain, with the branches represented by their impedances (or admittances) and the circuit variables (voltages and currents) represented by their equivalent phasors. KVL, KCL, and Ohm's law for each branch, and for the circuit as a whole, always apply. In fact, the analysis of ac circuits is performed in exactly the same

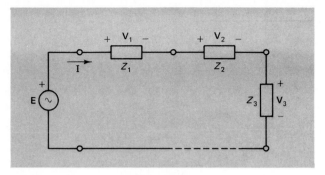

Figure 12.64

manner as the dc circuit analysis, except that all manipulations are performed using complex-number algebra.

The current $\mathbf{I}$ is the same in all the series-connected impedances shown in Fig. 12.64. Here

$$Z_T = Z_1 + Z_2 + Z_3 + \cdots \qquad (12.74)$$

and KVL can be expressed as

$$\mathbf{E} = \mathbf{V}_1 + \mathbf{V}_2 + \mathbf{V}_3 + \cdots \qquad (12.75)$$

while Ohm's law for the circuit as a whole and for each component provides

$$\mathbf{I} = \frac{\mathbf{E}}{Z_T} = \frac{\mathbf{V}_1}{Z_1} = \frac{\mathbf{V}_2}{Z_2} = \frac{\mathbf{V}_3}{Z_3} = \cdots \qquad (12.76)$$

Equation (12.76) can also be manipulated to produce the voltage-division principle, where

$$\mathbf{V}_1 = \mathbf{I}Z_1 = \frac{\mathbf{E}}{Z_T} Z_1 = \mathbf{E} \frac{Z_1}{Z_1 + Z_2 + Z_3 + \cdots} \qquad (12.77a)$$

Similarly,

$$\mathbf{V}_2 = \mathbf{E} \frac{Z_2}{Z_1 + Z_2 + Z_3 + \cdots} \quad \text{etc.} \qquad (12.77b)$$

The voltage $\mathbf{V}$ across each branch in the general parallel circuit shown in Fig. 12.65 is the same. Each branch impedance is by itself a series or a parallel connection of R, L, and C elements. Here

$$\frac{1}{Z_T} = \frac{1}{Z_1} + \frac{1}{Z_2} + \frac{1}{Z_3} + \cdots \qquad (12.78a)$$

or

$$Y_T = Y_1 + Y_2 + Y_3 + \cdots \qquad (12.78b)$$

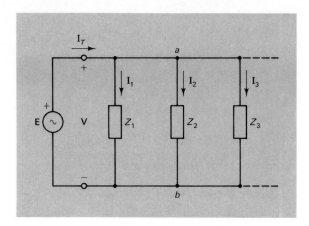

Figure 12.65

and KCL at nodes *a* or *b* provides

$$\mathbf{I}_T = \mathbf{I}_1 + \mathbf{I}_2 + \mathbf{I}_3 + \cdots \qquad (12.79)$$

Ohm's law for each branch and for the circuit as a whole can be expressed as follows:

$$\mathbf{E} = \mathbf{V} = \mathbf{I}_T Z_T = \mathbf{I}_1 Z_1 = \mathbf{I}_2 Z_2 = \mathbf{I}_3 Z_3 = \cdots \qquad (12.80)$$

The current-division principle can be derived from the manipulation of Eq. (12.80), where

$$\mathbf{I}_1 = \mathbf{I}_T \frac{Z_T}{Z_1} = \mathbf{I}_T \frac{Y_1}{Y_T} = \mathbf{I}_T \frac{Y_1}{Y_1 + Y_2 + Y_3 + \cdots} \qquad (12.81a)$$

Similarly,

$$\mathbf{I}_2 = \mathbf{I}_T \frac{Y_2}{Y_1 + Y_2 + Y_3 + \cdots} \quad \text{etc.} \qquad (12.81b)$$

Note the duality between the series and parallel circuit relationships (Y replaces Z and $\mathbf{I}$ replaces $\mathbf{V}$, and vice versa). In the special case of the two parallel branches in Fig. 12.66,

$$Z_T = \frac{1}{Y_T} = \frac{1}{Y_1 + Y_2} = \frac{1}{1/Z_1 + 1/Z_2} = \frac{Z_1 Z_2}{Z_1 + Z_2} \qquad (12.82)$$

while

$$\mathbf{V} = \mathbf{I}_1 Z_1 = \mathbf{I}_2 Z_2 = \mathbf{I}_T Z_T$$

and

$$\mathbf{I}_1 = \mathbf{I}_T \frac{Z_T}{Z_1} = \mathbf{I}_T \frac{Z_2}{Z_1 + Z_2} \qquad (12.83a)$$

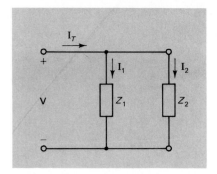

Figure 12.66

Also,

$$\mathbf{I}_2 = \mathbf{I}_T \frac{Z_1}{Z_1 + Z_2}$$ (12.83b)

The current-division principle expressed by Eqs. (12.83) indicates how the current $\mathbf{I}_T$ is split between the two parallel branch impedances, Z_1 and Z_2. This is a very useful tool in network analysis.

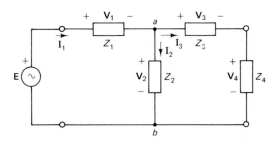

Figure 12.67

The circuit shown in Fig. 12.67 is an example of a simple single-source series–parallel ac network. Here also, KVL, KCL, and Ohm's law can be applied, resulting in

$$\text{KVL:} \quad \mathbf{E} = \mathbf{V}_1 + \mathbf{V}_2$$ (12.84)

with

$$\mathbf{V}_2 = \mathbf{V}_3 + \mathbf{V}_4$$ (12.85)

and

$$\text{KCL at node } a \text{ (or } b\text{):} \quad \mathbf{I}_1 = \mathbf{I}_2 + \mathbf{I}_3$$ (12.86)

Noting that Z_3 and Z_4 are in series, while the combined impedance $(Z_3 + Z_4)$ is in parallel with Z_2; the overall total impedance Z_T of the network, as seen by the source $\mathbf{E}$, is

$$Z_T = Z_1 + [Z_2 \| (Z_3 + Z_4)] = Z_1 + \frac{Z_2(Z_3 + Z_4)}{Z_2 + (Z_3 + Z_4)}$$ (12.87)

From Ohm's law for each branch impedance, we have

$$\mathbf{V}_1 = \mathbf{I}_1 Z_1 \tag{12.88}$$

$$\mathbf{V}_2 = \mathbf{I}_2 Z_2 = \mathbf{I}_3(Z_3 + Z_4) \tag{12.89}$$

$$\mathbf{V}_3 = \mathbf{I}_3 Z_3 \tag{12.90}$$

$$\mathbf{V}_4 = \mathbf{I}_3 Z_4 \tag{12.91}$$

Manipulation of the relationships above allows one to analyze such series-parallel circuits and obtain any required unknown network quantity. The power dissipated in any ac circuit is obtained from the general relationship

$$\boxed{P_{av} = VI \cos \phi_{\mathbf{VI}} \quad W} \tag{12.92a}$$

where $\mathbf{V}$ is the phasor voltage across the impedance Z and $\mathbf{I}$ is the phasor current flowing into it, while $\phi_{\mathbf{VI}}$ is the phase shift between the two phasors $\mathbf{V}$ and $\mathbf{I}$.

For a network as a whole, with the source $\mathbf{E}$ driving a current $\mathbf{I}_T$ into the network, whose impedance is Z_T,

$$\boxed{P_{av} = EI_T \cos \phi_{\mathbf{EI}_T} \quad W} \tag{12.92b}$$

Power in ac circuits is discussed in more detail in Chapter 13.

■ **EXAMPLE 12.28**

In the circuit shown in Fig. 12.68, $e(t) = 14.14 \sin(10t)$ volts. Find i_1, i_2, and v in the time domain.

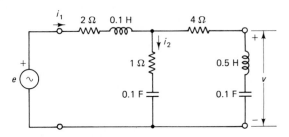

Figure 12.68

Solution With $\omega = 10\,\text{rad/s}$ and $\mathbf{E} = 10\underline{/0°}\,\text{V}$, the circuit can be redrawn in the frequency domain as shown in Fig. 12.69. The elements are represented by

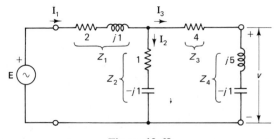

Figure 12.69

their impedances, and denoted

$$Z_1 = 2 + j1 = 2.236\underline{/26.6°}\ \Omega$$
$$Z_2 = 1 - j1 = 1.414\underline{/-45°}\ \Omega$$
$$Z_3 = 4\ \Omega$$
$$Z_4 = j5 - j1 = j4 = 4\underline{/90°}\ \Omega$$

According to the analysis of the similar circuit in Fig. 12.67, as described above,

$$Z_T = Z_1 + \frac{Z_2(Z_3 + Z_4)}{Z_2 + Z_3 + Z_4} = 2 + j1 + \frac{(1 - j1)(4 + j4)}{1 - j1 + 4 + j4}$$

$$= 2 + j1 + \frac{1.414\underline{/-45°} \times 5.657\underline{/45°}}{5.831\underline{/31°}} = 2 + j1 + 1.372\underline{/-31°}$$

$$= 2 + j1 + 1.18 - j0.707 = 3.18 + j0.293 = 3.19\underline{/5.3°}\ \Omega$$

Therefore,

$$\mathbf{I}_1 = \frac{\mathbf{E}}{Z_T} = \frac{10\underline{/0°}}{3.19\underline{/5.3°}} = 3.13\underline{/-5.3°}\ A$$

and

$$i_1(t) = 3.13\sqrt{2}\sin(10t - 5.3°)\ A$$

Now using the current-division principle for the two parallel branches, Z_2 and $(Z_3 + Z_4)$,

$$\mathbf{I}_2 = \mathbf{I}_1 \frac{Z_3 + Z_4}{Z_2 + Z_3 + Z_4} = 3.13\underline{/-5.3°} \times \frac{4 + j4}{5 + j3} = 3.13\underline{/-5.3°} \times \frac{5.657\underline{/45°}}{5.831\underline{/31°}}$$

$$= 3.04\underline{/8.7°}\ A$$

and

$$i_2(t) = 3.04\sqrt{2}\sin(10t + 8.7°)\ A$$

Also,

$$\mathbf{I}_3 = \mathbf{I}_1 \frac{Z_2}{Z_2 + Z_3 + Z_4} = 3.13\underline{/-5.3°} \times \frac{1 - j1}{5 + j3}$$

$$= 3.13\underline{/-5.3°} \times \frac{1.414\underline{/-45°}}{5.831\underline{/31°}} = 0.76\underline{/-81.3°}\ A$$

Thus from Ohm's law,

$$\mathbf{V} = \mathbf{I}_3 Z_4 = 0.76\underline{/-81.3°} \times 4\underline{/90°} = 3.04\underline{/8.7°}\ V$$

Therefore,

$$v(t) = 3.04\sqrt{2}\sin(10t + 8.7°)\ V$$

EXAMPLE 12.29

For the circuit shown in Fig. 12.70, find
(a) $\mathbf{I}$.
(b) $\mathbf{V}_D$.
(c) The phase angle between the Y_A and the Y_B waveforms.
(d) The equivalent simple series circuit at this frequency.

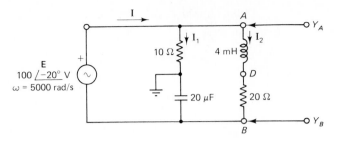

Figure 12.70

Solution (a) The impedances of the two branches are

$$\mathbf{Z}_1 = 10 - j\frac{1}{5000 \times 20 \times 10^{-6}} = 10 - j10 = 14.14\underline{/-45°}\ \Omega$$

and

$$Z_2 = 20 + j5000 \times 4 \times 10^{-3} = 20 + j20 = 28.28\underline{/45°}\ \Omega$$

Therefore, from Ohm's law,

$$\mathbf{I}_1 = \frac{\mathbf{E}}{\mathbf{Z}_1} = \frac{100\underline{/-20°}}{14.14\underline{/-45°}} = 7.07\underline{/25°} = 6.41 + j2.99\ \text{A}$$

and,

$$\mathbf{I}_2 = \frac{\mathbf{E}}{\mathbf{Z}_2} = \frac{100\underline{/-20°}}{28.28\underline{/45°}} = 3.536\underline{/-65°} = 1.49 - j3.20\ \text{A}$$

Applying KCL yields

$$\mathbf{I} = \mathbf{I}_1 + \mathbf{I}_2 = 7.9 - j0.21 = 7.903\underline{/-1.5°}\ \text{A}$$

(b) The voltage between points A and D is

$$\mathbf{V}_{AD} = \mathbf{V}_A - \mathbf{V}_D$$

where $\mathbf{V}_A$ and $\mathbf{V}_D$ are the voltages at points A and D with respect to ground. Applying Ohm's law for $\mathbf{V}_A$ and $\mathbf{V}_{AD}$; $\mathbf{V}_D$ can be expressed as

$$\begin{aligned}
\mathbf{V}_D &= \mathbf{V}_A - \mathbf{V}_{AD} \\
&= 10\underline{/0°}\ \mathbf{I}_1 - 20\underline{/90°}\ \mathbf{I}_2 \\
&= 10\underline{/0°} \times 7.07\underline{/25°} - 20\underline{/90°} \times 3.536\underline{/-65°} \\
&= 70.7\underline{/25°} - 70.7\underline{/25°} = 0\ \text{V}
\end{aligned}$$

(i.e., a voltmeter connected between point D and ground will read 0 V).

(c) Y_A monitors v_A (i.e., $\mathbf{V}_A$), where

$$\mathbf{V}_A = 10\underline{/0°}\ \mathbf{I}_1 = 70.7\underline{/25°}\ \text{V}$$

and Y_B monitors v_B (i.e., $\mathbf{V}_B$), where

$$\mathbf{V}_B = -Z_C\mathbf{I}_1 = -10\underline{/-90°} \times 7.07\underline{/25°} = 70.7\underline{/115°}\ \text{V}$$

(the number $-1 = 1\underline{/180°}$). Thus the Y_B waveform leads the Y_A waveform by 90°.

(d) To find the equivalent series circuit, the total impedance of the circuit must be calculated. This is easy to obtain from Ohm's law since $\mathbf{E}$ and $\mathbf{I}$ are known. Thus

$$Z_T = \frac{\mathbf{E}}{\mathbf{I}} = \frac{100\underline{/-20°}}{7.903\underline{/-1.5°}} = 12.66\underline{/-18.5°}\ \Omega$$

$$= 12 - j4\ \Omega$$

$$= R_{eq} - jX_{C eq} = R_{eq} - j\frac{1}{\omega C_{eq}}$$

Therefore,

$$R_{eq} = 12\ \Omega$$

and

$$C_{eq} = \frac{1}{\omega X_{C eq}} = \frac{1}{5000 \times 4} = 50\ \mu F$$

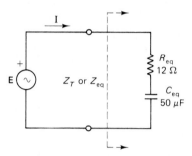

Figure 12.71

Z_T and its series components as obtained above are shown in Fig. 12.71. The equivalent series circuit is capacitive since $\mathbf{I}$ leads $\mathbf{E}$ at this frequency.

EXAMPLE 12.30

Both Z_A and Z_B in the circuit shown in Fig. 12.72 are composed of a single element each. The applied source has a voltage of 70.7 V and $\mathbf{E}$ leads $\mathbf{I}_3$. Also, $i_1(t) = 11.18 \sin(1000t + 26.5°)$ A and $i_3(t) = 7.07 \cos(1000t - 45°)$ A.

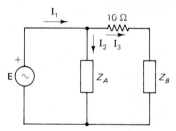

Figure 12.72

Find
(a) $\mathbf{I}_2$.
(b) The phase angle between $\mathbf{E}$ and $\mathbf{I}_3$.
(c) The values of the elements constituting Z_A and Z_B.

Solution (a) From the given waveforms, and with the cosine function changed into sine,

$$\mathbf{I}_1 = 0.707 \times 11.18\underline{/26.5°} = 7.905\underline{/26.5°}\ A$$

and
$$\mathbf{I}_3 = 0.707 \times 7.07\underline{/45°} = 5.0\underline{/45°} \text{ A}$$
Using KCL gives us
$$\mathbf{I}_1 = \mathbf{I}_2 + \mathbf{I}_3$$
Therefore,
$$\mathbf{I}_2 = \mathbf{I}_1 - \mathbf{I}_3 = (7.074 + j3.527) - (3.536 + j3.536)$$
$$\cong 3.538\underline{/0°} \text{ A}$$

(b) The three current phasors are shown in Fig. 12.73. Since $\mathbf{E}$ leads $\mathbf{I}_3$, then it also leads $\mathbf{I}_1$ and $\mathbf{I}_2$. Also, as $\mathbf{E}$ leads $\mathbf{I}_2$ (being the voltage across and the current through the single element Z_A), the phase shift between them must be 90°, and thus the element Z_A must be an inductor. Hence, as indicated in the phasor diagram,

$$\mathbf{E} = 70.7\underline{/90°} \text{ V}$$

The phase shift between $\mathbf{E}$ and $\mathbf{I}_3$ is thus 45°.

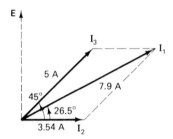

Figure 12.73

(c) Applying Ohm's law to each branch,
$$Z_A = \frac{\mathbf{E}}{\mathbf{I}_2} = \frac{70.7\underline{/90°}}{3.538\underline{/0°}} = 20\underline{/90°} \text{ Ω} = X_{LA}\underline{/90°} \text{ Ω}$$
Therefore,
$$L_A = \frac{X_{LA}}{\omega} = \frac{20}{1000} = 20 \text{ mH}$$
For the second branch,
$$\frac{\mathbf{E}}{\mathbf{I}_3} = 10 + Z_B$$
Therefore,
$$Z_B = \frac{70.7\underline{/90°}}{5\underline{/45°}} - 10 = 14.14\underline{/45°} - 10 = 10 + j10 - 10$$
$$= j10 = jX_{LB}$$
and
$$L_B = \frac{X_{LB}}{\omega} = \frac{10}{1000} = 10 \text{ mH}$$

EXAMPLE 12.31

In the circuit shown in Fig. 12.74, $i_T(t) = 0.158 \cos(1000t + 26.6°)$ A and $i_1(t) = 0.1 \sin(1000t + 135°)$ A. Find
(a) The voltage $e(t)$ of the applied source.

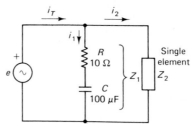

Figure 12.74

(b) The value of the unknown element.
(c) The phasors representing the voltages across R and C.

Solution (a) From the time-domain expression for the currents given,

$$\mathbf{I}_T = 0.158 \times 0.707\underline{/26.6° + 90°} = 0.112\underline{/116.6°} \text{ A}$$
$$= -0.05 + j0.1 \text{ A}$$

and

$$\mathbf{I}_1 = 0.0707\underline{/135°} \text{ A} = -0.05 + j0.05 \text{ A}$$

Thus, from KCL

$$\mathbf{I}_2 = \mathbf{I}_T - \mathbf{I}_1 = -0.05 + j0.1 - (-0.05 + j0.05) = j0.05$$
$$= 0.05\underline{/90°} \text{ A}$$

The impedance of the first branch, Z_1, is

$$Z_1 = R - j\frac{1}{\omega C} = 10 - j\frac{1}{1000 \times 100 \times 10^{-6}} = 10 - j10 \text{ }\Omega$$
$$= 14.14\underline{/-45°} \text{ }\Omega$$

Thus, from Ohm's law,

$$\mathbf{E} = \mathbf{I}_1 Z_1 = 0.0707\underline{/135°} \times 14.14\underline{/-45°} = 1\underline{/90°} \text{ V}$$

Therefore,

$$e(t) = 1.414 \sin(1000t + 90°) \text{ V}$$

(b) The impedance Z_2 of the unknown element can be obtained from Ohm's law:

$$Z_2 = \frac{\mathbf{E}}{\mathbf{I}_2} = \frac{1\underline{/90°}}{0.05\underline{/90°}} = 20\underline{/0°} \text{ }\Omega$$

Thus Z_2 is a resistance of value 20 Ω.

(c) Applying Ohm's law to the elements of the first branch gives us

$$\mathbf{V}_R = \mathbf{I}_1 Z_R = 0.0707\underline{/135°} \times 10\underline{/0°} = 0.707\underline{/135°} \text{ V}$$

and

$$\mathbf{V}_C = \mathbf{I}_1 Z_C = 0.0707\underline{/135°} \times 10\underline{/-90°} = 0.07\underline{/45°} \text{ V}$$

EXAMPLE 12.32

The phase angle between the two branch currents, i_1 and i_2, in Fig. 12.75 is 120°. Also, E is 100 V, I_1 is 5.0 A, $\omega = 1000$ rad/s, and C is 25 μF. Find
(a) The source current and draw the phasor diagram.
(b) The values of L and R.

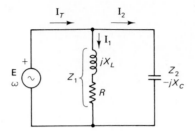

Figure 12.75

Solution (a) Let $\mathbf{E}$ be the reference phasor

$$\mathbf{E} = 100\underline{/0°}\ \text{V}$$

$$Z_2 = X_C\underline{/-90°} = \frac{1}{1000 \times 25 \times 10^{-6}}\underline{/-90°} = 40\underline{/-90°}\ \Omega$$

Therefore,

$$\mathbf{I}_2 = \frac{100\underline{/0°}}{40\underline{/-90°}} = 2.5\underline{/90°}\ \text{A}$$

Since $\mathbf{I}_1$ must lag $\mathbf{E}$ because Z_1 is an RL impedance, and the phase shift between $\mathbf{I}_1$ and $\mathbf{I}_2$ is 120°, then as the phasor diagram in Fig. 12.76 indicates,

$$\mathbf{I}_1 = 5\underline{/-30°}\ \text{A}$$

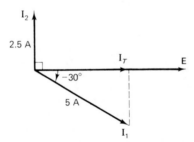

Figure 12.76

Now using KCL, the source current $\mathbf{I}_T$ is

$$\mathbf{I}_T = \mathbf{I}_1 + \mathbf{I}_2 = 5\underline{/-30°} + 2.5\underline{/90°}$$
$$= 4.33 - j2.5 + j2.5 = 4.33\underline{/0°}\ \text{A}$$

as shown in the phasor diagram.

(b) From Ohm's law for the impedance of the first branch, Z_1,

$$Z_1 = \frac{\mathbf{E}}{\mathbf{I}_1} = \frac{100\underline{/0°}}{5\underline{/-30°}} = 20\underline{/30°}\ \Omega$$
$$= 17.32 + j10 = R + jX_L$$

Therefore,

$$R = 17.32\ \Omega$$

and

$$X_L = 10\ \Omega = \omega L = 1000L$$

Therefore,

$$L = 10\ \text{mH}$$

EXAMPLE 12.33

Based on the assumed directions of current flow in the different branches of the lattice circuit shown in Fig. 12.77, find

(a) $\mathbf{V}_1$, $\mathbf{V}_2$, and $\mathbf{V}_3$ across each of the impedances Z_1, Z_2, and Z_3.

(b) $\mathbf{I}_1$, $\mathbf{I}_2$, and $\mathbf{I}_3$ flowing in each of the three impedances.

(c) The currents $\mathbf{I}_4$, $\mathbf{I}_5$, and $\mathbf{I}_6$ flowing in the three sources.

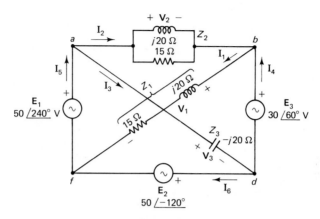

Figure 12.77

Solution   (a, b) Applying KVL to the *bdfb* loop, we have

$$-\mathbf{E}_2 - \mathbf{E}_3 + \mathbf{V}_1 = 0$$

Therefore,

$$\mathbf{V}_1 = \mathbf{E}_2 + \mathbf{E}_3 = 50\underline{/-120°} + 30\underline{/60°}$$
$$= (-25 - j43.3) + (15 + j25.98)$$
$$= -10 - j17.32 = 20\underline{/-120°} \text{ V}$$

Using Ohm's law for the impedance Z_1, where $Z_1 = 15 + j20 = 25\underline{/53.1°}\ \Omega$, we obtain

$$\mathbf{I}_1 = \frac{\mathbf{V}_1}{Z_1} = \frac{20\underline{/-120°}}{25\underline{/53.1°}} = 0.8\underline{/-173.1°} \text{ A}$$
$$= -0.794 - j0.096 \text{ A}$$

Applying KVL to the outside loop *abdfa*,

$$\mathbf{E}_3 + \mathbf{E}_2 - \mathbf{E}_1 + \mathbf{V}_2 = 0$$

Therefore,

$$\mathbf{V}_2 = \mathbf{E}_1 - (\mathbf{E}_2 + \mathbf{E}_3) = 50\underline{/240°} - 20\underline{/-120°} = 30\underline{/-120°} \text{ V}$$

Notice that the phase angle 240° is the same as the phase angle −120°. The admittance Y_2 of the branch containing Z_2 is

$$Y_2 = \frac{1}{15} + \frac{1}{j20} = 0.0667 - j0.05 = 0.0833\underline{/-36.9°} \text{ S}$$

From Ohm's law for this branch,

$$\mathbf{I}_2 = \frac{\mathbf{V}_2}{Z_2} = \mathbf{V}_2 Y_2 = 30\underline{/-120°} \times 0.0833\underline{/-36.9°}$$

$$= 2.5\underline{/-156.9°}\,\text{A}$$

$$= -2.3 - j0.981\,\text{A}$$

Applying KVL to the *adfa* loop yields

$$\mathbf{V}_3 + \mathbf{E}_2 - \mathbf{E}_1 = 0$$

Therefore,

$$\mathbf{V}_3 = \mathbf{E}_1 - \mathbf{E}_2 = 50\underline{/240°} - 50\underline{/-120°} = 0\,\text{V}$$

with

$$Z_3 = -j20\,\Omega$$

$$\mathbf{I}_3 = \frac{\mathbf{V}_3}{Z_3} = 0\,\text{A}$$

(c) Applying KCL at node *a*, we have

$$\mathbf{I}_5 = \mathbf{I}_2 + \mathbf{I}_3 = \mathbf{I}_2 = 2.5\underline{/-156.9°}\,\text{A}$$

Applying KCL at node *b* gives us

$$\mathbf{I}_4 + \mathbf{I}_2 = \mathbf{I}_1$$

Therefore,

$$\mathbf{I}_4 = \mathbf{I}_1 - \mathbf{I}_2 = (-0.794 - j0.096) - (-2.3 - j0.981)$$

$$= 1.506 + j0.885 = 1.747\underline{/30.4°}\,\text{A}$$

Applying KCL at node *d* yields

$$\mathbf{I}_3 = \mathbf{I}_4 + \mathbf{I}_6 = 0$$

Therefore,

$$\mathbf{I}_6 = -\mathbf{I}_4 = 1.747\underline{/-149.6°}\,\text{A}$$

As a check, at node *f* KCL provides

$$\mathbf{I}_5 = \mathbf{I}_1 + \mathbf{I}_6 = (-0.794 - j0.096) + (-1.506 - j0.885)$$

$$= -2.3 - j0.981 = 2.5\underline{/-156.9°}\,\text{A}$$

12.6 | IMPEDANCE MEASUREMENTS: THE IMPEDANCE BRIDGE

To measure an unknown impedance, Z_X (usually a capacitor or an inductor), a bridge circuit similar to Wheatstone bridge, as shown in Fig. 12.78, is used. Here the source $\mathbf{E}$ is an ac voltage source, whose frequency ω is usually fixed. Z_1, Z_2, and Z_3 are standard impedances which could be varied. Any two of these impedances could be used for bridge adjustment to achieve the balance condition. The variable components in any of these impedances are usually variable standard resistors because they are cheaper, more accurate, and of better quality than the variable standard capacitors or inductors. In actual practical bridges, the adjustable variable resistors are provided with dials calibrated to indicate the

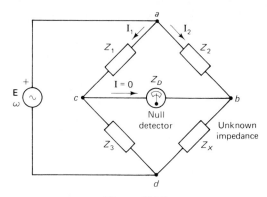

Figure 12.78

value of the unknown component (or components) directly once bridge balance is achieved.

The balance condition corresponds to zero deflection the sensitive null-detector (galvanometer); that is, the detector current $\mathbf{I}$ is zero. In this case

$$\mathbf{V}_{cb} = \mathbf{I}Z_D = 0 \qquad (12.93)$$

(i.e., points b and c are at the same potential level). Also, $\mathbf{I}_1$ flows in both Z_1 and Z_3, while $\mathbf{I}_2$ flows in both Z_2 and Z_X. It can thus be easily concluded that

$$\mathbf{V}_{ac} = \mathbf{V}_{ab} \Rightarrow \mathbf{I}_1 Z_1 = \mathbf{I}_2 Z_2 \qquad (12.94)$$

and

$$\mathbf{V}_{cd} = \mathbf{V}_{bd} \Rightarrow \mathbf{I}_1 Z_3 = \mathbf{I}_2 Z_X \qquad (12.95)$$

Taking the ratio of these two equations above we find that

$$\boxed{\frac{Z_1}{Z_3} = \frac{Z_2}{Z_X} \quad \text{(touching-arms ratio formula)}} \qquad (12.96a)$$

or

$$\boxed{Z_1 Z_X = Z_2 Z_3 \quad \text{(opposite-arms multiplication formula)}} \qquad (12.96b)$$

Equations (12.96) are the two ways in which the bridge balance condition can be expressed. Z_X can then be easily obtained in terms of the known Z_1, Z_2, and Z_3 impedances.

▮ EXAMPLE 12.34

The capacitance bridge is shown in Fig. 12.79. The model of a practical capacitor C_X usually includes a large parallel resistance R_D, called the dissipation or leakage resistance. At the balance condition of the bridge in Fig. 12.79,

$$\frac{Z_1}{Z_3} = \frac{Z_2}{Z_X}$$

or

$$Z_1 Y_3 = Z_2 Y_X$$

Therefore,

$$R_1\left(\frac{1}{R_3} + j\omega C_S\right) = R_2\left(\frac{1}{R_D} + j\omega C_X\right) \qquad (12.97)$$

C_S is a standard capacitor. The two complex numbers on each side of Eq. (12.97) are equal only if their real parts are equal and their imaginary parts are also equal.

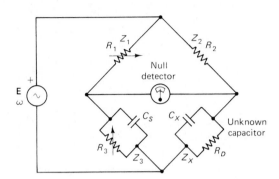

Figure 12.79

Here

$$\left.\begin{array}{c} \dfrac{R_1}{R_2} = \dfrac{R_2}{R_D} \Rightarrow R_D = R_2\dfrac{R_3}{R_1} \\[2em] \omega R_1 C_S = \omega R_2 C_X \Rightarrow C_X = C_S\dfrac{R_1}{R_2} \end{array}\right\} \text{ bridge equations} \qquad \begin{array}{c} (12.98) \\[2em] (12.99) \end{array}$$

R_1 and R_3 are successively and alternatively adjusted until the condition of the bridge balance (i.e., zero deflection in the null detector) is attained. If the dials on R_1 and R_3 indicate their respective values, the bridge equations (12.98) and (12.99) can be used to calculate C_X and R_D. Alternatively, the dial on R_1 can be calibrated to read C_X directly, while the dial on R_3 can be calibrated to indicate the value of R_D directly.

EXAMPLE 12.35

The inductance (Maxwell's) bridge is shown in Fig. 12.80. The model of a practical inductor L_X usually includes a small series resistance R_X to account for the resistance of the coil. At the condition of balance of the bridge in Fig. 12.80,

$$Z_1 Z_X = Z_2 Z_3$$

or

$$Z_X = Z_2 Z_3 Y_1$$

Therefore,

$$R_X + j\omega L_X = R_2 R_3\left(\frac{1}{R_1} + j\omega C_S\right) \qquad (12.100)$$

Equating the real and imaginary parts, respectively, on each side of Eq.

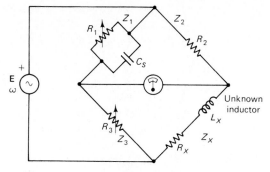

Figure 12.80

(12.100) above yields

$$R_X = R_2 \dfrac{R_3}{R_1} \left.\vphantom{\dfrac{R_3}{R_1}}\right\} \quad \text{bridge equations} \qquad (12.101)$$

$$L_X = R_3 R_2 C_S \qquad (12.102)$$

Also here, if the dials on R_3 and R_1 indicate the value of the corresponding resistance, the bridge equations above can be used to calculate R_X and L_X. The dial on R_3 can be calibrated to indicate the value of L_X directly, while the dial on R_1 can be calibrated to indicate the value of R_X. Many coils have low value for R_X (they are called high-Q coils; see Chapter 14). Because of the inverse proportionality between R_X and R_1, as indicated by Eq. (12.101), the corresponding value for R_1 to produce bridge balance could be quite large. Other bridge arrangements (e.g., the Hay bridge) are more suitable for high-Q inductor measurements.

DRILL PROBLEMS

Section 12.1

12.1. In the circuit shown in Fig. 12.81, $R = 10\,\Omega$, $L = 2\,\text{mH}$, and $e(t) = 14.142 \sin 10,000t$ V. Find

   (a) Z_T, the total impedance of the circuit, and draw the impedance triangle.
   (b) $\mathbf{I}$, $\mathbf{V}_R$, and $\mathbf{V}_L$ and draw the phasor diagram.
   (c) $i(t)$.
   (d) The power dissipated in this circuit.

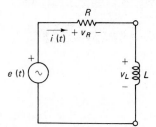

Figure 12.81

12.2. Repeat Drill Problem 12.1 for $R = 200\,\Omega$, $L = 0.1\,\text{H}$, and $e(t) = 7.071 \cos(1000t - 45°)$ V.

12.3. If a source whose voltage is $120\underline{/75°}$ V results in a current flow of $2.4\underline{/15°}$ A in a given series circuit, what are the values of the elements of this circuit? The frequency of the source is 200 Hz.

Section 12.2

12.4. In the circuit shown in Fig. 12.82, $R = 30\,\Omega$, $C = 10\,\mu$F, and $e(t) = 28.284 \sin(2500t + 45°)$ V. Find
   (a) Z_T, the total impedance of the circuit, and draw the impedance triangle.
   (b) $\mathbf{I}$, $\mathbf{V}_R$, and $\mathbf{V}_C$ and draw the phasor diagram.
   (c) $i(t)$.
   (d) The power dissipated in this network.

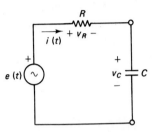

Figure 12.82

12.5. Repeat Drill Problem 12.2 for $R = 2\,\Omega$, $C = 50\,\mu$F, and $e(t) = 15 \cos(10,000t)$ V.

12.6. A 600 Hz source whose voltage is $50\underline{/20°}$ V results in a current flow of $2.5\underline{/50°}$ A in a given series circuit. What are the values of the elements of this circuit?

Section 12.3

12.7. In the circuit shown in Fig. 12.83, $R = 40\,\Omega$, $L = 20\,$mH, $C = 5\,\mu$F, and $e(t) = 14.142 \sin(4000t)$ V. Find
   (a) Z_T, the total impedance of the circuit, and draw the impedance triangle.
   (b) $\mathbf{I}$, $\mathbf{V}_R$, $\mathbf{V}_L$, and $\mathbf{V}_C$ and draw the phasor diagram.
   (c) $\mathbf{V}$ and verify KVL.
   (d) $i(t)$ and $v(t)$.
   (e) The power dissipated in this circuit.

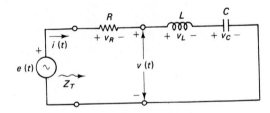

Figure 12.83

12.8. Repeat Drill Problem 12.7, but with the frequency of the source changed to 2500 rad/s.

12.9. What is the resonance frequency, f_0, of the circuit shown in Fig. 12.83 when the elements values are as given in Drill Problem 12.7? If the frequency of the source, $e(t)$, is adjusted to be the resonant frequency, find
   (a) $\mathbf{I}$, $\mathbf{V}_R$, $\mathbf{V}_L$, $\mathbf{V}_C$, and $\mathbf{V}$ in this case.
   (b) The power dissipated in this circuit.

Section 12.4

12.10. In the circuit shown in Fig. 12.84, $R = 10\,\Omega$, $L = 5\,\text{mH}$, and $C = 100\,\mu\text{F}$. If $i_S(t) = 0.2828 \sin(1000t)$ A:
  (a) Find Y_T of this circuit, and draw the admittance triangle.
  (b) Find $\mathbf{V}$, $\mathbf{I}_R$, $\mathbf{I}_L$, $\mathbf{I}_C$, and $\mathbf{I}_X$, and draw the phasor diagram,
  (c) Find $v(t)$ and $i_X(t)$.
  (d) Find the power dissipated in this circuit.
  (e) Verify KCL for this circuit.

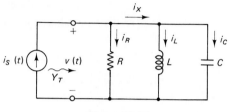

Figure 12.84

12.11. Repeat Drill Problem 12.10 if the frequency of the current source is doubled.

12.12. What is the resonance frequency, f_0, of the circuit shown in Fig. 12.84 when the elements values are as given in Drill Problem 12.10? If the frequency of the current source i_S is adjusted to f_0, find
  (a) $\mathbf{V}$, $\mathbf{I}_R$, $\mathbf{I}_L$, $\mathbf{I}_C$, and $\mathbf{I}_X$ in this case.
  (b) The power dissipated in this circuit.

12.13. At the frequency of the source given in Drill Problem 12.7, what is the equivalent parallel circuit that consists of single-element branches?

12.14. Find the equivalent parallel circuit that consists of single-element branches for the circuit examined in Drill Problem 12.8.

12.15. Find the simple series equivalent for the network examined in Drill Problem 12.10.

12.16. Repeat Drill Problem 12.15 but for the network examined in Drill Problem 12.11.

Section 12.5

12.17. In the network shown in Fig. 12.85, $\mathbf{I} = 2/\underline{30°}$ A. Find
  (a) $\mathbf{I}_1$ and $\mathbf{I}_2$ using the current-division principle.
  (b) Z_T and $\mathbf{E}$.
  (c) The elements of the simple series equivalent circuit to this network if the frequency of the source is 1600 Hz.

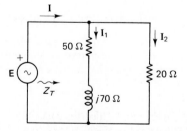

Figure 12.85

12.18. Repeat Drill Problem 12.17 but for the network shown in Fig. 12.86. Here $\mathbf{I} = 0.5/\underline{-45°}$ A, and the frequency of the source is 2500 Hz.

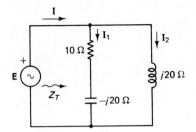

Figure 12.86

12.19. Analyze the circuit in Fig. 12.87 to obtain
   (a) Z_T.
   (b) I_1, I_2, and I_3.
   (c) V.
   (d) The power dissipated in this network.

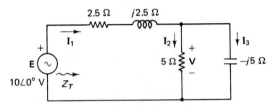

Figure 12.87

12.20. Analyze the circuit in Fig. 12.88 to obtain
   (a) Z_T.
   (b) V, I_1, and I_2.
   (c) V_{ab}.
   (d) The power dissipated in this circuit.

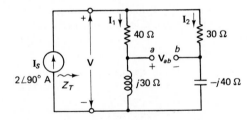

Figure 12.88

PROBLEMS

Note: The problems below that are marked with an asterisk (*) can be used as exercises in computer-aided network analysis or as programming exercises.

Section 12.1

(i) **12.1.** Find the frequency for which the phase angle ϕ between the applied voltage and the resulting current in the circuit of Fig. 12.89 is
   (a) $\phi = 30°$
   (b) $\phi = 45°$
   (c) $\phi = 60°$

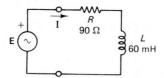

Figure 12.89

(i) **12.2.** The source voltage $e(t)$ in the circuit shown in Fig. 12.90 is $e(t) = 100 \sin(5000t)$ V. The oscilloscope controls are set as follows:

$$\text{time/div} = 0.1 \, \text{ms}, \quad \times \text{mag.:} \times 1$$
$$\text{Trigger: } Y_B, \text{ negative, auto. level}$$
$$\text{V/div} (Y_A) = 50 \, \text{V}$$
$$\text{V/div} (Y_B) = 20 \, \text{V}$$

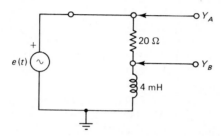

Figure 12.90

Draw the oscillogram thus obtained. Exchange the Y_B and the ground connections and then draw the corresponding oscillogram.

(i) **12.3.** An impedance consisting of two elements in series is connected to a $100\underline{/0°}$ V source whose frequency is 1000 rad/s. The resulting current is $(10 - j10)$ A. Find
  (a) The readings of a voltmeter connected across the source and an ammeter connected in series with the source and the impedance.
  (b) The value of the two elements of the impedance.
  (c) The current phasor if the source frequency is doubled.

(i) **12.4.** The voltage of a source $e(t)$ is $e(t) = 14.14 \sin(400t)$ V. This voltage produced the current $i(t) = -5 \cos(400t + 45°)$ A flowing in a black box. Find the equivalent simple series circuit of the black box. What is the power dissipated in this black box?

(i) **12.5.** A certain practical coil (solenoid) draws 5 A when connected to a 120 V, 25 Hz source. When the frequency of the source is changed to 60 Hz, the coil draws 3 A. Find the resistance and the inductance of the coil.

(i) **12.6.** In the circuit shown in Fig. 12.91, the voltage across the resistance is $30\underline{/120°}$ V. Find
  (a) The current **I**.
  (b) The source voltage **E** and the power dissipated in this circuit.
  (c) The voltage across L, $\mathbf{V}_L$. Draw the phasor diagram.

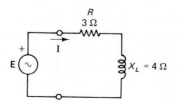

Figure 12.91

Problems

529

Section 12.2

(i) **12.7.** In the circuit shown in Fig. 12.92, the voltage across the resistance is $50\underline{/90°}$ V. Find

    (a) The current **I**.

    (b) The source voltage **E** and the power dissipated in this circuit.

    (c) The voltage across C, $\mathbf{V}_C$. Draw the phasor diagram.

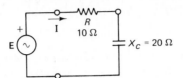

Figure 12.92

(i) **12.8.** A load consisting of two elements in series is connected to a $100\underline{/0°}$ V source whose frequency is 2000 rad/s. The resulting current is $(3 + j4)$ A.

    (a) What are the readings of a voltmeter connected across the source and an ammeter connected in series with the source and the load?

    (b) Find the value of the two elements of the load.

    (c) If the source frequency is doubled, find the new current phasor. Draw the complete phasor diagram in this case, showing the voltage across each element.

(i) **12.9.** The voltage across a certain black box and the current through this box are given by $v(t) = -10\sin(800t + 30°)$ V and $i(t) = -3.536\cos(800t - 15°)$ A. Find the equivalent simple series circuit of the black box and the power dissipated in it.

(ii) **12.10.** The oscillogram shown in Fig. 12.93(b) was obtained while performing some measurements on the circuit of Fig. 12.93(a). The scope settings were

$$\text{time/div.} = 5 \text{ ms}$$

$$\text{V/div } (Y_A) = \text{unknown}$$

$$\text{V/div } (Y_B) = 1 \text{ V}$$

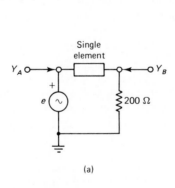

(a)

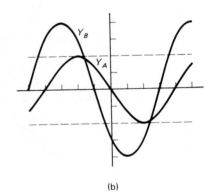

(b)

Figure 12.93

    (a) What is the value of the single element in the box?

    (b) What is the magnitude of the voltage source and the V/div for beam A?

(i) **12.11.** In the circuit shown in Fig. 12.94, $\mathbf{E} = 10\underline{/0°}$ V, $\mathbf{V}_C = 10\underline{/-30°}$ V, $X_C = 5\,\Omega$, and $\omega = 1000$ rad/s. Find the values of the elements constituting the impedance Z.

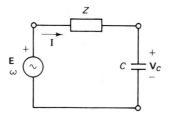

Figure 12.94

Section 12.3

(i) **12.12.** A practical inductor draws 30 A when connected to a 120 V dc source. It draws 24 A when connected to a 120 V, 60 Hz source. Find
  (a) The resistance and inductance of the inductor.
  (b) The value of the capacitor required, to be connected in series with the inductor so that the current from the 120 V 60 Hz source becomes 30 A.
  (c) The voltage across the capacitor and across the practical inductor in part (b). Draw the complete phase diagram.

(i) **12.13.** The black box in Fig. 12.95 contains a number of resistors, inductors, and capacitors, while $e(t) = 14.14 \sin(400t)$ V and $i(t) = -5 \cos(400t + 45°)$ A. Find the element required to be placed in series with the black box, without changing the ammeter's reading.

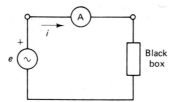

Figure 12.95

(i) **12.14.** In the circuit shown in Fig. 12.96, the voltage across the inductance is 20 V, while $\omega = 3.73 \times 10^5$ rad/s. Find the time-domain expression for $e(t)$ if at $t = 0$ s, the current $i(t)$ is zero. Draw the complete phasor diagram and calculate the power dissipated in this circuit.

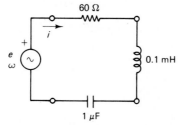

Figure 12.96

(ii) **12.15.** The current in a series RLC circuit lags the applied voltage by 30°. The voltage across the inductance is twice that of the voltage across the capacitance. The applied source voltage is 20 V and the value of the resistance is 17.3 Ω. Find
  (a) The current.
  (b) The values of L and C, assuming that $\omega = 1000$/rad/s.

(ii) **12.16.** Consider the circuit shown in Fig. 12.97, where **E** is 7.07 $\underline{/0°}$ V and both the Y_A and Y_B waveforms have 7.07 V peak voltages.
  (a) Find the voltage drop across R.
  (b) Find the values of L and C.

Problems

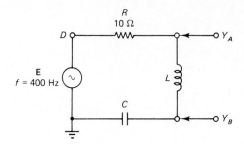

Figure 12.97

(c) Explain by logical sequential reasoning what happens to the current if the frequency of the source is increased above its present value of 400 Hz.

(d) Draw an oscillogram if the ground connection is moved to point D and the oscilloscope controls are adjusted as follows:

$$\text{Trigger: } Y_B, \text{ negative, auto. level}$$

$$\text{time/div} = 0.2 \text{ ms}$$

$$\text{V/div} (Y_A) = 2 \text{ V}$$

$$\text{V/div} (Y_B) = 5 \text{ V}$$

(ii) **12.17.** In the circuit shown in Fig. 12.98, $e(t) = 200 \sin(628t + 30°)$ V, $I = 1.0$ A, $V_L = 314$ V, and $C = 7.44 \, \mu\text{F}$. Find

(a) L.

(b) R.

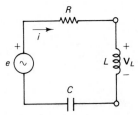

Figure 12.98

(i) **12.18.** The current in the circuit shown in Fig. 12.99 is $\mathbf{I} = (5\sqrt{3} + j5)$ A. Find the readings of the ideal voltmeter and the ideal ammeter connected as indicated. What is the value of the impedance Z_2?

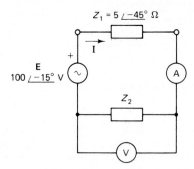

Figure 12.99

(i) **12.19.** The box in the circuit shown in Fig. 12.100(a) contains a single element. The corresponding oscillogram is shown in Fig. 12.100(b) with the scope settings:

$$\text{V/div} (Y_A) = 10 \text{ V}$$

$$\text{V/div} (Y_B) = 10 \text{ V}$$

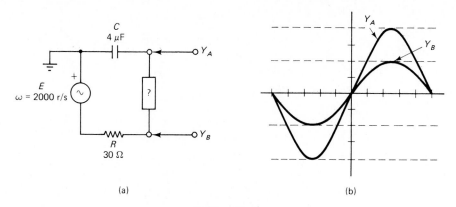

(a)

(b)

Figure 12.100

Find

(a) The value of the element.

(c) The value of E.

(c) The time/div setting.

(ii) **12.20.** The oscillogram obtained for the circuit shown in Fig. 12.101 indicates that Y_A has a peak value of 20 V, Y_B has a peak value of 16 V, Y_B lags Y_A, and the period of one cycle is 6.28 ms.

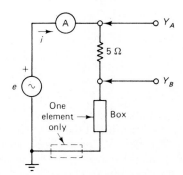

Figure 12.101

(a) Find the reading of the ammeter A.

(b) Find the nature and magnitude of the element in the box.

(c) What is the nature and magnitude of the element to be added in the dashed location so that the ammeter's reading would be the highest possible in this circuit?

(ii) **12.21.** The oscillogram corresponding to the circuit in Fig. 12.102(a) is shown in Fig. 12.102(b), with $e(t) = 50 \sin (6283t)$ V. The scope settings are

$$\text{time/div: unknown} \quad \times \text{mag.: } \times 5$$

$$\text{V/div } (Y_A)\text{: } 20 \text{ V}, \quad \text{V/div } (Y_B)\text{: unknown}$$

(a) Label the two traces of the oscillogram with the help of the phasor diagram.

(b) Find the time/div setting and the V/div setting for Y_B.

(c) Find the values of L and C.

Problems

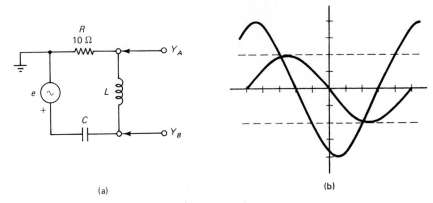

(a)   (b)

Figure 12.102

(i) **12.22.** The black box in the circuit shown in Fig. 12.103 contains two series-connected elements. By inspecting the oscillogram obtained, it was found that Y_A corresponds to $100 \sin (1000t)$ V and Y_B corresponds to $141.4 \cos (1000t - 30°)$ V.

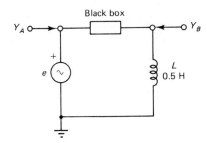

Figure 12.103

Find
(a) The current phasor $\mathbf{I}$.
(b) The nature and value of the two elements inside the black box.
(c) Draw a complete phasor diagram showing the current and voltages in this circuit. What is the average power dissipated?
(d) If L is changed, find its new value such that the current magnitude remains unchanged.

(ii) **12.23.** With the source shown in Fig. 12.104 operating at a radian frequency of ω_1, it was found that $\mathbf{I} = 10/\underline{0°}$ A. When the source frequency was doubled, the current in the circuit became $\mathbf{I} = 7.07/\underline{-45°}$ A. Find
(a) ω_1.
(b) The value of L.
(c) X_L and X_C at $2\omega_1$.

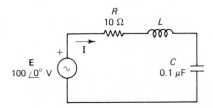

Figure 12.104

Section 12.4

(i) **12.24.** The black box in Fig. 12.105 contains two pure parallel-connected circuit elements. Here $e(t) = 141.4 \sin(300t)$ V and $i(t) = 1.414 \sin(300t + 30°)$ A.

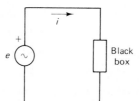

Figure 12.105

(a) What are the values of the two elements?
(b) Find the RMS current in each element. Draw the complete phasor diagram. What is the power dissipated in this circuit?
(c) What are the values of the elements of the equivalent simple series circuit?
(d) If the frequency is continuously variable in the original parallel circuit, what is the highest possible impedance of the black box?

(i) **12.25.** In a parallel RLC circuit, L is 10 mH. When the applied voltage is $e(t) = -353.6 \sin(3000t - 100°)$ V, the resulting source current is $i(t) = 12.5 \cos(3000t - 55°)$ A. Find the values of R and C in this circuit.

(i) **12.26.** In the circuit of Fig. 12.106, A_1 reads 5 A, A_2 reads 4 A, A_3 reads 3 A, and the voltmeter V reads 120 V (all meters are assumed to be ideal). Find
(a) The reading of ammeter A_4.
(b) The angle between **E** and $\mathbf{I}_T$. Draw the phasor diagram and find the total power dissipated in this circuit.
(c) The values of R, L, and C.

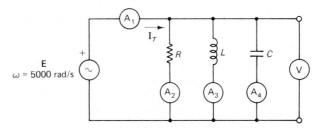

Figure 12.106

(ii) **12.27.** In the circuit of Fig. 12.107, the black box contains two parallel elements. When **E** is $100\underline{/0°}$ V, **I** is $40\underline{/37°}$ A. Find
(a) $\mathbf{I}_2$ and draw the phasor diagram.
(b) The elements in the black box.
(c) The values of the elements of the equivalent simple series circuit.

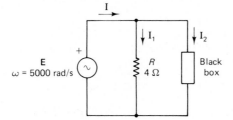

Figure 12.107

(ii) **12.28.** The black box in the circuit shown in Fig. 12.108(a) contains two elements in parallel. The oscillogram obtained is shown in Fig. 12.108(b), with the scope settings adjusted to

$$\text{time/div} = 5 \text{ ms}$$

$$\text{X-mag.: } \times 5$$

$$\text{V/div } (Y_A) = 50 \text{ V}$$

$$\text{V/div } (Y_B) = 0.2 \text{ V}$$

(The effect of the 3 Ω sense resistor on the current is negligible.)
(a) Find the values of the elements in the black box.
(b) What is the value of the extra element that can be connected in parallel with the black box without affecting the current magnitude?
(c) For the circuit in part (b), what are the values of the elements of the equivalent series connected load?

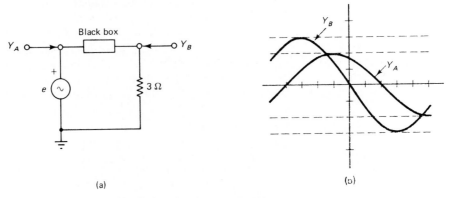

(a)

(b)

Figure 12.108

(ii) **12.29.** In the circuit shown in Fig. 12.109, E is 120 V, I_T is 15 A, $i_1(t) = 9\sqrt{2} \sin (1000t + 60°)$ A, and $i_1(t)$ leads $i_T(t)$. Find
(a) The values of R and the single element Z.
(b) The applied source voltage $e(t)$ in the time domain.
(c) The value of the extra element that can be added in parallel to this circuit without changing the magnitude of I_T.

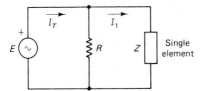

Figure 12.109

(ii) **12.30.** The oscillogram shown in Fig. 12.110(a) was obtained for the circuit shown in Fig. 12.110(b), while the scope settings were adjusted as follows:

$$\text{time/div} = 0.5 \text{ ms}$$

$$\text{V/div } (Y_A) = 1 \text{ V}$$

$$\text{V/div } (Y_B) = 50 \text{ V}$$

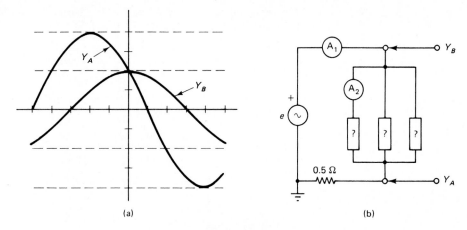

(a)   (b)

Figure 12.110

Each branch contains one pure element and the effect of the 0.5 Ω resistor on the current is negligible. Ammeter A_2 reads 8 A, which is leading the applied voltage $e(t)$ by 90°. Find

(a) The reading of ammeter A_1.

(b) The nature and value of the elements in the three branches.

Section 12.5

(i) ***12.31.** Find the input impedance Z_{in} as seen by the source in the circuit of Fig. 12.111. From this result, obtain the phasor form for $\mathbf{I}_1$, $\mathbf{I}_2$, and $\mathbf{V}$.

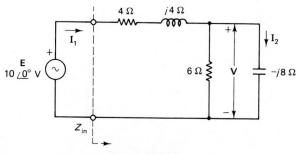

Figure 12.111

(ii) ***12.32.** In the circuit shown in Fig. 12.112, find $\mathbf{I}_1$, $\mathbf{I}_2$, and $\mathbf{V}$. What is the total power dissipated in this circuit?

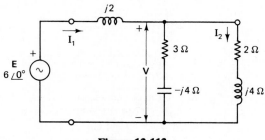

Figure 12.112

(i) ***12.33.** For the circuit shown in Fig. 12.113, find $\mathbf{I}_1$, $\mathbf{I}_2$, and $\mathbf{V}$. Calculate the total power dissipated in this circuit.

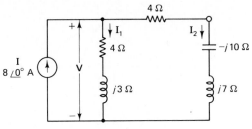

Figure 12.113

(ii) **12.34.** In the circuit shown in Fig. 12.114, $E = 1000$ V a $R = 10\,\Omega$. The current i_1 leads e, while $i_1 = 158.1 \cos(1000t + 26.6°)$ A and $i_3 = 100 \sin(1000t + 135°)$ A. Z_A and Z_B are the impedances of single elements. Find
  (a) The phase angle between e and i_2.
  (b) The RMS value of i_2.
  (c) The values of the elements constituting Z_A and Z_B.

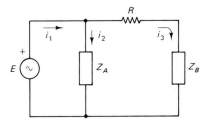

Figure 12.114

(ii) **12.35.** The black box in the circuit shown in Fig. 12.115 contains a single element, while $R = 10\,\Omega$ and $L = 10\,\text{mH}$. The currents i_1 and i_2 are given by $i_1(t) = 5\sqrt{5} \sin(1000t + 26.5°)$ A and $i_2(t) = 5\sqrt{2} \cos(1000t - 45°)$ A. Find
  (a) The time-domain and phasor representation of the applied source voltage $e(t)$.
  (b) The value of the element in the black box.

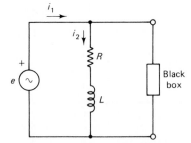

Figure 12.115

(i) ***12.36.** In the circuit shown in Fig. 12.116, $e = 10 \sin(1000t)$ V.
  (a) What is the total impedance seen by the source?
  (b) Find the currents $i(t)$, $i_1(t)$, and $i_2(t)$.
  (c) Draw the complete phasor diagram, showing all the voltages and currents.

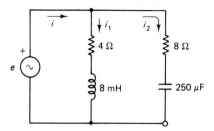

Figure 12.116

(i) *12.37. Find **I** and $\mathbf{V}_{AB}$ in the circuit shown in Fig. 12.117. Obtain the average power dissipated in this circuit, and the equivalent simple series and simple parallel forms for this circuit.

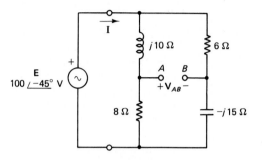

Figure 12.117

(ii) *12.38. Find the currents $\mathbf{I}_1$, $\mathbf{I}_2$, $\mathbf{I}_3$, $\mathbf{I}_4$, $\mathbf{I}_5$, and $\mathbf{I}_6$, in the directions indicated, in the circuit shown in Fig. 12.118.

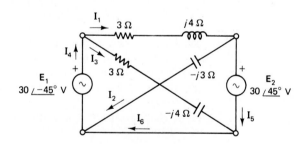

Figure 12.118

(ii) 12.39. The impedance Z_3 in the circuit shown in Fig. 12.119 is composed of two series

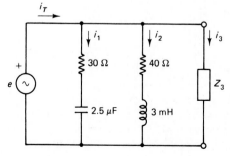

Figure 12.119

connected elements, while $i_T(t) = 4\sin(10{,}000t + 45°)\,\mathrm{A}$ and $i_1(t) = 1.414\cos(10{,}000t)\,\mathrm{A}$. Find

(a) $\mathbf{E}$, the voltage of the applied source in phasor form.

(b) $\mathbf{I}_3$.

(c) The values of the elements comprising the unknown impedance Z_3.

(d) The total power dissipated in this circuit.

(e) The equivalent simple series circuit of the overall network.

(ii) **12.40.** The oscillogram shown in Fig. 12.120(b) is obtained for the circuit shown in Fig. 12.120(a). The scope settings were

$$\text{time/div} = 0.1\ \text{ms}$$

$$\text{V/div }(Y_A) = 1\ \text{V}$$

$$\text{V/div }(Y_B) = 2\ \text{V}$$

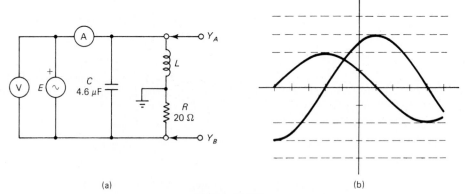

(a)   (b)

Figure 12.120

(a) With the help of the phasor diagram for this circuit, label the two waveforms.

(b) Find the value of L and the reading of the voltmeter.

(c) Taking $\mathbf{E}$ as the reference phasor, draw the complete phasor diagram for this circuit, showing all the currents and voltages.

(d) Express in phasor form the current in each branch and find the reading of the ammeter.

(e) Determine the values of the components of the simple series equivalent circuit.

Section 12.6

(ii) **12.41.** Find the balance conditions for the Hay bridge shown in Fig. 12.121. Using these conditions, obtain the expressions for R_X and L_X in terms of the other standard elements values and ω, the frequency of the source.

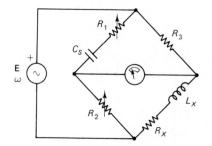

Figure 12.121

GLOSSARY

Admittance Diagram (or Triangle): A vector diagram displaying the magnitude and the direction of the parallel branch admittances and the total admittance of the network. In the case of ideal single element parallel branches, the admittance diagram depicts the conductance and the capacitive and inductive susceptances of the branches, and the total admittance of such a circuit.

Equivalent Circuits: At a specific frequency, a parallel load can be replaced by an equivalent series load (and vice versa) such that a given voltage source produces the same current flow in both the original and the equivalent load impedances.

Impedance Bridge: A circuit configuration similar to the Wheatstone bridge. Each branch consists of fixed and/or variable accurate R, L, and C components. Such circuits can be used in conjunction with a null detector in the bridge arm to measure the values of unknown impedance elements.

Impedance Diagram (or Triangle): A vector diagram displaying the magnitude and the direction of the series connected impedances and the total impedance of the network. In the case of simple *RLC* series circuits, the impedance diagram depicts the resistance and the inductive and capacitive reactances of the elements, and the total impedance of such a circuit.

ANSWERS TO DRILL PROBLEMS

12.1. (a) $Z_T = 10 + j20 = 22.3607\underline{/63.43°}\ \Omega$

(b) $I = 0.4472\underline{/-63.43°}\ A$,
$V_R = 4.472\underline{/-63.43°}\ V$,
$V_L = 8.944\underline{/26.57°}\ V$

(c) $i(t) = 0.6324 \sin (10{,}000t - 63.43°)\ A$

(d) $P_{av} = 2\ W$

12.3. $Z_T = 50\underline{/60°}\ \Omega = 25 + j43.3\ \Omega$,
$R = 25\ \Omega,\ L = 34.46\ mH$

12.5. (a) $Z_T = 2 - j2 = 2.8284\underline{/-45°}\ \Omega$

(b) $I = 3.75\underline{/135°}\ A$,
$V_R = 7.5\underline{/135°}\ V$,
$V_C = 7.5\underline{/45°}\ V$

(c) $i(t) = 5.3033 \sin (10{,}000t + 135°)\ A$

(d) $P_{av} = 28.125\ W$

12.7. (a) $Z_T = 40 + j30 = 50\underline{/36.87°}\ \Omega$

(b) $I = 0.2\underline{/-36.87°}\ A$,
$V_R = 8\underline{/-36.87°}\ V$,
$V_L = 16\underline{/53.13°}\ V$,
$V_C = 10\underline{/-126.87°}\ V$

(c) $V = V_L + V_C = 6\underline{/53.13°}\ V$,
$V_R + V_L + V_C = 10 + j0 = 10\underline{/0°}\ V = E$

(d) $i(t) = 0.28284 \sin (4000t - 36.87°)\ A,\ v(t) = 8.4853 \sin (4000t + 53.13°)\ V$

(e) $P_{av} = 1.6\ W$

12.9. $f = \dfrac{\omega_0}{2\pi} = 503.29\ Hz$

(a) $I = 0.25\underline{/0°}\ A$,
$V_R = 10\underline{/0°}\ V = E$,
$V_L = 15.811\underline{/90°}\ V$,
$V_C = 15.811\underline{/-90°}\ V$,
$V = V_L + V_C = 0\ V$

(b) $P_{av} = 2.5\ W$

12.11. (a) $Y_T = 0.1 + j0.1 = 0.14142\underline{/45°}\ S$

(b) $V = 1.4142\underline{/-45°}\ V$,
$I_R = 0.14142\underline{/-45°}\ A$,
$I_L = 0.14142\underline{/-135°}\ A$,
$I_C = 0.28284\underline{/45°}\ A$,
$I_X = 0.14142\underline{/45°}\ A$

(c) $v(t) = 2 \sin (2000t - 45°)$ V, $i_X(t) = 0.2 \sin (2000t + 45°)$ A

(d) $P_{av} = 0.2$ W

(e) $\mathbf{I}_R + \mathbf{I}_L + \mathbf{I}_C = 0.2\underline{/0°} = \mathbf{I}_S$

12.13. $Y_T = 0.02\underline{/-36.87°}$ S $= 0.016 - j0.012$ S, $R = 62.5\,\Omega$, $L = 20.83$ mH

12.15. $Z_T = 7.071\underline{/45°}\,\Omega = 5 + j5\,\Omega$, $R = 5\,\Omega$, $L = 5$ mH

12.17. (a) $\mathbf{I}_1 = 0.4041\underline{/-15°}$ A, $\mathbf{I}_2 = 1.7379\underline{/39.46°}$ A

(b) $Z_T = 17.379\underline{/9.46°}\,\Omega$, $\mathbf{E} = 34.758\underline{/39.46°}$ V

(c) $R = 17.143\,\Omega$, $L = 0.284$ mH

12.19. (a) $Z_T = 5 + j0 = 5\underline{/0°}\,\Omega$

(b) $\mathbf{I}_1 = 2\underline{/0°}$ A, $\mathbf{I}_2 = 1.4142\underline{/-45°}$ A, $\mathbf{I}_3 = 1.4142\underline{/45°}$ A

(c) $\mathbf{V} = 7.071\underline{/-45°}$ V

(d) $P_{av} = 20$ W

13

POWER
IN AC CIRCUITS

OBJECTIVES

- Ability to calculate the real power dissipated in any ac load and the power factor of the load.
- Understanding the concepts of the apparent power, the reactive power, the power triangle and the complex power and the ability to calculate these quantities for a given ac network and/or load.
- Understanding the techniques used to improve the power factor of a given load and being able to perform the required network analysis.
- Understanding the operation of the wattmeter and its circuit interconnection for power measurement.

13.1 | REAL POWER DISSIPATED IN AN AC LOAD

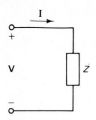

Figure 13.1

Any ac impedance Z, as in Fig. 13.1, composed of two or more elements of different nature, is either predominantly inductive [Fig. 13.2(a)] or predominantly capacitive [Fig. 13.2(b)]. The same can, of course, be stated for the total (or input) impedance of any ac circuit.

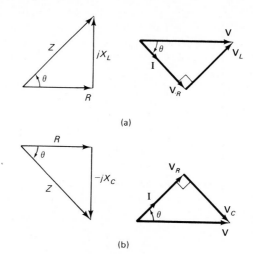

(a)

(b)

Figure 13.2

For a predominantly inductive load, as in Fig. 13.2(a),

$$Z = R + jX_L = |Z|\underline{/\theta°} \quad \Omega \tag{13.1}$$

If the voltage **V** across this impedance is considered to be the reference phasor,

$$\mathbf{V} = V\underline{/0°}\ \mathrm{V} \Leftrightarrow v(t) = \sqrt{2}\,V \sin \omega t \tag{13.2}$$

then the current **I** in the load Z is

$$\mathbf{I} = \frac{\mathbf{V}}{Z} = \frac{V\underline{/0°}}{|Z|\underline{/\theta°}} = \frac{V}{|Z|}\underline{/-\theta°}\,\mathrm{A} = I\underline{/-\theta°}\quad \mathrm{A} \tag{13.3a}$$

and

$$i(t) = \sqrt{2}\,I \sin(\omega t - \theta°) \quad \mathrm{A} \tag{13.3b}$$

The current **I** lags the voltage **V** by the phase angle $\theta°$. Note that

$$I = \frac{V}{|Z|} \quad \text{and} \quad \tan \theta = \frac{X_L}{R} = \frac{\omega L}{R} \tag{13.4}$$

For a predominantly capacitive load, as in Fig. 13.2(b)

$$Z = R - jX_C = |Z|\underline{/-\theta°} \quad \Omega \tag{13.5}$$

Also here, considering the voltage **V** across the load to be the reference phasor, as in Eq. (13.2), the current **I** flowing in the load can be determined from Ohm's law:

$$\mathbf{I} = \frac{\mathbf{V}}{Z} = \frac{V\underline{/0°}}{|Z|\underline{/-\theta°}} = \frac{V}{|Z|}\underline{/\theta°} = I\underline{/\theta°}\quad \mathrm{A} \tag{13.6a}$$

and

$$i(t) = \sqrt{2}\, I \sin{(\omega t + \theta°)} \quad \text{A} \tag{13.6b}$$

where

$$I = \frac{V}{|Z|} \quad \text{and} \quad \tan{\theta} = \frac{X_C}{R} = \frac{1}{\omega CR} \tag{13.7}$$

Here the current $\mathbf{I}$ leads the voltage $\mathbf{V}$ by the phase angle θ. In both cases, θ is the angle of the complex impedance Z. It is also the phase shift between the voltage $\mathbf{V}$ and the current $\mathbf{I}$ phasors.

As shown in Chapter 12 and repeated here, the instantaneous power in the load $p(t)$ is given by

$$
\begin{aligned}
p(t) &= v(t)i(t) \\
&= \sqrt{2}\, V \sin{(\omega t)}\sqrt{2}\, I \sin{(\omega t \pm \theta°)} \\
&= 2\, VI \sin{(\omega t)} \sin{(\omega t \pm \theta°)} \tag{13.8a}
\end{aligned}
$$

Using the trigonometric identity:

$$2 \sin A \sin B = \cos{(A - B)} - \cos{(A + B)}$$

Then

$$p(t) = VI[\cos{(\pm\theta°)} - \cos{(2\omega t \pm \theta°)}] \tag{13.8b}$$

Since the average value of any sinusoidal (or cosinusoidal) function over one complete cycle (or period) is zero, and since

$$\cos{(\pm\theta°)} = \cos{\theta°}$$

the real average power dissipated in the load P is given by

$$P = \overline{p(t)} = VI \cos{\theta°} \quad \text{W} \tag{13.9}$$

Thus, whether the load impedance Z is predominantly inductive or capacitive in nature, Eq. (13.9) provides the magnitude of the real average power dissipated in either type of load. This equation expresses the real power P in terms of the magnitudes of the voltage and current phasors (V and I) and the phase shift between them, $\theta°$.

To differentiate between the possible different natures of the load impedance, the quantity $\cos{\theta°}$ is called the *power factor* of the load, PF (sometimes denoted F_P). For *inductive loads*, since $\mathbf{I}$ lags $\mathbf{V}$, we say that the load has a *lagging power factor*. While for *capacitive loads*, since $\mathbf{I}$ leads $\mathbf{V}$, we say that the load has a *leading power factor*. Hence Eq. (13.9) can be rewritten in the form

$$\boxed{P = VI \cdot \text{PF} \quad \text{W}} \tag{13.10}$$

where the power factor PF is always

$$\boxed{\text{PF} = \cos{\theta°} \leq 1} \tag{13.11}$$

If the load is purely resistive, $Z = R/0°\ \Omega$, then

$$\text{PF} = \cos{0°} = 1 \tag{13.12a}$$

while if the load is purely inductive or capacitive, $Z = \pm jX = X/\underline{\pm 90°}\ \Omega$, then

$$PF = \cos{(\pm 90°)} = 0 \qquad (13.12b)$$

In this case, as Eqs. (13.12b) and (13.10) indicate, the power dissipated in the load is zero. This remark stresses the fact that no power is dissipated in either an inductor L or a capacitor C.

Referring to Fig. 13.2, we see that

$$V\cos\theta = V_R = IR \quad \text{V} \qquad (13.13)$$

where V_R is the voltage across the resistive component of the series equivalent form of the load impedance. Thus, from Eq. (13.9),

$$\boxed{P = IV\cos\theta = IV_R = I^2R \quad \text{W}} \qquad (13.14)$$

Thus all the real average power is dissipated only in the resistive component of the load impedance. This is true whether the load is composed of series- or parallel-connected elements (see Section 12.4).

EXAMPLE 13.1

Find the power delivered by the source (i.e., dissipated by the load) in the circuit shown in Fig. 13.3, given that the source voltage is $e(t) = 28.28 \sin{(10,000t)}$ V.

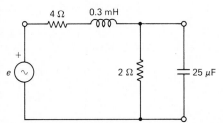

Figure 13.3

Solution   Here the source voltage phasor is:

$$\mathbf{E} = 20/\underline{0°}\ \text{V} \qquad \text{and} \qquad \omega = 10,000\ \text{rad/s}$$

The frequency-domain equivalent circuit is shown in Fig. 13.4. The total impedance of this circuit is

$$Z_T = 4 + j3 + \frac{2(-j4)}{2 - j4} = 4 + j3 + \frac{(-j8)(2 + j4)}{(2 - j4)(2 + j4)}$$

$$= 4 + j3 + \frac{1}{4 + 16}(-j16 + 32)$$

$$= 4 + j3 + 1.6 - j0.8 = 5.6 + j2.2 = 6.017/\underline{21.4°}\ \Omega$$

From Ohm's law for the entire circuit, we obtain

$$\mathbf{I} = \frac{\mathbf{E}}{Z_T} = \frac{20/\underline{0°}}{6.017/\underline{21.4°}} = 3.324/\underline{-21.4°}\ \text{A}$$

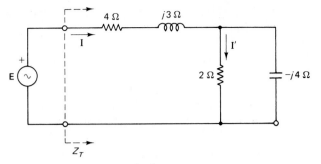

Figure 13.4

The total impedance is predominantly inductive in nature, and **I** lags **E**,

$$\theta^\circ = 21.4^\circ$$

$$\text{PF} = \cos 21.4^\circ = 0.931 \text{ lagging}$$

Therefore, the total real power dissipated in the circuit is

$$P = EI \cdot \text{PF} = 20 \times 3.324 \times 0.931 = 61.89 \text{ W}$$

Another method Since the equivalent series form of Z_T has a resistive component (R_{eq}) of 5.6 Ω, as obtained above, and the current flowing in it is **I**, then

$$P = I^2 R_{eq} = (3.324)^2 \times 5.6 = 61.87 \text{ W} \quad \text{(as above)}$$

A third method The current **I'** flowing in the 2 Ω resistor can be obtained from the current-division principle,

$$\mathbf{I'} = \mathbf{I} \frac{-j4}{2 - j4} = 3.324\underline{/-21.4^\circ} \times \frac{4\underline{/-90^\circ}}{4.472\underline{/-63.4^\circ}} = 2.973\underline{/-48^\circ} \text{ A}$$

Thus the total power dissipated in this circuit is

$$P = I^2 \times 4 + I'^2 \times 2 = (3.324)^2 \times 4 + (2.973)^2 \times 2$$
$$= 61.87 \text{ W} \quad \text{(as above)}$$

EXAMPLE 13.2

Given a source whose voltage is $7.07 \sin \omega t$ volts, find the PF and the power dissipated in each of the following cases.
(a) The resulting current is $i = 1.414 \sin(\omega t - 30^\circ)$ A.
(b) The resulting current is $i = 0.3535 \sin(\omega t + 60^\circ)$ A.
(c) The load impedance is $Z = 3 + j4$.

Solution The phasor of the source voltage is $\mathbf{V} = 5\underline{/0^\circ}$ V.
   (a) Here $\mathbf{I} = 1\underline{/-30^\circ}$ A. Thus **I** lags **V** by 30° (i.e., the load is inductive). Therefore,

$$\text{PF} = \cos 30^\circ = 0.866 \text{ lagging}$$

and

$$P = VI \cdot \text{PF} = 5 \times 1 \times 0.866 = 4.33 \text{ W}$$

(b) Here $\mathbf{I} = 0.25\underline{/60°}$ A. Thus $\mathbf{I}$ leads $\mathbf{V}$ by 60° (i.e., the load is capacitive). Therefore,

$$\text{PF} = \cos 60° = 0.5 \text{ leading}$$

and

$$P = VI \cdot \text{PF} = 5 \times 0.25 \times 0.5 = 0.625 \text{ W}$$

(c) $Z = 3 + j4 = 5\underline{/53.1°} \, \Omega$ (inductive load)

$$\mathbf{I} = \frac{\mathbf{V}}{Z} = \frac{5\underline{/0°}}{5\underline{/53.1°}} = 1\underline{/-53.1°} \text{ A}$$

$\mathbf{I}$ lags $\mathbf{V}$ by 53.1°; therefore,

$$\text{PF} = \cos 53.1° = 0.6 \text{ lagging}$$

and

$$P = VI \cdot \text{PF} = 5 \times 1 \times 0.6 = 3 \text{ W}$$

EXAMPLE 13.3

A 1000 Hz, 50 V source delivers 300 W to a load. The current in the load was measured to be 8 A. Find the impedance of the load if it is known that the load has a leading PF.

Solution   Since

$$P = VI \cdot \text{PF}$$

$$300 = 50 \times 8 \times \text{PF}$$

$$\text{PF} = \cos \theta_Z = \frac{300}{400} = 0.75 \text{ leading}$$

Thus

$$\theta_Z = 41.4°$$

If the source voltage is considered to be the reference phasor,

$$\mathbf{V} = 50\underline{/0°} \text{ V}$$

then $\mathbf{I}$ would lead $\mathbf{V}$ by 41.4°, that is,

$$\mathbf{I} = 8\underline{/41.4°} \text{ A}$$

and

$$Z = \frac{\mathbf{V}}{\mathbf{I}} = \frac{50\underline{/0°}}{8\underline{/41.4°}} = 6.25\underline{/-41.4°} \, \Omega$$

$$= 4.688 - j4.133 = R - jX_C$$

Therefore the load has a resistive component of 4.688 Ω, in series with a capacitor C, where

$$C = \frac{1}{\omega X_C} = \frac{1}{2\pi \times 1000 \times 4.133} = 38.51 \, \mu\text{F}$$

Check:

$$P = I^2 R = (8)^2 \times 4.688 = 300 \text{ W} \qquad \text{(as given)}$$

13.2 POWER TRIANGLE; APPARENT AND REACTIVE POWER

For the impedance in Fig. 13.1, the quantity

$$\boxed{S = VI \quad \text{Volt} \cdot \text{Ampere (or VA)}} \qquad (13.15)$$

is called the *apparent power* of an ac load. This is because in comparison to the case of dc resistive loads, it seems apparent that the power is the voltage–current product. Note that V and I in Eq. (13.15) are the RMS values of the corresponding voltage and current phasors. Since

$$|Z| = \frac{V}{I}$$

Eq. (13.15) can also be written in the form

$$\boxed{S = I^2 |Z| = \frac{V^2}{|Z|} \quad \text{VA}} \qquad (13.16)$$

The unit volt $\cdot$ ampere used here is chosen to distinguish the apparent power from the real power P, where

$$P = I^2 R = VI \cos \theta$$

whose unit is the watt. From the above, it can easily be concluded that

$$\boxed{P = S \cos \theta = S \cdot \text{PF} \quad \text{W}} \qquad (13.17)$$

Also, the quantity

$$\boxed{Q = S \sin \theta \quad \text{volt} \cdot \text{amperes reactive (VAR)}} \qquad (13.18)$$

is called the *reactive power* of the ac load, having the unit VAR. Q does not represent any measurable power dissipation. It is called reactive power because it represents power that could have been dissipated in the reactive component of Z (i.e., in X_L or X_C) had this component been resistive. It is therefore an imaginary power. It is sometimes referred to as *quadrature power*, for reasons to be made clear shortly.

Refer to Fig. 13.2:

$$V \sin \theta = V_X \text{ (voltage across the reactive component, } X_L \text{ or } X_C\text{)}$$

$$= IX \text{ (where } X \text{ is either } X_L \text{ or } X_C\text{)} \qquad (13.19)$$

Therefore, the reactive power Q can also be expressed in any of the forms

$$\boxed{\begin{aligned} Q &= VI \sin \theta \\ &= IV_X = I^2 X = \frac{V_X^2}{X} \quad \text{VAR} \end{aligned}} \qquad (13.20)$$

The trigonometric relations described in Eqs. (13.17) and (13.18) indicate that it is possible to represent the quantities S, P, and Q in the form of a triangle, called the *power triangle*. Two distinct power triangles are shown in Fig. 13.5, corresponding to the two cases: inductive load and capacitive load. The power triangle is clearly an exact replica of the corresponding impedance triangle, in particular since

$$P = I^2 R \qquad Q = I^2 X \qquad S = I^2 |Z|$$

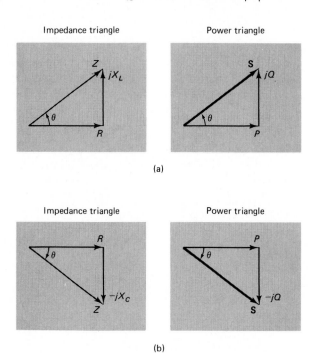

(a)

(b)

Figure 13.5

that is, each side of the power triangle is $I^2 \times$ the corresponding side of the impedance triangle. One can also refer to **S** as the *complex power*. It is a vector, represented as a complex number in polar or in rectangular forms as follows:

$$\boxed{\mathbf{S} = P + jQ = S\underline{/\theta^\circ} \quad \text{VA} \qquad \text{(inductive load, lagging PF)}} \qquad (13.21)$$

or

$$\boxed{\mathbf{S} = P - jQ = S\underline{/-\theta^\circ} \quad \text{VA} \qquad \text{(capacitive load, leading PF)}} \qquad (13.22)$$

where, as usual,

$$S = VI = \sqrt{P^2 + Q^2}, \qquad \tan \theta = \frac{Q}{P} \qquad \left[\text{or } \tan(-\theta) = \frac{-Q}{P} \right]$$

and

$$P = S \cos \theta, \qquad Q = S \sin \theta$$

Q is perpendicular to P, hence the name "quadrature power."

13.2.1 Relationship between Complex Power and Voltage and Current Phasors

The complete expression for the complex power **S** in polar form (i.e., its magnitude and phase) can be obtained directly from the voltage and current phasors for the given load Z. **S** is then the complex power of this load. For an inductive load,

$$Z = |Z|\underline{/\theta°} \quad \Omega \quad (\theta° \text{ positive, lagging PF})$$

let the current flowing in this load be represented by the phasor **I**, where

$$\mathbf{I} = I\underline{/\phi°} \quad A$$

Then the voltage phasor **V** across the load impedance Z is

$$\mathbf{V} = \mathbf{I}Z = I\underline{/\phi°}|Z|\underline{/\theta°} = I|Z|\underline{/\phi° + \theta°} = V\underline{/\phi° + \theta°} \quad V$$

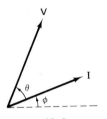

Figure 13.6

The corresponding phasor diagram is shown in Fig. 13.6.

The conjugate of the current phasor is

$$\mathbf{I}^* = I\underline{/-\phi°} \quad A$$

Therefore,

$$\mathbf{VI}^* = V\underline{/\phi° + \theta°} \, I\underline{/-\phi°} = VI\underline{/\theta°}$$
$$= S\underline{/\theta°} = \mathbf{S} \text{ (the complex power)}$$

and

$$\boxed{\mathbf{S} = \mathbf{VI}^* \quad VA} \tag{13.23}$$

Similar results, as above, are obtained for the case of the capacitive load, except that the angle $\theta°$ of Z (and of **S**) is negative, as is appropriate to represent this case.

In the expression for the complex power **S**, a positive imaginary part $(+jQ)$ corresponds to a positive phase angle $\theta°$, and represents the reactive power in an inductive load with lagging PF, while if the imaginary part of the complex power **S** is negative $(-jQ)$, it corresponds to a negative phase angle, $-\theta°$, and represents the reactive power in a capacitive load with leading PF.

In conclusion, the importance of the complex power **S**, as obtained from Eqs. (13.21), (13.22), or (13.23), is that it contains all the information pertaining to power in the given load. Its magnitude is the apparent power $(S = VI)$. Its phase angle, $\theta°$, is the power factor angle and is also the phase angle of the load impedance Z. Thus the power factor is:

$$PF = \cos\theta$$

lagging if $\theta°$ is positive (for inductive loads) or leading if $\theta°$ is negative (for capacitive loads). The real part of the complex power **S** is the real power dissipated in the load (P), while the imaginary part of the complex power is the reactive power in the load (Q), also indicating the nature of the load (inductive or capacitive) depending on the sign, as stated above.

13.2.2 Overall Complex Power for Series- and Parallel-Connected Loads

When two load impedances, Z_1 and Z_2, are connected in series as shown in Fig. 13.7, the same current phasor, $\mathbf{I}$, is flowing in each. Here

$$Z_T = Z_1 + Z_2$$

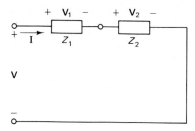

Figure 13.7

and from KVL,

$$\mathbf{V} = \mathbf{V}_1 + \mathbf{V}_2$$

The overall complex power in Z_T is

$$\begin{aligned}
\mathbf{S} &= \mathbf{VI}^* \\
&= (\mathbf{V}_1 + \mathbf{V}_2)\mathbf{I}^* \\
&= \mathbf{V}_1\mathbf{I}^* + \mathbf{V}_2\mathbf{I}^* \\
&= \mathbf{S}_1 + \mathbf{S}_2
\end{aligned} \tag{13.24}$$

where $\mathbf{S}_1$ and $\mathbf{S}_2$ are the complex powers for each of the load impedances Z_1 and Z_2. Clearly, this relationship can be extended for any number of series-connected loads.

When the two load impedances Z_1 and Z_2 are connected in parallel as shown in Fig. 13.8, the same voltage phasor, $\mathbf{V}$, exists across each. Here

$$Z_T = \frac{Z_1 Z_2}{Z_1 + Z_2}$$

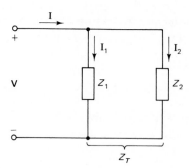

Figure 13.8

and from KCL,

$$\mathbf{I} = \mathbf{I}_1 + \mathbf{I}_2$$

Therefore,

$$\mathbf{I}^* = \mathbf{I}_1^* + \mathbf{I}_2^*$$

The overall complex power in Z_T is

$$\mathbf{S} = \mathbf{V}\mathbf{I}^*$$
$$= \mathbf{V}(\mathbf{I}_1^* + \mathbf{I}_2^*)$$
$$= \mathbf{V}\mathbf{I}_1^* + \mathbf{V}\mathbf{I}_2^*$$
$$= \mathbf{S}_1 + \mathbf{S}_2 \qquad (13.24a)$$

where $\mathbf{S}_1$ and $\mathbf{S}_2$ are the complex powers for each of the load impedances Z_1 and Z_2.

Thus whether the loads are connected in series or in parallel, or in general in any series–parallel combination, the total overall complex power of a network is the vectorial sum of the complex power of each of the loads constituting the network.

▮ EXAMPLE 13.4

A load whose impedance is $Z = 10\underline{/36.9°}\ \Omega$ is supplied by a source whose voltage is $\mathbf{V} = 30\underline{/0°}$ V. Find the complex power of this load and all the power information regarding this load.

Solution   Here

$$\mathbf{I} = \frac{\mathbf{V}}{Z} = \frac{30\underline{/0°}}{10\underline{/36.9°}} = 3\underline{/-36.9°}\ \text{A}$$

(using Ohm's law for the load) and

$$\mathbf{I}^* = 3\underline{/36.9°}\ \text{A}$$

Therefore,

$$\mathbf{S} = \mathbf{V}\mathbf{I}^* = 30\underline{/0°} \times 3\underline{/36.9°} = 90\underline{/36.9°}\ \text{VA}$$
$$= 72 + j54\ \text{VA}$$
$$= P + jQ$$

and

$$\text{apparent power} = 90\ \text{VA}\ (= VI)$$
$$\text{PF} = \cos(36.9°) = 0.8\ \text{lagging (inductive load)}$$

Note that

$$Z = 10\underline{/36.9°}\ \Omega = 8 + j6\ \Omega = R + jX_L$$
$$\text{real power dissipated} = P = 72\ \text{W}\ [= I^2 R = (3)^2 \times 8]$$
$$\text{reactive power} = Q = 54\ \text{VAR}\ [= I^2 X_L = (3)^2 \times 6]\ \text{(inductive)}$$

▮ EXAMPLE 13.5

A load absorbs an average power of 1 kW. Find the power factor of the load if
(a) $Q = 0$.
(b) $Q = 500$ VAR (inductive).
(c) $Q = 200$ VAR (capacitive).

Solution The real power in the load $= P = 1000\,\text{W}$.

(a) $\mathbf{S} = $ complex power $= P + jQ = 1000 + j0 = 1000\underline{/0°}\,\text{VA}$. Therefore,

$$PF = \cos 0° = 1$$

This is a purely resistive load.

(b) $\mathbf{S} = P + jQ = 1000 + j500 = 1118.0\underline{/26.6°}\,\text{VA}$. Therefore,

$$PF = \cos 26.6° = 0.894 \text{ lagging (inductive load)}$$

(c) $\mathbf{S} = P - jQ = 1000 - j200 = 1019.8\underline{/-11.3°}\,\text{VA}$. Therefore,

$$PF = \cos 11.3° = 0.981 \text{ leading (capacitive load)}$$

EXAMPLE 13.6

A load absorbs 2 kW at 0.9 PF lagging. Find the complex power of this load.

Solution Since

$$PF = \cos \theta° = 0.9 \quad (\text{lagging, i.e., inductive load})$$

then

$$\theta = 25.84°$$

Also,

$$\frac{Q}{P} = \tan \theta = \tan 25.84° = 0.4843$$

Therefore,

$$Q = 2000 \times 0.4843 = 968.6\,\text{VAR (inductive)}$$
$$\mathbf{S} = P + jQ = 2000 + j968.6 = 2222.2\underline{/25.84°}\,\text{VA}$$

EXAMPLE 13.7

A load absorbs 4 kW of power and has a reactive power of $Q = 3\,\text{kVAR}$ (capacitive). The voltage across the load is 500 V. Find the apparent power, PF, the load current, and the load impedance.

Solution The complex power $\mathbf{S}$ for this load is

$$\mathbf{S} = P - jQ = 4000 - j3000 = 5000\underline{/-36.9°}\,\text{VA}$$

Therefore,

$$\text{apparent power} = S = 5000\,\text{VA}$$
$$= VI$$
$$I = \text{load current} = \frac{S}{V} = \frac{5000}{500} = 10\,\text{A}$$

and

$$\text{power factor} = PF = \cos(-36.9) = 0.8 \text{ leading (capacitive load)}$$

To find the impedance, let

$$\mathbf{V} = 500\underline{/0°}\,\text{V} \quad (\text{i.e., the reference phasor})$$

Then

$$\mathbf{I} = 10\underline{/36.9°}\text{ A} \quad (\mathbf{I} \text{ leads } \mathbf{V} \text{ because the load is capacitive})$$

$$Z = \frac{\mathbf{V}}{\mathbf{I}} = \frac{500\underline{/0°}}{10\underline{/36.9°}} = 50\underline{/-36.9°}\,\Omega = 40 - j30\,\Omega = R - jX_C$$

$$R = 40\,\Omega \quad \text{and} \quad X_C = 30\,\Omega$$

Also (check),

$$P = I^2R \quad \text{that is, } R = \frac{P}{I^2} = \frac{4000}{(10)^2} = 40\,\Omega$$

and

$$Q = I^2X_C \quad \text{that is, } X_C = \frac{Q}{I^2} = \frac{3000}{(10)^2} = 30\,\Omega \quad \text{(as above)}$$

EXAMPLE 13.8

The circuit shown in Fig. 13.9 consists of three loads in a series–parallel connection, with each of the loads defined as indicated. Find the overall complex power of the circuit, its overall PF, and the phasor form of the source current $\mathbf{I}$.

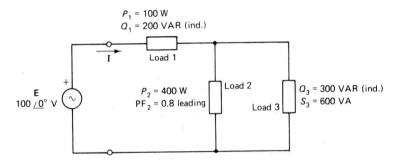

Figure 13.9

Solution The overall complex power $\mathbf{S}$ of this circuit is obtained using the vectorial addition expression of the complex power of each load, that is,

$$\mathbf{S} = \mathbf{S}_1 + \mathbf{S}_2 + \mathbf{S}_3$$

Here

$$\mathbf{S}_1 = 100 + j200 = 223.6\underline{/63.4°}\text{ VA}$$

For load 2,

$$\text{PF}_2 = 0.8 \text{ (leading, i.e., capacitive load)} = \frac{P_2}{S_2} = \cos\theta_2$$

Therefore,

$$S_2 = \frac{P_2}{\text{PF}_2} = \frac{400}{0.8} = 500 \text{ VA}$$

and

$$\theta_2 = -36.9° \quad \text{(negative angle since this load is capacitive)}$$

Thus

$$S_2 = 500\underline{/-36.9°}\text{ VA} = 400 - j300\text{ VA}$$

For load 3,

$$Q_3 = S_3 \sin\theta_3$$

Therefore,

$$\sin\theta_3 = \frac{Q_3}{S_3} = \frac{300}{600} = 0.5$$

$$\theta_3 = 30° \quad \text{(positive angle since this load is inductive)}$$
$$S_3 = 600\underline{/30°} = 519.6 + j300\text{ VA}$$

and

$$S = (100 + j200) + (400 - j300) + (519.6 + j300)$$
$$= 1019.6 + j200 = 1039\underline{/11.1°}\text{ VA}$$

The overall load is thus inductive, with PF $= \cos 11.1° = 0.981$ lagging. Since

$$S = 1039 = EI$$

then

$$I = \frac{1039}{100} = 10.39\text{ A}$$

The current phasor lags the voltage phasor $\mathbf{E}$ for an inductive circuit; therefore,

$$\mathbf{I} = 10.39\underline{/-11.1°}\text{ A}$$

EXAMPLE 13.9

Find the complex power for each of the three parallel-connected loads shown in Fig. 13.10. Then obtain the overall complex power of this network, its overall PF, and the phasor form of the source current $\mathbf{I}$.

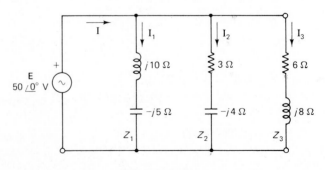

Figure 13.10

Solution   For load 1,

$$Z_1 = j10 - j5 = j5 = 5\underline{/90°}\ \Omega \quad \text{(purely inductive)}$$

Therefore,

$$\mathbf{I}_1 = \frac{\mathbf{E}}{Z_1} = \frac{50\underline{/0°}}{5\underline{/90°}} = 10\underline{/-90°}\ A$$

and

$$\mathbf{S}_1 = \mathbf{E}\mathbf{I}_1^* = 50\underline{/0°} \times 10\underline{/90°} = 500\underline{/90°}\ \text{VA} = j500\ \text{VA}$$

Note that no power ($P = 0$) is dissipated in this load, as it has no resistive component.

For load 2:

$$Z_2 = 3 - j4 = 5\underline{/-53.1°}\ \Omega \quad \text{(capacitive load)}$$

Therefore,

$$\mathbf{I}_2 = \frac{\mathbf{E}}{Z_2} = \frac{50\underline{/0°}}{5\underline{/-53.1°}} = 10\underline{/53.1°}\ A$$

and

$$\mathbf{S}_2 = \mathbf{E}\mathbf{I}_2^* = 50\underline{/0°} \times 10\underline{/-53.1°} = 500\underline{/-53.1°}\ \text{VA}$$
$$= 300 - j400\ \text{VA}$$

For load 3:

$$Z_3 = 6 + j8 = 10\underline{/53.1°}\ \Omega \quad \text{(inductive load)}$$

Therefore,

$$\mathbf{I}_3 = \frac{\mathbf{E}}{Z_3} = \frac{50\underline{/0°}}{10\underline{/53.1°}} = 5\underline{/-53.1°}\ A$$

and

$$\mathbf{S}_3 = \mathbf{E}\mathbf{I}_3^* = 50\underline{/0°} \times 5\underline{/53.1°} = 250\underline{/53.1°}\ \text{VA}$$
$$= 150 + j200\ \text{VA}$$

Thus the overall complex power of this network is

$$\mathbf{S} = \mathbf{S}_1 + \mathbf{S}_2 + \mathbf{S}_3 = (j500) + (300 - j400) + (150 + j200)$$
$$= 450 + j300 = 540.8\underline{/33.7°}\ \text{VA}$$

The overall complex power is inductive. The overall PF is then

$$\text{PF} = \cos 33.7° = 0.832 \text{ lagging}$$

To find the source current $\mathbf{I}$, one can use

$$\mathbf{S} = \mathbf{E}\mathbf{I}^*$$

Therefore,

$$\mathbf{I}^* = \frac{\mathbf{S}}{\mathbf{E}} = \frac{540.8\underline{/33.7°}}{50\underline{/0°}} = 10.82\underline{/33.7°}\ A$$

and

$$\mathbf{I} = 10.82\underline{/-33.7°}\ \text{A}$$

Check:

$$\mathbf{I} = \mathbf{I}_1 + \mathbf{I}_2 + \mathbf{I}_3 \ \text{(from KCL)}$$
$$= (-j10) + (6 + j8) + (3 - j4) = 9 - j6$$
$$= 10.82\underline{/-33.7°}\ \text{A} \qquad \text{(as above)}$$

13.3 POWER FACTOR CORRECTION

The generators and transformers used in power distribution systems to industrial and domestic customers are rated in apparent power ($S = VI$, i.e., in kVA). At a fixed operating voltage, the loading on the system increases as higher current flow in the distribution lines. Power companies thus charge customers that require high operating current an extra premium.

Most industrial and domestic electric loads operate at a required fixed amount of real power P. The power factor thus becomes increasingly important, since according to Eq. (13.9),

$$I = \text{load current} = \frac{P}{V \cos \theta} = \frac{P}{V \cdot \text{PF}} \tag{13.25}$$

As stated above, usually the load voltage V and the load power P are fixed constant quantities. Thus to reduce the line current I, the power factor must be increased. The ultimate value of the PF is unity (i.e., $= \cos 0°$). This situation corresponds to the overall load being equivalent to a purely resistive load, with the reactive power Q being zero, that is,

$$\mathbf{S} = P + j0 = P\underline{/0°}\ \text{VA} \tag{13.26}$$

Obviously, operating at a low PF requires high load current and is costly to the customer. For this reason, the power factor should be corrected (i.e., improved) to approach unity.

To correct (or improve) the power factor, without affecting the real power dissipated in the load impedance Z_L, a reactive impedance Z' (either a pure L or a pure C) is connected in parallel with the load, as shown in Fig. 13.11. Here,

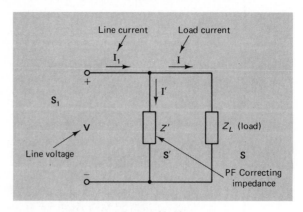

Figure 13.11

according to the properties of parallel circuits, the load voltage $\mathbf{V}$ and the load current $\mathbf{I}$ remain unchanged. The reactive impedance Z' does not dissipate any power (i.e., $P' = 0$ and $PF' = \cos 90° = 0$). Thus the overall power dissipated in the system remains the same, P.

By choosing Z' so that the new line current,

$$\mathbf{I}_1 = \mathbf{I} + \mathbf{I}'$$

would have a minimal phase shift from $\mathbf{V}$ (i.e., θ_1), the overall power factor of the system, $PF_1 = \cos \theta_1$, will improve. Ideally, θ_1 should approach zero; hence the overall power factor, PF_1, would approach unity. In this situation,

$$I_1 = \frac{P}{V \cos \theta_1} = \frac{P}{V} \tag{13.27}$$

that is, the magnitude of the line current I_1 is reduced to its minimum possible value and the overall reactive power of the system, Q_1, is reduced to approach zero. It is very convenient to understand and analyze the power factor correction technique through the consideration of the power triangle and the complex power, as shown in Fig. 13.12.

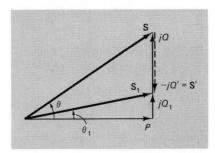

Figure 13.12

As developed in Section 13.2.2, the complex power of the overall system, $\mathbf{S}_1$ is obtained as the vectorial sum of $\mathbf{S}'$ (for impedance Z') and $\mathbf{S}$ (for load Z_L), that is,

$$\mathbf{S}_1 = \mathbf{S} + \mathbf{S}'$$
$$= (P + jQ) + (-jQ') = P + j(Q - Q') = P + jQ_1 \tag{13.28}$$

The objective behind power factor correction is to reduce the overall reactive power, Q_1. Thus if the original load Z_L is inductive ($P + jQ$), as is usually the case in practice, the added impedance Z' must be purely capacitive ($= -jX_C'$), with complex power $\mathbf{S}' = -jQ'$, as shown in Fig. 13.12 and Eq. (13.28).

In the special cases where the original load is capacitive, the added impedance Z' must be a pure inductance ($= jX_L'$), exactly the opposite of the situation above, to achieve the same objective of reducing Q_1. These cases are treated further in Chapter 14 as examples of general series–parallel resonance (see Section 14.3).

Thus to find the nature and magnitude of the required added impedance Z', whose complex power is $\mathbf{S}'$, obtain

$$\boxed{\begin{aligned} \mathbf{S}' &= \mathbf{S}_1 \text{ (overall complex power of the system)} \\ &\quad - \mathbf{S} \text{ (complex power of the original load)} \end{aligned}} \tag{13.28a}$$

If $\mathbf{S}'$ is a negative reactive power (i.e., $\mathbf{S}' = -jQ'$), Z' must be a capacitance, where

$$\boxed{Q' = \frac{V^2}{X_C'} = V^2 \omega C'} \qquad (13.29)$$

C' can then be easily calculated from Eq. (13.29).

On the other hand, if $\mathbf{S}'$ is a positive reactive power (i.e., $\mathbf{S}' = jQ'$), then Z' must be an inductance, where

$$Q' = \frac{V^2}{X_L'} = \frac{V^2}{\omega L'} \qquad (13.30)$$

The value of L' can then easily be obtained from Eq. (13.30).

▌ EXAMPLE 13.10

A load with 0.8 lagging PF absorbs 60 W from a 100 V, 60 Hz power line. It is required to correct the power factor to 0.9 lagging. Find
(a) The original and the final line currents.
(b) The value of the element to be added to achieve the required PF correction.

Solution   (a) Since the absorbed power remains the same before and after the PF correction, and

$$P = VI \cos \theta \qquad = VI \cdot PF \qquad = VI_1 \cdot PF_1$$
$$\text{(before correction)} \quad \text{(after correction)}$$

then

$$I \text{ (original line current)} = \frac{P}{V \cdot PF} = \frac{60}{100 \times 0.8} = 0.75 \text{ A}$$

and

$$I_1 \text{ (final line current after correction)} = \frac{60}{100 \times 0.9} = 0.667 \text{ A}$$

Thus the line current is reduced due to the power factor correction, as expected.

(b) The load is inductive since it has a lagging PF. To achieve the power factor correction (i.e., reduce the net reactive power), the impedance Z' added in parallel must be a capacitor C'.

For the original load,

$$P = 60 \text{ W}$$

$$PF = 0.8 = \cos \theta \Rightarrow \theta = 36.9°$$

$$S = \frac{P}{\cos \theta} = \frac{60}{0.8} = 75 \text{ VA}$$

Therefore,

$$\mathbf{S} = 75\underline{/36.9°} = 60 + j45 \text{ VA}$$

For the required situation after the PF correction,

$$P = 60 \text{ W}$$

$$\text{PF}_1 = 0.9 \text{ (lagging)} = \cos \theta_1 \Rightarrow \theta_1 = 25.84°$$

$$S_1 = \frac{P}{\cos \theta_1} = \frac{60}{0.9} = 66.67 \text{ VA}$$

Therefore,

$$\mathbf{S}_1 = 66.67\underline{/25.84°} = 60 + j29.06 \text{ VA}$$

Thus

$$\mathbf{S}' = \text{complex power of the added element, } Z'$$
$$= \mathbf{S}_1 - \mathbf{S}$$
$$= (60 + j29.06) - (60 + j45)$$
$$= -j15.94 = -jQ' \quad \text{(i.e., a capacitive impedance, as expected)}$$

and

$$Q' = 15.94 \text{ VAR} = \frac{V^2}{X'_C} = V^2 \omega C'$$
$$= (100)^2 \times 2\pi \times 60 \times C'$$
$$C' = \frac{15.94}{2\pi \times 60 \times 10^4} = 4.23 \ \mu F$$

EXAMPLE 13.11

The two series-connected loads in the circuit shown in Fig. 13.13 are specified as follows:

$$Z_a: \quad 250 \text{ VA, } 0.5 \text{ lagging PF}$$
$$Z_b: \quad 180 \text{ W, } 0.8 \text{ leading PF}$$

(a) Find the overall complex power and PF of this combined load.
(b) Find the value and the nature of the element required to correct the overall PF to unity.
(c) What is the percentage reduction in the line current, from before PF correction to after PF correction?

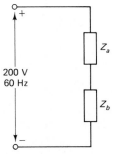

Figure 13.13

200 V
60 Hz

Solution (a) For Z_a:

$$S_a = 250 \text{ VA} \qquad \text{PF} = \cos \theta_a = 0.5 \text{ lagging} \Rightarrow \theta_a = 60°$$

Therefore,

$$\mathbf{S}_a = 250\underline{/60°} = 125 + j216.5 \text{ VA}$$

For Z_b:

$$P = 180 \text{ W}$$

$$\text{PF} = 0.8 \text{ leading} = \cos \theta_b \Rightarrow \theta_b = -36.9° \quad \text{(capacitive load)}$$

$$S_b = \frac{P}{PF} = \frac{180}{0.8} = 225 \text{ VA}$$

Therefore,

$$\mathbf{S}_b = 225\underline{/-36.9°} = 180 - j135.1 \text{ VA}$$

and

$$\mathbf{S} \text{ (overall load)} = \mathbf{S}_a + \mathbf{S}_b$$
$$= (125 + j216.5) + (180 - j135.1) = 305 + j81.4$$
$$= 315.7\underline{/15°} \text{ VA}$$

The overall load is inductive:

$$PF = \cos 15° = 0.966 \text{ lagging}$$

The line current in this case is

$$\frac{S}{V} \left(\text{or } \frac{P}{V \cdot PF} \right) = \frac{315.7}{200} = 1.58 \text{ A} \left(\text{or } = \frac{305}{200 \times 0.966} = 1.58 \text{ A} \right)$$

(b) Since it is required to correct the overall PF to unity,

$$P_{\text{overall}} = 305 \text{ W} \qquad \cos \theta_1 = 1.0 \Rightarrow \theta_1 = 0°$$

then

$$S_1 = \frac{P_{\text{overall}}}{\cos \theta_1} = 305 \text{ VA}$$

and

$$\mathbf{S}_1 = 305\underline{/0°} = 305 + j0 \text{ Va} \quad (\text{i.e., } Q_1 = 0 \text{ VAR})$$

Thus the complex power of the added impedance element, $\mathbf{S}'$, is

$$\mathbf{S}' = \mathbf{S}_1 - \mathbf{S}$$
$$= (305 + j0) - (305 + j81.4)$$
$$= -j81.4 = -jQ'$$

The power-factor-corrected system is shown in Fig. 13.14. Thus the added impedance, Z', must be a capacitor, with

$$Q' = 81.4 \text{ VAR} = \frac{V^2}{X'_C} = V^2 \omega C'$$

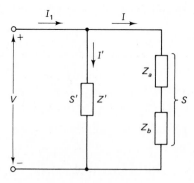

Figure 13.14

Therefore,

$$C' = \frac{81.4}{(200)^2 \times 2\pi \times 60} = 5.4 \, \mu F$$

Here the new line current, I_1, is

$$I_1 = \frac{S_1}{V} = \frac{305}{200} = 1.525 \text{ A}$$

(c) The reduction in the line current is

$$\frac{I - I_1}{I} \times 100\% = \frac{1.58 - 1.525}{1.58} \times 100 = 3.48\%$$

Note: In this particular example, the power factor correction is not really worthwhile. In practice, if the overall load has a PF of 0.9 (or higher), PF correction is not usually necessary.

13.4 POWER MEASUREMENT AND THE WATTMETER

The wattmeter, shown in Fig. 13.15, is the instrument used to measure the average power dissipated in a given load. It contains two coils, with a pair of terminals for each coil. One terminal of each of the coils, designated as the voltage and the current coils, is marked with ±polarity.

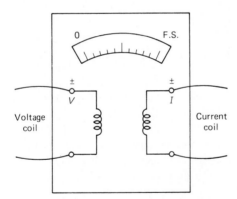

Figure 13.15

A schematic diagram of the construction of the wattmeter is shown in Fig. 13.16(a). The circuit representation of the wattmeter, as it may be connected to measure the average power in an impedance Z_L, is shown in Fig. 13.16(b).

The current coil is stationary, split into two halves as shown in Fig. 13.16(a). It has a very small impedance in order not to affect the value of the load current, which is monitored by flowing through this current coil. The current coil is connected in series with the load impedance as shown in Fig. 13.16(b). The magnetic field produced by the current coil is a function of time and is proportional to the load current i. Hence

$$\phi_i \propto i(t) \tag{13.31}$$

The voltage coil is a rotating high-impedance coil. The pointer and the restraining springs are attached to this coil. A high resistance, R_V, is already

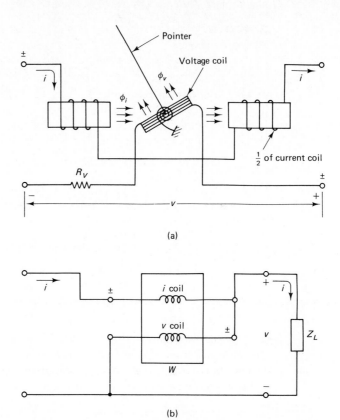

(a)

(b)

Figure 13.16

connected in series with the voltage coil inside the wattmeter, to reduce the loading effect of this coil as it is used to monitor the voltage v across the load. This is similar to the multiplier resistance R_{mult} used in dc voltmeters. The voltage coil must be connected across the load, as shown in Fig. 13.16(b), to monitor the load voltage v. Notice that usually one terminal of each of the current and voltage coils are connected together (to the same node in the circuit), as indicated in Fig. 13.16(b). The magnetic field produced by the rotating voltage coil is also a function of time and is proportional to the load voltage v. Hence

$$\phi_v \propto v(t) \tag{13.32}$$

A torque is produced as a result of the interaction of the two magnetic fields ϕ_i and ϕ_v. Due to the inertia of the moving coil, the deflection resulting from this torque is proportional to the average value of the torque. The torque itself is proportional to $\phi_i \phi_v$. Thus the pointer's deflection (i.e., the wattmeter reading)

$$\propto \overline{\phi_i \phi_v} \propto \overline{i(t)v(t)} \propto \overline{p(t)}(= P) \tag{13.33a}$$

where

$$P = \overline{p(t)} = VI \cos \phi_{vi}$$
$$= \text{average power dissipated in } Z_L \tag{13.33b}$$

as derived in Section 13.1. Wattmeter operation as described above is called the *electrodynamometer principle*. In many situations, the loading effects of the two coils, which are usually very small, are taken into account in calibrating the wattmeter.

Proper connections of the terminals of the wattmeter's coils are very important in order to obtain an upscale deflection (i.e., reading). The polarities of the load (Z_L) voltage and the direction of the current flow in the load must be marked in accordance with the V–I polarity convention. Current must enter (or leave) the current coil polarity marked ± at the same time that the voltage coil polarity marked ± is connected to the higher (or lower) potential point of the load. This is shown clearly in Fig. 13.16(b). Note that the simultaneous reversal of the two coil connections also results in proper upscale deflection and is thus allowed. However, polarity mixing, or reversal of one of the two coil connections and not the other, will result in no wattmeter reading (downscale deflection). The importance of the correct wattmeter connections will be more apparent in three-phase networks, where incorrect connections could still produce upscale wattmeter readings but are completely erroneous.

The presence of an ideal wattmeter in a circuit does not influence the average load power to be measured. The reading of the wattmeter will be the average power dissipated in the load (or by the source) to which the meter is connected, as given by Eq. (13.33b).

■ EXAMPLE 13.12

Find the reading of the two ideal wattmeters connected as shown in the circuit of Fig. 13.17.

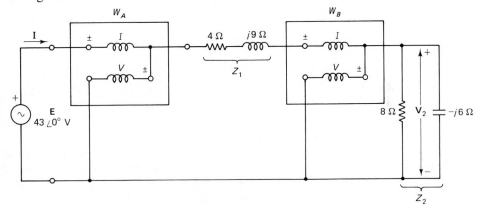

Figure 13.17

Solution The circuit should first be analyzed, ignoring the wattmeters. Here

$$Z_1 = 4 + j9 \ \Omega$$

$$Z_2 = \frac{8(-j6)}{8 - j6} = \frac{48 / -90°}{10 / -36.9°} = 4.8 / -53.1° \ \Omega$$

$$= 2.88 - j3.84 \ \Omega$$

The total circuit's impedance as seen by the source is

$$Z_T = Z_1 + Z_2 = 4 + j9 + 2.88 - j3.84$$

$$= 6.88 + j5.16 = 8.60 / 36.9° \ \Omega$$

Thus

$$\mathbf{I} = \frac{\mathbf{E}}{Z_T} = \frac{43\underline{/0^\circ}}{8.6\underline{/36.9^\circ}} = 5\underline{/-36.9^\circ}\,\text{A}$$

The voltage across Z_2, $\mathbf{V}_2$, is obtained from Ohm's law:

$$\mathbf{V}_2 = \mathbf{I}Z_2 = 5\underline{/-36.9^\circ} \times 4.8\underline{/-53.1^\circ} = 24.0\underline{/-90^\circ}\,\text{V}$$

The wattmeter W_A is connected such that its current coil monitors $\mathbf{I}$ and its voltage coil monitors $\mathbf{E}$. Hence its reading indicates the total power delivered by the source:

$$W_A \text{ reading} = EI \cos\theta_{EI}$$
$$= 43 \times 5 \times \cos(-36.9^\circ) = 172\,\text{W}$$

The wattmeter W_B is connected such that its current coil monitors $\mathbf{I}$ and its voltage coil monitors $\mathbf{V}_2$. Hence its reading indicates the average power dissipated in Z_2:

$$W_B \text{ reading} = V_2 I \cos\theta_{V_2 I}$$
$$= 24 \times 5 \times \cos(53.1^\circ) = 72\,\text{W}$$

Check:

$$P_2 = \frac{V_2^2}{8} = \frac{(24)^2}{8} = 72\,\text{W}$$
$$P_T = P_2 + I^2 \times 4 = 72 + (5)^2 \times 4 = 172\,\text{W} \qquad \text{(as above)}$$

DRILL PROBLEMS

Section 13.1

13.1. Find the PF and the real power dissipated by a load whose impedances is $Z = 20 + j30\,\Omega$ if the applied source voltage is $e(t) = 14.142\cos 1000t$ V.

13.2. If the voltage source $e(t) = 70.71\sin(2000t - 30^\circ)$ V drives the current $i(t) = 0.35355\sin(2000t + 30^\circ)$ A in a load, what are the power dissipated by this load and the power factor of this load? Find the values of the elements that constitute this series-connected load.

13.3. A 60 Hz, 120 V source delivers 1 kW of power to a load. The current flowing in the load is 15 A. Find the PF of the load and the impedance of this inductive load.

Section 13.2

13.4. Obtain the complex power and all the power parameters relating to the load examined in Drill Problem 13.1.

13.5. Obtain the complex power and all the power parameters relating to the load examined in Drill Problem 13.2.

13.6. Obtain the complex power and all the power parameters relating to the load examined in Drill Problem 13.3.

13.7. A 3 kW load has a 1 kVAR capacitive reactive power. If the voltage applied to this load is 250 V, find the apparent power, PF, load current, and load impedance.

13.8. In a circuit similar to that shown in Fig. 13.9:

> load 1: 200 W, $PF_1 = 0.75$ lagging
>
> load 2: 500 VAR (inductive), $S_2 = 700$ VA
>
> load 3: 300 W, 400 VAR capacitive)

The applied voltage source is $\mathbf{E} = 200\underline{/30°}$ V. Find the overall complex power of the system, its overall PF, and the current $\mathbf{I}$.

13.9. The two loads

$$Z_1 = 200\underline{/36.9°}\ \Omega$$
$$Z_2 = 70.71\underline{/-45°}\ \Omega$$

are connected in parallel to a voltage source whose voltage is $10\underline{/0°}$ V. Find the current in each of the impedances and in the source. Obtain the complex power of each load and of the overall system. What is the PF of the overall system?

Section 13.3

13.10. Correct the power factor of the system described in problem 8 to unity. What will be the value of the element to be added in parallel with the source if the frequency of the source is 120 Hz? What is the new value of the source's current?

13.11. Correct the power factor of the system described in problem 9 to 0.98 lagging. What will be the value of the element to be added in parallel with the source if the frequency of the source is 400 Hz? What will be the new value of the source's current?

PROBLEMS

Section 13.1

(i) **13.1.** In the circuit shown in Fig. 13.18, $\mathbf{E} = 10\underline{/0°}$ V. Find the power factor of the entire circuit. What is the average power delivered by the source? Determine the reactive and apparent power of this circuit.

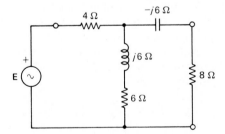

Figure 13.18

(i) **13.2.** Find the average power and the power factor of a load that allows a current $i(t) = 0.1414 \cos{(3000t - 30°)}$ A if the applied source voltage is $v(t) = 28.28 \sin{(3000t)}$ V.

(i) **13.3.** A 10 V source is applied to each of the following loads:
  (a) $Z_{La} = 3 + j4\ \Omega$.
  (b) $Z_{Lb} = 14 - j14\ \Omega$.
  (c) $Z_{Lc} = 50\underline{/30°}\ \Omega$.
  Find the power factor and the average power dissipated in each of these loads.

(i) **13.4.** A load dissipates 40 W when the current flowing in it is 5 A. The power factor of the load is 0.8 lagging. Find the impedance of the load if the frequency of the source is 2 kHz.

(i) **13.5.** A source whose voltage is $v(t) = 100 \sin{(5000t + 45°)}$ V delivers 500 W of power to an inductive load. The resistance of the load is 5 Ω. Find the PF and the inductance of this load.

(ii) **13.6.** A 70.7 V, 1000 Hz source provides 141.4 W to a load that has a leading PF. The source current is 4 A. Find the PF and the impedance of this load. Determine the elements of this load:
    (a) If the elements are connected in series.
    (b) If the elements are connected in parallel.

Section 13.2.

(i) **13.7.** Obtain the complete power triangle and the related power information if a $100/{-90°}$ V source is applied to a load whose impedance is
    (a) $Z_{La} = 8 - j6$ Ω.
    (b) $Z_{Lb} = 10 + j20$ Ω.

(i) **13.8.** What is the apparent power and the PF (hence determine **S**, the complex power) for each of the following loads.
    (a) $P_a = 100$ W, $Q_a = 300$ VAR (inductive).
    (b) $P_b = 400$ W, $Q_b = 200$ VAR (capacitive).

(i) **13.9.** Find the complex power of the load in each of the following situations.
    (a) $P = 1000$ W, PF = 0.707 leading.
    (b) $Q = 500$ VAR, PF = 0.8 lagging.
    (c) $S = 800$ VA, $P = 700$ W (inductive load).
    (d) $S = 200$ VA, $Q = 160$ VAR (capacitive load).

(i) **13.10.** Obtain the complete power triangle for a load whose current is $i = 1.414 \sin{(3000t - 60°)}$ A when the applied source voltage is $v = 70.7 \sin{(3000t + 10°)}$ V. What is the impedance of this load?

(ii) **13.11.** Determine the overall impedance of each of the following circuit conditions.
    (a) $\mathbf{S} = 3200/{-30°}$ VA, $\mathbf{V} = 800/30°$V.
    (b) $P = 2000$ W, PF = 0.707 lagging, and $I = 10$ A.
    (c) $Q = 1000$ VAR (capacitive), $P = 1000$ W, and $V = 100$ V.

(i) **13.12.** Find the complete power triangle for a load whose PF is 0.5 lagging if it draws 2.5 A from a 50 V, 2000 rad/s source. What is the impedance of this load?

(i) **13.13.** Find the complex power for each of Z_1 and Z_2 in the network shown in Fig. 13.19. What is the overall complex power of the network? Check that such quantity is equivalent to the complex power of the source.

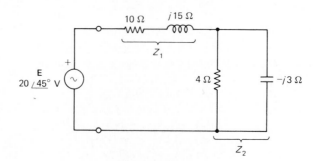

Figure 13.19

(ii) **13.14.** If the power dissipated in the 4 Ω resistor in the circuit shown in Fig. 13.20 is 400 W, find **E** if $\mathbf{I}_2$ is taken to be the reference phasor. Obtain the complex power of each branch of this circuit and the overall complex power.

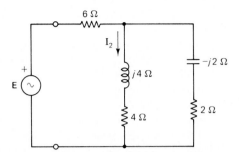

Figure 13.20

(ii) **13.15.** Find the overall complex power of the network shown in Fig. 13.21. If the applied source voltage is $\mathbf{V} = 100\underline{/0°}\,\text{V}$, find the source current **I**.

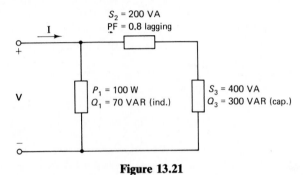

Figure 13.21

(ii) **13.16.** The overall power dissipated in the circuit shown in Fig. 13.22 is 200 W. Find the complete power triangle of this circuit.

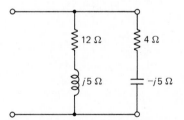

Figure 13.22

(ii) **13.17.** If the circuit shown in Fig. 13.23 has an overall apparent power of 500 VA and its overall PF is 0.8 lagging, find the unknown impedance Z.

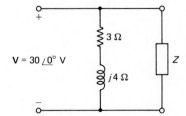

Figure 13.23

Problems

(i) **13.18.** Determine the complex power in each of the branches of the circuit shown in Fig. 13.23.

Section 13.3

(i) **13.19.** A load whose impedance is $(3 + j4)\ \Omega$ is connected across a 10 V, 1000 Hz source. Determine the PF of the load, the source current, and the complex power of the load. If it is required to correct the power factor to 0.95 lagging, what is the nature and magnitude of the element required to be connected in parallel with the load? What is the source current after the PF correction?

(i) **13.20.** Find the required parallel connected capacitor needed to improve the PF of a 100 kVA, 0.6 PF lagging motor to unity. The source is a 10 kV, 500 Hz source.

(ii) **13.21.** A 10 kW, 0.7 PF lagging load is connected in parallel with a capacitor to a 2400 V, 60 Hz source. Find the value of the capacitor to obtain an overall power factor of 0.95 leading. What is the source current with and without the parallel-connected capacitor?

(ii) **13.22.** The addition of 220 μF in parallel with a 200 kW load across a 2400 V, 60 Hz line improved the overall power factor to 0.95 lagging. What was the original power factor of the load? What is the line current before and after the power factor correction?

(ii) **13.23.** A 5 kW, 0.8 PF lagging induction motor is connected in parallel with a 2 kVA synchronous motor which operates with a leading PF. Find the the PF of the synchronous motor if the overall PF is 0.9 lagging.

Section 13.4

(i) **13.24.** Find the reading of the wattmeter when connected first as shown in Fig. 13.24(a) and then when connected as shown in Fig. (13.24(b)).

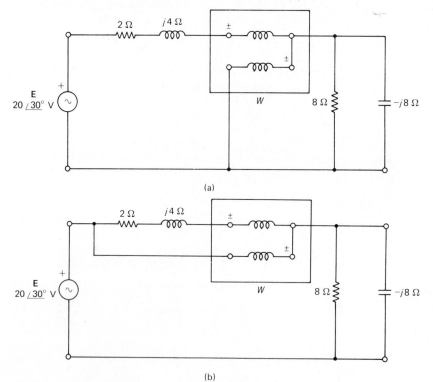

(a)

(b)

Figure 13.24

GLOSSARY

Apparent Power: The product of the magnitude of the ac voltage across a load and the magnitude of the ac current flowing in the load, without consideration of the phase shift between the two phasors (or waveforms). This quantity is not the real power in the load, but seems to correspond to the similar VI product in the case of dc circuit analysis.

Complex Power (S): A vector representing the power parameters of a load (or a system). Its real component is the real power, its imaginary component is the reactive (or quadrature) power, its magnitude is the apparent power, and its phase angle is the phase angle of the load (or the system).

Power Factor (PF or F_p): The ratio of the real ac power dissipated in a load to the apparent power of that load. It is the cosine of the phase shift between the voltage and current waveforms (or phasors) relating to this load.

Power Factor Correction: The addition of a reactive component (usually a capacitor) in parallel with an electrical load (or system) in order to establish a new system power factor closer to unity. Power factor correction does not affect the real power dissipated by the load (or the system).

Power Triangle: The vectorial representation of the power parameters, including the real power, reactive power, apparent power, and the phase shift of the load impedance. For a system of electrical loads, the power triangle can be extended to include the power parameters of each of the loads and the overall system's power parameters.

Reactive (or Quadrature) Power: The imaginary power associated with the reactive component(s) of a load (or a system). It is calculated in a manner similar to the calculation of the real power, as if the inductive or the capacitive reactance (or susceptance) were a resistance (or a conductance). It provides a measure of the energy storage property of the reactive component(s).

Real Power: The actual power dissipated in an electric load. Only resistive components of a load (or a system) do dissipate real power.

ANSWERS TO DRILL PROBLEMS

13.1. PF = 0.5547 lagging, P = 1.538 W

13.3. PF = 0.5556 lagging, Z = $8\underline{/56.25°}$ Ω, R = 4.4446 Ω, L = 17.64 mH

13.5. S = $12.5\underline{/-60°}$ VA, PF = cos(-60°) = 0.5 leading, P = 6.25 W, Q = 10.825 VAR (capacitive), S = 12.5 VA

13.7. S = 3162.28 VA, PF = 0.9487 leading, I = 12.649 A, Z = $19.764\underline{/-18.43°}$ Ω

13.9. I_1 = $0.05\underline{/-36.9°}$ A, I_2 = $0.14142\underline{/45°}$ A, I_T = $0.1565\underline{/26.56°}$ A, S_1 = $0.5\underline{/36.9°}$ VA, S_2 = $1.4142\underline{/-45°}$ VA, S_T = $1.565\underline{/-26.56°}$ VA, PF (overall system) = 0.894 leading

13.11. Connect a parallel inductor, L = 40.42 mH, I = 0.14286 A

14 | RESONANCE IN AC CIRCUITS

OBJECTIVES

- Understanding the concepts of the resonance phenomenon in *RLC* ac circuits and the conditions required to produce resonance.

- Ability to calculate the resonance frequency, the bandwidth and the cutoff frequencies, the quality factor and the value of the circuit's impedance at resonance in *RLC* series and parallel ac circuits.

- Understanding the major differences between circuit conditions in *RLC* series resonance versus *RLC* parallel resonance, and hence the various applications of both types of resonant circuits.

- In-depth understanding of the practical *RLC* tank circuit and the special resonance conditions and parameters associated with it.

Any ac *RLC* circuit, as in Fig. 14.1, experiences what is known as the *resonance phenomenon*. The circuit could be series, parallel, or general series–parallel. The resonance phenomenon is observed when the element values and the source frequency satisfy a condition known as the *resonance condition*.

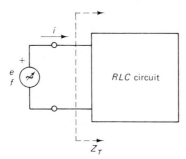

Figure 14.1

There are two general definitions of the resonance condition:

1. It is the condition at which the source voltage waveform $e(t)$ and the source current waveform $i(t)$ are in phase. This condition implies that Z_T (or Y_T), the total impedance (or admittance) of the *RLC* circuit, is a purely real number (i.e., the overall circuit appears as if it were equivalent to a single resistance). This condition is usually stated in the form

$$\text{Im}(Z_T) = 0 \quad \text{or} \quad \text{Im}(Y_T) = 0 \tag{14.1}$$

    Im $(\cdot)$ means the imaginary part of the complex number in parentheses.

2. It is the condition at which $|Z_T|$ (or $|Y_T|$) is a maximum or a minimum, excluding zero and infinite frequencies. The condition implies that

    a. If the circuit is driven by a voltage source, the current magnitude I will be observed to attain a maximum (or a minimum) at this condition.

    b. If the circuit is driven by a current source, the voltage magnitude V across the current source will be observed to attain a maximum (or a minimum) at this condition.

    This condition is usually expressed mathematically as

$$\frac{\partial}{\partial x}(|Z_T|) = 0 \quad \text{or} \quad \frac{\partial}{\partial x}(|Y_T|) = 0 \tag{14.2}$$

    where x is one of the circuit variables, either f, R, L, or C. This means that it is the condition at which the magnitude of the total circuit impedance (or admittance), when plotted versus the circuit variable, will have zero slope.

For the series *RLC* circuit and the three branch parallel *RLC*, where each branch contains only one ideal element, the two definitions for resonance will result in the same condition. However, for the general series–parallel *RLC* circuit, the two definitions could lead to two different resonance conditions.

In practice, resonant circuits are used to build electric filters. These are two-port networks, as shown in Fig. 14.2. The input signal could be either a voltage or a current source, with signal components having different frequencies, for example,

$$v_{in} = v_1(\omega_1) + v_2(\omega_2) + v_3(\omega_3) + \cdots$$

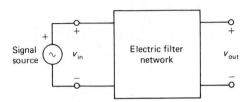

Figure 14.2

The function of the electric filter is to select one (or a range) of those signal components, depending on the element values and the circuit topology. These selected signal components will appear at the output port, while all the unselected signal components will be prevented from appearing at the output port (i.e., severely attenuated), because the network exhibits very high opposition (impedance) to those unselected signal frequency components.

14.1 RESONANCE IN A SERIES *RLC* CIRCUIT

A typical *RLC* series circuit is shown in Fig. 14.3 in both the time-domain and the equivalent frequency-domain model. The total impedance of this circuit is

$$Z_T = R + j(X_L - X_C) = R + jX$$

$$= R + j\left(\omega L - \frac{1}{\omega C}\right) = |Z_T|\underline{/\theta^\circ} \quad \Omega \tag{14.3}$$

X_L and X_C, and also $X = X_L - X_C$, are functions of frequency. This implies that the magnitude of the total impedance, $|Z_T|$, and its phase angle, θ, are also

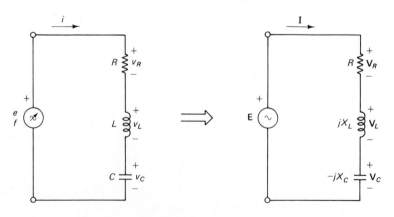

Figure 14.3

functions of the frequency. The graph in Fig. 14.4 shows the variation of

$$X(\omega) = X_L - X_C = \omega L - \frac{1}{\omega C} \tag{14.4}$$

as a function of frequency.

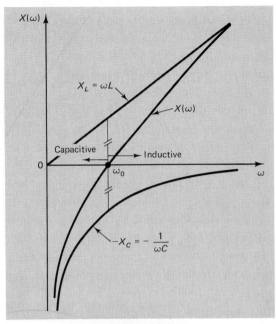

Figure 14.4

The first definition for resonance [Eq. (14.1)] results in the following condition:

$$\text{Im}(Z_T) = X(\omega) = \omega L - \frac{1}{\omega C} = 0 \tag{14.5}$$

The solution to this equation [i.e., the value of $\omega(\omega_0)$ that satisfies it] produces what is referred to as the *resonance frequency*. Here

$$\omega_0 L = \frac{1}{\omega_0 C} \tag{14.6}$$

$$\omega_0 = \frac{1}{\sqrt{LC}} \quad \text{rad/s} \tag{14.6a}$$

That is,

$$f_0 = \frac{\omega_0}{2\pi} = \frac{1}{2\pi\sqrt{LC}} \quad \text{Hz} \tag{14.6b}$$

Equation (14.6) is the resonance condition. Its graphical interpretation is shown clearly in Fig. 14.4. When condition (14.6) is satisfied (i.e., when the source is operating at this resonance frequency, f_0), then from Eq. (14.3),

$$Z_T\big|_{\omega_0} = R \tag{14.7}$$

indicating that Z_T will be purely real, equal to the circuit's resistive component only. Figure 14.5 shows the impedance triangle of this series circuit, corresponding to three different values of frequency. Also from Eq. (14.3),

$$\tan \theta = \frac{X}{R} = \frac{\omega L - 1/\omega C}{R} \tag{14.8}$$

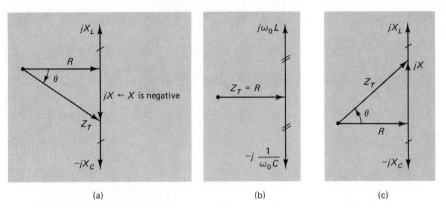

Figure 14.5 (a) $\omega < \omega_0$, below resonance; (b) $\omega = \omega_0$, at resonance; (c) $\omega > \omega_0$, above resonance.

The following are clear from Figs. 14.4 and 14.5:

1. When $\omega < \omega_0$, $X_C > X_L$; then X is a negative quantity. The circuit is predominantly capacitive. θ is a negative angle. Also, as ω decreases toward zero, X becomes an increasingly larger negative number, and θ tends to $-90°$.
2. When $\omega = \omega_0$, $X_C = X_L$; then X is zero. The circuit is resistive in nature. θ is zero at this resonance condition.
3. When $\omega > \omega_0$, $X_L > X_C$; then X is a positive quantity. The circuit is predominantly inductive. θ is a positive angle. As ω increases toward ∞, X becomes an increasingly larger positive number, and θ tends to $+90°$.

The graph of the phase variation θ with ω is shown in Fig. 14.6. The magnitude of $Z_T(\omega)$ can also be obtained from Eq. (14.3):

$$|Z_T(\omega)| = \sqrt{R^2 + X^2} = \sqrt{R^2 + \left(\omega L - \frac{1}{\omega C}\right)^2} \tag{14.9}$$

$|Z_T(\omega)|$ is plotted as a function of ω in Fig. 14.7. When ω is zero (i.e., dc), X_C is ∞ (i.e., open circuit). This means that $|Z_T|$ will be ∞ at this frequency. Similarly, when ω tends to ∞, X_L is ∞, and $|Z_T|$ also tends to ∞ at this frequency. Only at $\omega = \omega_0$, the resonance frequency, X is zero and $|Z_T|$ reaches a minimum:

$$|Z_T|_{\min} = R$$

Using the second definition of resonance [Eq. (14.2)], leading to the condition

$$\frac{\partial}{\partial \omega}|Z_T(\omega)| = 0 \tag{14.10}$$

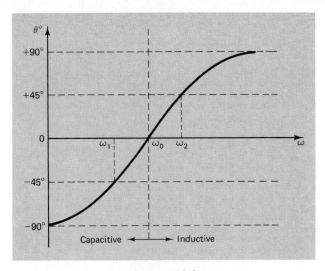

Figure 14.6

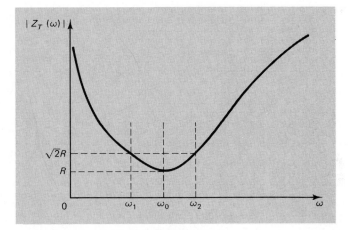

Figure 14.7

the same result as outlined above would have been obtained. As a result of manipulation of Eq. (14.10), we obtain

$$\omega L - \frac{1}{\omega C} = 0$$

whose solution is

$$\omega_0 = \frac{1}{\sqrt{LC}} \quad \text{rad/s}$$

as obtained in Eq. (14.6a). Here both definitions of resonance result in the same condition.

Two frequencies, ω_1 and ω_2, are indicated specifically in Figs. 14.6 and 14.7. Special importance is attached to these frequencies, as will be explained in the following discussions. Referring to Eq. (14.8) and Fig. 14.6, the angle θ is

$-45°$ at the specific frequency ω_1. Here

$$\tan(-45°) = -1 = \frac{1}{R}\left(\omega_1 L - \frac{1}{\omega_1 C}\right)$$

Therefore,

$$\omega_1 L - \frac{1}{\omega_1 C} = -R \qquad (14.11)$$

From Eq. (14.9),

$$|Z_T|_{\omega_1} = \sqrt{R^2 + \left(\omega_1 L - \frac{1}{\omega_1 C}\right)^2} = \sqrt{R^2 + R^2} = \sqrt{2}\,R \quad \Omega$$

as indicated in Fig. 14.7. Equation (14.11) can be solved to obtain the value of ω_1. First rewrite this equation in the form

$$\omega_1^2 + \frac{R}{L}\omega_1 - \frac{1}{LC} = 0$$

Then

$$\omega_1 = -\frac{R}{2L} + \sqrt{\frac{R^2}{4L^2} + \frac{1}{LC}}$$

The other root is rejected because it corresponds to a negative frequency. ω_1 can then be expressed in the form

$$\boxed{\omega_1 = \frac{R}{2L}\left(\sqrt{1 + \frac{4L}{CR^2}} - 1\right) \quad \text{rad/s}} \qquad (14.12)$$

Also with reference to Eq. (14.8) and Fig. 14.6, the angle θ is $+45°$ at the specific frequency ω_2. Here

$$\tan(45°) = 1 = \frac{1}{R}\left(\omega_2 L - \frac{1}{\omega_2 C}\right)$$

Therefore,

$$\omega_2 L - \frac{1}{\omega_2 C} = R \qquad (14.13)$$

Then from Eq. (14.9),

$$|Z_T|_{\omega_2} = \sqrt{R^2 + \left(\omega_2 L - \frac{1}{\omega_2 C}\right)^2} = \sqrt{R^2 + R^2} = \sqrt{2}\,R$$

as indicated in Fig. 14.7. Solving Eq. (14.13) to obtain ω_2 gives us

$$\omega_2^2 - \frac{R}{L}\omega_2 - \frac{1}{LC} = 0$$

Therefore,

$$\omega_2 = \frac{R}{2L} + \sqrt{\frac{R^2}{4L^2} + \frac{1}{LC}}$$

or

$$\omega_2 = \frac{R}{2L}\left(\sqrt{1 + \frac{4L}{CR^2}} + 1\right) \quad \text{rad/s} \tag{14.14}$$

Also here, the other root is rejected because it corresponds to a negative frequency.

The *RLC* series circuit is redrawn in Fig. 14.8 as a two-port network. The input is a variable-frequency voltage source, while the output port is across the

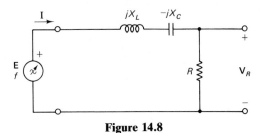

Figure 14.8

resistor *R*. The output voltage is $\mathbf{V}_R$. Since

$$I = \frac{E}{|Z_T|}$$

and,

$$V_R = IR$$

the magnitude of the current flowing in the circuit will be a function of the frequency. Its behavior is the inverse of the behavior of $|Z_T|$ with frequency (Fig. 14.7). When $|Z_T|$ is ∞, at dc and at infinite frequency, the current will be zero. When $|Z_T|$ is minimum $(= R)$ at ω_0, the current will be maximum $(= E/R)$ at this resonance frequency. At both ω_1 and ω_2, $|Z_T| = \sqrt{2}\,R$; the current will be $0.707(E/R)$. The output voltage V_R varies in exactly the same manner as *I* except for the scale factor *R*. Figure 14.9 shows such variations of *I* and V_R versus the variable frequency. Such plots are referred to as *frequency response* curves.

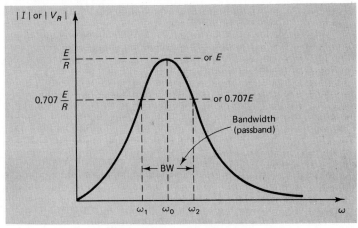

Figure 14.9   Selectivity curve

The shape of the graph in Fig. 14.9 is also called the *selectivity curve*. It shows that the circuit provides a discrimination function. Sources (signals) whose frequency is ω_0 or close to it are allowed to drive the largest current in the circuit; hence the output voltage will also be the highest. Other signals with frequencies much larger or much smaller than ω_0 will be allowed to drive only a very small current because the circuit represents quite a large impedance, $|Z_T|$, at these frequencies. The output voltage V_R will then be very small.

The frequencies ω_1 and ω_2 are used to define the frequency range in which the output voltage is higher than a certain limit. This limit was chosen to relate to the power dissipated in the *RLC* circuit. In general,

$$P(\omega) = I^2 R \quad \text{(as only the resistive component dissipates power)}$$

$$P(\omega_0) = \frac{E^2}{R} \quad \text{(the highest power dissipated)}$$

$$P(\omega_1) = P(\omega_2) = \left(\frac{E}{\sqrt{2}\,R}\right)^2 R = 0.5\frac{E^2}{R}$$

ω_1 and ω_2 are therefore called the *half-power points*. Since the factor 0.5 reduction in power corresponds to -3 dB, these two frequencies are also referred to as the 3-*dB points*. In terms of voltage, for $\omega_1 \le \omega \le \omega_2$,

$$0.707E < V_R < E$$

The factor 0.707 in voltage or current ratios also corresponds to -3 dB. The range of frequencies between ω_1 and ω_2 is called the *passband*. The circuit of Fig. 14.8 is, in fact, an example of what is called a *bandpass filter,* for obvious reasons. The width of this frequency range is called the *bandwidth* (BW):

$$\text{BW} = \omega_2 - \omega_1 \quad \text{rad/s} \tag{14.15}$$

or,

$$\text{BW} = f_2 - f_1 \quad \text{Hz}$$

ω_2 is called the *upper cutoff frequency* and ω_1 is called the *lower cutoff frequency*.

Substituting Eqs. (14.12) and (14.14) into Eq. (14.15), one can also define bandwidth as

$$\boxed{\text{BW} = \frac{R}{L} \quad \text{rad/s}} \tag{14.16}$$

In practice, a measure is used to quantify how selective a given bandpass *RLC* circuit is. This is called the *quality factor, Q*. It is defined as

$$\boxed{Q = \frac{\omega_0}{\text{BW}} = \frac{\omega_0}{\omega_2 - \omega_1} = \frac{f_0}{f_2 - f_1}} \tag{14.17}$$

The smaller the bandwidth is with respect to the center or resonance frequency, the higher the Q (i.e., the more selective the circuit is). Q is a dimensionless quantity. It could assume any numerical value (e.g., $0.1, 0.2, \ldots, 1, 2, \ldots, 10, \ldots, 100, \ldots, 1000, \ldots$). One usually says that the circuit has a high Q when $Q \ge 10$.

There are many useful forms in which Q can be expressed. Using Eqs.

(14.16), (14.17), and (14.6), we have

$$Q = \frac{\omega_0}{\text{BW}} = \frac{\omega_0 L}{R} = \frac{1}{\omega_0 C R} \qquad (14.18)$$

Since

$$\omega_0 = \frac{1}{\sqrt{LC}}$$

then

$$Q = \frac{1}{R}\sqrt{\frac{L}{C}} \quad \text{or} \quad Q^2 = \frac{L}{R^2 C} \qquad (14.19)$$

Also,

$$Q = \frac{(E/R)^2}{(E/R)^2} \frac{\omega_0 L}{R} = \frac{I|_{\omega_0}^2 \; X_L|_{\omega_0}}{I|_{\omega_0}^2 \; R} = \frac{\text{inductive VAR (at } \omega_0)}{\text{average power (at } \omega_0)} \qquad (14.20)$$

The expressions for the cutoff frequencies [Eqs. (14.12) and (14.14)] can also be written in terms of ω_0, BW, and Q as shown below.

$$\omega_1 = \frac{R}{2L}\left(\sqrt{1 + \frac{4L}{R^2 C}} - 1\right) = \frac{\text{BW}}{2}(\sqrt{1 + 4Q^2} - 1)$$

$$= \frac{\text{BW}}{2} 2Q \sqrt{1 + \frac{1}{4Q^2}} - \frac{\text{BW}}{2}$$

or

$$\omega_1 = \omega_0 \sqrt{1 + \frac{1}{4Q^2}} - \frac{\text{BW}}{2} \qquad (14.21)$$

Similarly,

$$\omega_2 = \omega_0 \sqrt{1 + \frac{1}{4Q^2}} + \frac{\text{BW}}{2} \qquad (14.22)$$

For circuits that have a high Q (i.e., $Q \geq 10$),

$$\omega_1 \cong \omega_0 - \frac{\text{BW}}{2} = \omega_0 - \frac{\omega_0}{2Q} \qquad (14.21\text{a})$$

and

$$\omega_2 \cong \omega_0 + \frac{\text{BW}}{2} = \omega_0 + \frac{\omega_0}{2Q} \qquad (14.22\text{a})$$

Thus ω_1 and ω_2 are in general not equally distant from ω_0 (i.e., they are not symmetrical about ω_0). However, for high-Q circuits, ω_0 is midway between ω_1 and ω_2 (to a very good approximation).

Through an examination of Eqs. (14.6a), (14.16), and (14.19), the following conclusions can be made:

    1. Keeping the center (resonance) frequency fixed, one can vary the

bandwidth (and Q) by varying only the value of the resistive component. The higher R is, the larger BW will be and the smaller Q will become.

2. Q can be kept constant while the resonance frequency may be varied if the values of L and C are increased or decreased by the same factor.

3. The bandwidth can be kept fixed while the center frequency may be varied, if only the value of C is varied. This is usually how a radio or television receiver is tuned (i.e., adjusting the center frequency to a required value).

Another very important observation has to be examined. If the voltage source in Fig. 14.3 is operating at a frequency equal to the circuit's resonance frequency ω_0,

$$e(t) = \sqrt{2}\,E \sin(\omega_0 t + \phi°) \Leftrightarrow \mathbf{E} = E\underline{/\phi°} \quad \text{V}$$

then

$$Z_T = R + j\left(\omega_0 L - \frac{1}{\omega_0 C}\right) = R\underline{/0°} \quad \Omega$$

$$\mathbf{I} = \frac{\mathbf{E}}{Z_T} = \frac{E}{R}\underline{/\phi°} \quad \text{A}$$

and

$$\mathbf{V}_R = \mathbf{I}Z_R = E\underline{/\phi°} = \mathbf{E} \quad \text{V}$$

$$\mathbf{V}_L = \mathbf{I}Z_L = \frac{E\omega_0 L}{R}\underline{/\phi° + 90°}$$

$$= QE\underline{/\phi° + 90°} \quad \text{V}$$

$$\mathbf{V}_C = \mathbf{I}Z_C = \frac{E}{\omega_0 CR}\underline{/\phi° - 90°}$$

$$= QE\underline{/\phi° - 90°} \quad \text{V}$$

These phasors are shown in the phasor diagram of Fig. 14.10. It is thus clear that

$$\boxed{\frac{V_L|_{\omega_0}}{E} = \frac{V_C|_{\omega_0}}{E} = Q} \tag{14.23}$$

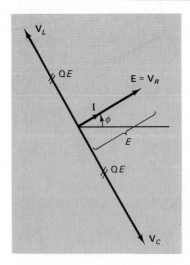

Figure 14.10

Thus at resonance, the voltage across both L and C are equal and each is Q times the source voltage. If Q is, say, 100, it is obvious that V_L and V_C could be quite large when the circuit is operated at its resonance frequency. However, $\mathbf{V}_L$ and $\mathbf{V}_C$ are always $180°$ out of phase, and hence their sum $(\mathbf{V}_{LC} = \mathbf{V}_X)$ is zero. The voltage across the resistive component, $\mathbf{V}_R$, is exactly equal to the source voltage $\mathbf{E}$. Clearly, KVL: $\mathbf{E} = \mathbf{V}_R + \mathbf{V}_L + \mathbf{V}_C$ is satisfied.

■ EXAMPLE 14.1

The signal voltage in the circuit shown in Fig. 14.11 is

$$e(t) = 0.01 \sin (2\pi \times 455 \times 10^3 t) \text{ V}$$

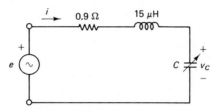

Figure 14.11

What should be the value of C in order that the circuit would resonate at this signal frequency? At this condition, find
(a) The values of I and V_C.
(b) The values of Q and BW of this circuit.

Solution   The circuit is required to resonate at

$$\omega_0 = 2\pi \times 455 \times 10^3 \text{ rad/s} = \frac{1}{\sqrt{LC}}$$

Therefore,

$$C = \frac{1}{\omega_0^2 L} = \frac{1}{(2\pi \times 455 \times 10^3)^2 \times 15 \times 10^{-6}} = 8.16 \text{ nF}$$

(a) Since at resonance

$$Z_T = R = 0.9\underline{/0°} \ \Omega$$

and

$$E = 0.01 \times 0.707 = 7.07 \text{ mV}$$

$$I = \frac{E}{|Z_T|} = \frac{7.07 \text{ m}}{0.9} = 7.86 \text{ mA}$$

$$V_C = IX_C|_{\omega_0} = IX_L|_{\omega_0}$$
$$= 7.86 \times 10^{-3} \times 2\pi \times 455 \times 10^3 \times 15 \times 10^{-6} = 0.337 \text{ V}$$

(b) Q can be obtained from any of its forms [e.g., Eq. (14.18) or (14.19)]:

$$Q = \frac{\omega_0 L}{R} = \frac{42.88}{0.9} = 47.65$$

As a check:

$$\frac{V_C}{E} = \frac{0.337}{0.00707} = 47.66 = Q \quad \text{(as expected)}$$

From Eq. (14.17),

$$BW = \frac{\omega_0}{Q} = \frac{2\pi \times 455 \times 10^3}{47.65} = 60.0 \, \text{krad/s} \quad (= 9.55 \, \text{kHz})$$

EXAMPLE 14.2

A series RLC circuit has a resonance frequency of 10 krad/s and a bandwidth of 400 rad/s.
(a) Find the cutoff (3-dB) frequencies.
(b) If the inductance of this circuit is 1 mH, find C and R.

Solution   Since ω_0 and BW are given, Q of this circuit can be easily found:

$$Q = \frac{\omega_0}{BW} = \frac{10 \times 10^3}{400} = 25$$

This is obviously a high-Q circuit ($Q > 10$).

(a) To obtain the cutoff frequencies, one has to use the forms for ω_1 and ω_2 in terms of ω_0 and Q, since these are the parameters given here. From Eqs. (14.21) and (14.22),

$$\omega_1 = \omega_0 \sqrt{1 + \frac{1}{4Q^2}} - \frac{BW}{2} = 10,002 - 200 = 9802 \, \text{rad/s}$$

(i.e., $f_1 = 1.56 \, \text{kHz}$) and

$$\omega_2 = \omega_0 \sqrt{1 + \frac{1}{4Q^2}} + \frac{BW}{2} = 10,002 + 200 = 10,202 \, \text{rad/s}$$

(i.e., $f_2 = 1.62 \, \text{kHz}$). These exact forms show that the cutoff frequencies are symmetrical around 10,002 rad/s and not around ω_0 (= 10,000 rad/s). However, the approximate forms [Eqs. (14.21a) and (14.22a)] are quite sufficient in this case, as $Q > 10$. Here

$$\omega_1 = \omega_0 - \frac{BW}{2} = 9.8 \, \text{krad/s}$$

$$\omega_2 = \omega_0 + \frac{BW}{2} = 10.2 \, \text{krad/s}$$

(b) As $\omega_0^2 = 1/LC$, then

$$C = \frac{1}{\omega_0^2 L} = \frac{1}{(10 \times 10^3)^2 \times 1 \times 10^{-3}} = 10 \, \mu\text{F}$$

Also, from $Q = \omega_0 L / R$, then

$$R = \frac{\omega_0 L}{Q} = \frac{10 \times 10^3 \times 1 \times 10^{-3}}{25} = 0.4 \, \Omega$$

EXAMPLE 14.3

A certain RLC series circuit has a maximum current of 2.5 A when the frequency of a 10 V source is adjusted to obtain this condition. The capacitor in this circuit is a 0.1 μF and the circuit's inductive reactance at this frequency is 50 Ω. Find

(a) R, Q, and L.
(b) ω_0 (the resonance frequency) and the circuit's bandwidth.

Solution (a) The frequency of the source is adjusted such that the circuit current reaches its maximum value. Clearly, this is the resonance frequency (still unknown), but at this frequency

$$Z_T = R\underline{/0^\circ} \ \ \Omega$$

As

$$I = \frac{E}{|Z_T|} = \frac{E}{R}$$

then

$$R = \frac{E}{I} = \frac{10}{2.5} = 4 \ \Omega$$

Now that R and $X_L|_{\omega_0}$ are known,

$$Q = \frac{\omega_0 L}{R} = \frac{50}{4} = 12.5$$

To find L, one can use the definition of Q in terms of the element values (as R and C are known). From Eq. (14.19),

$$Q = \frac{1}{R}\sqrt{\frac{L}{C}}$$

Therefore,

$$L = Q^2 R^2 C = (12.5 \times 4)^2 \times 0.1 \times 10^{-6} = 0.25 \text{ mH}$$

(b) Since the inductive reactance at resonance is given:

$$X_L|_{\omega_0} = \omega_0 L$$

then

$$\omega_0 = \frac{50}{0.25 \times 10^{-3}} = 200 \text{ krad/s} \qquad \text{that is, } f_0 = 31.85 \text{ kHz}$$

(Also check: $\omega_0 = 1/\sqrt{LC} = 200$ krad/s.)

$$\text{BW} = \frac{\omega_0}{Q} = \frac{200 \times 10^3}{12.5} = 16 \text{ krad/s or } 2.55 \text{ kHz.}$$

EXAMPLE 14.4

The source in the RLC circuit shown in Fig. 14.12(a) has a frequency of 1 kHz, being the resonance frequency of this circuit. The oscillogram obtained with the scope connected as indicated is shown in Fig. 14.12(b). Here

$$Y_A: \ \text{V/div} = 20 \qquad Y_B: \ \text{V/div} = \text{unknown}$$

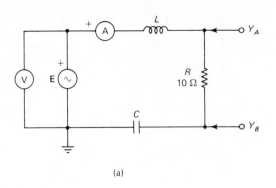

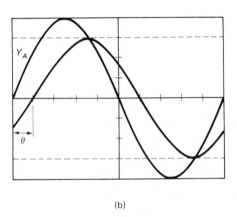

(a)

(b)

Figure 14.12

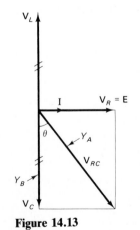

Figure 14.13

(a) Label the two waveforms.

(b) Find Y_B (V/div) and the readings of the voltmeter and ammeter.

(c) Find the values of L and C.

Solution   (a) It is usually easier in series circuits to take **I** as the reference phasor if no phasor is defined. The phasor diagram is then sketched as shown in Fig. 14.13 for the RLC series circuit at resonance. Y_B displays v_C, and Y_A displays v_{RC} according to the circuit connections. Based on the phasor diagram $Y_A(v_{RC})$ is leading $Y_B(v_C)$ by an angle $\theta°$. One can then easily label the waveform display in the oscillogram, the leading waveform is Y_A.

(b) From the oscillogram,

$$V_{RC} = 4 \times 20 \times 0.707 = 56.57 \text{ V}$$

$$\theta° = \frac{180°}{5 \text{ div}} \times 1 \text{ div} = 36° \text{ (since 1/2 cycle occupies five divisions)}$$

Using the geometrical relations from the phasor diagram gives us

$$V_C = V_{RC} \cos \theta = 56.57 \cos 36° = 45.76 \text{ V}$$

and

$$V_R = E = V_{RC} \sin \theta = 56.57 \sin 36° = 33.25 \text{ V}$$

Thus

$$\text{V/div } (Y_B) = \frac{45.76 \times 1.414}{3} = 21.57$$

Voltmeter reading: $E = 33.25$ V

Ammeter reading: $I = \dfrac{V_R}{R} = \dfrac{33.25}{10} = 3.325$ A

(c) At resonance,

$$X_L = X_C = \frac{V_C}{I} = \frac{45.76}{3.325} = 13.76 \ \Omega$$

Therefore,

$$L = \frac{X_L}{\omega} = \frac{13.76}{2\pi \times 10^3} = 2.19 \text{ mH}$$

and

$$C = \frac{1}{X_C\omega} = \frac{1}{13.76 \times 2\pi \times 10^3} = 11.57 \ \mu\text{F}$$

14.2 RESONANCE IN A THREE-BRANCH RLC PARALLEL CIRCUIT

Each of the three branches of the parallel circuit shown in Fig. 14.14 consists of a single ideal element. Even though this circuit is not practical, in particular because inductors usually include a finite resistance in their model, this ideal circuit is easier to analyze than the more practical form. Also, practical parallel circuits can always be converted to this ideal equivalent-circuit form, as shown in Section 14.3. This circuit is usually called the *ideal tank circuit* because of its ability to store a constant finite amount of energy at its resonance frequency.

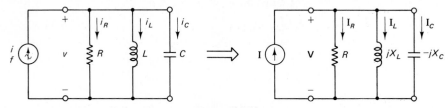

Figure 14.14

Parallel circuits are easier to analyze through their admittances. This circuit has a total admittance

$$Y_T = G + jB_C - jB_L = \frac{1}{R} + j\left(\omega C - \frac{1}{\omega L}\right) = |Y_T| \underline{/\theta^\circ} \text{ S} \qquad (14.24)$$

with

$$Y_T = \frac{\mathbf{I}}{\mathbf{V}} = \frac{1}{Z_T} \qquad \text{or} \qquad \mathbf{V} = \mathbf{I}Z_T = \mathbf{I}\frac{1}{Y_T} \qquad (14.25)$$

According to the first definition of resonance, [Eq. (14.1)], the voltage across the tank circuit $\mathbf{V}$, and the current flowing into it $\mathbf{I}$, are in phase when

$$\text{Im}\,(Y_T) = 0$$

that is,

$$\omega C - \frac{1}{\omega L} = 0$$

The value of $\omega(\omega_0)$ that satisfies this equation is the circuit's resonance frequency. Here

$$\omega_0 C = \frac{1}{\omega_0 L} \tag{14.26}$$

$$\omega_0 = \frac{1}{\sqrt{LC}} \quad \text{rad/s} \tag{14.26a}$$

or

$$f_0 = \frac{\omega_0}{2\pi} = \frac{1}{2\pi\sqrt{LC}} \quad \text{Hz} \tag{14.26b}$$

This is the same as was obtained for the series RLC case in Eq. (14.6).

The admittance triangles for the circuit are shown in Fig. 14.15, corresponding to three different values of frequency. By examining these diagrams and Eq. (14.24), where

$$|Y_T| = \sqrt{\frac{1}{R^2} + \left(\omega C - \frac{1}{\omega L}\right)^2} \tag{14.27}$$

$$\tan \theta = R\left(\omega C - \frac{1}{\omega L}\right) \tag{14.28}$$

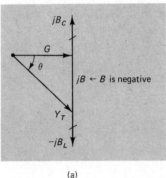 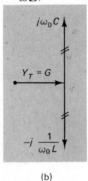 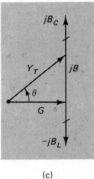

(a)   (b)   (c)

Figure 14.15   (a) $\omega < \omega_0$, below resonance; (b) $\omega = \omega_0$, at resonance; (c) $\omega > \omega_0$, above resonance.

it is clear that at resonance the total admittance of the circuit is

$$Y_T = \frac{1}{R}\underline{/0^\circ} \quad \text{S} \quad \text{(its minimum magnitude)}$$

that is,

$$|Y_T|_{\min} = |Y_T|_{\omega_0} = \frac{1}{R} \tag{14.29}$$

or

$$|Z_T|_{\max} = |Z_T|_{\omega_0} = R \tag{14.29a}$$

Based on the second definition of resonance, Eq. (14.2) can be expressed in this case as

$$\frac{\partial}{\partial\omega}|Y_T(\omega)| = 0$$

The solution of this equation also shows that the condition at which $|Y_T|$ is minimum corresponds to

$$\omega_0 C = \frac{1}{\omega_0 L}$$

resulting in

$$|Y_T|_{\min} = \frac{1}{R}$$

As in the case of the series RLC circuit both definitions of resonance result in the same resonance condition [Eq. (14.26)].

Figure 14.16 shows the behavior of the magnitude of the total circuit

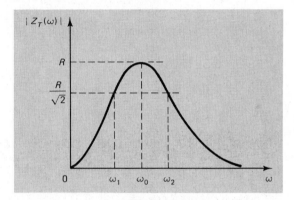

Figure 14.16

impedance $|Z_T|$ as the source's frequency is varied. The variation of the phase angle of Z_T, θ_Z, where

$$\theta_Z = -\theta_Y$$

is plotted as a function of the circuit's operating frequency in Fig. 14.17.

At $\omega = 0$ (dc), the inductor is equivalent to a short circuit, hence $|Z_T|$ is zero. In the frequency ·range $\omega < \omega_0$, the inductive branch has the smaller reactance and thus the larger current. The circuit current **I** will then be mostly inductive in nature. Here θ_Z is positive, approaching $+90°$ as ω tends to zero. At $\omega = \infty$ (very high frequency), the capacitor is equivalent to a short circuit, hence $|Z_T|$ is also zero. In the frequency range $\omega > \omega_0$, the capacitive branch has the smaller reactance and thus the larger current. The circuit's current **I** will then be mostly capacitive in nature. Here θ_Z is negative, approaching $-90°$ as ω tends to infinity.

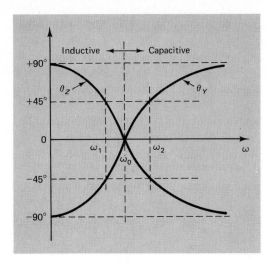

Figure 14.17

At resonance, $\omega = \omega_0$, the voltage and current waveforms are in phase (i.e., $\theta_Z = -\theta_Y = 0$), while the circuit's impedance will reach its highest magnitude, being equal to the value of the resistive branch. In theory, if the resistive branch is removed (i.e., $R = \infty$), the circuit's impedance at resonance could reach infinite magnitude (open circuit). This is not exactly correct in practice since the resistance of the inductive branch will modify this result.

It is important to note that even though the resonance frequency, ω_0, is determined in exactly the same manner in the series and the parallel RLC circuits, their behavior with frequency variations is quite different. In the series circuit the impedance of the circuit is minimum at resonance, whereas in the parallel circuit it is a maximum at ω_0. As the frequency is increased from below to above resonance, the nature of the series circuit changes from capacitive to inductive, whereas for the parallel circuit it changes from inductive to capacitive, exactly the opposite.

The two frequencies ω_1 and ω_2 (see Figs. 14.16 and 14.17) are also called the 3-dB cutoff, or half-power frequencies. By examining these two figures and Eqs. (14.27) and (14.28), it is clear that at ω_1,

$$\tan \theta_Y = \tan(-45°) = -1 = R\left(\omega_1 C - \frac{1}{\omega_1 L}\right)$$

Therefore,

$$\omega_1 C - \frac{1}{\omega_1 L} = -\frac{1}{R} \tag{14.30}$$

$$Y_T\big|_{\omega_1} = \frac{1}{R} + j\left(\omega_1 C - \frac{1}{\omega_1 L}\right) = \frac{1}{R} - j\frac{1}{R} = \frac{\sqrt{2}}{R}\underline{/-45°} \ \ \text{S}$$

that is,

$$Z_T\big|_{\omega_1} = \frac{R}{\sqrt{2}}\underline{/45°} \ \Omega \tag{14.31}$$

At ω_2,

$$\tan \theta_Y = \tan (45°) = 1 = R\left(\omega_2 C - \frac{1}{\omega_2 L}\right)$$

Therefore,

$$\omega_2 C - \frac{1}{\omega_2 L} = \frac{1}{R} \tag{14.32}$$

Here

$$Y_T\big|_{\omega_2} = \frac{\sqrt{2}}{R} \underline{/45°} \quad S$$

that is,

$$Z_T\big|_{\omega_2} = \frac{R}{\sqrt{2}} \underline{/-45°} \quad \Omega \tag{14.33}$$

Equations (14.30) and (14.32) can be solved in exactly the same manner as in the case of the series circuit, to find the values of ω_1 and ω_2, respectively. Considering only the positive frequency roots, we obtain

$$\omega_1 = \frac{1}{2RC}\left(\sqrt{1 + \frac{4R^2 C}{L}} - 1\right) \quad \text{rad/s} \tag{14.34}$$

and

$$\omega_2 = \frac{1}{2RC}\left(\sqrt{1 + \frac{4R^2 C}{L}} + 1\right) \quad \text{rad/s} \tag{14.35}$$

Here the bandwidth

$$\text{BW} = \omega_2 - \omega_1 = \frac{1}{RC} \quad \text{rad/s} \tag{14.36}$$

When the parallel *RLC* circuit is driven by a voltage source, the source current will be a minimum at the resonance frequency since $|Z_T|$ is a maximum there. As mentioned above, if the resistive element is removed, Z_T becomes ideally ∞ (an open circuit) at ω_0. This indicates that such tank circuits can considerably oppose (or completely prevent) the flow of signal current at some specific unwanted frequency, being equal to the resonance frequency of the tank circuit. Such an application is called a *frequency trap*.

In other practical applications (e.g., tuned amplifiers) the parallel *RLC* tank circuit is driven by a current source. The voltage across the tank circuit, being

$$V = I\,|Z_T|$$

will behave in exactly the same manner as the variation of $|Z_T|$ with frequency.

The graph in Fig. 14.18 is a replica of that in Fig. 14.16, except for the scale factor *I*. This shows that parallel *RLC* circuits can be used as selective circuits, allowing current signals with frequencies within the passband (BW) to exhibit high-voltage amplitudes, whereas signals with frequencies outside this band will produce very low voltage amplitudes (attenuated) across the tank circuit.

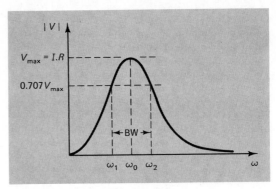

Figure 14.18

The quality factor, Q, is also used here as a measure of the circuit's selectivity. Similar to the series circuit case, Q is defined as

$$Q = \frac{\omega_0}{\text{BW}} \qquad (14.37)$$

The smaller (i.e., narrower) the bandwidth is, the higher the Q, indicating that the circuit is more selective.

Substituting for the bandwidth from Eq. (14.36) and using Eq. (14.26), Q may be expressed in many forms:

$$Q = \omega_0 CR = \frac{R}{\omega_0 L} = R\sqrt{\frac{C}{L}} \qquad (14.38)$$

Note that all these forms of Q [Eq. (14.38)] are exactly the opposite of those obtained for the series circuit case. Even so, in practice, parallel RLC selective circuits also reflect a large value for Q. A circuit is said to be a high-Q circuit when $Q \geq 10$.

The cutoff frequencies ω_1 and ω_2, which were expressed in terms of the element values in Eqs. (14.34) and (14.35), can also be expressed in terms of ω_0 and Q, using Eqs. (14.36) and (14.38):

$$\begin{aligned}
\omega_1 &= \frac{1}{2RC}\left(\sqrt{1 + \frac{4R^2 C}{L}} - 1\right) \\
&= \frac{\text{BW}}{2}\left(\sqrt{1 + 4Q^2} - 1\right) \\
&= \text{BW} \times Q\sqrt{1 + \frac{1}{4Q^2}} - \frac{\text{BW}}{2} \\
&= \omega_0\sqrt{1 + \frac{1}{4Q^2}} - \frac{\text{BW}}{2} \qquad (14.39)
\end{aligned}$$

Similarly,

$$\omega_2 = \omega_0\sqrt{1 + \frac{1}{4Q^2}} + \frac{\text{BW}}{2} \qquad (14.40)$$

These are exactly the same expressions as those obtained for the series circuit case in Eqs. (14.21) and (14.22). For high-Q circuits ($Q \geq 10$),

$$\omega_1 \cong \omega_0 - \frac{\text{BW}}{2} = \omega_0 - \frac{\omega_0}{2Q} \qquad (14.39a)$$

and

$$\omega_2 \cong \omega_0 + \frac{\text{BW}}{2} = \omega_0 + \frac{\omega_0}{2Q} \qquad (14.40a)$$

Again, note that ω_1 and ω_2 are not symmetrical around ω_0, except in high-Q circuits.

When the parallel RLC circuit is driven by a current source whose frequency is the resonance frequency of the circuit,

$$i(t) = \sqrt{2}\, I \sin{(\omega_0 t + \phi°)} \qquad \text{that is, } \mathbf{I} = I\underline{/\phi°} \quad \text{A}$$

The circuit's admittance is

$$Y_T = \frac{1}{R}\underline{/0°} \quad \text{S}$$

Therefore,

$$Z_T = \frac{1}{Y_T} = R\underline{/0°} \quad \Omega$$

Then, referring to Fig. 14.19, we see that

$$\mathbf{V} = \mathbf{I}Z_T = IR\underline{/\phi°} \quad \text{V}$$

$$\mathbf{I}_R = \frac{\mathbf{V}}{Z_R} = I\underline{/\phi°} = \mathbf{I}$$

$$\mathbf{I}_L = \frac{\mathbf{V}}{Z_L} = \frac{IR\underline{/\phi°}}{\omega_0 L\underline{/90°}} = I\frac{R}{\omega_0 L}\underline{/\phi° - 90°} = QI\underline{/\phi° - 90°} \quad \text{A}$$

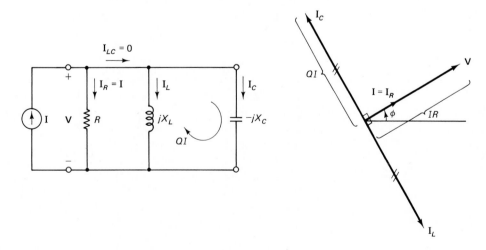

Figure 14.19

and

$$\mathbf{I}_C = \frac{\mathbf{V}}{Z_C} = \frac{IR\underline{/\phi°}}{(1/\omega_0 C)\underline{/-90°}} = I\omega_0 CR\underline{/\phi° + 90°}$$
$$= QI\underline{/\phi° + 90°} \quad \text{A}$$

Note that from KCL,

$$\mathbf{I}_{LC} = \mathbf{I}_L + \mathbf{I}_C = 0$$

as $\mathbf{I}_L$ and $\mathbf{I}_C$ have equal magnitudes and are 180° out of phase with each other. Also,

$$\mathbf{I} = \mathbf{I}_R + \mathbf{I}_L + \mathbf{I}_C = \mathbf{I}_R + \mathbf{I}_{LC} = \mathbf{I}_R$$

(i.e., KCL is satisfied by the derivations above).

It is clear from these derivations that at the resonance condition in a parallel *RLC* circuit,

$$\boxed{\frac{I_L|_{\omega_0}}{I} = \frac{I_C|_{\omega_0}}{I} = Q} \tag{14.41}$$

All the source current will flow in the resistive branch, while the *L* and *C* parallel branches exhibit an open circuit as far as the current source is concerned. However, a large current of magnitude $Q \times I$ is now circulating in *LC* loop, as shown in Fig. 14.19. This current is responsible for the energy storage in the tank circuit. Every quarter-cycle of the current source's period, energy will continuously oscillate between being stored in the magnetic field of the inductor and the electric field of the capacitor. This amount of energy is constant and no energy transfer takes place between the current source and the *LC* parallel branches.

Note on the Q factor of inductors.   A practical inductor (coil) possesses a finite resistance, R_L, due to the wire conductor from which the coil is made, as shown in Fig. 14.20. In practice, inductors used in high-frequency applications are usually specified in terms of their *Q* factor. This does not refer to any resonance phenomenon occurring in the coil. However, such a *Q* factor is defined as

$$Q = \frac{\omega L}{R_L} \tag{14.42}$$

One might wonder how a *Q* factor can be specified in this case, without any reference to frequency, especially since this *Q* is a function of frequency. To answer this question, the skin effect has to be explained. When current flows in a conducting wire, it creates a magnetic field around the conductor and in its material. This field is strongest in the center of the conductor. Due to the variation of the current (hence magnetic flux), especially in high-frequency applications, the electromagnetic effect induces a voltage in the conductor. This voltage is higher in the center than at the surface of the conductor. Such voltage opposes the current flow and forces the current to flow mainly in the conductor

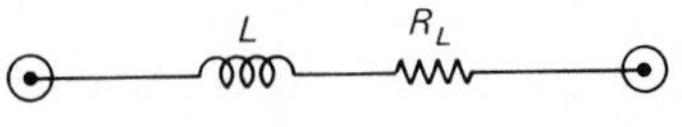

Figure 14.20   Model of a practical inductor.

area near the surface. This skin effect results in a reduction in the effective cross-sectional area of the conductor, as the operating frequency is increased. Hence the resistance of the wire, R_L $[= \rho(l/A)]$, increases with frequency.

Now referring to Eq. (14.42), since the numerator and denominator increase with frequency, Q will be almost constant $(= Q_L)$ for operating frequencies above a certain value (ω_r). This is the value (Q_L) that is specified for such coils. Figure 14.21 shows the variation of Q with frequency. For low-frequency operation, R_L is almost constant and Q rises linearly with ω. But as the frequency is increased and the skin effect becomes prominent, Q will reach its constant value, Q_L.

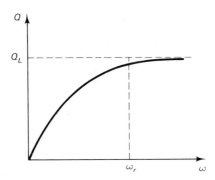

Figure 14.21

EXAMPLE 14.5

In a parallel RLC circuit, R is 250 Ω, L is 2 mH, and C is 20 μF. The circuit is driven by a 0.1 A current source operating at the circuit's resonance frequency, ω_0. Find

(a) ω_0, Q, BW, and the cutoff frequencies ω_1 and ω_2.
(b) The voltage across the tank circuit and the current in each branch.

Solution   (a) All the element values are given; therefore,

$$\omega_0 = \frac{1}{\sqrt{LC}} = 5000 \text{ rad/s} \qquad \text{that is, } f_0 = \frac{\omega_0}{2\pi} = 796.2 \text{ Hz}$$

and

$$Q = \frac{R}{\omega_0 L} = \omega_0 CR = R\sqrt{\frac{C}{L}} = 25 \qquad \text{BW} = \frac{\omega_0}{Q} = \frac{5000}{25} = 200 \text{ rad/s}$$

Since $Q > 10$, then

$$\omega_1 \cong \omega_0 - \frac{\text{BW}}{2} = 4900 \text{ rad/s} \qquad \text{and} \qquad \omega_2 \cong 5100 \text{ rad/s}$$

Using the exact expressions (14.39) and (14.40) yields

$$\omega_1 = \omega_0 \sqrt{1 + \frac{1}{4Q^2}} - \frac{\text{BW}}{2} = 4901 \text{ rad/s}$$

and

$$\omega_2 = 5101 \text{ rad/s}$$

Clearly, the approximate results are quite sufficient.

(b) Since the circuit is at resonance,

$$Z_T = R = 250\underline{/0°}\ \Omega$$

$$V = I\,|Z_T| = IR = 25\ \text{V}$$

$$I_R = \frac{V}{R} = I = 0.1\ \text{A}$$

and

$$I_L = I_C = QI = \frac{V}{\omega_0 L} = V\omega_0 C = 2.5\ \text{A}$$

EXAMPLE 14.6

If the current source in Example 14.5 is now operating at a frequency of 500 rad/s (i.e., $0.1\ \omega_0$), find the voltage across the tank circuit and the current in each branch.

Solution Using the general form of Y_T, [Eq. (14.24)], we obtain

$$Y_T = \frac{1}{R} + j\left(\omega C - \frac{1}{\omega L}\right) = 0.004 + j(0.01 - 1)$$

$$= 0.004 - j0.99 = 0.99\underline{/-89.8°}\ \text{S}$$

Since no phasor is defined as the reference, any can be used. Let

$$\mathbf{I} = 0.1\underline{/0°}\ \text{A}$$

$$\mathbf{V} = \mathbf{I}Z_T = \frac{\mathbf{I}}{Y_T} = \frac{0.1\underline{/0°}}{0.99\underline{/-89.8°}} = 0.101\underline{/89.8°}\ \text{V}$$

$$\mathbf{I}_R = \frac{\mathbf{V}}{Z_R} = \frac{0.101\underline{/89.8°}}{250\underline{/0°}} = 0.404\underline{/89.8°}\ \text{mA}$$

$$\mathbf{I}_C = \frac{\mathbf{V}}{Z_C} = \frac{0.101\underline{/89.8°}}{100\underline{/-90°}} = 1.01\underline{/179.8°}\ \text{mA}$$

and

$$\mathbf{I}_L = \frac{\mathbf{V}}{Z_L} = \frac{0.101\underline{/89.8°}}{1\underline{/90°}} = 101\underline{/-0.2°}\ \text{mA}$$

Comparing these results with those obtained in Example 14.5, it is clear that here $V\ (= 0.101\ \text{V})$ is much smaller than the voltage across the tank circuit at resonance $(= 25\ \text{V})$. The inductor current here is almost equal to the source current, since the inductive reactance is the smallest. The other branch currents are comparatively quite small. The voltage $\mathbf{V}$ leads the source current $\mathbf{I}$ by 89.8° (i.e., the circuit is inductive in nature).

EXAMPLE 14.7

When the frequency of a 10 V source is varied, it was observed that the minimum source current is 0.1 A, with the load consisting of a three-branch *RLC* parallel circuit (trap). The resonance frequency of this circuit is 2 MHz and its bandwidth is 200 kHz. Find the element values of the load components.

Solution The source current is minimum at the circuit's resonance frequency as the circuit's impedance reaches its maximum R; therefore,

$$R = \frac{V}{I} = \frac{10}{0.1} = 100\ \Omega$$

The circuit's quality factor is

$$Q = \frac{\omega_0}{\text{BW}} = \frac{2\pi \times 2 \times 10^6}{2\pi \times 200 \times 10^3} = 10$$

But

$$Q = \omega_0 CR \quad \text{(using Eq. (14.38)]}$$

Therefore,

$$C = \frac{Q}{\omega_0 R} = \frac{10}{2\pi \times 2 \times 10^6 \times 100} = 7.96\ \text{nF}$$

Now from $\omega_0^2 = 1/LC$,

$$L = \frac{1}{\omega_0^2 C} = 0.796\ \mu\text{H}$$

14.3 RESONANCE IN SERIES–PARALLEL *RLC* CIRCUITS

In this general type of circuit, the two definitions for resonance usually result in two different resonance conditions. Examples of such circuits will be examined here and the analysis technique used applies to any other series–parallel circuit.

14.3.1 Practical Tank Circuit

Since practical inductors do possess a finite resistance, a more realistic model of the *LC* tank circuit is shown in Fig. 14.22. The resistance R could be a separate element, the coil's resistance, or both.

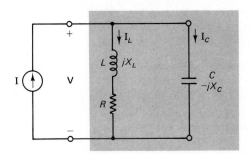

Figure 14.22

Based on the first definition of resonance and Eq. (14.1), for the current source **I** and the voltage across the tank circuit **V** to be in phase, the total admittance of the circuit $Y_T(= 1/Z_T)$ must be a purely real number at the resonance condition. Here

$$Y_T = \frac{1}{R + j\omega L} + j\omega C = \frac{R - j\omega L}{R^2 + \omega^2 L^2} + j\omega C$$

$$= \frac{R}{R^2 + \omega^2 L^2} + j\left(\omega C - \frac{1}{\omega L_{eq}}\right) \tag{14.43}$$

$$= \frac{1}{R_{eq}} + j\left(\omega C - \frac{1}{\omega L_{eq}}\right) \tag{14.43a}$$

where

$$R_{eq} = \frac{1}{R}(R^2 + \omega^2 L^2) \tag{14.44}$$

and

$$L_{eq} = \frac{1}{\omega^2 L}(R^2 + \omega^2 L^2) \tag{14.45}$$

The reasonance condition is obtained by equating the imaginary part of Y_T to zero, producing

$$\omega_0 C = \frac{\omega_0 L}{R^2 + \omega_0^2 L^2}$$

Therefore,

$$R^2 + \omega_0^2 L^2 = \frac{L}{C} \tag{14.46}$$

The resonance frequency ω_0 is the solution of Eq. (14.46):

$$\boxed{\omega_0 = \sqrt{\frac{1}{LC} - \left(\frac{R}{L}\right)^2} \quad \text{rad/s}} \tag{14.47}$$

The form in which Y_T is expressed in Eq. (14.43a) indicates that the general *RLC* circuit of Fig. 14.22 can be represented as being equivalent to a three-branch *RLC* circuit as in Fig. 14.14. The elements in this three-branch equivalent circuit are C, L_{eq}, and R_{eq}. Both R_{eq} and L_{eq} are functions of frequency, as shown in Eqs. (14.44) and (14.45), respectively. Obviously, this is not a real circuit but a mathematically equivalent model to facilitate easier determination of other important circuit parameters.

At the resonance frequency, ω_0, the total impedance of the circuit is purely real:

$$Z_T\big|_{\omega_0} = \frac{1}{Y_T\big|_{\omega_0}} = R_{eq}\big|_{\omega_0} = \frac{R^2 + \omega_0^2 L^2}{R}$$

Using the resonance condition [Eq. (14.46)] yields

$$\boxed{Z_T\big|_{\omega_0} = \frac{L}{RC} \,\Omega} \tag{14.48}$$

As explained above, the circuit is equivalent to a three-branch *RLC* parallel circuit. Thus, using the definition of Q as obtained in Section 14.2 [Eq. (14.38)], we have

$$Q = \frac{\omega_0}{BW} = \omega_0 C R_{eq} = \omega_0 C \frac{L}{RC} = \frac{\omega_0 L}{R} \qquad (14.49)$$

The circuit's quality factor is thus the same as Q of the inductive branch alone (or the coil's Q factor). The cutoff frequencies ω_1 and ω_2 can be obtained from the same general equations derived previously [Eqs. (14.39) and (14.40)] and rewritten below:

$$\omega_1 = \omega_0 \sqrt{1 + \frac{1}{4Q_2}} - \frac{BW}{2} \qquad \omega_2 = \omega_0 \sqrt{1 + \frac{1}{4Q^2}} + \frac{BW}{2}$$

Further simplification is possible for high-Q circuits ($Q \geq 10$). Here

$$(\omega_0 L)^2 \gg R^2 \qquad \left(\text{since } Q = \frac{\omega_0 L}{R} \right)$$

The resonance condition [Eq. (14.46)] can be rewritten as

$$R^2 + \omega_0^2 L^2 \approx \omega_0^2 L^2 = \frac{L}{C}$$

Therefore,

$$\omega_0 L \cong \frac{1}{\omega_0 C}$$

$$\omega_0 \cong \frac{1}{\sqrt{LC}} \quad \text{rad/s} \qquad (14.50)$$

This result is the same as obtained in the previous circuits. Using this approximation gives us

$$Z_T\big|_{\omega_0} = \frac{L}{RC} = \frac{\omega_0 L}{R} \frac{1}{\omega_0 C} \frac{R}{R}$$

$$\cong \frac{\omega_0 L}{R} \frac{\omega_0 L}{R} R \cong Q^2 R \quad \Omega \qquad (14.51)$$

This is a very important expression for the tank circuit's impedance at the resonance frequency. It shows that at resonance the tank circuit is equivalent to a resistance, being Q^2 times the resistance of the inductive branch. This can be, in fact, quite a large number. In the discussion of the ideal three-branch RLC parallel circuit, it was mentioned that if R is removed (i.e., ideal LC parallel circuit), then theoretically, the tank circuit is equivalent to an open circuit at resonance. Now, one can realize that practically the highest impedance attainable at resonance is $Q^2 R$.

The total circuit's impedance Z_T does exhibit a peak, or a maximum magnitude, as the frequency is varied. Practical applications of this tank circuit make use of such selectivity characteristics, as discussed before. The frequency at which this maximum magnitude of Z_T occurs is obtained from the second definition of resonance [Eq. (14.2)]. Here

$$\frac{\partial}{\partial \omega}|Y_T| = 0$$

where $|Y_T|$ is as defined in Eq. (14.43). This results in

$$R^2 + \omega_0^2 L^2 = \frac{L}{C}\sqrt{1 + \frac{2CR^2}{L}} \qquad (14.52)$$

In general, then, ω_0 obtained from Eq. (14.52) is not the same as ω_0 obtained from Eq. (14.46). However, for high-Q circuits,

$$Q^2 \cong \frac{L}{CR^2} \quad \text{[from Eq. (14.51)]}$$

Thus, Eq. (14.52) can be approximated as

$$R^2 + \omega_0^2 L^2 \cong \frac{L}{C}\sqrt{1 + \frac{2}{Q^2}} \cong \frac{L}{C}$$

as in Eq. (14.46). Hence, in high-Q circuits, the two definitions for resonance result in approximately the same resonance condition and resonance frequency.

A final remark on this circuit concerns the magnitude of the two branch currents at the resonance frequency. Here

$$I_C = \frac{V}{X_C} = I\frac{L}{RC}\omega_0 C = QI$$

and

$$I_L = \frac{V}{\sqrt{R^2 + \omega_0^2 L^2}} \cong I\frac{L}{RC}\frac{1}{\omega_0 L} \cong I\frac{L}{RC}\omega_0 C = QI$$

This is the same result as obtained in the ideal three-branch RLC parallel circuit.

■ EXAMPLE 14.8

A practical tank circuit has a 1 pF capacitor. The resonance frequency of the circuit is 1 MHz and its bandwidth is 50 kHz.
(a) What are the values of L and R?
(b) What is the tank circuit's impedance at resonance?

Solution (a) Since the resonance frequency and the bandwidth are given, then

$$Q = \frac{\omega_0}{\text{BW}} = \frac{f_0}{\text{BW (Hz)}} = \frac{10^6}{50 \times 10^3} = 20$$

This is a high-Q circuit. Using the approximation in Eq. (14.50) yields

$$\omega_0^2 \cong \frac{1}{LC}$$

$$L = \frac{1}{\omega_0^2 C} = \frac{1}{(2\pi \times 10^6)^2 \times 10^{-12}} = 25.33 \text{ mH}$$

As $Q = \omega_0 L/R$,

$$R = \frac{\omega_0 L}{Q} = \frac{2\pi \times 10^6 \times 25.33 \times 10^{-3}}{20} = 7.958 \text{ k}\Omega$$

(b) At the resonance frequency,

$$Z_T = Q^2 R = (20)^2 \times 7.958 \text{ k}\Omega = 3.183 \text{ M}\Omega$$

Also, using the exact form of Z_T [Eq. (14.48)], we have

$$Z_T = \frac{L}{RC} = \frac{25.33 \times 10^{-3}}{7.958 \times 10^3 \times 10^{-12}} = 3.183 \text{ M}\Omega \quad \text{(check)}$$

If one uses the exact expression for ω_0 [Eq. (14.47)] as another check on the element values:

$$\omega_0 = \sqrt{\frac{1}{LC} - \left(\frac{R}{L}\right)^2} = \sqrt{39.4789 \times 10^{12} - 0.0987 \times 10^{12}}$$

$$= 6.275 \text{ Mrad/s}$$

that is,

$$f_0 = \frac{\omega_0}{2\pi} = 0.9988 \text{ MHz}$$

as expected.

EXAMPLE 14.9

In the circuit shown in Fig. 14.23, $\mathbf{E}$ and $\mathbf{I}$ are in phase, with $E = 60$ V and $I = 0.72$ A.

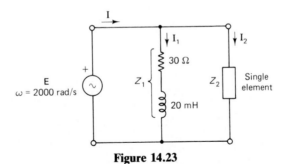

Figure 14.23

(a) Find the value of the element in the impedance Z_2.
(b) Find the circuit's quality factor and bandwidth.
(c) Draw a complete phasor diagram, taking $\mathbf{E}$ as the reference.

Solution (a) Let $\mathbf{E} = 60\underline{/0°}$ V; then

$$\mathbf{I} = 0.72\underline{/0°} \text{ A}$$

$$Z_1 = 30 + j20 \times 10^{-3} \times 2000 = 30 + j40 = 50\underline{/53.1°} \ \Omega$$

$$Y_1 = \frac{1}{Z_1} = 0.02\underline{/-53.1°} \text{ S}$$

At the given frequency ($\omega = 2000\,\text{rad/s}$), $\mathbf{E}$ and $\mathbf{I}$ are in phase. This is, therefore, the resonance frequency. Here Y_T is purely real.

$$Y_T = \frac{\mathbf{I}}{\mathbf{E}} = Y_1 + Y_2$$

Therefore,

$$\frac{0.72\underline{/0°}}{60\underline{/0°}} = 0.02\underline{/-53.1°} + Y_2$$

that is,

$$Y_2 = 0.012 - (0.012 - j0.016) = j0.016 = jB_C = j\omega C$$

Thus the element constituting the impedance Z_2 is a capacitor:

$$C = \frac{0.016}{2000} = 8\,\mu\text{F}$$

(b) $Q = \dfrac{\omega_0 L}{R} = \dfrac{2000 \times 20 \times 10^{-3}}{30} = 1.333$ and

$$\text{BW} = \frac{\omega_0}{Q} = \frac{2000}{1.333} = 1500\,\text{rad/s}$$

This is not a high-Q circuit. As a check on the results obtained, let us recalculate ω_0 from Eq. (14.47):

$$\omega_0 = \sqrt{\frac{1}{LC} - \left(\frac{R}{L}\right)^2} = \sqrt{\frac{1}{20 \times 10^{-3} \times 8 \times 10^{-6}} - \left(\frac{30}{20 \times 10^{-3}}\right)^2}$$

$$= 2000\,\text{rad/s} \quad \text{(as expected)}$$

Also,

$$Z_T\big|_{\omega_0} = \frac{L}{RC} = \frac{20 \times 10^{-3}}{30 \times 8 \times 10^{-6}} = 83.33\,\Omega$$

This agrees with the given values ($Z_T = \mathbf{E}/\mathbf{I} = 60\underline{/0°}/0.72\underline{/0°} = 83.33\underline{/0°}\,\Omega$). Note that $Q^2 R$ is only $53.33\,\Omega$, which is not the same as $Z_T\big|_{\omega_0}$. This is true here since the high-Q approximation used in developing Eq. (14.51) is not valid for this circuit.

(c) The branch currents can easily be calculated:

$$\mathbf{I}_1 = \frac{\mathbf{E}}{Z_1} = \frac{60\underline{/0°}}{50\underline{/53.1°}} = 1.2\underline{/-53.1°}\,\text{A}$$

and

$$\mathbf{I}_2 = \frac{\mathbf{E}}{Z_2} = \mathbf{E}Y_2 = 60\underline{/0°} \times 0.016\underline{/90°} = 0.96\underline{/90°}\,\text{A}$$

The phasor diagram is then as shown in Fig. 14.24, noting that KCL:

$$\mathbf{I} = \mathbf{I}_1 + \mathbf{I}_2$$

is satisfied.

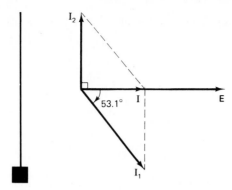

Figure 14.24

14.3.2 Inductor in Parallel with *RC* Branch

As another example of the resonance phenomena in series–parallel *RLC* circuits, the circuit shown in Fig. 14.25 will be considered here, to illustrate the technique usually used in obtaining the resonance frequency and the quality factor of such circuits. The resonance condition based on the first definition (i.e., **V** and **I** are in phase) is obtained by applying Eq. (14.1). Here

$$
\begin{aligned}
Y_T &= \frac{1}{R - j(1/\omega C)} - j\frac{1}{\omega L} \\[2mm]
&= \frac{R + j(1 + \omega C)}{R^2 + 1/\omega^2 C^2} - j\frac{1}{\omega L} \\[2mm]
&= \frac{R}{R^2 + 1/\omega^2 C^2} + j\left(\frac{1/\omega C}{R^2 + 1/\omega^2 C^2} - \frac{1}{\omega L}\right) \\[2mm]
&= \frac{1}{R_{eq}} + j\left(\omega C_{eq} - \frac{1}{\omega L}\right)
\end{aligned}
\tag{14.53}
$$

Figure 14.25

Equation (14.53) indicates that the given circuit can be transformed to the mathematically equivalent three-branch *RLC* parallel circuit, composed of R_{eq}, L, and C_{eq} in parallel, where

$$
R_{eq} = \frac{1}{R}\left(R^2 + \frac{1}{\omega^2 C^2}\right)
\tag{14.54}
$$

and

$$
C_{eq} = \frac{1}{\omega^2 C(R^2 + 1/\omega^2 C^2)} = \frac{C}{1 + \omega^2 C^2 R^2}
\tag{14.55}
$$

Both R_{eq} and C_{eq} are frequency dependent, as indicated above.

The resonance condition is obtained from

$$\text{Im}\,(Y_T) = 0 \qquad \text{when} \qquad \omega = \omega_0$$

Thus

$$R^2 + \frac{1}{\omega_0^2 C^2} = \frac{L}{C} \tag{14.56}$$

The solution of Eq. (14.56) will provide the resonance frequency, ω_0:

$$\omega_0 = \frac{1}{C\sqrt{(L/C) - R^2}} = \frac{1}{\sqrt{LC - R^2 C^2}} \quad \text{rad/s} \tag{14.57}$$

At this resonance frequency and using condition (14.56), we have

$$Z_T\big|_{\omega_0} = \frac{1}{Y_T\big|_{\omega_0}} = R_{\text{eq}}\big|_{\omega_0} = \frac{1}{R}\left(R^2 + \frac{1}{\omega_0^2 C^2}\right)$$

$$= \frac{L}{RC} \tag{14.58}$$

indicating that Z_T will be purely real, equivalent to a resistance whose value is L/RC, as was the case of the practical tank circuit.

Using the definition of the circuit's quality factor, based on the equivalent three-branch RLC parallel circuit [Eq. (14.38)]:

$$Q = \frac{\omega_0}{\text{BW}} = \frac{R_{\text{eq}}}{\omega_0 L} = \frac{L}{RC}\frac{1}{\omega_0 L} = \frac{1}{\omega_0 CR} \tag{14.59}$$

In the case of high-Q circuits ($Q \geq 10$), we can use the approximation

$$\omega_0^2 C^2 R^2 \ll 1 \qquad \text{that is,}\ \frac{1}{\omega_0^2 C^2} \gg R^2$$

The resonance condition [Eq. (14.56)] can then be rewritten in the form

$$R^2 + \frac{1}{\omega_0^2 C^2} \cong \frac{1}{\omega_0^2 C^2} \cong \frac{L}{C} \qquad \text{that is,}\ \omega_0 L \cong \frac{1}{\omega_0 C}$$

Therefore,

$$\omega_0 \cong \frac{1}{\sqrt{LC}} \quad \text{rad/s} \tag{14.60}$$

This is the same result as obtained in the previous cases. However, one has to keep in mind that ω_0 as obtained from Eq. (14.60) is only correct for high-Q circuits. The higher the circuit's Q is, the better is the approximation.

The circuit's impedance at resonance can also be expressed in the familiar form developed in the previous case. Here, for high-Q circuits,

$$Z_T\big|_{\omega_0} = \frac{L}{RC} = \frac{\omega_0 L}{R}\frac{1}{\omega_0 CR}R \cong \frac{1}{(\omega_0 CR)^2}R \cong Q^2 R \tag{14.61}$$

Also here, $|Z_T|$ peaks at resonance. Even though the resonance condition corresponding to the second definition of resonance results in a resonance frequency different from that obtained in Eq. (14.57), both resonance frequencies are approximately the same for high-Q circuits. As in the previous case, ω_1 and ω_2, the cutoff frequencies, can be obtained from the general forms [Eqs. (14.39) and (14.40)].

EXAMPLE 14.10

For the circuit shown in Fig. 14.26 find the resonance frequency ω_0 at which $\mathbf{E}$ and $\mathbf{I}$ are in phase. What is the value of Z_T at this frequency? Derive an expression for the quality factor Q of this circuit.

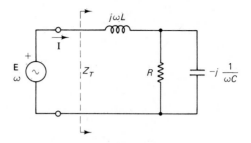

Figure 14.26

Solution In general (at any frequency ω), the total impedance of this circuit is

$$
\begin{aligned}
Z_T &= j\omega L + \frac{R[-j(1/\omega C)]}{R - j(1/\omega C)} \frac{j\omega C}{j\omega C} \\[2mm]
&= j\omega L + \frac{R}{1 + j\omega CR} \cdot \frac{1 - j\omega CR}{1 - j\omega CR} \\[2mm]
&= j\omega L + \frac{R - j\omega CR^2}{1 + \omega^2 C^2 R^2} = \frac{R}{1 + \omega^2 C^2 R^2} + j\omega L - j\frac{\omega CR^2}{1 + \omega^2 C^2 R^2} \\[2mm]
&= R_{eq} + j\omega L - j\frac{1}{\omega C_{eq}} = R_{eq} + j\left(\omega L - \frac{1}{\omega C_{eq}}\right)
\end{aligned}
$$

that is, the original circuit is equivalent (mathematically) to the series RLC circuit in Fig. 14.27, where

$$
R_{eq} = \frac{R}{1 + \omega^2 C^2 R^2}
$$

and

$$
C_{eq} = \frac{1 + \omega^2 C^2 R^2}{\omega^2 CR^2} = C\left(1 + \frac{1}{\omega^2 C^2 R^2}\right)
$$

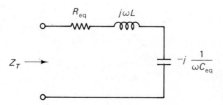

Figure 14.27

At the resonance frequency ω_0, Z_T must be a purely real number for **E** and **I** to be in phase [i.e., Im $(Z_T) = 0$]. Here

$$\omega_0 L = \frac{1}{\omega_0 C_{eq}} = \frac{\omega_0 C R^2}{1 + \omega_0^2 C^2 R^2}$$

$$1 + \omega_0^2 C^2 R^2 = \frac{C R^2}{L}$$

Therefore,

$$\omega_0 = \sqrt{\frac{1}{LC} - \frac{1}{C^2 R^2}} \quad \text{rad/s}$$

At this frequency,

$$Z_T = R_{eq}\big|_{\omega_0} = \frac{R}{1 + \omega_0^2 C^2 R^2}$$

Substituting from the resonance condition above gives us

$$Z_T = R \frac{L}{C R^2} = \frac{L}{CR} \quad \Omega \quad \text{(a purely real number)}$$

The quality factor Q of this circuit can easily be derived using the mathematically equivalent RLC series circuit in Fig. 14.27. Here

$$Q = \frac{\omega_0 L}{R_{eq}} = \frac{\omega_0 L}{R}(1 + \omega_0^2 C^2 R^2)$$

$$= \frac{\omega_0 L}{R} \frac{C R^2}{L} = \omega_0 C R$$

14.4 CONCLUDING REMARKS

A study of the resonance phenomena, defined as (1) the condition at which the source's voltage and current are in phase, or (2) the condition at which the magnitude of the total circuit's impedance (or admittance) is a maximum or a minimum, is conducted in this chapter. It is shown that both definitions result in the same resonance condition in the case of the series and the ideal three-branch parallel RLC circuits. Comprehensive analysis of these two circuits is presented, making use of impedance (or admittance) triangles and phasor diagrams. Throughout, comparison of the two circuits is made and similarities (or opposition) are pointed out. Also, the concepts of selectivity, the circuit's Q factor, bandwidth, and cutoff frequencies are developed and used in specific design examples.

Section 14.3 examines two cases of the general series–parallel RLC circuits, including the practical tank circuit. Here the difference between the two definitions of resonance becomes clearer. Simplifications based on the high-Q approximation are made, showing that in such cases the resonance conditions converge to the simple rule of

$$\omega_0 L \cong \frac{1}{\omega_0 C}$$

DRILL PROBLEMS

Section 14.1

14.1. Find the resonance frequency (f_0), quality factor (Q), bandwidth (BW), and the cutoff frequencies of an RLC series circuit whose resistance is $20\,\Omega$, inductance is $0.1\,H$, and capacitance is $0.4\,\mu F$.

14.2. Repeat Drill Problem 14.1 for the RLC series circuit that has the following elements values:
(a) $R = 50\,\Omega$, $L = 10\,mH$, and $C = 1\,\mu F$.
(b) $R = 300\,\Omega$, $L = 0.09\,H$, and $C = 10\,nF$.

14.3. An RLC series circuit is driven by the voltage source $\mathbf{E} = 20\underline{/0°}\,V$. The frequency of the source is adjusted to be the resonance frequency of the circuit, which is $750\,Hz$. If R is $10\,\Omega$ and X_L is $40\,\Omega$, find
(a) X_C at resonance.
(b) The values of L and C.
(c) Q, BW, and the cutoff frequencies of this circuit.
(d) $\mathbf{I}$, $\mathbf{V}_R$, $\mathbf{V}_L$, and $\mathbf{V}_C$ at resonance.

14.4. An RLC series circuit has a $2\,kHz$ resonance frequency and a quality factor of 8.
(a) Find the bandwidth of the circuit and its cutoff frequencies.
(b) If C is $0.2\,\mu F$, find the values of R and L.

14.5. A series RLC circuit has a resonance frequency of $5\,kHz$. Its capacitive reactance at resonance is $25\,\Omega$. The amount of current measured in this circuit was $0.1\,A$ when a $2\,V$ source, at the resonance frequency, is applied to the circuit. Find
(a) The values of R, L, and C.
(b) The BW and Q.
(c) The cutoff frequencies.
(d) The magnitude of the voltage across the capacitor under these conditions.

Section 14.2.

14.6. In an ideal parallel RLC circuit, R is $2\,k\Omega$, L is $5\,mH$, and C is $20\,nF$. Find
(a) ω_0, BW, Q, and the cutoff frequencies.
(b) The voltage across this circuit at resonance if it is driven by an ideal $0.01\,A$ current source.

14.7. Design an ideal tank circuit to allow $0.002\,A$ to flow from a $1\,V$ source applied to the circuit at the resonance frequency of the circuit, which is $10\,kHz$. The bandwidth of the circuit should be $0.5\,kHz$.

Section 14.3

14.8. In a practical tank circuit (see Fig. 14.22), R is $10\,\Omega$, L is $1\,mH$, and C is $10\,nF$.
(a) Find ω_0, Q, and BW.
(b) Find Z_T, the total impedance at resonance.
(c) If the applied current source has a value of $1.0\underline{/0°}\,mA$, find $\mathbf{V}$, $\mathbf{I}_L$, and $\mathbf{I}_C$.

14.9. In a practical tank circuit, the inductive branch has a $20\,\Omega$ resistance and a $100\,\Omega$ reactance at the resonance frequency of $7\,kHz$. At this frequency, the impedance is purely real. Find
(a) The value of X_C at resonance.
(b) The value of L and C.
(c) Q, BW, and the cutoff frequencies of this network.
(d) The impedance of the tank circuit at resonance.

14.10. Design a practical tank circuit whose resonance frequency is $20\,kHz$, BW is $1\,kHz$, and whose total impedance at resonance is $20\,k\Omega$.

PROBLEMS

Section 14.1

(i) **14.1.** In an RLC series circuit, the applied source voltage is $e(t) = 2 \sin (10{,}000t)$ V. The value of R is $10\,\Omega$ and the value of C is $0.1\,\mu$F. What should be the value of L so that the circuit would resonate at this frequency of the source? Find
   (a) The time-domain expression for $i(t)$.
   (b) The RMS voltages across L and C.
   (c) The quality factor and the bandwidth of this circuit.

(i) **14.2.** The components in Problem 14.2 are connected as three single-element branches in parallel. The applied source is the current source $i(t) = 1.414 \sin (10{,}000t)$ A. Find
   (a) Y_T of the circuit and the time-domain expression for $v(t)$, the voltage across the current source.
   (b) The RMS currents in the inductive and capacitive branches.
   (c) The quality factor and the bandwidth of this circuit.

(i) **14.3.** A series resonant circuit has a resonance frequency of 500 kHz and a bandwidth of 25 kHz. What is its quality factor? Determine the exact and approximate values of its 3-dB cutoff frequencies. If the value of the capacitor in this circuit is $0.1\,\mu$F, find the values of R and L.

(i) **14.4.** A series resonant circuit has a bandwidth of 20 kHz and a Q of 40. The resistor value is $10\,\text{k}\Omega$. Find the values of L and C in this circuit.

(ii) **14.5.** The resistor and inductor in a series RLC circuit have the values of $10\,\Omega$ and 1 mH, respectively. If Q is to be 50, find the resonance frequency, the value of the capacitor C, and the circuit's bandwidth.

(ii) **14.6.** A series RLC circuit carries a current of 2 A from a 16 V source at resonance. If the capacitive reactance at this resonance frequency is $100\,\Omega$, find the quality factor of this circuit. If this circuit is to have a bandwidth of 1 kHz, find the values of the required inductor and capacitor. What is the resonance frequency of this circuit?

(ii) **14.7.** Design a series RLC circuit that resonates at 2000 krad/s, has a quality factor of 10, and carries 1.5 A in response to a 9 V signal source. Find the bandwidth and the 3-dB cutoff frequencies of this circuit

(ii) **14.8.** The reading of the ideal ammeter in the circuit of Fig. 14.28 is 3.54 A. It does not change if the source frequency is doubled. The amplitude of the Y_A signal is 30 V (peak), and the amplitude of the Y_B signal is 40 V (peak).

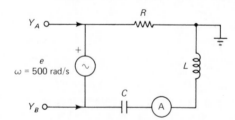

Figure 14.28

   (a) Explain which signal (Y_A or Y_B) is leading and by how much.
   (b) Explain which signal (Y_A or Y_B) is leading and by how much, when the frequency is doubled.
   (c) Find the values of R, L, and C.
   (d) What is the RMS value of the source voltage?
   (e) Find the resonance frequency and Q of this circuit. Explain what happens to the magnitudes of the Y_A and the Y_B signals at this frequency.

(ii) **14.9.** The oscillogram shown in Fig. 14.29(a) was obtained for the series *RLC* circuit shown in Fig. 14.29(b) with the scope settings adjusted as follows:

$$\text{time/div} = 1 \text{ ms}$$
$$\text{V/div } (Y_A) = 5 \text{ V}$$
$$\text{V/div } (Y_B) = \text{unknown}$$

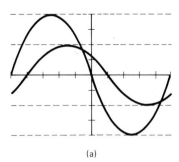

(a)

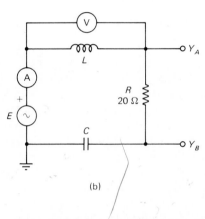

(b)

Figure 14.29

(a) Label the two waveforms and find the V/div of Y_B.
(b) Find the reading of the ammeter.
(c) If the voltmeter reading is 17 V, find the values of *L*, *C*, and *E*.
(d) Find the resonance frequency, quality factor, and bandwidth of this circuit.

Section 14.2

(ii) **14.10.** A three-branch ideal *RLC* parallel circuit is driven by a 10 V source. Its resonance frequency is 1 Mrad/s and its *Q* is 20. The source current at this frequency is 0.2 A. Find the value of each of the elements in the three parallel branches. Also determine the bandwidth and the 3-dB cutoff frequencies of this circuit.

(ii) **14.11.** An ideal three-branch *RLC* parallel circuit has an inductive branch with $L = 0.1$ mH. Find the value of *R* and *C* so that the circuit's bandwidth is 1 kHz and its *Q* is 40. What is the resonance frequency of this circuit?

(i) **14.12.** The source $e(t) = 1.414 \sin \omega t$ V is applied to a three-branch ideal *RLC* parallel circuit, where $R = 100 \, \Omega$, $L = 0.2$ mH, and $C = 0.22 \, \mu$F. Find
(a) The resonance frequency, *Q* and the bandwidth of this circuit.
(b) The branch currents and the source current at the resonance frequency in the time domain. Express these currents in phasor form and draw the phasor diagram.

Section 14.3

(i) **14.13.** A practical tank circuit has a 0.1 μF capacitive branch in parallel with a coil whose inductance is 0.5 mH and the coil's resistance is 1 Ω. Find the resonance frequency, the quality factor, and the bandwidth of this circuit. What is the impedance of this tank circuit at the resonance frequency?

(ii) **14.14.** A practical tank circuit should be deisgned to resonate at 1 MHz and have a bandwidth of 2 kHz. When this tank circuit is driven by a current source of value 0.01 A, at the resonance frequency, the voltage across the tank circuit should be 4 V. Find the values of the elements required.

(i) **14.15.** For the practical tank circuit shown in Fig. 14.30, find the frequency at which **E** and **I** are in phase. Calculate the value of Q and BW of this circuit. If **E** is $5\underline{/0°}$ V, draw a complete phasor diagram for this circuit showing the currents in the different branches. What is the value of the circuit's impedance at resonance?

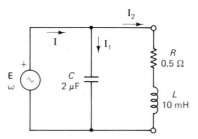

Figure 14.30

(ii) **14.16.** If in the circuit shown in Fig. 14.30, the inductive branch is removed, what should the value of L and R of the replacement branch be so that the tank circuit would have a Q of 50 and a bandwidth of 2 kHz? Recalculate the value of the impedance at resonance and the values of the different branch currents.

(ii) **14.17.** A tank circuit has a 5 mH coil and a 0.2 μF capacitor. Its impedance at resonance is 50 kΩ. Find the resistance of the coil, the resonance frequency, the Q, and the BW of this tank circuit.

(ii) **14.18.** For the circuit shown in Fig. 14.31, derive an expression for ω_0, the resonance frequency at which Z_T is real. What is the value of Z_T at this frequency? Also obtain the expression for the quality factor of this circuit.

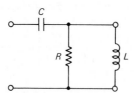

Figure 14.31

(ii) **14.19.** By converting each branch in the circuit shown in Fig. 14.32 into its single-element parallel equivalent, derive expressions for the resonance frequency and the quality factor of this circuit.

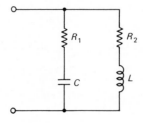

Figure 14.32

(i) **14.20.** In the circuit shown in Fig. 14.33, what is the value of X_C for the circuit to be at resonance? At this condition, find the currents $\mathbf{I}$, $\mathbf{I}_L$, and $\mathbf{I}_C$. Draw the phasor diagram. What is the value of Q of this circuit?

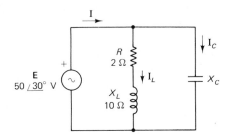

Figure 14.33

GLOSSARY

Bandpass Filter: An electrical network designed to allow signals whose frequencies are within a prespecified range to pass through with relatively little or no reduction in magnitude (i.e., low attenuation), while signals whose frequencies are outside this range are severely attenuated.

***Bandwidth* (BW):** The range of frequencies between the cutoff frequencies, or the range of frequencies between the cutoff frequency and a specific reference frequency.

Cutoff (Half-Power or −3 dB) Frequency: The frequency at which the response (voltage or current) drops to 70.71% of its peak value. Also, at this frequency the power dissipated in the network is half its maximum value.

Electric Filter: A network designed to pass (or reject) the transfer of electrical signals to a load when the signal frequencies are within a prespecified range.

Frequency Response: The behavior of the response (current or voltage) of an electrical network as a function of the variation in the source frequency. The response is usually displayed in the form of a graph showing the variation of the magnitude and/or phase of the response versus the frequency.

Frequency Trap: A parallel *LC* circuit, connected in series with the source (or load) to prevent the transfer of signals at a specific frequency from reaching the load. At this frequency the *LC* parallel circuit is at resonance and will be equivalent to a very high (ideally infinite) resistance.

Practical Tank Circuit: A parallel *LC* circuit in which the finite resistance of the practical inductor has a measurable impact on the properties and performance of the circuit, particularly at resonance.

Quality Factor (Q): A dimensionless quantity, being the ratio of the resonance frequency to the bandwidth of the circuit. It is a measure of the sharpness, or selectivity, of the resonant circuit.

Q Factor of an Inductor: A constant parameter of an inductor that provides the ratio of its reactance to its resistance when the coil is operated at frequencies above a specified critical frequency. The resistance of the inductor at such high frequencies is approximately linearly proportional to the frequency due to the skin effect.

Resonance Condition: A condition that establishes a relationship between the elements values (R, L, and C) of a circuit and the frequency of the applied source (the resonance frequency). At this condition, one or more of the following situations is satisfied:

(a) The voltage applied to the resonant circuit is in phase with the resulting current flow in the circuit.

(b) The magnitude of the impedance (or admittance) of the resonant circuit is either a maximum or a minimum.

(c) Maximum power is transferred to the circuit.

Selectivity: A characteristic of resonant circuits that is directly related to the bandwidth of the circuit. The Q factor is used as a measure of the selectivity. A highly selective resonant circuit would have a high value of Q and indicates that the bandwidth of the circuit is very narrow, and vice versa.

Skin Effect: A phenomenon that occurs in conductors when operated at high frequencies. Due to the electromagnetic induction, the electric current will tend to flow mainly in the area closest to the surface of the conductor, resulting in a reduction in the effective area of the conductor, hence a corresponding increase in the resistance of the conductor.

ANSWERS TO DRILL PROBLEMS

14.1. $f_0 = 795.77$ Hz, $Q = 25$, BW = 31.83 Hz, $f_1 = 780.01$ Hz, $f_2 = 811.84$ Hz

14.3.
(a) $X_C = 40\,\Omega$
(b) $L = 8.488$ mH, $C = 5.305\,\mu F$
(c) $Q = 4$, BW = 187.5 Hz, $f_1 = 662.09$ Hz, $f_2 = 849.59$ Hz
(d) $\mathbf{I} = 2\underline{/0°}$ A, $\mathbf{V}_R = 20\underline{/0°}$ V, $\mathbf{V}_L = 80\underline{/90°}$ V, $\mathbf{V}_C = 80\underline{/-90°}$ V

14.5.
(a) $R = 20$, Ω, $L = 0.796$ mH, $C = 1.273\,\mu F$

(b) $Q = 1.25$, BW = 4000 Hz
(c) $f_1 = 3385.16$ Hz, $f_2 = 7385.16$ Hz
(d) $V_C = 2.5$ V

14.7. $R = 500\,\Omega$, $Q = 20$, $L = 0.398$ mH, $C = 0.6366\,\mu F$

14.9.
(a) X_C (at ω_0) = 104 Ω
(b) $L = 2.274$ mH, $C = 0.2186\,\mu F$
(c) $Q = 5$, BW = 1400 Hz, $f_1 = 6334.9$ Hz, $f_2 = 7734.9$ Hz
(d) Z_T (at ω_0) = 520 Ω

15 | LOOP AND NODE ANALYSIS OF GENERAL AC CIRCUITS

OBJECTIVES

- Familiarity with topological terms and conditions used in network analysis.
- Ability to use the generalized loop analysis procedure to solve for a required unknown circuit quantity.
- Ability to use the generalized node analysis procedure to solve for a required unknown circuit quantity.
- Ability to apply the generalized network analysis techniques to networks containing dependent voltage and current sources.
- Familiarity with the $\Delta-Y$ and $Y-\Delta$ network conversions.

The loop and node analysis techniques have been introduced in Chapter 5 to analyze the general type of resistive networks. These methods are the most general analysis techniques that can be used to obtain any required response: voltage, current, or power, in almost any general electric circuit. As such, they are also used to analyze any general ac network. Here, however, impedances and admittances are the network quantities involved, instead of pure resistances and conductances. Also, the network variables are the phasors representing the sinusoidal waveforms. Hence the analysis here involves basically complex number manipulations. Because of its importance, a complete treatment of this subject will be represented here.

A typical general ac circuit is shown in Fig. 15.1. It contains current and voltage sources and impedances. A *node* is a point of interconnection of two or more elements in the network (e.g., points a, 1, 2, 3, 4, and b are nodes in the given network). A *principal node* is a point of interconnection of three or more elements in a network (e.g., points 1, 2, 3, and 4 are principal nodes in the given network). However, nodes a and b are not principal nodes.

A *branch* is the combination of elements forming a singular series path between any two principal nodes. In the network shown in Fig. 15.1, $\mathbf{E}_1$ and Z_1 form one branch; Z_2, $\mathbf{I}_{S_1}$, Z_3, and so on, are the other branches. In all, this network has eight branches. Note that this definition of a branch is more general than that used in Chapters 3 and 5 due to the introduction of principal nodes.

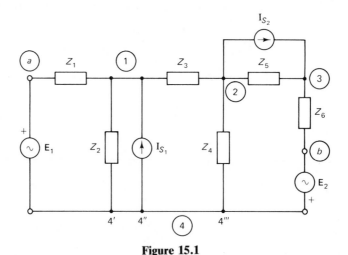

Figure 15.1

A *loop* is any closed path in a given network. On the other hand, a *mesh* is any closed path in a given network that does not have any element (or branch) inside it. Meshes are therefore special types of loops. For example, the path $a\,14'\,a$ is a loop or a mesh, but the path $a\,14''4'\,a$ is a loop but not a mesh because the branch Z_2 is inside this loop. Meshes can therefore be thought of as resembling window partitions (or window panes).

For a given network, using the definitions above, the solution requires either:

$$\text{number of independent mesh (or loop) equations}$$

$$= B \text{ (number of branches)} - N \text{ (number of principal nodes)} + 1 \quad (15.1)$$

or

$$\text{number of independent node equations}$$

$$= N \text{ (number of principal nodes)} - 1 \quad (15.2)$$

The method that needs the least number of equations is usually the preferred method of analysis because the solution required is obviously easier to carry out. In the network shown in Fig. 15.1,

$$\text{required number of mesh equations} = 8 - 4 + 1 = 5$$

while

$$\text{required number of node equations} = 4 - 1 = 3$$

As will be shown in the following discussions, source conversions can be quite useful and could result in reducing the necessary number of equations to solve the network.

15.2 LOOP (OR MESH) ANALYSIS

Loop analysis is based on writing KVL for the necessary number of loops (or meshes) in a given network in order to solve the network (i.e., determine all the values of the electrical variables). The question now arises: What are the circuit variables involved, and how many loop (or mesh) equations are necessary?

Consider the circuit in Fig. 15.2; it has two principal nodes, 1 and 2. It has three branches and two meshes, marked 1 and 2. It also has three loops, the two meshes mentioned above and the loop consisting of the circuit along the outside path of the network.

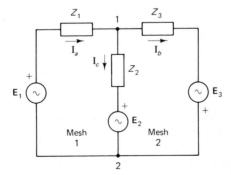

Figure 15.2

If one were able to determine $\mathbf{I}_a$, $\mathbf{I}_b$, and $\mathbf{I}_c$, as shown in the Fig. 15.2, all the voltages across the components could be found, and also the power in any element. Thus everything about the network would be known. However, we do not really have to find all these three currents (phasor quantities) initially, because if $\mathbf{I}_a$ and $\mathbf{I}_b$ (for example) are known, then by applying KCL at node 1,

$$\mathbf{I}_a = \mathbf{I}_b + \mathbf{I}_c \quad (15.3a)$$

or

$$\mathbf{I}_c = \mathbf{I}_a - \mathbf{I}_b \qquad\qquad (15.3b)$$

(i.e., $\mathbf{I}_c$ can easily be derived). All that is needed, then, are the values of two circuit variables—the two currents $\mathbf{I}_a$ and $\mathbf{I}_b$ in this case.

It is well known from algebra that *to determine the values of n unknowns, n independent equations (or conditions) are required.* The simultaneous solution of these equations will result in unique and finite values of these unknowns. If we write only $(n - 1)$ equations, or fewer, no solution can be found. Also, if we write $(n + 1)$ equations, or more, no unique solution can be found. More than one answer may be obtained or we may encounter quantities such as 0/0. When this situation occurs, we conclude that a redundant equation was written, which is in fact algebraically obtained from the other n equations by addition and/or subtraction. The conclusions is: *To determine n unknowns, n and exactly n independent equations are necessary and sufficient.*

Going back to the network of Fig. 15.2, two equations are thus needed to determine the two principal unknowns, $\mathbf{I}_a$ and $\mathbf{I}_b$. Any two loops can be considered and KVL can be written for each in terms of the two unknown currents. By the way, these currents are referred to as *branch currents.*

To develop a systematic approach to the analysis technique, it is usually easier to consider the meshes constituting the network because they are clearer to identify. This is recommended but not necessary. In the case of planar networks (with no lines or wires crossing) such as the one considered in Fig. 15.2, it will be shown that mesh equations can be written in an easy and systematic manner. However, for nonplanar networks (as for example the lattice network shown in Fig. 15.3), loops must be considered. In fact, node analysis may be the preferred technique in this case.

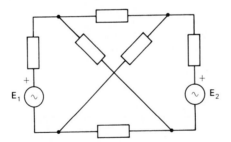

Figure 15.3

Through network topology considerations, the network unknowns in the loop analysis technique, and hence the number of sufficient and necessary equations, are obtained through the application of Eq. (15.1).

For the circuit in Fig. 15.2, $B = 3$, $N = 2$; therefore, the number of necessary loop (or mesh) equations is

$$n = B - N + 1 = 3 - 2 + 1 = 2$$

as predicted above. Note that this network has two meshes. It was also pointed out in relation to the circuit of Fig. 15.1 that five mesh equations were needed, while the network itself contains five meshes. Thus one can conclude that the number of equations needed in mesh or loop analysis (and hence also the number of unknowns) is the same as the number of meshes in the given planar network.

For the circuit in Fig. 15.2, KVL can be written for the two meshes as follows:

$$\mathbf{I}_a Z_1 + \mathbf{I}_c Z_2 + \mathbf{E}_2 - \mathbf{E}_1 = 0 \quad \text{(for mesh 1)}$$

and

$$\mathbf{I}_b Z_3 - \mathbf{E}_3 - \mathbf{E}_2 - \mathbf{I}_c Z_c = 0 \quad \text{(for mesh 2)}$$

Replacing $\mathbf{I}_c$ by $(\mathbf{I}_a - \mathbf{I}_b)$, the two equations above can be rewritten as follows:

$$\mathbf{I}_a Z_1 + (\mathbf{I}_a - \mathbf{I}_b)Z_2 = \mathbf{E}_1 - \mathbf{E}_2$$

and

$$\mathbf{I}_b Z_3 - (\mathbf{I}_a - \mathbf{I}_b)Z_2 = \mathbf{E}_2 + \mathbf{E}_3$$

Rearranging the terms, we have

$$\mathbf{I}_a(Z_1 + Z_2) - \mathbf{I}_b Z_2 = \mathbf{E}_1 - \mathbf{E}_2 \tag{15.4}$$

$$-\mathbf{I}_a Z_2 + \mathbf{I}_b(Z_2 + Z_3) = \mathbf{E}_2 + \mathbf{E}_3 \tag{15.5}$$

To systemize and generalize the loop (or mesh) analysis technique, one can think in terms of loop (or mesh) currents, instead of some specific branch currents. It will be assumed that each mesh (or loop) is carrying a mesh current flowing in these closed paths as indicated in Fig. 15.4. This is actually the same circuit of Fig. 15.2. These currents are fictitious since they suggest the existence of two streams of current carriers flowing in the impedance Z_c, common between the two meshes. This is, of course, not true. However, one can think of the net current flow in this common branch as being made up of two components: one due to the sources in mesh 1 and the other component due to the sources in mesh 2. The net current in this common branch is the algebraic sum of its component currents.

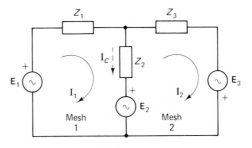

Figure 15.4

This assumption, or suggestion, makes the choice of the unknown variables quite systematic. Since in any network containing n meshes we need n equations (i.e., KVL) in terms of n unknowns, it is obvious that the simplest choice to make is to choose these unknowns to be the n mesh currents.

In the network of Fig. 15.2 it was suggested that the two unknowns be the two branch currents $\mathbf{I}_a$ and $\mathbf{I}_b$. We could have easily taken $\mathbf{I}_b$ and $\mathbf{I}_c$ or $\mathbf{I}_a$ and $\mathbf{I}_c$ instead. In simple circuits like this one, the choice of the unknowns is not very critical, but in more complicated networks, the choice of unknowns could be critical and could result in unnecessarily more complicated solutions. For this reason mesh (or loop) currents are the simplest choice.

The direction of the loop (or mesh) currents is arbitrary; some could be assumed to flow clockwise and some counterclockwise or all in one direction or the other. Again it is recommended that one choose all such mesh currents to flow clockwise. The solution of the network equations will provide the proper current directions. A negative current only means that current is actually flowing in the direction opposite to that assumed. Also, a net current in a certain branch is the algebraic sum of its component currents, including signs. The recommended clockwise direction only makes the solution more systematic.

From a comparison of the current choices made in Figs. 15.2 and 15.4, it is clear that

$$\mathbf{I}_1 = \mathbf{I}_a \quad \text{(current flowing in } \mathbf{E}_1 \text{ and } Z_1) \tag{15.6}$$

$$\mathbf{I}_2 = \mathbf{I}_b \quad \text{(current flowing in } Z_3 \text{ and } \mathbf{E}_3) \tag{15.7}$$

and

$$\mathbf{I}_c = \mathbf{I}_1 - \mathbf{I}_2 \quad \text{(in the downward direction, flowing in } Z_2 \text{ and } \mathbf{E}_2)$$

$$= \mathbf{I}_a - \mathbf{I}_b \quad \text{[as in Eq. (15.3b)]} \tag{15.8}$$

Hence Eqs. (15.4) and (15.5) can now be written in terms of the mesh currents $\mathbf{I}_1$ and $\mathbf{I}_2$, as follows:

$$(Z_1 + Z_2)\mathbf{I}_1 - Z_2\mathbf{I}_2 = \mathbf{E}_1 - \mathbf{E}_2 \tag{15.9}$$

$$-Z_2\mathbf{I}_1 + (Z_2 + Z_3)\mathbf{I}_2 = \mathbf{E}_2 + \mathbf{E}_3 \tag{15.10}$$

Now consider the network in Fig. 15.5. This network obviously has three meshes. Thus three mesh equations have to be written in terms of the three unknown mesh currents. Also as a check, this network has four principal nodes and six branches; that is, the number of mesh equations required is

$$B - N + 1 = 6 - 4 + 1 = 3$$

as above.

The three mesh currents are chosen in a clockwise direction as shown in Fig. 15.5. Writing KVL for each of the meshes yields

$$-\mathbf{E}_1 + \mathbf{I}_1Z_1 + (\mathbf{I}_1 - \mathbf{I}_3)Z_3 + (\mathbf{I}_1 - \mathbf{I}_2)Z_2 = 0 \tag{15.11}$$

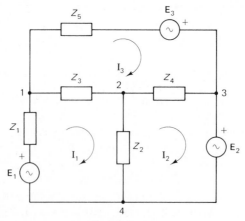

Figure 15.5

Here $(\mathbf{I}_1 - \mathbf{I}_3)$ is the net current in Z_3 and $(\mathbf{I}_1 - \mathbf{I}_2)$ is the net current in Z_2, both in the direction of $\mathbf{I}_1$. Similarly, for the second mesh,

$$(\mathbf{I}_2 - \mathbf{I}_1)Z_2 + (\mathbf{I}_2 - \mathbf{I}_3)Z_4 + \mathbf{E}_2 = 0 \qquad (15.12)$$

and for the third mesh,

$$\mathbf{I}_3 Z_5 - \mathbf{E}_3 + (\mathbf{I}_3 - \mathbf{I}_2)Z_4 + (\mathbf{I}_3 - \mathbf{I}_1)Z_3 = 0 \qquad (15.13)$$

Rearranging the terms and combining the coefficients of the different currents, the three equations above can be rewritten in the form

$$(Z_1 + Z_2 + Z_3)\mathbf{I}_1 - Z_2\mathbf{I}_2 - Z_3\mathbf{I}_3 = \mathbf{E}_1 \qquad (15.14)$$
$$-Z_2\mathbf{I}_1 + (Z_2 + Z_4)\mathbf{I}_2 - Z_4\mathbf{I}_3 = -\mathbf{E}_2 \qquad (15.15)$$
$$-Z_3\mathbf{I}_1 - Z_4\mathbf{I}_2 + (Z_3 + Z_4 + Z_5)\mathbf{I}_3 = \mathbf{E}_3 \qquad (15.16)$$

Writing the mesh (or loop) equations in the form above shows the systematic approach used in developing the mesh (or loop) analysis technique. Examination of the mesh equations (15.9) and (15.10) for the circuit of Fig. 15.4 and the mesh equations (15.14), (15.15), and (15.16) for the circuit of Fig. 15.5 reveals that in general the mesh equations can be written in the form

$$Z_{11}I_1 \quad \longrightarrow \quad Z_{12}I_2 = E_{11} \qquad (15.17)$$
$$-Z_{21}I_1 \quad + \quad Z_{22}I_2 = E_{22} \qquad (15.18)$$

for the two-mesh circuit, and

$$Z_{11}I_1 \quad - \quad Z_{12}I_2 \quad - \quad Z_{13}I_3 \quad = \quad E_{11} \qquad (15.19)$$
$$-Z_{21}I_1 \quad + \quad Z_{22}I_2 \quad - \quad Z_{23}I_3 \quad = \quad E_{22} \qquad (15.20)$$
$$-Z_{31}I_1 \quad - \quad Z_{32}I_2 \quad + \quad Z_{33}I_3 \quad = \quad E_{33} \qquad (15.21)$$

Principal diagonal

for the three-mesh circuit, where
Z_{11} = the total or self-impedance of mesh 1, carrying the current $\mathbf{I}_1$
Z_{22} = the total or self-impedance of mesh 2, carrying the current $\mathbf{I}_2$
Z_{33} = the total or self-impedance of mesh 3, carrying the current $\mathbf{I}_3$
$Z_{12} = Z_{21}$ = common or mutal impedance between meshes 1 and 2
$Z_{13} = Z_{31}$ = common or mutual impedance between meshes 1 and 3
$Z_{23} = Z_{32}$ = common or mutual impedance between meshes 2 and 3
$\mathbf{E}_{11}$ = the net algebraic sum of voltage sources in mesh 1, driving the current $\mathbf{I}_1$ in the assumed direction (clockwise)
$\mathbf{E}_{22}$ = the net algebraic sum of voltage sources in mesh 2, driving the current $\mathbf{I}_2$ in the assumed direction (clockwise)
$\mathbf{E}_{33}$ = the net algebraic sum of voltage sources in mesh 3, driving the current $\mathbf{I}_3$ in the assumed direction (clockwise),
etc.

Any set of linear algebraic equations, such as Eqs. (15.17) and (15.18) or Eqs. (15.19), (15.20), and (15.21), can be expressed in a matrix form as

$$\boxed{[Z][\mathbf{I}] = [\mathbf{E}]} \qquad (15.22)$$

where $[Z]$, called the system matrix, is an $n \times n$ arrangement of elements consisting of the characteristic impedances of the network listed above. It has the form

$$[Z] = \begin{bmatrix} Z_{11} & -Z_{12} & -Z_{13} \\ -Z_{21} & Z_{22} & -Z_{23} \\ -Z_{31} & -Z_{32} & Z_{33} \end{bmatrix} \qquad (15.23)$$

$[\mathbf{I}]$ is an n-column matrix, consisting of the n mesh currents, and taking the form

$$[\mathbf{I}] = \begin{bmatrix} \mathbf{I}_1 \\ \mathbf{I}_2 \\ \mathbf{I}_3 \end{bmatrix} \qquad (15.24)$$

and $[\mathbf{E}]$ is also an n-column matrix, consisting of the net voltage sources in the n meshes. It takes the form

$$[\mathbf{E}] = \begin{bmatrix} \mathbf{E}_{11} \\ \mathbf{E}_{22} \\ \mathbf{E}_{33} \end{bmatrix} \qquad (15.25)$$

This "shorthand" matrix notation is very useful. The rules of matrix algebra can be applied to such matrix equations.

15.2.1   Important Remarks

The following observations on the mesh (or loop) analysis techniques are quite useful and important, to minimize the chances of errors and facilitate the required solution.

1. The impedance terms in the principal diagonal (above) represent the self- or total impedance of each of the meshes. They are the coefficients of the corresponding mesh currents and they are always expressed with a positive sign.
2. The off-diagonal impedance terms represent the common or mutual impedances between each pair of meshes. If there is no common impedance between two meshes, its value is obviously zero. It is important to note the symmetry of these off-diagonal terms, since as observed above:

$$Z_{ij} = Z_{ji} = \text{impedance common between meshes}$$

$$\text{(or loops) } i \text{ and } j \text{ (or } j \text{ and } i)$$

The signs of these terms are always negative if all the mesh currents are chosen clockwise, or all chosen counterclockwise. This is because in such cases the currents in any two adjacent meshes always flow in opposite directions through the common impedance between the two meshes. However, if some of the mesh (or loop) current directions are clockwise and some are counterclockwise, we have to be very careful, and each common impedance must be examined separately. If the two adjacent mesh currents flow in a common impedance in the same direction, the sign of the corresponding mutual impedance term is positive; otherwise, it is negative. The symmetry observation also applies to these algebraic signs.

3. The right-hand terms in the mesh (or loop) equations always represent the net algebraic sum of the source voltages in each of these meshes for each of the corresponding mesh equations. To obtain this sum, sources that aid (or help) the current to flow in the assumed direction are taken with a positive sign, while those sources that try to drive the mesh current in the direction opposite to that assumed are taken with a negative sign. If there are no sources in a certain mesh (or loop), this mesh equation will have a zero term on the right-hand side of the equation.

4. Always check that each element, whether an impedance or a source, in the given network has at least one mesh current flowing in it. Otherwise, the choice of mesh currents in such a case is inappropriate and no solution will be obtained. Also, never write more equations than the number necessary for the given network, as specified in Eq. (15.1).

5. *Source Conversion*: Current sources are difficult but not impossible to deal with when using the mesh (or loop) analysis technique for a network containing such sources. The reason is that all the terms in the mesh (or loop) equations are voltages, since these equations are KVL for each mesh. There is no relationship that allows us to express the voltage across an ideal current source. This voltage can have any value and it is, in fact, determined by the elements and their connections, external to the current source.

   The easiest way to overcome this difficulty is to consider the impedance in parallel with the current source as the internal impedance of this source and convert this practical current source into its equivalent voltage source model. The other alternative is to allow only one mesh (or loop) current to flow in a mesh (or loop) containing the current source. This second alternative is the only one possible if there are no impedances in parallel with the ideal current source.

Whenever possible it is recommended to convert practical current sources into their equivalent voltage source models because this allows one to follow the same systematic approach outlined above. Source conversion also reduces the number of meshes in a given network.

As an example, consider the network shown in Fig. 15.6. It contains the current source I_S. It has three meshes, five branches, and three principal nodes.

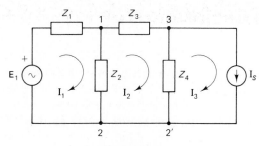

Figure 15.6

Therefore, the number of mesh equations required is 3 (= 5 − 3 + 1). According to the second alternative above, Fig. 15.6 indicates the three mesh currents assumed, noting that I_3 is the only mesh (or loop) current flowing in the current source. Therefore,

$$I_3 = I_S \tag{15.26}$$

while KVL for the other two meshes produce

$$I_1 Z_1 + (I_1 - I_2)Z_2 = E_1$$

or

$$(Z_1 + Z_2)I_1 - Z_2 I_2 = E_1 \tag{15.27}$$

and

$$(I_2 - I_1)Z_2 + I_2 Z_3 + (I_2 - I_3)Z_4 = 0$$

or

$$-Z_2 I_1 + (Z_2 + Z_3 + Z_4)I_2 - Z_4 I_3 = 0 \tag{15.28}$$

Note that Eq. (15.26) is the only equation that can be written for the third mesh. Substituting from Eq. (15.26) in Eq. (15.28) for I_3 gives us

$$-Z_2 I_1 + (Z_2 + Z_3 + Z_4)I_2 = Z_4 I_S \tag{15.29}$$

Now Eqs. (15.27) and (15.29) can be solved to obtain the two unknown currents I_1 and I_2.

The current source I_S in parallel with Z_4 forms a practical current source between points 3 and 2'. Using source conversion, the equivalent voltage source model is shown in the equivalent circuit in Fig. 15.7, with $E_S = Z_4 I_S$. The polarities of E_S are as indicated. The network now has two meshes. The two mesh currents I_1 and I_2 are assumed to flow clockwise in the two meshes as indicated in Fig. 15.7. The two mesh equations can then be written by inspection following the systematic procedure outlined above. This results in Eqs. (15.27) and (15.29) directly. In more complicated circuits, the use of source conversion is less prone to error.

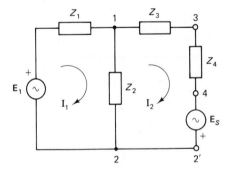

Figure 15.7

If it was required to find the current in Z_4 in the original circuit of Fig. 15.6, this current is not I_2 as the equivalent circuit in Fig. 15.7 suggests. This is because the equivalent voltage source model in Fig. 15.7 is mathematically equivalent to the current source as far as the external network is concerned, but not with respect to its internal components. Note that the circuit in Fig. 15.7 has an extra node, 4, that did not exist in the original network of Fig. 15.6. Once I_2 is obtained from the solution of the network in Fig. 15.7, one can go back to the original network in Fig. 15.6 and apply KCL at node 3, giving

$$\mathbf{I}_2 = \mathbf{I}_{Z_4} + \mathbf{I}_S$$

Therefore, the current in Z_4 is

$$\mathbf{I}_{Z_4} = \mathbf{I}_2 - \mathbf{I}_S \qquad (15.30)$$

15.2.2 Solution of the Mesh (or Loop) Equations

A system of linear algebraic equations such as, for example, Eqs. (15.19), (15.20), and (15.21), can be solved either through direct algebraic manipulations if they are simple enough, or by applying Cramer's rule using determinants. For the set of equations mentioned here, the system determinant Δ is

$$\Delta = \begin{vmatrix} Z_{11} & -Z_{12} & -Z_{13} \\ -Z_{21} & Z_{22} & -Z_{23} \\ -Z_{31} & -Z_{32} & Z_{33} \end{vmatrix} \begin{matrix} Z_{11} & -Z_{12} \\ -Z_{21} & Z_{22} \\ -Z_{31} & -Z_{32} \end{matrix}$$

(check symmetry around the principle diagonal)

$$\begin{aligned} = Z_{11}Z_{22}Z_{33} &- Z_{12}Z_{23}Z_{31} - Z_{13}Z_{12}Z_{32} \\ &-Z_{31}Z_{22}Z_{13} - Z_{32}Z_{23}Z_{11} - Z_{33}Z_{21}Z_{12} \end{aligned} \qquad (15.31)$$

Expanding a second- or third-order determinant can be done systematically as shown above and according to the rules introduced in Chapter 5. In general, determinants can be expanded in terms of a column (or a row) and the corresponding cofactors. For example, using the first column, the determinant Δ above can be expanded as follows:

$$\Delta = Z_{11}\Delta_{11} - Z_{21}\Delta_{21} - Z_{31}\Delta_{31} \qquad (15.32)$$

where Δ_{11}, Δ_{21}, and Δ_{31} are the cofactors of the corresponding terms of the column used for the expansion. Any cofactor of a term in a determinant (e.g., a_{ij}) is obtained from the original determinant by canceling the row i and the column j in which this element exists, and multiplying by $(-1)^{i+j}$. For example,

$$\Delta_{11} = (-)^2 \begin{vmatrix} Z_{22} & -Z_{23} \\ -Z_{32} & Z_{33} \end{vmatrix} = Z_{22}Z_{33} - Z_{23}Z_{32} \qquad (15.33)$$

$$\Delta_{21} = (-)^3 \begin{vmatrix} -Z_{12} & -Z_{13} \\ -Z_{32} & Z_{33} \end{vmatrix} = -(-Z_{12}Z_{33} - Z_{13}Z_{32}) \qquad (15.34)$$

and

$$\Delta_{31} = (-)^4 \begin{vmatrix} -Z_{12} & -Z_{13} \\ Z_{22} & -Z_{23} \end{vmatrix} = (Z_{12}Z_{23} + Z_{13}Z_{22}) \qquad (15.35)$$

Substituting these cofactors into Eq. (15.32) will result in the same expansion as indicated in Eq. (15.31). This technique, however, can be extended to any higher-order determinants, resulting in a progressive and systematic reduction of the order of the determinants to be expanded. As shown above, expanding a third-order determinant reduces to expanding second-order determinants through the cofactors.

Reviewing Cramer's rule for solving the system of equations (15.19), (15.20), and (15.21) to obtain the unknown mesh currents $\mathbf{I}_1$, $\mathbf{I}_2$, and $\mathbf{I}_3$, we see that

$$\mathbf{I}_1 = \frac{\begin{vmatrix} \mathbf{E}_{11} & -Z_{12} & -Z_{13} \\ \mathbf{E}_{22} & Z_{22} & -Z_{23} \\ \mathbf{E}_{33} & -Z_{32} & Z_{33} \end{vmatrix}}{\Delta} = \mathbf{E}_{11}\frac{\Delta_{11}}{\Delta} + \mathbf{E}_{22}\frac{\Delta_{21}}{\Delta} + \mathbf{E}_{33}\frac{\Delta_{31}}{\Delta} \qquad (15.36)$$

$$\mathbf{I}_2 = \frac{\begin{vmatrix} Z_{11} & \mathbf{E}_{11} & -Z_{13} \\ -Z_{21} & \mathbf{E}_{22} & -Z_{23} \\ -Z_{31} & \mathbf{E}_{33} & Z_{33} \end{vmatrix}}{\Delta} = \mathbf{E}_{11}\frac{\Delta_{12}}{\Delta} + \mathbf{E}_{22}\frac{\Delta_{22}}{\Delta} + \mathbf{E}_{33}\frac{\Delta_{32}}{\Delta} \qquad (15.37)$$

and

$$\mathbf{I}_3 = \frac{\begin{vmatrix} Z_{11} & -Z_{12} & \mathbf{E}_{11} \\ -Z_{21} & Z_{22} & \mathbf{E}_{22} \\ -Z_{31} & -Z_{32} & \mathbf{E}_{33} \end{vmatrix}}{\Delta} = \mathbf{E}_{11}\frac{\Delta_{13}}{\Delta} + \mathbf{E}_{22}\frac{\Delta_{23}}{\Delta} + \mathbf{E}_{33}\frac{\Delta_{33}}{\Delta} \qquad (15.38)$$

Note that to obtain the unknown current, $\mathbf{I}_n$, the numerator determinant is obtained from the system's determinant by replacing column n by the right-hand-side terms of the system of equations. All the numerator determinants above are expanded in terms of the column containing the right-hand-side terms [i.e., the net source voltages in each mesh (or loop)]. Note also that the cofactors are those corresponding to the elimination of the row and column in which these net source voltage terms exist.

EXAMPLE 15.1

Find the output power of the voltage source in the circuit shown in Fig. 15.8 and also the power dissipated in each resistance in this circuit.

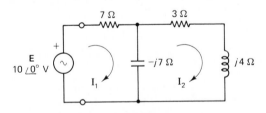

Figure 15.8

Solution This circuit has two meshes. The mesh currents are assumed to flow as indicated in Fig. 15.8. Clearly, the values of these mesh currents need to be determined before power calculations can be made.

For this circuit the two mesh equations can easily be written following the systematic approach described above.

$$\underset{\substack{\uparrow \\ \text{total impedance} \\ \text{of mesh 1}}}{(7 - j7)\mathbf{I}_1} \quad \underset{\substack{\uparrow \\ \text{common impedance} \\ \text{between meshes 1 and 2}}}{-\,(-j7)\mathbf{I}_2} \quad \underset{\substack{\uparrow \\ \text{net source voltage} \\ \text{aiding } \mathbf{I}_1}}{= 10}$$

and similarly,

$$-(-j7)\mathbf{I}_1 + (3 + j4 - j7)\mathbf{I}_2 = 0$$

Rewriting these two equations yields

$$(7 - j7)\mathbf{I}_1 + j7\mathbf{I}_2 = 10$$
$$j7\mathbf{I}_1 + (3 - j3)\mathbf{I}_2 = 0$$

Here

$$\Delta = \begin{vmatrix} (7 - j7) & j7 \\ j7 & (3 - j3) \end{vmatrix} = (7 - j7)(3 - j3) - (j7)(j7)$$

$$= 21 - 21 - j21 - j21 + 49$$

$$= 49 - j42 = 64.54\underline{/-40.6^\circ}$$

(Notice the symmetry around the principal diagonal in the mesh equations above.) Now, applying Cramer's rule, we have

$$\mathbf{I}_1 = \frac{\begin{vmatrix} 10 & j7 \\ 0 & 3 - j3 \end{vmatrix}}{\Delta} = \frac{10(3 - j3)}{\Delta} = \frac{42.43\underline{/-45^\circ}}{64.54\underline{/-40.6^\circ}} = 0.657\underline{/-4.4^\circ}\ \text{A}$$

and

$$\mathbf{I}_2 = \frac{\begin{vmatrix} 7 - j7 & 10 \\ j7 & 0 \end{vmatrix}}{\Delta} = \frac{-10 \times j7}{\Delta} = \frac{70\underline{/-90^\circ}}{64.54\underline{/-40.6^\circ}} = 1.085\underline{/-49.4^\circ}\ \text{A}$$

Hence

$$\text{source power} = EI_1 \cos\theta_{EI_1} = 10 \times 0.657 \cos(-4.4^\circ)$$
$$= 6.55\ \text{W}$$
$$P_{7\Omega} = |I_1|^2 \times 7 = (0.657)^2 \times 7 = 3.02\ \text{W}$$
$$P_{3\Omega} = |I_2|^2 \times 3 = (1.085)^2 \times 3 = 3.53\ \text{W}$$

As a check, all the power delivered by the source is dissipated by the two resistors in the network ($P_{\text{source}} = P_{7\Omega} + P_{3\Omega}$), as the inductor and the capacitor do not dissipate any power.

EXAMPLE 15.2

Write the mesh equations for the circuit shown in Fig. 15.9 and then calculate the current in the 10 Ω resistor.

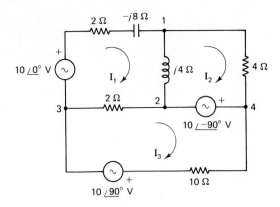

Figure 15.9

Solution Obviously, this network contains three meshes. Also, it has four principal nodes and six branches (i.e., $n = 6 - 4 + 1 = 3$, as expected). The mesh currents are assumed to flow in the clockwise direction, as indicated.

Following the systematic procedure of writing the mesh equations, we have

$$(2 + 2 + j4 - j9)\mathbf{I}_1 - j4\mathbf{I}_2 - 2\mathbf{I}_3 = 10$$
$$-j4\mathbf{I}_1 + (4 + j4)\mathbf{I}_2 - 0\mathbf{I}_3 = -10\underline{/-90°} = j10$$
$$-2\mathbf{I}_1 - 0\mathbf{I}_2 + (2 + 10)\mathbf{I}_3 = 10\underline{/-90°} - 10\underline{/90°}$$
$$= -j10 - j10 = -j20$$

Note that there are no impedances common between meshes 2 and 3; that is why $Z_{23} = Z_{32} = 0$. Simplifiying the equations above yields

$$(4 - j4)\mathbf{I}_1 - j4\mathbf{I}_2 - 2\mathbf{I}_3 = 10$$
$$-j4\mathbf{I}_1 + (4 + j4)\mathbf{I}_2 - 0\mathbf{I}_3 = j10$$
$$-2\mathbf{I}_1 - 0\mathbf{I}_2 + 12\mathbf{I}_3 = -j20$$

Observe the symmetry around the principal diagonal. Only the current $\mathbf{I}_3$ needs to be determined, since it is the current in the 10 Ω resistor. Using Cramer's Rule, one can write:

$$\mathbf{I}_3 = \frac{\begin{vmatrix} 4 - j4 & -j4 & 10 \\ -j4 & 4 + j4 & j10 \\ -2 & 0 & -j20 \end{vmatrix}}{\begin{vmatrix} 4 - j4 & -j4 & -2 \\ -j4 & 4 + j4 & 0 \\ -2 & 0 & 12 \end{vmatrix}}$$

$$= \frac{10[2(4 + j4)] - j10[-j8] - j20[(4 - j4)(4 + j4) + 16]}{-2[2(4 + j4)] + 12[(4 - j4)(4 + j4) + 16]}$$

$$= \frac{80 + j80 - 80 - j20 \times 48}{-16 - j16 + 12 \times 48} = \frac{-j880}{560 - j16}$$

$$= \frac{880/-90°}{560.23/-1.6°} = 1.57/-88.4° \text{ A}$$

The numerator and denominator determinants are both expanded in terms of the elements of the third column and their corresponding cofactors.

EXAMPLE 15.3

The circuit shown in Fig. 15.10 is a typical three-phase system feeding a Δ-connected load. Calculate the three line currents I_A, I_B, and I_C.

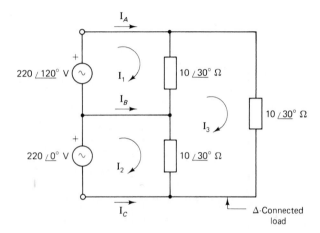

Figure 15.10

Solution For the three meshes of this network, the three mesh currents are chosen as indicated. The three mesh equations can then be written as below, noting that $Z_{12} = Z_{21} = 0$ Ω.

$$10/30° \, I_1 - 0I_2 - 10/30° \, I_3 = 220/120°$$
$$-0I_1 + 10/30° \, I_2 - 10/30° \, I_3 = 220/0°$$
$$-10/30° \, I_1 - 10/30° \, I_2 + 30/30° \, I_3 = 0$$

Here the system determinant is

$$\Delta = \begin{vmatrix} 10/30° & 0 & -10/30° \\ 0 & 10/30° & -10/30° \\ -10/30° & -10/30° & 30/30° \end{vmatrix}$$

$$= 10/30° \, [300/60° - 100/60°] - 10/30° \, [100/60°]$$

$$= 2000/90° - 1000/90° = 1000/90°$$

To determine the line currents, I_1 and I_2 need to be calculated. However, I_3 is not really needed here unless we also have to calculate the currents in the load impedances. Using Cramer's rule yields

$$\mathbf{I}_1 = \frac{\begin{vmatrix} 220\underline{/120^\circ} & 0 & -10\underline{/30^\circ} \\ 220 & 10\underline{/30^\circ} & -10\underline{/30^\circ} \\ 0 & -10\underline{/30^\circ} & 30\underline{/30^\circ} \end{vmatrix}}{\Delta}$$

$$= \frac{220\underline{/120^\circ}\,[300\underline{/60^\circ} - 100\underline{/60^\circ}] - 220[-100\underline{/60^\circ}]}{1000\underline{/90^\circ}}$$

$$= \frac{44{,}000\underline{/180^\circ} + 22{,}000\underline{/60^\circ}}{1000\underline{/90^\circ}} = 44\underline{/90^\circ} + 22\underline{/-30^\circ}$$

$$= j44 + (19.05 - j11) = 19.05 + j33 = 38.1\underline{/60^\circ}\ \text{A}$$

and

$$\mathbf{I}_2 = \frac{\begin{vmatrix} 10\underline{/30^\circ} & 220\underline{/120^\circ} & -10\underline{/30^\circ} \\ 0 & 220 & -10\underline{/30^\circ} \\ -10\underline{/30^\circ} & 0 & 30\underline{/30^\circ} \end{vmatrix}}{\Delta}$$

$$= \frac{-220\underline{/120^\circ}\,[-100\underline{/60^\circ}] + 220[300\underline{/60^\circ} - 100\underline{/60^\circ}]}{1000\underline{/90^\circ}}$$

$$= \frac{22{,}000\underline{/180^\circ} + 44{,}000\underline{/60^\circ}}{1000\underline{/90^\circ}} = 22\underline{/90^\circ} + 44\underline{/-30^\circ}$$

$$= j22 + (38.1 - j22) = 38.1\underline{/0^\circ}\ \text{A}$$

From Fig. 15.10, it is clear that

$$\mathbf{I}_A = \mathbf{I}_1 = 38.1\underline{/60^\circ}\ \text{A}$$
$$\mathbf{I}_B = \mathbf{I}_2 - \mathbf{I}_1 = 38.1 - (19.05 + j33)$$
$$= 19.05 - j33 = 38.1\underline{/-60^\circ}\ \text{A}$$

and

$$\mathbf{I}_C = -\mathbf{I}_2 = -38.1\underline{/0^\circ} = 38.1\underline{/180^\circ}\ \text{A}$$

EXAMPLE 15.4

In the circuit shown in Fig. 15.11, find the value of $\mathbf{V}_2$ such that the current in the $(3 - j2)\ \Omega$ impedance is zero.

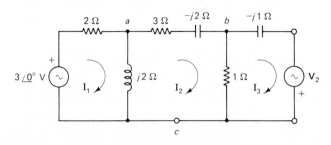

Figure 15.11

Solution Assuming the three mesh currents in the clockwise direction as indicated in Fig. 15.11, the three mesh equations can then be written as follows:

$$(2 + j2)\mathbf{I}_1 - j2\mathbf{I}_2 + 0\mathbf{I}_3 = 3$$

$$-j2\mathbf{I}_1 + (3 + 1 + j2 - j2)\mathbf{I}_2 - 1\mathbf{I}_3 = 0$$

$$-0\mathbf{I}_1 - 1\mathbf{I}_2 + (1 - j1)\mathbf{I}_3 = \mathbf{V}_2$$

It is required here that the current in the $(3 - j2)$ Ω impedance, which is $\mathbf{I}_2$, be zero. Using Cramer's rule, we have

$$\mathbf{I}_2 = \frac{\begin{vmatrix} 2 + j2 & 3 & 0 \\ -j2 & 0 & -1 \\ 0 & \mathbf{V}_2 & 1 - j1 \end{vmatrix}}{\Delta} = 0$$

Thus

$$-3[-j2(1 - j1)] - \mathbf{V}_2[-1(2 + j2)] = 0$$

and

$$\mathbf{V}_2 = \frac{j6(1 - j1)}{-(2 + j2)} = \frac{6 + j6}{-(2 + j2)} = -3 = 3\underline{/180°}\ \text{V}$$

As a check, this result could also be obtained by observing that when the current in the $(3 - j2)$ Ω impedance is zero,

$$\mathbf{V}_{ab} = 0 = \mathbf{V}_{ac} - \mathbf{V}_{bc}$$

This also indicates that there is effectively an open circuit between the first and the third mesh (i.e., they are independent of one another). Thus

$$\mathbf{V}_{ac} = \mathbf{V}_{bc}$$

Using the voltage-division principle, we have

$$3\frac{j2}{2 + j2} = -\mathbf{V}_2\frac{1}{1 - j1}$$

or

$$\mathbf{V}_2 = -3 \times \frac{j2(1 - j1)}{2 + j2} = -3 \times \frac{2 + j2}{2 + j2} = -3 = 3\underline{/180°}\ \text{V}$$

as obtained above.

EXAMPLE 15.5

Find the current in the 2 Ω resistor in the circuit shown in Fig. 15.12.

Solution As can easily be observed, this circuit has five meshes. It has five principal nodes, as marked, and nine branches (i.e., $n = 9 - 5 + 1 = 5$). This means that five loop (or mesh) equations are required to solve it. However, the two current sources can be converted to their equivalent voltage source

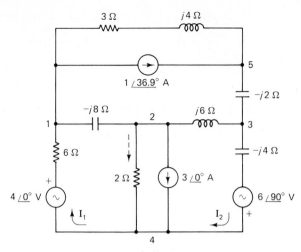

Figure 15.12

models to reduce the number of meshes, as shown in Fig. 15.13. Here

$$\mathbf{E}_1 = 1\underline{/36.9°} \times (3 + j4)$$
$$= 1\underline{/36.9°} \times 5\underline{/53.1°} = 5\underline{/90°} \text{ V}$$

and

$$\mathbf{E}_2 = 3\underline{/0°} \times 2 = 6\underline{/0°} \text{ V}$$

while the polarities of these sources are as indicated in Fig. 15.13.

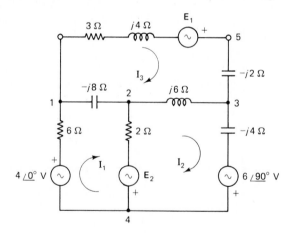

Figure 15.13

The three mesh currents for the equivalent circuit in Fig. 15.13 are assumed in the clockwise direction, as shown. By inspection, the three required mesh equations can easily be written following the systematic procedure developed above. Thus

$$(6 - j8 + 2)\mathbf{I}_1 - 2\mathbf{I}_2 - (-j8)\mathbf{I}_3 = 4 + \mathbf{E}_2 = 4 + 6 = 10$$
$$-2\mathbf{I}_1 + (2 + j6 - j4)\mathbf{I}_2 - j6\mathbf{I}_3 = 6\underline{/90°} - \mathbf{E}_2 = -6 + j6$$
$$-(-j8)\mathbf{I}_1 - j6\mathbf{I}_2 + (3 + j4 - j2 + j6 - j8)\mathbf{I}_3 = \mathbf{E}_1 = 5\underline{/90°} = j5$$

Simplifying the equations above, we obtain:

$$(8 - j8)\mathbf{I}_1 - 2\mathbf{I}_2 + j8\mathbf{I}_3 = 10$$
$$-2\mathbf{I}_1 + (2 + j2)\mathbf{I}_2 - j6\mathbf{I}_3 = -6 + j6$$
$$j8\mathbf{I}_1 - j6\mathbf{I}_2 + 3\mathbf{I}_3 = j5$$

The symmetry around the principal diagonal is clearly observed in the set of equations above. Here

$$\Delta = \begin{vmatrix} 8 - j8 & -2 & j8 \\ -2 & (2 + j2) & -j6 \\ j8 & -j6 & 3 \end{vmatrix}$$
$$= (8 - j8)(6 + j6 + 36) + 2(-6 - 48) + j8(j12 - j16 + 16)$$
$$= (8 - j8)(42 + j6) - 108 + j8(16 - j4)$$
$$= 336 + 48 - j336 + j48 - 108 + j128 + 32 = 308 - j160$$
$$= 374.08\underline{/-27.5°}$$

Applying Cramer's rule, we have

$$\mathbf{I}_1 = \frac{\begin{vmatrix} 10 & -2 & j8 \\ -6 + j6 & 2 + j2 & -j6 \\ j5 & -j6 & 3 \end{vmatrix}}{\Delta}$$

$$= \frac{10(6 + j6 + 36) + 2(-18 + j18 - 30) + j8(j36 + 36 - j10 + 10)}{\Delta}$$

$$= \frac{420 + j60 - 96 + j36 - 208 + j368}{\Delta} = \frac{116 + j464}{\Delta}$$

$$= \frac{478.28\underline{/76°}}{347.08\underline{/-27.5°}} = 1.378\underline{/103.5°} \text{ A}$$

and

$$\mathbf{I}_2 = \frac{\begin{vmatrix} 8 - j8 & 10 & j8 \\ -2 & -6 + j6 & -j6 \\ j8 & j5 & 3 \end{vmatrix}}{\Delta}$$

$$= \frac{(8 - j8)(-18 + j18 - 30) + 2(30 + 40) + j8(-j60 + j48 + 48)}{\Delta}$$

$$= \frac{-384 + j384 + j144 + 144 + 140 + 96 + j384}{\Delta} = \frac{-4 + j912}{\Delta}$$

$$= \frac{912.01\underline{/90.2°}}{347.08\underline{/-27.5°}} = 2.628\underline{/117.7°} \text{ A}$$

$\mathbf{I}_3$ is not required to find the current in the 2 Ω resistor. This current cannot be found from the equivalent circuit in Fig. 15.13 (i.e., it is not equal to $\mathbf{I}_1 - \mathbf{I}_2$) because of the source conversion involving this 2 Ω resistor. Referring to the original circuit in Fig. 15.12, it is clear that $\mathbf{I}_1$ is the current in the $4\underline{/0°}$ V source and $\mathbf{I}_2$ is the current in the $6\underline{/90°}$ V source. Now applying KCL at node 4, we have

$$\mathbf{I}_{2\Omega} + 3\underline{/0°} + \mathbf{I}_2 = \mathbf{I}_1 \quad (\mathbf{I}_{2\Omega} \text{ is assumed to flow downward})$$

Therefore,

$$\begin{aligned}
\mathbf{I}_{2\Omega} &= \mathbf{I}_1 - \mathbf{I}_2 - 3 \\
&= -0.322 + j1.34 - (-1.222 + j2.333) - 3 \\
&= -2.1 - j0.99 = 2.32\underline{/-154.8°} \text{ A}
\end{aligned}$$

15.3 NODE VOLTAGE ANALYSIS

This technique of network analysis is based on writing KCL at each of the principal nodes of a given network. The unknowns in such cases are the voltages at each of the nodes with respect to a reference node (datum or ground). This reference node is usually chosen to be the one that has the largest number of branches connected to it. The network shown in Fig. 15.14 has three principal nodes as indicated, while nodes a and b are not principal nodes. Each of nodes 1 and 2 has three branches connected to it, while node 3 has four branches connected to it. It is also the common node between the two voltage sources. Thus node 3 is chosen to be the reference node. The ground symbol is usually connected to this reference node, even though it might not be physically connected to ground. By so doing, one forces the voltage at this reference node to be zero, thus simplifying the solution. Once the voltages at the other nodes, 1 and 2, are obtained with respect to this reference, the network is said to be solved. All the currents in, voltages across and power dissipated in any of the elements of the network can then easily be calculated, simply by making use of Ohm's law.

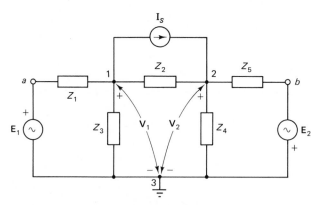

Figure 15.14

The node voltages obtained by this analysis technique (e.g., $\mathbf{V}_1$ and $\mathbf{V}_2$ in Fig. 15.14) are in fact the potential differences between the nodes in question and the reference node. Thus even if the reference node is not in reality at the ground potential (but at a finite voltage $\mathbf{V}_3$), these potential differences do not change. They are the exact and correct solutions. The only change that should be considered if the reference node is not at the ground potential is that the proper voltages at the different nodes (with respect to the real ground) are obtained by adding algebraically the node potential differences (e.g., V_1 or V_2) to the real voltage of the reference node (e.g., V_3). This, however, does not change any current in or any potential difference across any element in the network.

Since the reference node is considered to be at 0 V in the nodal analysis, the number of unknowns and also the number of necessary node voltage equations is $(N-1)$, where N is the number of principal nodes. This result was expressed in Eq. (15.2). In the network shown in Fig. 15.14, N is 3. Thus two node voltage equations (using KCL) need to be written in terms of the two unknown voltages V_1 and V_2, being the voltages at the two principal nodes with respect to the reference (ground voltage) node.

Before writing KCL for each of the principal nodes, it is recommended that each voltage source and its series-connected impedance (if any) be converted to an equivalent practical current source model. Even though this step is not necessary, it has the advantage of facilitating the development of an easy, systematic, and shortcut method of writing the required node equations by inspection; similar to the technique developed for the loop current method (mesh analysis).

As developed in Chapter 4, any practical voltage source model (i.e., an ideal voltage source in series with an impedance) can be converted to a mathematically equivalent practical current source model, as far as the rest of the network is concerned.

In the network of Fig. 15.14, E_1 and Z_1 and E_2 and Z_5, being the practical voltage sources, can be converted to the corresponding practical current sources, as shown in the equivalent circuit of Fig. 15.15.

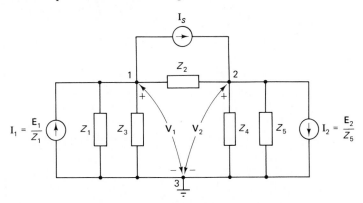

Figure 15.15

I_1 ($= E_1/Z_1$) in parallel with Z_1 replaces the first voltage source between nodes 1 and 3; similarly, I_2 ($= E_2/Z_5$) in parallel with Z_5 replaces the second voltage source between nodes 2 and 3. Careful consideration must be exercised to determine the correct directions of the equivalent current source, since this could be the cause of many errors in the network solution. Notice also that nodes a and b are eliminated from the equivalent circuit of Fig. 15.15.

In writing KCL at any particular node, we will encounter situations such as that illustrated in Fig. 15.16. Since the current in the impedance Z is unknown, it is equally correct to assume that it will flow from node 1 to node 2 (I) or in the

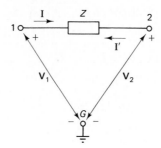

Figure 15.16

reverse direction ($\mathbf{I}'$). Clearly,

$$\mathbf{I}' = -\mathbf{I} \tag{15.39}$$

The correct result depends on the sign of the actual answer obtained. Writing KVL for the closed path 12G or 21G, we have

$$\mathbf{I}Z + \mathbf{V}_2 - \mathbf{V}_1 = 0 \tag{15.40a}$$

or

$$\mathbf{I}'Z + \mathbf{V}_1 - \mathbf{V}_2 = 0 \tag{15.40b}$$

Therefore:

$$\mathbf{I} = \text{current leaving node 1}$$

$$= \text{current entering node 2} = \frac{\mathbf{V}_1 - \mathbf{V}_2}{Z} \tag{15.41a}$$

or

$$\mathbf{I}' = \text{current entering node 1}$$

$$= \text{current leaving node 2} = \frac{\mathbf{V}_2 - \mathbf{V}_1}{Z} \tag{15.41b}$$

Either way of expressing currents at a node (entering or leaving) can be used in writing KCL at a node; that is, both forms of Eq. (15.41) are correct and algebraically equivalent, since Eq. (15.39) is satisfied. This is clear from KCL, as

algebraic sum of currents leaving a node

$$= \text{algebraic sum of currents entering this node} \tag{15.42}$$

A term in the left-hand side (LHS) of Eq. (15.42) expressed as leaving the node will acquire a negative sign when transferred to the RHS of the equation, equivalent to expressing this same current as entering this node.

To develop the systematic shortcut procedure for nodal analysis, all the *currents flowing in elements* connected to a node are expressed as *leaving* the node and their sum is equated to the algebraic sum of the *current sources entering* this node; the current source is taken with a positive sign when entering the node and with a negative sign when leaving the node (obviously, a zero value is used when there are no current sources connected to the node in question). Thus for nodes 1 and 2 in Fig. 15.15,

$$\frac{\mathbf{V}_1}{Z_1} + \frac{\mathbf{V}_1}{Z_3} + \frac{\mathbf{V}_1 - \mathbf{V}_2}{Z_2} = \mathbf{I}_1 - \mathbf{I}_S \tag{15.43}$$

and

$$\frac{V_2}{Z_5} + \frac{V_2}{Z_4} + \frac{V_2 - V_1}{Z_2} = I_S - I_2 \tag{15.44}$$

Combining terms and using the admittances of elements, $Y\ (= 1/Z)$, instead of the impedances, Eqs. (15.43) and (15.44) can be rewritten as follows:

$$(Y_1 + Y_2 + Y_3)V_1 - Y_2 V_2 = I_1 - I_S \tag{15.45}$$

and

$$-Y_2 V_1 + (Y_2 + Y_4 + Y_5)V_2 = I_S - I_2 \tag{15.46}$$

Before going on with the analysis procedure, it is interesting to confirm that Eqs. (15.43) and (15.44) could have been also obtained from the original circuit of Fig. 15.14 without source conversions. Here KCL applied at each of nodes 1 and 2 results in

$$\frac{V_1 - E_1}{Z_1} + \frac{V_1}{Z_3} + \frac{V_1 - V_2}{Z_2} + I_S = 0$$

that is,

$$\frac{V_1}{Z_1} + \frac{V_1}{Z_3} + \frac{V_1 - V_2}{Z_2} = \frac{E_1}{Z_1} - I_S = I_1 - I_S$$

as in Eq. (15.43). Similarly,

$$\frac{V_2 - (-E_2)}{Z_5} + \frac{V_2}{Z_4} + \frac{V_2 - V_1}{Z_2} = I_S$$

that is,

$$\frac{V_2}{Z_5} + \frac{V_2}{Z_4} + \frac{V_2 - V_1}{Z_2} = I_S - \frac{E_2}{Z_5} = I_S - I_2$$

as in Eq. (15.44). Note that

$\quad$ V_a (voltage at node a with respect to node 3, or ground) = E_1

$\quad$ V_b (voltage at node b with respect to node 3, or ground) = $-E_2$

In general, then, Eqs. (15.45) and (15.46) can be written in the form

$$Y_{11}V_1 - Y_{12}V_2 = I_{11} \tag{15.47}$$
$$-Y_{21}V_1 + Y_{22}V_2 = I_{22} \tag{15.48}$$

It is easy to extend this general form to a network having four principal nodes (i.e., $n = N - 1 = 3$) and to write the three node equations for the three node voltages, V_1, V_2, and V_3 as

$$Y_{11}V_1 \quad - \quad Y_{12}V_2 \quad - \quad Y_{13}V_3 \quad = \quad I_{11} \tag{15.49}$$
$$-Y_{21}V_1 \quad + \quad Y_{22}V_2 \quad - \quad Y_{23}V_3 \quad = \quad I_{22} \tag{15.50}$$
$$-Y_{31}V_1 \quad - \quad Y_{32}V_2 \quad + \quad Y_{33}V_3 \quad = \quad I_{33} \tag{15.51}$$

Principal diagonal

where Y_{ii} = total or self-admittance of all the branches connected to node i.

$\quad Y_{ij} = Y_{ji}$ = common or mutual admittance of the branch connected between nodes i and j

$\quad \mathbf{V}_i$ = node voltage at node i with respect to the reference node

$\quad \mathbf{I}_{ii}$ = algebraic sum of all the current sources entering node i

Either set of linear node voltage equations, (15.47) and (15.48) or (15.49), (15.50), and (15.51) can be expressed by the general matrix equation

$$[Y][\mathbf{V}] = [\mathbf{I}] \qquad (15.52)$$

where

$$[Y] = \begin{bmatrix} Y_{11} & -Y_{12} & -Y_{13} \\ -Y_{21} & Y_{22} & -Y_{23} \\ -Y_{31} & -Y_{32} & Y_{33} \end{bmatrix}$$

an $n \times n$ system matrix; n is the number of equations

$$[\mathbf{V}] = \begin{bmatrix} \mathbf{V}_1 \\ \mathbf{V}_2 \\ \mathbf{V}_3 \end{bmatrix}$$

an n-column matrix representing the node voltages

and

$$[\mathbf{I}] = \begin{bmatrix} \mathbf{I}_{11} \\ \mathbf{I}_{22} \\ \mathbf{I}_{33} \end{bmatrix}$$

an n-column matrix representing the algebraic sum or the source currents entering each of the nodes

15.3.1 Important Remarks

These remarks are intended to clarify the understanding of how the systematic procedure for writing the node voltage equations is developed and to check for possible errors in writing down these equations.

1. Similar to the mesh analysis technique, the nodal analysis results in a set of linear algebraic equations in terms of the n unknown node voltages. The coefficients of the voltages in this set of equations are admittances. They are always symmetrically equivalent around the principal diagonal, i.e., $Y_{12} = Y_{21}$ (admittance of the branch connected between nodes 1 and 2). Similarly, $Y_{13} = Y_{31}$, $Y_{23} = Y_{32}$, and so on. Here all these mutual admittance terms have negative signs.

2. All the terms constituting the principal diagonal are always expressed with positive signs. Each of these terms is the coefficient of the node voltage $\mathbf{V}_i$ in the node voltage equation for node i. They represent the total or self-admittances of each of the nodes. To obtain these terms correctly, the simple systematic procedure described below can be followed. The self-admittance of node i, Y_{ii}, is obtained by first eliminating the sources in the network [i.e., making them dead (voltage sources replaced by a short circuit and current sources replaced by an open circuit)]. Then assume that all the other nodes in the network, except node i in question of course, are temporarily at the ground or reference node potential. Add all the admittances of all the branches

connected between node i and all the other "temporary" grounds. This sum is the self- or total admittance of node i. Notice that the admittance of a branch between two "temporary" grounds does not contribute to the node's self-admittance because it is equivalent to a short circuit when the voltage across it is zero.

3. The RHS of the node equations always represents the algebraic sum of all the current sources entering the node in question. If a current source is actually in a dirrection leaving the node in question, it must be taken with a negative sign.

4. It is always preferable to convert all the practical voltage sources in a given network to their equivalent current source model before starting the nodal analysis. If this is not possible for a certain ideal voltage source, as for example source $\mathbf{E}$ in Fig. 15.17 because there is no impedance in series with the source between node i and j, the solution is slightly more complicated. Here node i has a voltage $\mathbf{V}_i$ and node j has a voltage $(\mathbf{V}_i + \mathbf{E})$. Write only one node equation for these two nodes in terms of $\mathbf{V}_i$. Such a KCL (for both nodes combined) would be in the form

$$\text{sum of currents leaving node } i = \mathbf{I}'$$
$$= -\text{sum of currents leaving node } j$$

or

$$\text{sum of currents leaving node } i$$
$$+ \text{ sum of currents leaving node } j = 0 \qquad (15.53)$$

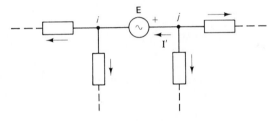

Figure 15.17

5. The nodal equations obtained following the systematic procedure outlined above can be solved in general using Cramer's rule and determinants, in exactly the same manner as explained for the mesh analysis method. Once the solutions for the different node voltages are obtained, any of the elements' currents or powers can easily be derived. It is important to note that a negative answer for a certain node voltage only means that this node is lower in potential than the voltage of the reference node, which is considered to be the ground or zero voltage datum.

■ EXAMPLE 15.6

Write the node voltage equations for the network shown in Fig. 15.18; then calculate the value of the current in the 10 Ω resistor.

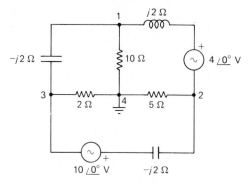

Figure 15.18

Solution This network has four principal nodes, as marked in the figure. Thus three node equations are required in terms of three node voltages. Node 4 is chosen to be the reference (or ground) node because this will facilitate calculation of the current in the 10 Ω resistor, since all that is then needed is the value of $\mathbf{V}_1$, the voltage at node 1, with respect to node 4 (ground).

The two voltage sources are first converted to their equivalent current source models as shown in Fig. 15.19. Following the procedure described above, the three node equations can be written as follows:
For node 1,

$$\left(\frac{1}{10} - \frac{1}{-j2} + \frac{1}{j2}\right)\mathbf{V}_1 \quad - \quad \left(\frac{1}{j2}\right)\mathbf{V}_2 \quad - \quad \left(\frac{1}{-j2}\right)\mathbf{V}_3 = -j2$$

- ↑ self-admittance of node 1
- ↑ common admittance between nodes 1 and 2
- ↑ common admittance between nodes 1 and 3
- ↖ $\mathbf{I}_2$, current source entering node 1

Similarly, for nodes 2 and 3,

$$-\left(\frac{1}{j2}\right)\mathbf{V}_1 + \left(\frac{1}{5} + \frac{1}{j2} + \frac{1}{-j2}\right)\mathbf{V}_2 - \left(\frac{1}{-j2}\right)\mathbf{V}_3 = \mathbf{I}_1 - \mathbf{I}_2$$

$$= j5 - (-j2)$$

Figure 15.19

and

$$-\left(\frac{1}{-j2}\right)\mathbf{V}_1 - \left(\frac{1}{-j2}\right)\mathbf{V}_2 + \left(\frac{1}{2} + \frac{1}{-j2} + \frac{1}{-j2}\right)\mathbf{V}_3 = -\mathbf{I}_1 = -j5$$

Simplifying the equations above yields

$$0.1\mathbf{V}_1 + j0.5\mathbf{V}_2 - j0.5\mathbf{V}_3 = -j2$$
$$j0.5\mathbf{V}_1 + 0.2\mathbf{V}_2 - j0.5\mathbf{V}_3 = j7$$
$$-j0.5\mathbf{V}_1 - j0.5\mathbf{V}_2 + (0.5 + j1)\mathbf{V}_3 = -j5$$

Note the symmetry around the principal diagonal. Cramer's rule can now be used to calculate $\mathbf{V}_1$:

$$\mathbf{V}_1 = \frac{\begin{vmatrix} -j2 & j0.5 & -j0.5 \\ j7 & 0.2 & -j0.5 \\ -j5 & -j0.5 & (0.5 + j1) \end{vmatrix}}{\begin{vmatrix} 0.1 & j0.5 & -j0.5 \\ j0.5 & 0.2 & -j0.5 \\ -j0.5 & -j0.5 & (0.5 + j1) \end{vmatrix}}$$

$$= \frac{-j2[0.1 + j0.2 + 0.25] - j7[j0.25 - 0.5 + 0.25] - j5[0.25 + j0.1]}{0.1[0.1 + j0.2 + 0.25] - j0.5[j0.25 - 0.5 + 0.25] - j0.5[0.25 + j0.1]}$$

$$= \frac{-j0.7 + 0.4 + 1.75 + j1.75 - j1.25 + 0.5}{0.035 + j0.02 + 0.125 + j0.125 - j0.125 + 0.05} = \frac{2.65 - j0.2}{0.21 + j0.02}$$

$$= \frac{2.658\underline{/-4.3°}}{0.211\underline{/5.4°}} = 12.6\underline{/-9.7°}\ \text{V}$$

From Fig. 15.18, using Ohm's law, we have

$$\mathbf{I}_{10\,\Omega} = \frac{\mathbf{V}_1}{10} = 1.26\underline{/-9.7°}\ \text{A}$$

EXAMPLE 15.7

The two equal capacitors and the resistor in the bridged-T network shown in Fig. 15.20 are varied until the current in the detector, Z_D, is zero. For this

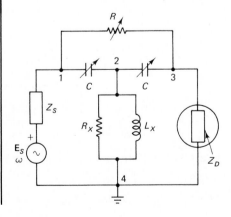

Figure 15.20

balance condition, determine the values of the unknowns L_X and R_X in terms of the other circuit components.

Solution This network has four principal nodes, marked as shown in Fig. 15.20. Node 4 is chosen to be the reference (or ground) node since the largest number of branches are connected to it. Three node equations should then be written in terms of the three node voltages, $\mathbf{V}_1$, $\mathbf{V}_2$, and $\mathbf{V}_3$. At the balance condition, the voltage across Z_D, $\mathbf{V}_3$, should be zero since the current in Z_D is zero. The frequency of the source is ω.

The circuit is redrawn in Fig. 15.21, with the voltage source converted into its equivalent current source model and the elements expressed in terms of their admittances. Following the systematic procedure, the three node voltage equations can be written by inspection as follows:

$$\left(\frac{1}{Z_S} + \frac{1}{R} + j\omega C\right)\mathbf{V}_1 - j\omega C\mathbf{V}_2 - \frac{1}{R}\mathbf{V}_3 = \frac{\mathbf{E}_S}{Z_S}$$

$$-j\omega C\mathbf{V}_1 + \left(\frac{1}{R_X} + \frac{1}{j\omega L_X} + 2j\omega C\right)\mathbf{V}_2 - j\omega C\mathbf{V}_3 = 0$$

$$-\frac{1}{R}\mathbf{V}_1 - j\omega C\mathbf{V}_2 + \left(\frac{1}{Z_D} + \frac{1}{R} + j\omega C\right)\mathbf{V}_3 = 0$$

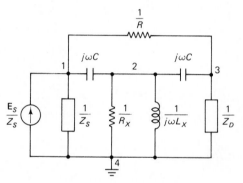

Figure 15.21

Note that there are no current sources connected to nodes 2 and 3. Here $\mathbf{V}_3$ has to be determined and equated to zero to achieve the balance condition.

$$\mathbf{V}_3 = \frac{\begin{vmatrix} \dfrac{1}{Z_S} + \dfrac{1}{R} + j\omega C & -j\omega C & \dfrac{\mathbf{E}_S}{Z_S} \\[2mm] -j\omega C & \dfrac{1}{R_X} - j\dfrac{1}{\omega L_X} + j2\omega C & 0 \\[2mm] -\dfrac{1}{R} & -j\omega C & 0 \end{vmatrix}}{\Delta} = 0$$

Expanding the numerator determinant in terms of the third column yields

$$\frac{\mathbf{E}_S}{Z_S}\left[-\omega^2 C^2 + \frac{1}{R}\left(\frac{1}{R_X} - j\frac{1}{\omega L_X} + j2\omega C\right)\right] = 0$$

Therefore,

$$\left(-\omega^2 C^2 + \frac{1}{RR_X}\right) - j\left(\frac{1}{\omega L_X R} - \frac{2\omega C}{R}\right) = 0$$

This complex number is zero when both its real and imaginary parts are zero.

$$-\omega^2 C^2 + \frac{1}{RR_X} = 0$$

and

$$\frac{1}{\omega L_X R} - \frac{2\omega C}{R} = 0$$

Therefore,

$$R_X = \frac{1}{\omega^2 C^2 R} \ \Omega \quad \text{and} \quad L_X = \frac{1}{2\omega^2 C} \ \text{H}$$

EXAMPLE 15.8

Find the voltage $\mathbf{V}_{AB}$ in the circuit shown in Fig. 15.22.

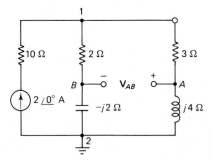

Figure 15.22

Solution This circuit has only two principal nodes. Node 2 is chosen to be the reference node. By writing the KCL at node 1, the node voltage $\mathbf{V}_1$ can be determined. Here

$$\frac{\mathbf{V}_1}{2 - j2} + \frac{\mathbf{V}_1}{3 + j4} = 2\underline{/0°}$$

The RHS is the value of the current source entering node 1. Note that the 10 Ω resistor has no effect on the value of this current source. It has no effect on the self (or total)-admittance of node 1, because when the current source is dead (i.e., an open circuit) this branch has an infinite impedance (i.e., zero admittance). Notice also that there are no mutual admittance terms. Thus

$$\mathbf{V}_1\left(\frac{1}{2.828\underline{/-45°}} + \frac{1}{5\underline{/53.1°}}\right) = 2$$

$$\mathbf{V}_1(0.25 + j0.25 + 0.12 - j0.16) = 2$$

and

$$\mathbf{V}_1 = \frac{2}{0.37 + j0.09} = \frac{2}{0.381\underline{/13.7°}} = 5.25\underline{/-13.7°} \ V$$

Therefore, using the voltage-division principle, we have

$$\mathbf{V}_{AB} = \mathbf{V}_{A2} - \mathbf{V}_{B2} = \mathbf{V}_1 \frac{j4}{3 + j4} - \mathbf{V}_1 \frac{-j2}{2 - j2}$$

$$= \mathbf{V}_1(0.8\underline{/36.9°} + 0.7072\underline{/135°})$$

$$= 5.25\underline{/-13.7°}\,(0.64 + j0.48 - 0.5 + j0.5)$$

$$= 5.25\underline{/-13.7°}\,(0.14 + j0.98)$$

$$= 5.25\underline{/-13.7°} \times 0.99\underline{/81.9°} = 5.2\underline{/68.2°}\ \text{V}$$

EXAMPLE 15.9

Write the nodal equations for the network shown in Fig. 15.23. What should be the value of **E** such that the current in the 4 Ω resistor be zero? What is the value of $\mathbf{V}_1$ in this case?

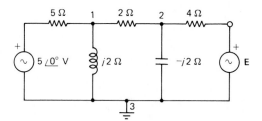

Figure 15.23

Solution   The two practical voltage sources are first converted into their equivalent current source models, resulting in the circuit shown in Fig. 15.24. The two node voltage equations can then easily be written by inspection as follows:

$$\left(\frac{1}{5} + \frac{1}{j2} + \frac{1}{2}\right)\mathbf{V}_1 - \frac{1}{2}\mathbf{V}_2 = 1$$

$$-\frac{1}{2}\mathbf{V}_1 + \left(\frac{1}{2} + \frac{1}{4} + \frac{1}{-j2}\right)\mathbf{V}_2 = \frac{\mathbf{E}}{4}$$

Simplifying the two equations above yields

$$(0.7 - j0.5)\mathbf{V}_1 - 0.5\mathbf{V}_2 = 1$$

$$-0.5\mathbf{V}_1 + (0.75 + j0.5)\mathbf{V}_2 = \frac{\mathbf{E}}{4}$$

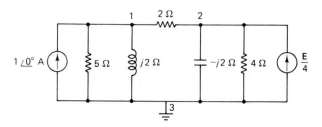

Figure 15.24

Reexamining the original circuit of Fig. 15.23, we see that the current in the $4\,\Omega$ resistor would be zero if $\mathbf{V}_2$ were equal to $\mathbf{E}$ (to produce $\mathbf{V}_{4\Omega} = \mathbf{V}_2 - \mathbf{E} = 0$). Therefore, $\mathbf{V}_2$ should be evaluated and equated to $\mathbf{E}$. From Cramer's rule,

$$\mathbf{V}_2 = \frac{\begin{vmatrix} 0.7 - j0.5 & 1 \\ -0.5 & \dfrac{\mathbf{E}}{4} \end{vmatrix}}{\begin{vmatrix} 0.7 - j0.5 & -0.5 \\ -0.5 & 0.75 + j0.5 \end{vmatrix}} = \mathbf{E}$$

Thus

$$\frac{\mathbf{E}(0.175 - j0.125) + 0.5}{0.525 + 0.25 - j0.375 + j0.35 - 0.25} = \mathbf{E}$$

$$\mathbf{E}(0.175 - j0.125) + 0.5 = \mathbf{E}(0.525 - j0.025)$$

Therefore,

$$\mathbf{E} = \frac{0.5}{0.35 + j0.1} = \frac{0.5}{0.364 \underline{/15.9°}} = 1.374 \underline{/-15.9°}\ \text{V}$$

As $\mathbf{E}$ is now determined, then $\mathbf{E}/4\,\Omega = 0.3435 \underline{/-15.9°} = (0.33 - j0.094)$ A. Again applying Cramer's rule to obtain $\mathbf{V}_1$ from the nodal equations above we obtain

$$\mathbf{V}_1 = \frac{\begin{vmatrix} 1 & -0.5 \\ 0.33 - j0.094 & 0.75 + j0.5 \end{vmatrix}}{\Delta}$$

$$= \frac{0.75 + j0.5 + 0.165 - j0.047}{0.525 - j0.025}$$

$$= \frac{0.915 + j0.453}{0.525 - j0.025} = \frac{1.021 \underline{/26.3°}}{0.5256 \underline{/-2.7°}} = 1.943 \underline{/29°}\ \text{V}$$

EXAMPLE 15.10

Using the systematic shortcut procedure, write the nodal equations for the circuit shown in Fig. 15.25. Find the voltage across the $1 \underline{/0°}$ A current source.

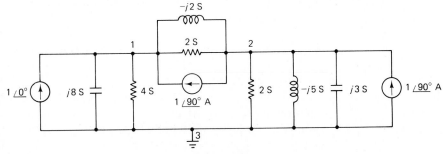

Figure 15.25

Solution The three principal nodes are marked as shown in Fig. 15.25, with node 3 chosen as the reference (ground) node. $\mathbf{V}_1$ is the required voltage across the $1\underline{/0°}$ A current source. Following the systematic shortcut nodal analysis procedure, the two node equations below can be written by inspection (notice that the elements are specified by their admittances):

$$(4 + j8 + 2 - j2)\mathbf{V}_1 - (2 - j2)\mathbf{V}_2 = 1\underline{/0°} + 1\underline{/90°} = 1 + j1$$

$$-(2 - j2)\mathbf{V}_1 + (2 - j2 + 2 - j5 + j3)\mathbf{V}_2 = 1\underline{/90°} - 1\underline{/90°} = 0$$

Simplifying, we have

$$(6 + j6)\mathbf{V}_1 - (2 - j2)\mathbf{V}_2 = 1 + j1$$

$$-(2 - j2)\mathbf{V}_1 + (4 - j4)\mathbf{V}_2 = 0$$

Using Cramer's rule to solve for $\mathbf{V}_1$ gives us

$$\mathbf{V}_1 = \frac{\begin{vmatrix} (1 + j1) & -(2 - j2) \\ 0 & (4 - j4) \end{vmatrix}}{\begin{vmatrix} (6 + j6) & -(2 - j2) \\ -(2 - j2) & (4 - j4) \end{vmatrix}} = \frac{(1 + j1)(4 - j4)}{(6 + j6)(4 - j4) - (2 - j2)(2 - j2)}$$

$$= \frac{8}{48 - (4 - 4 - j4 - j4)} = \frac{8}{48 + j8} = \frac{1}{6 + j1} = 0.164\underline{/-9.5°}\ \text{V}$$

15.4 CHOICE OF THE METHOD OF ANALYSIS

For nonplanar networks, it is usually recommended that the nodal analysis technique be used. This is not because the loop analysis method does not work, but because it is sometimes difficult and confusing to select the proper independent loops.

For planar networks it is usually recommended to choose the technique that results in the least number of system equations (n), as this obviously simplifies and reduces the amount of manipulations required to obtain the solution. Remember that in nodal analysis

$$n = N - 1$$

and in loop analysis

$$n = B - N + 1$$

as examined above.

If both techniques require the same number of equations, the choice of the appropriate analysis method depends on whether the network has mostly voltage sources, in which case loop analysis is more appropriate, or it has mostly current sources, in which case nodal analysis is more appropriate. This choice is made because the required amount of source conversion manipulation is reduced.

EXAMPLE 15.11

Find the current **I** in the network shown in Fig. 15.26.

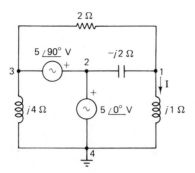

Figure 15.26

Solution   This network has four principal nodes, as marked in Fig. 15.26. It also has six branches. Therefore, the solution requires

$$n = 4 - 1 = 3$$

node voltage equations, or

$$n = 6 - 4 + 1 = 3$$

loop (or mesh) equations. This is clear since the network actually has three meshes.

Initially, it seems that mesh analysis is more appropriate since both methods require the same number of equations and the network contains two voltage sources (actually both are ideal). However, in the nodal analysis, with node 4 chosen as the reference (or ground) point, the voltages at nodes 2 and 3 are actually known:

$$\mathbf{V}_2 = 5\underline{/0°}\ \text{V} = 5\ \text{V}$$

and

$$\mathbf{V}_3 = 5\underline{/0°} - 5\underline{/90°} = 5 - j5 = 7.072\underline{/-45°}\ \text{V}$$

Therefore, there is no need to write nodal equations for these two nodes. All that is needed, then, to solve this network is to write one node equation for node 1. Using KCL at node 1 whose voltage is $\mathbf{V}_1$,

$$\frac{\mathbf{V}_1}{j1} + \frac{\mathbf{V}_1 - \mathbf{V}_2}{-j2} + \frac{\mathbf{V}_1 - \mathbf{V}_3}{2} = 0$$

Substituting for $\mathbf{V}_2$ and $\mathbf{V}_3$ and simplifying gives us

$$-j\mathbf{V}_1 + j0.5(\mathbf{V}_1 - 5) + 0.5(\mathbf{V}_1 - 5 + j5) = 0$$
$$(-j1 + j0.5 + 0.5)\mathbf{V}_1 = j2.5 + 2.5 - j2.5 = 2.5$$

Therefore,

$$\mathbf{V}_1 = \frac{2.5}{0.5 - j0.5} = \frac{5}{1 - j1} = 3.536\underline{/45°}\ \text{V}$$

Hence

$$\mathbf{I} = \frac{\mathbf{V}_1}{j1} = \frac{3.536\underline{/45°}}{1\underline{/90°}} = 3.536\underline{/-45°} \text{ A}$$

15.5 | DEPENDENT SOURCES IN NETWORK ANALYSIS

The voltage and current sources (whether ideal or practical) that have been dealt with so far are called *independent sources*. The values of these sources, that is, the generated voltage $\mathbf{E}_S$ or the generated current $\mathbf{I}_S$ [see Fig. 15.27(a)], do not depend on any other current or voltage in or across any other element in the network external to the source.

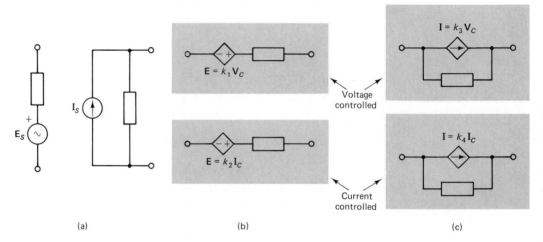

Figure 15.27 (a) Independent sources; (b) dependent voltage sources; (c) dependent current sources.

Many equivalent-circuit models of electrical or electronic devices and transducers (e.g., BJT, FET, etc.) contain what is referred to as *dependent or controlled sources*. They can be voltage or current sources as shown in Fig. 15.27(b) and (c). These dependent or controlled sources are only mathematical models. The voltage or current generated by such sources is controlled by a voltage across some element $\mathbf{V}_C$, or controlled by the current flowing in some element $\mathbf{I}_C$.

When the value of the source is controlled by a voltage ($\mathbf{V}_C$, referred to as the *controlling voltage*) somewhere else in the network, the source is said to be *voltage controlled*. When the value of the source is controlled by a current ($\mathbf{I}_C$, referred to as the *controlling current*) somewhere else in the network, the source is said to be current controlled. This is indicated in Fig. 15.27.

Loop (or mesh) and node analysis techniques also apply to networks containing such sources. Here, however, we have to apply these methods based on an element-by-element consideration, using KVL and KCL. The shortcut systematic procedures and the symmetry observations discussed above do not usually apply in these cases.

Source conversions, from dependent voltage source to dependent current source models, or vice versa, can be used as needed to simplify the networks, *provided that* the controlling element (which provides $\mathbf{V}_C$ or $\mathbf{I}_C$) *is not* part of the source transformation.

■ EXAMPLE 15.12

Find the voltage at node 2, $\mathbf{V}_2$, in the circuit shown in Fig. 15.28.

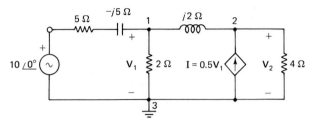

Figure 15.28

Solution This circuit contains a voltage-dependent current source $\mathbf{I}$, where

$$\mathbf{I} = 0.5\mathbf{V}_1$$

There are three nodes and three meshes in this network. Thus nodal analysis should be the preferred method since only two equations are needed.

The practical independent voltage source can first be converted into its current source model, as shown in Fig. 15.29. Here

$$\mathbf{I}_S = \frac{10\underline{/0^\circ}}{5 - j5} = \frac{10\underline{/0^\circ}}{7.072\underline{/-45^\circ}} = 1.414\underline{/45^\circ}\ \mathrm{A}$$

$$= 1 + j1\ \mathrm{A}$$

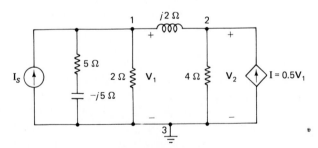

Figure 15.29

Writing KCL at node 1, we have

$$\frac{\mathbf{V}_1}{2} + \frac{\mathbf{V}_1}{5 - j5} + \frac{\mathbf{V}_1 - \mathbf{V}_2}{j2} = 1 + j1$$

while KCL at node 2 results in

$$\frac{\mathbf{V}_2}{4} + \frac{\mathbf{V}_2 - \mathbf{V}_1}{j2} = \mathbf{I} = 0.5\,\mathbf{V}_1$$

Combining the coefficients of the voltages in the two equations above and rearranging yields

$$\mathbf{V}_1(0.5 + 0.1 + j0.1 - j0.5) + \mathbf{V}_2(j0.5) = 1 + j1$$
$$\mathbf{V}_1(-0.5 + j0.5) + \mathbf{V}_2(0.25 - j0.5) = 0$$

No symmetry is observed around the diagonal axis in these two equations. Now applying Cramer's rule to determine $\mathbf{V}_2$, we get

$$\mathbf{V}_2 = \frac{\begin{vmatrix} 0.6 - j0.4 & 1 + j1 \\ -0.5 + j0.5 & 0 \end{vmatrix}}{\begin{vmatrix} 0.6 - j0.4 & j0.5 \\ -0.5 + j0.5 & 0.25 - j0.5 \end{vmatrix}}$$

$$= \frac{-(1 + j1)(-0.5 + j0.5)}{(0.6 - j0.4)(0.25 - j0.5) - j0.5(-0.5 + j0.5)}$$

$$= \frac{0.5 + 0.5 + j0.5 - j0.5}{0.15 - 0.2 - j0.1 - j0.3 + j0.25 + 0.25} = \frac{1}{0.2 - j0.15}$$

$$= 4\underline{/36.9°} \text{ V}$$

EXAMPLE 15.13

The circuit shown in Fig. 15.30 contains a current-dependent voltage source $\mathbf{E}$. Find the value of the current $\mathbf{I}_0$.

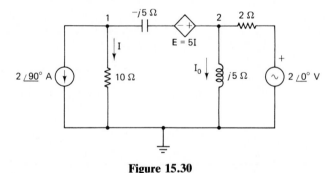

Figure 15.30

Solution If mesh analysis is to be used, the current source $2\underline{/90°}$ A should first be converted into a voltage source. However, this transformation is *not allowable* because the source's parallel resistor (the 10 Ω) is the controlling element of the dependent voltage source $\mathbf{E}$, and should not therefore be part of any source conversion.

Nodal analysis can be applied, starting with the conversion of the two voltage sources into their current source equivalent models, as shown in Fig. 15.31. Notice that the dependent voltage source becomes a dependent current source, $\mathbf{I}_S$:

$$\mathbf{I}_S = \frac{\mathbf{E}}{-j5} = \frac{5\mathbf{I}}{-j5} = j\mathbf{I} = j\frac{\mathbf{V}_1}{10} = j0.1\mathbf{V}_1$$

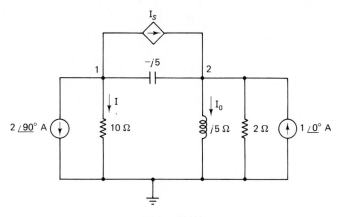

Figure 15.31

At node 1, KCL results in

$$\frac{\mathbf{V}_1}{10} + \frac{\mathbf{V}_1 - \mathbf{V}_2}{-j5} = -2\underline{/90°} - \mathbf{I}_S = -j2 - j0.1\mathbf{V}_1$$

and at node 2, KCL results in

$$\frac{\mathbf{V}_2}{2} + \frac{\mathbf{V}_2}{j5} + \frac{\mathbf{V}_2 - \mathbf{V}_1}{-j5} = 1\underline{/0°} + \mathbf{I}_S = 1 + j0.1\mathbf{V}_1$$

Combining terms and rearranging, the two equations above can be expressed as follows:

$$(0.1 + j0.2 + j0.1)\mathbf{V}_1 - j0.2\mathbf{V}_2 = -j2$$
$$(-j0.2 - j0.1)\mathbf{V}_1 + (0.5 - j0.2 + j0.2)\mathbf{V}_2 = 1$$

Hence

$$(0.1 + j0.3)\mathbf{V}_1 - j0.2\mathbf{V}_2 = -j2$$
$$-j0.3\mathbf{V}_1 + 0.5\mathbf{V}_2 = 1$$

$\mathbf{V}_2$ can then be determined from Cramer's rule:

$$\mathbf{V}_2 = \frac{\begin{vmatrix} 0.1 + j0.3 & -j2 \\ -j0.3 & 1 \end{vmatrix}}{\begin{vmatrix} 0.1 + j0.3 & -j0.2 \\ -j0.3 & 0.5 \end{vmatrix}} = \frac{0.1 + j0.3 + 0.6}{0.05 + j0.15 + 0.06} = \frac{0.7 + j0.3}{0.11 + j0.15}$$

$$= \frac{0.7615\underline{/23.2°}}{0.186\underline{/53.7°}} = 4.094\underline{/-30.5°} \text{ V}$$

Thus

$$\mathbf{I}_0 = \frac{\mathbf{V}_2}{j5} = \frac{4.094\underline{/-30.5°}}{5\underline{/90°}} = 0.819\underline{/-120.5°} \text{ A}$$

15.6 Δ–Y AND Y–Δ CONVERSIONS

Many connections (or arrangements) of electrical loads form the shape of a Δ, which is also called a Π *network* because when the network is inverted it has the shape of a Π, as shown in Fig. 15.32(a). These connections involve three loads, Z_a, Z_b, and Z_c, between three nodes, 1, 2, and 3.

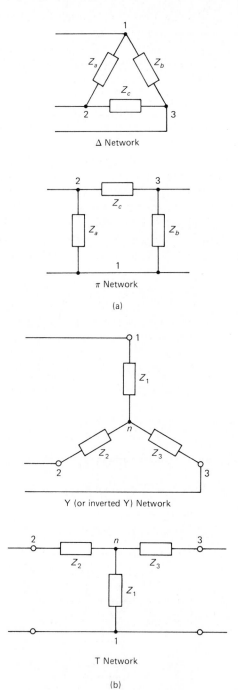

Δ Network

π Network

(a)

Y (or inverted Y) Network

T Network

(b)

Figure 15.32

Similarly, another popular connection of three loads (e.g., Z_1, Z_2, and Z_3 between three nodes, such as nodes 1, 2, and 3) can take the shape of a Y (or an inverted Y) and is called a *Y network*. This type of connection (or arrangement) of loads is also referred to as a *T network* (also due to its shape), as shown in Fig. 15.32(b).

These loads exist quite often in three-phase networks and also in electrical filters, attenuation pads, and matching networks.

As was the objective in many of the topics discussed in network analysis, the intention here is to find a technique to convert from one form of these load connections (Δ or Π) to the other (Y or T), and vice versa, so as to simplify the network and make the task of network analysis easier. Obviously, the need for such conversions depends on the rest of the network's topology and possibly also the connections of the sources in the network.

Thus, if the Δ network is to be equivalent to the Y network, between the same three nodes (or terminals), replacing one by the other should not disturb or change any voltage or current anywhere else in the remainder of the network, external to the converted load. This equivalency, however, does not hold for voltages and currents internal to the Δ or Y networks. In fact, as can be observed from Fig. 15.32, the Δ-to-Y conversion creates an extra internal node, n. The reverse conversion obviously removes this internal node.

This equivalency then requires that the impedance between each pair of nodes (three such pairs) in the Δ network must be the same as that between the same pair of nodes in the Y network. Thus between nodes 1 and 2,

$$Z_1 + Z_2 = \frac{Z_a(Z_b + Z_c)}{Z_a + Z_b + Z_c} \tag{15.54}$$

between nodes 2 and 3,

$$Z_2 + Z_3 = \frac{Z_c(Z_a + Z_b)}{Z_a + Z_b + Z_c} \tag{15.55}$$

and between nodes 3 and 1,

$$Z_1 + Z_3 = \frac{Z_b(Z_a + Z_c)}{Z_a + Z_b + Z_c} \tag{15.56}$$

Subtracting Eq. (15.55) from Eq. (15.56) gives us

$$Z_1 - Z_2 = \frac{Z_a(Z_b - Z_c)}{Z_a + Z_b + Z_c} \tag{15.57}$$

Adding Eqs. (15.54) and (15.57), we obtain

$$Z_1 = \frac{Z_a Z_b}{Z_a + Z_b + Z_c} \tag{15.58}$$

Subtracting Eq. (15.57) from Eq. (15.54), we obtain

$$Z_2 = \frac{Z_a Z_c}{Z_a + Z_b + Z_c} \tag{15.59}$$

Then from Eq. (15.56),

$$Z_3 = -Z_1 + \frac{Z_b(Z_a + Z_c)}{Z_a + Z_b + Z_c} = \frac{Z_b Z_c}{Z_a + Z_b + Z_c} \qquad (15.60)$$

Equations (15.58), (15.59), and (15.60) then show the conversion equations required to convert a Δ-connected load to its equivalent Y-connected form of load connection. A pictorial summary of this conversion process is shown in Fig. 15.33. *Any of the impedances in the equivalent Y-connected load is equal to the product of the two adjacent Δ-connected impedances, divided by the sum of all the three Δ-connected impedances.* For example, if Z_1 is to be calculated, the numerator is the product of Z_a and Z_b, the adjacent impedances in Fig. 15.33, and similarly for the other two elements.

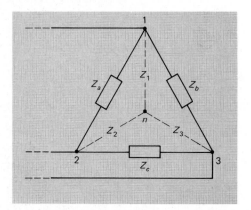

Figure 15.33   Δ-to-Y conversion.

When the three Δ-connected impedances are equal, each is Z_Δ; then the equivalent Y-connected impedances are also equal and each is Z_Y, where

$$Z_Y = \frac{Z_\Delta}{3} \qquad (15.61)$$

as is evident from any of the three conversion equations above.

To develop the relationships required for the conversion of the Y-connected load to its equivalent Δ-connected load, Eqs. (15.58), (15.59), and (15.60) will first be used to construct to the following expression:

$$Z_1 Z_2 + Z_2 Z_3 + Z_3 Z_1 = \frac{Z_a^2 Z_b Z_c + Z_c^2 Z_a Z_b + Z_b^2 Z_a Z_c}{(Z_a + Z_b + Z_c)^2}$$

$$= \frac{Z_a Z_b Z_c (Z_a + Z_b + Z_c)}{(Z_a + Z_b + Z_c)^2} = \frac{Z_a Z_b Z_c}{Z_a + Z_b + Z_c} \qquad (15.62)$$

Now, dividing Eq. (15.62) by each of Eqs. (15.58), (15.59), and (15.60) in turn results in

$$Z_a = \frac{Z_1Z_2 + Z_2Z_3 + Z_3Z_1}{Z_3} \qquad (15.63)$$

$$Z_b = \frac{Z_1Z_2 + Z_2Z_3 + Z_3Z_1}{Z_2} \qquad (15.64)$$

$$Z_c = \frac{Z_1Z_2 + Z_2Z_3 + Z_3Z_1}{Z_1} \qquad (15.65)$$

Equations (15.63), (15.64), and (15.65) then show the conversion equations required to convert a Y-connected load to its equivalent Δ-connected load. A pictorial summary of this conversion process is shown in Fig. 15.34. *Any of the impedances in the equivalent Δ-connected load is equal to the sum of the products of all the impedances of the Y-connected load taken two at a time, divided by the opposite impedance in the Y-connected load.* For example, if Z_a is required, we divide by Z_3, the opposite impedance in the Y-connected load in Fig. 15.34, and similarly for the other two elements.

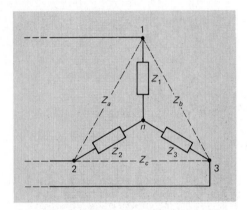

Figure 15.34 Y-to-Δ conversion.

When all three impedances in the Y-connected load are equal, each is Z_Y, then the equivalent Δ-connected impedances are also equal and each is Z_Δ, where:

$$Z_\Delta = 3Z_Y$$

as is evident from any of the conversion equations above. This equation confirms the result obtained in Eq. (15.61).

The conversion equations developed here also apply when converting a Π section into an equivalent T section, similar to the Δ-to-Y conversion, and vice versa.

■ EXAMPLE 15.14

Find the input impedance Z_{in} of the network shown in Fig. 15.35.

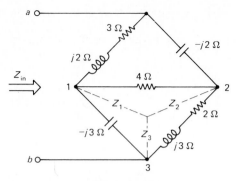

Figure 15.35

Solution To find the overall impedance between terminals a and b (i.e., Z_{in}) in the network shown, we notice that series and/or parallel equivalents of combinations of impedances cannot be applied. However, using the Δ-to-Y conversion for either the top three impedances (top Δ) or the bottom three impedances (bottom Δ) should make the overall network easier to handle. Here, for the bottom Δ load,

$$Z_1 = \frac{-j3 \times 4}{4 - j3 + 2 + j3} = \frac{-j12}{6} = -j2 \, \Omega$$

$$Z_2 = \frac{4(2 + j3)}{6} = 1.333 + j2 \, \Omega$$

and

$$Z_3 = \frac{-j3(2 + j3)}{6} = 1.5 - j1 \, \Omega$$

The network then becomes as shown in the equivalent circuit of Fig. 15.36, which is clearly a simpler series–parallel type of network. Thus:

$$Z_{in} = 1.5 - j1 + \frac{(3 + j2 - j2)(1.333 + j2 - j2)}{3 + j2 - j2 + 1.333 + j2 - j2} = 1.5 - j1 + \frac{4}{4.333}$$

$$= 2.423 - j1 = 2.621\underline{/-22.4°} \, \Omega$$

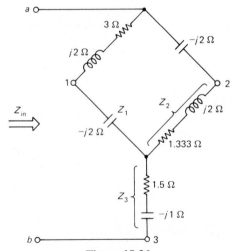

Figure 15.36

EXAMPLE 15.15

Convert the network shown in Fig. 15.37 into a Π-connected equivalent circuit.

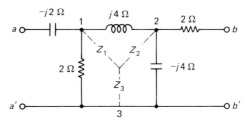

Figure 15.37

Solution This network is neither a Δ(Π) nor a Y(T) network, but its middle part (between nodes 1, 2, and 3) is a Π section which can be converted into a T section, as indicated. Here

$$Z_1 = \frac{j4 \times 2}{2 + j4 - j4} = j4 \ \Omega$$

$$Z_2 = \frac{j4 \times -j4}{2} = 8 \ \Omega$$

$$Z_3 = \frac{-j4 \times 2}{2} = -j4 \ \Omega$$

The overall network is then redrawn as shown in the equivalent form of Fig. 15.38(a). This is now a T (or a Y) network. It can then be easily converted

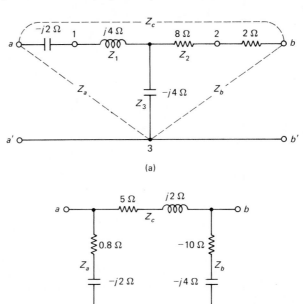

(a)

(b)

Figure 15.38

into a Π section, as required. This is indicated by the dashed impedances in Fig. 15.38(a). Here

$$Z_a = \frac{j2(-j4) + j2 \times 10 + 10(-j4)}{10} = 0.8 - j2 \; \Omega$$

$$Z_b = \frac{8 - j20}{j2} = -10 - j4 \; \Omega$$

and

$$Z_c = \frac{8 - j20}{-j4} = 5 + j2 \; \Omega$$

The final equivalent Π-connected network is shown in Fig. 15.38(b). Notice the negative resistance in Z_b. This should not worry us because such conversions produce only mathematical equivalencies.

DRILL PROBLEMS

Section 15.2

15.1. Write the mesh equations for the network shown in Fig. 15.39 and then determine the current in the 3 Ω resistor.

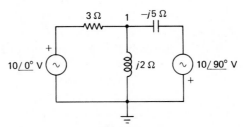

Figure 15.39

15.2. Convert the current source in the network shown in Fig. 15.40 into its equivalent voltage source model. Write the mesh equations and determine the current in the 4 Ω resistor.

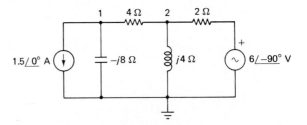

Figure 15.40

15.3. Convert the current sources in the network shown in Fig. 15.41 into their equivalent voltage source models. Write the corresponding mesh equations and determine the current in the 1 Ω resistor.

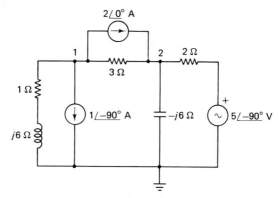

Figure 15.41

Section 15.3

15.4. Convert the voltage sources in the network shown in Fig. 15.39 into their equivalent current source models. Obtain, using nodal analysis, the voltage at node 1.

15.5. Convert the voltage source in the network shown in Fig. 15.40 into the equivalent current source model. Using nodal analysis, obtain the voltages at nodes 1 and 2.

15.6. Using the node voltage analysis method, obtain the voltages at nodes 1 and 2 in the network shown in Fig. 15.41.

Section 15.5

15.7. Using nodal analysis, obtain the voltage across the $j2\,\Omega$ inductor in the network shown in Fig. 15.42.

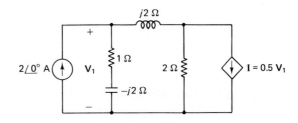

Figure 15.42

15.8. Analyze the network shown in Fig. 15.42 using the mesh current technique to obtain the current in the $j2\,\Omega$ inductor.

Section 15.6

15.9. Referring to Fig. 15.33, obtain the equivalent Y-connected impedances to each of the following Δ-connected loads:
   (a) $Z_a = j2\,\Omega$, $Z_b = 2 - j4\,\Omega$, $Z_c = 3 + j2\,\Omega$.
   (b) $Z_a = 4\,\Omega$, $Z_b = -j5\,\Omega$, $Z_c = j8\,\Omega$.

15.10. Referring to Fig. 15.34, obtain the equivalent Δ-connected impedances to each of the following Y-connected loads:
   (a) $Z_1 = 3\,\Omega$, $Z_2 = j2\,\Omega$, $Z_3 = -j2\,\Omega$.
   (b) $Z_1 = 1 + j2\,\Omega$, $Z_2 = 2 - j4\,\Omega$, $Z_3 = 5\,\Omega$.

PROBLEMS

Note: The problems below that are marked with an asterisk (*) can be used as exercises in computer-aided network analysis or as programming exercises.

Section 15.2

(i) ***15.1.** Find the power provided by each of the sources in the circuit shown in Fig. 15.43 using the mesh analysis technique.

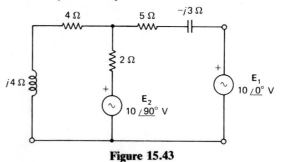

Figure 15.43

(i) ***15.2.** Find the values of the three currents I_1, I_2, and I_3 in the circuit shown in Fig. 15.44.

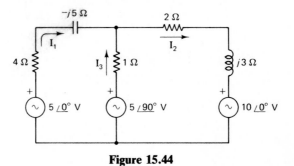

Figure 15.44

(i) ***15.3.** Find the three line currents I_A, I_B, and I_C in the three-phase circuit shown in Fig. 15.45.

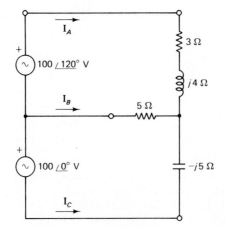

Figure 15.45

(i) ***15.4.** Find the current in the 4 Ω resistor in the circuit shown in Fig. 15.46.

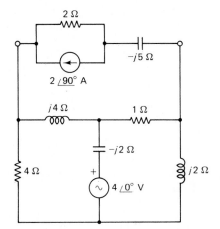

Figure 15.46

(i) ***15.5.** Find the current **I** in the circuit shown in Fig. 15.47.

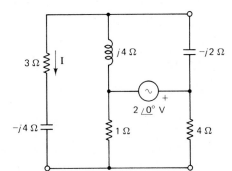

Figure 15.47

(i) ***15.6.** In the network shown in Fig. 15.48, calculate the power in each of the three resistors.

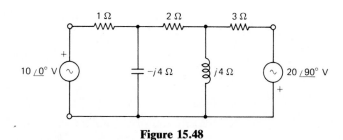

Figure 15.48

(ii) **15.7.** At the condition of balance in the bridge circuit shown in Fig. 15.49, no current flows in the detector whose impedance is Z_D. Using loop analysis, find the values of R_X and L_X in terms of the other bridge elements. The frequency of the source is ω rad/s.

Problems **659**

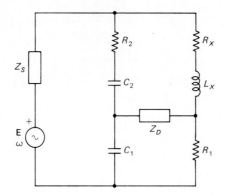

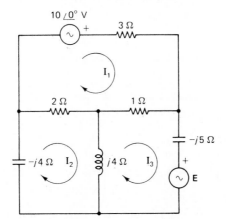

Figure 15.49

(i)   **15.8.** Find the value of the source **E** in the circuit shown in Fig. 15.50 that causes the mesh current $\mathbf{I}_1$ to be zero.

Figure 15.50

(i)   **15.9.** Calculate the value of the source **E** in the circuit shown in Fig. 15.51 that causes the current in the 4 Ω resistor to be zero.

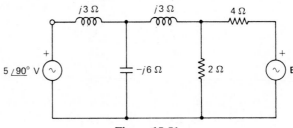

Figure 15.51

Section 15.3

(i)   *15.10. Find the node voltages $\mathbf{V}_1$ and $\mathbf{V}_2$ in the circuit shown in Fig. 15.52.

(i)   15.11   Replace the whole top branch in the circuit shown in Fig. 15.46, including the current source and the $-j5\,\Omega$ capacitor, by a single equivalent current source. Use nodal analysis to find the voltage across the 4 Ω resistor.

(i)   *15.12. Find the node voltage $\mathbf{V}_1$ in the circuit shown in Fig. 15.53.

(i)   15.13. In the circuit of Fig. 15.45, consider the common node between the two sources to be the ground node and consider the common point between the loads to be node 1. Find the voltage $\mathbf{V}_1$ at this node. Use this value of $\mathbf{V}_1$ to recalculate the three line currents $\mathbf{I}_A$, $\mathbf{I}_B$, and $\mathbf{I}_C$.

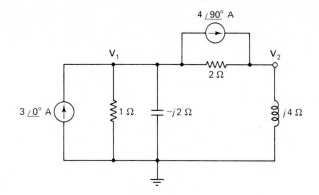

Figure 15.52

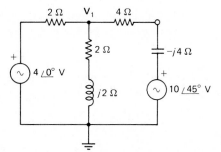

Figure 15.53

(i) **15.14.** Using nodal analysis, calculate the value of the source voltage **E** that results in the current **I** becoming zero in the circuit shown in Fig. 15.54.

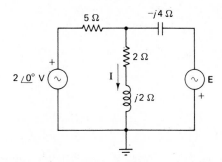

Figure 15.54

(i) ***15.15.** Find the power dissipated in the 5 Ω resistor in the circuit shown in Fig. 15.55 using nodal analysis.

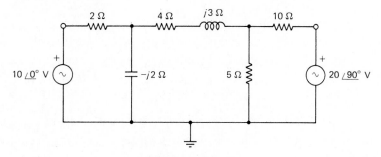

Figure 15.55

(ii) **15.16.** In the circuit of Fig. 15.56, what should be the value of the source **V** so that no current flows in the $-j4\,\Omega$ capacitor? Use nodal analysis.

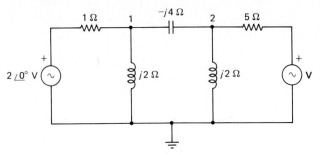

Figure 15.56

(ii) **15.17.** Again using nodal analysis in the circuit of Fig. 15.56, recalculate the value of the source **V** so that the current in the $5\,\Omega$ resistor becomes zero.

Section 15.5

(ii) ***15.18.** The circuit shown in Fig. 15.57 contains a dependent current source $\mathbf{I}_S$. Calculate the voltage $\mathbf{V}_2$.

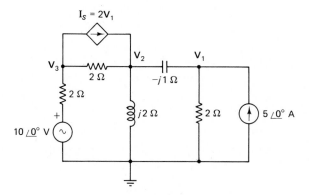

Figure 15.57

(ii) ***15.19.** Calculate the node voltages $\mathbf{V}_1$ and $\mathbf{V}_2$ in the network shown in Fig. 15.58. Note that this network contains the dependent voltage source $\mathbf{E}_S$.

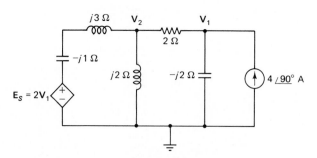

Figure 15.58

Section 15.6

(i) **15.20.** Obtain the Y (or T)-connected equivalent circuit for the Π-connected impedances shown in Fig. 15.59.

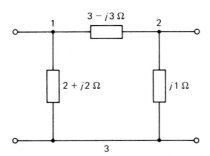

Figure 15.59

(i) *15.21.** Use Δ–Y conversion to simplify the circuit shown in Fig. 15.60. Calculate the value of the source current **I**.

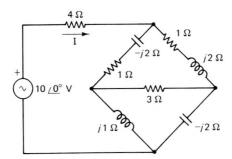

Figure 15.60

(ii) *15.22.** Use Δ-to-Y or Y-to-Δ conversions to simplify the bridged circuit shown in Fig. 15.61. Calculate **I**.

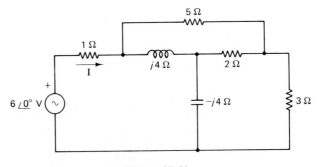

Figure 15.61

(i) **15.23.** The Π and the Y impedances connected in parallel as shown in Fig. 15.62 are both balanced (i.e., each consists of equal loads). Find the single equivalent Π section and then convert it to find the single equivalent T section.

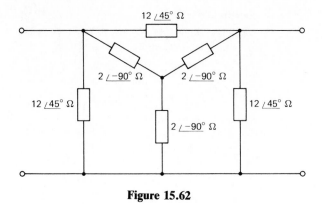

Figure 15.62

(i) **15.24.** Replace the network shown in Fig. 15.63 by an equivalent Y-connected circuit.

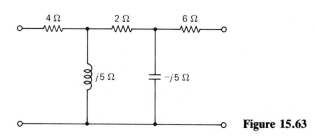

Figure 15.63

(i) **15.25.** Obtain the Δ (or Π)-connected equivalent circuit for the network shown in Fig. 15.64.

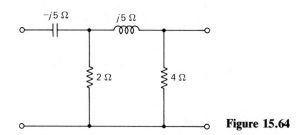

Figure 15.64

GLOSSARY

Delta (Δ) Connection: A connection of any three general impedance branches to form a network that has the shape of the capital Greek letter delta (Δ), or an inverted delta (∇). This configuration is also called the Π connection.

Dependent (or Controlled) Source: A current or a voltage source whose magnitude and phase are determined (i.e., controlled) by a current or a voltage somewhere else in the network.

Independent Source: A current or a voltage source whose magnitude and phase are independent of the network to which it is connected. In other words, the magnitude and phase of such a source are independent of the value of the load impedance connected to the source.

Loop (or Mesh) Analysis: A method of network analysis that relies on expressing KVL around the closed paths (i.e., loops or meshes) of the network, in terms of artificially assumed loop (or mesh) currents.

Nodal Analysis: A method of network analysis that relies on expressing KCL at each of the principal nodes of the network, in terms of the assumed voltages at each of the nodes with respect to a common reference (datum or ground) node.

Source Conversion: The method through which a practical voltage source is converted into a practical current source, or vice versa, so that the terminal behavior of the source remains the same as far as the rest of the network is concerned.

Wye (Y) Connection: A connection of any three general impedance branches to form a network that has the shape of the capital letter Y, or an inverted Y. The three branches have a common node in the middle of the Y configuration. This configuration is also called the T connection.

ANSWERS TO DRILL PROBLEMS

15.1. $I_{3\,\Omega} = 2.68 \underline{/-81.7^\circ}$ A
(from left to right)

15.3. $I_{1\,\Omega} = 0.8715 \underline{/134.56^\circ}$ A
(downward)

15.5. $V_1 = 7.4423 \underline{/-172.88^\circ}$ V,
$V_2 = 4.7069 \underline{/-101.31^\circ}$ V

15.7. $V_1 = 2.8284 \underline{/8.13^\circ}$ V,
$V_2 = 2.8284 \underline{/-81.87^\circ}$ V,
V (voltage across the $j2\ \Omega$

inductor) $= V_1 - V_2 =$
$4 \underline{/53.13^\circ}$ V

15.9. **(a)** $Z_1 = 1.6 + j0.8\ \Omega$,
$Z_2 = -0.8 + j1.2\ \Omega$,
$Z_3 = 2.8 - j1.6\ \Omega$
(b) $Z_1 = -2.4 - j3.2\ \Omega$,
$Z_2 = 3.84 + j5.12\ \Omega$,
$Z_3 = 6.4 - j4.8\ \Omega$

16

NETWORK THEOREMS FOR AC CIRCUITS

OBJECTIVES

- To understand the additional network analysis tools afforded through the Superposition, Thevenin's, Norton's, and the maximum power transfer theorems.
- Ability to use the superposition technique to analyze any ac network in general, and those containing multi-sources with multi-frequencies in particular.
- Ability to apply Thevenin's and Norton's theorems to simplify any ac network in general, including those containing dependent sources.
- Understanding the applications of the general expressions of the maximum power transfer theorem.

Thévenin's and Norton's theorems, and also the superposition theorem, constitute alternative methods for solving ac electric circuits, instead of the general loop and node analysis techniques. Sometimes these methods are much easier to use and hence save time and effort, but this is not true in general. However, as discussed here, these methods do have their special importance because in certain network situations one or the other of these theorems is the most appropriate, if not the only one, to use. The maximum power transfer theorem is treated in general terms as it relates to ac circuits.

16.1 THE SUPERPOSITION THEOREM

This theorem states that *the response (voltage* **V** *or current* **I**) *in any element of a linear network containing many sources is the algebraic phasor sum of the separate individual responses in this element when each source in turn acts alone in the network, while all the other sources in the network are killed* (*i.e., set to produce zero output*). Note that dependent sources *should not* be killed.

To kill a voltage source (i.e., make the voltage across it zero), it should be replaced temporarily by a short circuit, while to kill a current source (i.e., make the current from it zero), it should be replaced temporarily by an open circuit. The internal impedances of practical voltage and current sources remain connected in the network even after the corresponding source is killed.

The network shown in Fig. 16.1 has two independent voltage sources and one independent current source. The response **I** in the impedance Z_2, for example, is obtained from

$$\mathbf{I} = \mathbf{I}' \big|_{\text{due to } \mathbf{E}_1 \text{ alone}} + \mathbf{I}'' \big|_{\text{due to } \mathbf{E}_2 \text{ alone}} + \mathbf{I}'' \big|_{\text{due to } \mathbf{I}_3 \text{ alone}} \qquad (16.1)$$

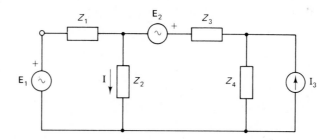

Figure 16.1

according to the application of the superposition theorem. Of course, when each source acts alone, the other sources are killed. This clearly results in simplifying the circuit and the required analysis to obtain the partial responses (**I**′, **I**″, and **I**‴), but at the same time the analysis has to be performed as many times as there are sources in the particular network. The circuit is treated as a phasor circuit, with all the elements represented by their impedances or admittances.

This theorem is actually a restatement of the general loop (or node) analysis techniques. The currents in the different loops were obtained through the expansion of the corresponding determinants, as was expressed by Eqs. (15.36), (15.37), and (15.38). In general, then, the current $\mathbf{I}_n$ in loop n of a given network can be expressed as

$$\mathbf{I}_n = \mathbf{E}_{11}\frac{\Delta_{1n}}{\Delta} + \mathbf{E}_{22}\frac{\Delta_{2n}}{\Delta} + \mathbf{E}_{33}\frac{\Delta_{3n}}{\Delta} + \cdots \qquad (16.2a)$$

$$= \mathbf{I}'\,\big|_{\text{due to } \mathbf{E}_{11}} + \mathbf{I}''\,\big|_{\text{due to } \mathbf{E}_{22}} + \mathbf{I}'''\,\big|_{\text{due to } \mathbf{E}_{33}} + \cdots \qquad (16.2b)$$

where $\mathbf{E}_{11}$, $\mathbf{E}_{22}$, $\mathbf{E}_{33}$, and so on, are the net voltage sources in each of the loops. By proper choice of these loops, it is possible in many cases that these net voltage sources could be only a single source at a time.

In some networks, where this theorem is most applicable, the simplification resulting from killing the sources to obtain the partial response due to each source at a time reduces the required analysis to simply the application of Ohm's law and/or voltage- or current-division rules. These partial responses can then be easily obtained by inspection. However, in more complicated networks, the superposition theorem might require similar or more effort than in the loop or node analysis methods. Here, of course, superposition is not the appropriate analysis technique.

It is important to note that the application of the superposition theorem through Eq. (16.1) implies that all the sources in the network have the same operating frequency in order that the phasor representations involved become possible (i.e., are mathematically appropriate). The direct application of Eq. (16.1) is not correct, however, when the sources in the given network are operating at different frequencies (f or ω).

The superposition theorem is *the only analysis tool* available for networks that contain *sources with different operating frequencies*; loop and node analysis techniques cannot be applied because they involve phasor and impedance representations, which are appropriate only at a single operating frequency. Because the superposition theorem implies analysis of the network with each source acting alone at its specific operating frequency, then for each source, ac network analysis involving phasor and impedance representations can be performed.

In such cases all the impedances in the network (which are, in general, frequency dependent, e.g., jX_L and $-jX_C$) will assume different values when each of the sources is considered alone. The circuit is then solved for each source independently and the response is expressed as a time function, for example,

$$i' = \sqrt{2}\,I'\sin(\omega't + \phi')\,\text{A}$$

corresponding to $\mathbf{I}' = I'\underline{/\phi'}\,\text{A}$ at ω',

$$i'' = \sqrt{2}\,I''\sin(\omega''t + \phi'')\,\text{A}$$

corresponding to $\mathbf{I}'' = I''\underline{/\phi''}\,\text{A}$ at ω'', and so on. The overall resultant response is therefore

$$i(t) = i' + i'' + i''' + \cdots$$
$$= \sqrt{2}\,I'\sin(\omega't + \phi') + \sqrt{2}\,I''\sin(\omega''t + \phi'') + \cdots \qquad (16.3)$$

The RMS value of the response is obtained from

$$I = \sqrt{I'^2 + I''^2 + I'''^2 + \cdots} \qquad (16.4)$$

This follows from the definition of the RMS based on average power.

Electronic circuits (e.g., amplifiers) usually contain dc sources ($\omega = 0\,\text{rad/s}$)

for biasing and signal sources. They are therefore analyzed using superposition, once at dc and once at the signal's operating frequency. The equivalent circuit in each case is clearly quite different, as inductors and capacitors are short circuits and open circuits, respectively, when dc analysis is performed. Also, networks with nonsinusoidal sources (e.g., square wave, triangular, etc., waveforms) are analyzed using superposition and Fourier series.

EXAMPLE 16.1

Using superposition, find the voltage **V** in the network shown in Fig. 16.2.

Solution The two sources here are operating at the same frequency; hence it is possible to represent the circuit in the phasor-impedance form shown in Fig. 16.2.

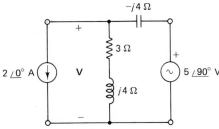

Figure 16.2

With the voltage source acting alone while the current source is open circuited, the circuit becomes a simple series one. Using the voltage-division principle, we have

$$\mathbf{V}' = j5 \cdot \frac{3 + j4}{3 + j4 - j4} = \frac{5}{4}(-4 + j3) = -6.67 + j5 \text{ V}$$

Now, with the current source acting alone while the voltage source is short circuited, the circuit becomes a simple parallel one. Using Ohm's law and noting the polarities gives us

$$\mathbf{V}'' = -2 \times \frac{-j4(3 + j4)}{3 + j4 - j4} = \frac{8}{3}(-4 + j3) = -10.67 + j8 \text{ V}$$

Thus

$$\mathbf{V} = \mathbf{V}' + \mathbf{V}'' = -17.34 + j13 = 21.67\underline{/143.1°} \text{ V}$$

EXAMPLE 16.2

Find the current **I** in the circuit shown in Fig. 16.3 using superposition.

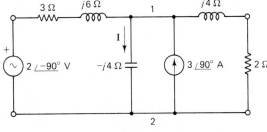

Figure 16.3

Solution   The circuit is redrawn in Fig. 16.4(a), with the voltage source acting alone while the current source is open circuited. Here

$$Z_T = 3 + j6 + \frac{-j4(2 + j4)}{2 + j4 - j4}$$

$$= 3 + j6 + 8 - j4$$

$$= 11 + j2 = 11.18\underline{/10.3°}\ \Omega$$

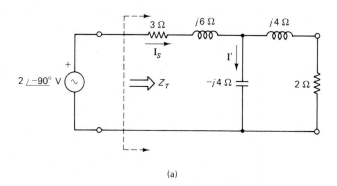

(a)

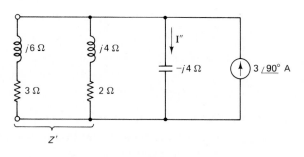

(b)

Figure 16.4

Therefore,

$$\mathbf{I}_S = \frac{2\underline{/-90°}}{11.18\underline{/10.3°}}$$

$$= 0.179\underline{/-100.3°}\ A$$

Using current division, we have

$$\mathbf{I}' = \mathbf{I}_S\frac{2 + j4}{2 + j4 - j4} = 0.179\underline{/-100.3°} \times 2.236\underline{/63.4°}$$

$$= 0.4\underline{/-36.9°} = 0.32 - j0.24\ A$$

The circuit is redrawn in Fig. 16.4(b) with the current source acting alone while the voltage source is short circuited. Here

$$Z' = \frac{(2 + j4)(3 + j6)}{5 + j10} = 1.2 + j2.4 \ \Omega$$

Again applying the current-division principle, we have

$$\mathbf{I}'' = 3\underline{/90°} \times \frac{Z'}{Z' - j4} = j3 \times \frac{1.2 + j2.4}{1.2 + j2.4 - j4}$$

$$= 3\underline{/90°} \times \frac{2.683\underline{/63.4°}}{2\underline{/-53.1°}} = 4.025\underline{/206.5°} \ \mathrm{A}$$

$$= -3.6 - j1.8 \ \mathrm{A}$$

Therefore,

$$\mathbf{I} = \mathbf{I}' + \mathbf{I}'' = 0.32 - j0.24 - 3.6 - j1.8 = -3.28 - j2.04$$

$$= 3.86\underline{/-148.1°} \ \mathrm{A}$$

Note that the original circuit in Fig. 16.3 has two principal nodes, as marked. Therefore, one node equation is necessary and sufficient, while the voltage source can easily be converted to a current source. The circuit's solution using the node analysis method would have been the easier alternative here. Try it!

EXAMPLE 16.3

Find the waveform of the voltage $v(t)$ across the current source in the circuit shown in Fig. 16.5 given that $e(t) = 2.828 \ \sin(5000t) \ \mathrm{V}$ and $i_S(t) = 1.414 \cos(10,000t) \ \mathrm{A}$.

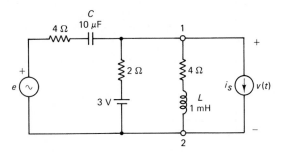

Figure 16.5

Solution As can be observed in Fig. 16.5, this circuit has three sources, each at a different operating frequency. Therefore, the application of the superposition theorem is the only analysis method that can be used to find the required response.

The circuit is redrawn in Fig. 16.6(a) with the dc source ($\omega = 0 \ \mathrm{rad/s}$) acting alone. Here L is equivalent to a short circuit, while C is equivalent to an open circuit. Therefore,

$$\mathbf{V}' = 3 \times \frac{4}{4 + 2} = 2 \ \mathrm{V}$$

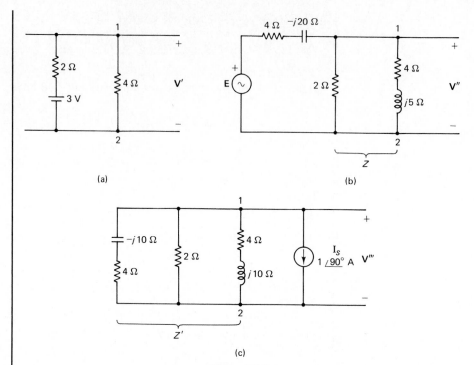

Figure 16.6

The voltage source $e(t)$ is represented by the phasor

$$\mathbf{E} = 2\underline{/0^\circ}\,\text{V} \qquad \text{at} \qquad \omega_1 = 5000\,\text{rad/s}$$

Figure 16.6(b) shows the circuit operating at this frequency, with the source $\mathbf{E}$ acting alone while the other two sources are killed. Note that here

$$Z_C = -j\frac{1}{\omega_1 c} = -j\frac{1}{5000 \times 10 \times 10^{-6}} = -j20\,\Omega$$

$$Z_L = j\omega_1 L = j5000 \times 1 \times 10^{-3} = j5\,\Omega$$

In this circuit

$$Z = \frac{2(4 + j5)}{2 + 4 + j5} = \frac{12.81\underline{/51.3^\circ}}{7.81\underline{/39.8^\circ}} = 1.64\underline{/11.5^\circ}\,\Omega = 1.607 + j0.327\,\Omega$$

Using the voltage-division principle gives us

$$\mathbf{V''} = 2\underline{/0^\circ} \times \frac{1.607 + j0.327}{4 - j20 + 1.607 + j0.327} = 2\underline{/0^\circ} \times \frac{1.64\underline{/11.5^\circ}}{20.456\underline{/-74.1^\circ}}$$

$$= 0.16\underline{/85.6^\circ}\,\text{V}$$

In the time domain this voltage phasor represents the waveform

$$v''(t) = 0.227\sin(5000t + 85.6^\circ)\,\text{V}$$

The current source $i_S(t)$ is represented by the phasor:

$$\mathbf{I}_S = 1\underline{/90^\circ}\,\text{A} \qquad \text{at} \qquad \omega_2 = 10{,}000\,\text{rad/s}$$

Here Z_C is $-j10\ \Omega$ and Z_L is $j10\ \Omega$. The circuit is redrawn as indicated in Fig. 16.6(c), operating at the frequency ω_2, with the current source $\mathbf{I}_S$ acting alone while the two voltage sources are killed. Here

$$\frac{1}{Z'} = \frac{1}{2} + \frac{1}{4 - j10} + \frac{1}{4 + j10}$$

$$= \frac{(4 - j10)(4 + j10) + 8 + j20 + 8 - j20}{2(4 - j10)(4 + j10)}$$

$$= \frac{16 + 100 + 16}{2(16 + 100)} = 0.569\ \text{S}$$

Therefore,

$$Z' = 1.758\ \Omega$$

and

$$\mathbf{V}''' = -\mathbf{I}_S Z' = -1\underline{/90^\circ} \times 1.758$$

$$= 1.758\underline{/-90^\circ}\ \text{V}$$

In the time domain, this voltage phasor represents the waveform

$$v'''(t) = 2.486 \sin(10{,}000t - 90^\circ)\ \text{V}$$

Thus, the required overall response is

$$v(t) = V' + v''(t) + v'''(t)$$

$$= 2 + 0.227 \sin(5000t + 85.6^\circ)$$

$$+ 2.486 \sin(10{,}000t - 90^\circ)\ \text{V}$$

16.2 THÉVENIN'S AND NORTON'S THEOREMS

In many practical situations, a network may contain a variable component or element (e.g., a transducer), while the rest of the network is composed of fixed elements. If it is required to find the response ($\mathbf{V}$ or $\mathbf{I}$) in this variable component, the whole network should be analyzed each time this component varies. Thévenin's and/or Norton's theorems provide a method through which the fixed part of such networks could be converted into a very simple equivalent circuit, either in the form of a practical voltage source (Thévenin's equivalent) or in the form of a practical current source (Norton's equivalent). Obviously, this conversion will result in considerable simplification of the analysis, as the variable component changes.

In Fig. 16.7(a), part A of the network consists of fixed components, including impedances and independent and dependent sources, while part B of the network usually consists of the variable component or load. Also, part B of the network does not contain any voltages or currents that control any of the dependent sources that may exist in part A of the network. Thévenin's theorem states that *as far as the response in part B of the network is concerned, part A of the network is equivalent to a single voltage source, $\mathbf{V}_{\text{th}}$, in series with a single impedance, Z_{th}.*

The representation of the Thévenin equivalent-circuit model is shown in

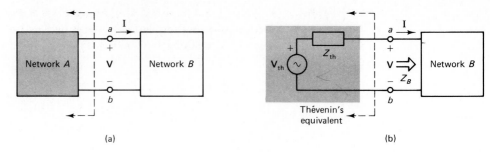

(a)   (b)

Figure 16.7

Fig. 16.7(b). It should be noted, however, that currents and voltages inside part
A of the network are not the same as in the Thévenin equivalent-circuit model.
The equivalency is correct only for voltages and currents in part B of the
network. This means that $\mathbf{V}$ and $\mathbf{I}$ obtained from the Thévenin equivalent circuit
in Fig. 16.7(b) are exactly the same as $\mathbf{V}$ and $\mathbf{I}$ calculated through the analysis of
the original circuit in Fig. 16.7(a).

To find the parameters $\mathbf{V}_{th}$ and Z_{th} of the Thévenin equivalent circuit, the
following procedure is performed:

1. $\mathbf{V}_{th}$ is obtained by measuring or calculating the open-circuit voltage
    between terminals a and b of part A of the network. The open-circuit
    condition means that part B of the network should be disconnected (or
    removed), as shown in Fig. 16.8. Thus

$$\mathbf{V}_{th} = \mathbf{V}_{ab}\big|_{\text{o.c.}}$$ (16.5)

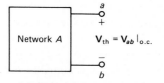

Figure 16.8

The polarities of the $\mathbf{V}_{th}$ source are the same as the polarities of the
calculated (or measured) voltage phasor $\mathbf{V}_{ab}\big|_{\text{o.c.}}$. Notice that $\mathbf{V}_{ab}$ means
the potential of point a with respect to point b (i.e., b is assumed to be
the reference); then in the equivalent circuit of Fig. 16.7(b), $\mathbf{V}_{th}$ has its
positive terminal toward point a.

2. Z_{th} is obtained through one of the following methods; in either case
    part B of the network should still be disconnected.
    a. When network A does not contain any dependent sources, all the
        independent sources in this part of the network should be killed.
        Ideal voltage sources are dead when short circuited and ideal
        current sources are dead when open circuited. The input impedance
        of network A, as seen between terminals a and b, is then calculated
        using the simple methods of series, parallel, and/or Δ–Y (or Y–Δ)

equivalent impedances. As indicated in Fig. 16.9(a), here

$$Z_{th} = Z_{in} \quad \text{(between points } a \text{ and } b\text{)} \tag{16.6}$$

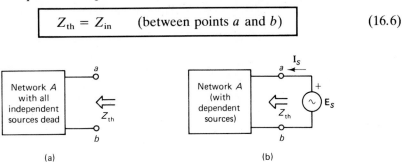

(a)　　　　　　　　　　　　　(b)

Figure 16.9

b. The input impedance of network A, between points a and b, can also be calculated or measured by inserting a voltage source $\mathbf{E}_S$ between terminals a and b and finding the current $\mathbf{I}_S$ that this source will drive into part A of the network, as shown in Fig. 16.9(b).

This method is used particularly if part A of the network contains dependent sources, which should not be killed. However, the independent sources in network A should always be killed when Z_{th} is calculated. As indicated in Fig. 16.9(b), here

$$Z_{th} = Z_{in} = \frac{\mathbf{E}_S}{\mathbf{I}_S} \tag{16.7}$$

Norton's theorem states that *as far as the response in part B of the network is concerned, part A of the network is equivalent to a single current source, $\mathbf{I}_N$, in parallel with a single impedance Z_N.*

The representation of the Norton equivalent–circuit model is shown in Fig. 16.10.

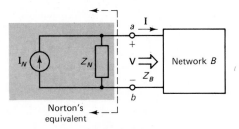

Figure 16.10

In the Norton equivalent–circuit model, Z_N is found in exactly the same way as Z_{th}, following the procedure described above. Thus

$$Z_N = Z_{th} \tag{16.8}$$

The other parameter in the Norton equivalent–circuit model, which is the value of the current source $\mathbf{I}_N$, is obtained as follows. Remove part B from the original circuit and replace it by a short circuit between points *a* and *b,* as shown in Fig. 16.11. The current flowing in this short-circuit wire, from *a* to *b,* due to the sources in network *A* is then calculated, or measured. Here

$$\boxed{\mathbf{I}_N = \mathbf{I}_{ab}\big|_{\text{s.c.}}} \tag{16.9}$$

Notice that the direction of the current source arrow (or polarities) of $\mathbf{I}_N$ is such that it drives current from *a* to *b,* as the current in Eq. (16.9) was calculated.

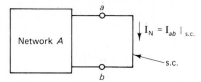

Figure 16.11

Remark. Since both the Thévenin and Norton equivalent-circuit models can replace the original network *A* and produce the same voltage and current $\mathbf{V}$ and $\mathbf{I}$, in network *B* when it is connected between the terminals *a* and *b,* these circuit models are also equivalent to each other [i.e., Fig. 16.7(b) and Fig. 16.10 are equivalent]. This must, of course, be true no matter what the value of the input impedance of network *B,* Z_B, is.

From the network in Fig. 16.7(b),

$$\mathbf{I} = \frac{\mathbf{V}_{\text{th}}}{Z_{\text{th}} + Z_B} \tag{16.10}$$

and

$$\mathbf{V} = \mathbf{I}Z_B = \mathbf{V}_{\text{th}}\frac{Z_B}{Z_{\text{th}} + Z_B} \tag{16.11}$$

From the network in Fig. 16.10, using the current-division principle, we have

$$\mathbf{I} = \mathbf{I}_N\frac{Z_N}{Z_N + Z_B} = \frac{\mathbf{I}_N Z_{\text{th}}}{Z_{\text{th}} + Z_B} \quad (\text{as } Z_{\text{th}} = Z_N) \tag{16.12}$$

and

$$\mathbf{V} = \mathbf{I}Z_B = \mathbf{I}_N Z_{\text{th}}\frac{Z_B}{Z_{\text{th}} + Z_B} \tag{16.13}$$

Comparing Eqs. (16.10) and (16.12) describing the same current value, and Eqs. (16.11) and (16.13) describing the same voltage value, we can conclude that

$$\mathbf{V}_{\text{th}} \text{ must be equal to } \mathbf{I}_N Z_{\text{th}}$$

Thus

$$Z_N = Z_{\text{th}} \tag{16.8}$$

and

$$\mathbf{V}_{\text{th}} = \mathbf{I}_N Z_N \tag{16.14}$$

or

$$\boxed{\mathbf{I}_N = \frac{\mathbf{V}_{\text{th}}}{Z_{\text{th}}}} \tag{16.15}$$

This is, in fact, another development (or proof) for the voltage-to-current source conversion, or vice-versa.

EXAMPLE 16.4

For the circuit shown in Fig. 16.12, find the Thévenin and Norton equivalent-circuit models between terminals a and b. Check the source conversion equivalency. Show that both circuit models produce the same current in a load Z_L of $(3 + j7)\ \Omega$ when connected between terminals a and b.

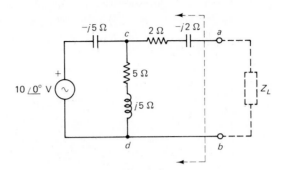

Figure 16.12

Solution $Z_{th}(=Z_N)$ can easily be obtained, with Z_L removed and the voltage source short circuited, by looking between terminals a and b. Here,

$$Z_{th} = 2 - j2 + \frac{-j5(5 + j5)}{-j5 + 5 + j5} = 2 - j2 + 5 - j5$$

$$= 7 - j7\ \Omega = 9.9\underline{/-45^\circ}\ \Omega$$

In determining $\mathbf{V}_{th}(=\mathbf{V}_{ab}|_{o.c.})$, no current flows in the $(2 - j2)\ \Omega$ branch since Z_L is removed, and thus the voltage across this branch is zero. Thus

$$\mathbf{V}_{th} = \mathbf{V}_{ab}|_{o.c.} = \mathbf{V}_{cd}$$

$$= 10\underline{/0^\circ} \times \frac{5 + j5}{5 + j5 - j5}$$

$$= 10\underline{/0^\circ} \times 1.414\underline{/45^\circ}$$

$$= 14.14\underline{/45^\circ}\ \text{V}$$

The Thévenin equivalent circuit model is then as shown in Fig. 16.13(a).

When the load $Z_L(=3 + j7\ \Omega)$ is connected to terminals a and b, then

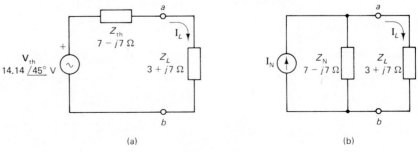

(a)   (b)

Figure 16.13

from Ohm's law,

$$\mathbf{I}_L = \frac{\mathbf{V}_{th}}{Z_{th} + Z_L} = \frac{14.14\underline{/45°}}{7 - j7 + 3 + j7}$$

$$= 1.414\underline{/45°}\ \text{A}$$

To obtain Norton's current source $\mathbf{I}_N$, terminals a and b of the original circuit in Fig. 16.12 are short circuited (notice that Z_L is removed). Using Ohm's law and then current division, $\mathbf{I}_{ab}$ can be calculated in one step as follows:

$$\mathbf{I}_N = \mathbf{I}_{ab}\big|_{s.c.} = \frac{10\underline{/0°}}{-j5 + \dfrac{(5 + j5)(2 - j2)}{5 + j5 + 2 - j2}} \times \frac{5 + j5}{5 + j5 + 2 - j2}$$

$$= \frac{10\underline{/0°} \times 7.071\underline{/45°}}{-j5(7 + j3) + (5 + j5)(2 - j2)}$$

$$= \frac{70.71\underline{/45°}}{15 - j35 + 20} = \frac{70.71\underline{/45°}}{49.5\underline{/-45°}} = 1.428\underline{/90°}\ \text{A}$$

To check the source conversion equivalency, Eq. (16.15) must be satisfied, that is,

$$\frac{\mathbf{V}_{th}}{Z_{th}} = \frac{14.14\underline{/45°}}{9.9\underline{/-45°}} = 1.428\underline{/90°} = \mathbf{I}_N \quad \text{(check!)}$$

Then Norton's equivalent-circuit model is as shown in Fig. 16.13(b). When the load Z_L is connected, using the current-division principle results in

$$\mathbf{I}_L = \mathbf{I}_N \frac{Z_N}{Z_N + Z_L} = 1.428\underline{/90°} \times \frac{7 - j7}{7 - j7 + 3 + j7}$$

$$= 1.428\underline{/90°} \times \frac{9.9}{10}\underline{/-45°} = 1.414\underline{/45°}\ \text{A}$$

Thus using either the Thévenin or the Norton equivalent-circuit model, the load current $\mathbf{I}_L$ is exactly the same. Thus both circuit models produce the same response.

EXAMPLE 16.5

Find the Thévenin equivalent for the network shown in Fig. 16.14, between terminals 1 and 2. Notice that this network contains a dependent voltage source $\mathbf{V}_S = 3\mathbf{V}_X$.

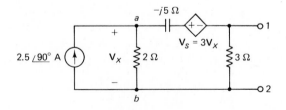

Figure 16.14

Solution To obtain $\mathbf{V}_{th}(=\mathbf{V}_{12}|_{o.c.})$, the practical current source is first converted into its equivalent voltage source model, as shown in Fig. 16.15(a). This conversion is acceptable here because the controlling voltage $\mathbf{V}_X$ is across the source terminals a and b, which still exist in the equivalent circuit, being also the terminals of the equivalent practical voltage source.

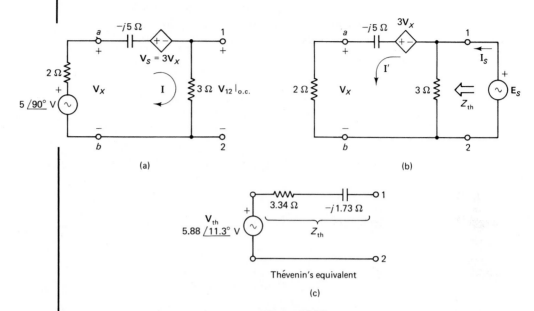

(a)

(b)

Thévenin's equivalent

(c)

Figure 16.15

Assume that the current in the closed loop of Fig. 16.15(a) is $\mathbf{I}$, as shown. Here, using Ohm's law, we have

$$\mathbf{I} = \frac{j5 - 3\mathbf{V}_X}{2 - j5 + 3} = \frac{j5 - 3\mathbf{V}_X}{5 - j5}$$

but

$$\mathbf{V}_X = j5 - 2\mathbf{I}$$

($2\mathbf{I}$ is the voltage drop across the $2\,\Omega$ source resistance); therefore,

$$\mathbf{I}(5 - j5) = j5 - 3(j5 - 2\mathbf{I}) = j5 - j15 + 6\mathbf{I}$$

$$\mathbf{I} = \frac{-j10}{-1 - j5} = \frac{10/90°}{5.1/78.7°} = 1.96/11.3°\ \text{A}$$

Thus

$$\mathbf{V}_{th} = \mathbf{V}_{12}|_{o.c.} = 3\mathbf{I} = 5.88/11.3°\ \text{V}$$

To find Z_{th}, the independent current source is killed by open circuiting it. The dependent source cannot be killed. The circuit is then redrawn as shown in Fig. 16.15(b) driven by the source $\mathbf{E}_S$ across terminals 1 and 2, resulting in the source current $\mathbf{I}_S$. The input impedance of the network between terminals 1 and 2 is Z_{th}, which is in this case

$$Z_{th} = \frac{\mathbf{E}_S}{\mathbf{I}_S}$$

Here

$$\mathbf{I}' = \frac{\mathbf{V}_X}{2} = \frac{\mathbf{E}_S + 3\mathbf{V}_X}{2 - j5}$$

or

$$2\mathbf{E}_S + 6\mathbf{V}_X = (2 - j5)\mathbf{V}_X$$

$$\mathbf{V}_X = \frac{2\mathbf{E}_S}{-4 - j5}$$

Therefore

$$\mathbf{I}' = \frac{\mathbf{V}_X}{2} = \frac{-\mathbf{E}_S}{4 + j5}$$

But from KCL at node 1,

$$\mathbf{I}_S = \frac{\mathbf{E}_S}{3} + \mathbf{I}' = \frac{\mathbf{E}_S}{3} - \frac{\mathbf{E}_S}{4 + j5} = \mathbf{E}_S \frac{4 + j5 - 3}{3(4 + j5)}$$

Thus

$$Z_{\text{th}} = \frac{\mathbf{E}_S}{\mathbf{I}_S} = \frac{3(4 + j5)}{1 + j5} = \frac{19.21\underline{/51.3°}}{5.1\underline{/78.7°}} = 3.767\underline{/-27.4°}\ \Omega$$

$$= 3.34 - j1.733\ \Omega$$

The Thévenin equivalent circuit between terminals 1 and 2 is thus as shown in Fig. 16.15(c).

EXAMPLE 16.6

Find the Thévenin equivalent circuit for the network shown in Fig. 16.16 between terminals a and b.

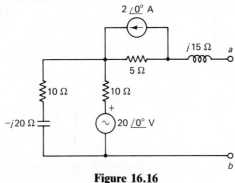

Figure 16.16

Solution   To calculate Z_{th}, the two independent sources are killed by open circuiting the ideal current source and short circuiting the ideal voltage source. Looking at the network between terminals a and b, we see that

$$Z_{\text{th}} = Z_{\text{in}}$$

$$= 5 + j15 + \frac{10(10 - j20)}{10 + 10 - j20}$$

$$= 5 + j15 + \frac{(10 - j20)}{2 - j2}\frac{(2 + j2)}{(2 + j2)} = 5 + j15 + \frac{20 + 40 + j20 - j40}{8}$$

$$= 5 + j15 + 7.5 - j2.5$$

$$= 12.5 + j12.5 = 17.68\underline{/45°}\ \Omega$$

The circuit is redrawn as in Fig. 16.17(a), with the current source converted to a voltage source. Here

$$\mathbf{I}' = \frac{20\underline{/0^\circ}}{10 + 10 - j20} = \frac{1\underline{/0^\circ}}{1.414\underline{/-45^\circ}} = 0.707\underline{/45^\circ}\,\text{A}$$

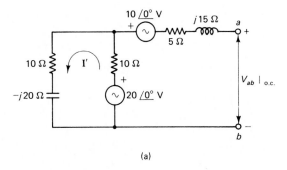

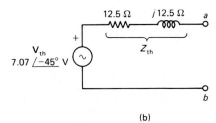

(a)

(b)

Figure 16.17

Thus, noting that no current flows in the $(5 + j15)\,\Omega$ impedance and that the voltage drop across the $10\,\Omega$ resistor has polarities opposing the $20\underline{/0^\circ}$ V source, $\mathbf{V}_{\text{th}}$ can be calculated as follows:

$$\begin{aligned}
\mathbf{V}_{\text{th}} = \mathbf{V}_{ab}\big|_{\text{o.c.}} &= 20\underline{/0^\circ} - 10 \times \mathbf{I}' - 10\underline{/0^\circ} \\
&= 10 - 10(0.5 + j0.5) \\
&= 10 - 5 - j5 = 5 - j5 \\
&= 7.07\underline{/-45^\circ}\,\text{V}
\end{aligned}$$

The Thévenin equivalent circuit is thus as shown in Fig. 16.17(b).

EXAMPLE 16.7

Find the Thévenin equivalent circuit for the bridge network shown in Fig. 16.18 between terminals a and b.

Solution   Noting that the voltage source $\mathbf{E}$ is across the two parallel branches $(Z_1 + Z_3)$ and $(Z_2 + Z_4)$, $\mathbf{V}_{\text{th}}$ can be calculated as follows:

$$\begin{aligned}
\mathbf{V}_{\text{th}} = \mathbf{V}_{ab}\big|_{\text{o.c.}} &= \mathbf{V}_{ao} - \mathbf{V}_{bo} \\
&= \mathbf{E}\frac{Z_3}{Z_1 + Z_3} - \mathbf{E}\frac{Z_4}{Z_2 + Z_4} = \mathbf{E}\frac{Z_2 Z_3 - Z_1 Z_4}{(Z_1 + Z_3)(Z_2 + Z_4)}\quad\text{V}
\end{aligned}$$

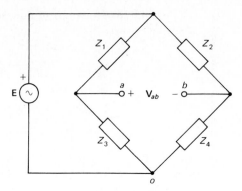

Figure 16.18

The network is redrawn as shown in Fig. 16.19, after short circuiting the independent voltage source. Thus

$$Z_{\text{th}} = \frac{Z_1 Z_3}{Z_1 + Z_3} + \frac{Z_2 Z_4}{Z_2 + Z_4}$$

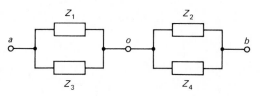

Figure 16.19

EXAMPLE 16.8

Using Thévenin's equivalent circuit, find the value of $\mathbf{E}_2$ in the circuit shown in Fig. 16.20 such that the current in the $(3 + j4)$ Ω impedance is $1\underline{/0°}$ A, flowing from a to b.

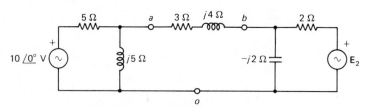

Figure 16.20

Solution   The impedance $(3 + j4)$ Ω will be considered as the load impedance Z_L (or part B of the network). Removing it, one can then find the Thévenin equivalent for the remainder of the network between points a and b.

With the two independent voltage sources killed (i.e., short circuited),

$$Z_{th} = Z_{in} \quad \text{(between points } a \text{ and } b)$$

$$= \frac{5(j5)}{5 + j5} + \frac{2(-j2)}{2 - j2} = \frac{j5(1 - j1)}{(1 + j1)(1 - j1)} + \frac{-j2(1 + j1)}{(1 - j1)(1 + j1)}$$

$$= \frac{j5 + 5 - j2 + 2}{2} = 3.5 + j1.5 = 3.81\underline{/23.2°}\ \Omega$$

With Z_L removed, it is now easy to calculate $\mathbf{V}_{ab}|_{o.c.}$ since the circuit is then made of two separate independent loops. Using the voltage-division principle, we have

$$\mathbf{V}_{th} = \mathbf{V}_{ab}|_{o.c.} = \mathbf{V}_{ao} - \mathbf{V}_{bo} = 10\underline{/0°} \times \frac{j5}{5 + j5} - \mathbf{E}_2 \frac{-j2}{2 - j2}$$

$$= \frac{j10}{(1 + j1)}\frac{(1 - j1)}{(1 - j1)} - \mathbf{E}_2\frac{-j1(1 + j1)}{(1 - j1)(1 + j1)}$$

$$= 5 + j5 - (0.5 - j0.5)\mathbf{E}_2$$

The Thévenin equivalent circuit is shown in Fig. 16.21 with the load Z_L reconnected between terminals a and b. The load current is also indicated. From Ohm's law,

$$\mathbf{I}_L = \frac{\mathbf{V}_{th}}{Z_{th} + Z_L}$$

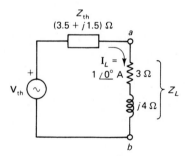

Figure 16.21

Then

$$1\underline{/0°} = \frac{5 + j5 - (0.5 - j0.5)\mathbf{E}_2}{6.5 + j5.5}$$

or

$$6.5 + j5.5 = 5 + j5 - (0.5 - j0.5)\mathbf{E}_2$$

Therefore,

$$\mathbf{E}_2 = \frac{-1.5 - j0.5}{0.5 - j0.5} = \frac{-3 - j1}{1 - j1}\frac{1 + j1}{1 + j1} = \frac{-3 + 1 - j1 - j3}{2}$$

$$= -1 - j2 = 2.236\underline{/-116.6°}\ \text{V}$$

As discussed above, any network with any number of sources and elements can be converted between two terminals (say, a and b) into its Thévenin equivalent-circuit model, as shown in Fig. 16.22. Thévenin's equivalent circuit is, in fact, the practical voltage source model (i.e., $\mathbf{E}$ could be $\mathbf{V}_{th}$ and Z_S could be Z_{th}). The question then arises as to how to choose a variable load impedance Z_L such that maximum power is transferred from the network (or the source) to the load when this load is connected to terminals a and b.

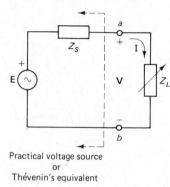

Practical voltage source
or
Thévenin's equivalent

Figure 16.22

The answer to this question is provided through the maximum power transfer theorem. Two distinct possibilities are considered through this theorem.

1. If Z_L is a completely arbitrary impedance,

$$Z_L = R + jX$$

where both R and X (with its sign, i.e., inductive or capacitive) can be chosen or varied to be any value required. The ac source, in general, has an impedance

$$Z_S = R_S + jX_S$$

where X_S is positive when the impedance of the source is inductive and X_S is negative when the impedance of the source is capacitive. In the circuit of Fig. 16.22,

$$\mathbf{I} = \frac{\mathbf{E}}{Z_S + Z_L} = \frac{\mathbf{E}}{(R_S + R) + j(X_S + X)} \tag{16.16}$$

This current is a phasor quantity. The load power is only dissipated in the resistive component of the load, R, as the reactive component X (whether L or C) does not dissipate any power. To calculate the value of the power dissipated in the load, P_L, only the magnitude of the current phasor is required.

$$P_L = |I|^2 R = \frac{E^2 R}{(R_S + R)^2 + (X_S + X)^2} \tag{16.17}$$

Here P_L is a function of two variables, R and X. P_L reaches its maximum value when, according to calculus, the two following conditions (involving partial derivatives) are satisfied. (The results will be given directly without the intermediate mathematical steps.)

$$\frac{\partial P_L}{\partial X} = 0 \qquad (16.18)$$

and

$$\frac{\partial P_L}{\partial R} = 0 \qquad (16.19)$$

With P_L as expressed in Eq. (16.17), condition (16.18) results in

$$X_S + X = 0$$

or

$$X = -X_S \qquad (16.20)$$

The meaning of this result will be explained below. With P_L as expressed in Eq. (16.17), condition (16.19) results in

$$(R_S + R)^2 + (X_S + X)^2 - 2R(R_S + R) = 0$$

that is,

$$R^2 = R_S^2 + (X_S + X)^2$$

or

$$R = \sqrt{R_S^2 + (X_S + X)^2} \qquad (16.21)$$

However, since $X = -X_S$, from Eq. (16.20), Eq. (16.21) reduces to

$$R = R_S \qquad (16.22)$$

Thus to produce maximum power in the load (i.e., maximum power transfer), the load impedance must be chosen to be

$$\boxed{Z_L = R + jX = R_S - jX_S = Z_S^*} \qquad (16.23)$$

This means that the load impedance must be the complex conjugate of the source impedance.

It is important to understand the meaning of this maximum power transfer condition. Figure 16.23 represents the components of the impedances of the circuit in Fig. 16.22, to help us understand this concept.

When the impedance of the source is inductive (as shown in Fig. 16.23), the load impedance must be chosen to be capacitive, and vice versa, according to Eq.

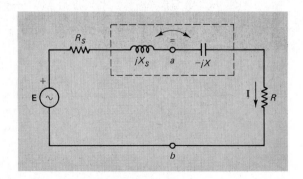

Figure 16.23

(16.20), such that

$$|X| = |X_S|$$

This is, in fact, the resonace condition in a simple series circuit. As discussed in Chapter 14, the total circuit's impedance is minimum at resonance and thus $|\mathbf{I}|$ is maximum. This obviously results in the highest power to be developed in the load. With $X_S + X = 0$, the dashed box in Fig. 16.23, representing the overall reactive component of the total impedance of the circuit, is equivalent to a short circuit. The circuit then becomes similar to the all resistive dc case, again requiring that R be equal to R_S for maximum power to develop in the load.

Note that if the impedance of the source is resistive only $(X_S = 0)$, this requires that X be also zero and Eq. (16.23) reduces to

$$R = R_S \tag{16.24}$$

exactly as was obtained for dc circuits.

2. If the load Z_L consists of only a varaible (arbitrary) resistive component, R, while its reactive component, X, is fixed (or zero), only condition (16.19) above is valid. As was examined above, maximum power transfer is obtained in this case when according to Eq. (16.21),

$$\boxed{R = \sqrt{R_S^2 + (X_S + X)^2}} \tag{16.21}$$

and when X is zero,

$$R = \sqrt{R_S^2 + X_S^2} = |Z_S| \tag{16.25}$$

Therefore, when only the resistive component of the load impedance is variable, it must be chosen to be equal to the magnitude of the total fixed impedance of the Thévenin equivalent circuit, to achieve maximum power transfer.

In general, Eq. (16.17) is used to calculate the maximum load power. This equation becomes quite simple when the load impedance Z_L is equal to the complex-conjugate source impedance, Z_S^*, as in Eq. (16.23) discussed in case 1 above. Here Eq. (16.17) reduces to

$$\boxed{P_L|_{max} = \frac{E^2 R}{(R_S + R)^2 + (X_S + X)^2} = \frac{E^2 R_S}{(2R_S)^2} = \frac{E^2}{4R_S}} \tag{16.26}$$

■ EXAMPLE 16.9

For the circuit in Example 16.4, the load impedance is varied. Find the values of Z_L required to produce maximum power in the load in each of the following cases, and then calculate the value of this maximum load power.
(a) Z_L is completely arbitrary.
(b) Z_L consists of a variable resistance only.
(c) Z_L is $R + j2\,\Omega$, where only R is a variable resistance.

Solution   Thévenin's equivalent of the circuit in Example 16.4 is redrawn in Fig. 16.24.
(a) When Z_L is completely arbitrary,

$$Z_L = Z_S^* = Z_{th}^* = 7 + j7\,\Omega$$

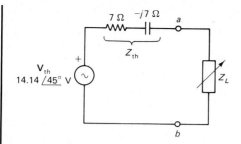

Figure 16.24

for maximum power transfer to the load. Here

$$P_L\big|_{max} = \frac{E^2}{4R_S} = \frac{V_{th}^2}{4R_{th}} = \frac{(14.14)^2}{4 \times 7} = 7.14 \text{ W}$$

(b) When Z_L is a variable resistance only, R, it should be chosen to be

$$R = \sqrt{R_{th}^2 + X_{th}^2} = \sqrt{49 + 49} = 9.9 \ \Omega$$

Here, from Eq. (16.17),

$$P_L\big|_{max} = \frac{V_{th}^2 R}{(R_{th} + R)^2 + (X_{th})^2} = \frac{(14.14)^2 \times 9.9}{(7 + 9.9)^2 + (-7)^2} = 5.92 \text{ W}$$

(c) Here the required value of R can be obtained from Eq. (16.21):

$$R = \sqrt{R_{th}^2 + (X + X_{th})^2} = \sqrt{(7)^2 + (2 - 7)^2} = 8.6 \ \Omega$$

that is,

$$Z_L = R + j2 = 8.6 + j2 \ \Omega$$

In this case,

$$P_L\big|_{max} = \frac{V_{th}^2 R}{(R_{th} + R)^2 + (X + X_{th})^2} = \frac{(14.14)^2 \times 8.6}{(7 + 8.6)^2 + (2 - 7)^2} = 6.41 \text{ W}$$

Notice that case 1, when the load is completely arbitrary, always results in the highest possible load power, delivered from a given circuit.

■ **EXAMPLE 16.10**

Repeat Example 16.9 for the circuit discussed in Example 16.6.

Solution The Thévenin equivalent of the circuit in Example 16.6 is redrawn in Fig. 16.25.

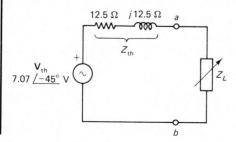

Figure 16.25

(a) When Z_L is completely arbitrary,

$$Z_L = Z_{th}^* = 12.5 - j12.5 \ \Omega$$

for maximum power transfer to the load. Here

$$P_L|_{max} = \frac{V_{th}^2}{4R_{th}} = \frac{(7.07)^2}{4 \times 12.5} = 1.0 \ \text{W}$$

(b) When Z_L is a variable resistance only, R, it should be chosen such that

$$R = \sqrt{R_{th}^2 + X_{th}^2} = \sqrt{(12.5)^2 + (12.5)^2} = 17.68 \ \Omega$$

Here,

$$P_L|_{max} = \frac{V_{th}^2 R}{(R + R_{th})^2 + (X_{th})^2} = \frac{(7.07)^2 \times 17.68}{(12.5 + 17.8)^2 + (12.5)^2} = 0.828 \ \text{W}$$

(c) Here the required value of R is obtained from Eq. (16.21):

$$R = \sqrt{R_{th}^2 + (X + X_{th})^2}$$
$$= \sqrt{(12.5)^2 + (2 + 12.5)^2} = 19.14 \ \Omega$$

that is,

$$Z_L = R + j2 = 19.14 + j2 \ \Omega$$

In this case,

$$P_L|_{max} = \frac{V_{th}^2 R}{(R + R_{th})^2 + (X + X_{th})^2} = \frac{(7.07)^2 \times 19.14}{(12.5 + 19.14)^2 + (2 + 12.5)^2}$$
$$= 0.79 \ \text{W}$$

Again notice that case 1 results in the highest possible load power.

EXAMPLE 16.11

For the circuit in Fig. 16.26, find the Thévenin equivalent circuit between terminals a and b. Choose Z_L for maximum power transfer to the load and calculate the value of this maximum load power.

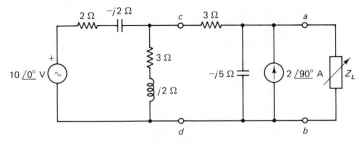

Figure 16.26

Solution　It is frequently possible to apply Thévenin's theorem many times to different parts in a given network. For example, in the circuit shown here, Thévenin's theorem can be applied first to the part of the network to the left of points c and d. With the rest of the circuit removed,

$$\mathbf{V}'_{th} = \mathbf{V}_{cd}\big|_{o.c.} = 10\underline{/0^\circ} \times \frac{3 + j2}{3 + j2 + 2 - j2} = 6 + j4 = 7.21\underline{/33.7^\circ}\, V$$

and

$$Z'_{th} = \frac{(2 - j2)(3 + j2)}{2 - j2 + 3 + j2} = \frac{6 + 4 - j6 + j4}{5} = 2 - j0.4\,\Omega$$

The circuit can then be redrawn as shown in Fig. 16.27(a). The current source between points a and b has been converted to its voltage source model with $\mathbf{E}_S = -j5 \times 2\underline{/90^\circ} = 5\underline{/-90^\circ} \times 2\underline{/90^\circ} = 10\underline{/0^\circ}\, V$ as shown. With the load Z_L removed, the circuit becomes a simple series one. Here

$$\mathbf{I}' = \frac{6 + j4 - 10}{2 - j0.4 + 3 - j5} = \frac{-4 + j4}{5 - j5.4}$$

$$= \frac{5.657\underline{/135^\circ}}{7.359\underline{/-47.2^\circ}} = 0.769\underline{/182.2^\circ}\, A$$

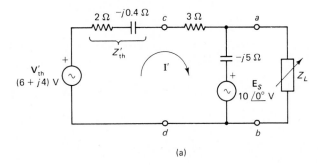

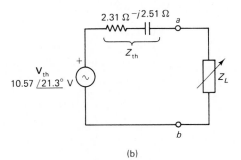

(a)

(b)

Figure 16.27

Hence,

$$\mathbf{V}_{th} = \mathbf{V}_{ab}\big|_{o.c.} = 10\underline{/0^\circ} + (-j5)\mathbf{I}'$$

$$= 10 + 5\underline{/-90^\circ} \times 0.769\underline{/182.2^\circ} = 10 + 3.845\underline{/92.2^\circ}$$

$$= 10 + (-0.148 + j3.842) = 9.852 + j3.842$$

$$= 10.57\underline{/21.3^\circ}\, V$$

With the two voltage sources in Fig. 16.27(a) short circuited, to kill them, Z_{th}

can easily be calculated, looking to the network between points a and b:

$$Z_{th} = \frac{-j5(5 - j0.4)}{5 - j0.4 - j5} = \frac{5\underline{/-90°} \times 5.016\underline{/-4.6°}}{7.359\underline{/-47.2°}} = 3.408\underline{/-47.4°}\ \Omega$$

$$= 2.31 - j2.51\ \Omega$$

The Thévenin equivalent circuit is then as shown in Fig. 16.27(b), with the load Z_L reconnected. As there are no restrictions on the choice of Z_L, one can choose it according to case 1 above, to be

$$Z_L = Z_{th}^* = 2.31 + j2.51\ \Omega$$

to achieve the highest possible load power. Here

$$P_L|_{max} = \frac{V_{th}^2}{4R_{th}} = \frac{(10.57)^2}{4 \times 2.31} = 12.09\ W$$

DRILL PROBLEMS

Section 16.1

16.1. Using the superposition theorem, find the voltage V in the network shown in Fig. 16.28.

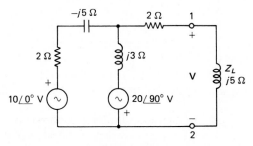

Figure 16.28

16.2. Apply the superposition theorem to determine the current I in the network shown in Fig. 16.29.

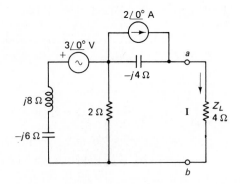

Figure 16.29

16.3. Determine the voltage V in the network shown in Fig. 16.30 using the superposition theorem.

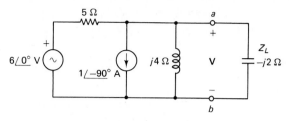

Figure 16.30

16.4. Determine the current **I** in the network shown in Fig. 16.31 using the superposition theorem.

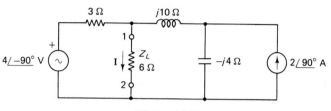

Figure 16.31

Section 16.2

16.5. Obtain the Thévenin's equivalent circuit for the network shown in Fig. 16.28 between terminals 1 and 2. Use this equivalent circuit to determine the voltage **V** across the load impedance Z_L.

16.6. Obtain the Thévenin's equivalent circuit for the network shown in Fig. 16.29 between terminals a and b. Use this equivalent circuit to determine the current **I** in the load impedance Z_L.

16.7. For the same circuit shown in Fig. 16.29, determine the Norton's equivalent between terminals a and b. Prove the validity of the source equivalency in comparison with the results obtained in Drill Problem 16.6.

16.8. Obtain the Thévenin's equivalent circuit for the network shown in Fig. 16.30 between terminals a and b. Using this model, determine the voltage **V** across the load impedance Z_L.

16.9. Obtain the Norton's equivalent circuit for the network shown in Fig. 16.31 between terminals 1 and 2. Use this equivalent circuit to determine the current **I** in the load impedance Z_L.

16.10. Find the Thévenin's equivalent circuit for the network shown in Fig. 16.32 between terminals a and b.

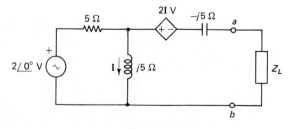

Figure 16.32

16.11. Using the results of problem 5, determine the value that Z_L should have in order to obtain maximum power transfer to the load. Z_L is completely arbitrary. What is the value of this maximum power?

16.12. If Z_L in the network shown in Fig. 16.29 is a variable resistance, what should be its value to obtain maximum power transfer to this load? What is the value of this maximum power? Use the Thévenin's equivalent circuit obtained in Drill Problem 16.6.

16.13. If Z_L in the network shown in Fig. 16.30 is completely arbitrary, what should be its value in order to obtain the maximum possible power in this load? What is the value of this maximum power? Use the Thévenin's equivalent circuit obtained in Drill Problem 16.8.

16.14. Convert the Norton's equivalent of the network shown in Fig. 16.31, as obtained in Drill Problem 16.9, into its equivalent Thévenin's model. If Z_L is a variable resistance, what should be its value for maximum power transfer? Determine the value of this maximum load power.

16.15. Using the Thévenin's equivalent circuit obtained in Drill Problem 16.10, determine the value that Z_L should have, if it is completely arbitrary, to obtain maximum power in the load impedance shown in Fig. 16.32. What is the value of this maximum load power?

PROBLEMS

Note: The problems below that are marked with an asterisk (*) can be used as exercises in computer-aided network analysis or as programming exercises.

Section 16.1

(i)   *16.1. Using the superposition theorem, find the current **I** in the network shown in Fig. 16.33.

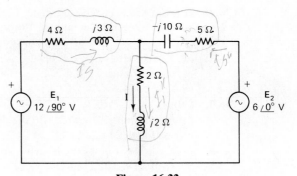

Figure 16.33

(i)   *16.2. Find the current **I** in the $3\,\Omega$ resistor in the network shown in Fig. 16.34, using the superposition theorem.

(i)   16.3. In the network shown in Fig. 16.35, the $\mathbf{V}_1$ and $\mathbf{V}_2$ sources produce the same value of voltage **V** across the $5\,\Omega$ resistor, when each of the sources is acting alone. Find the ratio $\mathbf{V}_1/\mathbf{V}_2$.

(i)   16.4. In the circuit shown in Fig. 16.36, $i = 1.414 \sin (5000t + 90°)\,\text{A}$, and $e = 7.071 \sin (10{,}000t - 45°)\,\text{V}$. Find the time-domain expression for the waveform of the voltage v.

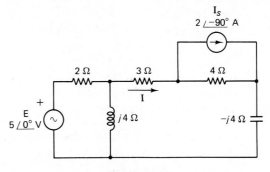

Figure 16.34

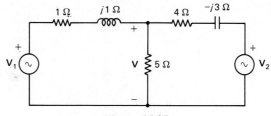

Figure 16.35

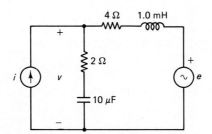

Figure 16.36

(i)   **16.5.**   Find the time-domain waveform for i_L in the network shown in Fig. 16.37, if $e = 28.28 \cos (1000t)$ V and, $i = 5\sqrt{2} \sin (2000t)$ A.

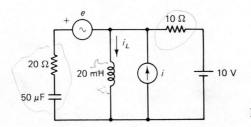

$X_L = wL.$
$X_L = w\, 70 \times 10^{-3}$
$z_L = j X_L -$

Figure 16.37

Section 16.2

(i)   **16.6.**   Obtain the Thévenin and Norton equivalents for the network shown in Fig. 16.38. Check the source equivalency property of these circuit models.

(i)   ***16.7.**   Find the Thévenin and Norton equivalent circuits for the network shown in Fig. 16.39 between terminals a and b. Connect a 20 Ω load resistance between these

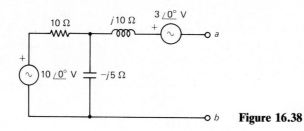

Figure 16.38

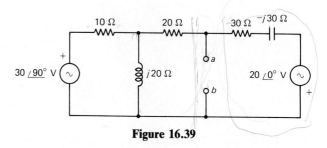

Figure 16.39

terminals. Calculate the current in this load using each of the equivalent circuit models.

(i)   **16.8.**   Obtain the Thévenin and Norton equivalent circuits for the network shown in Fig. 16.40, between the two terminals *a* and *b*. The current source **I** is zero in this case.

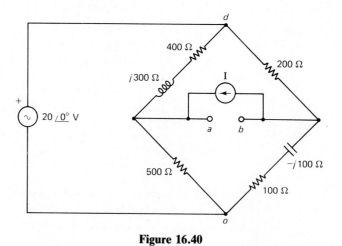

Figure 16.40

(i)   **16.9.**   Repeat Problem 16.8 if the current source **I** now has the value $1/90°$ A.

(i)   ***16.10.**   Find the Thévenin equivalent for the network shown in Fig. 16.41 between the

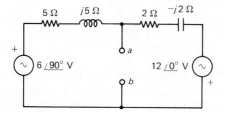

Figure 16.41

terminals a and b. Calculate the power dissipated in an impedance of $(3 + j4)\ \Omega$ when connected to terminals a and b.

(ii) ***16.11.** Find the Thévenin equivalent between terminals a and b for the network shown in Fig. 16.42. Calculate the voltage across a $10\ \Omega$ resistive load when connected between the a and b terminals.

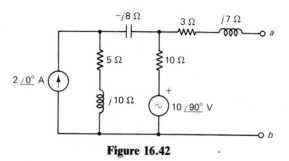

Figure 16.42

(i) ***16.12.** Find the Thévenin and Norton equivalents for the network shown in Fig. 16.43, between terminals a and b. Calculate the current in a load impedance $Z_L = 10/60°\ \Omega$ when connected between the terminals a and b.

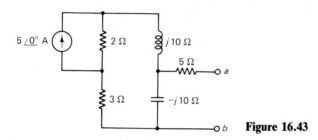

Figure 16.43

(i) **16.13.** Obtain the Thévenin and Norton equivalents for the network shown in Fig. 16.44 between terminals a and b. Check the source equivalency condition for the two circuit models obtained.

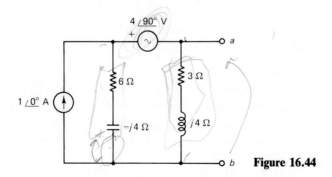

Figure 16.44

(ii) ***16.14.** Find the Thévenin equivalent for the network shown in Fig. 16.45 between terminals a and b. Calculate the power dissipated in a $(5 - j5)\ \Omega$ impedance when connected between the a and b terminals.

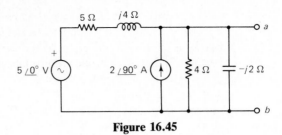

Figure 16.45

(ii) **16.15.** Find the Thévenin equivalent between terminals a and b of the network shown in Fig. 16.46. Notice that this network contains a dependent current source.

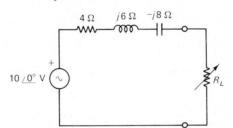

Figure 16.46

Section 16.3

(i) **16.16.** In the circuit shown in Fig. 16.47, find the value of R_L which results in maximum power transfer to the load? What is the value of this maximum load power?

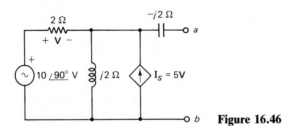

Figure 16.47

(i) **16.17.** If the load in problem 16.16 is completely arbitrary, what should be its value Z_L to achieve maximum power transfer to this load? What is the value of this maximum power?

(i) **16.18.** For the circuit considered in Problem 16.6, find the value of Z_L necessary to produce maximum power transfer to the load Z_L when connected between the terminals a and b. What is the value of this maximum load power?
  (a) Z_L is completely arbitrary.
  (b) Z_L consists of a variable resistance only.

(i) **16.19.** Repeat Problem 16.18 for the network considered in Problem 16.7.

(i) **16.20.** Repeat Problem 16.18 for the network considered in Problem 16.10.

(i) **16.21.** Repeat Problem 16.18 for the network considered in Problem 16.11.

(ii) **16.22.** Find the load Z_L required to achieve maximum power transfer to the load when it is connected to terminals a and b of the network considered in Problem 16.12 if
  (a) Z_L is completely arbitrary.
  (b) $Z_L = R + j5$, where R is the only variable element.
  Find the maximum load power in each case.

(i) **16.23.** Repeat Problem 16.22 for the network considered in Problem 16.14.

GLOSSARY

Maximum Power Transfer Theorem: The theorem that establishes the condition(s) necessary to ensure that maximum power is produced in a load connected to a given network or source.

Norton's Theorem: The theorem through which any given two-terminal linear network can be replaced between those two terminals by an equivalent practical current source model consisting of an ideal current source in parallel with a single impedance branch. The equivalent model produces the same response in the load connected between the two terminals as that produced by the original network.

Superposition Theorem: A method of network analysis that permits the determination of the response (current or voltage) in an element as an algebraic phasor (or time-domain) sum of the partial responses due to the influence of each of the sources in the network acting alone and independently, while the other sources are temporarily killed.

Thévenin's Theorem: The theorem through which any given two-terminal linear network can be replaced between those two terminals by an equivalent practical voltage source model consisting of an ideal voltage source in series with a single impedance. The equivalent model produces the same response in the load connected between the two terminals as that produced by the original network.

ANSWERS TO DRILL PROBLEMS

16.1. $\mathbf{V} = 16.01\underline{/-106.77°}$ V

16.3. $\mathbf{V} = 4.879\underline{/-11.53°}$ V

16.5. $\mathbf{V}_{th} = 35.5317\underline{/-129.29°}$ V,
$Z_{th} = 4.25 + j5.25\ \Omega$,
$\mathbf{V} = 16.01\underline{/-106.77°}$ V

16.7. $\mathbf{I}_N = 2.1095\underline{/-31.42°}$ A,
$Z_N = 1 - j3$
$\quad = 3.1623\underline{/-71.57°}\ \Omega$,
$\mathbf{I}_N \cdot \mathbf{Z_N} = 6.6708\underline{/-103°} = \mathbf{V}_{th}$

16.9. $\mathbf{I}_N = 2.6667\underline{/-90°}$ A,
$Z_N = 2.4 + j1.2\ \Omega$,
$\mathbf{I} = 0.8433\underline{/-71.56°}$ A

16.11. $Z_L = 4.25 - j5.25\ \Omega$,
$P_{Lmax} = 74.2648$ W

16.13. $Z_L = 1.951 - j2.439\ \Omega$,
$P_{Lmax} = 3.0503$ W

16.15. $Z_L = 1.5 + j1.5\ \Omega$,
$P_{Lmax} = 0.3867$ W

17 | MUTUALLY COUPLED CIRCUITS AND TRANSFORMERS

OBJECTIVES

- Understanding the circuit concepts of mutual inductance as a consequence of the electromagnetic induction phenomena, and the basic interrelationships involving the induced voltage and its polarities.

- Understanding the interrelationships between self and mutual inductances of magnetically coupled coils, the coupling coefficient and the dot convention.

- Ability to analyze mutually coupled circuits using KVL, the loop-current equations, and the conductively coupled equivalent circuits.

- Understanding the operations of, the assumptions implied by, and the inter-relationships involving the ideal transformer.

- Understanding and using the concepts of impedance transformation for both the ideal and the practical transformers.

- Ability to analyze circuits containing autotransformers.

The multiloop or multinode networks considered thus far are called *conductively coupled circuits*. A loop affects the loop beside it through the current conduction in the mutual (or common) impedance between the loops. The same current conduction in the common impedance between two nodes is the cause of one node voltage affecting the node beside it.

A different type (or method) of coupling between adjacent circuits will be discussed in this chapter. Instead of the coupling occurring through the contact of the common element, here the coupling occurs without contact between the adjacent circuits. Here the coupling occurs through the magnetic fields generated by one or more of the adjacent circuits. Such circuits are then said to be *magnetically*, or *mutually*, or *inductively coupled*.

17.1 SELF- AND MUTUAL INDUCTANCE

Consider the two coils in Fig. 17.1. They are in close physical proximity of each other. Coil 1 has N_1 turns and is referred to as the *primary coil* (or winding) and coil 2 has N_2 turns and is referred to as the *secondary coil*. The space between the coils could be air, or the coils could be wound on a common core made of iron, ferrite, or even plastic.

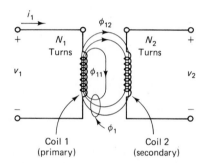

Coil 1 (primary)   Coil 2 (secondary)   **Figure 17.1**

When a time-varying current i_1 flows in coil 1, it generates a time-varying magnetic flux ϕ_1 in the space (or field) surrounding this coil. Part of this flux, ϕ_{12}, will link with coil 2. This part is called the *mutual flux*. Another part of this flux, ϕ_{11}, will not link with coil 2. This last part is then called the *leakage flux* of coil 1. This is clearly shown in Fig. 17.1. Here

$$\phi_1 = \phi_{11} + \phi_{12} \tag{17.1}$$

According to Faraday's law, the varying magnetic fields will cause the generation of induced voltages, as the varying flux lines cut across the conductor from which the coil is made. Such induced voltages are proportional to the rate of change of the flux linkage with the coil. In the case of the coils in Fig. 17.1,

$$v_1 = N_1 \frac{d\phi_1}{dt} \tag{17.2}$$

as all of the flux ϕ_1 links coil 1, while

$$v_2 = \pm N_2 \frac{d\phi_{12}}{dt} \tag{17.3}$$

because only ϕ_{12} links with coil 2. The polarity of v_2 depends on the winding sense of the two coils. This induced voltage polarity (of v_2) is determined through the application of Lenz's law, as will be discussed later.

From the study of the electromagnetic induction phenomena in Chapters 9 and 10, it is well known that the flux is proportional to the current producing it. Thus

$$\phi_1 = \frac{N_1 i_1}{\mathcal{R}_1} \tag{17.4}$$

and

$$\phi_{12} = \frac{N_1 i_1}{\mathcal{R}_{12}} \tag{17.5}$$

The two relationships above represent Ohm's law for the magnetic circuits, with

$$\mathcal{R} = \text{reluctance of the magnetic path of } \phi_1 = \frac{l_1}{\mu A_1}$$

and

$$\mathcal{R}_{12} = \text{reluctance of the magnetic path of } \phi_{12} = \frac{l_2}{\mu A_{12}}$$

The length and cross-sectional dimensions are effective (or average) values.

Substituting from Eqs. (17.4) and (17.5) into Eqs. (17.2) and (17.3), respectively, produces

$$\boxed{v_1 = \frac{N_1^2}{\mathcal{R}_1} \frac{di_1}{dt} = L_1 \frac{di_1}{dt}} \tag{17.6}$$

and

$$\boxed{v_2 = \pm \frac{N_1 N_2}{\mathcal{R}_{12}} \frac{di_1}{dt} = \pm M \frac{di_1}{dt}} \tag{17.7}$$

In Eqs. (17.6) and (17.7), the constants of proportionalities are

$$L_1 = \text{self-inductance of coil 1} = \frac{N_1^2}{\mathcal{R}_1} \text{ H} \tag{17.8}$$

and

$$M = \text{mutual inductance between the two coils}$$

$$= \frac{N_1 N_2}{\mathcal{R}_{12}} \text{ H} \tag{17.9}$$

Both L_1 and M obviously have the same physical units (the henry), and both are constants, depending on the physical parameters and dimensions of the magnetic flux paths.

In terms of the self-inductance L_1 and the mutual inductance M, Eqs. (17.4) and (17.5) can be rewritten as follows:

$$N_1 \phi_1 = \frac{N_1^2}{\mathcal{R}_1} i_1 = L_1 i_1 \tag{17.10a}$$

or

$$L_1 = \frac{N_1\phi_1}{i_1} \tag{17.10b}$$

and

$$N_2\phi_{12} = \frac{N_1 N_2}{\mathscr{R}_{12}} i_1 = Mi_1 \tag{17.11a}$$

or

$$M = \frac{N_2\phi_{12}}{i_1} \tag{17.11b}$$

It is clear from Eqs. (17.6) and (17.7) that if the current i does not vary with time, as is the case in dc source excitation, di_1/dt is zero. Thus the voltage drop across coil 1, v_1, and the mutually induced voltage drop across coil 2, v_2 (which resembles a source), are both zero. Mutual coupling is therefore evident only in ac circuits, or circuits that are driven by time-varying sources.

The process analyzed above is reversible as shown in Fig. 17.2. Here a current i_2 is flowing in coil 2, producing the magnetic flux ϕ_2. Part of this flux will link the two coils, ϕ_{21}, and part will not link coil 1, ϕ_{22}. Again here ϕ_{21} is called the mutual flux and ϕ_{22} is called the leakage flux. Here

$$\phi_2 = \phi_{21} + \phi_{22} \tag{17.12}$$

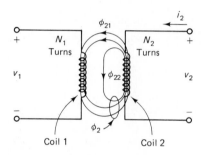

Coil 1   Coil 2   **Figure 17.2**

From Faraday's law

$$v_1 = \pm N_1 \frac{d\phi_{21}}{dt} \tag{17.13}$$

and

$$v_2 = N_2 \frac{d\phi_2}{dt} \tag{17.14}$$

Also here the polarity of the mutually induced voltage v_1 (which resembles a source) will be determined through the application of Lenz's law. As mentioned above, such polarity depends on the winding sense of the two coils.

The magnetic fluxes ϕ_2 and ϕ_{21} are proportional to the current i_2 that produced them. Here

$$\phi_2 = \frac{N_2 i_2}{\mathscr{R}_2} \tag{17.15}$$

and

$$\phi_{21} = \frac{N_2 i_2}{\mathscr{R}_{21}} \tag{17.16}$$

17.1   Self- and Mutual Inductance

$\mathcal{R}_2$ is the reluctance of the magnetic path of ϕ_2, and $\mathcal{R}_{21}$ is the reluctance of the path of the mutual flux ϕ_{21}. Since the path of the mutual flux between the two coils is physically the same for ϕ_{21} as well as for ϕ_{12}, then

$$\mathcal{R}_{21} = \mathcal{R}_{12} \qquad (17.17)$$

This quantity depends only on the physical dimensions and parameters of the magnetic material constituting the flux path between the two coils.

Substituting from Eqs. (17.15) and (17.16) into Eqs. (17.14) and (17.13), respectively, results in

$$v_1 = \pm \frac{N_1 N_2}{\mathcal{R}_{21}} \frac{di_2}{dt} = \pm M \frac{di_2}{dt} \qquad (17.18)$$

and

$$v_2 = \frac{N_2^2}{\mathcal{R}_2} \frac{di_2}{dt} = L_2 \frac{di_2}{dt} \qquad (17.19)$$

The mutual inductance M is exactly the same as before, since

$$M = \frac{N_1 N_2}{\mathcal{R}_{12}} = \frac{N_1 N_2}{\mathcal{R}_{21}} \qquad (17.20)$$

while the self-inductance of the second coil, L_2 is given by

$$L_2 = \frac{N_2^2}{\mathcal{R}_2} \qquad (17.21)$$

As in the previous case, Eqs. (17.15) and (17.16) can be expressed in terms of L_2 and M as follows:

$$N_2 \phi_2 = \frac{N_2^2}{\mathcal{R}_2} i_2 = L_2 i_2 \qquad (17.22a)$$

or

$$L_2 = \frac{N_2 \phi_2}{i_2} \qquad (17.22b)$$

and

$$N_1 \phi_{21} = \frac{N_1 N_2}{\mathcal{R}_{21}} i_2 = M i_2 \qquad (17.23a)$$

or

$$M = \frac{N_1 \phi_{21}}{i_2} \qquad (17.23b)$$

17.2 COUPLING COEFFICIENTS

The *coefficient of coupling k* is defined as the fraction of the total flux generated by either coil that mutually links, or couples, the other coil. This quantity depends on the closeness, or proximity, and the axis or orientation of the two coils. Therefore,

$$k = \frac{\phi_{12}}{\phi_1} = \frac{\phi_{12}}{\phi_{11} + \phi_{12}}$$

or

$$\phi_{12} = k\phi_1 \tag{17.24}$$

and also

$$k = \frac{\phi_{21}}{\phi_2} = \frac{\phi_{21}}{\phi_{22} + \phi_{21}}$$

or

$$\phi_{21} = k\phi_2 \tag{17.25}$$

It is clear that

$$0 \leq k \leq 1 \tag{17.26}$$

When the flux linkage between the two coils is a very small fraction of the total flux, that is,

$$\phi_{12} \ll \phi_1 \quad \text{or} \quad \phi_{21} \ll \phi_2$$

then k is very small (≈ 0). The coils are then said to be *loosely coupled*.

On the other hand, when the leakage flux is very small compared to the mutual (or linkage) flux, that is,

$$\phi_{11} \ll \phi_{12} \quad \text{or} \quad \phi_{22} \ll \phi_{21}$$

then k is approximately 1, its maximum value, indicating that almost all the flux generated by either coil links with the other coil. The coils are then said to be *closely coupled*.

From Eqs. (17.10b) and (17.11b),

$$\frac{M}{L_1} = \frac{N_2}{N_1}\frac{\phi_{12}}{\phi_1} = \frac{N_2}{N_1}k \tag{17.27}$$

and from Eqs. (17.22b) and (17.23b),

$$\frac{M}{L_2} = \frac{N_1}{N_2}\frac{\phi_{21}}{\phi_2} = \frac{N_1}{N_2}k \tag{17.28}$$

Therefore,

$$\frac{M}{L_1}\frac{M}{L_2} = \frac{N_2}{N_1}k\frac{N_1}{N_2}k = k^2$$

and

$$\boxed{M = k\sqrt{L_1 L_2}} \tag{17.29}$$

Based on the discussions above, M is very small (≈ 0) for the case of loosely coupled coils, while for the case of closely coupled coils, with $k \approx 1$, M is very close to its maximum value: $\sqrt{L_1 L_2}$.

A very important observation can be deduced from Eqs. (17.27) and (17.28):

$$k = \frac{N_1 M}{N_2 L_1} = \frac{N_2 M}{N_1 L_2} \tag{17.30}$$

Therefore,

$$\boxed{\dfrac{L_1}{L_2} = \left(\dfrac{N_1}{N_2}\right)^2}$$

(17.31)

In all the derivations above, the magnetic material between the two coils is assumed to be linear (i.e., B versus H or ϕ versus Ni is a straight-line relationship). This, of course, makes the use of the linear form of Ohm's law for the magnetic circuits valid [e.g., Eqs. (17.4), (17.5), (17.15), and (17.16)]. In most practical applications, this assumption is approximately correct, at least in a limited range of the B–H curve of the magnetic material, which could be air, iron core, or ferrite core.

17.3 DOT CONVENTION: POLARITY OF THE MUTUALLY INDUCED VOLTAGE

When the current i_1 flows in the primary winding (coil 1), it causes a voltage drop across the self-inductance L_1 of this coil, given by Eq. (17.6):

$$v_{1L} = L_1 \dfrac{di_1}{dt}$$

with the normal voltage-drop polarity convention. As explained above, this current also results in a mutually induced voltage (i.e., a generated source) in the magnetically coupled secondary winding (coil 2). This voltage is given by Eq. (17.7):

$$v_{2M} = \pm M \dfrac{di_1}{dt}$$

Once the secondary circuit is closed, including either a load impedance Z_2 or another independent voltage source e_2, or both, the current i_2 will flow in this secondary coil. The flow of this current, i_2, will also simultaneously produce:

1. A voltage drop across the self-inductance L_2, given by Eq. (17.19):

$$v_{2L} = L_2 \dfrac{di_2}{dt}$$

   with polarities consistent with the normal voltage drop convention.

2. A mutually induced voltage (i.e., a generated source) in the primary coil, given by Eq. (17.18):

$$v_{1M} = \pm M \dfrac{di_2}{dt}$$

The combined effects of these voltage drops should be accounted for when writing the loop equations (KVL) for both the primary and the secondary circuits. This will be discussed in the following section. However, before writing the proper loop equations, the correct polarities of the mutually induced voltages in each of the two coils must be determined.

The principles discussed above are illustrated in Fig. 17.3. This figure shows

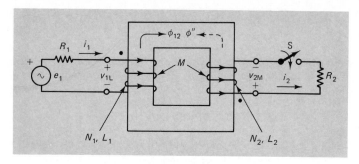

Figure 17.3

the two coils wound around an iron core. It is an example of an iron-core transformer. The source e_1 drives the current i_1 in coil 1, the primary winding. With the winding directions for the two coils as indicated in the figure, application of the RHR shows that the mutual flux coupling the two coils, ϕ_{12}, is clockwise as shown.

Based on Lenz's law, a flux ϕ'' should be produced by coil 2, to oppose the buildup of the flux ϕ_{12}. The application of the RHR to coil 2 indicates that ϕ'' would be produced if the current in the secondary coil flows in the direction shown in the figure. This current, i_2 is called the *natural current*. It is due to the mutually induced (or generated) voltage in the secondary winding, which in this case is $v_{2M} = M(di_1/dt)$.

As can be observed from Fig. 17.3, the polarity of this mutually induced voltage, v_{2M}, is now determined. Its determination was aided by the application of the RHR and Lenz's law, with the knowledge of the direction of winding of the two coils. Note that v_{2M} will always exist, while i_2 will flow only when the switch S is closed. The equivalent circuit of the secondary network is then as shown in Fig. 17.4. The loop equation for this secondary circuit can thus be written as follows:

$$i_2 R_2 + L_2 \frac{di_2}{dt} \quad - \quad M\frac{di_1}{dt} = 0 \tag{17.32}$$

$$\underset{\text{voltage drop } v_{2L}}{\text{self-inductance}} \qquad \underset{\text{generated voltage } v_{2M}}{\text{mutual inductance}}$$

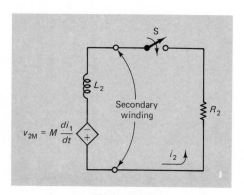

Figure 17.4

The overall voltage across the secondary winding is thus made up of two components, one due to its self-inductance L_2, and the other due to the mutual inductance M caused by the current flow in coil 1.

Now, assuming that the secondary circuit is closed, the current i_2 would be allowed to flow. This current will generate a coupling flux ϕ_{21} as shown in Fig. 17.5 (this figure is only the core and windings in Fig. 17.3 redrawn to show only the mutual inductance effects). According to the RHR, ϕ_{21} is counterclockwise as shown. Also here, Lenz's law indicates that a flux ϕ' should be generated in coil 1 to oppose the buildup of ϕ_{21}, as shown in Fig. 17.5. Through the application of the RHR, ϕ' will result when the current i_1 flows in the primary winding in the direction indicated. This current is considered to be the result of the mutually generated voltage v_{1M} $[= M(di_2/dt)]$. The polarities of this voltage are thus determined as indicated in Fig. 17.5.

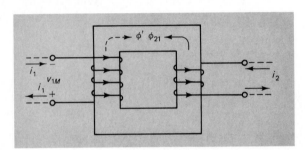

Figure 17.5

The complete equivalent circuit of the primary network is shown in Fig. 17.6, taking into account the mutually induced voltage due to the flow of the current i_2 in coil 2. The loop equation for this primary circuit can thus be written as follows:

$$i_1 R_1 + L_1 \frac{di_1}{dt} \quad - \quad M\frac{di_2}{dt} = e_1 \tag{17.33}$$

$$\underbrace{\phantom{i_1 R_1 + L_1 \frac{di_1}{dt}}}_{\substack{\text{self-inductance}\\\text{voltage drop}}} \qquad \underbrace{\phantom{M\frac{di_2}{dt}}}_{\substack{\text{mutual inductance}\\\text{generated voltage}}}$$

$$v_{1L} \qquad\qquad\qquad v_{1M}$$

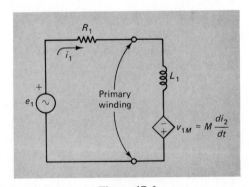

Figure 17.6

One should also note here that the total voltage across the primary winding is made up of the two components: v_{1L}, the self-inductance voltage drop, and v_{1M}, the mutually induced voltage caused by the current flow in coil 2.

It is thus clear that the polarities of the mutually induced voltage terms can be completely determined through the use of the RHR and Lenz's law. Such polarities are also dependent on the direction of winding of the two coils. Obviously, if the direction of winding of the secondary (or the primary) coil is reversed, the polarities of v_{2M} and the direction of i_2 will reverse.

Figure 17.7 shows the same two coupled coils network examined above (refer to Fig. 17.3), except that the direction of winding of the secondary coil is reversed. Following a similar discussion to that carried out above, one can easily see that the direction of the natural current i_2 and therefore the mutually induced voltage in the secondary coil, v_{2M}, are both reversed compared to the previous case. However, the complete analysis of the situation illustrated in Fig. 17.7, in a manner parallel to that outlined above, reveals that the loop equations for the primary and the secondary circuits will be exactly the same as obtained in Eqs. (17.32) and (17.33).

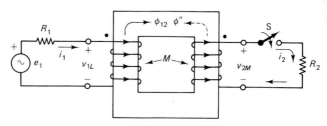

Figure 17.7

To simplify the analysis of mutually coupled circuits and to eliminate the need for showing and examining the winding directions of the two coils, a method is devised to determine the polarity of the mutually induced voltages, v_{1M} and v_{2M}, from the circuit diagrams. This method is referred to as the *dot convention*. It is explained below.

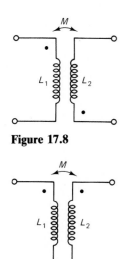

Figure 17.8

Figure 17.9

1. *Dot determination.* One terminal of each of the mutually coupled coils, the primary and the secondary, is marked with a dot on the corresponding circuit diagrams. Such dotted terminals are indicated in Figs. 17.3 and 17.7. The symbolic representations of these two cases in circuit diagrams are shown in Figs. 17.8 and 17.9, respectively. The determination of which of the terminals of the two coils is to be marked with a dot, and the meaning of such a convention, are discussed below.

Consider only the core and the two coils. Take one terminal of one coil as the starting point, and mark a dot at this terminal. Allow a hypothetical current to flow into this dotted terminal (i.e., this terminal has a positive voltage drop polarity). Using the RHR and Lenz's law (as explained above), determine the direction of the natural current in the other coil (or coils if there are more than two coupled coils). The terminal from which this *natural current leaves* the other coil(s) is then also dotted (i.e., it also has a positive voltage polarity due to the mutually induced voltage). Both (or all) of the dotted terminals, one for each of the coils, will then have the same instantaneous voltage polarity. This means that the voltage waveforms at these dot-marked terminals are in phase.

2. *The Dot Rule.* Once the dotted terminals are determined, or given on a circuit diagram (e.g., provided by the manufacturer), they are used in the circuit analysis of mutually coupled networks to determine the polarities of the mutually induced voltages. For each of the circuits in which the coupled coils are connected, a certain direction of current flow is assumed. Then:

a. When one current enters a coil through its dotted terminal, while the other current(s) leaves the other coil(s) from its dotted terminal, the signs of the mutually induced voltage terms in the loop equations are opposite to the signs of the self-inductance voltage-drop terms. This is the situation in the analysis above for both of Figs. 17.3 and 17.7, resulting in the corresponding positive and negative signs in Eqs. (17.32) and (17.33).

b. When both of the assumed currents enter the coupled coils, or both leave the coupled coils, through the dotted terminals, the signs of the mutually induced voltage terms are the same as the signs of the self-inductance voltage-drop terms in the loop equations. This situation may be encountered when each of the coupled circuits contains an independent voltage source.

The dot rule is quite general and its use in any circuit situation will result in the proper identification and determination of the polarities of the mutually induced voltage terms. Its application in practice will be shown in the examples discussed below.

17.4 ANALYSIS OF CIRCUITS CONTAINING COUPLED COILS

Consider the two-mutually coupled coils, or transformer, circuit shown in Fig. 17.10. The currents in the primary and the secondary circuits are assumed to flow as indicated. Since both i_1 and i_2 enter the two dotted terminals of the two coils, then according to case b of the dot rule, the signs of the mutually induced voltages will be the same as the signs of the self-inductance voltage-drop terms in both loops.

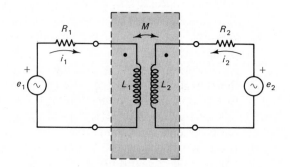

Figure 17.10

Thus the loop equation for the primary circuit can be written as follows:

$$e_1 = i_1 R_1 + L_1 \frac{di_1}{dt} + M \frac{di_2}{dt} \tag{17.34}$$

and the loop equation for the secondary circuit is, similarly,

$$e_2 = i_2 R_2 + L_2 \frac{di_2}{dt} + M \frac{di_1}{dt} \tag{17.35}$$

When the sources generate sinusoidal waveforms of the same frequency, the two loop equations above can be rewritten in terms of phasors and impedances, noting that the derivative d/dt in the time domain corresponds to $j\omega$ in the frequency domain. Thus

$$\boxed{\mathbf{E}_1 = (R_1 + j\omega L_1)\mathbf{I}_1 + j\omega M \mathbf{I}_2} \tag{17.36}$$

and

$$\boxed{\mathbf{E}_2 = j\omega M \mathbf{I}_1 + (R_2 + j\omega L_2)\mathbf{I}_2} \tag{17.37}$$

These two loop equations can then be solved to determine the two loop currents. A different situation is shown in the circuit of Fig. 17.11, where the primary

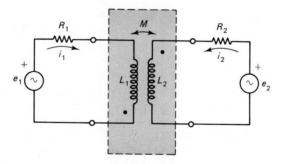

Figure 17.11

winding has been reversed. The directions of the currents flowing in both the primary and the secondary circuits are assumed to be as before. Here the current i_2 enters the secondary coil from its dotted terminal, while the current i_1 leaves the primary coil through its dotted terminal. Thus, according to case a of the dot rule, the mutually induced voltage terms have opposite signs in comparison with the self-inductance voltage-drop terms. The phasor-impedance forms of the corresponding loop equations (in the frequency domain) are therefore

$$\boxed{\mathbf{E}_1 = (R_1 + j\omega L_1)\mathbf{I}_1 - j\omega M \mathbf{I}_2} \tag{17.38}$$

and

$$\boxed{\mathbf{E}_2 = -j\omega M \mathbf{I}_1 + (R_2 + j\omega L_2)\mathbf{I}_2}$$ (17.39)

The solution of these two loop equations will then determine the values of the two loop currents in this case.

17.4.1 Conductively Coupled Equivalent Circuits

Other mutually coupled circuits can also be analyzed in a manner similar to that outlined above. The loop equations obtained for the cases considered in Figs. 17.10 and 17.11 could have resulted from the analysis of a conductively coupled two-loop circuit as shown in the general network of Fig. 17.12. Such a circuit is called the *conductively coupled equivalent circuit* to the original mutually (or magnetically) coupled circuit. In many cases, as will be apparent shortly, such equivalent circuits could contain negative elements! In other words, they may be physically unrealizable, but they are still valid mathematically equivalent models.

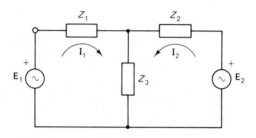

Figure 17.12

The two loop equations for the network of Fig. 17.12 are

$$\mathbf{E}_1 = (Z_1 + Z_3)\mathbf{I}_1 + Z_3\mathbf{I}_2$$ (17.40)

and

$$\mathbf{E}_2 = Z_3\mathbf{I}_1 + (Z_2 + Z_3)\mathbf{I}_2$$ (17.41)

Note that the common impedance term, Z_3, has a positive sign, as both loop currents $\mathbf{I}_1$ and $\mathbf{I}_2$ flow in it in the same direction, aiding each other.

Through comparing Eqs. (17.36) and (17.37) to Eqs. (17.40) and (17.41), respectively, the following conclusions can be made:

$$Z_3 = j\omega M$$
$$Z_1 + Z_3 = R_1 + j\omega L_1$$ (17.42)

Therefore,

$$Z_1 = R_1 + j\omega(L_1 - M)$$ (17.43)

Similarly,

$$Z_2 + Z_3 = R_2 + j\omega L_2$$

and

$$Z_2 = R_2 + j\omega(L_2 - M) \tag{17.44}$$

Using the impedance values obtained above, the circuit shown in Fig. 17.13 is therefore the conductively coupled equivalent circuit to the mutually coupled circuit of Fig. 17.10. The mutually coupled coils in the dashed box of Fig. 17.10 are thus equivalent to the three inductor arrangement in the dashed box of Fig. 17.13. M could be larger (or smaller) than either L_1 or L_2. If it is larger than either self-inductance, negative elements may result.

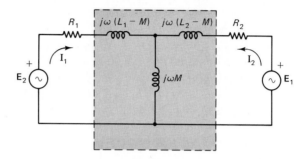

Figure 17.13

Similarly, by comparing Eqs. (17.38) and (17.39) to Eqs. (17.40) and (17.41), the following conclusions can be made:

$$Z_3 = -j\omega M$$
$$Z_1 + Z_3 = R_1 + j\omega L_1 \tag{17.45}$$

Therefore,

$$Z_1 = R_1 + j\omega(L_1 + M) \tag{17.46}$$

Similarly,

$$Z_2 + Z_3 = R_2 + j\omega L_2$$

and

$$Z_2 = R_2 + j\omega(L_2 + M) \tag{17.47}$$

These impedance values thus result in the circuit shown in Fig. 17.14. It is therefore the conductively coupled equivalent circuit to the mutually coupled

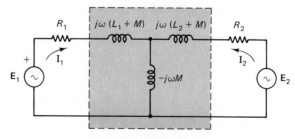

Figure 17.14

coils inside the dashed box of Fig. 17.11. Here the two mutually coupled coils inside the dashed box of Fig. 17.11 are equivalent to the three inductor arrangement inside the dashed box of Fig. 17.14. It is obvious here that the common impedance branch, Z_3, has a negative inductor element $(-M)$. It should also be noted that the reversal of the dotted terminal does account for the change in the sign of the M terms of the conductively coupled equivalent circuit. In terms of reactances,

$$X_M = \omega M = k\sqrt{\omega^2 L_1 L_2} = k\sqrt{X_{L_1} X_{L_2}} \tag{17.48}$$

EXAMPLE 17.1

Two coupled coils have self-inductances of $L_1 = 100\,mH$ and $L_2 = 400\,mH$. The coupling coefficient k is 0.8. Find M. If N_1 is 1000 turns, what is the value of N_2? The current $i_1 = 2\sin(500t)$ A enters the coil 1, find the flux ϕ_1 and the mutually induced voltage v_{2M}.

Solution   Since

$$M = k\sqrt{L_1 L_2}$$

then

$$M = 0.8\sqrt{0.1 \times 0.4} = 0.16\,H = 160\,mH$$

Note that

$$L_1 < M < L_2$$

Also,

$$\frac{L_1}{L_2} = \left(\frac{N_1}{N_2}\right)^2$$

Therefore,

$$N_2 = N_1\sqrt{\frac{L_2}{L_1}} = 1000\sqrt{\frac{0.4}{0.1}} = 2000\ \text{turns}$$

From Eq. (17.7),

$$v_{2M} = M\frac{di_1}{dt} = 0.16\frac{d}{dt}(2\sin 500t)$$

$$= 0.16 \times 2 \times 500\cos 500t$$

$$= 160\sin(500t + 90°)\ \text{V}$$

(The determination of the polarity of v_{2M} is not required in this example.) Now

$$\phi_1 = \frac{1}{k}\phi_{12} = \frac{1}{k}\frac{Mi_1}{N_2} \quad \text{[using Eq. (17.11a)]}$$

$$= \frac{L_1 i_1}{N_1} \quad\quad\quad \text{[using Eq. (17.10a)]}$$

$$= \frac{0.1}{1000} \times 2\sin(500t) = 2 \times 10^{-4}\sin(500t)\ \text{Wb}$$

(Either equality gives the same result.)

EXAMPLE 17.2

Determine the dotted terminals of the coupled coils shown in Fig. 17.15, then find the expression for the circuit's current, **I**.

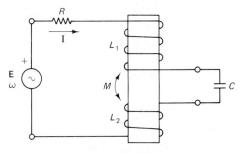

Figure 17.15

Solution   The two coils are considered separately, as shown in Fig. 17.16(a). Here i_1 enters coil 1 from the first dotted terminal. The application of the RHR results in the coupling flux ϕ_{12} shown. From Lenz's law, ϕ' is opposing the buildup of the flux ϕ_{12}. Finally, applying the RHR to coil 2 shows that the natural current i_2 (causing the flux ϕ') should leave from that terminal of coil 2, which is then dotted, as Fig. 17.16(a) indicates. The dotted terminals are thus determined. The circuit is then redrawn as shown in Fig. 17.16(b).

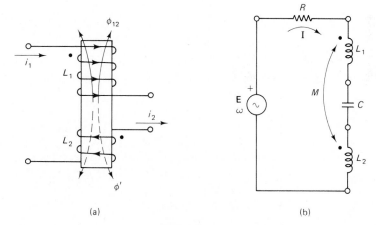

(a)   (b)

Figure 17.16

The current **I** enters each of the two coils from its dotted terminal. Thus each coil generates a mutually induced voltage in the other coil, with the same polarity as the self-inductance voltage-drop terms. Therefore,

$$\mathbf{E} = R\mathbf{I} + j\omega L_1\mathbf{I} + j\omega M\mathbf{I} + \frac{1}{j\omega C}\mathbf{I} + j\omega L_2\mathbf{I} + j\omega M\mathbf{I}$$

$$= \mathbf{I}\left(R + j\omega L_{\text{eq}} - j\frac{1}{\omega C}\right) = \mathbf{I}Z_T$$

where
$$L_{eq} = L_1 + L_2 + 2M$$

Thus
$$\mathbf{I} = \frac{\mathbf{E}}{Z_T} = \frac{\mathbf{E}}{R + j\omega L_{eq} - j(1/\omega C)} \ \text{A}$$

EXAMPLE 17.3

Figure 17.17

Two mutually coupled inductors L_1 and L_2 are shown in Fig. 17.17.
(a) Determine M considering series connections.
(b) Find the equivalent inductance of the two coils when connected in parallel, with the two dotted terminals connected together.

Solution The two dotted terminals are determined following an analysis similar to that discussed in Example 17.2. Figure 17.17 indicates the results of such dot determination.

(a) When the coils are connected in series as shown in Fig. 17.18(a), the current $\mathbf{I}$ enters each of the coils from its dotted terminal, as in Example 17.2. Therefore,

$$L_A = L_1 + L_2 + 2M \qquad \text{(series-aiding connection)}$$

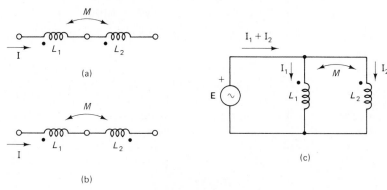

(a)

(b)

(c)

Figure 17.18

When the coils are connected in series as shown in Fig. 17.18(b), the current $\mathbf{I}$ will enter one coil from its dotted terminal and leave the other coil from its dotted terminal. Thus the mutually induced voltages have opposite signs as compared with the self-inductance voltage-drop terms. Thus here

$$L_B = L_1 + L_2 - 2M \quad \text{(series-opposing connection)}$$

Thus
$$L_A - L_B = 4M$$

or

$$M = \frac{L_A - L_B}{4}$$

(b) The two coils are connected in parallel as shown schematically in Fig. 17.18(c). Consider the source $\mathbf{E}$ connected in parallel with the two coils. As is clear from the circuit connection,

$$\text{source current} = \mathbf{I}_1 + \mathbf{I}_2$$

and

$$Z_{eq} = \frac{\mathbf{E}}{\mathbf{I}_1 + \mathbf{I}_2}$$

Each of the two branch currents enters the corresponding coil from the dotted terminal. Thus the mutually induced voltages have the same signs as the self-inductance voltage-drop terms. Therefore, using KVL for each of the two branches yields

$$\mathbf{E} = j\omega L_1 \mathbf{I}_1 + j\omega M \mathbf{I}_2$$

and

$$\mathbf{E} = j\omega M \mathbf{I}_1 + j\omega L_2 \mathbf{I}_2$$

Cramer's rule can be used to solve these two equations to determine $\mathbf{I}_1$ and $\mathbf{I}_2$:

$$\mathbf{I}_1 = \frac{\begin{vmatrix} \mathbf{E} & j\omega M \\ \mathbf{E} & j\omega L_2 \end{vmatrix}}{\begin{vmatrix} j\omega L_1 & j\omega M \\ j\omega M & j\omega L_2 \end{vmatrix}} = \frac{\mathbf{E}j\omega(L_2 - M)}{-\omega^2 L_1 L_2 + \omega^2 M^2}$$

and

$$\mathbf{I}_2 = \frac{\begin{vmatrix} j\omega L_1 & \mathbf{E} \\ j\omega M & \mathbf{E} \end{vmatrix}}{\Delta} = \frac{\mathbf{E}j\omega(L_1 - M)}{-\omega^2 L_1 L_2 + \omega^2 M^2}$$

$$\mathbf{I}_1 + \mathbf{I}_2 = \mathbf{E}\frac{j\omega(L_1 + L_2 - 2M)}{-\omega^2(L_1 L_2 - M^2)}$$

Therefore,

$$Z_{eq} = \frac{\mathbf{E}}{\mathbf{I}_1 + \mathbf{I}_2} = \frac{j^2\omega^2(L_1 L_2 - M^2)}{j\omega(L_1 + L_2 - 2M)} = j\omega L_{eq}$$

where

$$L_{eq} = \frac{L_1 L_2 - M^2}{L_1 + L_2 - 2M}$$

EXAMPLE 17.4

Obtain the dotted equivalent circuit for the coupled coils in the network shown in Fig. 17.19. Use this circuit to determine the voltage across the capacitor. Note that the circuit diagram given indicates Z_{L_1}, Z_{L_2}, and Z_M directly.

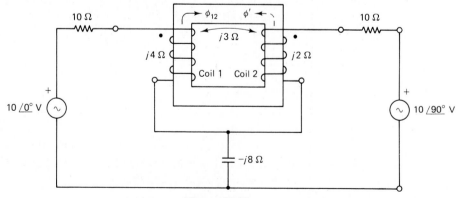

Figure 17.19

Solution By considering only the two coils in Fig. 17.19, and assuming that a current enters the top terminal of coil 1 (dotted), it is easy to apply the RHR and Lenz's law to the two coils. In the figure, the mutual flux ϕ_{12} and the opposing flux ϕ' are indicated. Finally, when the RHR is applied to coil 2, the natural current that causes ϕ' will be seen to leave coil 2 from its top terminal, which is then also dotted.

The complete circuit is then redrawn as shown in Fig. 17.20. The two loop currents are assumed to flow in the directions indicated in the figure. One loop current enters one dotted terminal (coil 1) while the other loop current leaves the other dotted terminal (coil 2). The mutually induced voltages will therefore have the opposite polarities compared to the self-inductance voltage-drop terms. The two loop equations can thus be written as follows:

$$10 = (10 + j4 - j8)\mathbf{I}_1 - j3\mathbf{I}_2 - (-j8)\mathbf{I}_2$$
$$= (10 - j4)\mathbf{I}_1 + j5\mathbf{I}_2$$

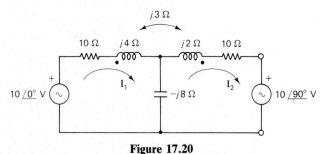

Figure 17.20

Note that $-j3\mathbf{I}_2$ is the mutually induced voltage term, and

$$-j10 = \mathbf{I}_2(10 + j2 - j8) - j3\mathbf{I}_1 - (-j8)\mathbf{I}_1$$
$$= j5\mathbf{I}_1 + (10 - j6)\mathbf{I}_2$$

Notice the symmetry in the impedance determinant (i.e., $Z_{12} = Z_{21}$). Using Cramer's rule to solve for the two currents, we have

$$\mathbf{I}_1 = \frac{\begin{vmatrix} 10 & j5 \\ -j10 & 10 - j6 \end{vmatrix}}{\begin{vmatrix} 10 - j4 & j5 \\ j5 & 10 - j6 \end{vmatrix}} = \frac{100 - j60 - 50}{100 - 24 - j40 - j60 + 25} = \frac{50 - j60}{101 - j100}$$

$$= \frac{78.1/-50.2°}{142.1/-44.7°} = 0.549/-5.5° \text{ A}$$

$$= 0.547 - j0.053 \text{ A}$$

and

$$\mathbf{I}_2 = \frac{\begin{vmatrix} 10 - j4 & 10 \\ j5 & -j10 \end{vmatrix}}{\Delta} = \frac{-j100 - 40 - j50}{101 - j100} = \frac{-40 - j150}{101 - j100}$$

$$= \frac{155.2/-104.9°}{142.1/-44.7°} = 1.092/-60.2° \text{ A}$$

$$= 0.543 - j0.948 \text{ A}$$

Thus the voltage across the capacitor is

$$\mathbf{V}_C = -j8(\mathbf{I}_1 - \mathbf{I}_2) = -j8(0.004 + j0.895)$$
$$= 8\underline{/-90°} \times 0.895\underline{/89.7°} = 7.16\underline{/-0.3°} \text{ V}$$

EXAMPLE 17.5

Find the voltage across coil 2 in the circuit shown in Fig. 17.21.

Solution   The two loop currents are assumed to flow in a clockwise direction as shown in Fig. 17.21. The voltage across the capacitor is the same as the voltage across coil 2, since these two elements are in parallel. Thus only the value of $\mathbf{I}_2$ needs to be determined. The two loop equations can be written as follows:

$$50\underline{/45°} = (3 + j4 + j5 + 2 \times j3)\mathbf{I}_1 - (j5)\mathbf{I}_2 - (j3)\mathbf{I}_2$$

and

$$0 = (j5 - j8)\mathbf{I}_2 - (j5)\mathbf{I}_1 - (j3)\mathbf{I}_1$$

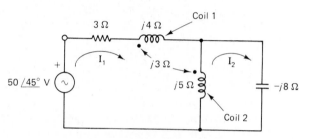

Figure 17.21

As $\mathbf{I}_1$ flows in both of the coupled coils in the first loop, entering each of the loops from its dotted terminal, the mutually induced voltages, $2 \times j3\mathbf{I}_1$, have positive polarities, similar to the voltage drops across the self-inductances. However, the loop current $\mathbf{I}_1$ enters the dotted terminal of coil 1, while the loop current $\mathbf{I}_2$ leaves the dotted terminal of coil 2; therefore, the mutually induced voltages, $j3\mathbf{I}_2$ and $j3\mathbf{I}_1$, have negative signs compared with the self-inductance voltage drops.

Rewriting the loop equations gives us

$$(3 + j15)\mathbf{I}_1 - j8\mathbf{I}_2 = 5\underline{/45°}$$
$$-j8\mathbf{I}_1 + (-j3)\mathbf{I}_1 = 0$$

Using Cramer's Rule to solve for $\mathbf{I}_2$ yields

$$\mathbf{I}_2 = \frac{\begin{vmatrix} 3 + j15 & 5\underline{/45°} \\ -j8 & 0 \end{vmatrix}}{\begin{vmatrix} 3 + j15 & -j8 \\ -j8 & -j3 \end{vmatrix}} = \frac{8\underline{/90°} \times 5\underline{/45°}}{-j9 + 45 + 64} = \frac{400\underline{/135°}}{109 - j9}$$
$$= 3.657\underline{/139.7°} \text{ A}$$

Therefore,

$$\mathbf{V}_{\text{Coil 2}} = \mathbf{V}_C = \mathbf{I}_2(-j8) = 3.657\underline{/139.7°} \times 8\underline{/-90°}$$
$$= 29.26\underline{/49.7°} \text{ V}$$

EXAMPLE 17.6

Obtain the Thévenin equivalent circuit between terminals a and b for the network shown in Fig. 17.22.

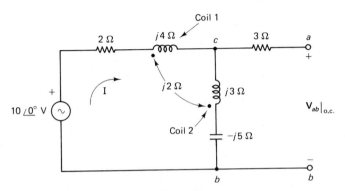

Figure 17.22

Solution   With terminals a and b open circuited, only the current $\mathbf{I}$ flows in the network as shown in Fig. 17.22. Because $\mathbf{I}$ enters one coil from its dotted terminal and leaves the other coil from its dotted terminal, the mutually induced voltages between the two coils will have the opposite polarity with respect to the self-inductance voltage drops. Here

$$10 = (2 + j4 + j3 - 2 \times j2 - j5)\mathbf{I}$$
$$= (2 - j2)\mathbf{I}$$

Therefore,

$$\mathbf{I} = \frac{10}{2 - j2} = \frac{10}{2.828\underline{/-45°}} = 3.536\underline{/45°}$$
$$= 2.5 + j2.5 \text{ A}$$

Thus

$$\mathbf{V}_{\text{th}} = \mathbf{V}_{ab}|_{oc} = \mathbf{V}_{cb} = \mathbf{I}(j3 - j5) - \mathbf{I}(j2)$$

(the last term is the mutually induced voltage, being the effect of coil 1 on coil 2). Therefore,

$$\mathbf{V}_{\text{th}} = \mathbf{I}(-j4) = 3.536\underline{/45°} \times 4\underline{/-90°} = 14.14\underline{/-45°} \text{ V}$$
$$= 10 - j10 \text{ V}$$

To find Z_{th}, the independent voltage source is short circuited. A source $\mathbf{E}$ is applied across terminals a and b, driving the current $\mathbf{I}_2$; hence

$$Z_{\text{th}} = \frac{\mathbf{E}}{\mathbf{I}_2}$$

To analyze the circuit in Fig. 17.23(a), the two loop currents $\mathbf{I}_1$ and $\mathbf{I}_2$ are assumed to flow as shown in the figure. The loop current $\mathbf{I}_1$ flows in both coils, entering one and leaving the other from the dotted terminals. Thus the mutually induced voltages due to this current, $2 \times j2\mathbf{I}_1$, will have opposite polarities compared with the self-inductance terms. On the other hand, $\mathbf{I}_1$ and $\mathbf{I}_2$ both leave the corresponding coils from the dotted terminals; thus the mutually induced voltages $j2\mathbf{I}_1$ and $j2\mathbf{I}_2$ have the same polarities as the voltage drops across the self-inductances.

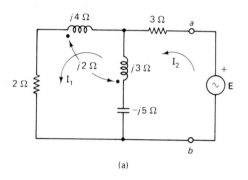

(a)

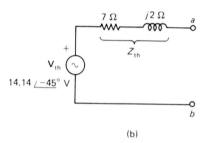

(b)

Figure 17.23

The two loop equations can then be written as

$$(2 + j4 + j3 - j5 - 2 \times j2)\mathbf{I}_1 - (j3 - j5)\mathbf{I}_2 + j2\mathbf{I}_2 = 0$$

and

$$(3 + j3 - j5)\mathbf{I}_2 - (j3 - j5)\mathbf{I}_1 + j2\mathbf{I}_1 = \mathbf{E}$$

By collecting terms, the two loop equations above can be simplified to

$$(2 - j2)\mathbf{I}_1 + j4\mathbf{I}_2 = 0$$
$$j4\mathbf{I}_1 + (3 - j2)\mathbf{I}_2 = \mathbf{E}$$

Thus

$$\mathbf{I}_2 = \frac{\begin{vmatrix} 2 - j2 & 0 \\ j4 & \mathbf{E} \end{vmatrix}}{\begin{vmatrix} 2 - j2 & j4 \\ j4 & 3 - j2 \end{vmatrix}} = \frac{\mathbf{E}(2 - j2)}{6 - 4 - j6 - j4 + 16} = \frac{\mathbf{E}(2 - j2)}{18 - j10}$$

Therefore,

$$Z_{th} = \frac{\mathbf{E}}{\mathbf{I}_2} = \frac{18 - j10}{2 - j2} = \frac{20.591\underline{/-29°}}{2.828\underline{/-45°}} = 7.28\underline{/16°} = 7 + j2 \; \Omega$$

The Thévenin equivalent circuit is thus as shown in Fig. 17.23(b).

EXAMPLE 17.7

For the mutually coupled circuit shown in Fig. 17.24, obtain the equivalent conductively coupled circuit as seen by the source.

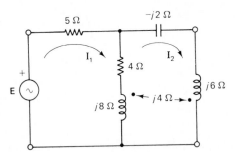

Figure 17.24

Solution Assuming the two loop currents $\mathbf{I}_1$ and $\mathbf{I}_2$, flowing clockwise as indicated, the two loop equations for the circuit in Fig. 17.24 can be written as

$$\mathbf{E} = (5 + 4 + j8)\mathbf{I}_1 - (4 + j8)\mathbf{I}_2 - j4\mathbf{I}_2$$

and

$$0 = (4 - j2 + j6 + j8 + 2 \times j4)\mathbf{I}_2 - (4 + j8)\mathbf{I}_1 - j4\mathbf{I}_1$$

The mutually induced voltage terms and their polarities are determined following the dot rule, in a manner similar to that discussed in the previous examples. The two loop equations above can be simplified as follows:

$$\mathbf{E} = (9 + j8)\mathbf{I}_1 - (4 + j12)\mathbf{I}_2$$

and

$$0 = -(4 + j12)\mathbf{I}_1 + (4 + j20)\mathbf{I}_2$$

As the topology of the original circuit suggests, the source would see a conductively coupled circuit made of a series branch, Z_1, and two parallel branches, Z_2 and Z_3, as indicated in Fig. 17.25(a). In this conductively coupled

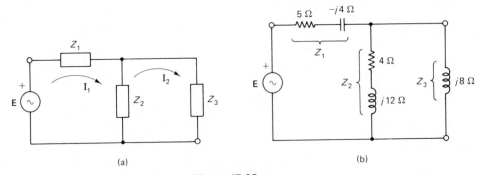

(a)   (b)

Figure 17.25

circuit no mutual inductance exists. Here

$$\mathbf{E} = (Z_1 + Z_2)\mathbf{I}_1 - Z_2\mathbf{I}_2$$

and

$$0 = -Z_2\mathbf{I}_1 + (Z_2 + Z_3)\mathbf{I}_2$$

By comparing the two sets of loop equations above, for the mutually coupled circuit and for the equivalent conductively coupled circuit, the following conclusions can be made:

$$Z_2 = 4 + j12 \ \Omega$$

$$Z_1 + Z_2 = 9 + j8 \ \Omega$$

Therefore,

and

$$Z_1 = 9 + j8 - (4 + j12) = 5 - j4 \ \Omega$$

$$Z_2 + Z_3 = 4 + j20 \ \Omega$$

Thus

$$Z_3 = 4 + j20 - (4 + j12)$$

$$= j8 \ \Omega$$

The conductively coupled equivalent circuit can thus be represented as shown in Fig. 17.25(b).

17.5 IDEAL TRANSFORMER

Consider the transformer circuit shown in Fig. 17.26. It consists of two mutually coupled coils. Later, the remainder of the network connections, (e.g., source, load, etc.) will be taken into account. The primary and secondary currents $\mathbf{I}_1$ and $\mathbf{I}_2$ are considered to flow as indicated, while the primary and the secondary voltages $\mathbf{V}_1$ and $\mathbf{V}_2$ have the polarities shown in Fig. 17.26. Accordingly, the two loop (or KVL) equations for the primary and the secondary coils can be expressed as follows:

$$\boxed{\mathbf{V}_1 = j\omega L_1\mathbf{I}_1 - j\omega M\mathbf{I}_2} \tag{17.49}$$

and

$$\boxed{\mathbf{V}_2 = j\omega M\mathbf{I}_1 - j\omega L_2\mathbf{I}_2} \tag{17.50}$$

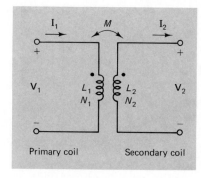

Figure 17.26

Notice that the mutually induced voltages have opposite polarities with respect to the self-inductance voltage drops, since $\mathbf{I}_1$ enters coil 1 from its dotted terminal, while $\mathbf{I}_2$ leaves coil 2 from its dotted terminal. This is in accordance with the dot rule. However, in the secondary coil, the self-inductance voltage drop has a negative sign due to the assumed direction of $\mathbf{I}_2$, according to the voltage–current polarity convention for passive elements.

The ratio of these two coil voltages is therefore

$$\frac{\mathbf{V}_1}{\mathbf{V}_2} = \frac{L_1\mathbf{I}_1 - M\mathbf{I}_2}{M\mathbf{I}_1 - L_2\mathbf{I}_2} \tag{17.51}$$

The first and most important characteristic of an ideal transformer is that its primary and secondary fluxes have no leakage components. Thus all the flux produced due to the flow of $\mathbf{I}_1$ links with the secondary coil, while all the flux produced due to the flow of $\mathbf{I}_2$ links with the primary coil. Therefore,

$$\phi_1 = \phi_{12}$$

and

$$\phi_2 = \phi_{21}$$

This means that the two coils are perfectly (closely) coupled. The coupling coefficient k as defined by Eqs. (17.24) and (17.25) is therefore

$$k = \frac{\phi_{12}}{\phi_1} = \frac{\phi_{21}}{\phi_2} = 1 \tag{17.52}$$

As a result, the mutual inductance M, from Eq. (17.29), is

$$M = k\sqrt{L_1 L_2} = \sqrt{L_1 L_2} \tag{17.53}$$

Substituting this value of M for the perfectly coupled coils in Eq. (17.51), the voltage ratio becomes

$$\frac{\mathbf{V}_1}{\mathbf{V}_2} = \frac{L_1\mathbf{I}_1 - \sqrt{L_1 L_2}\,\mathbf{I}_2}{\sqrt{L_1 L_2}\,\mathbf{I}_1 - L_2\mathbf{I}_2} = \frac{\sqrt{L_1}}{\sqrt{L_2}}\frac{\sqrt{L_1}\,\mathbf{I}_1 - \sqrt{L_2}\,\mathbf{I}_2}{\sqrt{L_1}\,\mathbf{I}_1 - \sqrt{L_2}\,\mathbf{I}_2} = \sqrt{\frac{L_1}{L_2}} \tag{17.54}$$

But from Eq. (17.31),

$$\frac{L_1}{L_2} = \left(\frac{N_1}{N_2}\right)^2$$

Therefore,

$$\frac{\mathbf{V}_1}{\mathbf{V}_2} = \frac{N_1}{N_2} = a \tag{17.55}$$

a is a constant real number, called the *turns ratio*. Equation (17.55) therefore indicates that $\mathbf{V}_1$ and $\mathbf{V}_2$ are in phase. This equation can also be written in the form

$$\frac{\mathbf{V}_1}{N_1} = \frac{\mathbf{V}_2}{N_2} = \text{transformer's volt/turn parameter} \tag{17.56}$$

Thus the ideal transformer is characterized by the fact that it has a unique constant volt/turn ratio for both the primary and secondary windings.

It is important to note that Eq. (17.55) could easily have been obtained from Eqs. (17.2) and (17.3) or from Eqs. (17.13) and (17.14). These equations described the primary and secondary voltages in the time domain in terms of the magnetic flux. However, in these equations only one current was assumed to flow at a time, either in the primary or secondary windings. Thus the derivation above is the most general since it considers the usual general case, where currents are flowing in both the primary and secondary circuits.

Another important characteristic of the ideal transformer is implied in the derivations above. Both coils are considered to be ideal inductors (i.e., each of the coils has a zero resistance). For this reason, Eqs. (17.49) and (17.50) did not account for any resistive voltage drop.

The third important property of the ideal transformer is that it does not dissipate any power. Since the two coil windings are assumed to have zero resistance, there is no electric power loss in the two coils. There is, however, another form of power loss in mutually coupled circuits. It is due to magnetic core losses, referred to as *iron losses,* as most transformers have iron cores. Two causes contribute to iron losses:

1. *Eddy-current losses.* The time-varying magnetic flux in the iron core produces an induced voltage in the magnetic core itself. Current will thus flow in the iron core, which conducts electricity. These currents are called *eddy currents.* They do therefore introduce an extra source of power loss. Losses due to eddy currents can be considerably reduced by properly designing the iron core. It is usually constructed using thin laminated magnetic sheets with thin insulating sheets between them.

2. *Hysteresis losses.* The magnetic material of the core can be considered as made up from a large number of infinitesimally small magnets. These magnets continuously change their orientations in response to the varying magnetic field. Such molecular motions encounter friction and result in power loss.

Ideal transformers have zero magnetic core losses. This implies also that the reluctance $\mathcal{R}$ of the magnetic circuit model of the iron core is zero, resembling a magnetic short circuit. Since Ohm's law for a magnetic circuit takes the form (see Chapter 9)

$$Ni = \mathcal{R}\phi$$

a zero reluctance means that no ampere turns are required to establish the magnetic flux in the core. Also, since

$$\mathcal{R} = \frac{l}{\mu A}$$

a zero reluctance implies an infinite permeability ($\mu = \infty$).

As can be observed from Eqs. (17.8) and (17.21), the zero reluctance of the magnetic circuit implies also that both L_1 and L_2 are infinite. However, the ratio L_1/L_2 will still be finite [$= (N_1/N_2)^2$], as the reluctance terms $\mathcal{R}_1$ and $\mathcal{R}_2$ cancel out at the limit of this ratio.

From Eqs. (17.49) and (17.50),

$$\left. \frac{\mathbf{V}_1}{j\omega L_1} \right|_{L_1 \to \infty} = 0 = \mathbf{I}_1 - \mathbf{I}_2 \frac{M}{L_1} = \mathbf{I}_1 - \mathbf{I}_2 \sqrt{\frac{L_2}{L_1}}$$

and

$$\frac{\mathbf{V}_2}{j\omega L_2}\bigg|_{L_2 \to \infty} = 0 = \frac{M}{L_2}\mathbf{I}_1 - \mathbf{I}_2 = \sqrt{\frac{L_1}{L_2}}\mathbf{I}_1 - \mathbf{I}_2$$

Both of these relations result in the following current ratio expression:

$$\frac{\mathbf{I}_1}{\mathbf{I}_2} = \sqrt{\frac{L_2}{L_1}} = \frac{N_2}{N_1} = \frac{1}{a} = \frac{1}{\text{turns ratio}} \tag{17.57}$$

Equation (17.57) thus indicates that the primary and secondary currents, flowing in the directions assumed in Fig. 17.26, are in phase, since their ratio is a real number. This equation can also be written in the form

$$N_1\mathbf{I}_1 = \text{primary ampere turn}$$
$$= N_2\mathbf{I}_2 = \text{secondary ampere turn} \tag{17.58}$$

Since the primary current and voltage are in phase with the secondary current and voltage, respectively, the phase shift between the current and voltage phasors in the primary winding is the same as the phase shift between the current and voltage phasors in the secondary winding. Using this fact and Eqs. (17.56) and (17.58), the following conclusion can be made:

$$\text{input power} = V_1 I_1 \cos \theta_{V_1 I_1} = \frac{N_1}{N_2} V_2 \left(\frac{N_2}{N_1}\right) I_2 \cos \theta_{V_2 I_2}$$
$$= V_2 I_2 \cos \theta_{V_2 I_2} = \text{output power} \tag{17.59}$$

This equation justifies the assumption of zero power loss in the ideal transformer. One can combine Eqs. (17.56) and (17.58) in the form

$$a = \frac{N_1}{N_2} \text{(turns ratio)} = \frac{\mathbf{V}_1}{\mathbf{V}_2} = \frac{\mathbf{I}_2}{\mathbf{I}_1} \tag{17.60}$$

When the turns ratio a is larger than 1, the primary voltage is larger than the secondary voltage, while the primary current is smaller than the secondary current. This is called a *step-down transformer*, referring to the voltage. On the other hand, if a is less than 1, the secondary voltage is larger than the primary voltage, while the secondary current is less than the primary current. Again with reference to the voltage, this is called a *step-up transformer*. Many applications exist for both types of transformers. In fact, when the primary and the secondary windings of a step-up transformer are exchanged, the transformer becomes a step-down transformer. A unity turns ratio transformer ($a = 1$) is used primarily to electrically isolate the primary and the secondary circuits, for grounding purposes. In this case both the primary and the secondary voltages and currents are the same.

High-power iron-core transformers have very closely coupled coils and can be considered to be approximately ideal. Two basic parameters are usually defined for a given transformer: its turn ratio a and its VA (or power) rating.

Figure 17.27 shows the typical circuit diagram symbol of an ideal iron-core transformer. In comparison with Fig. 17.26, the secondary winding has been reversed in this case. To keep the phase relationships between the voltages and currents the same as in Fig. 17.26, the secondary voltage and current are reversed as indicated in Fig. 17.27. Thus, also in Fig. 17.27, $\mathbf{V}_1$ and $\mathbf{V}_2$ are in phase and $\mathbf{I}_1$

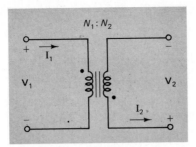

Figure 17.27

and $\mathbf{I}_2$ are in phase. Such an in-phase relationship can always be made to apply if the voltages are referenced such that both have the same polarity at the corresponding dotted terminal. Also, the current directions should be taken such that when one current enters from a dotted terminal, the other current leaves from the other dotted terminal.

17.5.1 Equivalent Circuits and Impedance Transformation

Normally, the source is connected to the primary winding of a transformer. The primary circuit is said to be active. The load (e.g., Z_2) is usually connected to the secondary winding of the transformer. The secondary circuit is said to be passive. Figure 17.28(a) shows such connections. Here

$$\begin{aligned}
\mathbf{E} &= Z_1\mathbf{I}_1 + \mathbf{V}_1 \\
&= Z_1\mathbf{I}_1 + a\mathbf{V}_2 \quad \text{[using Eq. (17.55)]} \\
&= Z_1\mathbf{I}_1 + aZ_2\mathbf{I}_2 \\
&= Z_1\mathbf{I}_1 + aZ_2a\mathbf{I}_1 \quad \text{[using Eq. (17.57)]} \\
&= (Z_1 + a^2Z_2)\mathbf{I}_1 \\
&= (Z_1 + Z_r)\mathbf{I}_1 \quad\quad\quad\quad\quad\quad (17.61)
\end{aligned}$$

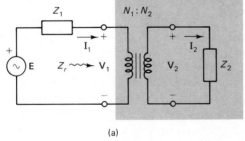
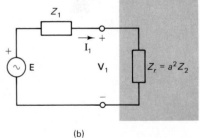

(a)   (b)

Figure 17.28

Equation (17.61) also describes the equivalent circuit shown in Fig. 17.28(b). Z_r is called the reflected impedance:

$$\boxed{\begin{aligned}
Z_r &= \text{secondary impedance reflected into the primary circuit} \\
&= \text{impedance looking between the terminals of the primary coil} \\
&= \frac{\mathbf{V}_1}{\mathbf{I}_1} = \frac{a\mathbf{V}_2}{(1/a)\mathbf{I}_2} = a^2\frac{\mathbf{V}_2}{\mathbf{I}_2} = a^2Z_2 = \left(\frac{N_1}{N_2}\right)^2 Z_2
\end{aligned}} \quad (17.62)$$

as obtained above in Eq. (17.61).

17.5   Ideal Transformer

725

If the dotted terminals of the transformer are not assigned, it is possible that the assumed polarity of the secondary voltage and the assumed direction of the secondary current be opposite to those indicated in Figs. 17.26 and 17.27. Therefore, in such cases, it is possible that a 180° phase shift actually exists between $\mathbf{I}_1$ and $\mathbf{I}_2$ and between $\mathbf{V}_1$ and $\mathbf{V}_2$. Here

$$\frac{\mathbf{V}_1}{\mathbf{V}_2} = -\frac{N_1}{N_2} = -a$$

and

$$\frac{\mathbf{I}_1}{\mathbf{I}_2} = -\frac{N_2}{N_1} = -\frac{1}{a}$$

Following a derivation similar to that above, it can be shown that even in such cases the reflected impedance is still $a^2 Z_2$. Thus as far as the determination of $\mathbf{I}_1$ is concerned [Eq. (17.61)], the dotted terminals of the transformer are of no consequence.

One of the important applications of transformers is in situations where the load impedance Z_2 is to be matched to the source impedance Z_1 to achieve maximum power transfer conditions. In such situations, it is possible to choose $a(= N_1/N_2)$ such that

$$Z_r = a^2 Z_2 = Z_1^* \tag{17.63}$$

This is especially easy to achieve if the circuit impedances are all resistive, since

$$R_r = a^2 R_2 = R_1$$

Therefore, the required turns ratio a is given by

$$a = \frac{N_1}{N_2} = \sqrt{\frac{R_1}{R_2}} \tag{17.64}$$

EXAMPLE 17.8

Using the concept of reflected impedances, find the values of $\mathbf{I}_1$, $\mathbf{I}_2$, $\mathbf{I}_3$, $\mathbf{V}_1$, and $\mathbf{V}_2$ in the ideal transformer circuit shown in Fig. 17.29.

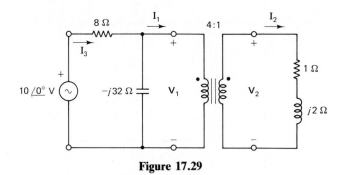

Figure 17.29

Solution   Here

$$a = \frac{N_1}{N_2} = 4$$

that is,

$$a^2 = 16$$

and as

$$Z_2 = 1 + j2 \ \Omega$$

then

$$Z_r = a^2 Z_2$$
$$= 16(1 + j2)$$
$$= 16 + j32 \ \Omega$$
$$= 35.78\underline{/63.4°} \ \Omega$$

The equivalent circuit is thus redrawn as shown in Fig. 17.30. The total impedance of the circuit as seen by the source, Z_T, is

$$Z_T = 8 + \frac{-j32(16 + j32)}{-j32 + 16 + j32}$$
$$= 8 + -j2(16 + j32) = 8 - j32 + 64 = 72 - j32$$
$$= 78.79\underline{/-24°} \ \Omega$$

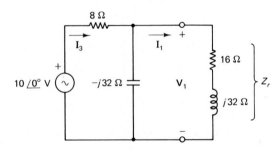

Figure 17.30

Thus

$$\mathbf{I}_3 = \frac{10\underline{/0°}}{Z_T} = \frac{10\underline{/0°}}{78.79\underline{/-24°}} = 0.1269\underline{/24°} \ A$$

Using the current-division principle yields

$$\mathbf{I}_1 = \mathbf{I}_3 \frac{-j32}{16 + j32 - j32} = 0.1269\underline{/24°} \times 2\underline{/-90°}$$
$$= 0.2538\underline{/-66°} \ A$$

Hence

$$\mathbf{V}_1 = \mathbf{I}_1 Z_r = 0.2538\underline{/-66°} \times 35.78\underline{/63.4°} = 9.08\underline{/-2.6°} \ V$$

As the secondary current direction and the secondary voltage polarities are in accordance with those indicated in Fig. 17.26, then from Eq. (17.55),

$$\mathbf{V}_2 = \frac{1}{a} \mathbf{V}_1 = \frac{1}{4} \times 9.08\underline{/-2.6°} = 2.27\underline{/-2.6°} \ V$$

and from Eq. (17.47),

$$\mathbf{I}_2 = a\mathbf{I}_1 = 4 \times 0.2538\underline{/-66°} = 1.015\underline{/-66°} \ A$$

It is clear that this is a step-down transformer. Notice that P_{in}, the input primary winding power, is

$$V_1 I_1 \cos\theta_{V_1 I_1} = 9.08 \times 0.2538\cos 63.4° = 1.032 \text{ W}$$

while P_{out}, the output power from the secondary winding, is

$$V_2 I_2 \cos\theta_{V_2 I_2} = 2.27 \times 1.015\cos 63.4° = 1.032 \text{ W}$$

as expected in the case of ideal transformers.

17.6 PRACTICAL TRANSFORMERS: THE TRANSFORMER CIRCUIT MODEL

Air-core transformers have small (fractional) coefficients of coupling k. Also, practical iron-core transformers have coupling coefficients which are less than unity, typically 0.95. Such transformers do not satisfy the conditions for the ideal transformer discussed in Section 17.5. Besides the nonunity coupling coefficients due to the magnetic flux leakage, there is always a finite amount of power loss in the coil resistances. Also, the magnetic losses in the core are finite. The primary and secondary inductances L_1 and L_2 could be large, but not infinite.

Circuits containing such practical transformers can always be analyzed using the techniques developed above for mutually coupled circuits. In some applications, a circuit model for the transformer could be useful. This model is developed through the addition of extra elements to the ideal transformer model, to account for the transformer's imperfections.

Consider the transformer circuit shown in Fig. 17.31. The secondary circuit is passive, containing the load Z_L, as is usually the case. The loop equations for the primary and secondary circuits can be written as follows:

$$\mathbf{E} = \mathbf{I}_p Z_1 + \mathbf{I}_p j\omega L_p \mp \mathbf{I}_s j\omega M \tag{17.65}$$

and

$$0 = \mp \mathbf{I}_p j\omega M + \mathbf{I}_s Z_L + \mathbf{I}_s j\omega L_s \tag{17.66}$$

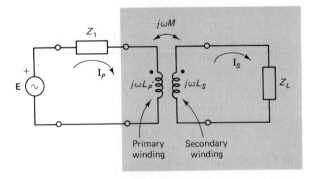

Figure 17.31

The negative signs associated with the mutually induced voltage terms in Eqs. (17.65) and (17.66) are the proper signs according to the dot rule when applied to the circuit of Fig. 17.31. However, the positive signs are included to account for either reversing the direction of $\mathbf{I}_s$ or reversing the dotted terminal in the secondary winding. From Eq. (17.66),

$$\mathbf{I}_s = \pm\mathbf{I}_p \frac{j\omega M}{Z_L + j\omega L_s} \tag{17.67}$$

Substituting in Eq. (17.65) yields

$$\begin{aligned}
\mathbf{E} &= \mathbf{I}_p(Z_1 + j\omega L_p) - \mathbf{I}_p \frac{j^2\omega^2 M^2}{Z_L + j\omega L_s} \\
&= \mathbf{I}_p\left(Z_1 + j\omega L_p + \frac{\omega^2 M^2}{Z_L + j\omega L_s}\right) \\
&= \mathbf{I}_p(Z_1 + j\omega L_p + Z_r) \tag{17.68}
\end{aligned}$$

Here

$$M = k\sqrt{L_p L_s}$$

or

$$X_M = \omega M = k\sqrt{\omega^2 L_p L_s} = k\sqrt{X_p X_s}$$

and

$$\boxed{\text{reflected impedance } Z_r = \frac{\omega^2 M^2}{Z_L + j\omega L_s}} \tag{17.69}$$

Equation (17.68) thus describes the circuit in Fig. 17.32. It is called the equivalent circuit of the transformer referred to the primary. It is clear from this analysis that due to the magnetic coupling, the impedance of the secondary circuit affects the value of the primary current, through the addition of the reflected impedance Z_r in series with the primary winding impedance.

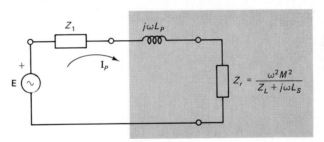

Figure 17.32

As the results of Eq. (17.68) indicate, whether the signs used in Eqs. (17.65) and (17.66) were the positive or the negative signs, the same expression is obtained. Thus as far as the primary current is concerned, the location of the dotted terminals or the assumed direction of the secondary current has no effect, nor do they have any effect on the value of Z_r. This confirms the similar remark made earlier concerning the reflected impedance using the ideal transformer model.

There is, however, a basic difference between the two reflected impedances. In the case of the ideal transformer, the reflected impedance $a^2 Z_L$ preserves the nature of the load impedance Z_L, and it replaces the primary winding. On the ot!.... hand, in the case of the practical transformer the reflected impedance indicates an inversion process since Z_L appears in the denominator of Eq.

(17.69). Thus if the load was capacitive, the reflected impedance could be inductive, and vice versa. Also, the reflected impedance in the practical transformer case is in series with the primary winding impedance and does not replace it.

EXAMPLE 17.9

Using the reflected impedance concept, find $\mathbf{I}_p$ and $\mathbf{I}_s$ in the circuit shown in Fig. 17.33.

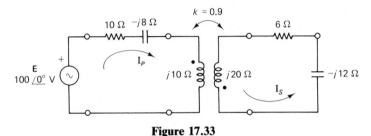

Figure 17.33

Solution Here

$$X_M = \omega M = k\sqrt{X_p X_s}$$
$$= 0.9\sqrt{10 \times 20}$$
$$= 12.728 \ \Omega$$

Therefore, according to Eq. (17.69),

$$Z_r = \frac{\omega^2 M^2}{Z_L + j\omega L_s} = \frac{(12.728)^2}{6 - j12 + j20} = \frac{162}{6 + j8} = \frac{162}{10\underline{/53.1°}}$$
$$= 16.2\underline{/-53.1°} = 9.72 - j12.96 \ \Omega$$

Notice that the total secondary impedance of $6 + j8 \ \Omega$, being inductive, reflects as a capacitive load (Z_r) in the primary circuit. The total impedance seen by the source is thus

$$Z_T = 10 - j8 + j10 + Z_r = 10 + j2 + 9.72 - j12.96$$
$$= 19.72 - j10.96 = 22.56\underline{/-29.1°} \ \Omega$$

Hence

$$\mathbf{I}_p = \frac{\mathbf{E}}{Z_T} = \frac{100\underline{/0°}}{22.56\underline{/-29.1°}} = 4.43\underline{/29.1°} \ \text{A}$$

The direction of the secondary current $\mathbf{I}_s$ is correct according to the dot convention. Its value can thus be calculated using Eq. (17.67), with the positive sign

$$\mathbf{I}_s = \mathbf{I}_p \frac{j\omega M}{Z_L + j\omega L_s} = 4.43\underline{/29.1°} \times \frac{12.728\underline{/90°}}{6 + j8}$$
$$= 5.64\underline{/66°} \ \text{A}$$

A circuit model used to approximate the actual behavior of a practical transformer is shown in Fig. 17.34. In this model R_p and R_s are the finite

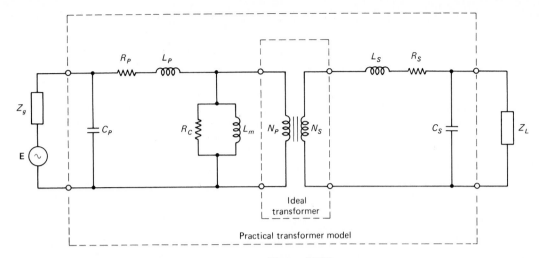

Figure 17.34

resistances of the primary and secondary windings. C_p and C_s are the stray capacitances of each of the two coils, as each coil is made of conducting wire turns separated by the wire insulator. L_p and L_s are the leakage inductances of the two coils, where

$$L_p = \frac{N_p \phi_{11}}{i_p} \qquad \text{and} \qquad L_s = \frac{N_s \phi_{22}}{i_s}$$

R_c is a resistance included in the model to account for the core losses. L_m is called the *magnetizing inductance*. The current flow in L_m is that required to establish the mutual core flux (i.e., the current required to overcome the reluctance of the magnetic core). This current is called the *magnetizing* or *exciting current*. In good iron-core transformers, this magnetizing current is usually very small, representing only about 5% of the full-load current. In ideal transformers, with infinite core permeability, no magnetizing current is required and L_m is infinite.

In the middle of the model, the ideal transformer is included, converting the current and voltage reaching it from the primary to the secondary circuits. The elements added to the ideal transformer, as discussed above and as shown in Fig. 17.34, account for the transformer's imperfections.

17.7 | AUTOTRANSFORMERS

The *autotransformer* is a single continuous coil winding on a magnetic core as shown in Fig. 17.35. The input voltage $\mathbf{V}_1$ and the output voltage $\mathbf{V}_2$ have a common terminal which is the bottom terminal of the winding. The output voltage(s) is obtained from a fixed or a sliding tap(s).

When the current $\mathbf{I}_1$ enters from the top terminal of the coil, as shown in Fig. 17.35, this terminal would be dotted. The figure shows the magnetic field ϕ_1 created by the current $\mathbf{I}_1$. According to Lenz's law, the opposing flux ϕ' is produced. The application of the RHR indicates that the natural current $\mathbf{I}_n$ (shown with dashed lines in Fig. 17.35) will flow out of the tap when the output circuit is closed. This terminal would then be also dotted. Due to the tightly

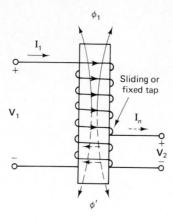

Figure 17.35

wound single coil construction of the autotransformer, the leakage flux is minimal and the autotransformer can be considered as an ideal transformer, to a very good approximation.

Figure 17.36 shows the two possible circuit connections for using the autotransformer. In such transformers, two relationships can always be applied. They are the key to analyzing circuits containing autotransformers.

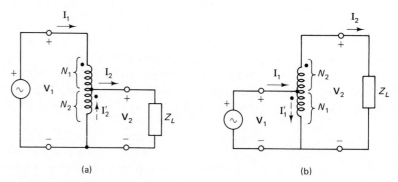

Figure 17.36 (a) Step-down autotransformer; (b) step-up autotransformer.

First, the volts/turn ratio is a constant parameter for a given autotransformer. In the circuit of Fig. 17.36(a),

$$\frac{\mathbf{V}_1}{N_1 + N_2} = \frac{\mathbf{V}_2}{N_2}$$

or

$$\frac{\mathbf{V}_1}{\mathbf{V}_2} = \frac{N_1 + N_2}{N_2} = 1 + \frac{N_1}{N_2} \tag{17.70}$$

and in the circuit of Fig. 17.36(b),

$$\frac{\mathbf{V}_1}{N_1} = \frac{\mathbf{V}_2}{N_1 + N_2}$$

or

$$\frac{\mathbf{V}_1}{\mathbf{V}_2} = \frac{N_1}{N_1 + N_2} \tag{17.71}$$

The second relationship concerns the power conditions in the autotransformer. Since it is assumed to be an ideal transformer, there is no power loss in it. Therefore, in general terms, the input complex power $\mathbf{S}_1$ is equal to the output complex power $\mathbf{S}_2$, that is,

$$\mathbf{S}_1 = \mathbf{V}_1 \mathbf{I}_1^* = \mathbf{S}_2 = \mathbf{V}_2 \mathbf{I}_2^* \tag{17.72}$$

This complex-number relationship means that the real parts are equal and the imaginary parts are also equal, in both sides of the equation. When there is only one tap and a single load impedance Z_L, the phase shift between $\mathbf{V}_2$ and $\mathbf{I}_2$, $\theta_{V_2 I_2}$, is the same as the phase shift between $\mathbf{V}_1$ and $\mathbf{I}_1$, $\theta_{V_1 I_1}$.

In such a case, Eq. (17.72) can be expressed as

$$V_1 I_1 = V_2 I_2$$

or

$$\frac{I_1}{I_2} = \frac{V_2}{V_1}$$

But since the voltage relationships above [Eqs. (17.70) and (17.71)] indicate that $\mathbf{V}_1$ and $\mathbf{V}_2$ are in phase (their ratio is a constant real number), then

$$\frac{\mathbf{I}_1}{\mathbf{I}_2} = \frac{\mathbf{V}_2}{\mathbf{V}_1} \tag{17.73}$$

In the case of the circuit in Fig. 17.36(a),

$$\frac{\mathbf{I}_1}{\mathbf{I}_2} = \frac{N_2}{N_1 + N_2} \tag{17.74}$$

and in the case of the circuit in Fig. 17.36(b),

$$\frac{\mathbf{I}_1}{\mathbf{I}_2} = \frac{N_1 + N_2}{N_1} = 1 + \frac{N_2}{N_1} \tag{17.75}$$

The main difference between autotransformers and ordinary transformers is that the autotransformer exhibits not only magnetic coupling but also electrical conduction coupling. KCL can be applied at the tap node, providing the following conditions.

In the case of the circuit in Fig. 17.36(a),

$$\mathbf{I}_1 + \mathbf{I}_2' = \mathbf{I}_2$$

Therefore,

$$\frac{\mathbf{I}_2'}{\mathbf{I}_1} = \frac{\mathbf{I}_2}{\mathbf{I}_1} - 1 \qquad \text{and using Eq. (17.74)}$$

$$= \frac{N_1 + N_2}{N_2} - 1 = \frac{N_1}{N_2} \tag{17.76}$$

This is the same equation as that obtained for the ideal transformer. Here $\mathbf{I}_2'$ is the magnetically induced current due to the mutual coupling. The output current $\mathbf{I}_2$ is thus made up of two components: $\mathbf{I}_2'$, due to the magnetic coupling, and $\mathbf{I}_1$, due to the electrical conduction coupling.

In the case of the circuit in Fig. 17.36(b), KCL applied at the tap node provides

$$\mathbf{I}_1 = \mathbf{I}_1' + \mathbf{I}_2$$

Therefore,

$$\frac{\mathbf{I}_1'}{\mathbf{I}_2} = \frac{\mathbf{I}_1}{\mathbf{I}_2} - 1 \quad \text{and using Eq. (17.75)}$$

$$= 1 + \frac{N_2}{N_1} - 1 = \frac{N_2}{N_1} \tag{17.77}$$

Also here, this equation is the same as that obtained for the ideal transformer. Here $\mathbf{I}_1'$ is the current that produces the magnetic coupling. In this case, therefore, the input current $\mathbf{I}_1$ is the sum of two components: $\mathbf{I}_1'$, which causes the magnetic coupling, and $\mathbf{I}_2$, the electrically coupled output current.

It is clear that autotransformers cannot be used in cases where electrical isolation is required, for example when the input and output circuits need separate grounds. However, they have an important advantage over the comparably constructed ideal transformers, with electrically isolated primary and secondary circuits. This advantage is due to the added dimension of extra coupling due to the electrical conduction. They can have a much higher power-handling capability than the equivalently constructed isolated ideal transformer.

■ EXAMPLE 17.10

Find $\mathbf{V}_2$, $\mathbf{I}_2$, $\mathbf{I}_1$, and $\mathbf{I}_2'$ for the ideal autotransformer circuit shown in Fig. 17.37. Calculate the circuit's power.

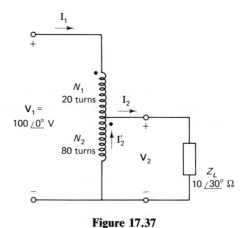

Figure 17.37

Solution Using the volts/turn condition gives us

$$\frac{\mathbf{V}_1}{N_1 + N_2} = \frac{\mathbf{V}_2}{N_2}$$

Therefore,

$$\mathbf{V}_2 = \mathbf{V}_1 \frac{N_2}{N_1 + N_2}$$

$$= 100\underline{/0^\circ} \times \frac{80}{20 + 80}$$

$$= 80\underline{/0^\circ} \text{ V}$$

Thus

$$\mathbf{I}_2 = \frac{\mathbf{V}_2}{Z_L} = \frac{80\underline{/0^\circ}}{10\underline{/30^\circ}}$$

$$= 8\underline{/-30^\circ} \text{ A}$$

Since there is only one load in the secondary, the phase shift between $\mathbf{V}_1$ and $\mathbf{I}_1$ is the same as the phase shift between $\mathbf{V}_2$ and $\mathbf{I}_2$. Here $\mathbf{I}_2$ lags $\mathbf{V}_2$ by 30°, as indicated above. From Eq. (17.74), which applies in this case,

$$\frac{\mathbf{I}_1}{\mathbf{I}_2} = \frac{N_2}{N_1 + N_2} = \frac{80}{100} = 0.8$$

Therefore,

$$\mathbf{I}_1 = 0.8\mathbf{I}_2 = 6.4\underline{/-30^\circ} \text{ A}$$

As KCL at the tap node provides,

$$\mathbf{I}_1 + \mathbf{I}_2' = \mathbf{I}_2$$

Therefore,

$$\mathbf{I}_2' = \mathbf{I}_2 - \mathbf{I}_1 = 8\underline{/-30^\circ} - 6.4\underline{/-30^\circ} = 1.6\underline{/-30^\circ} \text{ A}$$

The input power to the autotransformer is the same as the output power, as the transformer is assumed to be ideal. Therefore,

$$P_{\text{in}} = P_{\text{out}} = V_1 I_1 \cos\theta_{V_1 I_1} = V_2 I_2 \cos\theta_{V_2 I_2}$$

$$= 100 \times 6.4 \times \cos 30^\circ = 554.26 \text{ W}$$

EXAMPLE 17.11

In the ideal transformer circuit shown in Fig. 17.38, $\mathbf{I}_2$ is $50\underline{/-36.9^\circ}$ A and $\mathbf{I}_3$ is $16\underline{/0^\circ}$ A. Find $\mathbf{I}_1$.

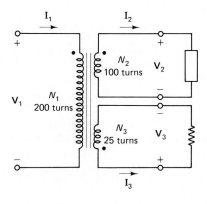

Figure 17.38

Solution From the volts/turn condition for this multisecondary ideal transformer:

$$\frac{\mathbf{V}_1}{N_1} = \frac{\mathbf{V}_2}{N_2} = \frac{\mathbf{V}_3}{N_3}$$

$$\frac{\mathbf{V}_1}{\mathbf{V}_2} = \frac{N_1}{N_2} = \frac{200}{100} = 2$$

and

$$\frac{\mathbf{V}_1}{\mathbf{V}_3} = \frac{N_1}{N_3} = \frac{200}{25} = 8$$

Note that all the three voltages are in phase, in accordance with the dot rule.

The second condition that could be applied in this case is the power condition:

input complex power = output complex power

that is

$$\mathbf{S}_1 = \mathbf{S}_2 + \mathbf{S}_3 \quad \text{or} \quad \mathbf{V}_1\mathbf{I}_1^* = \mathbf{V}_2\mathbf{I}_2^* + \mathbf{V}_3\mathbf{I}_3^*$$

Equating the real parts and the imaginary parts for each side of this equation,

$$V_1 I_1 \cos \theta_1 = V_2 I_2 \cos \theta_2 + V_3 I_3 \cos \theta_3$$

(real input power = real output power)

and

$$V_1 I_1 \sin \theta_1 = V_2 I_2 \sin \theta_2 + V_3 I_3 \sin \theta_3$$

(reactive input power = reactive output power)

Noting that the load in the secondary loop (N_3) is resistive, then $\mathbf{V}_3$ and $\mathbf{I}_3$ are in phase (i.e., $\theta_3 = 0°$). All the three voltage phasors have therefore zero phase angle, while θ_1 and θ_2 are the phase shifts of the corresponding currents and voltages. The ratio of the two power relationships above provides

$$\tan \theta_1 = \frac{V_2 I_2 \sin \theta_2 + V_3 I_3 \sin \theta_3}{V_2 I_2 \cos \theta_2 + V_3 I_3 \cos \theta_3}$$

$$= \frac{\dfrac{V_2}{V_1} I_2 \sin \theta_2 + \dfrac{V_3}{V_1} I_3 \sin \theta_3}{\dfrac{V_2}{V_1} I_2 \cos \theta_2 + \dfrac{V_3}{V_1} I_3 \cos \theta_3}$$

Substituting in this relationship, using the voltage ratios obtained above, while $\theta_2 = 36.9°$ and $\theta_3 = 0°$, we have

$$\tan \theta_1 = \frac{0.5 \times 50 \sin (36.9°) + 0}{0.5 \times 50 \cos (36.9°) + \frac{1}{8} \times 16 \cos 0°} = \frac{15}{22} = 0.6818$$

Therefore,

$$\theta_1 = 34.3°$$

Now, to obtain the magnitude of I_1, either of the power equalities above can be used. From the reactive power equation,

$$I_1 \sin \theta_1 = \frac{V_2}{V_1} I_2 \sin \theta_2 + \frac{V_3}{V_1} I_3 \sin \theta_3$$

then

$$I_1 \sin (34.3°) = \tfrac{1}{2} \times 50 \times \sin (36.9°) = 15$$

and

$$I_1 = 26.62 \text{ A}$$

Therefore,

$$\mathbf{I}_1 = 26.62\underline{/-34.3°} \text{ A}$$

Note that the conjugation of the current phasors in the complex power equation reverses the actual phases with respect to the voltage phasors.

DRILL PROBLEMS

Sections 17.1 to 17.3

17.1. If the coupling coefficient between two magnetically coupled coils is 0.9, and the self-inductance (L_1) of the first coil is 80 mH while the mutual inductance between the coils is 36 mH, find the self-inductance of the second coil (L_2).

17.2. For the two mutually coupled inductors in Drill Problem 17.1, N_1 is 200 turns. Find N_2.

17.3. For the same two coils examined in Drill Problems 17.1 and 17.2, the current flowing in coil 1 is

$$i_1(t) = 2.5 \times 10^{-3} \sin 2000t \text{ A}$$

Find
   (a) The flux ϕ_1 and ϕ_{12}.
   (b) The voltages v_1 and v_2.

17.4. The two inductors $L_1 = 20$ mH and $L_2 = 50$ mH are connected in series as shown in Fig. 17.18(a). The total inductance was found to be 100 mH. Find
   (a) The value of the mutual inductance M.
   (b) The value of the coupling coefficient k.
   (c) The total inductance if these two coils are connected as shown in Fig. 17.18(b).

Section 17.4

17.5. If the winding sence of coil 2 in the circuit shown in Fig. 17.19 is reversed, rewrite the loop equations for this network and obtain the voltage across the capacitor in this case.

17.6. If the winding sence of coil 2 in the network shown in Fig. 17.21 is reversed, rewrite the loop equations for this network and calculate the voltage across coil 2 in this case.

17.7. Obtain the conductively coupled equivalent network for the circuit analyzed in Drill Problem 17.5.

17.8. Obtain the conductively coupled equivalent network for the network analyzed in Drill Problem 17.6.

Section 17.5

17.9. What should be the turns ratio of an ideal transformer if it is to be used in matching a 4 Ω load resistance to a source whose internal resistance is 100 Ω? If the source voltage is 8 V, what will be the load voltage?

17.10. An ideal transformer whose turns ratio is 3:1 is used in a circuit similar to that shown in Fig. 17.28. What should be the load impedance Z_2 if maximum power transfer is to be achieved, given that Z_1 is $36 + j54\,\Omega$? If the source voltage is $10\underline{/0°}\,\text{V}$, calculate $\mathbf{I}_1$, $\mathbf{I}_2$, $\mathbf{V}_1$, and $\mathbf{V}_2$ in this case. What is the power delivered by the source and the power dissipated in the load?

Section 17.6

17.11. Write the loop equations for the practical transformer circuit shown in Fig. 17.33. Use these equations to calculate the values of $\mathbf{I}_p$ and $\mathbf{I}_s$ and compare these results to the answers obtained in Example 17.9.

17.12. Obtain the conductively coupled equivalent circuit to the practical transformer circuit examined in Drill Problem 17.11.

17.13. If the terminals of the primary winding of the transformer shown in Fig. 17.33 are reversed, calculate the new value of the current $\mathbf{I}_s$ using the reflected impedance concept.

17.14. Recalculate the secondary current $\mathbf{I}_s$ using the loop equations for the transformer circuit discussed in Drill Problem 17.13.

PROBLEMS

Note: The problems below that are marked with an asterisk (*) can be used as exercises in computer-aided network analysis or as programming exercises.

Sections 17.1 to 17.3

(i)   **17.1.** Find the mutual inductance and the turns ratio for two coils, one with an inductance of $0.9\,\text{H}$ and the other has an inductance of $0.4\,\text{H}$. The coefficient of coupling between the coils is 0.8.

(i)   **17.2.** Two coils have a coupling coefficient of 0.9. One has 100 turns and the other has 400 turns. A total flux ϕ_1 of $0.5\,\text{mWb}$ exists in the 100-turn coil when the current flowing in it is $0.2\,\text{A}$. Find the values of L_1, L_2, and M.

(ii)   **17.3.** Two coils, one having double the value of the other, have a total inductance of $130\,\text{mH}$ when connected in series aiding, and have a total inductance of $50\,\text{mH}$ when connected in series opposing (refer to Example 17.3). What are the values of L_1, L_2, M, and k?

(i)   **17.4.** The two inductors of Problem 17.1 are connected in parallel opposing (refer to Example 17.3). What is the equivalent inductance of such a connection?

(i)   **17.5.** Identify the other dotted terminal in the three coupled coils arrangements in Fig. 17.39.

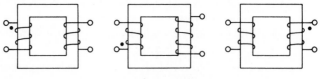

Figure 17.39

Section 17.4

(i)   ***17.6.** Identify the dotted terminals for the coupled coils shown in Fig. 17.40, then find the value of X_M. Write the loop equation for this circuit and solve it to obtain the value of the current $\mathbf{I}$.

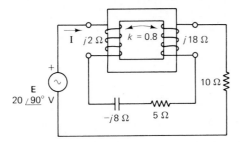

Figure 17.40

(i) ***17.7.** Write the loop equation for the circuit shown in Fig. 17.41, after identifying the dotted terminals. What is the value of the current **I**? What is the voltage across the two points a and b?

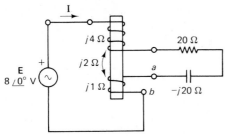

Figure 17.41

(ii) ***17.8.** Calculate the equivalent inductance between terminals a and b for the three series inductors in Fig. 17.42. Notice that each inductor is mutually coupled with the other two.

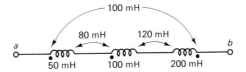

Figure 17.42

(ii) ***17.9.** Calculate the input impedance of the network shown in Fig. 17.43, taking into account the appropriate mutual coupling effect.

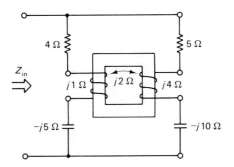

Figure 17.43

(ii) ***17.10.** Analyze the circuit in Fig. 17.44, including the mutual coupling effect indicated, to obtain the voltage across the $j5\ \Omega$ inductor.

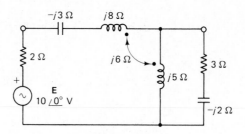

Figure 17.44

(ii) ***17.11.** Write the appropriate loop equations for the circuit shown in Fig. 17.45 and obtain the current in the $-j8\ \Omega$ impedance.

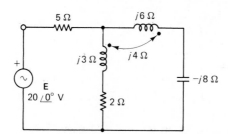

Figure 17.45

(ii) ***17.12.** Obtain Thévenin and Norton equivalents for the network shown in Fig. 17.46 between the terminals a and b.

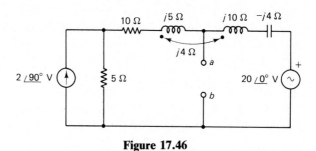

Figure 17.46

(ii) ***17.13.** Find the source's current in the circuit shown in Fig. 17.47.

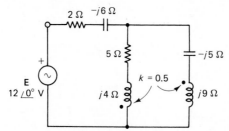

Figure 17.47

(i) **17.14.** Obtain the conductively coupled equivalent circuit for the network shown in Fig. 17.44.

(i) **17.15.** Obtain the conductively coupled equivalent circuit for the network shown in Fig. 17.45.

(i)　**17.16.**　Obtain the conductively coupled equivalent circuit for the network shown in Fig. 17.47.

Section 17.5

(i)　***17.17.**　Calculate I_1, I_2, V_1, and V_2 for the ideal transformer circuit shown in Fig. 17.48, when

　　(a)　$Z_2 = 2 + j1\ \Omega$.

　　(b)　$Z_2 = 1 - j1\ \Omega$.

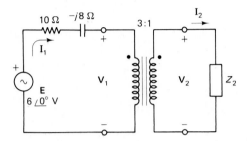

Figure 17.48

(i)　***17.18.**　Find I_1, I_2, I_3, V_1, and V_2 for the ideal transformer circuit shown in Fig. 17.49.

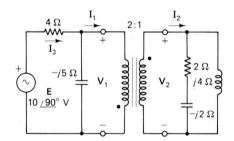

Figure 17.49

(ii)　***17.19.**　Find the currents I_1, I_2, and I_3 in the ideal transformer circuit shown in Fig. 17.50.

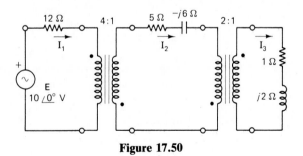

Figure 17.50

Section 17.6

(i)　***17.20.**　In the circuit shown in Fig. 17.51, X_p is $8\ \Omega$ and X_s is $2\ \Omega$. Write the appropriate loop equations to calculate I_1, I_2, V_1, and V_2.

(i)　***17.21.**　Calculate the same circuit variables in Problem 17.20 if the source E_2 is replaced by a short circuit.

(i)　***17.22.**　Write the appropriate loop equations for the circuit shown in Fig. 17.52 to obtain V_1, V_2, I_1, and I_2.

(ii)　**17.23.**　Resolve Problem 17.22 using the reflected impedance concept. What will be the

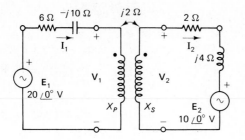

Figure 17.51

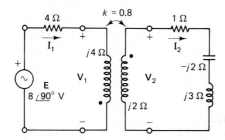

Figure 17.52

values of the required circuit variables if the dotted terminal in the primary winding is reversed?

(ii) ***17.24.** In the circuit of Fig. 17.53, I_2 is $2\underline{/90°}$ A. Find V_1, V_2, I_1, and I_2' for this autotransformer.

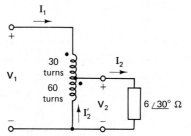

Figure 17.53

(ii) ***17.25.** Using the appropriate voltage and current relationships for the autotransformer circuit shown in Fig. 17.54, calculate I_1, I_1', and I_2.

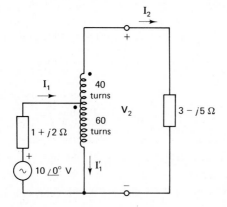

Figure 17.54

GLOSSARY

Autotransformer: A transformer that consists of a single continuous winding on a magnetic core. This coil is common between the primary and the secondary circuits of the transformer. Because this type of transformer uses both conductive and magnetic coupling, it has a larger power-handling capability than that of an equivalently constructed two winding transformer. On the other hand, both the primary and the secondary circuits must have the same ground terminal.

Conductively Coupled Equivalent Circuit: A circuit that is equivalent to another circuit containing mutually coupled coils. The conductivity coupled equivalent circuit has no mutual inductance terms, but produces the same response as the original mutually coupled circuit, due only to the current conduction in the common branches between the different loops.

Coupling Coefficient (k):   The fraction of the total flux generated by one coil that links with another, closeby coil. Its value ranges between zero and unity, depending on the closeness and the orientation angle of the axes of the two coils.

Dot Convention: A technique for marking (or labeling) the in-phase voltage terminals of magnetically coupled coils. It is also used to determine the polarities of the mutually induced voltages of the magnetically coupled coils.

Ideal Transformer: Two closely coupled coils, the primary and the secondary, whose coupling coefficient is unity. It has zero copper losses (no coil resistances) and zero iron losses (no hysteresis or eddy current losses in the iron core).

Leakage Flux: The portion of the magnetic flux lines, generated by one coil, that does not link with (pass through the magnetic circuit of) a second, closeby coil.

Loosely Coupled Coils: The situation corresponding to two coils whose coupling coefficient is very low.

Mutual Coupling: The situation that exists when the magnetic field generated in one coil causes an induced voltage in another, closeby coil, and vice-versa. The two (or more) coils are said to be mutually coupled, or said to form a transformer.

Mutual Inductance: A quantity that relates the magnetically induced voltage in one coil due to the time variation of the current flow (or magnetic flux produced) in another coil.

Primary Coil (or Winding):   The coil or winding to which the source of electrical energy is normally connected.

Reflected Impedance: The impedance appearing in the primary circuit of a transformer, equivalent to the effects of the flow of the secondary current in the load impedance connected in the secondary circuit (i.e., the reflected impedance accounts for the complete secondary circuit).

Secondary coil (or Winding):   The coil or winding to which the electrical load is normally connected.

Step-Down Transformer: A transformer whose secondary voltage is smaller than its primary voltage (i.e., whose turns ratio is larger than 1).

Step-Up Transformer: A transformer whose secondary voltage is larger than its primary voltage (i.e., whose turns ratio is smaller than 1).

Turns Ratio (a):   The ratio of the number of turns in the primary winding (or coil) to the number of turns in the secondary winding (or coil) of a transformer.

ANSWERS TO DRILL PROBLEMS

17.1. $L_2 = 20\,\text{mH}$

17.3. (a) $\phi_1 = 10^{-6} \sin 2000t$ Wb,
$\phi_{12} = 0.9 \times 10^{-6} \sin 2000t$ Wb

   (b) $v_1 = 0.4 \cos 2000t$ V,
$v_2 = 0.18 \cos 2000t$ V

17.5. $(10 - j4)\mathbf{I}_1 + j11\mathbf{I}_2 = 10$,
$j11\mathbf{I}_1 + (10 - j6)\mathbf{I}_2 = -j10$,
$\mathbf{V}_C = 5.539\underline{/15.6°}$ V

17.7. The conductively coupled equivalent circuit is similar to that shown in Fig. 17.12 with the direction of $\mathbf{I}_2$ reversed. Here $Z_1 = 10 + j7\,\Omega$, $Z_2 = 10 + j5\,\Omega$, and $Z_3 = -j11\,\Omega$.

17.9. $N_1/N_2 = a = 5:1$, $V_2 = 0.8$ V

17.11. $(10 + j2)\mathbf{I}_p - j12.728\mathbf{I}_s = 100$,
$-j12.728\mathbf{I}_p + (6 + j8)\mathbf{I}_s = 0$,
$\mathbf{I}_p = 4.4324\underline{/29.06°}$ A, $\mathbf{I}_s = 5.6416\underline{/65.93°}$ A (as in Example 17.9)

17.13. $\mathbf{I}_s = 5.64\underline{/-114°}$A

18 | THREE-PHASE CIRCUITS

OBJECTIVES

- Understanding how the two sequences of the three-phase voltage generation are produced, and the corresponding phasor subscripting and phasor diagram representations.

- Familiarity with the properties of the Y- and Δ-winding connections of the three-phase generators, using three or four wire systems.

- Familiarity with the special network conditions resulting from the Y- and Δ-connection of loads; balanced and unbalanced.

- Ability to analyze three-phase networks for any type of load connection, to obtain the required line or phase currents and power.

- Understanding the techniques used to calculate and measure power in three-phase networks, particularly using the two-wattmeters method.

Almost all electric power generation (using alternators), distribution, and industrial usage (through motors) is done using three-phase systems. A three-phase generator produces three sinusoidal voltage waveforms, that is, it is equivalent to three separate ac single-phase voltage sources, as will be illustrated below. Even though multiphase generators and alternators producing more phases are available for some special applications, they are not as practical as three-phase systems. As will be shown in this chapter, three-phase systems have the advantage of a uniform power dissipation versus pulsating power in a single-phase ac system. This results in less vibrations in the generators and motors, and better transmission efficiency.

18.1 TWO-PHASE SYSTEMS

The schematic diagram in Fig. 18.1 shows a simplified construction of a two-phase generator. The two coils, A and B, are placed perpendicular to each other on the rotor as shown. As the rotor is moved counterclockwise, each coil generates a sinusoidal waveform due to electromagnetic induction. With each coil having the same number of turns, moving with the same angular speed (ω) in the same magnetic field, each coil will generate a voltage waveform having the same magnitude $E(0.707E_{max})$. The voltage generated by coil B lags that generated by coil A by 90°, as shown in Fig. 18.2.

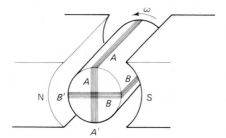

Figure 18.1

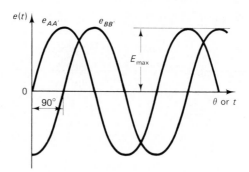

Figure 18.2

Therefore, taking the waveform $e_{AA'}(t)$ as the reference, we have

$$e_{AA'}(t) = E_{max} \sin \omega t \Rightarrow \mathbf{E}_{AA'} = E\underline{/0°} \text{ V} \qquad (18.1\text{a})$$

and

$$e_{BB'}(t) = E_{max} \sin(\omega t - 90°) \Rightarrow \mathbf{E}_{BB'} = E\underline{/-90°} \text{ V} \qquad (18.1\text{b})$$

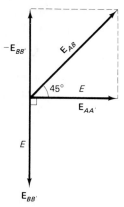

E

$E_{BB'}$

Figure 18.3

The corresponding phasors are shown in Fig. 18.3. Both waveforms have the same angular frequency ω $(=2\pi f)$. In North America, power is generated with a frequency of 60 Hz (i.e., $\omega = 377$ rad/s). The generator coils are shown schematically in Fig. 18.4. If one terminal of each coil is connected together, terminals A' and B' in the diagram, the new node thus formed is called the *neutral*, N. Conductors (wires) used to connect these sources to the loads are referred to as *lines*. Figure 18.4 also shows the circuit diagram of such a two-phase three-line generating system. The line-to-neutral voltages are called the *phase voltages*:

$$\mathbf{V}_{AN} = \mathbf{E}_{AN} = \mathbf{E}_{AA'} = E\underline{/0°}\text{ V}$$
$$\mathbf{V}_{BN} = \mathbf{E}_{BN} = \mathbf{E}_{BB'} = E\underline{/-90°}\text{ V}$$

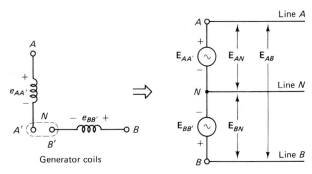

Figure 18.4

Also, the line-to-line voltage is referred to simply as the *line voltage*. Here

$$\mathbf{V}_{AB} = \mathbf{V}_{AN} - \mathbf{V}_{BN} = \mathbf{E}_{AB} = E - (-jE) = E + jE = \sqrt{2}\,E\underline{/45°}\text{ V}$$

as shown in the phasor diagram of Fig. 18.3.

18.2 | THREE-PHASE VOLTAGE GENERATION

Expanding the two-phase generator, through the addition of a third coil on the rotor such that the coils are equally spaced by 120° from each other as shown in Fig. 18.5, will result in the generation of the three-phase voltage waveforms shown in Fig. 18.6. It is assumed here that each coil has the same number of turns, that the rotation is counterclockwise, and that the location at which coil A is, is the zero reference of time (or angle), at which instant $e_{AA'}(t)$ is zero and positive going.

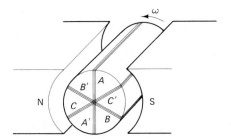

Figure 18.5

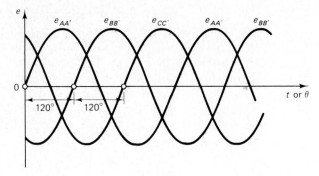

Figure 18.6

The two terminals of each coil are denoted A, A', B, B', and C, C'. The voltage generated by each of the three coils is referenced as indicated in Fig. 18.6. As can be observed from the simplified generator of Fig. 18.5, $e_{AA'}$ is zero, positive going first. Then $e_{BB'}$ is zero and positive going 120° later. Finally, $e_{CC'}$ is zero and positive going after another 120°. These *starting points* are circled in Fig. 18.6. Thus waveform $e_{AA'}$ leads $e_{BB'}$ by 120° and waveform $e_{BB'}$ leads $e_{CC'}$ by 120°. Hence

$$e_{AA'}(t) = E_{max} \sin \omega t \Rightarrow \mathbf{E}_{AA'} = E\underline{/0°}\ \mathrm{V} \tag{18.2a}$$

where

$$E = 0.707 E_{max}$$

$$e_{BB'}(t) = E_{max} \sin (\omega t - 120°) \Rightarrow \mathbf{E}_{BB'} = E\underline{/-120°}\ \mathrm{V} \tag{18.2b}$$

$$e_{CC'}(t) = E_{max} \sin (\omega t - 240°) = E_{max} \sin (\omega t + 120°)$$

$$\Rightarrow \mathbf{E}_{CC'} = E\underline{/120°}\ \mathrm{V} \tag{18.2c}$$

These are called the *phase voltages*. Figure 18.7 represents the corresponding phasor diagram. It is easy to see that

$$\mathbf{E}_{AA'} + \mathbf{E}_{BB'} + \mathbf{E}_{CC'} = E + (-0.5E - j0.866E)$$
$$+ (-0.5E + j0.866E) = 0 \tag{18.3}$$

The sequence in which the voltage waveforms are generated, or appear with respect to time, can be observed from Figs. 18.5 and 18.6 to be

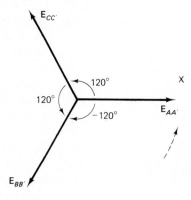

Figure 18.7

$\overline{ABC}\overline{ABC}AB$.... This is called the ABC (or BCA or CAB) or positive sequence.

If the rotor of the simple generator in Fig. 18.5 is rotated clockwise, or if the rotation is counterclockwise but coils B and C are interchanged, it is clear that the sequence will then become $\overline{ACB}\overline{ACB}AC$.... This is called the ACB (or CBA or BAC) or negative sequence. Here $e_{AA'}(t)$ is still the reference waveform, but it now leads $e_{CC'}(t)$ by 120°, while $e_{CC'}(t)$ leads $e_{BB'}(t)$ by another 120°. Thus in the ACB or negative sequence, the phase voltages are

$$\mathbf{E}_{AA'} = E\underline{/0°}\text{ V} \tag{18.4a}$$

$$\mathbf{E}_{BB'} = E\underline{/+120°}\text{ V} \tag{18.4b}$$

$$\mathbf{E}_{CC'} = E\underline{/-120°}\text{ V} \tag{18.4c}$$

as shown in Fig. 18.8.

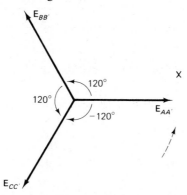

Figure 18.8

An easy way to identify the sequence of the phasor diagram (or remember it) is to assume a fixed point such as point x in Figs. 18.7 and 18.8. Then imagining that the phasor diagram is rotated counterclockwise, the vector subscripts A, B, and C will pass by this fixed point in the ABC (positive) or ACB (negative) sequence.

These simple representations of the generated three-phase voltages also apply to the modern practical generators, which are much more involved than the simplified diagram of Fig. 18.5 indicates. In such generators, the magnetic field usually rotates while the voltage coils are stationary.

18.3 THREE-PHASE GENERATOR WINDING CONNECTIONS

The three windings (coils) of the generator could be connected together in one of two topologies. If terminal A' is connected to B, and terminal B' is connected to C and terminal C' is connected to A, as shown in Fig. 18.9, this results in the delta connection (Δ-connected generator), as the shape of the topology indicates. Here three lines are used to connect the three-phase generator to its loads.

The line-to-line voltages are the same as the corresponding coil (or phase) voltages. Thus

$$\mathbf{E}_{AB} = \mathbf{E}_{AA'} = E\underline{/0°}\text{ V} \left.\begin{array}{l}\\ \end{array}\right\}ABC \quad \text{or} \quad = E\underline{/0°}\text{ V}\left.\begin{array}{l}\\ \end{array}\right\}ACB \tag{18.5a}$$

$$\mathbf{E}_{BC} = \mathbf{E}_{BB'} = E\underline{/-120°}\text{ V} \quad \text{or} \quad = E\underline{/120°}\text{ V} \tag{18.5b}$$

$$\mathbf{E}_{CA} = \mathbf{E}_{CC'} = E\underline{/120°}\text{ V} \quad\text{sequence} \quad \text{or} \quad = E\underline{/-120°}\text{ V} \quad\text{sequence} \tag{18.5c}$$

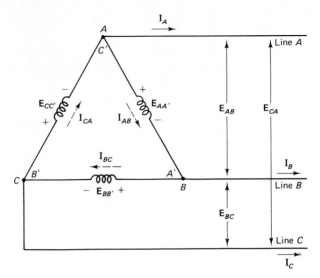

Figure 18.9

The phasor diagrams in Fig. 18.10 show these line voltages. Notice that the sequence can still be identified through the subscripts (considering the first-occurring subscripts alone, or the second-occurring subscripts alone) and the counterclockwise rotation past a fixed point, as noted above.

The phase currents (coil currents) are not the same as the line currents in the Δ-connected source. We discuss this point further later. This generator connection, however, is seldom used in practice because any slight imbalance in magnitude or phase of the three-phase voltages will not result in a zero sum, as in Eq. (18.3). The result will be a large circulating current in the coils, which could be damaging and is undesirable.

The second topology in which the three generator windings could be connected is shown in Fig. 18.11. This is called the Y-connected generator. The three coil terminals A', B', and C' are tied together, forming the node N (neutral). This node could be used for common ground connection.

To feed loads connected to this source, either a four-line system, as shown in Fig. 18.11, can be used, or a three-line system when the neutral wire (line) is removed. As will be clear later, balanced loads result in no current flow in the neutral line. Also, Δ-connected loads do not need this neutral line.

This generator system voltages are then defined as line-to-neutral voltages (e.g., $\mathbf{E}_{AN}$) or as line-to-line voltages (e.g., $\mathbf{E}_{AB}$). When either definition is given, the other can be found from it as will be shown below. Clearly, this also depends on the phase sequence, being either ABC or ACB.

It is clear from Fig. 18.11 that the line-to-neutral voltages are the same as the phase (or coil) voltages. Thus for an ABC sequence,

$$\mathbf{E}_{AN} = E\underline{/0°} \text{ V} \tag{18.6a}$$

$$\mathbf{E}_{BN} = E\underline{/-120°} \text{ V} \tag{18.6b}$$

$$\mathbf{E}_{CN} = E\underline{/120°} \text{ V} \tag{18.6c}$$

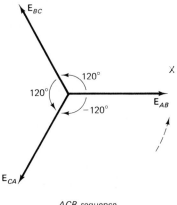

ACB sequence

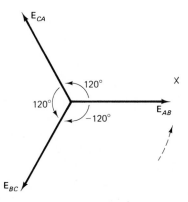

ABC sequence

Figure 18.10

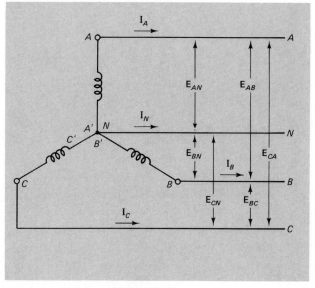

Figure 18.11

The line-to-line voltages are

$$\mathbf{E}_{AB} = \mathbf{E}_{AN} - \mathbf{E}_{BN} \qquad 30° = \sqrt{3}\,E\underline{/30°}\ \text{V}$$

$$= E - (-0.5E - j0.866E) = 1.5E + j0.866E$$

or

$$\boxed{\mathbf{E}_{AB} = 1.5E + j0.866E = 1.732E\underline{/30°} = \sqrt{3}\,E\underline{/30°}\ \text{V}} \qquad (18.7a)$$

The line-to-line voltage is thus $\sqrt{3}$ times the phase voltage. Notice that $\mathbf{E}_{AB}$ leads $\mathbf{E}_{AN}$ by 30°. Similarly,

$$\boxed{\mathbf{E}_{BC} = \mathbf{E}_{BN} - \mathbf{E}_{CN} = \sqrt{3}\,E\underline{/-90°}\ \text{V}} \qquad (18.7b)$$

$\mathbf{E}_{BC}$ leads $\mathbf{E}_{BN}$ by 30°; and

$$\boxed{\mathbf{E}_{CA} = \mathbf{E}_{CN} - \mathbf{E}_{AN} = \sqrt{3}\,E\underline{/+150°}\ \text{V}} \qquad (18.7c)$$

$\mathbf{E}_{CA}$ leads $\mathbf{E}_{CN}$ by 30°.

The resulting line-to-line phasor voltages also form a balanced three-phase generator system. Their sum is

$$\boxed{\mathbf{E}_{AB} + \mathbf{E}_{BC} + \mathbf{E}_{CA} = 0} \qquad (18.8)$$

Notice again here that if the phasor diagram is imagined to rotate past a fixed point x in the plane, in a counterclockwise direction, the first and second subscripts of the line-to-line phasor voltages go past this point in the proper sequence (here it is the ABC or positive sequence). This is shown in Fig. 18.12. Similarly, for the ACB sequence (see Fig. 18.13),

$$\boxed{\mathbf{E}_{AN} = E\underline{/0°}\ \text{V}} \qquad (18.9a)$$

$$\boxed{\mathbf{E}_{BN} = E\underline{/120°}\ \text{V}} \qquad (18.9b)$$

$$\boxed{\mathbf{E}_{CN} = E\underline{/-120°}\ \text{V}} \qquad (18.9c)$$

Then

$$\mathbf{E}_{AB} = \mathbf{E}_{AN} - \mathbf{E}_{BN}$$

$$= E - (-0.5E + j0.866E) = 1.5E - j0.866E = 1.732E\underline{/-30°}$$

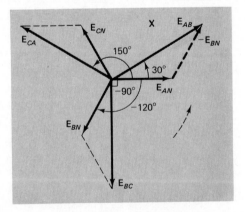

Figure 18.12 ABC sequence.

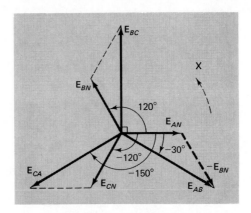

Figure 18.13 *ACB* sequence.

or

$$\mathbf{E}_{AB} = \sqrt{3}\,E\underline{/-30°}\ \text{V} \tag{18.10a}$$

and

$$\mathbf{E}_{BC} = \sqrt{3}\,E\underline{/90°}\ \text{V} \tag{18.10b}$$

$$\mathbf{E}_{CA} = \sqrt{3}\,E\underline{/-150°}\ \text{V} \tag{18.10c}$$

Also here, the line-to-line voltages are $\sqrt{3}$ times the phase voltages but lag the corresponding phase voltage by 30°. They also form a balanced three-phase generator system (each is 120° out of phase with respect to the others). Notice that the phasor diagram in Fig. 18.13 is exactly that in Fig. 18.12, inverted upside down with respect to the horizontal axis.

In the Y-connected generator (Fig. 18.11), the line currents ($\mathbf{I}_A$, $\mathbf{I}_B$, and $\mathbf{I}_C$) are the same as the phase (coil) currents.

18.4 BALANCED THREE-PHASE Y-CONNECTED LOADS

The balanced three-phase Y-connected load has three equal phase impedances (Z) as shown in Fig. 18.14. The connection is obviously similar to the source

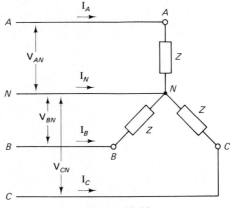

Figure 18.14

connection of Fig. 18.11. Clearly,

$$\mathbf{V}_{AN} = \mathbf{E}_{AN} \quad \text{etc.}$$

and

$$\mathbf{V}_{AB} = \mathbf{E}_{AB} \quad \text{etc.}$$

that is, using $\mathbf{V}$ or $\mathbf{E}$ to represent phasor voltages is equivalent.

Here the line currents ($\mathbf{I}_A$, $\mathbf{I}_B$, and $\mathbf{I}_C$) are also the same as the phase currents, as is clear from Fig. 18.14.

$$\mathbf{I}_A = \frac{\mathbf{V}_{AN}}{Z} = \frac{E\underline{/0°}}{|Z|\underline{/\theta°}} = \frac{E}{|Z|}\underline{/-\theta°} \qquad \text{A} \left.\vphantom{\begin{array}{c}1\\1\\1\end{array}}\right\} \qquad (18.11a)$$

$$\mathbf{I}_B = \frac{\mathbf{V}_{BN}}{Z} = \frac{E\underline{/-120°}}{|Z|\underline{/\theta°}} = \frac{E}{|Z|}\underline{/-120° - \theta°} \quad \text{A} \left.\vphantom{\begin{array}{c}1\\1\\1\end{array}}\right\} \quad (ABC \text{ sequence}) \quad (18.11b)$$

$$\mathbf{I}_C = \frac{\mathbf{V}_{CN}}{Z} = \frac{E\underline{/120°}}{|Z|\underline{/\theta°}} = \frac{E}{|Z|}\underline{/120° - \theta°} \qquad \text{A} \left.\vphantom{\begin{array}{c}1\\1\\1\end{array}}\right\} \qquad (18.11c)$$

Figure 18.15 shows the corresponding phasor diagram.

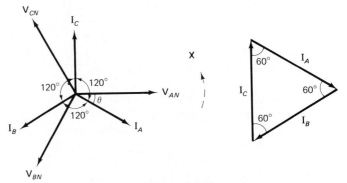

Figure 18.15

The three line currents have equal magnitudes, each being $E/|Z|$. Each leads or lags the corresponding voltage-to-neutral phasor by the phase angle of Z (i.e., $\theta°$). The three currents are shifted by 120° from each other. Hence the three line currents also form a balanced set of variables. Their algebraic (or vectorial) sum will be zero, as is clear from the phasor sum shown in Fig. 18.15. Also, one can easily notice that these currents do have the same sequence as the driving voltage sequence. Therefore, the remarks above are also valid for the ACB sequence.

Since KCL applies at the node N of the load,

$$\mathbf{I}_N + \mathbf{I}_A + \mathbf{I}_B + \mathbf{I}_C = 0$$

$$\boxed{\mathbf{I}_N = -(\mathbf{I}_A + \mathbf{I}_B + \mathbf{I}_C) = 0} \qquad (18.12)$$

The neutral wire does not carry any current when the load is balanced, as explained here. It can thus be removed without affecting any of the voltages and currents in the circuit of Fig. 18.14. However, when the Y-connected load is not balanced (i.e., different phase impedances), a net current will flow in this neutral line. It is useful in such cases to keep the phase voltages (line-to-neutral voltages) balanced because node N in the generator and in the load will always be at the same potential only if the neutral line is connected.

18.5 ONE-LINE EQUIVALENT CIRCUIT FOR BALANCED LOADS

Based on the circuit analysis of the balanced Y-connected load above, it is clear that all the line currents can be obtained when only one of them is calculated and the phase sequence is known.

From the one-line (e.g., line A) equivalent circuit of Fig. 18.16,

$$\boxed{\mathbf{I}_A = \frac{\mathbf{V}_{AN}}{Z_Y}} \quad (Z_Y \text{ is the phase impedance}) \qquad (18.13)$$

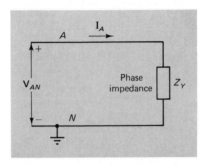

Figure 18.16

Then $\mathbf{I}_B$ and $\mathbf{I}_C$ are 120° out of phase with $\mathbf{I}_A$. The one that leads and the one that lags can easily be determined from the given sequence.

■ EXAMPLE 18.1

A three-phase four-wire ACB system with a line-to-neutral voltage of 120 V feeds a balanced Y-connected load with a phase impedance of $15\underline{/-20°}\ \Omega$. Find the line currents and draw the phasor diagram. Check the value of $\mathbf{I}_N$.

Solution   Using the one-line equivalent circuit with

$$\mathbf{V}_{AN} = 120\underline{/0°}\ \text{V}$$

$$\mathbf{I}_A = \frac{\mathbf{V}_{AN}}{Z_Y} = \frac{120\underline{/0°}}{15\underline{/-20°}} = 8\underline{/20°}\ \text{A}$$

The phasor diagram for this ACB system is shown in Fig. 18.17. Here

$$\mathbf{I}_B = 8\underline{/140°}\ \text{A}$$

and

$$\mathbf{I}_C = 8\underline{/-100°}\ \text{A}$$

Therefore,

$$\mathbf{I}_N = -(\mathbf{I}_A + \mathbf{I}_B + \mathbf{I}_C)$$
$$= -(7.517 + j2.736 - 6.128 + j5.142 - 1.389 - j7.878) = 0\ \text{A}$$

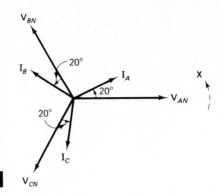

Figure 18.17

EXAMPLE 18.2

A three-phase three-wire 199 V *ABC* system feeds a balanced Y-connected load with a phase impedance of $23\underline{/30°}\ \Omega$. Find the line currents and draw the phasor diagram. The network connection is shown in Fig. 18.18.

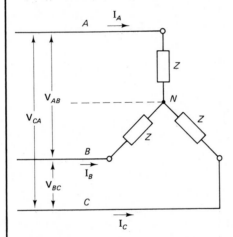

Figure 18.18

Solution Here the line-to-line voltage (V_L) is 199 V. As the load is balanced, node *N* of the load and the source are at the same potential; thus the phase voltages are also balanced with

$$V_{ph} = \frac{V_L}{\sqrt{3}} = \frac{199}{1.732} = 114.9\ \text{V}$$

Let $\mathbf{V}_{AN} = 114.9\underline{/0°}$ V (with $\mathbf{V}_{AB} = 199\underline{/30°}$ V). Now using the one-line equivalent circuit, we have

$$\mathbf{I}_A = \frac{\mathbf{V}_{AN}}{Z} = \frac{114.9\underline{/0°}}{23\underline{/30°}} = 5\underline{/-30°}\ \text{A}$$

Because this is an *ABC* system, $\mathbf{I}_A$ leads $\mathbf{I}_B$ by 120° and lags $\mathbf{I}_C$ by 120°. Thus

$$\mathbf{I}_B = 5\underline{/-150°}\ A$$

and

$$\mathbf{I}_C = 5\underline{/90°}\ \text{A}$$

The complete phasor diagram is as shown in Fig. 18.19.

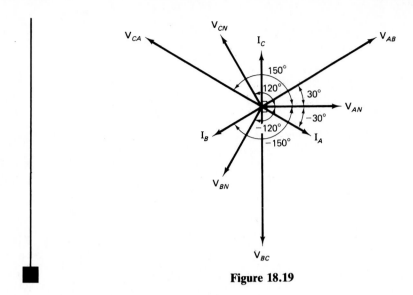

Figure 18.19

18.6 | BALANCED THREE-PHASE Δ-CONNECTED LOADS

This type of load connection requires a three-wire system. Here the line-to-line voltage is equal to the phase voltage of the load, as Fig. 18.20 indicates. Thus the phase currents can easily be calculated:

$$\mathbf{I}_{AB} = \frac{\mathbf{V}_{AB}}{Z_\Delta} \tag{18.14a}$$

$$\mathbf{I}_{BC} = \frac{\mathbf{V}_{BC}}{Z_\Delta} \tag{18.14b}$$

$$\mathbf{I}_{CA} = \frac{\mathbf{V}_{CA}}{Z_\Delta} \tag{18.14c}$$

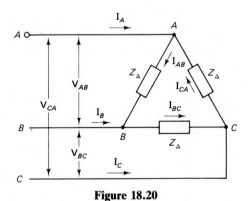

Figure 18.20

These phase currents also form a balanced set of phasors. They have equal magnitudes and are out of phase with each other by 120°. The phase sequence determines the lead and lag conditions. The line currents are obtained by applying KCL at each of the Δ nodes:

$$\mathbf{I}_A = \mathbf{I}_{AB} - \mathbf{I}_{CA} \tag{18.15a}$$

$$\mathbf{I}_B = \mathbf{I}_{BC} - \mathbf{I}_{AB} \tag{18.15b}$$

$$\mathbf{I}_C = \mathbf{I}_{CA} - \mathbf{I}_{BC} \tag{18.15c}$$

Therefore,

$$\mathbf{I}_A + \mathbf{I}_B + \mathbf{I}_C = \mathbf{I}_{AB} - \mathbf{I}_{CA} + \mathbf{I}_{BC} - \mathbf{I}_{AB} + \mathbf{I}_{CA} - \mathbf{I}_{BC} = 0 \tag{18.15d}$$

Let $\mathbf{I}_{AB} = I\underline{/\phi^\circ}$ A; then $\mathbf{I}_{CA} = I\underline{/\phi^\circ + 120^\circ}$ A in the ABC sequence. Therefore, using Eq. (18.15a), we have

$$
\begin{aligned}
\mathbf{I}_A &= I \cos \phi + jI \sin \phi - I \cos (\phi + 120^\circ) - jI \sin (\phi + 120^\circ) \\
&= -2I \sin (\phi + 60^\circ) \sin (-60^\circ) + j2I \cos (\phi + 60^\circ) \sin (-60^\circ) \\
&= \sqrt{3}\, I[\sin (\phi + 60^\circ) - j \cos (\phi + 60^\circ)] \\
&= \sqrt{3}\, I\{\sin [90^\circ + (\phi - 30^\circ)] - j \cos [90^\circ + (\phi - 30^\circ)]\} \\
&= \sqrt{3}\, I[\cos (\phi - 30^\circ) + j \sin (\phi - 30^\circ)] \\
&= \sqrt{3}\, I\underline{/\phi - 30^\circ}\ \text{A} \tag{18.16}
\end{aligned}
$$

and

$$|\mathbf{I}_L| = \sqrt{3}\,|\mathbf{I}_{\text{ph}}| \tag{18.16a}$$

The two trigonometric relations

$$\cos A - \cos B = -2 \sin \left(\frac{A + B}{2}\right) \sin \left(\frac{A - B}{2}\right)$$

and

$$\sin A - \sin B = 2 \cos \left(\frac{A + B}{2}\right) \sin \left(\frac{A - B}{2}\right)$$

were used in the derivation above. The phasor diagram schematic of Fig. 18.21 verifies this result geometrically.

Thus in a balanced three-phase Δ-connected load, the line currents are $\sqrt{3}$ times the phase current in magnitude. The line currents lag the corresponding phase currents by 30° in the ABC system ($\mathbf{I}_A$ lags $\mathbf{I}_{AB}$, $\mathbf{I}_B$ lags $\mathbf{I}_{BC}$, and $\mathbf{I}_C$ lags $\mathbf{I}_{CA}$), but lead by 30° in the ACB system.

If the Δ-connected load is transformed to an equivalent Y-connected load as shown in Fig. 18.22, the one-line equivalent circuit method can be used to obtain the line currents, using the phase voltage $\mathbf{V}_{AN}$ as before. Here

$$\mathbf{I}_A = \frac{\mathbf{V}_{AN}}{Z_Y} = 3\,\frac{\mathbf{V}_{AN}}{Z_\Delta}$$

Figure 18.21

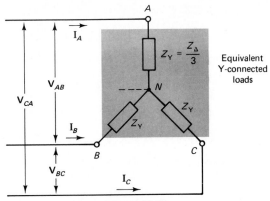

Figure 18.22

Therefore,

$$|\mathbf{I}_L| = 3\frac{|V_{ph}|}{|Z_\Delta|} = \sqrt{3}\frac{|V_L|}{|Z_\Delta|} = \sqrt{3}\,|\mathbf{I}_{ph}|$$

as above.

EXAMPLE 18.3

A three-phase three-wire 120 V ABC system feeds a Δ-connected load whose phase impedance is $30\underline{/45°}\,\Omega$. Find the phase and line currents in this system and draw the phasor diagram.

Solution Referring to Fig. 18.20, and for an ABC sequence with $\mathbf{V}_{AN}$ being the reference,

$$\mathbf{V}_{AB} = 120\underline{/30°}\,\text{V} \qquad \mathbf{V}_{BC} = 120\underline{/-90°}\,\text{V} \qquad \mathbf{V}_{CA} = 120\underline{/150°}\,\text{V}$$

Then

$$
\begin{aligned}
\mathbf{I}_{AB} &= \frac{\mathbf{V}_{AB}}{Z_\Delta} = \frac{120\underline{/30°}}{30\underline{/45°}}\\
&= 4\underline{/-15°}\,\text{A}\\
\mathbf{I}_{BC} &= \frac{\mathbf{V}_{BC}}{Z_\Delta} = \frac{120\underline{/-90°}}{30\underline{/45°}}\\
&= 4\underline{/-135°}\,\text{A}\\
\mathbf{I}_{CA} &= \frac{\mathbf{V}_{CA}}{Z_\Delta} = \frac{120\underline{/150°}}{30\underline{/45°}}\\
&= 4\underline{/105°}\,\text{A}
\end{aligned}
$$

Therefore,

$$
\begin{aligned}
\mathbf{I}_A = \mathbf{I}_{AB} - \mathbf{I}_{CA} &= (3.864 - j1.035) - (-1.305 + j3.864)\\
&= 4.9 - j4.9 = 6.928\underline{/-45°}\,\text{A}
\end{aligned}
$$

Notice that $\mathbf{I}_A$ has a magnitude $\sqrt{3} \times 4$ A and lags $\mathbf{I}_{AB}$ by 30°, as expected.

Similarly,

$$\mathbf{I}_B = 6.928\underline{/-165°} \text{ A}$$
$$\mathbf{I}_C = 6.928\underline{/75°} \text{ A}$$

Also, using the one-line equivalent circuit, as shown in Fig. 18.23(a), with

$$\mathbf{V}_{AN} = \frac{120}{\sqrt{3}}\underline{/0°} = 69.28\underline{/0°} \text{ V}$$

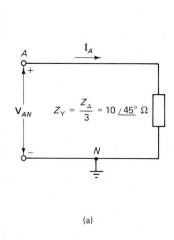

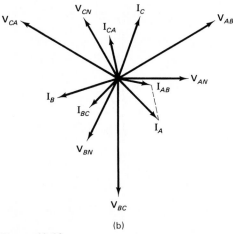

(a)   (b)

Figure 18.23

then

$$\mathbf{I}_A = \frac{\mathbf{V}_{AN}}{Z_Y} = \frac{69.28\underline{/0°}}{10\underline{/45°}} = 6.928\underline{/-45°} \text{ A}$$

as obtained above. $\mathbf{I}_B$ lags $\mathbf{I}_A$ by 120° and $\mathbf{I}_C$ leads $\mathbf{I}_A$ by 120°. The complete phasor diagram is shown in Fig. 18.23(b).

18.7 UNBALANCED Δ-CONNECTED LOADS

In general, the phase impedances of the Δ-connected load are not equal as shown in Fig. 18.24. Given the line voltages and the phase sequence, it is fairly simple to

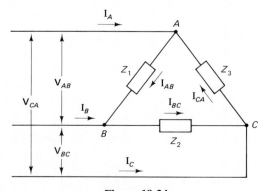

Figure 18.24

determine the phase and line currents. Here

$$\mathbf{I}_{AB} = \frac{\mathbf{V}_{AB}}{Z_1} \qquad \mathbf{I}_{BC} = \frac{\mathbf{V}_{BC}}{Z_2} \qquad \mathbf{I}_{CA} = \frac{\mathbf{V}_{CA}}{Z_3}$$

The line currents $\mathbf{I}_A$, $\mathbf{I}_B$, and $\mathbf{I}_C$ are obtained through applying KCL at the Δ nodes. Thus Eqs. (18.15) are still applicable here. Because of the imbalance of the load, the single-line equivalent-circuit method cannot be used in such situations.

■ EXAMPLE 18.4

A three-phase three-wire 240 V *ACB* system feeds the Δ-connected load shown in Fig. 18.25. Find the phase and line currents and draw the phasor diagram.

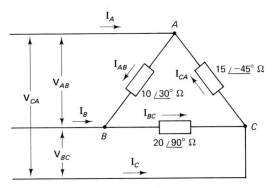

Figure 18.25

Solution With $\mathbf{V}_{AN}$ being the reference voltage phasor*, even though it is not used here [refer to Fig. 18.13 and Eqs. (18.10)],

$$\mathbf{V}_{AB} = 240\underline{/-30°}\text{ V}$$
$$\mathbf{V}_{BC} = 240\underline{/90°}\text{ V}$$
$$\mathbf{V}_{CA} = 240\underline{/-150°}\text{ V}$$

Thus

$$\mathbf{I}_{AB} = \frac{240\underline{/-30°}}{10\underline{/30°}} = 24\underline{/-60°}\text{ A}$$

$$\mathbf{I}_{BC} = \frac{240\underline{/90°}}{20\underline{/90°}} = 12\underline{/0°}\text{ A}$$

$$\mathbf{I}_{CA} = \frac{240\underline{/-150°}}{15\underline{/-45°}} = 16\underline{/-105°}\text{ A}$$

---

* Obviously, any other voltage phasor could have been taken as a reference, reflecting as a constant rotation of the complete phasor diagram. For the sake of consistency, this choice is as good as any!

From KCL [Eqs. (18.15)],

$$\mathbf{I}_A = \mathbf{I}_{AB} - \mathbf{I}_{CA}$$
$$= (12 - j20.78) - (-4.14 - j15.45)$$
$$= 16.14 - j5.33 = 17\underline{/-18.3°}\text{ A}$$

$$\mathbf{I}_B = \mathbf{I}_{BC} - \mathbf{I}_{AB}$$
$$= 12 - (12 - j20.78)$$
$$= 20.78\underline{/90°}\text{ A}$$

$$\mathbf{I}_C = \mathbf{I}_{CA} - \mathbf{I}_{BC}$$
$$= (-4.14 - j15.45) - 12$$
$$= -16.14 - j15.45 = 22.34\underline{/-136.3°}\text{ A}$$

The complete phasor diagram is shown in Fig. 18.26.

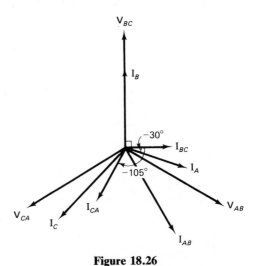

Figure 18.26

Two important observations can be made from this example:

1. The phase currents $\mathbf{I}_{AB}$, $\mathbf{I}_{BC}$, and $\mathbf{I}_{CA}$ do not form a balanced set of phasors. They have unequal magnitudes and are not separated by 120°. Their sum is not zero.

2. The line currents $\mathbf{I}_A$, $\mathbf{I}_B$, and $\mathbf{I}_C$ also do not form a balanced set of phasors; however, their sum $(\mathbf{I}_A + \mathbf{I}_B + \mathbf{I}_C = 0)$ is zero. This is true algebraically [refer to Eqs. (18.15)] and numerically from Example 18.4. This is also true from KCL, which can be stated in general: *The algebraic (or vectorial) sum of all currents entering a closed surface in an electrical circuit is zero.*

18.8 | UNBALANCED FOUR-WIRE Y-CONNECTED LOADS

The phase impedances (Z_A, Z_B, and Z_C) of the Y-connected load are generally unequal. Because this system (Fig. 18.27) includes the neutral wire connecting the neutral nodes of the source and the load, the balanced line-to-neutral voltages

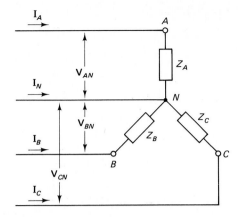

Figure 18.27

appear across the corresponding phase impedances. Here the line currents ($\mathbf{I}_A$, $\mathbf{I}_B$, and $\mathbf{I}_C$), being equal to the phase currents, are determined through application of Ohm's law:

$$\mathbf{I}_A = \frac{\mathbf{V}_{AN}}{Z_A} \qquad \mathbf{I}_B = \frac{\mathbf{V}_{BN}}{Z_B} \qquad \mathbf{I}_C = \frac{\mathbf{V}_{CN}}{Z_C} \tag{18.17}$$

They form an unbalanced set of phasors. Their sum is not zero. However, from KCL applied at node N of the load,

$$\mathbf{I}_N + \mathbf{I}_A + \mathbf{I}_B + \mathbf{I}_C = 0$$

Therefore,

$$\mathbf{I}_N = -(\mathbf{I}_A + \mathbf{I}_B + \mathbf{I}_C) \tag{18.18}$$

that is, the neutral wire carries the net sum of the unbalanced line currents.

EXAMPLE 18.5

A three-phase four-wire 208 V* ABC system feeds the Y-connected unbalanced load shown in Fig. 18.28. Obtain the line currents in this system, and draw the phasor diagram.

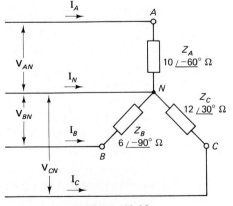

Figure 18.28

* Usually the voltage specified (208 V in this problem statement) refers to the line-to-line voltage, unless specifically defined as line-to-neutral voltage.

Solution For this ABC system, the line-to-neutral (or phase) voltages are

$$\mathbf{V}_{AN} = \frac{208}{\sqrt{3}}\,\underline{/0°} = 120\underline{/0°}\ \text{V}$$

$$\mathbf{V}_{BN} = 120\underline{/-120°}\ \text{V}$$

$$\mathbf{V}_{CN} = 120\underline{/120°}\ \text{V}$$

Therefore,

$$\mathbf{I}_A = \frac{120\underline{/0°}}{10\underline{/-60°}} = 12\underline{/60°}\ \text{A}$$

$$\mathbf{I}_B = \frac{120\underline{/-120°}}{6\underline{/-90°}} = 20\underline{/-30°}\ \text{A}$$

$$\mathbf{I}_C = \frac{120\underline{/120°}}{12\underline{/30°}} = 10\underline{/90°}\ \text{A}$$

The current in the neutral wire is

$$\mathbf{I}_N = -(\mathbf{I}_A + \mathbf{I}_B + \mathbf{I}_C)$$
$$= -(6 + j10.39 + 17.32 - j10 + j10)$$
$$= -23.32 - j10.39 = 25.53\underline{/-156°}\ \text{A}$$

The complete phasor diagram is shown in Fig. 18.29.

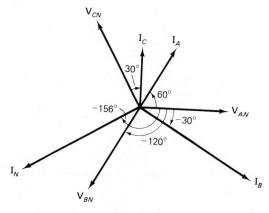

Figure 18.29

18.9 | UNBALANCED THREE-WIRE Y-CONNECTED LOADS

With only three wires connected to the unbalanced Y-connected load, the circuit is a two-loop network which can easily be solved using loop analysis to determine the values of $\mathbf{I}_1$ and $\mathbf{I}_2$. From Fig. 18.30,

$$\mathbf{V}_{AB} = \mathbf{I}_1(Z_A + Z_B) - \mathbf{I}_2(Z_B) \tag{18.19}$$

$$\mathbf{V}_{BC} = -\mathbf{I}_1(Z_B) + \mathbf{I}_2(Z_B + Z_C) \tag{18.20}$$

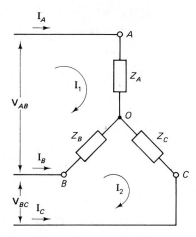

Figure 18.30

Equations (18.19) and (18.20) are then solved to obtain $\mathbf{I}_1$ and $\mathbf{I}_2$. Then

$$\mathbf{I}_A = \mathbf{I}_1 \tag{18.21a}$$

$$\mathbf{I}_B = \mathbf{I}_2 - \mathbf{I}_1 \tag{18.21b}$$

$$\mathbf{I}_C = -\mathbf{I}_2 \tag{18.21c}$$

Also,

$$\mathbf{V}_{AO} = \mathbf{I}_A Z_A \tag{18.22a}$$

$$\mathbf{V}_{BO} = \mathbf{I}_B Z_B \tag{18.22b}$$

$$\mathbf{V}_{CO} = \mathbf{I}_C Z_C \tag{18.22c}$$

The common node between the three impedances of the three-phase Y-connected load is called O, not N, since it could have a potential quite different from that of N, the neutral point of the source. In fact, the phase voltages $\mathbf{V}_{AO}$, $\mathbf{V}_{BO}$, and $\mathbf{V}_{CO}$ determined from Eqs. (18.22) form an unbalanced set of phasors which are quite different from the phase voltages of the source, $\mathbf{V}_{AN}$, $\mathbf{V}_{BN}$, and $\mathbf{V}_{CN}$, in general. The voltage $\mathbf{V}_{ON}$ is called the *displacement neutral voltage*. It can be obtained from:

$$\boxed{\begin{array}{c} \mathbf{V}_{ON} = \mathbf{V}_{OA} + \mathbf{V}_{AN} = \mathbf{V}_{AN} - \mathbf{V}_{AO} \\ \text{or} \\ \mathbf{V}_{ON} = \mathbf{V}_{BN} - \mathbf{V}_{BO} = \mathbf{V}_{CN} - \mathbf{V}_{CO} \end{array}} \tag{18.23}$$

When the fourth wire is connected, points O and N are forced to have the same potential (i.e., $\mathbf{V}_{ON}$ is zero), as is the situation in the four-wire systems discussed above.

Notice also that from KCL at node O,

$$\boxed{\mathbf{I}_A + \mathbf{I}_B + \mathbf{I}_C = 0} \tag{18.24}$$

Equations (18.21) confirm this fact. Refer also to the general form of KCL stated above. Thus Eq. (18.24) is valid for all three-phase three-wire systems, whether the load is Δ- or Y-connected, balanced or unbalanced.

EXAMPLE 18.6

A three-phase three-wire $300\,V$ ACB system feeds the unbalanced Y-connected load shown in Fig. 18.31. Find the line currents and the phase voltages of the load. Also determine the displacement neutral voltage $\mathbf{V}_{ON}$.

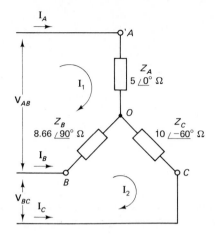

Figure 18.31

Solution   Here $\mathbf{V}_{AB} = 300\underline{/-30°}\,V$ and $\mathbf{V}_{BC} = 300\underline{/90°}\,V$ for an ACB system with $\mathbf{V}_{AN}$ as reference (see Fig. 18.13). Using loop analysis yields

$$\mathbf{V}_{AB} = 300\underline{/-30°} = \mathbf{I}_1(5 + j8.66) - \mathbf{I}_2(j8.66)$$
$$\mathbf{V}_{BC} = 300\underline{/90°} = -\mathbf{I}_1(j8.66) + \mathbf{I}_2(j8.66 + 5 - j8.66)$$

Thus

$$\mathbf{I}_1 = \frac{\begin{vmatrix} 300\underline{/-30°} & -j8.66 \\ 300\underline{/90°} & 5 \end{vmatrix}}{\begin{vmatrix} 5 + j8.66 & -j8.66 \\ -j8.66 & 5 \end{vmatrix}} = \frac{1500\underline{/-30°} - 2598}{25 + j43.3 + 75}$$

$$= \frac{1299 - j750 - 2598}{100 + j43.3} = \frac{1500\underline{/-150°}}{109\underline{/23.4°}} = 13.76\underline{/-173.4°}\,A$$

$$\mathbf{I}_2 = \frac{\begin{vmatrix} 10\underline{/60°} & 300\underline{/-30°} \\ 8.66\underline{/-90°} & 300\underline{/90°} \end{vmatrix}}{100 + j43.3} = \frac{3000\underline{/150°} - 2598\underline{/-120°}}{109\underline{/23.4°}}$$

$$= \frac{-2598 + j1500 - (-1299 - j2250)}{109\underline{/23.4°}} = \frac{-1299 + j3750}{109\underline{/23.4°}}$$

$$= \frac{3968.6\underline{/109.1°}}{109\underline{/23.4°}} = 36.41\underline{/85.7°}\,A$$

Therefore,

$$\mathbf{I}_A = \mathbf{I}_1 = 13.76\underline{/-173.4°}\,A \qquad \mathbf{V}_{AO} = \mathbf{I}_A Z_A = 68.8\underline{/-173.4°}\,V$$
$$\mathbf{I}_B = \mathbf{I}_2 - \mathbf{I}_1 = 2.73 + j36.31 - (-13.69 - j1.58)$$
$$= 16.42 + j37.89 = 41.29\underline{/66.6°}\,A$$

and
$$\mathbf{V}_{BO} = \mathbf{I}_B Z_B = 357.6\underline{/156.6°} \text{ V}$$
$$\mathbf{I}_C = -\mathbf{I}_2 = 36.41\underline{/-94.3°} \text{ A} \qquad \mathbf{V}_{CO} = \mathbf{I}_C Z_C = 364.1\underline{/-154.3°} \text{ V}$$

Notice the unbalanced nature of the line currents and the phase voltages of the load. The displacement neutral voltage is

$$\mathbf{V}_{ON} = \mathbf{V}_{AN} - \mathbf{V}_{AO} = 173.2\underline{/0°} - 68.8\underline{/-173.4°}$$
$$= 173.2 - (-68.3 - j7.91) = 241.5 + j7.91 = 241.6\underline{/1.9°} \text{ V}$$

Also,

$$\mathbf{V}_{ON} = \mathbf{V}_{BN} - \mathbf{V}_{BO} = 173.2\underline{/120°} - 357.6\underline{/156.6°}$$
$$= -86.6 + j150 + 328.2 - j142 = 241.6 + j8 = 241.7\underline{/1.9°} \text{ V}$$

(check)

18.10 POWER IN THREE-PHASE SYSTEMS

In the case of balanced three-phase systems, the average power dissipated in each phase of the load is the same (P_{ph}) and the total power dissipated by the three-phase balanced load is

$$P_T = 3P_{ph} \tag{18.25}$$

When the load is connected in a Δ form as in Fig. 18.32, the wattmeter, connected to any of the three load phases as shown, will read the average power in that phase. This power reading is

$$P_{ph} = I_{ph}V_{ph}\cos\theta_Z$$

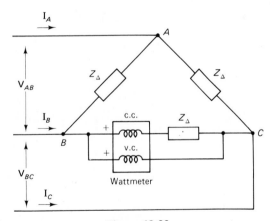

Figure 18.32

Here

$$I_{ph} = \text{phase current magnitude}$$
$$= \frac{1}{\sqrt{3}}I_L$$
$$= \frac{1}{\sqrt{3}} \times \text{line current magnitude}$$

as was proved in Eq. (18.16).

$$V_{\text{ph}} = \text{phase voltage magnitude}$$
$$= V_L = \text{line voltage magnitude}$$

θ_Z is the phase angle of the impedance Z_Δ. Therefore,

$$P_{\text{ph}} = \frac{1}{\sqrt{3}} I_L V_L \cos \theta_Z \qquad (18.25a)$$

Also, when the load is connected in a Y form as in Fig. 18.33, the wattmeter, connected to any of the three phases as shown, will read the average power in that phase. This power reading is

$$P_{\text{ph}} = I_{\text{ph}} V_{\text{ph}} \cos \theta_Z$$

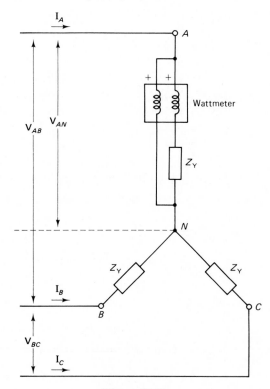

Figure 18.33

where

$$I_{\text{ph}} = \text{phase current magnitude} = I_L$$

$$V_{\text{ph}} = \text{phase voltage magnitude} = \frac{1}{\sqrt{3}} V_L \quad [\text{from Eqs. (18.7)}]$$

$$\theta_Z = \text{phase angle of the impedance } Z_Y$$

Notice that from the Δ–Y transformation, $Z_Y = \frac{1}{3} Z_\Delta$ (i.e., the phase of Z_Y is the same as the phase of Z_Δ). Thus for the Y-connected load,

$$P_{\text{ph}} = \frac{1}{\sqrt{3}} I_L V_L \cos \theta_Z \qquad (18.25b)$$

Hence for both the Δ-connected and the Y-connected balanced three-phase loads, the total power in the system is obtained from the same expression:

$$P_T = 3P_{\text{ph}} = \sqrt{3}\,I_L V_L \cos\theta_Z \qquad (18.26)$$

Remark. Remember that θ_Z is *not* the phase angle between any particular I_L and any other particular V_L. It is always the angle of the load impedance.

If we transform a Δ-connected load into the equivalent Y-connected load, or vice versa, the phase power read by a wattmeter connected to any of the phases of the Δ load or of the Y load will be exactly the same, as Eqs. (18.25) indicate.

It is easy to extend the result of Eq. (18.26) above to the complete power triangle of the load system, with

$$S_T = \text{total apparent power} = \sqrt{3}\,I_L V_L \quad \text{VA} \qquad (18.26a)$$
$$Q_T = \text{total reactive power} = \sqrt{3}\,I_L V_L \sin\theta_Z \quad \text{VAR} \qquad (18.26b)$$

The consideration of the instantaneous power in the balanced three-phase load system is of special interest. Analysis based on Δ-connected or Y-connected loads lead to the same results, similar to the argument above. Then, for a Y-connected balanced three-phase load, let

$$v_{AN} = \sqrt{2}\,V_{\text{ph}} \sin\omega t \qquad \text{then} \qquad i_{AN} = \sqrt{2}\,I_L \sin(\omega t - \theta_Z)$$
$$v_{BN} = \sqrt{2}\,V_{\text{ph}} \sin(\omega t - 120°) \qquad \text{then} \qquad i_{BN} = \sqrt{2}\,I_L \sin(\omega t - 120° - \theta_Z)$$
$$v_{CN} = \sqrt{2}\,V_{\text{ph}} \sin(\omega t + 120°) \qquad \text{then} \qquad i_{CN} = \sqrt{2}\,I_L \sin(\omega t + 120° - \theta_Z)$$

for an *ABC* sequence (similar results would be obtained for an *ACB* sequence). Thus

$$p_T(t) = 2V_{\text{ph}}I_L[\sin\omega t \sin(\omega t - \theta_Z) + \sin(\omega t - 120°)\sin(\omega t - 120° - \theta_Z)$$
$$+ \sin(\omega t + 120°)\sin(\omega t + 120° - \theta_Z)]$$

Using the trigonometric relationship:

$$\sin A \sin B = \tfrac{1}{2}[\cos(A - B) - \cos(A + B)]$$

Then

$$p_T(t) = V_{\text{ph}}I_L\{3\cos\theta_Z - [\cos(2\omega t - \theta_Z) + \cos(2\omega t - \theta_Z - 120°)$$
$$+ \cos(2\omega t - \theta_Z + 120°)]\}$$

$$= \frac{1}{\sqrt{3}}V_L I_L\{3\cos\theta_Z - [\cos\psi + \cos\psi\cos 120° + \sin\psi\sin 120°$$

$$+ \cos\psi\cos 120° - \sin\psi\sin 120°]\} \quad \text{where } \psi = 2\omega t - \theta_Z$$

$$= \frac{1}{\sqrt{3}}V_L I_L\{3\cos\theta_Z - [\cos\psi + 2\cos\psi(-\tfrac{1}{2})]\}$$

$$= \sqrt{3}\,V_L I_L \cos\theta_Z \qquad (18.27)$$

We can conclude from the result above that the total instantaneous power in a balanced three-phase system is a constant (independent of time) equal to the total average power of the system. This is true whether the load is Δ or Y connected or whether its sequence is *ABC* or *ACB*. This is one of the major advantages of three-phase transmission and operation. Even though the instantaneous power in each phase (single-phase) is pulsating (i.e., continuously varying with time), the total instantaneous power of the whole load system is a constant. This obviously results in minimizing the vibrations associated with power variations with time in both the generating system (alternators) and the load system (e.g., motors).

In the case of unbalanced three-phase loads, clearly Eqs. (18.26) and (18.27) do not apply. Here each phase dissipates a different amount of power. The average phase power has to be measured using a separate wattmeter for each phase, or by reconnecting the wattmeter to each phase in turn. Power calculations are performed for each phase of the load independently (as if it were a separate load), based on each of the different phase voltages and the corresponding phase currents. For the overall system,

$$P_T = P_A + P_B + P_C \qquad (18.28)$$

where P_A, P_B, and P_C are the average powers (or wattmeter readings) of the three different phases.

18.11 TWO-WATTMETER METHOD OF POWER MEASUREMENT

In many instances, the three-phase load is enclosed (e.g., a three-phase motor) and the three phases are unaccessible to connect the wattmeters required to measure the power in each of the phases. The only access is to the lines feeding such loads. Also, it is not usually convenient to interrupt the operation of the three-phase load to reconnect the wattmeter to each phase of the load in turn, particularly for the cases of unbalanced three-phase load. If the three phases of an unbalanced load are accessible and the interruption of the operation is acceptable, then obviously the power measurements can be performed by connecting one wattmeter to each of the three phases of the load.

As will be discussed here, only two wattmeters are in fact sufficient to measure the power in any three-phase system, whether the load is balanced or not, with the only access to the system being the three lines.

Consider the three wattmeters *A*, *B*, and *C* shown in Fig. 18.34. The current coil of each wattmeter is connected in series in the line, sensing the different line currents. The voltage coil (v.c.) of each wattmeter has one terminal connected to each of the lines, while the other terminal of each is connected to the common point *X*. Thus

$$\bar{p}_A = \overline{i_A v_{AX}} = W_A \quad (p_A, i_A, \text{ and } v_{AX} \text{ are the instantaneous functions of time, with } \bar{p}_A \text{ being the average power, i.e., the reading of wattmeter } A)$$

(18.29a)

Similarly,

$$\bar{p}_B = \overline{i_B v_{BX}} = W_B \quad (\text{reading of wattmeter } B) \qquad (18.29b)$$

and

$$\bar{p}_C = \overline{i_C v_{CX}} = W_C \quad (\text{reading of wattmeter } C) \qquad (18.29c)$$

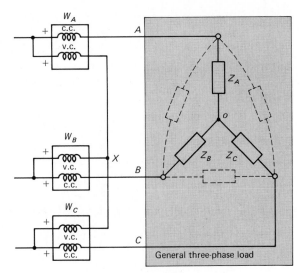

Figure 18.34

Adding these quantities (the readings of the three wattmeters) gives us

$$\bar{p}_A + \bar{p}_B + \bar{p}_C = \text{Av}(i_A v_{AX}) + \text{Av}(i_B v_{BX}) + \text{Av}(i_C v_{CX})$$
$$= \text{Av}(i_A v_{AX} + i_B v_{BX} + i_C v_{CX})$$
$$= \text{Av}[i_A(v_{AO} + v_{OX}) + i_B(v_{BO} + v_{OX}) + i_C(v_{CO} + v_{OX})]$$
$$= \text{Av}[i_A v_{AO} + i_B v_{BO} + i_C v_{CO} + v_{OX}(i_A + i_B + i_C)]$$

Whether the load is balanced or not, or whether it is Δ-connected or Y-connected, according to the general form of KCL [e.g., Eq. (18.24)],

$$i_A + i_B + i_C = 0$$

This relationship is true whether the variables are in the time domain or in the phasor equivalent form. Thus

$$W_A + W_B + W_C = \text{Av}\,[i_A v_{AO} + i_B v_{BO} + i_C v_{CO}]$$
$$= P_A + P_B + P_C = P_T \qquad (18.30)$$

In Eq. (18.30), clearly $\overline{i_A v_{AO}}$ is the average power in phase A of the load, as is the case for the other two terms. Thus the sum of the three wattmeters readings is, in fact, the total power of the system, independent of the actual potential of point X.

Even though point O is hypothetical, the result in Eq. (18.30) is correct whether the load is Δ or Y connected because power is the same in the Δ-connected load as it is in its equivalent Y-connected load. Also, this result is independent of the phase sequence or of whether the three-phase load is balanced or not.

If point X is taken to coincide with any of the three lines, the voltage coil of the wattmeter in this line will have zero voltage across it and thus this wattmeter reading will be zero, allowing this particular wattmeter to be removed. Figure 18.35 shows the case where point X is chosen to coincide with line B. The two other arrangements can be obtained similarly and used to measure the total

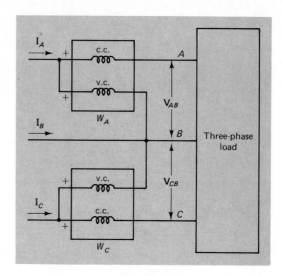

Figure 18.35

system power. Here

$$\boxed{P_T = W_A + W_C} \tag{18.31}$$

where

$$\boxed{\begin{aligned} W_A &= I_A V_{AB} \cos \theta_{\mathbf{I}_A,\, \mathbf{v}_{AB}} \\ W_C &= I_C V_{CB} \cos \theta_{\mathbf{I}_C,\, \mathbf{v}_{CB}} \end{aligned}} \tag{18.31a}$$
$$\tag{18.31b}$$

Notice that the voltage across the voltage coil of W_C is $\mathbf{V}_{CB}$ (not $\mathbf{V}_{BC}$) because of the wattmeter polarities. Also, the sum in Eq. (18.31) is algebraic since it is possible that one of the wattmeter readings may be negative. Similarly, for the other two-wattmeter arrangements, the relationship between the watt-meter polarities and the voltage phasor must be carefully examined.

Note. The negative wattmeter reading is obtained by reversing the polarities of one of its coils, but not both. When polarities are reversed compared with those shown in Fig. 18.35, to obtain an upscale deflection, the reading of the corresponding wattmeter is considered negative and algebraically added to the other wattmeter reading to obtain the total power.

Considering the case when *the two-wattmeter method* is applied to measure the total power of a *balanced three-phase load system* as shown in Fig. 18.36. This system is assumed to be an ABC system. The two wattmeters are connected to lines A and B as shown, that is, point X (the common voltage node) is taken to coincide with line C in this case. Thus

$$W_A = I_A V_{AC} \cos \theta_{\mathbf{I}_A,\, \mathbf{v}_{AC}} \tag{18.32a}$$
$$W_B = I_B V_{BC} \cos \theta_{\mathbf{I}_B,\, \mathbf{v}_{BC}} \tag{18.32b}$$

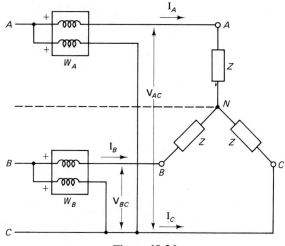

Figure 18.36

Taking $\mathbf{V}_{AN}$ to be the reference phasor, the phase and line voltages of this balanced three-phase ABC system are as shown in Fig. 18.37. If the impedance Z is assumed to be inductive with angle θ_Z, the line currents will lag the corresponding phase voltages by the angle θ_Z as shown. Thus

$$W_A = I_L V_L \cos(\theta_Z - 30°) \tag{18.33a}$$

$$W_B = I_L V_L \cos(\theta_Z + 30°) \tag{18.33b}$$

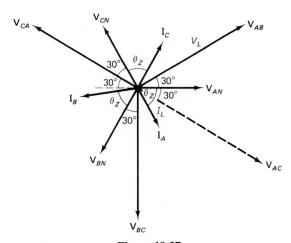

Figure 18.37

Therefore,

$$
\begin{aligned}
P_T = W_A + W_B &= I_L V_L[\cos(\theta_Z - 30°) + \cos(\theta_Z + 30°)] \\
&= I_L V_L(\cos\theta_Z \cos 30° + \sin\theta_Z \sin 30° + \cos\theta_Z \cos 30° - \sin\theta_Z \sin 30°) \\
&= \sqrt{3}\, V_L I_L \cos\theta_Z \tag{18.33c}
\end{aligned}
$$

as obtained in Eq. (18.27). This derivation, even though it is one case out of many possible situations for different sequences or load connections, also proves the validity of the two-wattmeter method. It is intended here as an example to show how one can obtain the readings of the two wattmeters.

Knowing the two wattmeter readings and the line voltage, the magnitude and phase of the balanced Y-connected impedances Z can be calculated. Referring to the example in Fig. 18.37, it can be easily proved that

$$W_A - W_B = V_L I_L \sin \theta_Z \qquad (18.33d)$$

Therefore,

$$\tan \theta_Z = \sqrt{3} \frac{W_A - W_B}{W_A + W_B} \qquad (18.34)$$

This is obtained by dividing Eq. (18.33d) by Eq. (18.33c). Hence the phase angle of the load impedance is obtained in both magnitude and sign. Then from

$$P_T = W_A + W_B = \sqrt{3} V_L I_L \cos \theta_Z$$

I_L can be obtained. Thus

$$\boxed{|Z_Y| = \frac{V_{ph}}{I_L} = \frac{1}{\sqrt{3}} \frac{V_L}{I_L}} \qquad (18.35)$$

For other wattmeter connections and other sequences, one can follow an analysis procedure similar to that above to obtain the phase angle of the Y-connected (or Δ-connected since $Z_\Delta = 3Z_Y$) balanced load impedance. Summarizing such results, we have, for the ABC sequence,

$$\boxed{\tan \theta_Z = \sqrt{3} \frac{W_A - W_B}{W_A + W_B} = \sqrt{3} \frac{W_B - W_C}{W_B + W_C} = \sqrt{3} \frac{W_C - W_A}{W_C + W_A}} \qquad (18.36)$$

and for the ACB sequence,

$$\boxed{\tan \theta_Z = \sqrt{3} \frac{W_B - W_A}{W_B + W_A} = \sqrt{3} \frac{W_C - W_B}{W_C + W_B} = \sqrt{3} \frac{W_A - W_C}{W_A + W_C}} \qquad (18.37)$$

Note that in all such balanced load situations, one wattmeter will read as in Eq. (18.33a), while the other will read as in Eq. (18.33b).

18.12 GENERAL EXAMPLES OF THREE-PHASE SYSTEMS

■ EXAMPLE 18.7

A three-phase three-wire ACB 120 V system supplies a Δ-connected balanced load with impedance of $30/{-60°}\ \Omega$ as shown in Fig. 18.38. Determine the line currents and draw the phasor diagram. Find the readings of the two wattmeters used to determine the total system power.

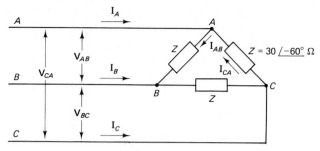

Figure 18.38

Solution Taking $\mathbf{V}_{AN}$ as the reference phasor:

$$\mathbf{V}_{AB} = 120\underline{/-30°}\ \text{V}$$
$$\mathbf{V}_{BC} = 120\underline{/90°}\ \text{V}$$
$$\mathbf{V}_{CA} = 120\underline{/-150°}\ \text{V}$$

(for an *ACB* system, with a 120 V line voltage; check with Fig. 18.13). The phase currents can then easily be found from Ohm's law:

$$\mathbf{I}_{AB} = \frac{\mathbf{V}_{AB}}{Z} = \frac{120\underline{/-30°}}{30\underline{/-60°}} = 4\underline{/30°}\ \text{A}$$

This is a balanced system. Thus, for the *ACB* sequence,

$$\mathbf{I}_{BC} = 4\underline{/150°}\ \text{A}$$
$$\mathbf{I}_{CA} = 4\underline{/-90°}\ \text{A}$$

(As a check:

$$\mathbf{I}_{CA} = \frac{\mathbf{V}_{CA}}{Z} = \frac{120\underline{/-150°}}{30\underline{/-60°}}$$
$$= 4\underline{/-90°}\ \text{A})$$

For the line current $\mathbf{I}_A$, applying KCL at node *A* of the load, we have

$$\mathbf{I}_A = \mathbf{I}_{AB} - \mathbf{I}_{CA} = 4\underline{/30°} - 4\underline{/-90°} = 3.464 + j2 - (-j4)$$
$$= 3.464 + j6 = 6.928\underline{/60°}\ \text{A}$$

(*Note:* $I_L = \sqrt{3}\,I_{\text{ph}} = \sqrt{3} \times 4$.)

The two other line currents can easily be determined for this balanced *ACB* system:

$$\mathbf{I}_B = 6.928\underline{/180°}\ \text{A}$$

and

$$\mathbf{I}_C = 6.928\underline{/-60°}\ \text{A}$$

The phasor diagram is as shown in Fig. 18.39.

The two wattmeters (since their location in the lines is undefined) will

18.12   General Examples of Three-Phase Systems

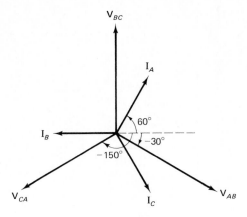

Figure 18.39

read

$$W_1 = V_L L_L \cos(\theta_Z - 30°) \quad (\theta_Z \text{ is the angle of } Z)$$
$$= 120 \times 6.928 \cos(-60° - 30°) = 0 \text{ W}$$
$$W_2 = V_L I_L \cos(\theta_Z + 30°)$$
$$= 120 \times 6.928 \cos(-60° + 30°) = 720 \text{ W}$$
$$\text{total system power} = W_1 + W_2 = 0 + 720 = 720 \text{ W}$$

As a check on the system's total power,

$$P_T = \sqrt{3} \, V_L I_L \cos \theta_Z = 1.732 \times 120 \times 6.928 \cos(-60°)$$
$$= 720 \text{ W} \quad \text{(check)}$$

or since $Z = 30\underline{/-60°} = 15 - j26 \, \Omega$, its resistive component is 15 Ω; then

$$P_T = 3P_{\text{ph}} = 3 \times I_{\text{ph}}^2 \times R_{\text{ph}} = 3 \times (4)^2 \times 15 = 720 \text{ W} \quad \text{(check)}$$

EXAMPLE 18.8

A three-phase three-wire 200 V ABC system feeds a balanced Y-connected load with impedance of $70.7\underline{/45°} \, \Omega$ as shown in Fig. 18.40. Find the line currents and draw the phasor diagram. Find the total system's power and the two wattmeter readings when they are connected to lines B and C.

Solution Here $V_L = 200$ V, $V_{\text{ph}} = 200/\sqrt{3} = 115.47$ V. For this ABC system with $\mathbf{V}_{AN}$ as reference,

$$\mathbf{V}_{AN} = 115.47\underline{/0°} \text{ V}$$
$$\mathbf{V}_{BN} = 115.47\underline{/-120°} \text{ V}$$
$$\mathbf{V}_{CN} = 115.47\underline{/120°} \text{ V}$$

and

$$\mathbf{V}_{AB} = 200\underline{/30°} \text{ V}$$
$$\mathbf{V}_{BC} = 200\underline{/-90°} \text{ V}$$
$$\mathbf{V}_{CA} = 200\underline{/150°} \text{ V} \quad \text{(check with Fig. 18.12)}$$

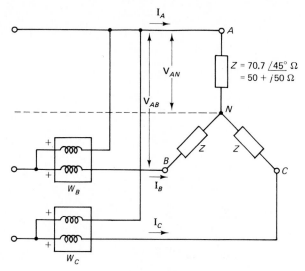

Figure 18.40

Here node N is the neutral because the load is balanced. From Ohm's law,

$$\mathbf{I}_A = \frac{\mathbf{V}_{AN}}{Z} = \frac{115.47\underline{/0°}}{70.7\underline{/45°}}$$

$$= 1.633\underline{/-45°}\ \text{A}$$

Thus the two other line currents are

$$\mathbf{I}_B = 1.633\underline{/-165°}\ \text{A}$$

and

$$\mathbf{I}_C = 1.633\underline{/75°}\ \text{A}$$

The phasor diagram is as shown in Fig. 18.41. The total system's power is

$$P_T = \sqrt{3}\ V_L I_L \cos\theta_Z \quad (\theta_Z \text{ is the angle of } Z)$$

$$= \sqrt{3} \times 200 \times 1.633 \cos 45°$$

$$= 400\ \text{W}$$

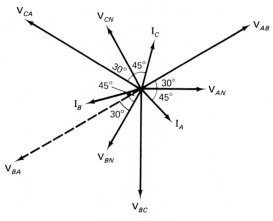

Figure 18.41

or

$$P_T = 3P_{ph} = 3 \times (1.633)^2 \times 50$$
$$= 400 \text{ W}$$

The readings of the two wattmeters will be

$$W_B = I_B V_{BA} \cos \theta_{\mathbf{I}_B, \mathbf{v}_{BA}}$$
$$= 1.633 \times 200 \cos (45° - 30°) = 315.47 \text{ W}$$

$$W_C = I_C V_{CA} \cos \theta_{\mathbf{I}_C, \mathbf{v}_{CA}}$$
$$= 1.633 \times 200 \cos (30° + 45°) = 84.53 \text{ W}$$

Therefore,

$$P_T = W_B + W_C = 315.47 + 84.53 = 400 \text{ W}$$

as above.

■ EXAMPLE 18.9

A Δ-connected balanced load with impedance $15\underline{/30°}\ \Omega$ and a Y-connected balanced load with impedance $10\underline{/-30°}\ \Omega$ are both connected to a three-phase three-wire ABC 208 V system. Find the line currents, the total system power, and the two wattmeter readings.

Solution For these type of problems, the single-line equivalent-circuit technique is the most appropriate method to analyze the system, as both loads are balanced. Converting the Δ load into its equivalent Y-connected form, we have

$$Z'_Y = \tfrac{1}{3}Z_\Delta = 5\underline{/30°}\ \Omega$$

The two load systems are now balanced Y connected, as shown in Fig. 18.42. The line-to-neutral voltages are the same across the corresponding phase impedances of the two loads. The phase loads are thus in parallel and one equivalent Y-connected circuit can be used. Its phase impedance is

$$Z_{Y\text{eq}} = \frac{10\underline{/-30°} \times 5\underline{/30°}}{10\underline{/-30°} + 5\underline{/30°}} = \frac{50\underline{/0°}}{13.228\underline{/-10.9°}} = 3.78\underline{/10.9°}\ \Omega$$

With

$$\mathbf{V}_{AN} = \frac{208}{\sqrt{3}} \underline{/0°} = 120\underline{/0°} \text{ V}$$

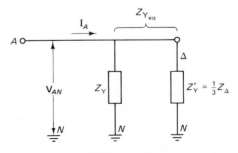

Figure 18.42

then,

$$\mathbf{I}_A = \frac{120\underline{/0^\circ}}{3.78\underline{/10.9^\circ}} = 31.75\underline{/-10.9^\circ}\ \text{A}$$

For this balanced ABC sequence system,

$$\mathbf{I}_B = 31.75\underline{/-10.9^\circ - 120^\circ} = 31.75\underline{/-130.9^\circ}\ \text{A}$$

and

$$\mathbf{I}_C = 31.75\underline{/-10.9^\circ + 120^\circ} = 31.75\underline{/109.1^\circ}\ \text{A}$$
$$P_T = \text{total system's power} = \sqrt{3}\,V_L I_L \cos\theta_{\text{Zeq}}$$
$$= \sqrt{3} \times 208 \times 31.75 \cos(10.9^\circ) = 11.232\ \text{kW}$$

The wattmeter readings will be

$$W_1 = V_L I_L \cos(\theta_{\text{Zeq}} - 30^\circ)$$
$$= 208 \times 31.75 \cos(10.9^\circ - 30^\circ) = 6.24\ \text{kW}$$
$$W_2 = V_L I_L \cos(\theta_{\text{Zeq}} + 30^\circ)$$
$$= 208 \times 31.75 \cos(10.9^\circ + 30^\circ) = 4.992\ \text{kW}$$

Clearly, $P_T = W_1 + W_2$.

EXAMPLE 18.10

A three-phase three-wire ACB 240 V system supplies the unbalanced Δ-connected load shown in Fig. 18.43. Find the three line currents and the total system's power. If the two wattmeters are connected to lines A and B, what will be their readings?

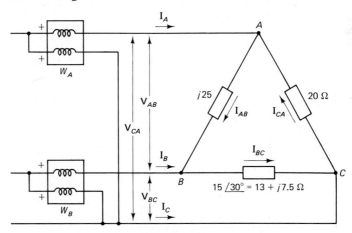

Figure 18.43

Solution   For the ACB sequence, with $\mathbf{V}_{AN}$ as reference phasor,

$$\mathbf{V}_{AB} = 240\underline{/-30^\circ}\ \text{V}$$
$$\mathbf{V}_{BC} = 240\underline{/90^\circ}\ \text{V}$$
$$\mathbf{V}_{CA} = 240\underline{/-150^\circ}\ \text{V}$$

Thus the three phase currents are

$$\mathbf{I}_{AB} = \frac{240\underline{/-30°}}{25\underline{/90°}} = 9.6\underline{/-120°}\,\text{A}$$

$$\mathbf{I}_{BC} = \frac{240\underline{/90°}}{15\underline{/30°}} = 16\underline{/60°}\,\text{A}$$

$$\mathbf{I}_{CA} = \frac{240\underline{/-150°}}{20\underline{/0°}} = 12\underline{/-150°}\,\text{A}$$

Applying KCL at the three nodes of the load, we have

$$\mathbf{I}_A = \mathbf{I}_{AB} - \mathbf{I}_{CA} = (-4.8 - j8.31) - (-10.39 - j6)$$
$$= 5.59 - j2.31 = 6.05\underline{/-22.5°}\,\text{A}$$
$$\mathbf{I}_B = \mathbf{I}_{BC} - \mathbf{I}_{AB} = (8 + j13.86) - (-4.8 - j8.31)$$
$$= 12.8 + j22.17 = 25.6\underline{/60°}\,\text{A}$$
$$\mathbf{I}_C = \mathbf{I}_{CA} - \mathbf{I}_{BC} = (-10.39 - j6) - (8 + j13.86)$$
$$= -18.39 - j19.86 = 27.07\underline{/-132.8°}\,\text{A}$$

The phasor diagram in Fig. 18.44 shows the line voltages and currents. The total system's power is the sum of the three phase powers. Here

$$P_T = P_{AB} + P_{BC} + P_{CA}$$
$$= 0 + (16)^2 \times 12.99 + (12)^2 \times 20 = 6205\,\text{W}$$

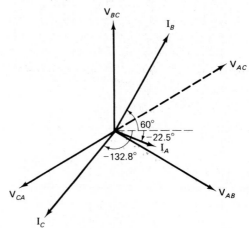

Figure 18.44

The two wattmeter readings are

$$W_A = V_{AC}I_A \cos\theta_{\mathbf{V}_{AC},\,\mathbf{I}_A}$$
$$= 240 \times 6.05 \cos(30° + 22.5°) = 884\,\text{W}$$
$$W_B = V_{BC}I_B \cos\theta_{\mathbf{V}_{BC},\,\mathbf{I}_B}$$
$$= 240 \times 25.6 \cos 30° = 5321\,\text{W}$$

As a check,

$$W_A + W_B = 884 + 5321 = 6205\,\text{W} = P_T$$

EXAMPLE 18.11

The two-wattmeter method is applied to a three-phase three-wire 120 V ABC system. With the meters connected to lines A and C, $W_A = 920$ W and $W_C = 460$ W. Find the impedance of the balanced Δ-connected load.

Solution   Here

$$P_T = W_A + W_C = 920 + 460 = 1380 \text{ W}$$

For the balanced load,

$$P_T = \sqrt{3}\, V_L I_L \cos \theta_Z$$

But from Eq. (18.36),

$$\tan \theta_Z = \sqrt{3}\, \frac{W_C - W_A}{W_C + W_A} = \sqrt{3} \times \frac{460 - 920}{460 + 920} = -0.577$$

$$\theta_Z = -30°$$

Thus

$$1380 = \sqrt{3} \times 120 \times I_L \cos(-30°)$$

Therefore,

$$I_L = 7.667 \text{ A}$$

As the load is Δ-connected, $V_{\text{ph}} = V_L = 120$ V. Also, for the balanced load:

$$I_{\text{ph}} = \frac{1}{\sqrt{3}} I_L = \frac{7.667}{1.732} = 4.426 \text{ A}$$

Therefore,

$$|Z_\Delta| = \frac{120}{4.426} = 27.11 \ \Omega$$

$$Z_\Delta = 27.11\underline{/-30°} \ \Omega$$

Note that for the equivalent Y-connected load,

$$|Z_Y| = \frac{V_{\text{ph}}}{I_L} = \frac{120}{\sqrt{3} \times 7.667} = 9.04 \ \Omega \qquad \text{that is, } Z_Y = 9.04\underline{/-30°} \ \Omega = \frac{1}{3} Z_\Delta$$

EXAMPLE 18.12

A 100 V three-phase three-wire ABC system feeds the unbalanced Y-connected load shown in Fig. 18.45. Find the line currents, the total system power, and the two wattmeter readings when they are connected to lines A and C.

Solution   For this ABC system; with $\mathbf{V}_{AN}$ as the reference phasor,

$$\mathbf{V}_{AB} = 100\underline{/30°} \text{ V}$$

$$\mathbf{V}_{BC} = 100\underline{/-90°} \text{ V}$$

$$\mathbf{V}_{CA} = 100\underline{/150°} \text{ V}$$

Applying loop analysis to the network yields

$$(20 + j17.33)\mathbf{I}_1 - 10\mathbf{I}_2 = 100\underline{/30°}$$

$$-10\mathbf{I}_1 + (10 - j10)\mathbf{I}_2 = 100\underline{/-90°}$$

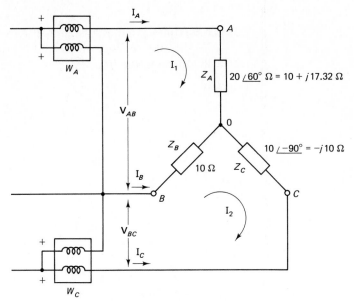

Figure 18.45

Solving for $\mathbf{I}_1$ and $\mathbf{I}_2$, we have

$$\mathbf{I}_1 = \frac{\begin{vmatrix} 100\underline{/30°} & -10 \\ 100\underline{/-90°} & 10 - j10 \end{vmatrix}}{\begin{vmatrix} 20 + j17.33 & -10 \\ -10 & 10 - j10 \end{vmatrix}} = \frac{1414\underline{/-15°} + 1000\underline{/-90°}}{374.25\underline{/-4.1°} - 100}$$

$$= \frac{1365.8 - j1366}{273.3 - j26.8} = \frac{1931.7\underline{/-45°}}{274.6\underline{/-5.6°}} = 7.03\underline{/-39.4°} \text{ A}$$

$$\mathbf{I}_2 = \frac{\begin{vmatrix} 20 + j17.33 & 100\underline{/30°} \\ -10 & 100\underline{/-90°} \end{vmatrix}}{274.6\underline{/-5.6°}} = \frac{2646.4\underline{/-49.1°} + 1000\underline{/30°}}{274.6\underline{/-5.6°}}$$

$$= \frac{2598.7 - j1500.3}{274.6\underline{/-5.6°}} = \frac{3000.7\underline{/-30°}}{274.6\underline{/-5.6°}} = 10.93\underline{/-24.4°} \text{ A}$$

Therefore,

$$\mathbf{I}_A = \mathbf{I}_1 = 7.03\underline{/-39.4°} \text{ A}$$
$$\mathbf{I}_B = \mathbf{I}_2 - \mathbf{I}_1 = 9.95 - j4.52 - (5.43 - j4.46) = 4.52 - j0.06$$
$$= 4.52\underline{/-0.8°} \text{ A}$$
$$\mathbf{I}_C = -\mathbf{I}_2 = 10.93\underline{/155.6°} \text{ A}$$

The phasor diagram is shown in Fig. 18.46. The total system power is

$$P_T = P_A + P_B + P_C = (7.03)^2 \times 10 + (4.52)^2 \times 10 + 0$$
$$= 698.5 \text{ W}$$

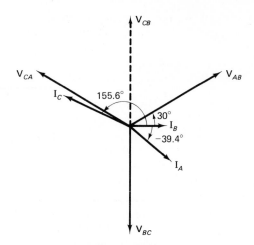

Figure 18.46

The two wattmeter readings are

$$W_A = I_A V_{AB} \cos \theta_{\mathbf{I}_A, \mathbf{v}_{AB}}$$
$$= 7.03 \times 100 \cos (69.4°) = 247.3 \text{ W}$$

$$W_C = I_C V_{CB} \cos \theta_{\mathbf{I}_C, \mathbf{v}_{CB}}$$
$$= 10.93 \times 100 \cos (155.6° - 90°) = 451.5 \text{ W}$$

(Check: $W_A + W_C = 698.8 \text{ W} = P_T$.)

DRILL PROBLEMS

Sections 18.4 to 18.6

18.1. A three-phase four-wire ABC system, whose line voltage is 346.4 V feeds a balanced Y-connected load whose impedance is $25/-30° \, \Omega$. Find the four line currents and draw the phasor diagram.

18.2. Using the one-line equivalent-circuit method, calculate the line currents in a three-phase three-wire ACB system having a 206 V line voltage feeding a balanced Y-connected load whose impedance is $12/45° \, \Omega$.

18.3. If the phase voltage of a Y-connected three-phase three-wire ABC generator is 100 V, and this system feeds a balanced Δ-connected load whose impedance is $10/60° \, \Omega$, find
  (a) The phase currents in the load impedances.
  (b) The line currents.
  Draw the phasor diagram of this system.

18.4. A balanced Δ-connected load whose impedance is $8/30° \, \Omega$ is fed by a 160 V three-phase three-wire ACB generator. Find the phase currents of the load and the line currents. Draw the phasor diagram.

18.5. By converting the load in Drill Problem 18.3 into an equivalent Y-connected load and using the one-line method, recalculate the line currents to confirm the results obtained in that problem.

18.6. Convert the load given in Drill Problem 18.4 into its equivalent Y-connected load and then use the one-line method to recalculate the line currents.

18.7. For the following balanced loads, determine the sequence and the third line current in each of the following cases:

    (a) $\mathbf{I}_A = 4\underline{/30°}$ A, $\mathbf{I}_B = 4\underline{/-90°}$ A.

    (b) $\mathbf{I}_C = 5\underline{/-120°}$ A, $\mathbf{I}_A = 5\underline{/0°}$ A.

18.8. If each of the cases examined in Drill Problem 18.7 results due to the application of a 200 V three-phase three-line system, having the determined sequence, find

    (a) The value of the impedance in case (a) of Drill Problem 18.7 if this load is Y-connected.

    (b) The value of the impedance in case (b) of Drill Problem 18.7 if this load is Δ-connected.

Sections 18.7 to 18.9

18.9. A three-phase three-wire 300 V ABC system feeds a Δ-connected load as shown in Fig. 18.24, where $Z_1 = 15\underline{/-60°}$ Ω, $Z_2 = 10\underline{/30°}$ Ω, and $Z_3 = 20\underline{/0°}$ Ω. Find the phase and line currents and draw the phasor diagram.

18.10. A three-phase four-wire 346.4 V ACB system feeds an unbalanced Y-connected load as shown in Fig. 18.27, where $Z_A = 10\underline{/0°}$ Ω, $Z_B = 20\underline{/-30°}$ Ω, and $Z_C = 10\underline{/60°}$ Ω. Calculate the values of the four line currents in this system and draw the phasor diagram.

18.11. If the neutral line in the system described in Drill Problem 18.10 is disconnected, calculate the new values of the three-line currents. Draw the phasor diagram. Determine the value of the displacement neutral voltage V_{ON}.

Sections 18.10 and 18.11

18.12. Use two methods to calculate the total power in each of the following cases:

    (a) The system examined in Drill Problem 18.1.

    (b) The system examined in Drill Problem 18.2.

    (c) The system examined in Drill Problem 18.3.

18.13. If the total power measured in a 200 V three-phase system feeding a balanced Δ-connected load is 5 kW, and the load impedance is known to have a 0.8 lagging power factor, find the magnitude of the line current and the value of the load impedance.

18.14. A 173.2 V three-phase three-wire ABC system feeds a balanced Y-connected load as shown in Fig. 18.36. The readings of the two wattmeters connected as shown in this figure were $W_A = 600$ W and $W_B = 800$ W. Find the line current and the load impedance.

18.15. Find the readings of the two wattmeters connected to lines A and C of the unbalanced system examined in Drill Problem 18.9.

18.16. Repeat Problem 18.15 if the two wattmeters are now connected to lines B and C.

18.17. Calculate the total power in the three-phase system examined in Drill Problem 18.9, by adding the three phase powers. Compare this total power to the results obtained in Drill Problems 18.15 and 18.16.

PROBLEMS

Note: All of the problems below can be used as exercises on computer-aided network analysis or as programming exercises.

Sections 18.4 to 18.6

(i)   **18.1.** A balanced Y-connected load, with an impedance of $10\underline{/30°}$ Ω is connected to a three-phase four-wire ACB system having an effective line voltage of 500 V. Obtain the line currents.

(i) **18.2.** A Δ-connected balanced load with an impedance of $45\underline{/60°}\,\Omega$ is connected to a three-phase three-wire 208 V ABC system. Calculate the phase and the line currents.

(i) **18.3.** Using the one-line equivalent-circuit method, recalculate the line currents in Problem 18.2.

(i) **18.4.** Determine the sequence in each of the following cases.
    (a) $\mathbf{I}_A = 10\underline{/60°}\,A$, $\mathbf{I}_C = 10\underline{/-60°}\,A$, balanced.
    (b) $\mathbf{I}_A = 5\underline{/45°}\,A$, $\mathbf{I}_B = 5\underline{/-75°}\,A$, balanced.
    Find the third line current in each case above.

(ii) **18.5.** The currents in Problem 18.4(a) were obtained from a 100 V three-wire three-phase system whose load is Y-connected. Find the load impedance.

(ii) **18.6.** The currents in Problem 18.4(b) were obtained from a 208 V three-wire three-phase system whose load is Δ-connected. Find the load impedance.

(i) **18.7.** Find the thrid line current in an unbalanced three-phase three-wire system when $\mathbf{I}_A = 6\underline{/45°}\,A$ and $\mathbf{I}_B = 10\underline{/-30°}\,A$.

(ii) **18.8.** A balanced Δ-connected load has one phase current $\mathbf{I}_{BC} = 2\underline{/-90°}\,A$. Find the other phase currents and the three line currents if the system is an ABC system. If the line voltage is 100 V, what is the load impedance?

(i) **18.9.** Find the line currents for a three-phase three-wire 200 V ABC system feeding
    (a) A balanced Δ-connected load with a load impedance of $15\underline{/45°}\,\Omega$
    (b) A balanced Y-connected load with a load impedance of $7.071\underline{/45°}\,\Omega$
    when each load is connected alone to the system, and when the loads are connected simultaneously.

(i) **18.10.** A balanced Δ-connected load, with $Z_\Delta = 9\underline{/90°}\,\Omega$ and a balanced Y-connected load with $Z_Y = 4\underline{/0°}\,\Omega$ are supplied by a three-phase three-wire ACB system with a 208 V line voltage. Calculate the line currents using the one-line equivalent-circuit technique.

Sections 18.7 to 18.9

(i) **18.11.** A three-phase three-wire 300 V ABC system feeds an unbalanced Δ-connected load with $Z_{AB} = 20\underline{/60°}\,\Omega$, $Z_{BC} = 10\underline{/0°}\,\Omega$, and $Z_{CA} = 10\underline{/90°}\,\Omega$. Obtain the phase and the line currents.

(i) **18.12.** A three-phase four-wire 120 V ABC system feeds an unbalanced Y-connected load with $Z_A = 5\underline{/0°}\,\Omega$, $Z_B = 10\underline{/30°}\,\Omega$, and $Z_C = 20\underline{/60°}\,\Omega$. Obtain the four line currents.

(i) **18.13.** A three-phase four-wire 120 V ABC system feeds an unbalanced Y-connected load with $Z_A = 6\underline{/90°}\,\Omega$, $Z_B = 4\underline{/30°}\,\Omega$, and $Z_C = 12\underline{/0°}\,\Omega$. Find the four line currents.

(ii) **18.14.** The unbalanced Y-connected load in Problem 18.12 is fed by a three-phase three-wire ABC system with a line voltage of 208 V. Obtain the line currents in this case. Calculate the phase voltages of the load and the displacement neutral voltage $\mathbf{V}_{ON}$.

(ii) **18.15.** The unbalanced Y-connected load in Problem 18.13 is fed by a three-phase three-wire ACB system with a line voltage of 100 V. Obtain the line currents in this case and calculate the phase voltages of the load. What is the value of the displacement neutral voltage, $\mathbf{V}_{ON}$?

Sections 18.10 to 18.12

(ii) **18.16.** Obtain the readings of two wattmeters connected to a three-phase three-wire 120 V system feeding a balanced Δ-connected load with a load impedance of $12\underline{/30°}\,\Omega$. Assume either phase sequence. Find the phase power and compare the total power to the sum of the wattmeter readings.

Problems

785

(i)   **18.17.**   Repeat Problem 18.16 but with the load connected as a balanced Y load, with an impedance of $10\underline{/70°}\,\Omega$.

(ii) **18.18.**   Two wattmeters are connected to lines A and B of a three-phase three-wire ABC system with a line voltage of 208 V feeding a balanced Y-connected load. Wattmeter W_A reads -200 W and wattmeter W_B reads 1000 W. Find the line current and the load impedance.

(ii) **18.19.**   Two wattmeters are connected to lines A and C of a three-phase three-wire ACB system with a line voltage of 500 V feeding a balanced Δ-connected load. The wattmeter readings were $W_A = 500$ W and $W_C = 1500$ W. Find the load impedance.

(i)   **18.20.**   Find the readings of two wattmeters connected to lines A and B of the unbalanced system in Problem 18.11. Calculate each phase power and compare the total power to the sum of the two wattmeter readings.

(i)   **18.21.**   Find the readings of two wattmeters connected to lines B and C of the unbalanced system in Problem 18.14. Calculate each phase power and compare the total power to the sum of the two wattmeter readings.

(i)   **18.22.**   In Problem 18.15, find the readings of the two wattmeters when connected
  (a)   To lines A and C.
  (b)   To lines B and C.
  Compare the sum of the two wattmeter readings in each of the cases above to the total power of the system, by summing the phase powers.

(ii) **18.23.**   A 100 V ABC system is connected to the load shown in Fig. 18.47. Find the readings of the meter
  (a)   If it is a high-impedance voltmeter.
  (b)   If it is a very low impedance ammeter.

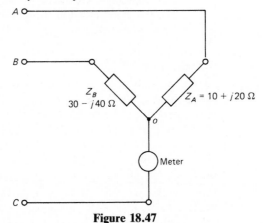

Figure 18.47

(i)   **18.24.**   A three-phase 200 V ACB system feeds the unbalanced Δ-connected load shown in Fig. 18.48. Find the phase and the line currents. Find each phase power and the total system power.

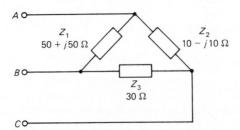

Figure 18.48

(i) **18.25.** Find the readings of the two wattmeters in Problem 18.24 when connected
  (a) To lines A and B.
  (b) To lines B and C.

GLOSSARY

***Balanced Load*:**   A multiphase load that consists of equal impedances in each of the phases.

***Delta (Δ)-Connected Generator*:**   A three-phase generator in which the three-phase voltage coils are connected together in the shape of the capital Greek letter delta (Δ).

***Lines*:**   The wires used to connect a three-phase (or a multiphase) generator to its load. A three-phase system has three active lines (A, B, and C) and may or may not have a fourth neutral line.

***Line Voltage*:**   The voltage difference between each pair of lines, excluding the neutral line if it exists, in a multiphase system.

***Neutral Line*:**   The wire that connects the neutral nodes of a multiphase generator and the corresponding multiphase load. When the multiphase load is balanced zero current flows in the neutral line.

***Neutral Point*:**   The common node in a Y-connected generator or a Y-connected load, which may or may not be connected to ground. It is the reference point for the phase voltages of a multiphase system.

***Phase Current*:**   The current that flows in each of the phases of a multiphase generator-load system.

***Phase Sequence*:**   The order in which the generated sinusoidal phase voltages of a multiphase generator occur with respect to time. In a three-phase system, the phases are referred to as A, B, and C. In the ABC (or positive) sequence, the phase A voltage leads the phase B voltage by $120°$, and the phase B voltage leads the phase C voltage by $120°$. In the ACB (or negative) sequence, the phase A voltage leads the phase C voltage by $120°$, while the phase C voltage leads the phase B voltage by $120°$.

***Phase Voltage*:**   The voltage generated by each of the separate phase coils in a multiphase generator. In a three-phase four-line Y-connected generator, the phase voltage will be the voltage difference appearing between each of the lines and the neutral line. In a three-phase, three-line Δ-connected generator, the phase voltage will be the voltage difference appearing between each of the line pairs.

***Three-Phase Generator*:**   An ac source that produces three sinusoidal voltage waveforms. These voltage waveforms are referred to as the phase voltages.

***Two-wattmeter Method*:**   A method for measuring (or determining) the total power dissipated in any three-phase load (whether or not it is balanced and whether it is Δ- or Y-connected) using only two wattmeters connected to the lines feeding the three-phase load.

***Unbalanced Load*:**   A multiphase load not having the same impedance in each phase.

Y-Connected Generator: A three-phase generator in which the three-phase voltage coils are connected in the shape ot the letter Y. The common terminal of the three coils is the neutral point of the generator.

ANSWERS TO DRILL PROBLEMS

18.1. $\mathbf{I}_A = 8\underline{/30°}\,A$, $\mathbf{I}_B = 8\underline{/-90°}\,A$, $\mathbf{I}_C = 8\underline{/150°}\,A$, $\mathbf{I}_N = 0\,A$

18.3. (a) $\mathbf{I}_{AB} = 17.32\underline{/-30°}\,A$, $\mathbf{I}_{BC} = 17.32\underline{/-150°}\,A$, $\mathbf{I}_{CA} = 17.32\underline{/90°}\,A$

(b) $\mathbf{I}_A = 30\underline{/-60°}\,A$, $\mathbf{I}_B = 30\underline{/180°}\,A$, $\mathbf{I}_C = 30\underline{/60°}\,A$

18.5. $\mathbf{I}_A = 30\underline{/-60°}\,A$, $\mathbf{I}_B = 30\underline{/180°}\,A$, $\mathbf{I}_C = 30\underline{/60°}\,A$

18.7. (a) $\mathbf{I}_C = 4\underline{/150°}\,A$, *ABC* sequence

(b) $\mathbf{I}_B = 5\underline{/120°}\,A$, *ACB* sequence

18.9. $\mathbf{I}_{AB} = 20\underline{/90°}\,A$, $\mathbf{I}_{BC} = 30\underline{/-120°}\,A$, $\mathbf{I}_{CA} = 15\underline{/150°}\,A$, $\mathbf{I}_A = 18.03\underline{/43.9°}\,A$, $\mathbf{I}_B = 48.36\underline{/-108.1°}\,A$, $\mathbf{I}_C = 33.54\underline{/86.6°}\,A$

18.11. $\mathbf{I}_A = 24.83\underline{/-2.46°}\,A$, $\mathbf{I}_B = 8.533\underline{/137.7°}\,A$, $\mathbf{I}_C = 19.078\underline{/-165.81°}\,A$, $\mathbf{V}_{ON} = 49.24\underline{/167.5°}\,V$

18.13. $I_L = 18.04\,A$, $Z_\Delta = 19.2\underline{/36.87°}\,\Omega$

18.15. $W_A = 5.25\,kW$, $W_B = 10.03\,kW$

18.17. $P_T = 15.29\,kW = W_A + W_B = W_B + W_C$

19

COMPUTER-AIDED CIRCUIT ANALYSIS USING SPICE

OBJECTIVES

- Familiarity with the specific guidelines used to compile a SPICE data file to describe a given electric circuit and a specific required type of analysis.
- Ability to perform dc, ac and transient analysis of any of the typical networks discussed in this textbook using the SPICE software package.

All the analysis procedures examined in this text are described in terms of algebraic, closed-form solutions, which are quite systematic in nature. This applies to the dc analysis, transient analysis, and ac analysis of all the linear time-invariant networks analyzed in previous chapters. In particular, the loop and node analysis procedures, which are generally described in terms of matrices and solved using determinants, were found to be quite general and apply to any kind of electical network. This conclusion leads one to reflect on the question: If the network can be described to a computer, can the computer be programmed to do the numerical calculations necessary to obtain the required solution?

Obviously, the answer is "yes." Computers are powerful machines that can perform many numerical computations in a very fast and accurate manner. Besides, writing a software package to describe an algebraic solution procedure is now an easy task.

Many such software packages (computer programs) have been developed in the past 20 years. Some are suitable for mainframe computers, some for minis, and some for micros (or personal computers). In fact, some of these programs are so powerful that they can be used to analyze networks that are 100 times more complex than any considered in this text. Clearly, such networks cannot be analyzed by hand. Also, most of these programs can analyze nonlinear and time-variant networks. Models for electronic components such as diodes, BJTs, JFETs, and MOSFETs can be incorporated in such computer programs.

One of the most general and powerful circuit analysis programs is SPICE. It is being used in many electronics industries, in research and development laboratories, and in universities and colleges.

In this chapter we examine how to prepare and describe an electric network to the computer using the SPICE program, how to ask the computer to perform a certain analysis procedure, and how to provide the answers in a suitable form (e.g., printed tables or graphs).

SPICE is a general-purpose circuit-simulation program for nonlinear dc, nonlinear transient, and linear ac analysis. The circuits it simulates may contain resistors, capacitors, inductors, mutual inductors, independent voltage and current sources, all four types of dependent sources, transmission lines, and the four most common semiconductor devices. It has built-in models for these nonlinear devices. Only those features of SPICE that apply to linear dc, transient, and ac analysis of the circuit types considered in this text are examined here. Those who are interested in the full range of capabilities of SPICE should refer to the User's Guide, which can be obtained from their computer centers or by contacting the University of California, Berkeley, California.

The data files necessary to describe to the computer the circuits to be analyzed are discussed here. However, the way in which SPICE is invoked using these data files depends on the computer's operating system. This varies from computer to computer and from place to place. The appropriate personnel in the computer center should be consulted about the relevant operating system commands to invoke SPICE. Under IBM's CMS operating system, for example, SPICE is run by entering "SPICE FILENAME FILETYPE." A version of this program, called PSICE, is now available for the IBM-PC.

With the availability of such powerful network analysis computer programs, the natural question that is often asked is: Why learn or study all these network analysis techniques if the computer can do the work for us, much faster, more accurately, and even more efficiently? Besides the philosophical answers to this question, the real answer is that knowledge of circuit analysis is important to check on the validity and reasonableness of the computer results, at least in an approximate manner. As the saying goes: "garbage in, garbage out." Also, in many situations one may not always have computer facilities available. In many other cases, simple circuits similar to those examined here are often encountered in practice. For such circuits it is as easy and as fast to analyze the network problem by hand, using a pocket calculator as an aid. Therefore, computer-aided circuit analyses have a very useful role and a wide range of applications, but they do not eliminate the need for the knowledge gained through the study of network analysis techniques.

19.2 CIRCUIT DESCRIPTION: THE DATA FILE

19.2.1 Input format

SPICE is designed to accept the free-format type of input. The input on every line always begins at column 1. If the amount of information needs more space than is available on one line, line continuation can be effected by entering a + (plus) in column 1 of the next, immediately following line. SPICE then continues reading the information beginning with column 2. Fields on a line are separated by one or more blanks, a comma, an equal sign (=), or a left or right parenthesis. Extra spaces are ignored.

A name field must begin with a letter (A through Z) and cannot contain delimiters. Only the first eight characters of the name are used.

A number field may be an integer field (e.g., 21, -75, etc.), a floating-point field (e.g., 6.143), either an integer or a floating-point number followed by an integer exponent (e.g., 1.3E-10, 7.95E3, etc.), or either an integer or a floating-point number followed by one of the following *scale factors*:

$T = 1E12 = 10^{12}$	$G = 1E9 = 10^9$	$MEG = 1E6 = 10^6$
$K = 1E3 = 10^3$	$M = 1E-3 = 10^{-3}$	$U = 1E-6 = 10^{-6}$
$N = 1E-9 = 10^{-9}$	$P = 1E-12 = 10^{-12}$	$F = 1E-15 = 10^{-15}$

Letters immediately following a number that are not scale factors are ignored. Thus 10, 10 V, 10 volts, and 10 Hz are all accepted by the program as the numerical value of 10. Also, 1000, 1000.0, 1E3, 1.0E3, 1 K, 1 KHz, and 1 KV are all accepted by the program as the numerical value 1000.

19.2.2 Data File (General Remarks)

The circuit to be analyzed is described to SPICE by a set of *element line records* which define the circuit's topology and element values. A set of *control lines* is used to define the model parameters and the execution (run) controls.

The first line in the input deck must be a title line, for example,

Power Amplifier Circuit

Its contents are printed verbatim as the heading for each section of the output. The last line in the input deck must be an

```
.END
```

line. The period is an integral part of the name. The order of the remaining lines is arbitrary, except that a continuation line must immediately follow the line being continued.

Sometimes it may be of value to insert comments in the input data file. Comment lines can be placed anywhere in the data file. They have no effect on the circuit description. Comment lines must start with an asterisk (*) in column 1, for example:

```
* This part represents the LP input filter section

* The gain of this amplifier section should be 100
```

Each element in the circuit is specified by an element line which contains the element name, the circuit nodes to which the element is connected, and the value of the parameters that determine the electrical characteristics of the element. The first letter of the element name specifies the element type. The format for the SPICE element types is an alphanumeric string (e.g., XXXX, YYYYYYY, ZZZZZZZZ) of one up to eight characters. For example, a resistor's name must begin with the letter R and can contain from one to eight characters. The following represent valid resistor names:

```
R, R12, ROUT, R3AC  etc.
```

In the following descriptions, all indicated punctuation (parentheses, equal signs, etc.) is required. However, data fields enclosed within the signs"⟨ ⟩" are optional.

It is very important to note that SPICE always uses the associated reference convention shown in Fig. 19.1. Positive current flows in the direction of the voltage drop, no matter what the nature of the element between the nodes. Thus whether the element in Fig. 19.1 is an active element (e.g., a source) or a passive element (e.g., R, L, or C), positive voltage is considered as the voltage of the first node specified with respect to the second node specified, while positive current is considered to flow from the first node specified, into the element, toward the second node specified.

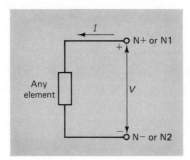

Figure 19.1

Nodes are numbered with nonnegative integers but need not be numbered sequentially. The *ground* (datum) *node* must be numbered *zero*. Loops containing only voltage sources and/or inductors are not allowed. Each node in the circuit must have a dc path to the ground, obviously not through capacitors. At least two elements must be connected to each node. An open-circuit condition between two nodes can be simulated by a resistance of very large magnitude (compared to the other elements in the circut, e.g., $10^{10}\,\Omega$).

19.2.3   Element Lines

Resistors.   General form:

```
RXXXXXXX N1 N2 VALUE <TC=TC1,TC2>
```

■ EXAMPLE 19.1

Refer to Fig. 19.2.

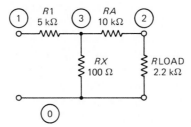

Figure 19.2

```
R1      1 3 5K
RX      3 0 100.0    TC=0.001,0.0002
RA      3 2 10.0E3
RLOAD   2 0 2.2K
```

Here, N1 and N2 are the two element nodes. VALUE is the value of the resistance in ohms; it may be positive or negative but not zero. TC1 and TC2 are optional temperature coefficients. If they are not specified, zero is assumed for both. The value of the resistance as a function of temperature is calculated by SPICE using the equation

$$value(TEMP)=value(TNOM)*(1+TC1*(TEMP-TNOM)$$
$$+TC2*(TEMP-TNOM)**2)$$

SPICE uses a nominal temperature (TNOM) of 27°C, or 300°K, for all the elements whose characteristics are temperature dependent. As will be shown later, a TEMP control line can be used to specify any other operating temperature, where the circuit may have to be simulated.

Capacitors and inductors.   General form:

```
CXXXXXXX N+ N- VALUE <IC=INTCOND>
LYYYYYYY N+ N- VALUE <IC=INTCOND>
```

■ EXAMPLE 19.2

Refer to Fig. 19.3.

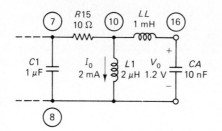

Figure 19.3

```
C1    7   8  1UF
R15   7  10  10.0
L1   10   8  2.0UH  IC=2MA
LL   10  16  1E-3
CA   16   8  10NF   IC=1.2V
```

Here, N+ and N− are the positive and negative element nodes, respectively, based on the expected current flow or voltage drop, or based on the initial condition (IC) so that such a value is numerically positive. VALUE is the value of the capacitance in farads or the value of the inductance in henrys. The initial condition is optional, indicating the initial (at zero time) voltage on the capacitor or the initial current flowing in the inductor in volts and amperes, respectively. Note that the initial conditions (if any) apply only if the UIC, "Use Initial Condition," option is specified on the .TRAN control card (for transient analysis).

Coupled (mutual) inductors. General form:

```
KXXXXXXX  LYYYYYYY  LZZZZZZZ  VALUE
```

EXAMPLE 19.3

Refer to Fig. 19.4.

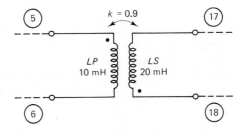

Figure 19.4

```
LP      5    6   10.0 M
LS     18   17   20.0 E−3
KTRANS LP   LS    0.90
```

Here LYYYYYYY and LZZZZZZZ are the names of the two coupled inductors. VALUE is the value of the coefficient of coupling k. Clearly, as was examined previously, $0 < k < 1$. The first nodes on each of the two coupled inductor lines are the dotted terminals, accounting for the dot convention.

Independent sources. General form:

```
VXXXXXXX N+ N- ⟨DC DC/TRAN VALUE⟩ ⟨AC ACMAG ACPHASE⟩
IYYYYYYY N+ N- ⟨DC DC/TRAN VALUE⟩ ⟨AC ACMAG ACPHASE⟩
```

■ EXAMPLE 19.4

Refer to Fig. 19.5.

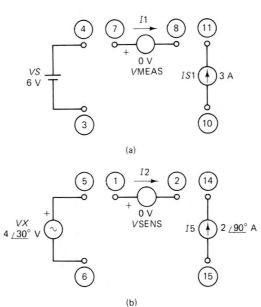

Figure 19.5 (a) Dc sources; (b) ac sources.

```
VS        4  3 DC  6.0
VMEAS     7  8
IS1      10 11 DC  3.0
```

or

```
IS1      11 10 DC  -3.0
VX        5  6 AC   4    30
VSENS     1  2
I5       15 14 AC   2.0 90
```

Here N+ and N− are the positive and negative nodes, respectively. Voltage and current sources need not be grounded. Positive current is assumed to flow from the positive node, N+, through the source, to the negative node. A current source of positive value will force current to flow out of node N+, through the source, and into node N−, as shown in Example 19.4 and Fig. 19.5.

It is important to note that in SPICE, voltage sources are used as "ammeters" to sense the current flowing in a certain branch, in addition to being used for circuit excitation. As shown in Example 19.4, the zero-valued sources VMEAS and VSENS may be inserted into the circuit for the purpose of measuring current. VMEAS will sense a positive dc current, I1, flowing from node 7 to node 8. Similarly VSENS will measure a positive ac current, I2, flowing from node 1 to node 2. These "ammeter"-type sources have no effect on circuit operation since they represent short circuits.

DC/TRAN VALUE is the dc and transient analysis value of the source. This value may be omitted if the source has a zero value for both the dc and the transient analysis. Constant power supply sources (time invariant) should have their values indicated and may be optionally preceded by the letters DC.

For ac analysis, ACMAG and ACPHASE are the magnitude and phase of the source, respectively. The source is set to this value in ac analysis. If ACMAG is omitted following the keyword AC, a value of unity is assumed. If ACPHASE is omitted a value of zero is assumed.

Any independent source can be assinged a time-dependent function for transient analysis. The value of the source at the zero time is the value used for the dc analysis. SPICE has five independent source time functions: pulse, exponential, sinusoidal, piecewise linear, and single-frequency FM. Only the pulse function will be examined here. It is specified with the parameters shown in parentheses, as follows:

```
PULSE (V1  V2  TD  TR  TF  PW  PER)
```

As an example, refer to Fig. 19.6.

```
VIN  3  0   PULSE (-1  1  2N  2N  2N  8N  20N)
```

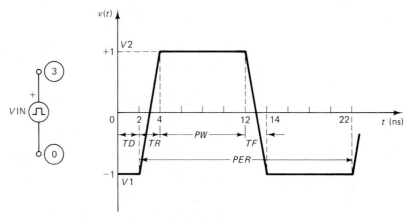

Figure 19.6

Here

V1   is the initial value (in volt or amperes).
V2   is the pulse value (in volts or amperes).
TD   is the delay time
TR   is the rise time.
TF   is the fall time.
PW   is the pulse width.
PER   is the period of the waveform.

For information regarding the other time functions available with SPICE and the default values if the parameters are undefined, the reader should consult the user's manual.

Linear dependent sources. SPICE allows circuits to contain linear dependent sources, characterized as one of the following:

1. *Voltage-controlled sources* [Fig. 19.7(a)]. These are sources whose values are obtained from the equation

$$IS = G * VC$$

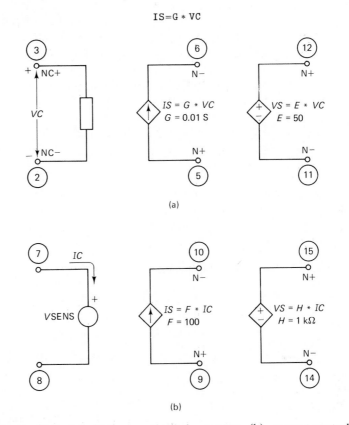

Figure 19.7 (a) Voltage-controlled sources; (b) current-controlled sources.

(i.e., voltage-controlled current source), where G is the transconductance in siemens, or from the equation

$$VS = E * VC$$

(i.e., voltage-controlled voltage source), where E is the voltage gain. Here VC is the controlling voltage.

The voltage-controlled current source is described by the general form

GXXXXXXX N+ N− NC+ NC− Value

In the example shown in Fig. 19.7(a),

G 5 6 3 2 0.01

The voltage-controlled voltage source is described by the general form

EXXXXXXX N+ N− NC+ NC− VALUE

In the example shown in Fig. 19.7(a),

```
          E 12 11  3  2  50.0
```

In the general forms above, N+ and N− are the positive and negative nodes, respectively, of the controlled source. Note that positive current flows in from N+, through the source, to N−. NC+ and NC− are the positive and negative nodes, respectively, of the controlling voltage. VALUE is either the value of G, the transconductance in siemens or mhos, or the value of the voltage gain, as appropriate for the type of source considered.

   2. *Current-controlled sources* [Fig. 19.7(b)]. These are sources whose values are obtained from the equation

```
            IS=F * IC
```

(i.e., current-controlled current source), where F is the current gain, or from the equation

```
            VS=H * IC
```

(i.e., current-controlled voltage source), where H is the transresistance in ohms. In the relationships above, IC is the controlling current. The current-controlled current source is described by the general form

```
      FXXXXXXX N+  N−  VNAME  VALUE
```

In the example shown in Fig. 19.7(b),

```
        VSENS   7  8
        F  9   10 VSENS  100
```

The current-controlled voltage source is described by the general form

```
      HXXXXXXX N+  N−  VNAME  VALUE
```

in the example shown in Fig. 19.7(b),

```
        VSENS   7  8
        H 15   14 VSENS  1.0K
```

In the general forms above, N+ and N− are the positive and negative nodes, respectively, of the controlled source. Also, positive current flow is from N+, through the source, to N−. VNAME is the name of the measuring (or sensing) voltage source, of zero value, through which the controlling current, IC, is flowing. The positive direction of flow of the controlling current is from the positive node of VNAME, through the source, to the negative node of VNAME, as indicated in Fig. 19.7(b) and in Fig. 19.5. VALUE is either the value of F, the current gain, or the value of H, the transresistance in ohms, as appropriate for the type of source considered.

19.3 TYPES OF ANALYSIS

19.3.1   DC Analysis

The dc analysis portion of SPICE determines the dc operating point of the circuit, with inductors shorted and capacitors opened. A dc analysis is automatically performed prior to a transient analysis to determine the initial conditions for the transient calculations. The dc analysis is also performed prior to an ac small-signal analysis to determine the linearized small-signal models for the nonlinear devices in the circuit being analyzed. The dc analysis can also be used to generate dc transfer curves: A specified independent voltage or current source is stepped over a user-specified range and the dc output variables are stored for each sequential source value. Such data can then be used to plot the required transfer characteristics.

Other features that can be obtained using SPICE's dc analysis, but are not considered here, are:

1. The dc small-signal value of a transfer function (ratio of output variable to input source), input resistance, and output resistance
2. The dc small-signal sensitivities of a specified output variable with respect to the different circuit parameters
3. Small-signal models for nonlinear devices.

19.3.2 AC (Small-Signal) Analysis

The ac small-signal portion of SPICE computes the ac output variables as functions of frequency. The program first computes the dc operating point of the circuit and determines the linearized small-signal models for all the nonlinear devices in the circuit. The resultant linear circuit is then analyzed over a user-specified range of frequencies. The desired output of an ac small-signal analysis is usually a transfer function (e.g., voltage gain, transimpedance, etc., as functions of the variable frequency). If the circuit has only one ac input, it is convenient to set that input to unity magnitude and zero phase angle, so that the output variables required have the same value as the transfer function of the output variables with respect to the input.

Other features that can be obtained using SPICE's ac analysis, but are not considered here, are:

1. The generation and simulation of white noise by resistors and semiconductor devices.
2. The distortion characteristics of a circuit, assuming one or two signal frequencies imposed at the input.

19.3.3 Transient Analysis

The transient analysis portion of SPICE computes the transient output variables as functions of time over a user-specified time interval. The initial conditions are automatically determined by a dc analysis, as mentioned above. All time-independent sources (e.g., power supplies) are set to their dc value.

Another feature that can be obtained through SPICE's transient analysis is the Fourier analysis of the output time-domain waveform, to determine the first nine coefficients of the Fourier expansion. This option, however, is not considered here.

19.3.4 Analysis at Different Temperatures

All the input data for SPICE are assumed to have been measured at room temperature, specified to be 27°C (or 300°K). This is also the nominal temperature at which all temperature-dependent component parameters are simulated for the different types of circuit analysis. The circuit can be simulated at other temperatures by using the .TEMP control line. This feature is of prime importance for circuits containing nonlinear semiconductor devices since many of these device parameters are temperature dependent.

19.4 CONTROL LINES

After the circuit has been described to the computer through SPICE's elements lines, as discussed in Section 19.2, the rest of the input data file contains the program control lines. These lines ask the appropriate portion of SPICE to perform the required type of circuit analysis and direct the program to produce the output in the required format. Only those control lines that are used in conjunction with the simple dc, ac, and transient analysis discussed above are examined here.

.TEMP Line   The general form of this line is

```
.TEMP T1 ⟨T2⟩ ⟨T3⟩ ...
```

for example,

```
.TEMP 0.0 50.0 100.0
```

This line specifies the temperatures at which the circuit is to be stimulated. T1, T2, T3, and so on, are the different temperatures in degrees C. Temperatures of less than −223.0°C are ignored. As mentioned above, model data are specified at TNOM degrees C (TNOM = 27°C). If the .TEMP line is omitted, the simulation will be performed at a temperature equal to TNOM.

.WIDTH line.   The general form of this line is

```
.WIDTH   IN=COLNUM   OUT=COLNUM
```

for example,

```
.WIDTH   IN=72   OUT=133
```

COLNUM is the number of the last column to be read from each line of input. This setting takes effect with the next line read. The default value of COLNUM is 80. The OUT parameter specifies the output print width. Permissible values for the output print width are 80 and 133.

.OP line.   The general form of this line is

```
.OP
```

The inclusion of this line in an input data file will require SPICE to determine the dc operating point (OP) of the circuit with all the inductors shorted and all the capacitors opened. SPICE performs a dc operating point analysis if no other analyses are required. As mentioned before, a dc analysis is performed automatically prior to a transient analysis to determine the initial conditions for the transient calculations. Also, prior to an ac small-signal analysis, a dc analysis is performed to determine the linearized small-signal models for nonlinear devices.

.DC line. The general form of this line is

```
.DC SCRNAM VSTART VSTOP VINCR (SCR2 START2 STOP2 INCR2)
```

for example,

```
.DC  VIN   0.0   10.0  1.0
.DC  VS    1.0   5.0   0.5  VX   0.0  1.0  0.1
.DC  IS    2.0   4.0   0.5  VS1  1.0  5.0  1.0
.DC  VS1   15.0  15.0  0.1
```

This line defines the limits and the sweep values of the source (or sources) in order to obtain the dc transfer characteristics. SCRNAM is the name of an independent voltage or current source. VSTART, VSTOP, and VINCR are the starting, final, and incremental values of the source. In the first example, the value of the voltage source VIN will be swept from 0 V to 10 V in steps of 1 V. Each time the source VIN has a different value, the dc analysis will be performed to obtain a certain output variable, as will be shown below in the .PRINT and/or .PLOT lines. A second source, SCR2, may optionally be specified with its associated sweep parameters. In such cases, as in the second and third examples above, the first source will be swept over its range for each of the incremented values of the second source. This option can be very useful, in particular for obtaining semiconductor device output characteristics. In the fourth example above, the dc analysis is performed only at the given values of the sources (e.g., VS1 = 15 V), since VSTART = VSTOP. Note that VINCR should not be zero.

.IC line. The general form of this line is

```
.IC V(NODNUM1)=VALUE1  V(NODNUM2)=VALUE2 . . .
```

for example,

```
.IC V(5)=2  V(3)=-5  V(7)=3.5
```

This line is for setting the initial conditions (IC) for the transient analysis. It has two different interpretations depending on whether the UIC (use initial conditions) optional parameter is specified on the .TRAN line or not (see below):

1. When the UIC parameter is specified on the .TRAN line, the node voltages (with respect to the ground node) specified on the .IC line are used to compute the capacitor, inductor, and nonlinear element initial conditions. This is equivalent to specifying the IC = . . . parameter on each of the device lines, but is much more convenient. The IC = . . . parameters can still be specified and will take precedence over the .IC values. In this case (i.e., when the UIC parameter is specified) *no dc bias solution* (i.e., the operating point analysis) is computed before the transient analysis. One should then take care to specify all dc source voltages on the .IC card if they are to be used in computing the device's initial conditions.

2. When the UIC parameter is not specified on the .TRAN line, the dc bias (i.e., the operating point analysis) solution will be computed before the transient analysis. In this case, the node voltages specified on the .IC line will be forced to the desired initial values during the bias solution. During the transient analysis, the constraint on these node voltages is removed.

.TRAN line. The general form of this line is

```
.TRAN  TSTEP  TSTOP  ⟨TSTART⟩ ⟨TMAX⟩ ⟨UIC⟩
```

for example,

```
.TRAN    1NS    100NS
.TRAN    1NS    100NS  20NS
.TRAN    10NS   1US    UIC
```

Here TSTEP is the printing or plotting increment of time for line-printer output. TSTEP is also the suggested computing-time increment. TSTOP is the final value of time and TSTART is the initial value of time for computing the transient response. If TSTART is omitted, it is assumed to be zero. The transient analysis always begins at time zero. In the interval between the zero time and TSTART, the circuit is analyzed to reach a steady state, but no output is stored. In the interval between TSTART and TSTOP, the circuit is analyzed and outputs are stored. TMAX is the maximum step size that SPICE will use for computing the transient response. This is useful when one wishes to guarantee a computing interval which is smaller than the printer increment TSTEP. If TMAX is omitted, the program chooses either TSTEP or (TSTOP − TSTART)/50.0, whichever is smaller. In the three examples above, TSTEP is actually the computing interval taken.

The optional keyword UIC (i.e., use initial conditions) indicates that the user does not want SPICE to solve for the quiescent operating point before beginning the transient analysis. When this keyword is specified, SPICE uses the values specified using IC = . . . on the various elements lines as the initial conditions for the transient and proceeds with the analysis. If the .IC line has been specified, the node voltages on the .IC line are used to compute the initial conditions for the devices. When the UIC is not specified on the .TRAN line, an initial operating point analysis is performed, as described above.

.AC line. The general form of this line is one of the following:

```
.AC  DEC  ND  FSTART   FSTOP
.AC  OCT  NO  FSTART   FSTOP
.AC  LIN  NP  FSTART   FSTOP
```

for example,

```
.AC  DEC  10  1.   10KHz
.AC  LIN  100 1.0  100.0
.AC  LIN  1   1K   1K
```

This line instructs SPICE to perform an ac analysis of the circuit described at different frequencies in the range FSTART, the starting frequency to FSTOP, the final frequency. The frequency points at which the analysis is performed are determined as indicated below.

1. DEC stands for decade variation and ND is the number of points per decade. Thus in the first example above, there are four decades in the frequency range 1 Hz to 10 kHz. Each decade is divided into 10 points. The frequency multiplication factor x is obtained from $x^{ND} = 10$. Here x is 1.2589.

2. OCT stands for octave variation and NO is the number of points per octave. The frequency multiplication factor x is obtained from $x^{NO} = 2$.

3. LIN stands for linear variation and NP is the number of points in the frequency range. In the second example above, the frequency range from 1 to 100 Hz is divided into 100 points (i.e., frequncies of 1-Hz interval are used in the ac analysis).

Note that in order for this analysis to be meaningful, at least one independent source in the circuit must have been specified with an ac value. In the third example above, the ac analysis is required at only one frequency since FSTART = FSTOP (= 1.0 kHz).

.PRINT lines   The general form of this line is

```
.PRINT PRTYPE OV1 ⟨OV2) ⟨OV3) . . . ⟨OV8)
```

for example

```
.PRINT    DC      V(2)        I(VMEAS)    V(9,3)
.PRINT    TRAN    V(5)        I(VIN)
.PRINT    AC      VM(2,1)     VP(2,1)     VDB(7)
```

This line defines the contents of a tabular listing of one to eight output variables, OV1, OV2, . . . , OV8. PRTYPE is the type of analysis (DC, AC, or TRAN) for which the specified outputs are desired. The form for voltage output variables is as follows:

```
V(N1⟨,N2⟩)
```

This specifies the voltage difference between nodes N1 and N2 (i.e., the voltage at node N1 with respect to that at node N2). If N2 and the preceding comma are omitted, ground (node 0) is assumed. The specification then refers the voltage at the node indicated with respect to ground.

For AC analysis, five additional outputs can be accessed by replacing the letter V by

VR: real part of the voltage
VI: imaginary part of the voltage
VM: magnitude of the voltage
VP: phase of the voltage
VDB: $20 \log_{10}$ (voltage magnitude)

The form for current output variables is as follows:

```
I(VXXXXXXX)
```

This specifies the current flowing in the independent voltage source named VXXXXXXX. This voltage source could be a power supply, or a dead source, of value zero, inserted into the branch in which the current is to be measured (as an ammeter; see the element lines above). Note that positive current flows from the positive node, N+, through the source, to the negative node, N−.

For AC analysis, the corresponding replacements for the letter I may be made in the same way as described above for the AC voltage output. These are

IR, II, IM, IP, IDB

There is no limit on the number of .PRINT lines for each type of analysis.

.PLOT lines. The general form of this line is

```
.PLOT PLTYPE OV1 ⟨PLO1,PHI1⟩ ⟨OV2⟩ ⟨PLO2,PHI2⟩ ... ⟨OV8⟩
```

for example,

```
.PLOT  DC    V(2)    V(7)    (4,10)   V(3,1)
.PLOT  TRAN  V(5,1)  I(VIN)  V(4)     (0,5)    V(9)    (0,10)
.PLOT  AC    VM(3)   VP(3)   VM(5,2)  VDB(6)   VP(6)
```

This line defines the contents of one plot of one to eight output variables. PLTYPE is the type of analysis (DC, AC, TRAN) for which the specified outputs are required. The syntax for OV1, OV2, and so on, is the same as for the .PRINT line described above.

The optional vertical plot limits (PLO, PHI) may be specified after any of the output variables. All output variables to the left of a pair of plot limits will be plotted using the same lower and upper plot bounds. Thus in the first example above, V(2) and V(7) are plotted on a 4 unit-to-10 unit grid. If plot limits are not specified, as in the third example above, SPICE will automatically determine the minimum and maximum values of all output variables being plotted on the same graph and scale the plot to fit. More than one scale will be used if the output variable values warrant (i.e., mixing output variables with values which are orders of magnitude different still results in readable plots). The overlap of two or more traces on any one graph is indicated by the letter X.

When more than one output variable appears on the same plot, the first variable specified on the .PLOT line will be printed as well as plotted. If a printout of all variables is desired, a companion .PRINT line should be included. There is no limit on the number of .PLOT lines specified for each type of analysis.

19.5 GENERAL EXAMPLES

EXAMPLE 19.5

Prepare the circuit in Fig. 19.8 for computer-aided analysis using SPICE. Write a program to obtain the values of the voltages V_1, and V_3 and the currents I_2, I_3, and I_4, as indicated in the figure.

Solution The nodes are designated as shown in Fig. 19.9. Also, a 0 V (short circuit, current sensing) independent source VM1 is added to measure the

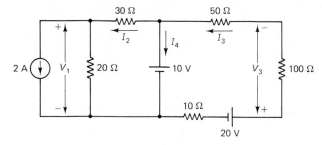

Figure 19.8

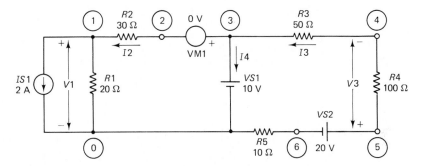

Figure 19.9

current I_2. The current I_3 is the current flowing in source VS2, with the reverse sign, while the current I_4 is the current flowing in source VS1. Here the input data file will be as follows.

```
SOLUTION TO EXAMPLE 19.5
R1    1   0    20.0
R2    1   2    30.0
R3    3   4    50.0
R4    4   5    100.0
R5    6   0    10.0
IS1   1   0    DC  2.0
VS1   3   0    DC  10.0
VS2   5   6    DC  20.0
VM1   3   2
.PRINT    DC  V(1)  V(5,4)  I(VM1)  I(VS2)  I(VS1)
.DC  IS1  2.0  2.0  0.1
.END
```

The printed output results are shown in Fig. 19.10. The inclusion of the line .DC . . . requires the computations of the outputs specified in the .PRINT line (see the .DC line in Section 19.4). This is what is needed in this problem. However, if the .DC . . . line is deleted or only the line .OP is used, as shown also in Fig. 19.10, the results of the dc analysis are printed automatically, in the form of all the node voltages and all the independent voltage source currents in the given circuit. As the results indicate, I_4 [$= I(VS1)$] actually flows in the opposite direction to that shown in Fig. 19.8, while I_3 [$= -I(VS2)$] is flowing in the correct direction. KCL is satisfied. The results obtained here can easily be checked through applying the familiar network analysis techniques.

```
******** 1/ 1/80******** PSpice 2.04 (Nov 1985) ******* 0:11:53********

SOLUTION TO EXAMPLE 19.5

**************************************************************************


****      CIRCUIT DESCRIPTION

R1 1 0 20.0
R2 1 2 30.0
R3 3 4 50.0
R4 4 5 100.0
R5 6 0 10.0
IS1 1 0 DC 2.0
VS1 3 0 DC 10.0
VS2 5 6 DC 20.0
VM1 3 2
.PRINT DC V(1) V(5,4) I(VM1) I(VS2) I(VS1)
.DC IS1 2.0 2.0 0.1
.END


****      DC TRANSFER CURVES              TEMPERATURE =   27.000 DEG C

**************************************************************************

 IS1          V(1)         V(5,4)       I(VM1)       I(VS2)       I(VS1)


2.000E+00    -2.000E+01   6.250E+00    1.000E+00    -6.250E-02   -9.375E-01


        JOB CONCLUDED

        TOTAL JOB TIME              7.25

**************************************************************************

****      CIRCUIT DESCRIPTION

R1 1 0 20.0
R2 1 2 30.0
R3 3 4 50.0
R4 4 5 100.0
R5 6 0 10.0
IS1 1 0 DC 2.0
VS1 3 0 DC 10.0
VS2 5 6 DC 20.0
VM1 3 2
.PRINT DC V(1) V(5,4) I(VM1) I(VS2) I(VS1)
.OP
.END


**************************************************************************


****      SMALL SIGNAL BIAS SOLUTION      TEMPERATURE =   27.000 DEG C

 NODE    VOLTAGE      NODE    VOLTAGE      NODE    VOLTAGE      NODE    VOLTAGE

(  1)   -20.0000    (  2)   10.0000     (  3)   10.0000     (  4)   13.1250

(  5)   19.3750     (  6)    -.6250


     VOLTAGE SOURCE CURRENTS
     NAME        CURRENT

     VS1       -9.375D-01

     VS2       -6.250D-02

     VM1        1.000D+00


     TOTAL POWER DISSIPATION   5.06D+01  WATTS

        JOB CONCLUDED

        TOTAL JOB TIME              7.15
```

Figure 19.10

EXAMPLE 19.6

Write a SPICE program to describe the circuit shown in Fig. 19.11 to the computer. It is required to obtain the values of I_1, I_2, V_{ab}, and I_x, and to plot the transfer characteristics for V_{ab} and I_x as V_{S1} varies between 15 and 25 V in steps of 1 V and I_S varies between 2.0 and 3.0 A in 0.5 A steps.

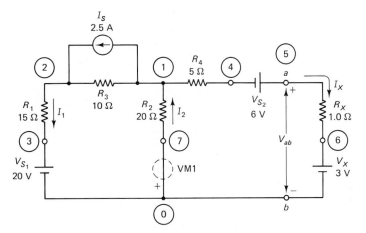

Figure 19.11

Solution   The elements are named and the nodes are designated as indicated in Fig. 19.11. Note that node 7 and the short-circuit source VM1 are added to measure the current I_2. I_1 is the current flowing in the independent source V_{S1}, while I_x is the current flowing in the independent source V_x, in the proper direction. Here the input data file will be as follows.

```
SOLUTION TO EXAMPLE 19.6
R1     3    2    15.0
R2     1    7    20.0
R3     1    2    10.0
R4     1    4    5.0
RX     5    6    1.0
IS     1    2    DC  2.5
VS1    3    0    DC  20.0
VS2    5    4    DC  6.0
VX     6    0    DC  3.0
VM1    0    7
.PRINT    DC    I(VS1)  I(VM1)  I(VX)    V(5)
.DC VS1   15.0     25.0  1.0  IS  2.0  3.0  0.5
.PLOT     DC    V(5)
.PLOT     DC    I(VX)
.END
```

The printout and the graphs obtained by running this program are shown in Fig. 19.12. Notice that the table indicates the values of the required outputs three times: as VS1 is stepped from 15 to 25 V for each of the stepped values of IS (2 A, 2.5 A, and 3.0 A). The arrow indicates the results of the dc circuit analysis corresponding to the values of the sources as shown in Fig. 19.11.

```
******** 1/ 1/80******** PSpice 2.04 (Nov 1985) ******* 0:33:21********


SOLUTION TO EXAMPLE 19.6


****************************************************************************

****     CIRCUIT DESCRIPTION

R1 3 2 15.0
R2 1 7 20.0
R3 1 2 10.0
R4 1 4 5.0
RX 5 6 1.0
IS 1 2 DC 2.5
VS1 3 0 DC 20.0
VS2 5 4 DC 6.0
VX 6 0 DC 3.0
VM1 0 7
.PRINT DC I(VS1) I(VM1) I(VX) V(5)
.DC VS1 15.0 25.0 1.0 IS 2.0 3.0 0.5
.PLOT DC V(5)
.PLOT DC I(VX)
.PROBE
.END


****************************************************************************

****     DC TRANSFER CURVES                 TEMPERATURE =   27.000 DEG C



      VS1           I(VS1)        I(VM1)        I(VX)         V(5)


   ┌─1.500E+01     9.091E-02     1.364E-01     4.545E-02     3.045E+00
   │  1.600E+01     5.714E-02     1.286E-01     7.143E-02     3.071E+00
   │  1.700E+01     2.338E-02     1.208E-01     9.740E-02     3.097E+00
I_S│  1.800E+01    -1.039E-02     1.130E-01     1.234E-01     3.123E+00
   │  1.900E+01    -4.416E-02     1.052E-01     1.494E-01     3.149E+00
   │  2.000E+01    -7.792E-02     9.740E-02     1.753E-01     3.175E+00
=2A┤  2.100E+01    -1.117E-01     8.961E-02     2.013E-01     3.201E+00
   │  2.200E+01    -1.455E-01     8.182E-02     2.273E-01     3.227E+00
   │  2.300E+01    -1.792E-01     7.403E-02     2.532E-01     3.253E+00
   │  2.400E+01    -2.130E-01     6.623E-02     2.792E-01     3.279E+00
   └─2.500E+01    -2.468E-01     5.844E-02     3.052E-01     3.305E+00
   ┌─1.500E+01     2.597E-01     1.753E-01    -8.442E-02     2.916E+00
   │  1.600E+01     2.260E-01     1.675E-01    -5.844E-02     2.942E+00
   │  1.700E+01     1.922E-01     1.597E-01    -3.247E-02     2.968E+00
I_S│  1.800E+01     1.584E-01     1.519E-01    -6.494E-03     2.994E+00
   │  1.900E+01     1.247E-01     1.442E-01     1.948E-02     3.019E+00
   │  2.000E+01  →  9.091E-02     1.364E-01     4.545E-02     3.045E+00
=2.5A  2.100E+01     5.714E-02     1.286E-01     7.143E-02     3.071E+00
   │  2.200E+01     2.338E-02     1.208E-01     9.740E-02     3.097E+00
   │  2.300E+01    -1.039E-02     1.130E-01     1.234E-01     3.123E+00
   │  2.400E+01    -4.416E-02     1.052E-01     1.494E-01     3.149E+00
   └─2.500E+01    -7.792E-02     9.740E-02     1.753E-01     3.175E+00


      VS1           I(VS1)        I(VM1)        I(VX)         V(5)


   ┌─1.500E+01     4.286E-01     2.143E-01    -2.143E-01     2.786E+00
   │  1.600E+01     3.948E-01     2.065E-01    -1.883E-01     2.812E+00
   │  1.700E+01     3.610E-01     1.987E-01    -1.623E-01     2.838E+00
I_S│  1.800E+01     3.273E-01     1.909E-01    -1.364E-01     2.864E+00
   │  1.900E+01     2.935E-01     1.831E-01    -1.104E-01     2.890E+00
   │  2.000E+01     2.597E-01     1.753E-01    -8.442E-02     2.916E+00
=3A┤  2.100E+01     2.260E-01     1.675E-01    -5.844E-02     2.942E+00
   │  2.200E+01     1.922E-01     1.597E-01    -3.247E-02     2.968E+00
   │  2.300E+01     1.584E-01     1.519E-01    -6.494E-03     2.994E+00
   │  2.400E+01     1.247E-01     1.442E-01     1.948E-02     3.019E+00
   └─2.500E+01     9.091E-02     1.364E-01     4.545E-02     3.045E+00
```

Figure 19.12

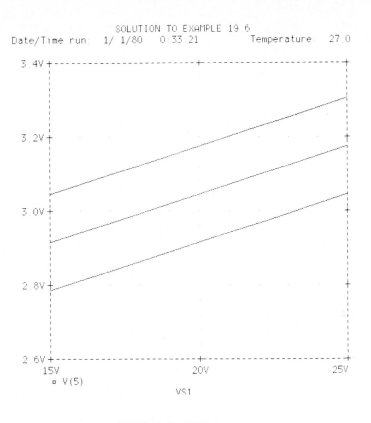

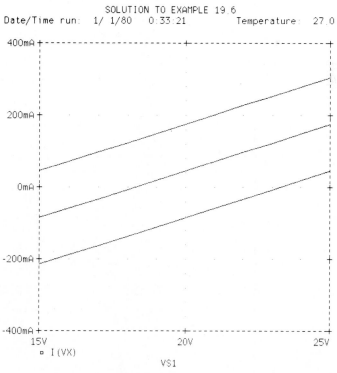

Figure 19.12—(*continued*)

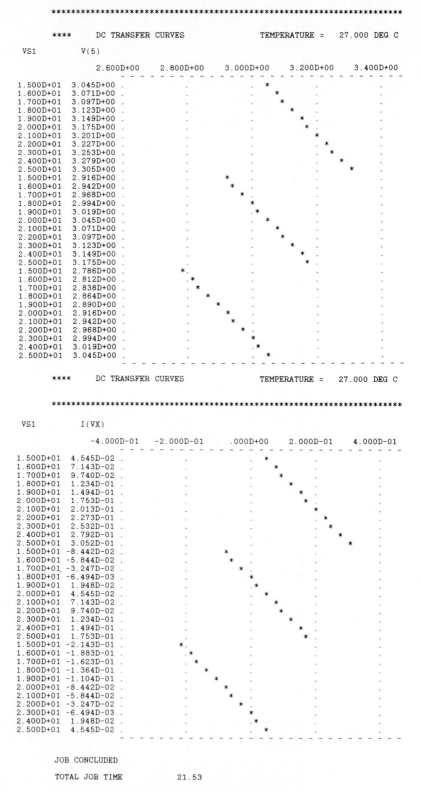

```
**************************************************************************

      ****      DC TRANSFER CURVES              TEMPERATURE =    27.000 DEG C

      VS1         V(5)

                  2.600D+00    2.800D+00    3.000D+00    3.200D+00    3.400D+00
                - - - - - - - - - - - - - - - - - - - - - - - - - - - - - - - -
  1.500D+01  3.045D+00 .           .           .     *     .          .
  1.600D+01  3.071D+00 .           .           .      *    .          .
  1.700D+01  3.097D+00 .           .           .       *   .          .
  1.800D+01  3.123D+00 .           .           .        *  .          .
  1.900D+01  3.149D+00 .           .           .         * .          .
  2.000D+01  3.175D+00 .           .           .          *.          .
  2.100D+01  3.201D+00 .           .           .          .*          .
  2.200D+01  3.227D+00 .           .           .          . *         .
  2.300D+01  3.253D+00 .           .           .          .  *        .
  2.400D+01  3.279D+00 .           .           .          .   *       .
  2.500D+01  3.305D+00 .           .           .          .    *      .
  1.500D+01  2.916D+00 .           .           .   *      .          .
  1.600D+01  2.942D+00 .           .           .  *       .          .
  1.700D+01  2.968D+00 .           .           . *        .          .
  1.800D+01  2.994D+00 .           .           .*         .          .
  1.900D+01  3.019D+00 .           .           .*         .          .
  2.000D+01  3.045D+00 .           .           . *        .          .
  2.100D+01  3.071D+00 .           .           .  *       .          .
  2.200D+01  3.097D+00 .           .           .  *       .          .
  2.300D+01  3.123D+00 .           .           .   *      .          .
  2.400D+01  3.149D+00 .           .           .    *     .          .
  2.500D+01  3.175D+00 .           .           .     *    .          .
  1.500D+01  2.786D+00 .           *.          .          .          .
  1.600D+01  2.812D+00 .           .*          .          .          .
  1.700D+01  2.838D+00 .           . *         .          .          .
  1.800D+01  2.864D+00 .           .   *       .          .          .
  1.900D+01  2.890D+00 .           .    *      .          .          .
  2.000D+01  2.916D+00 .           .     *     .          .          .
  2.100D+01  2.942D+00 .           .      *    .          .          .
  2.200D+01  2.968D+00 .           .       *   .          .          .
  2.300D+01  2.994D+00 .           .         * .          .          .
  2.400D+01  3.019D+00 .           .          .*          .          .
  2.500D+01  3.045D+00 .           .           .     *    .          .
                - - - - - - - - - - - - - - - - - - - - - - - - - - - - - - - -

      ****      DC TRANSFER CURVES              TEMPERATURE =    27.000 DEG C

**************************************************************************

      VS1         I(VX)

                 -4.000D-01   -2.000D-01    .000D+00     2.000D-01    4.000D-01
                - - - - - - - - - - - - - - - - - - - - - - - - - - - - - - - -
  1.500D+01  4.545D-02 .           .           .     *    .          .
  1.600D+01  7.143D-02 .           .           .      *   .          .
  1.700D+01  9.740D-02 .           .           .       *  .          .
  1.800D+01  1.234D-01 .           .           .        * .          .
  1.900D+01  1.494D-01 .           .           .         *.          .
  2.000D+01  1.753D-01 .           .           .          *          .
  2.100D+01  2.013D-01 .           .           .          .*         .
  2.200D+01  2.273D-01 .           .           .          . *        .
  2.300D+01  2.532D-01 .           .           .          .   *      .
  2.400D+01  2.792D-01 .           .           .          .    *     .
  2.500D+01  3.052D-01 .           .           .          .      *   .
  1.500D+01 -8.442D-02 .           .           *          .          .
  1.600D+01 -5.844D-02 .           .         * .          .          .
  1.700D+01 -3.247D-02 .           .        *  .          .          .
  1.800D+01 -6.494D-03 .           .          .*          .          .
  1.900D+01  1.948D-02 .           .          . *         .          .
  2.000D+01  4.545D-02 .           .           .*         .          .
  2.100D+01  7.143D-02 .           .           . *        .          .
  2.200D+01  9.740D-02 .           .           . *        .          .
  2.300D+01  1.234D-01 .           .           .   *      .          .
  2.400D+01  1.494D-01 .           .           .    *     .          .
  2.500D+01  1.753D-01 .           .           .     *    .          .
  1.500D+01 -2.143D-01 .           *.          .          .          .
  1.600D+01 -1.883D-01 .           .*          .          .          .
  1.700D+01 -1.623D-01 .           . *         .          .          .
  1.800D+01 -1.364D-01 .           .   *       .          .          .
  1.900D+01 -1.104D-01 .           .     *     .          .          .
  2.000D+01 -8.442D-02 .           .       *   .          .          .
  2.100D+01 -5.844D-02 .           .         * .          .          .
  2.200D+01 -3.247D-02 .           .        *  .          .          .
  2.300D+01 -6.494D-03 .           .          .*          .          .
  2.400D+01  1.948D-02 .           .          . *         .          .
  2.500D+01  4.545D-02 .           .           .*         .          .
                - - - - - - - - - - - - - - - - - - - - - - - - - - - - - - - -

      JOB CONCLUDED

      TOTAL JOB TIME            21.53
```

Figure 19.12—(*continued*)

EXAMPLE 19.7

Using SPICE, analyze the circuit shown in Fig. 19.13 to obtain the values of V_1, I_1, and I. Note that this network contains dependent sources.

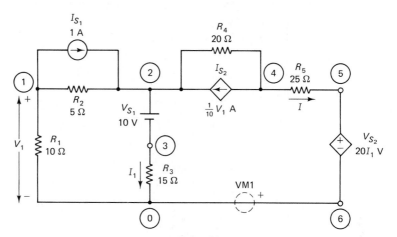

Figure 19.13

Solution   The nodes are designated as shown in Fig. 19.13. The short-circuit independent voltage source VM1 and node 6 are added to measure the current I because dependent voltage sources cannot be used to measure current. Here the input data file will be as follows:

```
SOLUTION TO EXAMPLE 19.7
R1      1       0       10.0
R2      1       2       5.0
R3      3       0       15.0
R4      2       4       20.0
R5      4       5       25.0
IS1     1       2       DC      1.0
VS1     2       3       DC      10.0
VM1     6       0
G       4       2       1       0       0.1     ← describing the voltage-
                                                  controlled current source I_S2
H       5       6       VS1     20.0            ← describing the current-
                                                  controlled voltage source V_S2
.PRINT      DC      V(1)    I(VS1)      I(VM1)
.DC     VS1     10.0    10.0                0.1     ← this line requires that the
                                                      dc analysis be performed at
                                                      the fixed values of the sources
                                                      in the given circuit
.END
```

Note that in describing the current-controlled voltage source, V_{S2},

$$V_{S2} = H * I_C = 20 \times I_1$$

there is no need to add a sensing short-circuit voltage source, as the controlling current I_1 is the current flowing in V_{S1} in the proper direction. Also, the required value of the current *I*, in the .PRINT line, is obtained as the current flowing in VM1. This current flows in the proper direction from node 6 to node 0. The description of the dependent source I_{S2} follows directly from Section 19.2.3.

The printout obtained by running this program is shown in Fig. 19.14. Try to use the loop analysis techniques to check the results obtained from this program!

```
******** 1/ 1/80******** PSpice 2.04 (Nov 1985) ******* 0:06:45********

SOLUTION TO EXAMPLE 19.7

****************************************************************************

****      CIRCUIT DESCRIPTION

R1 1 0 10.0
R2 1 2 5.0
R3 3 0 15.0
R4 2 4 20.0
R5 4 5 25.0
IS1 1 2 DC 1.0
VS1 2 3 DC 10.0
VM1 6 0
G 4 2 1 0 0.1
H 5 6 VS1 20.0
.PRINT DC V(1) I(VS1) I(VM1)
.DC VS1 10.0 10.0 0.1
.END

****      DC TRANSFER CURVES             TEMPERATURE =   27.000 DEG C

****************************************************************************

VS1          V(1)         I(VS1)       I(VM1)

1.000E+01    5.128E-01   -2.821E-01    2.308E-01

          JOB CONCLUDED

          TOTAL JOB TIME          8.56
```
Figure 19.14

EXAMPLE 19.8

The capacitor in the circuit shown in Fig. 19.15 was initially charged to a voltage of 20 V. At $t = 0$, the switch was closed. Using SPICE, calculate the initial values of i_1, i_2, and i_3. Plot the variation of i_1, i_2, i_3, and v_C in the time span from $t = 0$ to $t = 50$ ms.

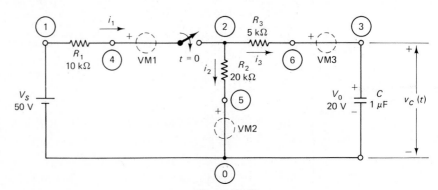

Figure 19.15

Solution The nodes are designated as shown in the circuit diagram of Fig. 19.15. The three 0-V sources VM1, VM2, and VM3 (and nodes 4, 5, and 6) are added to measure the required branch currents.

Here the input data file will be as follows.

```
SOLUTION TO EXAMPLE 19.8
R1      1     4        10.0K
R2      2     5        20.0K
R3      2     6        5.0K
C       3     0        1.0U
VS      1     0        DC    50.0
VM1     4     2
VM2     5     0
VM3     6     3
.IC     V(1)=50.0    V(3)=20.0
.TRAN       1M        50M
.PRINT      DC        I(VM1)      I(VM2)     I(VM3)
.PLOT       TRAN      V(3)
.PLOT       TRAN      I(VM1)      I(VM2)     I(VM3)
.END
```

The transient analysis starts at time zero. Using 1 ms computing and plotting incremental time, the transient analysis is performed until TSTOP (= 50 ms). Since the optional UIC keyword is not specified, the optional IC = ... on the element cards (here it is the initial voltage on the capacitor) cannot be used. However, the .IC card specifies the initial voltages at the nodes, including V_0 [= $V(3)$] and V_S [=V(1)]. In this case, an operating point solution will be computed before the transient analysis, using the initial conditions specified on the .IC card. The meaning of the .PRINT and .PLOT cards is quite clear.

The printout and the graphs obtained by running this program are shown in Fig. 19.16. Note that since the .DC ... line is omitted, the program automatically provides the results in the form of all the node voltages and all the currents flowing in the specified independent voltage sources.

```
******** 1/ 1/80******** PSpice 2.04 (Nov 1985) ******* 0:27:05********


SOLUTION TO EXAMPLE 19.8


*******************************************************************************

****      CIRCUIT DESCRIPTION

R1 1 4 10.0K
R2 2 5 20.0K
R3 2 6 5.0K
C 3 0 1.0U
VS 1 0 DC 50.0
VM1 4 2
VM2 5 0
VM3 6 3
.IC V(1)=50.0 V(3)=20.0
.TRAN 1M 50M
.PRINT DC I(VM1) I(VM2) I(VM3)
.PLOT TRAN I(VM1) I(VM2) I(VM3)
.PLOT TRAN V(3)
.PROBE
.END


*******************************************************************************
```

Figure 19.16

```
****      INITIAL TRANSIENT SOLUTION        TEMPERATURE =   27.000 DEG C


    NODE   VOLTAGE     NODE   VOLTAGE     NODE   VOLTAGE     NODE   VOLTAGE

    ( 1)   50.0000     ( 2)   25.7143     ( 3)   20.0000     ( 4)   25.7143

    ( 5)    .0000      ( 6)   20.0000


****      TRANSIENT ANALYSIS               TEMPERATURE =   27.000 DEG C

***************************************************************************


LEGEND:

*:  I(VM1)
+:  I(VM2)
=:  I(VM3)
    TIME      I(VM1)

(*)------------  1.000D-03   1.500D-03   2.000D-03   2.500D-03   3.000D-03
                -  -  -  -  -  -  -  -  -  -  -  -  -  -  -  -  -  -  -  -

(+)------------  1.200D-03   1.400D-03   1.600D-03   1.800D-03   2.000D-03
                -  -  -  -  -  -  -  -  -  -  -  -  -  -  -  -  -  -  -  -

(=)------------   .000D+00   5.000D-04   1.000D-03   1.500D-03   2.000D-03
                -  -  -  -  -  -  -  -  -  -  -  -  -  -  -  -  -  -  -  -
 .000D+00   2.429D-03 .        +        .           . =        * .         .
1.000D-03   2.366D-03 .         +       .          .=          * .         .
2.000D-03   2.309D-03 .          +      .         =.          * .          .
3.000D-03   2.256D-03 .           +     .        =  .        * .           .
4.000D-03   2.208D-03 .            +    .      =    .      * .             .
5.000D-03   2.163D-03 .            .+   .    =      .      * .             .
6.000D-03   2.122D-03 .          + =    .          .      * .              .
7.000D-03   2.085D-03 .          . =+   .          . .   * .               .
8.000D-03   2.051D-03 .          . =    +          . .*   .                .
9.000D-03   2.019D-03 .          .=     .    +     . *    .                .
1.000D-02   1.990D-03 .          =      .      +   .*     .                .
1.100D-02   1.964D-03 .         =.      .       + *.      .                .
1.200D-02   1.939D-03 .        = .      .       + *.      .                .
1.300D-02   1.917D-03 .       =  .      .       + * .     .                .
1.400D-02   1.896D-03 .       =  .      .          X      .                .
1.500D-02   1.877D-03 .      =   .      .          X      .                .
1.600D-02   1.860D-03 .      =   .      .       * +.      .                .
1.700D-02   1.844D-03 .      =   .      .       * +.      .                .
1.800D-02   1.830D-03 .     =    .      .       * +.      .                .
1.900D-02   1.816D-03 .     =    .      .       * +.      .                .
2.000D-02   1.804D-03 .    =     .      .       *  +      .                .
2.100D-02   1.793D-03 .    =     .      .       *  +      .                .
2.200D-02   1.782D-03 .    =     .      .      *   .+     .                .
2.300D-02   1.773D-03 .   =      .      .      *   .+     .                .
2.400D-02   1.764D-03 .   =      .      .      *   .+     .                .
2.500D-02   1.756D-03 .  =       .      .      *   .+     .                .
2.600D-02   1.749D-03 .  =       .      .     *    . +    .                .
2.700D-02   1.742D-03 .  =       .      .     *    . +    .                .
2.800D-02   1.736D-03 .  =       .      .     *    . +    .                .
2.900D-02   1.730D-03 . -        .      .     *    . +    .                .
3.000D-02   1.725D-03 . =        .      .     *    . +    .                .
3.100D-02   1.720D-03 . =        .      .     *    .  +   .                .
3.200D-02   1.716D-03 . =        .      .     *    .  +   .                .
3.300D-02   1.712D-03 . =        .      .     *    .  +   .                .
3.400D-02   1.708D-03 . =        .      .    *     .  +   .                .
3.500D-02   1.705D-03 .=         .      .    *     .  +   .                .
3.600D-02   1.701D-03 .=         .      .    *     .  +   .                .
3.700D-02   1.699D-03 .=         .      .    *     .  +   .                .
3.800D-02   1.696D-03 .=         .      .    *     .  +   .                .
3.900D-02   1.694D-03 .=         .      .    *     .  +   .                .
4.000D-02   1.691D-03 .=         .      .    *     .   +  .                .
4.100D-02   1.689D-03 .=         .      .    *     .   +  .                .
4.200D-02   1.687D-03 .=         .      .    *     .   +  .                .
4.300D-02   1.686D-03 .=         .      .    *     .   +  .                .
4.400D-02   1.684D-03 .=         .      .    *     .   +  .                .
4.500D-02   1.683D-03 .=         .      .    *     .   +  .                .
4.600D-02   1.681D-03 .=         .      .    *     .   +  .                .
4.700D-02   1.680D-03 .=         .      .    *     .   +  .                .
4.800D-02   1.679D-03 =          .      .    *     .   +  .                .
4.900D-02   1.678D-03 =          .      .    *     .   +  .                .
5.000D-02   1.677D-03 =          .      .    *     .   +  .                .
                -  -  -  -  -  -  -  -  -  -  -  -  -  -  -  -  -  -  -  -
```

Figure 19.16—(*continued*)

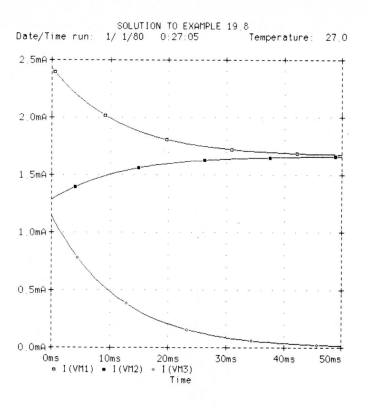

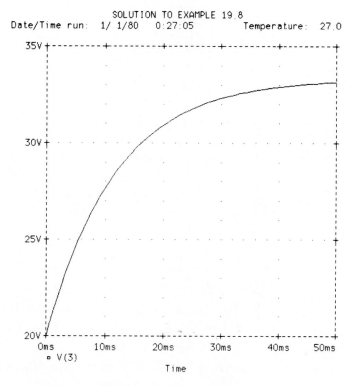

Figure 19.16—(*continued*)

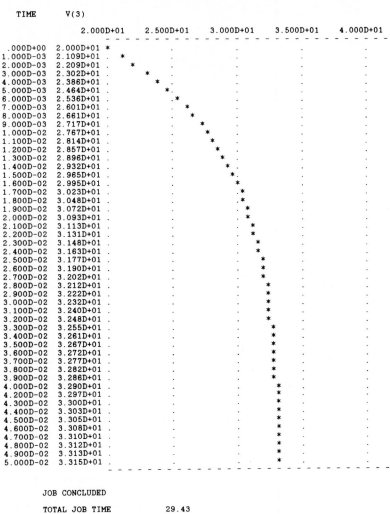

```
****        TRANSIENT ANALYSIS                 TEMPERATURE =    27.000 DEG C

***********************************************************************

        TIME        V(3)

                    2.000D+01     2.500D+01     3.000D+01     3.500D+01     4.000D+01
                - - - - - - - - - - - - - - - - - - - - - - - - - - - - - - - - - - -
       .000D+00   2.000D+01 *        .            .            .            .
      1.000D-03   2.109D+01 .    *   .            .            .            .
      2.000D-03   2.209D+01 .       *.            .            .            .
      3.000D-03   2.302D+01 .        .  *         .            .            .
      4.000D-03   2.386D+01 .        .     *.     .            .            .
      5.000D-03   2.464D+01 .        .       *.   .            .            .
      6.000D-03   2.536D+01 .        .       .*   .            .            .
      7.000D-03   2.601D+01 .        .        . * .            .            .
      8.000D-03   2.661D+01 .        .        .   *            .            .
      9.000D-03   2.717D+01 .        .        .    *           .            .
      1.000D-02   2.767D+01 .        .        .     *          .            .
      1.100D-02   2.814D+01 .        .        .      *         .            .
      1.200D-02   2.857D+01 .        .        .       *        .            .
      1.300D-02   2.896D+01 .        .        .        *       .            .
      1.400D-02   2.932D+01 .        .        .         *.     .            .
      1.500D-02   2.965D+01 .        .        .          *.    .            .
      1.600D-02   2.995D+01 .        .        .           *    .            .
      1.700D-02   3.023D+01 .        .        .           .*   .            .
      1.800D-02   3.048D+01 .        .        .           .*   .            .
      1.900D-02   3.072D+01 .        .        .           . *  .            .
      2.000D-02   3.093D+01 .        .        .           .  * .            .
      2.100D-02   3.113D+01 .        .        .           .   *.            .
      2.200D-02   3.131D+01 .        .        .           .   *.            .
      2.300D-02   3.148D+01 .        .        .           .    *            .
      2.400D-02   3.163D+01 .        .        .           .    *            .
      2.500D-02   3.177D+01 .        .        .           .    .*           .
      2.600D-02   3.190D+01 .        .        .           .    .*           .
      2.700D-02   3.202D+01 .        .        .           .    . *          .
      2.800D-02   3.212D+01 .        .        .           .    . *          .
      2.900D-02   3.222D+01 .        .        .           .    . *          .
      3.000D-02   3.232D+01 .        .        .           .    . *          .
      3.100D-02   3.240D+01 .        .        .           .    . *          .
      3.200D-02   3.248D+01 .        .        .           .    . *          .
      3.300D-02   3.255D+01 .        .        .           .    .  *         .
      3.400D-02   3.261D+01 .        .        .           .    .  *         .
      3.500D-02   3.267D+01 .        .        .           .    .  *         .
      3.600D-02   3.272D+01 .        .        .           .    .  *         .
      3.700D-02   3.277D+01 .        .        .           .    .  *         .
      3.800D-02   3.282D+01 .        .        .           .    .  *         .
      3.900D-02   3.286D+01 .        .        .           .    .  *         .
      4.000D-02   3.290D+01 .        .        .           .    .   *        .
      4.200D-02   3.297D+01 .        .        .           .    .   *        .
      4.300D-02   3.300D+01 .        .        .           .    .   *        .
      4.400D-02   3.303D+01 .        .        .           .    .   *        .
      4.500D-02   3.305D+01 .        .        .           .    .   *        .
      4.600D-02   3.308D+01 .        .        .           .    .   *        .
      4.700D-02   3.310D+01 .        .        .           .    .   *        .
      4.800D-02   3.312D+01 .        .        .           .    .   *        .
      4.900D-02   3.313D+01 .        .        .           .    .   *        .
      5.000D-02   3.315D+01 .        .        .           .    .   *        .
                - - - - - - - - - - - - - - - - - - - - - - - - - - - - - - - - - - -

        JOB CONCLUDED

        TOTAL JOB TIME            29.43
```

Figure 19.16—(*continued*)

EXAMPLE 19.9

The inductor in the circuit shown in Fig. 19.17 has no initial current flowing in it. Use SPICE to obtain i_L and v_L as functions of time for a period of 2.5 ms after closing the switch.

Solution   The nodes are designated as shown in Fig. 19.17. The 0-V source, VM, is added to measure the inductor's current, i_L. Here the initial conditions will be used in the transient analysis. The input data file is as follows.

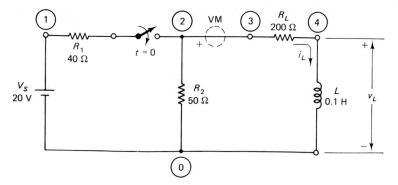

Figure 19.17

```
SOLUTION TO EXAMPLE 19.9
R1     1    2    40.0
R2     2    0    50.0
RL     3    4    200.0
L      4    0    0.1      IC=0.0
VS     1    0    DC       20.0
VM     2    3
.IC  V(1)=20.0
.TRAN     50.0U    2.5M    0.0    25.0U  UIC
.PLOT     TRAN     V(4)    I(VM)
.PRINT    TRAN     I(VM)
.END
```

The program performs the transient analysis using the specified initial condition. The plot indicates the variation of both v_L and i_L as functions of time for a period of 2.5 ms. By taking TSTEP = 50 μs, TMAX = 25 μs, 100 points are used in the calculation and 50 in the plots. It is easy to check the correctness of the initial and final conditions in this circuit, as obtained from the computer program.

The .PLOT line will also provide a table for the variation v_L [= V(4)]. If a printout table for the variation of i_L [= I(VM)] is also required, the .PRINT line is used as indicated in the program above. The output obtained by running this program is shown in Fig. 19.18.

```
******** 1/ 1/80******** PSpice 2.04 (Nov 1985) ******* 0:16:59********

SOLUTION TO EXAMPLE 19.9

*****************************************************************************
****      CIRCUIT DESCRIPTION

R1 1 2 40.0
R2 2 0 50.0
RL 3 4 200.0
L 4 0 0.1 IC=0.0
VS 1 0 DC 20.0
VM 2 3
.IC V(1)=20.0
.TRAN 50.0U 2.5M 0.0 25.0U UIC
.PLOT TRAN V(4) I(VM)
.PRINT TRAN I(VM)
.PROBE
.END


*****************************************************************************
```

Figure 19.18

TEMPERATURE = 27.000 DEG C

****   TRANSIENT ANALYSIS

TIME	I(VM)	TIME	I(VM)
.000E+00	1.542E-08	1.300E-03	4.722E-02
5.000E-05	5.245E-03	1.350E-03	4.751E-02
1.000E-04	9.952E-03	1.400E-03	4.777E-02
1.500E-04	1.416E-02	1.450E-03	4.801E-02
2.000E-04	1.793E-02	1.500E-03	4.822E-02
2.500E-04	2.131E-02	1.550E-03	4.840E-02
3.000E-04	2.433E-02	1.600E-03	4.857E-02
3.500E-04	2.703E-02	1.650E-03	4.872E-02
4.000E-04	2.944E-02	1.700E-03	4.886E-02
4.500E-04	3.160E-02	1.750E-03	4.898E-02
5.000E-04	3.354E-02	1.800E-03	4.908E-02
5.500E-04	3.527E-02	1.850E-03	4.918E-02
6.000E-04	3.682E-02	1.900E-03	4.927E-02
6.500E-04	3.821E-02	1.950E-03	4.934E-02
7.000E-04	3.945E-02	2.000E-03	4.941E-02
7.500E-04	4.056E-02	2.050E-03	4.948E-02
8.000E-04	4.155E-02	2.100E-03	4.953E-02
8.500E-04	4.244E-02	2.150E-03	4.958E-02
9.000E-04	4.323E-02	2.200E-03	4.962E-02
9.500E-04	4.395E-02	2.250E-03	4.966E-02
1.000E-03	4.458E-02	2.300E-03	4.970E-02
1.050E-03	4.515E-02	2.350E-03	4.973E-02
1.100E-03	4.566E-02	2.400E-03	4.976E-02
1.150E-03	4.612E-02	2.450E-03	4.978E-02
1.200E-03	4.653E-02	2.500E-03	4.981E-02
1.250E-03	4.689E-02		

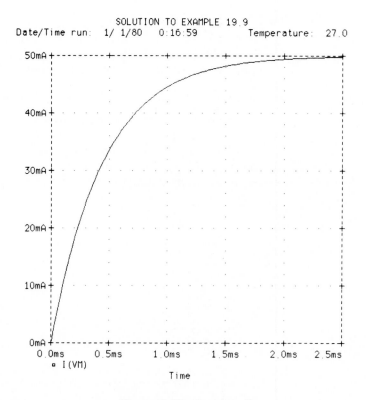

SOLUTION TO EXAMPLE 19.9
Date/Time run: 1/ 1/80   0:16:59   Temperature: 27.0

Figure 19.18—(*continued*)

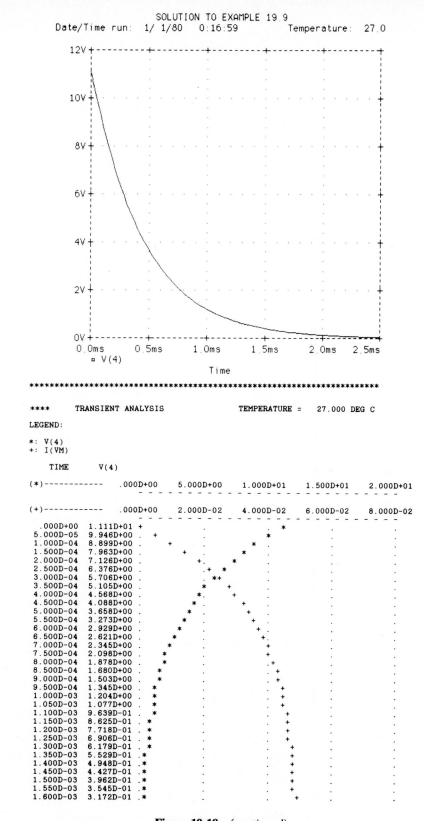

SOLUTION TO EXAMPLE 19.9
Date/Time run: 1/ 1/80   0:16:59   Temperature: 27.0

```
****************************************************************************

****      TRANSIENT ANALYSIS                    TEMPERATURE =   27.000 DEG C

LEGEND:

*: V(4)
+: I(VM)

     TIME        V(4)

(*)-----------    .000D+00    5.000D+00    1.000D+01    1.500D+01    2.000D+01
                                                                            
(+)-----------    .000D+00    2.000D-02    4.000D-02    6.000D-02    8.000D-02
- - - - - - - - - - - - - - - - - - - - - - - - - - - - - - - - - - - - - - -
  .000D+00   1.111D+01 +       .           .           .     *     .
 5.000D-05   9.946D+00 .   +   .           .           *     .     .
 1.000D-04   8.899D+00 .       +     .     .        *  .     .     .
 1.500D-04   7.963D+00 .       .  +  .     .      *  . .     .     .
 2.000D-04   7.126D+00 .       .     . +   .    *    . .     .     .
 2.500D-04   6.376D+00 .       .     .   .+ *  .     . .     .     .
 3.000D-04   5.706D+00 .       .     .     *+   .     . .     .     .
 3.500D-04   5.105D+00 .       .     .   * .  + .     . .     .     .
 4.000D-04   4.568D+00 .       .     . *.    .    +   . .     .     .
 4.500D-04   4.088D+00 .       .   *   .     .      + . .     .     .
 5.000D-04   3.658D+00 .       .  *    .     .       +. .     .     .
 5.500D-04   3.273D+00 .       . *     .     .       .+.     .     .
 6.000D-04   2.929D+00 .       .*      .     .       . +.     .     .
 6.500D-04   2.621D+00 .      *.       .     .       .  +     .     .
 7.000D-04   2.345D+00 .     * .       .     .       .   +    .     .
 7.500D-04   2.098D+00 .     *.        .     .       .    +   .     .
 8.000D-04   1.878D+00 .    * .        .     .       .      + .     .
 8.500D-04   1.680D+00 .   *  .        .     .       .       +.     .
 9.000D-04   1.503D+00 .   *  .        .     .       .       .+     .
 9.500D-04   1.345D+00 .  *   .        .     .       .       . +    .
 1.000D-03   1.204D+00 .  *   .        .     .       .       .  +   .
 1.050D-03   1.077D+00 . *    .        .     .       .       .  +   .
 1.100D-03   9.639D-01 . *    .        .     .       .       .   +  .
 1.150D-03   8.625D-01 .*     .        .     .       .       .   +  .
 1.200D-03   7.718D-01 .*     .        .     .       .       .   +  .
 1.250D-03   6.906D-01 .*     .        .     .       .       .   +  .
 1.300D-03   6.179D-01 .*     .        .     .       .       .    + .
 1.350D-03   5.529D-01 .*     .        .     .       .       .    + .
 1.400D-03   4.948D-01 .*     .        .     .       .       .    + .
 1.450D-03   4.427D-01 .*     .        .     .       .       .    + .
 1.500D-03   3.962D-01 .*     .        .     .       .       .    + .
 1.550D-03   3.545D-01 .*     .        .     .       .       .    + .
 1.600D-03   3.172D-01 .*     .        .     .       .       .     +.
```

Figure 19.18—(*continued*)

```
1.700D-03  2.540D-01 .*          .          .   +      .          .
1.750D-03  2.273D-01 .*          .          .   +      .          .
1.800D-03  2.034D-01 .*          .          .   +      .          .
1.850D-03  1.820D-01 *           .          .   +      .          .
1.900D-03  1.628D-01 *           .          .   +      .          .
1.950D-03  1.457D-01 *           .          .   +      .          .
2.000D-03  1.304D-01 *           .          .   +      .          .
2.050D-03  1.167D-01 *           .          .   +      .          .
2.100D-03  1.044D-01 *           .          .   +      .          .
2.150D-03  9.341D-02 *           .          .   +      .          .
2.200D-03  8.359D-02 *           .          .   +      .          .
2.250D-03  7.479D-02 *           .          .   +      .          .
2.300D-03  6.693D-02 *           .          .   +      .          .
2.350D-03  5.989D-02 *           .          .   +      .          .
2.400D-03  5.359D-02 *           .          .   +      .          .
2.450D-03  4.795D-02 *           .          .   +      .          .
2.500D-03  4.289D-02 *           .          .   +      .          .
                     - - - - - - - - - - - - - - - - - - - - - - - -


       JOB CONCLUDED

       TOTAL JOB TIME        32.79
```

Figure 19.18—(*continued*)

EXAMPLE 19.10

Analyze the circuit shown in Fig. 19.19 to obtain the values of $\mathbf{I}$, $\mathbf{V}$, $\mathbf{I}_0$, and $\mathbf{V}_0$. This circuit contains two dependent sources. The frequency of the sources is 200 Hz.

Solution The elements and the nodes in this ac circuit are designated as shown in Fig. 19.19. The 0 V source, VSENS, is added as shown, to measure the current $\mathbf{I}$ in the proper direction. The input data file is then as follows.

```
SOLUTION TO EXAMPLE 19.10
R1      2    3    50.0
VS1     2    1    AC  5.0    90.0
L1      3    4    0.1
R2      4    8    100.0
C       8    0    10.0E-6
VSENS   0    1
R3      5    6    200.0
L2      6    7    0.2
VS2     7    0    AC  10.0   0.0
F       3    4    VSENS      10.0      ← describing the source I₁
E       5    4    4    0     12.0      ← describing the source V₁
.AC  LIN 1    200.0         200.0
.PRINT  AC  VM(4)  VP(4)  VM(6)  VP(6)  IM(VSENS)  IP(VSENS)
+IM(VS2)  IP(VS2)
.END
```

The current-controlled current source $\mathbf{I}_1$, given by

$$\mathbf{I}_1 = 10\mathbf{I} \qquad \text{or} \qquad IS = F * IC$$

is described by the F line above. The controlling current $\mathbf{I}$ is measured in the proper direction by VSENS. Similarly, the voltage-controlled voltage source $\mathbf{V}_1$ given by

$$\mathbf{V}_1 = 12\mathbf{V} \qquad \text{or} \qquad VS = E * VC$$

is described by the E line shown above.

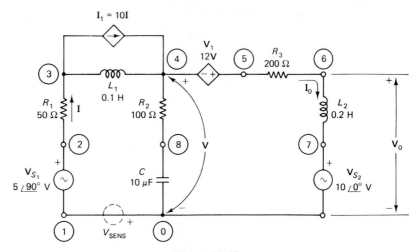

Figure 19.19

The .AC line specifies only one frequency, the operating frequency of 200.0 HZ. The .PRINT card has eight output variables, the magnitude and phase of each of the required ac quantities, $\mathbf{V}$, $\mathbf{V}_0$, $\mathbf{I}$, and $\mathbf{I}_0$, respectively. The output obtained by running this program is shown in Fig. 19.20. Notice that since there are no dc sources in this circuit, the results of the initial dc analysis of the circuit are all zeros, as indicated in the printout.

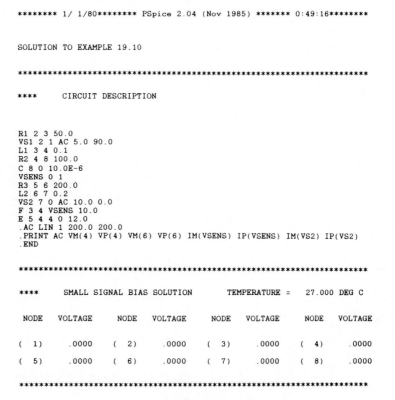

```
******** 1/ 1/80******** PSpice 2.04 (Nov 1985) ******* 0:49:16********


SOLUTION TO EXAMPLE 19.10


****************************************************************************

****      CIRCUIT DESCRIPTION


R1 2 3 50.0
VS1 2 1 AC 5.0 90.0
L1 3 4 0.1
R2 4 8 100.0
C 8 0 10.0E-6
VSENS 0 1
R3 5 6 200.0
L2 6 7 0.2
VS2 7 0 AC 10.0 0.0
F 3 4 VSENS 10.0
E 5 4 4 0 12.0
.AC LIN 1 200.0 200.0
.PRINT AC VM(4) VP(4) VM(6) VP(6) IM(VSENS) IP(VSENS) IM(VS2) IP(VS2)
.END


****************************************************************************

****      SMALL SIGNAL BIAS SOLUTION        TEMPERATURE =    27.000 DEG C

  NODE   VOLTAGE     NODE   VOLTAGE     NODE   VOLTAGE     NODE   VOLTAGE

  ( 1)    .0000     ( 2)    .0000     ( 3)    .0000     ( 4)    .0000

  ( 5)    .0000     ( 6)    .0000     ( 7)    .0000     ( 8)    .0000


****************************************************************************
```

Figure 19.20

```
****    AC ANALYSIS                        TEMPERATURE =   27.000 DEG C


   FREQ      VM(4)       VP(4)       VM(6)       VP(6)      IM(VSENS)

2.000E+02   6.988E-01   -1.865E+01   1.084E+01   -1.310E+01   4.651E-03


****************************************************************************
****    AC ANALYSIS                        TEMPERATURE =   27.000 DEG C


   FREQ      IP(VSENS)   IM(VS2)     IP(VS2)

2.000E+02   -1.753E+02   1.003E-02   -1.671E+02

               JOB CONCLUDED
               TOTAL JOB TIME          10.44
```

Figure 19.20—(*continued*)

EXAMPLE 19.11

Obtain a plot of the variation of the source current $\mathbf{I}_S$, the load current $\mathbf{I}_L$, and the load voltage $\mathbf{V}_L$ in the magnetically coupled circuit shown in Fig. 19.21. The frequency of the source varies from 100 Hz to 10 kHz.

Solution   In preparation for computer-aided circuit analysis, using SPICE, the circuit elements and nodes are designated as shown in Fig. 19.21. The two short-circuit sources, VM1 and VM2, are added to the circuit to monitor the required currents. SPICE requires that each node must have a dc path to ground. For this reason, the reference (ground) nodes (marked node 0) of the primary and secondary circuits are connected together. This does not affect the results of the analysis. The input data file is thus as follows.

```
SOLUTION TO EXAMPLE 19.11
RS     2     3     100.0
C1     3     4     2.0E-6
RP     4     0     200.0
LP     4     0     0.1
VS     1     0     AC  100.0  0.0
VM1    1     2
LS     0     5     0.2
C2     5     0     5.0U
K      LP    LS    0.80
L2     5     7     0.4
RL     7     8     50.0
VM2    8     0
.AC    DEC   20    100.0      10.0K
.PLOT        AC    IM(VM1)    IP(VM1)
.PLOT        AC    IM(VM2)    IP(VM2)
.PLOT        AC    VM(7)      VP(7)
.END
```

SPICE ignores any missing node number. Here the circuit does not have a node numbered 6. To account for the dot convention, notice that the first node in each of the LP and LS lines is the dotted terminal. The K line accounts

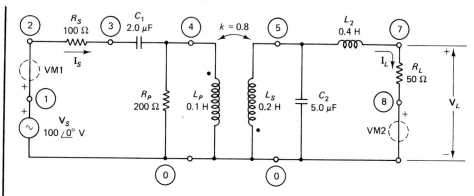

Figure 19.21

for the magnetic coupling. The frequency varies over two decades, with 20 analysis points required per decade, as indicated on the .AC analysis line. Each of the required output variables is plotted (magnitude and phase) on a separate graph. The output obtained by running this program is shown in Fig. 19.22.

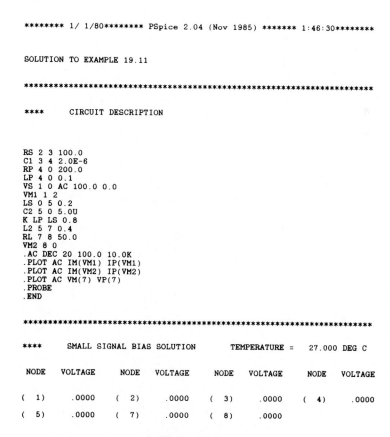

```
******** 1/ 1/80******** PSpice 2.04 (Nov 1985) ******* 1:46:30********


SOLUTION TO EXAMPLE 19.11


*******************************************************************************

****     CIRCUIT DESCRIPTION


RS 2 3 100.0
C1 3 4 2.0E-6
RP 4 0 200.0
LP 4 0 0.1
VS 1 0 AC 100.0 0.0
VM1 1 2
LS 0 5 0.2
C2 5 0 5.0U
K LP LS 0.8
L2 5 7 0.4
RL 7 8 50.0
VM2 8 0
.AC DEC 20 100.0 10.0K
.PLOT AC IM(VM1) IP(VM1)
.PLOT AC IM(VM2) IP(VM2)
.PLOT AC VM(7) VP(7)
.PROBE
.END


*******************************************************************************

****     SMALL SIGNAL BIAS SOLUTION      TEMPERATURE =   27.000 DEG C

  NODE   VOLTAGE     NODE   VOLTAGE     NODE   VOLTAGE     NODE   VOLTAGE

  (  1)    .0000    (  2)    .0000    (  3)    .0000    (  4)    .0000

  (  5)    .0000    (  7)    .0000    (  8)    .0000
```

Figure 19.22

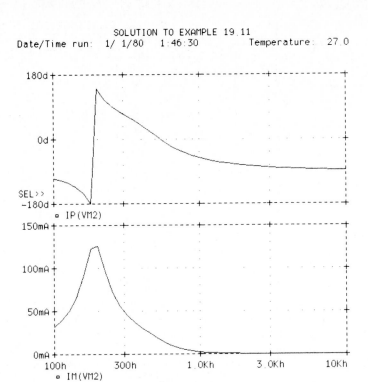

SOLUTION TO EXAMPLE 19.11
Date/Time run: 1/ 1/80 1:46:30   Temperature: 27.0

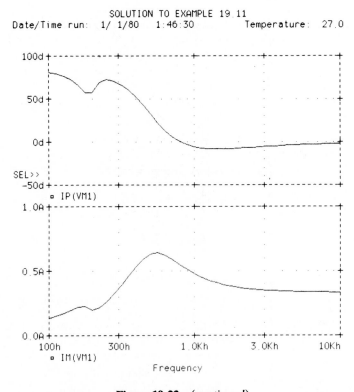

SOLUTION TO EXAMPLE 19.11
Date/Time run: 1/ 1/80 1:46:30   Temperature: 27.0

Figure 19.22—(*continued*)

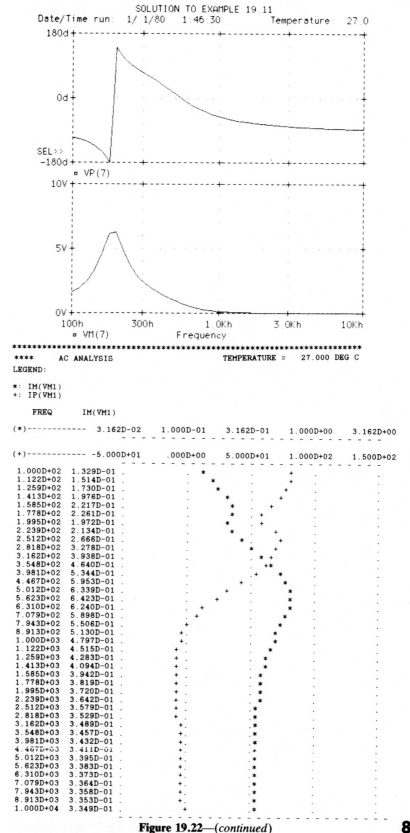

Figure 19.22—(continued)

```
**************************************************************************

****     AC ANALYSIS                      TEMPERATURE =    27.000 DEG C

LEGEND:

*:  IM(VM2)
+:  IP(VM2)

     FREQ       IM(VM2)

(*)------------  1.000D-06     1.000D-04     1.000D-02     1.000D+00     1.000D+02

                                  - - - - - - - - - - - - - - - - - - - - - -

(+)------------ -2.000D+02     -1.000D+02     .000D+00     1.000D+02     2.000D+02

                                  - - - - - - - - - - - - - - - - - - - - - -
 1.000D+02  3.195D-02 .                       + .            *            .
 1.122D+02  3.941D-02 .                      +              *             .
 1.259D+02  5.012D-02 .                   +  .              *             .
 1.413D+02  6.641D-02 .                 +    .             *              .
 1.585D+02  9.168D-02 .            +         .            *               .
 1.778D+02  1.227D-01 .      +               .           *                .
 1.995D+02  1.256D-01 .    +                 .          *              +   .
 2.239D+02  9.724D-02 .                      .          *           +      .
 2.512D+02  7.236D-02 .                      .          *         + .      .
 2.818D+02  5.600D-02 .                      .          *       +          .
 3.162D+02  4.500D-02 .                      .          *     +            .
 3.548D+02  3.698D-02 .                      .          *  +               .
 3.981D+02  3.051D-02 .                      .          *+                 .
 4.467D+02  2.477D-02 .                      .          X                  .
 5.012D+02  1.943D-02 .                      .       +*                    .
 5.623D+02  1.458D-02 .                      .      +.*                    .
 6.310D+02  1.052D-02 .                      .    +  *                     .
 7.079D+02  7.391D-03 .                      .   +  *.                     .
 7.943D+02  5.117D-03 .                      .  +  *                       .
 8.913D+02  3.525D-03 .                      . +  *                        .
 1.000D+03  2.429D-03 .                      .+   *                        .
 1.122D+03  1.678D-03 .                      + *                           .
 1.259D+03  1.163D-03 .                    + *                             .
 1.413D+03  8.084D-04 .                    .+*                             .
 1.585D+03  5.639D-04 .                    .+*                             .
 1.778D+03  3.944D-04 .                    X                               .
 1.995D+03  2.765D-04 .                   *+                               .
 2.239D+03  1.942D-04 .                 *+                                 .
 2.512D+03  1.366D-04 .             *  +                                   .
 2.818D+03  9.622D-05 .            *  +                                    .
 3.162D+03  6.785D-05 .           *.  +                                    .
 3.548D+03  4.788D-05 .           *   +                                    .
 3.981D+03  3.381D-05 .          *  . +                                    .
 4.467D+03  2.388D-05 .         *   . +                                    .
 5.012D+03  1.688D-05 .        *    . +                                    .
 5.623D+03  1.193D-05 .      *      . +                                    .
 6.310D+03  8.440D-06 .     *       . +                                    .
 7.079D+03  5.970D-06 .    *        . +                                    .
 7.943D+03  4.224D-06 .   *         . +                                    .
 8.913D+03  2.989D-06 . *           . +                                    .
 1.000D+04  2.115D-06 . *           . +                                    .

**************************************************************************

****     AC ANALYSIS                      TEMPERATURE =    27.000 DEG C

LEGEND:

*:  VM(7)
+:  VP(7)

     FREQ       VM(7)

(*)------------  1.000D-04     1.000D-02     1.000D+00     1.000D+02     1.000D+04

                                  - - - - - - - - - - - - - - - - - - - - - -

(+)------------ -2.000D+02     -1.000D+02     .000D+00     1.000D+02     2.000D+02

                                  - - - - - - - - - - - - - - - - - - - - - -
 1.000D+02  1.597D+00 .                       + .             *            .
 1.122D+02  1.970D+00 .                      +  .            *             .
 1.259D+02  2.506D+00 .                   +     .           *              .
 1.413D+02  3.320D+00 .                 +       .          .*              .
 1.585D+02  4.584D+00 .            +            .          *               .
 1.778D+02  6.133D+00 .      +                  .          *               .
 1.995D+02  6.279D+00 .                         .          *            +  .
 2.239D+02  4.862D+00 .                         .          *         + .    .
 2.512D+02  3.618D+00 .                         .          *       + .      .
 2.818D+02  2.800D+00 .                         .          *      +         .
 3.162D+02  2.250D+00 .                         .          *   +            .
 3.548D+02  1.849D+00 .                         .          *  +             .
```

Figure 19.22—(*continued*)

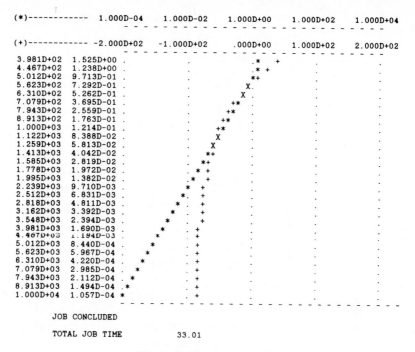

(*)------------ 1.000D-04   1.000D-02   1.000D+00   1.000D+02   1.000D+04

(+)------------ -2.000D+02   -1.000D+02   .000D+00   1.000D+02   2.000D+02

3.981D+02 1.525D+00 .   .   .* +
4.467D+02 1.238D+00 .   .   .* +
5.012D+02 9.713D-01 .   .   .*+
5.623D+02 7.292D-01 .   .   X.
6.310D+02 5.262D-01 .   .   X .
7.079D+02 3.695D-01 .   .   +*
7.943D+02 2.559D-01 .   .   +*
8.913D+02 1.763D-01 .   .   +*
1.000D+03 1.214D-01 .   .   +*
1.122D+03 8.388D-02 .   .   X
1.259D+03 5.813D-02 .   .   X
1.413D+03 4.042D-02 .   .   *+
1.585D+03 2.819D-02 .   . *+
1.778D+03 1.972D-02 .   . * +
1.995D+03 1.382D-02 .   .* +
2.239D+03 9.710D-03 .   .* +
2.512D+03 6.831D-03 .   *. +
2.818D+03 4.811D-03 .   *. . +
3.162D+03 3.392D-03 .   * .   +
3.548D+03 2.394D-03 .   *   .   +
3.981D+03 1.690D-03 .   *   .   +
4.467D+03 1.194D-03 .   *   .   +
5.012D+03 8.440D-04 .   *   .   +
5.623D+03 5.967D-04 .   *   .   +
6.310D+03 4.220D-04 .   *   .   +
7.079D+03 2.985D-04 .   *   .   +
7.943D+03 2.112D-04 . *   .   +
8.913D+03 1.494D-04 .*   .   +
1.000D+04 1.057D-04 *   .   +

JOB CONCLUDED

TOTAL JOB TIME   33.01

Figure 19.22—(*continued*)

EXAMPLE 19.12

Obtain the magnitude- and phase-frequency plots of the voltage transfer function $\mathbf{V}_2/\mathbf{V}_1$ and the input admittance $\mathbf{I}_1/\mathbf{V}_1$ of the filter circuit shown in Fig. 19.23 over the frequency range 10 Hz to 10 kHz.

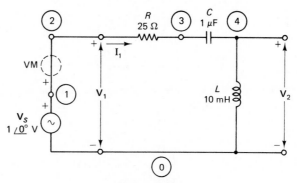

Figure 19.23

Solution   The source $\mathbf{V}_S$ is chosen with a value $1\underline{/0°}\,\text{V}$, so that

$$G_{21} = \text{voltage transfer function} = \frac{\mathbf{V}_2}{\mathbf{V}_1} = \mathbf{V}_2$$

while

$$Y_{11} = \text{input admittance} = \frac{\mathbf{I}_1}{\mathbf{V}_1} = \mathbf{I}_1$$

The short-circuit source, VM, is added to monitor I_1 in the proper direction. The nodes are designated as shown in Fig. 19.23. Here the input data file is as follows:

```
SOLUTION TO EXAMPLE 19.12
R      2    3    25.0
C      3    4    1.0E-6
L      4    0    10.0M
VS     1    0    AC  1.0  0.0
VM     1    2
.AC    DEC  10   10.0     10.0K
.PLOT  AC   IM(VM)        IP(VM)     ← equivalent to Y₁₁(ω) in S
.PLOT  AC   VDB(4)        VP(4)      ← equivalent to G₂₁(ω)
.END
```

The magnitude and phase of the input admittance Y_{11} are equivalent to the magnitude and phase of the current I_1. These are obtained in the first plot. The transfer voltage function is obtained directly from V_2 [$=V(4)$]. The second plot shows the magnitude, in decibels, and phase of this transfer function. the output obtained by running the program is shown in Fig. 19.24. As the results of the $G_{21}(\omega)$ plot indicate, this filter is a high-pass filter. Also notice that this circuit has a series resonance at the frequency

$$f_0 = \frac{1}{2\pi\sqrt{LC}} = 1.592 \text{ kHz}$$

At this frequency the current $|I_1|$ peaks, as can clearly be seen from the computer results.

```
******** 1/ 1/80******** PSpice 2.04 (Nov 1985) ******* 2:20:07********

SOLUTION TO EXAMPLE 19.12

************************************************************************

****      CIRCUIT DESCRIPTION

R 2 3 25.0
C 3 4 1.0E-6
L 4 0 10.0M
VS 1 0 AC 1.0 0.0
VM 1 2
.AC DEC 10 10.0 10.0K
.PLOT AC IM(VM) IP(VM)
.PLOT AC VDB(4) VP(4)
.PROBE
.END

************************************************************************

****      SMALL SIGNAL BIAS SOLUTION       TEMPERATURE =   27.000 DEG C

NODE   VOLTAGE     NODE   VOLTAGE     NODE   VOLTAGE     NODE    VOLTAGE

( 1)   .0000       ( 2)   .0000       ( 3)    .0000      ( 4)     .0000
```

Figure 19.24

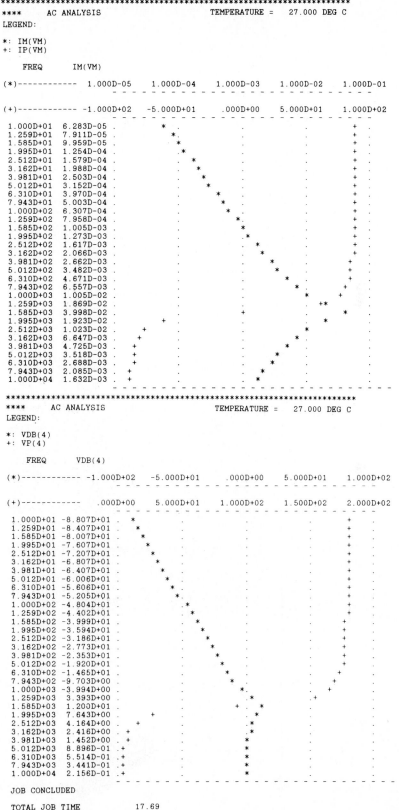

```
************************************************************************
****       AC ANALYSIS                    TEMPERATURE =    27.000 DEG C
LEGEND:

*: IM(VM)
+: IP(VM)

     FREQ        IM(VM)

(*)------------   1.000D-05   1.000D-04   1.000D-03   1.000D-02   1.000D-01


(+)------------  -1.000D+02  -5.000D+01    .000D+00   5.000D+01   1.000D+02
                - - - - - - - - - - - - - - - - - - - - - - - - - - - - -
 1.000D+01  6.283D-05 .            *   .         .         .         + .
 1.259D+01  7.911D-05 .           *.   .         .         .         + .
 1.585D+01  9.959D-05 .            *   .         .         .         + .
 1.995D+01  1.254D-04 .            .*  .         .         .         + .
 2.512D+01  1.579D-04 .            . *     .         .         .     + .
 3.162D+01  1.988D-04 .            .   *   .         .         .     + .
 3.981D+01  2.503D-04 .            .    *  .         .         .     + .
 5.012D+01  3.152D-04 .            .     * .         .         .     + .
 6.310D+01  3.970D-04 .            .   *   .         .         .     + .
 7.943D+01  5.003D-04 .            .    *  .         .         .     + .
 1.000D+02  6.307D-04 .            .     * .         .         .     + .
 1.259D+02  7.958D-04 .            .    *. .         .         .     + .
 1.585D+02  1.005D-03 .            .     * .         .         .     + .
 1.995D+02  1.273D-03 .            .    .* .         .         .     + .
 2.512D+02  1.617D-03 .            .    . *.        .         .      + .
 3.162D+02  2.066D-03 .            .    .  *        .         .      + .
 3.981D+02  2.662D-03 .            .    .   *       .         .     +  .
 5.012D+02  3.482D-03 .            .    .    *      .         .     +  .
 6.310D+02  4.671D-03 .            .    .     *     .         .     +  .
 7.943D+02  6.557D-03 .            .    .      *    .         .    +   .
 1.000D+03  1.005D-02 .            .    .       .   .         .   +    .
 1.259D+03  1.869D-02 .            .    .       .   .       +*        .
 1.585D+03  3.998D-02 .            .    .       .   +       .    *    .
 1.995D+03  1.923D-02 .            .  +     .       .       .    *    .
 2.512D+03  1.023D-02 .        +   .       .       .       .   *      .
 3.162D+03  6.647D-03 .     +  .       .       .       .   *         .
 3.981D+03  4.725D-03 .    +   .       .       .       .  *          .
 5.012D+03  3.518D-03 .    +   .       .       .      *  .          .
 6.310D+03  2.688D-03 .    +   .       .       .     *   .          .
 7.943D+03  2.085D-03 .  +     .       .       .    *    .          .
 1.000D+04  1.632D-03 .  +     .       .       .   *     .          .
                - - - - - - - - - - - - - - - - - - - - - - - - - - - - -

************************************************************************
****       AC ANALYSIS                    TEMPERATURE =    27.000 DEG C
LEGEND:

*: VDB(4)
+: VP(4)

     FREQ        VDB(4)

(*)------------  -1.000D+02  -5.000D+01    .000D+00   5.000D+01   1.000D+02
                - - - - - - - - - - - - - - - - - - - - - - - - - - - - -

(+)------------    .000D+00   5.000D+01   1.000D+02   1.500D+02   2.000D+02
                - - - - - - - - - - - - - - - - - - - - - - - - - - - - -
 1.000D+01 -8.807D+01 . *       .         .         .         .     +  .
 1.259D+01 -8.407D+01 .  *      .         .         .         .     +  .
 1.585D+01 -8.007D+01 .   *     .         .         .         .     +  .
 1.995D+01 -7.607D+01 .   *     .         .         .         .     +  .
 2.512D+01 -7.207D+01 .    *    .         .         .         .     +  .
 3.162D+01 -6.807D+01 .     *   .         .         .         .     +  .
 3.981D+01 -6.407D+01 .     *   .         .         .         .     +  .
 5.012D+01 -6.006D+01 .      *  .         .         .         .     +  .
 6.310D+01 -5.606D+01 .      *. .         .         .         .     +  .
 7.943D+01 -5.205D+01 .       *.         .         .         .     +  .
 1.000D+02 -4.804D+01 .       .*         .         .         .     +   .
 1.259D+02 -4.402D+01 .       . *        .         .         .     +   .
 1.585D+02 -3.999D+01 .       .  *       .         .         .    +    .
 1.995D+02 -3.594D+01 .       .   *      .         .         .    +    .
 2.512D+02 -3.186D+01 .       .    *     .         .         .    +    .
 3.162D+02 -2.773D+01 .       .     *    .         .         .    +    .
 3.981D+02 -2.353D+01 .       .      *   .         .         .    +    .
 5.012D+02 -1.920D+01 .       .       *  .         .         .    +    .
 6.310D+02 -1.465D+01 .       .        * .         .         .   +     .
 7.943D+02 -9.703D+00 .       .         *.         .         .  +      .
 1.000D+03 -3.994D+00 .       .         . *        .         . +       .
 1.259D+03  3.393D+00 .       .         .  *       .         .+        .
 1.585D+03  1.200D+01 .       .         .      +   . *       .         .
 1.995D+03  7.643D+00 .       .    +    .         . *       .         .
 2.512D+03  4.164D+00 .     + .         .         .*        .         .
 3.162D+03  2.416D+00 . +     .         .         .*        .         .
 3.981D+03  1.452D+00 . +     .         .         *         .         .
 5.012D+03  8.896D-01 .+      .         .         *         .         .
 6.310D+03  5.514D-01 .+      .         .         *         .         .
 7.943D+03  3.441D-01 .+      .         .         *         .         .
 1.000D+04  2.156D-01 .+      .         .         *         .         .
                - - - - - - - - - - - - - - - - - - - - - - - - - - - - -

JOB CONCLUDED

TOTAL JOB TIME              17.69
```

Figure 19.24—(*continued*)

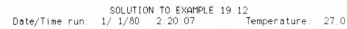

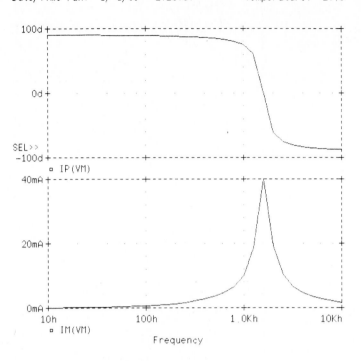

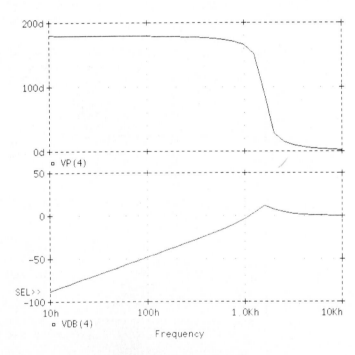

Figure 19.24—(*continued*)

PROBLEMS

(i) **19.1.** Use SPICE to obtain the voltage V and currents I_1 and I_2 in the circuit shown in Fig. 19.25.

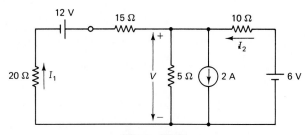

Figure 19.25

(ii) **19.2.** Analyze the circuit shown in Fig. 19.26 using SPICE. Determine the values of I_1, I_2, I_3, V_4, V_5, and V_6, as designated in the circuit diagram.

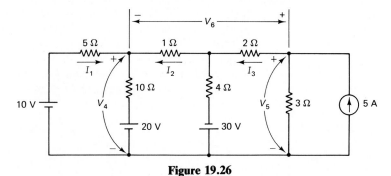

Figure 19.26

(ii) **19.3.** Analyze the circuit shown in Fig. 19.27 using SPICE. Determine the values of V', I, V_L, and I_L. Plot the transfer characteristics showing the variations of V_L and I_L as the source E varies between 0 and 20 V, in 1 V steps.

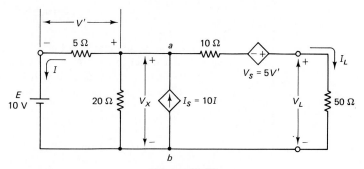

Figure 19.27

(i) **19.4.** Remove the voltage source V_S in the circuit shown in Fig. 19.27 (i.e., replace it with a short circuit). Reanalyze the circuit to obtain the transfer characteristics showing the variations of V_x and V_L as the source voltage E varies between 5 and 15 V, in 1 V steps.

(ii) **19.5.** Use SPICE to analyze the circuit shown in Fig. 19.28. Obtain the values of I, V', and I_x.

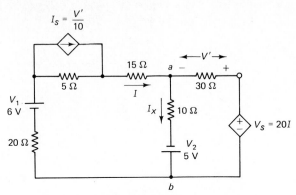

Figure 19.28

(i) **19.6.** Remove the branch between points a and b in the circuit shown in Fig. 19.28, (i.e., replace this branch by an open circuit). Reanalyze the circuit to obtain the transfer characteristics for V_{ab} as the source voltage V_1 varies between 0 and 12 V in steps of 2 V.

(i) **19.7.** In the circuit shown in Fig. 19.29, the switch has been in position 1 for a very long time. What is the initial capacitor's voltage V_0, $V_0 = v_C(0^-)$? Write a SPICE program to obtain the variations of i_C and v_C during the transient from $t = 0$ to $t = 300$ ms.

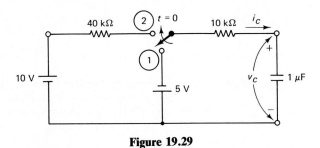

Figure 19.29

(i) **19.8.** If the switch in circuit shown in Fig. 19.30 has been in position 2 for a very long time, find the initial current in the inductor just before the moment of switching: $I_0 = i_L(0^-)$. Using SPICE, find the transient response of this circuit in the form of the variation of i_L and v_L as functions of time for $t = 0$ to $t = 0.6$ ms.

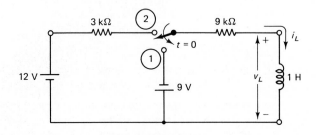

Figure 19.30

(i) **19.9.** Before the switch in the circuit shown in Fig. 19.31 was closed, the initial voltage on the capacitor C_1, V_{10}, was 3 V and the initial voltage on the capacitor C_2, V_{20}, was 6 V (notice the polarities). Use SPICE to plot the variation of the voltage across each of C_1 and C_2 as a function of time in the interval $t = 0$ to $t = 600$ ms.

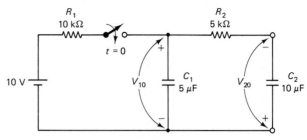

Figure 19.31

(i) **19.10.** Find the initial current in the inductor, I_0, in the circuit shown in Fig. 19.32. After closing the switch at $t = 0$, use SPICE to obtain the transient response of this circuit in the form of the variation of i_L, i, and v as functions of time for $t = 0$ to $t = 1.2$ ms.

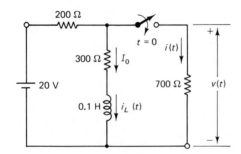

Figure 19.32

(i) **19.11.** There was no initial voltage on the capacitor and no initial current in the inductor in the circuit shown in Fig. 19.33. The switch was closed at $t = 0$. Obtain, using SPICE, the transient response of this network showing the variation of each of v_C, i_C, v_L, and i_L as functions of time for $t = 0$ to $t = 5$ ms.

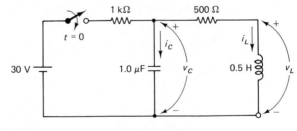

Figure 19.33

(i) **19.12.** Analyze the ac circuit shown in Fig. 19.34, using SPICE, to obtain the values of $\mathbf{I}_1$, $\mathbf{I}_2$, $\mathbf{I}_3$, and $\mathbf{V}$. All sources are operating at a frequency of 1 kHz.

(ii) **19.13.** The sources in the circuit shown in Fig. 19.35 operate at a frequency of 10 kHz. Use SPICE to obtain the values of $\mathbf{I}_1$, $\mathbf{I}_2$, and $\mathbf{V}_x$ in this circuit.

(i) **19.14.** In the circuit shown in Fig. 19.35, $\mathbf{V}_S$ is now changed to $10\mathbf{V}_x$ and $\mathbf{I}_S$ is now changed to $1/5\mathbf{V}_x$. Obtain, using SPICE, the variations of the magnitude and

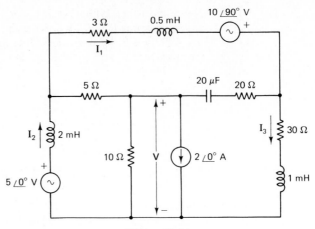

Figure 19.34

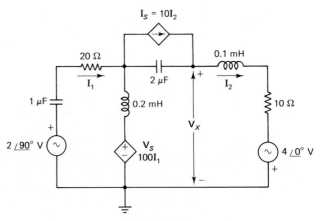

Figure 19.35

phase of $\mathbf{I}_1$, $\mathbf{I}_2$, and $\mathbf{V}_x$ as functions of frequency if the frequency of the sources is allowed to change from 1 kHz to 100 kHz.

(i) **19.15.** Analyze the mutually coupled circuit shown in Fig. 19.36, using SPICE, to obtain the value of each of the branch currents $\mathbf{I}_1$, $\mathbf{I}_2$, and $\mathbf{I}_3$.

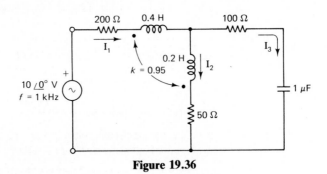

Figure 19.36

(ii) **19.16.** Using SPICE, analyze the transformer circuit shown in Fig. 19.37 to obtain $\mathbf{I}_1$, $\mathbf{V}_p$, $\mathbf{I}_2$, and $\mathbf{V}_s$ if the frequency of the sources is 500 Hz.

834   Chapter 19   Computer-Aided Circuit Analysis Using SPICE

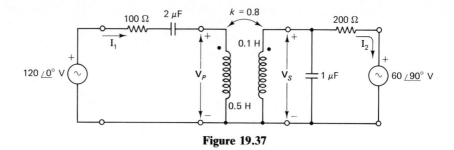

Figure 19.37

(i) **19.17.** Obtain the magnitude and phase variations with frequency for $\mathbf{V}_p$ and $\mathbf{V}_s$ in the circuit of Fig. 19.37 if the frequency is allowed to vary from 100 Hz to 2 kHz.

(i) **19.18.** Obtain the magnitude- and phase-frequency plots for the voltage transfer function $\mathbf{V}_2/\mathbf{V}_1$ of the circuit shown in Fig. 19.38, using SPICE. The frequency is allowed to vary over the range from 1 kHz to 20 kHz. Determine the 3-dB points of this filter network.

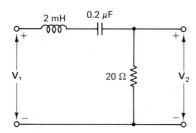

Figure 19.38

(i) **19.19.** Using SPICE, obtain the magnitude- and phase-frequency plots for the voltage transfer function $\mathbf{V}_2/\mathbf{V}_1$ of the circuit shown in Fig. 19.39. The range of variation of the frequency is from 1000 Hz to 100 kHz. Determine the 3-dB point of this filter network.

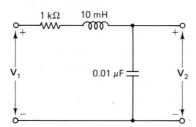

Figure 19.39

(i) **19.20.** Using SPICE, obtain the magnitude- and phase-frequency plots for the voltage transfer function $\mathbf{V}_2/\mathbf{V}_1$ of the circuit shown in Fig. 19.40 over the frequency range from 100 kHz to 1 MHz. Determine the 3-dB points of this filter network.

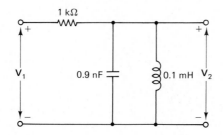

Figure 19.40

Problems

(i) **19.21.** Obtain the magnitude- and phase-frequency plots of the voltage transfer function V_2/V_1 for the network shown in Fig. 19.41, using SPICE. The frequency is allowed to vary over the range 5000 Hz to 500 kHz. Determine the 3-dB point of this filter network.

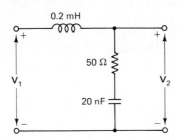

Figure 19.41

(i) **19.22.** Using SPICE, obtain the input admittance (magnitude and phase) for each of the networks in Figs. 19.40 and 19.41.

GLOSSARY

Control Lines: Lines in a computer software program whose function is to direct the program to perform a certain type of analysis (e.g., dc, ac, transient, etc.) at certain temperature(s), produce an output in the form of a table or graph (or both), and end the program. In SPICE, the control lines always start with a dot ".". These lines control the execution (run) of the program.

Data File: The computer software record that provides the input data information required to simulate and analyze a given circuit. This file includes the element lines, the control lines, and the comment lines.

Element Lines: Lines in a computer software program that contain the description of all the active and passive components of a given network. Each element line describes where the element is connected (i.e., between which nodes), its value, and any relevant initial condition. Special element line formats are used for controlled sources, coupled coils, and transmission lines. The record of the element lines defines the circuit's topology to the computer.

SPICE (or Pspice): A general-purpose circuit simulation software program package used in the analysis and design of analog and digital integrated circuits, to perform nonlinear dc, nonlinear transient, and linear ac analysis of the given circuit. Pspice is the version used for personal computers.

ANSWERS TO SELECTED PROBLEMS

Chapter 1

1.1. $-1.602 \, \text{mC}$

1.3. 2.809×10^{16} electrons

1.5. $0.1 \, \mu\text{C}$

1.7. $90 \, \text{J}$

1.9. $5 \, \text{C}$

1.11. $6 \, \text{W}$

1.13. $30 \, \text{W}$, $1800 \, \text{J}$

1.15. $3600 \, \text{C}$, $21{,}600 \, \text{J}$

1.17. $4 \, \text{A}$, $24 \, \text{W}$, $0.48 \, \text{kWh}$

1.19. $4973.33 \, \text{W}$, $41.44 \, \text{A}$

Chapter 2

2.1. $0.4156 \, \Omega$

2.3. $0.0319 \, \text{mm}$

2.5. **(a)** $0.6279 \, \Omega$, $18.736 \, \Omega$, $935 \, \Omega$, $15.083 \, \Omega$;
(b) $0.3666 \, \Omega$, $18.713 \, \Omega$, $1015 \, \Omega$, $8.827 \, \Omega$

2.7. $0.3 \, \text{V}$, $6 \, \text{mW}$

2.9. $352.94 \, \Omega$, $0.102 \, \text{W}$

2.11. $1.2 \, \text{mA}$, $14.4 \, \text{mW}$

2.13. $150 \, \text{V}$, $20 \, \text{mA}$

2.15. $0.8 \, \text{A}$, $156.25 \, \Omega$

2.17. $78.125 \, \text{s}$

2.19. (2) has maximum safe voltage,
(3) has maximum safe current

2.21. **(a)** $V_{\text{max}} = 5.5 \, \text{V}$, $I_{\text{max}} = 27.4 \, \text{mA}$;
(b) $V_{\text{max}} = 8.7 \, \text{V}$, $I_{\text{max}} = 17.3 \, \text{mA}$

2.23. **(a)** 5.4 to $6.6 \, \text{V}$ and 75.76 to $92.95 \, \text{mA}$;
(b) 128.25 to $141.75 \, \text{V}$ and 3.53 to $3.9 \, \text{mA}$

Chapter 3

3.1. **(a)** $I_1 = 2 \, \text{A}$, $I_2 = 7 \, \text{A}$;
(b) $I_1 = 3 \, \text{A}$, $I_2 = 1 \, \text{A}$, $I_3 = 2 \, \text{A}$, $I_4 = 5 \, \text{A}$

3.3. $V_1 = 2 \, \text{V}$, $V_2 = 9 \, \text{V}$, $V_3 = 13 \, \text{V}$

3.5. $R_T = 120 \, \Omega$, $I = 0.1 \, \text{A}$, $V_1 = 1 \, \text{V}$, $V_2 = 4$, $V_3 = 7 \, \text{V}$

3.7. $V_{ab} = 16.667\,\text{V}$,
$P_{6\Omega} = 0.1667\,\text{W}$,
$P_{15\Omega} = 0.4168\,\text{W}$,
$P_{12V} = 2.0\,\text{W}$

3.9. $V_{\max} = 50\,\text{V}$, the weaker link is the $300\,\Omega$

3.11. $R_3 = 5\,\Omega$, $R_T = 10\,\Omega$,
$P_{R_3} = 500\,\text{W}$

3.13. $R_2 = 40\,\Omega$ or $2.5\,\Omega$

3.15. $V = 211.76\,\text{V}$,
$I_1 = I_2 = 2.1175\,\text{A}$,
$I_3 = 1.059\,\text{A}$,
$I_4 = 0.706\,\text{A}$

3.17. $R_T = 11.43\,\Omega$

3.19. $V_{ab} = 12\,\text{V}$, $I_3 = 1.333\,\text{A}$,
$I_4 = 0.667\,\text{A}$, $I_5 = 1\,\text{A}$

3.21. (d)

3.25. $I_{BC} = 0\,\text{A}$, $I_{BA} = 2\,\text{A}$

3.27. $R_3 = 80\,\Omega$, $V = 8\,\text{V}$

3.29. $V_1 = 8\,\text{V}$, $V_2 = 2\,\text{V}$

3.31. $I_1 = 0.2\,\text{A}$, $I_2 = 0.15\,\text{A}$,
$V = 2\,\text{V}$

3.33. $I_1 = 0.8\,\text{A}$, $I_2 = 1.2\,\text{A}$,
$V_2 = 14.4\,\text{V}$, $V_3 = 4.8\,\text{V}$

3.35. (a) $E = 200\,\text{V}$;
(b) $P_{5\,k\Omega} = 2\,\text{W}$

3.37. $E_{\max} = 12\,\text{V}$, R_1 and R_2 are weakest

3.39. $R = 1\,\Omega$, $P_R = 25\,\text{W}$

3.41. $R = 15\,\Omega$

3.43. $R_x = 450\,\Omega$

3.45. (a) $V_A = 8\,\text{V}$; (b) $V_B = 4\,\text{V}$;
(c) $V_{AB} = 4\,\text{V}$

3.47. $E = 96\,\text{V}$

3.49. $E = 55\,\text{V}$, $I_T = 25\,\text{mA}$

3.51. (a) $E = 135\,\text{V}$;
(b) $V_{AB} = 20\,\text{V}$

3.53. $E = 90\,\text{V}$, $I = -5\,\text{A}$

3.55. $R = 10\,\Omega$

3.57. (a) $R = 6\,\Omega$, $I_{2\Omega} = 0\,\text{A}$;
(b) $3.097\,\Omega$

3.59. $R_2 = 4\,\Omega$, $R_4 = 5\,\Omega$,
$I_1 = 3\,\text{A}$
$E = 20\,\text{V}$

Chapter 4

4.1. (a) $E = 80\,\text{V}$, $R_{\text{int}} = 10\,\Omega$,
(b) $I = 1\,\text{A}$

4.3. $E = 60\,\text{V}$, $R_{\text{int}} = 100\,\Omega$,
$V = 50\,\text{V}$

4.5. $V = 133.33\,\text{V}$,
$R_L = 666.67\,\Omega$

4.7. $I = 0.9\,\text{mA}$, $V = 1.0\,\text{V}$

4.9. $I_S = 8\,\text{A}$, $R_{\text{int}} = 10\,\Omega$,
$I = 1\,\text{A}$

4.11. $V = 133.33\,\text{V}$,
$R_L = 666.67\,\Omega$

4.13. $P_{\max} = 15.625\,\text{W}$, $\eta = 50\%$

4.15. (a) $V = 27\,\text{V}$;
(b) $P_{\max} = 121.5\,\text{W}$;
(c) $P_L = 77.76\,\text{W}$;
(d) $\eta_1 = 66.67\%$
and $\eta_2 = 33.33\%$

4.17. (a) $E = 50\,\text{V}$,
$R_{\text{int}} = 12.5\,\Omega$;
(b) $P_L = 37.5\,\text{W}$;
(c) $P_{\max} = 50\,\text{W}$

4.19. $R_b = 150\,\Omega$, $R_{sd} = 75\,\Omega$,
V.R. $= 6.67\%$, $P_S = 2.88\,\text{W}$

4.21. $R_b = 666.67\,\Omega$, $R_{sd} = 400\,\Omega$,
V.R. $= 25\%$

4.23. $R_b = 4\,k\Omega$, $R_{sd} = 1.429\,k\Omega$

4.25. $R_b = 9.709\,k\Omega$, $R_{sd} = 291\,\Omega$

4.27. $R_1 = 1900\,\Omega$, $R_2 = 1000\,\Omega$,
$R_3 = 50\,\Omega$, $E = 300\,\text{V}$

4.29. $R_1 = 160\,\Omega$, $R_2 = 50\,\Omega$,
$P_S = 37.5\,\text{W}$

4.31. (a) $E = 270\,\text{V}$,
$R_1 = 2000\,\Omega$
$R_2 = 1000\,\Omega$,
$R_3 = 100\,\Omega$;
(b) $V_A = 259.62\,\text{V}$,
$V_B = 51.92\,\text{V}$,
$V_C = -5.2\,\text{V}$

Chapter 5

5.1. (a) 10; (b) 27; (c) 183

5.3. $x_1 = 12$, $x_2 = -14$,
$x_3 = -12$

5.5. $I_{5\Omega} = 0.548\,\text{A}$,
$I_{1\Omega} = 1.161\,\text{A}$,
$V_{3\Omega} = 1.839\,\text{V}$

5.6. $I_{5\Omega} = 1.263\,\text{A}$ downward

5.9. $I_{3\Omega} = 2.9655\,\text{A}$,
$V_{1\Omega} = -0.8276\,\text{V}$
$I_{4\Omega} = 0.7931\,\text{A}$

5.11. $V_1 = -2.727\,\text{V}$,
$V_2 = -0.545\,\text{V}$,
$I_{2\,\Omega} = 1.091\,\text{A}$ from 2 to 1

5.13. $V_{3\,\Omega} = 5.6315\,\text{V}$,
$V_{5\,\Omega} = -3.3521\,\text{V}$

5.15. $I_{2\,\Omega} = 0.312\,\text{A}$,
$I_{3\,\Omega} = 0.9677\,\text{A}$

5.17. (a) $I_{2\,\Omega} = 0.312\,\text{A}$;
(b) $I_{6\,\text{V}} = 0.161\,\text{A}$ from right to left;
(c) $I_{1\,\Omega} = 0.473\,\text{A}$ downward

5.19. $V_1 = 0.6127\,\text{V}$,
$V_2 = 0.7284\,\text{V}$

Chapter 6

6.1. $V = 36\,\text{V}$

6.3. $V_{25\,\Omega} = -9.139\,\text{V}$

6.5. $I = -1.25\,\text{A}$

6.7. $V_{\text{th}} = 90\,\text{V}$, $R_{\text{th}} = 30\,\Omega$,
$V = 36\,\text{V}$

6.9. $V_{\text{th}} = -10.6\,\text{V}$, $R_{\text{th}} = 4\,\Omega$,
$V = -9.138\,\text{V}$, $R_L = 4\,\Omega$,
$P_{L\,\text{max}} = 7.0225\,\text{W}$

6.11. $I_N = 3\,\text{A}$, $R_N = 30\,\Omega$,
$V = 36\,\text{V}$

6.13. $I_N = -2.65\,\text{A}$, $R_N = 4\,\Omega$,
$V = -9.138\,\text{V}$,
$I_N R_N = 10.6\,\text{V} = V_{\text{th}}$

6.15. $I_T = 4.5\,\text{A}$, $R_T = 1\,\Omega$,
$I = 0.9\,\text{A}$, $V = 2.7\,\text{V}$

6.17. $I_N = 1.667\,\text{A}$, $R_N = 2.4\,\Omega$,
$V = 2.054\,\text{V}$, $I = 0.811\,\text{A}$

6.19. $I_N = 1\,\text{A}$, $R_N = 4\,\Omega$

Chapter 7

7.1. $R_{\text{sh}} = 250\,\Omega$, $R_A = 200\,\Omega$

7.3. $R_{\text{sh}} = 2.041\,\Omega$, $R_A = 2\,\Omega$;
(a) $I_m = 0.1\,\text{mA}$;
(b) $I_{\text{sh}} = 4.9\,\text{mA}$;
(c) $I = 5\,\text{mA}$

7.5. $R_A = 5.4\,\Omega$, $I_{\text{max}} = 10\,\text{mA}$,
I (reading) $= 9.488\,\text{A}$,
loading effect error $= 5.12\%$

7.7. (a) $R_{\text{mult}} = 9.5\,\text{k}\Omega$;
(b) $R_{\text{mult}} = 19.5\,\text{k}\Omega$;
(c) $R_{\text{mult}} = 99.5\,\text{k}\Omega$

7.9. V_{R_1} (ideal) $= 60\,\text{V}$,
V_{R_2} (ideal) $= 30\,\text{V}$,
V_{R_1} (reading) $= 45\,\text{V}$,
V_{R_2} (reading) $= 22.5\,\text{V}$

7.11. (a) $R_{\text{mult}} = 19.9\,\text{k}\Omega$;
(b) $R_{\text{sh}} = 5.263\,\Omega$;
(c) $R_1 = 100\,\Omega$,
$R_2 = 1450\,\Omega$

7.13. $S = 5\,\text{k}\Omega/\text{V}$

7.15. $E = 20\,\text{V}$

7.17. (a) $R_1 = 16.95\,\Omega$,
$R_2 = 983.33\,\Omega$;
(b) $R_1 = 200\,\Omega$,
$R_2 = 9833.3\,\Omega$;
(c) 16.67% of F.S. deflection,
66.67% of F.S. deflection

7.19. (a) $T = 0.5\,\text{ms}$;
(b) $T_r = 1\,\text{ms}$;
(c) $V_{\text{pp}}(Y_A) = 20\,\text{V}$,
$V_{\text{pp}}(Y_B) = 12\,\text{V}$;
(d) $\phi = 180°$

Chapter 8

8.1. (a) $C = 88.5\,\text{pF}$;
(b) $C = 442.5\,\text{pF}$

8.3. (a) $Q = 50\,\mu\text{C}$,
$W = 0.625\,\text{mJ}$;
(b) $Q = 33\,\text{nC}$,
$W = 0.165\,\mu\text{J}$

8.5. $C_{\text{eq}} = 3.6\,\mu\text{F}$,
$V_1 = 7.2\,\text{V}$, $Q_1 = 4.32\,\mu\text{C}$,
$W_1 = 155.52\,\mu\text{J}$
$V_2 = 4.8\,\text{V}$, $Q_2 = 38.4\,\mu\text{C}$,
$W_2 = 92.16\,\mu\text{J}$
$V_3 = 2.4\,\text{V}$, $Q_3 = 4.8\,\mu\text{C}$,
$W_3 = 5.76\,\mu\text{J}$
$V_4 = 2.4\,\text{V}$, $Q_4 = 4.8\,\mu\text{C}$,
$W_4 = 5.76\,\mu\text{J}$

8.7. $v_C = 20e^{-1000t}\,\text{V}$,
$i = 2e^{-1000t}\,\text{mA}$;
(a) $v_C = 2.707\,\text{V}$,
$i = 0.2707\,\text{mA}$;
(b) $v_C = 0.3663\,\text{V}$,
$i = 0.03663\,\text{mA}$

8.9. $v_C = 10(1 - e^{-5t})\,\text{V}$,
$i = 0.1e^{-5t}\,\text{mA}$,

$v_C = 7.769\,\text{V},$

$i = 0.0223\,\text{mA}$

8.11. $v_C = 6 + 4e^{-t/80\,\text{ms}}\,\text{V},$

$i_S = 0.06 - 0.01e^{-t/80\,\text{ms}}\,\text{mA},$

$v_C(\infty) = 6\,\text{V},$

$i_S(\infty) = 0.06\,\text{mA}$

8.13. $v_C\,(0^+) = 12.5\,\text{V},$

$v_C(t) = 15 - 2.5e^{-t/12\,\text{ms}}\,\text{V}$

8.15. **(a)** $v_R = 3.68\,\text{V},$

$v_R = -4.71\,\text{V};$

(b) $\dfrac{dv_C}{dt} = 3.68\,\text{kV/s},$

$\dfrac{dv_C}{dt} = -4.71\,\text{kV/s};$

(c) $t = 1.248\,\text{ms}$

8.17. **(a)** $i_S = 21.84\,\text{mA};$

(b) $\dfrac{dv_C}{dt} = -17.28\,\text{V/s};$

(c) $v_{\text{switch}} = 195.65\,\text{V}$

8.19. $v_A\,(1^-\,\text{ms}) = 10\,\text{V},$

$v_A(1^+\,\text{ms}) = 5\,\text{V},$

$v_B\,(1\,\text{ms}) = 8.16\,\text{V},$

$v_{AB}\,(1^-\,\text{ms}) = 1.84\,\text{V},$

$v_{AB}\,(1^+\,\text{ms}) = -3.16\,\text{V}$

8.21. **(a)** $v_C = 10.38\,\text{V},$

$i = 0.812\,\text{mA};$

(b) $v_C = 1.613\,\text{V},$

$i = -0.81\,\text{mA};$

(c) $\dfrac{di}{dt} = 1.096\,\text{A/s}$

8.23. **(a)** $v_C = 5.7\,\text{V},$

$\dfrac{dv_C}{dt} = -142.5\,\text{V/s};$

(b) $v_{12} = 2.3\,\text{V}$

Chapter 9

9.1. **(a)** $\mathscr{R} = 1.989 \times 10^6\,\text{A/Wb};$

(b) $B = 3.75\,\text{WB/m}^2;$

(c) $H = 1.492 \times 10^4\,\text{At/m};$

(d) $n = 995\,\text{turns}$

9.3. **(a)** $A = 119.4\,\text{cm}^2$

(b) $H = 266.67\,\text{At/m};$

(c) $B = 3.35 \times 10^{-2}\,\text{Wb/m}^2$

9.5. $I = 5.71\,\text{A}$

9.7. $n_2 = 95.5\,\text{turns}$

9.9. $B = 68.54\,\text{mWb/m}^2$

9.11. The a terminal is positive

9.13. The a terminal is positive

9.15. Out of b, into a

9.17. Upward

Chapter 10

10.1. $L = 0.126\,\text{mH}$

10.3. $L = 0.1005\,\text{H}$

10.6. $L_{\text{eq}} = 12\,\text{mH}$

10.7. **(a)** $t = 0.778\,\text{s};$

(b) $v_{\text{switch}} = 138.08\,\text{V}$

10.9. **(a)** $t = 2.03\,\text{ms};$

(b) $\dfrac{di_L}{dt} = 14.94\,\text{A/s}$

10.11. $i_L = 0.08 - 0.04e^{-5000t}\,\text{A},$

$v_L(0.7\,\text{ms}) = 0.121\,\text{V}$

10.13. **(a)** $E = 100\,\text{V},$

$L = 250.9\,\text{mH};$

(b) $t = 6.231\,\text{ms};$

(c) $v_{\text{switch}} = 157.51\,\text{V}$

10.15. **(a)** $i_1 = -i_2 = 3.315\,\text{A};$

(b) $v_L = -33.15\,\text{V};$

(c) $v = -16.575\,\text{V};$

(d) $\dfrac{di_1}{dt} = -331.5\,\text{A/s};$

(e) $v_{\text{switch}} = 76.575\,\text{V}$

10.17. **(a)** $v_L = 119.2\,\text{V};$

(b) $t = 53.18\,\text{ms};$

(c) $v_{\text{switch}} = 317.75\,\text{V}$

Chapter 11

11.1. **(a)** $V_m = 50\,\text{V}, f = 50\,\text{HZ},$

$T = 20\,\text{ms}, \phi = 90°;$

(b) $V_m = 100\,\text{V},$

$f = 795.77\,\text{HZ},$

$T = 1.26\,\text{ms}, \phi = 135°;$

(c) $V_m = 70.7\,\text{V},$

$f = 159.15\,\text{HZ},$

$T = 6.28\,\text{ms}, \phi = -60°;$

(d) $I_m = 1.414\,\text{A},$

$f = 63.66\,\text{Hz},$

$T = 15.71\,\text{ms},$

$\phi = -120°;$

(e) $I_m = 0.5\,\text{A}, f = 60\,\text{HZ},$

$T = 16.67\,\text{ms}, \phi = 135°$

11.3. **(a)** 50 V, 49.385 V,

47.555 V;

(b) 70.7 V, −98.97 V,
87.86 V;

(c) −61.23 V, −36.78 V,
−3.336 V;

(d) −1.225 A, −1.341 A,
−1.403 A;

(e) 0.3535 A, 0.281 A,
0.199 A

11.5. **(a)** $V_{av} = \dfrac{V_m}{2} = 5$ V;

(b) $V_{av} = \dfrac{2V_m}{\pi} = 6.37$ V

11.7. $\mathbf{V}_1 = 141.4/90°$ V,
$\mathbf{V}_2 = 70.7/135°$ V,
$\mathbf{V}_3 = 100/{-30°}$ V,
t_{01} (starting point) =
−3.93 ms,
$t_{02} = -5.89$ ms,
$t_{03} = 1.31$ ms,
$\mathbf{V}_2$ leads $\mathbf{V}_1$ by 45°,
$\mathbf{V}_1$ leads $\mathbf{V}_3$ by 120°

11.9. **(a)** $5/53.1°$,
$3 - j4 = 5/{-53.1°}$;

(b) $10/126.87°$,
$-6 - j8 = 10/{-126.9°}$;

(c) $20/{-60°}$,
$10 + j17.32 = 20/60°$;

(d) $28.28/{-135°}$,
$-20 + j20 = 28.28/135°$;

(e) $10.77/111.8°$,
$-4 - j10$
$= 10.77/{-111.8°}$

11.11. **(a)** $17.07 + j2.93 =$
$17.32/9.7°$;

(b) $0.46 - j7.2 =$
$7.21/{-86.3°}$;

(c) $50/180°$;

(d) $0.707/{-285°} = 0.707/75°$;

(e) $-81.28 + j19.22 =$
$83.52/166.7°$

11.13. $v_L = 80 \sin (100t + 120°)$ V;
$\mathbf{I} = 2.828/30°$ A;
$\mathbf{V}_L = 56.57/120°$ V

11.15. **(a)** $Z_R = 1000/0°$ Ω at all
values of f;

(b) $Z_{C_1} = 3978.9/{-90°}$ Ω,
$Z_{C_2} = 530.5/{-90°}$ Ω;

(c) $Z_{L_1} = 2.26/90°$ Ω,
$Z_{L_2} = 16.96/90°$ Ω

11.17. **(a)** $\mathbf{V} = 50/30°$ V,
$\mathbf{I} = 5/120°$ A,
element is a capacitor,
$C = 100 \,\mu$F;

(b) $P_{av} = 0$ W;

(c) voltmeter's reading =
50 V,
ammeter's reading =
5 A

11.19. $v_1 = 100 \sin (523.6t - 150°)$,
$\mathbf{V}_1 = 70.7/{-150°}$ V,
$v_2 = 100 \sin (523.6t + 150°)$,
$\mathbf{V}_2 = 70.7/150°$ V

11.21. **(a)** $i(t) =$
$23.33 \sin (314t - 45°)$ A;

(b) $P_{av} = 2722.5$ W;

(c) $i(t) =$
$23.33 \sin (314t + 45°)$ A,
$P_{av} = 2722.5$ W

11.23. **(a)** $Z_C = 378.94/{-90°}$ Ω,
$Y_C = 2.64 \times 10^{-3}/90°$ S;

(b) $i =$
$1.49 \sin (120\pi t + 60°)$ A;
ammeter reading =
1.05 A;

(c) $Z_C = 94.74/{-90°}$ Ω,
$Y_C = 10.56 \times 10^{-3}/90°$ S
$i =$
$5.96 \sin (240\pi t + 60°)$ A,
ammeter reading $= 4.2$ A

Chapter 12

12.1. **(a)** 137.83 Hz;
(b) 238.73 Hz;
(c) 413.5 Hz

12.3. **(a)** Voltmeter reading =
100 V,
ammeter reading =
14.14 A;
(b) 5 Ω and 5 mH;
(c) $\mathbf{I} = 8.94/{-63.4°}$ A

12.5. $R = 19$ Ω, $L = 93.37$ mH
12.7. **(a)** $\mathbf{I} = 5/90°$ A;
(b) $\mathbf{E} = 111.8/26.6°$ V,
$P_{av} = 250$ W;
(c) $\mathbf{V}_C = 100/0°$ V

12.9. $R = 2$ Ω, $C = 625 \,\mu$F,
$P_{av} = 12.5$ W
12.11. $R = 2.5$ Ω, $L = 0.67$ mH

12.13. $C = 625\ \mu\text{F}$

12.15. (a) $\mathbf{I} = 1.0\underline{/0°}\ \text{A}$;
(b) $L = 20\ \text{mH}, C = 100\ \mu\text{F}$

12.17. (a) $L = 0.5\ \text{H}$;
(b) $R = 100\ \Omega$

12.19. (a) $L = 31.25\ \text{mH}$;
(b) $E = 15.7\ \text{V}$;
(c) time/div. = $0.314\ \text{ms/div}$

12.21. (a) Y_A is the smaller appearing waveform;
(b) V/div (Y_B) = $32.36\ \text{V/div}$, time/div = $0.5\ \text{ms}$;
(c) $L = 4.9\ \text{mH}$, $C = 6.84\ \mu\text{F}$ or $4.16\ \mu\text{F}$

12.23. (a) $\omega_1 = 1.5 \times 10^6\ \text{rad/s}$;
(b) $L = 4.45\ \mu\text{H}$;
(c) $X_L = 13.34\ \Omega$, $X_C = 3.34\ \Omega$

12.25. $R = 40\ \Omega, C = 2.77\ \mu\text{F}$

12.27. (a) $\mathbf{I}_2 = 25\underline{/73.7°}\ \text{A}$;
(b) $R = 14.29\ \Omega$, $C = 48.0\ \mu\text{F}$;
(c) $R = 2\ \Omega$, $C = 133.33\ \mu\text{F}$

12.29. (a) $R = 10\ \Omega, Z = Z_C$, $C = 75\ \mu\text{F}$;
(b) $e(t) = 120\sqrt{2}\sin(1000t - 30°)\ \text{V}$;
(c) $L = 6.67\ \text{mH}$

12.31. $Z_{\text{in}} = Z_T = 7.92\underline{/8.13°}\ \Omega$, $\mathbf{I}_1 = 1.26\underline{/-8.13°}\ \text{A}$, $\mathbf{I}_2 = 0.76\underline{/45°}\ \text{A}$, $\mathbf{V} = 6.06\underline{/-45°}\ \text{V}$

12.33. $\mathbf{I}_1 = 5\underline{/-36.9°}\ \text{A}$, $\mathbf{I}_2 = 5\underline{/36.9°}\ \text{A}$, $\mathbf{V} = 25\underline{/0°}\ \text{V}, P_{\text{av}} = 200\ \text{W}$

12.35. (a) $e(t) = 100\sin(1000t + 90°)\ \text{V}$;
(b) $L = 20\ \text{mH}$

12.37. $\mathbf{I} = 7.19\underline{/-47.8°}\ \text{A}$, $\mathbf{V}_{\text{AB}} = 49.3\underline{/151.9°}\ \text{V}$, $R_S = 13.9\ \Omega, X_{L_s} = 0.68\ \Omega$, $R_p = 13.93\ \Omega, X_{L_p} = 285\ \Omega$, $P_{\text{av}} = 717.7\ \text{W}$

12.39. (a) $\mathbf{E} = 50\underline{/36.9°}\ \text{V}$;
(b) $\mathbf{I}_3 = 1.414\underline{/45°}\ \text{A}$;

(c) $R_3 = 35.0\ \Omega$, $C_3 = 20\ \mu\text{F}$;
(d) $P_{\text{av}} = 140\ \text{W}$;
(e) $R_S = 17.5\ \Omega, C_s = 40.0\ \mu\text{F}$

12.40. (a) Y_B is the leading waveform;
(b) $L = 2.86\ \text{mH}$, voltmeter reading = $3.535\ \text{V}$;
(d) $\mathbf{I}_C = 0.085\underline{/90°}\ \text{A}$, $\mathbf{I}_2 = 0.1414\underline{/-36.9°}\ \text{A}$, $\mathbf{I}_T = 0.113\underline{/0°}\ \text{A}$, ammeter reading = $0.113\ \text{A}$
(e) $R_S = 31.25\ \Omega$

12.41. $R_X = \omega^2 R_1 R_2 R_3 C_S^2 / (1 + \omega^2 R_1^2 C_S^2)\ \Omega$, $L_X = R_2 R_3 C_S / (1 + \omega^2 R_1^2 C_S^2)\ \text{H}$

Chapter 13

13.1. $P = 9.926\ \text{W}$, $\mathbf{S} = 9.963\underline{/4.9°}\ \text{VA}$, apparent power = $9.963\ \text{VA}$, reactive power = $0.851\ \text{VAR}$

13.3. (a) $P = 12\ \text{W}$, PF = 0.6 lagging;
(b) $P = 3.57\ \text{W}$, PF = 0.707 leading;
(c) $P = 1.732\ \text{W}$, PF = 0.866 lagging

13.5. PF = 0.707 lagging, $L = 1\ \text{mH}$

13.7. (a) $\mathbf{S} = 800 - j600 = 1000\underline{/-36.9°}\ \text{VA}$, PF = 0.8 lead;
(b) $\mathbf{S} = 200 + j400 = 447.2\underline{/63.4°}\ \text{VA}$, PF = 0.448 lag

13.9. (a) $\mathbf{S} = 1414\underline{/-45°}\ \text{VA}$;
(b) $\mathbf{S} = 833.3\underline{/36.9°}\ \text{VA}$;
(c) $\mathbf{S} = 800\underline{/29°}\ \text{VA}$;
(d) $\mathbf{S} = 200\underline{/-53.1°}\ \text{VA}$

13.11. (a) $Z = 200\underline{/-30°}\ \Omega$;
(b) $Z = 28.28\underline{/45°}\ \Omega$;
(c) $Z = 7.07\underline{/-45°}\ \Omega$

13.13. $\mathbf{S}_T = 15.16 + j17.33$
$= 23.02\underline{/48.8°}$ VA

13.15. $\mathbf{S}_T = 524.6 - j110$
$= 536\underline{/-11.8°}$ VA,
$\mathbf{I} = 5.36\underline{/11.8°}$ A

13.17. $Z = 2.4 + j1.28 =$
$2.72\underline{/28.1°}$ Ω,
$\mathbf{S}_T = 500\underline{/36.9°}$ VA

13.19. PF = 0.6 lagging,
$\mathbf{I} = 2\underline{/-53.1°}$ A,
$\mathbf{S} = 20\underline{/53.1°}$ VA,
$C = 19.19\,\mu$F,
$\mathbf{I}$ (after correction) =
$1.263\underline{/-18.2°}$ A

13.21. $C = 6.21\,\mu$F,
I_S (without C) = 5.95 A,
I_S (with C) = 4.39 A

13.23. PF of syn. motor
$= \cos 10.9° = 0.982$ leading

Chapter 14

14.1. $L = 0.1$ H;
(a) $i(t) = 0.2 \sin 10{,}000t$ A;
(b) $|V_L| = |V_C| = 141.42$ V;
(c) $Q = 100$,
BW = 100 rad/s

14.3. $Q = 20$,
$f_1 \cong 487.5$ kHz,
$f_2 \cong 512.5$ kHz,
$f_1 = 487.66$ kHz,
$f_2 = 512.66$ kHz,
$L = 1.01\,\mu$H, $R = 0.159$ Ω

14.5. $\omega_0 = 5 \times 10^5$ rad/s,
$C = 4$ nF, $BW = 1591.5$ Hz

14.7. $R = 6$ Ω, $L = 30\,\mu$H,
$C = 8.33$ nF,
BW = 200 krad/s,
$\omega_1 = 1902.5$ krad/s,
$\omega_2 = 2102.5$ krad/s

14.9. (a) Y_A is the leading waveform, V/div $(Y_B) = 8.09$;
(b) $I = 0.416$ A;
(c) $L = 65$ mH,
$C = 57.9\,\mu$F,
$E = 10$ V;
(d) $\omega_0 = 515.5$ rad/s,
$Q = 1.68$,
BW = 307.7 rad/s

14.11. $R = 1007.3$ Ω,
$C = 0.158\,\mu$F,
$f = 40$ kHz

14.13. $\omega_0 = 141.407$ krad/s,
$Q = 70.7$,
BW = 2000 rad/s,
$Z_0 = 5000$ Ω

14.15. $\omega_0 = 7070.9$ rad/s,
$Q = 141.42$, BW = 50 rad/s,
$I_1 = 70.71\underline{/90°}$ mA,
$I_2 = 70.71\underline{/-89.6°}$ mA,
$I_T = 0.5$ mA,
Z_T at $\omega_0 = 10$ kΩ

14.17. $R = 0.5$ Ω,
$\omega_0 = 31622.6$ rad/s,
$Q = 316.2$, BW = 100 rad/s

14.19. $\omega_0 = \dfrac{1}{\sqrt{LC}}$
$\times \sqrt{\dfrac{1 - (C/L)R_2^2}{1 - (C/L)R_1^2}}$ rad/s,

$Q = \dfrac{1}{\omega_0 C(R_1 + R_2)}$
$\times \dfrac{1 - (C/L)R_2^2}{1 - (C/L)R_1 R_2}$

Chapter 15

15.1. $P_{E_1} = 19.23$ W,
$P_{E_2} = 11.53$ W

15.3. $\mathbf{I}_A = 23.96\underline{/74.74°}$ A,
$\mathbf{I}_B = 4.972\underline{/161.23°}$ A,
$\mathbf{I}_C = 24.76\underline{/-93.7°}$ A

15.5. $\mathbf{I} = 0.406\underline{/64.6°}$ A

15.7. $L_x = R_1 R_2 C_1$, $R_x = \dfrac{C_1}{C_2} R_1$

15.9. $\mathbf{E} = 2.169\underline{/12.5°}$ V

15.11. $\mathbf{V}_{4\Omega} = 1.596\underline{/37.2°}$ V

15.13. $\mathbf{V}_1 = 24.834\underline{/-18.7°}$ V,
$\mathbf{I}_A = 23.96\underline{/74.7°}$ A,
$\mathbf{I}_B = 4.976\underline{/161.3°}$ A,
$\mathbf{I}_C = 24.76\underline{/-93.7°}$ A

15.15. $P_{5\Omega} = 0.7866$ W

15.17. $\mathbf{V} = 2\underline{/180°}$ V

15.19. $\mathbf{V}_1 = \mathbf{V}_2 = 8\underline{/0°}$ V

15.21. $\mathbf{I} = 1.87\underline{/2.3°}$ A

15.23. $Z_{\Delta_T} = 8.143\underline{/-61.3°}$ Ω,
$Z_{Y_T} = 2.714\underline{/-61.3°}$ Ω

15.25. $Z_a = -j5\ \Omega$, $Z_b = 12.5\ \Omega$,
$Z_c = 3.123\underline{/38.7°}\ \Omega$

Chapter 16

16.1. $\mathbf{I} = 1.934\underline{/42.9°}$ A

16.3. $\dfrac{\mathbf{V}_1}{\mathbf{V}_2} = 0.283\underline{/81.9°}$

16.5. $i_L(t) = 1$
$+\ 0.392 \sin (1000t - 123.7°)$
$+\ 1.262 \sin (2000t - 88°)$ A

16.7. $Z_{\text{th}} = Z_{\text{N}}$
$= 18.878\underline{/-12.8°}\ \Omega$,
$\mathbf{V}_{\text{th}} = 16.05\underline{/125.5°}$ V,
$\mathbf{I}_{\text{N}} = 0.85\underline{/138.2°}$ A,
$\mathbf{I}_L = 0.415\underline{/131.78°}$ A,

16.9. $Z_{\text{th}} = Z_{\text{N}}$
$= 332.87\underline{/7.53°}\ \Omega$,
$\mathbf{V}_{\text{th}} = 333.3\underline{/97.2°}$ V,
$\mathbf{I}_{\text{N}} = 1.001\underline{/89.6°}$ A

16.11. $Z_{\text{th}} = 10.178\underline{/50.7°}\ \Omega$,
$\mathbf{V}_{\text{th}} = 17.35\underline{/64.7°}$ V,
$\mathbf{V}_L = 9.514\underline{/42.1°}$ V

16.13. $Z_{\text{th}} = Z_{\text{N}} = 4.006\underline{/19.4°}\ \Omega$,
$\mathbf{V}_{\text{th}} = 5.556\underline{/0°}$ V,
$\mathbf{I}_{\text{N}} = 1.387\underline{/-19.4°}$ A

16.15. $Z_{\text{th}} = 1.992\underline{/-84.8°}\ \Omega$,
$\mathbf{V}_{\text{th}} = 9.954\underline{/95.2°}$ V

16.17. $Z_L = 4.472\underline{/26.6°}\ \Omega$,
$P_{L\,\text{max}} = 6.25$ W

16.19. (a) $Z_L = 18.878\underline{/12.8°}\ \Omega$,
$P_{L\,\text{max}} = 3.498$ W;
(b) $Z_L = 18.878\ \Omega$,
$P_{L\,\text{max}} = 3.454$ W

16.21. (a) Z_L
$= 10.178\underline{/-50.7°}\ \Omega$,
$P_{L\,\text{max}} = 11.668$ W;
(b) $Z_L = 10.178\underline{/0°}\ \Omega$,
$P_{L\,\text{max}} = 9.051$ W

16.23. (a) $Z_L = 1.825\underline{/47.3°}\ \Omega$,
$P_{L\,\text{max}} = 1.788$ W;
(b) $R = 3.863\ \Omega$,
$P_{L\,\text{max}} = 0.868$ W

Chapter 17

17.1. $M = 0.48$, $\dfrac{N_1}{N_2} = 1.5$

17.3. $L_1 = 30\ \text{mH}$, $L_2 = 60\ \text{mH}$,
$M = 20\ \text{mH}$, $k = 0.471$

17.5. Bottom, top, top.

17.7. $\mathbf{I} = 0.35\underline{/28.8°}$ A,
$\mathbf{V}_{ab} = 1.051\underline{/118.8°}$ V

17.9. $Z_{\text{in}} = 4.068\underline{/-56.3°}\ \Omega$

17.11. $\mathbf{I} = 2.8\underline{/22.02°}$ A

17.13. $\mathbf{I}_S = 2.008\underline{/58.6°}$ A

17.15. Z_1 in series with $Z_2 \parallel Z_3$:
$Z_1 = 5 - j4\ \Omega$,
$Z_2 = 2 + j7\ \Omega$, $Z_3 = j2\ \Omega$

17.17. (a) $\mathbf{I}_1 = 0.214\underline{/-2°}$ A,
$\mathbf{I}_2 = 0.642\underline{/-2°}$ A,
$\mathbf{V}_1 = 4.307\underline{/24.6°}$ V,
$\mathbf{V}_2 = 1.436\underline{/24.6°}$ V;
(b) $\mathbf{I}_1 = 0.235\underline{/41.8°}$ A,
$\mathbf{I}_2 = 0.705\underline{/41.8°}$ A,
$\mathbf{V}_1 = 2.991\underline{/-3.2°}$ V,
$\mathbf{V}_2 = 0.997\underline{/-3.2°}$ V

17.19. $\mathbf{I}_1 = 0.0628\underline{/-11.6°}$ A,
$\mathbf{I}_2 = 0.2512\underline{/168.4°}$ A,
$\mathbf{I}_3 = 0.502\underline{/-11.1°}$ A

17.21. $\mathbf{I}_1 = 2.975\underline{/22.8°}$ A,
$\mathbf{I}_2 = 0.941\underline{/41.2°}$ A,
$\mathbf{V}_1 = 22\underline{/111.3°}$ V,
$\mathbf{V}_2 = 4.207\underline{/104.7°}$ V

17.23. $\mathbf{I}_1 = 1.556\underline{/61.4°}$ A,
$\mathbf{I}_2 = 1.114\underline{/-100.2°}$ A,
$\mathbf{V}_1 = 3.914\underline{/139.7°}$ V,
$\mathbf{V}_2 = 1.575\underline{/-55.2°}$ V
When the dotted terminal is reversed: $\mathbf{V}_1$ and $\mathbf{I}_1$ are as above,
$\mathbf{I}_2 = 1.114\underline{/79.8°}$ A,
$\mathbf{V}_2 = 1.575\underline{/124.8°}$ V

17.25. $\mathbf{I}_1 = 4.786\underline{/-5.5°}$ A,
$\mathbf{I}_1' = 1.914\underline{/-5.5°}$ A,
$\mathbf{I}_2 = 2.876\underline{/-5.5°}$ A

Chapter 18

18.1. $\mathbf{I}_A = 28.87\underline{/-30°}$ A,
$\mathbf{I}_B = 28.87\underline{/90°}$ A,
$\mathbf{I}_C = 28.87\underline{/-150°}$ A

18.3. $\mathbf{I}_{AB} = 4.622\underline{/-30°}$ A,
$\mathbf{I}_{BC} = 4.622\underline{/-150°}$ A,
$\mathbf{I}_{CA} = 4.622\underline{/90°}$ A,
$\mathbf{I}_A = 8.0\underline{/-60°}$ A,
$\mathbf{I}_B = 8\underline{/180°}$ A,
$\mathbf{I}_C = 8.0\underline{/60°}$ A

18.5. $Z_Y = 5.774\underline{/-60°}\,\Omega$

18.7. $\mathbf{I}_C = 12.925\underline{/176.6°}\,A$

18.9. **(a)** $\mathbf{I}_A = 23.094\underline{/-45°}\,A,$
$\mathbf{I}_B = 23.094\underline{/-165°}\,A,$
$\mathbf{I}_C = 23.094\underline{/75°}\,A;$

(b) $\mathbf{I}_A = 16.33\underline{/-45°}\,A,$
$\mathbf{I}_B = 16.33\underline{/-165°}\,A,$
$\mathbf{I}_C = 16.33\underline{/75°}\,A,$
With both loads
connected,
$\mathbf{I}_A = 39.424\underline{/-45°}\,A,$
$\mathbf{I}_B = 39.424\underline{/-165°}\,A,$
$\mathbf{I}_C = 39.424\underline{/75°}\,A$

18.11. $\mathbf{I}_{AB} = 15\underline{/-30°}\,A,$
$\mathbf{I}_{BC} = 30\underline{/-90°}\,A,$
$\mathbf{I}_{CA} = 30\underline{/60°}\,A,$
$\mathbf{I}_A = 33.54\underline{/-93.4°}\,A,$
$\mathbf{I}_B = 25.98\underline{/-120°}\,A,$
$\mathbf{I}_C = 57.95\underline{/75°}\,A$

18.13. $\mathbf{I}_A = 11.547\underline{/-90°}\,A,$
$\mathbf{I}_B = 17.32\underline{/-150°}\,A,$
$\mathbf{I}_C = 5.774\underline{/120°}\,A,$
$\mathbf{I}_N = 23.478\underline{/40.4°}\,A$

18.15. $\mathbf{I}_A = 10.5\underline{/-87.3°}\,A,$
$\mathbf{V}_{AO} = 63.0\underline{/27°}\,V,$
$\mathbf{I}_B = 14.488\underline{/84.1°}\,A,$
$\mathbf{V}_{BO} = 57.95\underline{/114.1°}\,V,$
$\mathbf{I}_C = 4.392\underline{/-116.7°}\,A,$
$\mathbf{V}_{CO} = 52.7\underline{/-116.7°}\,V,$
$\mathbf{V}_{ON} = 5.98\underline{/-150.3°}\,V$

18.17. $W_1 = 636.86\,W,$
$W_2 = -144.36\,W,$
$W_T = W_1 + W_2 = 492.5\,W.$
$= 3P_{\text{ph}} = 3 \times 164.16\,W$

18.19. $Z_\Delta = 283.45\underline{/-40.9°}\,\Omega$

18.21. $W_B = 3236.7\,W,$
$W_C = 511\,W,$
$P_A = 1022.45\,W,$
$P_B = 2214.25\,W,$
$P_C = 511.08\,W$

18.23. **(a)** Voltmeter
reading $= 145.37\,V;$
(b) ammeter reading
$= 5.819\,A$

18.25. **(a)** $W_A = 1853.7\,W,$
$W_B = 1879.6\,W;$
(b) $W_B = 1067.9\,W;$
$W_C = 2666.3\,W$

Chapter 19

19.1. $I_1 = 0.4348\,A,\ I_1 = 0.9217\,A$
$V = -3.217\,V$

19.3. $V' = 4.04\,V,\ I = 80.81\,mA$
$V_L = 10.35\,V,\ I_L = 0.2071\,A$
See figure.

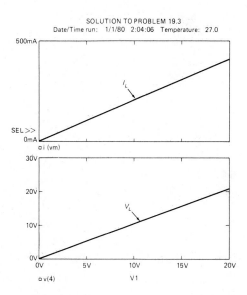

SOLUTION TO PROBLEM 19.3
Date/Time run: 1/1/80 2:04:06 Temperature: 27.0

19.5. $I = 0.2385\,A,\ V' = 5.538\,V$
$I_x = 0.4231\,A$

19.7. $V_0 = -5\,V;$ see figure.

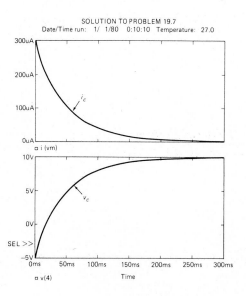

SOLUTION TO PROBLEM 19.7
Date/Time run: 1/1/80 0:10:10 Temperature: 27.0

19.9. See figure.

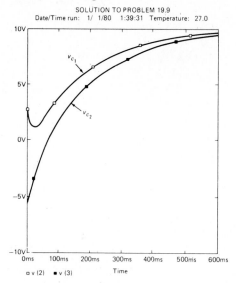

SOLUTION TO PROBLEM 19.9
Date/Time run: 1/ 1/80   1:39:31   Temperature: 27.0

□ v (2)   ■ v (3)

19.11. See figure.

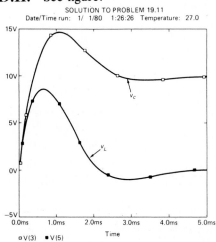

SOLUTION TO PROBLEM 19.11
Date/Time run: 1/ 1/80   1:26:26   Temperature: 27.0

□ V(3)   ■ V (5)

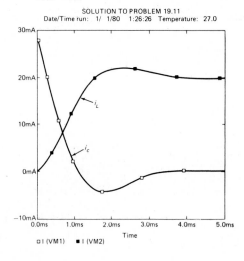

SOLUTION TO PROBLEM 19.11
Date/Time run: 1/ 1/80   1:26:26   Temperature: 27.0

□ I (VM1)   ■ I (VM2)

19.13. $I_1 = 0.02\underline{/78.63°}$ A,
$I_2 = 0.0506\underline{/-57.85°}$ A
$V_x = 3.471\underline{/-175.7°}$ V

19.15. $I_1 = 0.02568\underline{/-36.97°}$ A,
$I_2 = 0.03553\underline{/-32.93°}$ A
$I_3 = 0.01008\underline{/157.4°}$ A

19.17. See figures.

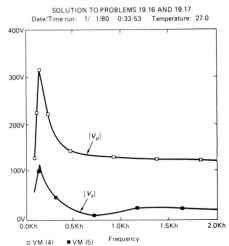

SOLUTION TO PROBLEMS 19.16 AND 19.17
Date/Time run: 1/ 1/80   0:33:53   Temperature: 27.0

□ VM (4)   ■ VM (5)

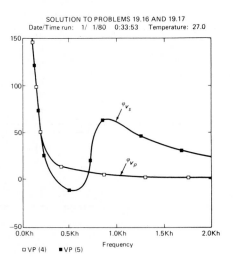

SOLUTION TO PROBLEMS 19.16 AND 19.17
Date/Time run: 1/ 1/80   0:33:53   Temperature: 27.0

□ VP (4)   ■ VP (5)

19.19. $f_{3\,dB} = 20.21$ kHz; see figure.

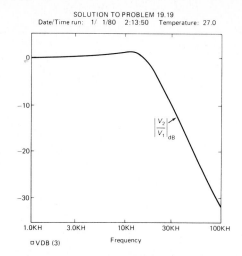

SOLUTION TO PROBLEM 19.19
Date/Time run: 1/ 1/80 2:13:50 Temperature: 27.0

□ VDB (3)   Frequency

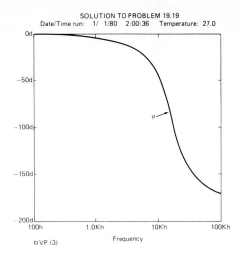

SOLUTION TO PROBLEM 19.19
Date/Time run: 1/ 1/80 2:00:36 Temperature: 27.0

□ VP (3)   Frequency

19.21. $f_{3\,dB} = 128.9\,\text{kHz}$; see figures.

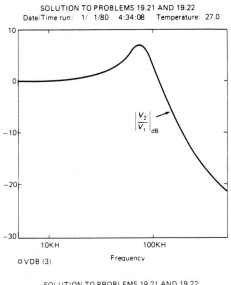

SOLUTION TO PROBLEMS 19.21 AND 19.22
Date/Time run: 1/ 1/80 4:34:08 Temperature: 27.0

□ VDB (3)   Frequency

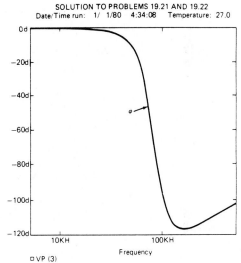

SOLUTION TO PROBLEMS 19.21 AND 19.22
Date/Time run: 1/ 1/80 4:34:08 Temperature: 27.0

□ VP (3)   Frequency

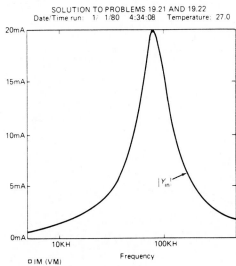

SOLUTION TO PROBLEMS 19.21 AND 19.22
Date/Time run: 1/ 1/80 4:34:08 Temperature: 27.0

□ IM (VM)   Frequency

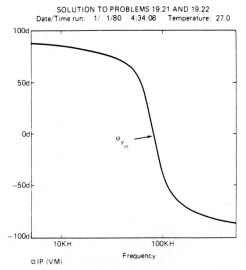

SOLUTION TO PROBLEMS 19.21 AND 19.22
Date/Time run: 1/ 1/80 4:34:08 Temperature: 27.0

□ IP (VM)   Frequency

INDEX

Dielectric, 274
  constant, 274
Differential equation, 291, 367
Digital reading instruments, 234
Diode, 30, 40
Discharging transients, 296, 376
Dot convention, 704–8
Double-subscripted notation, 71–73
Dynamic resistance, 41

E

Eddy current losses, 723
Effective resistance, 595
Effective value, 413
Efficiency, 22
Electric field, 274
Electrodynamometer principle, 564
Electrolytic capacitor, 279
Electromagnet, 333
Electromagnetic induction, 344
Electromagnetism, 236, 333
Electromotive force, 12
Electron, 7, 8
Electron gun, 257
Electron spin, 330
Electrostatic force, 258, 274, 330
Element lines, 791, 793–98
Energy, 12, 20
  storage, 272, 275, 362
Equivalent circuits, 83–85
EVM, 245
Exponential curve, 293–94

F

Farad, 273
Faraday's law, 344–46, 359
Ferrites, 340
Ferromagnetic materials, 336
Fields:
  electric, 274
  magnetic, 330
Filters, 574
  band-pass, 580
  high-pass, 828
Final value:
  of LC circuit, 368–70
  of RC circuit, 290, 301
Flux:
  density, 331
  electric, 274
  leakage, 699

linkage, 345, 699–702
  magnetic, 330
Flyback time, 260
Focusing anodes, 258
Force, 234
Form factor, 414
Fourier series, 669
Four-wire system, 762–63
Free electron, 8
Free format, 791
Frequency, 405
Frequency domain representation, 428–29
Frequency response, 579
Frequency trap, 591
Full-load current, 125
Full-scale current, 236

G

Galvanometer, 98
Generator, 347
  ac, 402
  three-phase, 747–49
  two-phase, 746–47
Germanium, 9, 30
Graphical analysis, 128–29
Graticule, 255
Ground, 71, 72, 632–33

H

Half-power frequency, 580
Hay bridge, 540
Henry, 344, 359
Hertz, 405
Holes, 8
Horsepower, 21
Hysteresis, 338
  loss, 339

I

ICs, 42
Ideal current source, 133
Ideal tank circuit, 587
Ideal transformer, 721–26
Ideal voltage source, 121
Igniton, 376
Impedance, 428
  bridge, 522
  of a capacitor, 430
  determinant, 623
  diagram, 456, 482